Fascinating Life Sciences

This interdisciplinary series brings together the most essential and captivating topics in the life sciences. They range from the plant sciences to zoology, from the microbiome to macrobiome, and from basic biology to biotechnology. The series not only highlights fascinating research; it also discusses major challenges associated with the life sciences and related disciplines and outlines future research directions. Individual volumes provide in-depth information, are richly illustrated with photographs, illustrations, and maps, and feature suggestions for further reading or glossaries where appropriate.

Interested researchers in all areas of the life sciences, as well as biology enthusiasts, will find the series' interdisciplinary focus and highly readable volumes especially appealing.

James S. Albert · Roberto E. Reis
Editors

Ecology and Evolution of Amazonian Fishes

Assembling the Most Diverse Continental Vertebrate Fauna

Springer

Editors
James S. Albert
University of Louisiana at Lafayette
Lafayette, LA, USA

Roberto E. Reis
Pontifical Catholic University of Rio Grande do Sul
Porto Alegre, RS, Brazil

ISSN 2509-6745 ISSN 2509-6753 (electronic)
Fascinating Life Sciences
ISBN 978-3-032-09327-1 ISBN 978-3-032-09328-8 (eBook)
https://doi.org/10.1007/978-3-032-09328-8

This Springer imprint is published by the registered company Springer Nature Switzerland AG
The registered company address is: Gewerbestrasse 11, 6330 Cham, Switzerland

Frontispiece

A sample of phenotypic diversity in Amazonian fishes. **Left column**: Oceolate river stingray *Potamotrygon motoro* (Potamotrygonidae), Common hatchetfish *Gasteropelecus sternicla* (Gasteropelecidae), Red-bellied piranha *Pygocentrus caribe* (Serrasalmidae), Amazon leaffish *Monocirrhus polyacanthus* (Polycentridae). **Right column**: Black ghost knifefish *Apteronotus albifrons* (Apteronotidae), Fitzcarraldo's banded knifefish *Gymnotus chaviro* (Gymnotidae), Bumblebee catfish *Pseudopimelodus bufonis* (Pseudopimelodidae), Bristlenose pleco *Ancistrus* sp. (Loricariidae), Pike cichlid *Crenicichla semicincta* (Cichlidae). Photos by Tiago P. Carvalho.

We dedicate this book to Gerald R. Smith to honor his many and profound contributions to the study of freshwater fishes. Jerry was a mentor to the editors and a perennial source of wisdom regarding the complementary roles of fossil and living fishes in unraveling the causes of evolution.
A brief anecdote says so much. It was Fall 1989 in the UMMZ Fish Division wet lab when Jerry rushed in with a new paper on the osteology of Luvarus imperialis by Tyler et al. (1989). The Louvar is the one pelagic species of Acanthuroidei, a clade of otherwise nearshore marine fishes. He passed the reprint around, noting the telltale surgeonfish characters and also features that resemble tunas and billfishes that swim in open water. Unity of form, diversity of function. Homology and homoplasy. Evolution of ecological traits and historical ecology.
Thanks, Jerry, we miss your boundless energy and many useful insights.

Dedication

Gerald R. Smith
Museum of Zoology
University of Michigan
1935 -2022

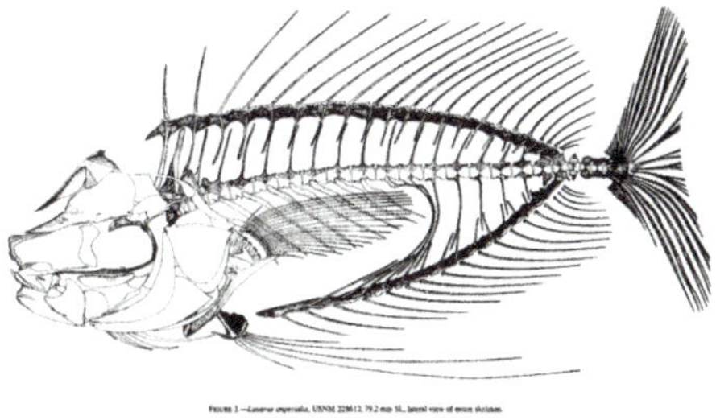

Luvarus imperialis
Tyler et al. (1989)

Foreword

To the best of our knowledge, Earth is the only place in the universe capable of sustaining life. Yet given what dominates its surface and depths, we might more appropriately have called it *Water*.

Even so, like most others, I have spent virtually all my life on land. But unlike most, I have been fortunate to study and work in extraordinary ecosystems—from pristine tropical rainforests to alpine meadows and grasslands. As incredible as these are to any biologist, it has been beneath the surface of the water—working as a young divemaster and free-diving at every opportunity—that I have been most astonished by life's complexity and beauty: its myriad forms, adaptations, and interactions.

Indeed, for anyone interested in biodiversity, the aquatic world demands our attention—especially the region covered in this book: the American tropics, or Neotropics. Across nearly all major groups of organisms we know reasonably well, this is where diversity reaches its zenith: the highest richness of vascular plants, butterflies, birds, amphibians, lizards, snakes, mammals, ants, and so on. And, as this book compellingly shows, the Neotropics are unmatched in their freshwater fish diversity, with a staggering 6345 species reported here. With new species described every year and ongoing research continuing to reveal new ecological insights and evolutionary adaptations, we are still just scratching the surface.

However, the thrill of studying such a remarkably diverse component of life on Earth is starkly contrasted by the scale of threats it faces. Never before in human history have so many species been at such imminent risk of extinction—driven by the profound harm we are inflicting on our planet.

The threats are numerous and complex: the deforestation of tropical rainforests, which amounted to 6.7 million hectares in 2024; the chemical pollution linked to oil extraction and gold and mineral mining; unsustainable hunting and fishing; the introduction—both accidental and deliberate—of non-native species; the illegal wildlife trade; and the increasing frequency of droughts, fires, and other extreme events associated with climate change. The list goes on.

The combined effect of these pressures is unambiguous. According to data compiled by the International Union for Conservation of Nature (version 2025-1), of the 5842 freshwater fishes assessed so far in the vast class Actinopterygii, 949 are officially classified as threatened—including 217 listed as critically endangered, meaning they face imminent extinction unless drastic, urgent measures are taken to secure their survival. The Living Planet Index, which tracks vertebrate population trends over time, reports a 91% decline across 533 populations of 84 species of migratory freshwater fishes in Latin America and the Caribbean since 1975—the most severe regional drop worldwide over this period.

It is fair to ask: in the face of such a crisis, do we really need more science? Shouldn't we all be focused exclusively on protecting biodiversity, rather than continuing to study it? As compelling as this might sound from an activist's standpoint, I strongly disagree. The reason is simple: we cannot effectively protect what we do not understand.

And understanding Neotropical freshwater fishes is precisely what this book sets out to achieve—across as many dimensions as one could hope for. From phylogenetic relationships to morphological diversity; from patterns of species richness and endemism to the processes of speciation, extinction, and dispersal; from habitat occupancy to sensory ecology—and much more. By bridging deep-time evolutionary histories with contemporary ecological dynamics—two perspectives of the same telescope with which we view nature—this collection of chapters offers a rich, multidimensional framework for understanding biodiversity: how it is formed, maintained, and—hopefully—safeguarded.

This book will no doubt become the standard reference in the field—an invaluable resource for students, seasoned professionals, and curious enthusiasts alike. But it should also be essential reading for anyone willing to have their mind blown by a world that is, quite literally, upside down from the one we inhabit daily.

Consider the subterranean fishes, which through evolution have lost their eyesight and pigmentation; or the peculiar algal grazers; or the wood-eating catfishes; or those that specialize in feeding on the fins or mucus of other fishes; or those adapted to seasonally hypoxic waters, and so on. The list of ecological marvels is endless—and I can't stop reading. In many ways, this book is not only a compendium of facts but also a celebration of natural history: combining the ultra-modern with the classical, the seemingly trivial with the truly extraordinary.

I applaud the 45 authors for sharing their expertise and insights—much of it acquired through years of strenuous, and at times dangerous, fieldwork. And I thank James Albert and Roberto Reis for their vision and leadership in bringing this remarkable volume to life, inspiring current and future generations of Neotropical fish lovers.

Now to the most important part: enjoy the read—and let it unveil the dazzling jewels that make our watery planet so uniquely special.

Alexandre Antonelli
Executive Director of Science
Royal Botanic Gardens
Kew, London, UK

Full Professor of Biodiversity and Systematics
Crown Princess Victoria Chair in Biodiversity
University of Gothenburg
Gothenburg, Sweden

Preface

The Richest Vertebrate Fauna

Fishing in South America is by far the most dangerous of all forms of scientific exploration. In order to maintain good health in the tropics my advice would be to eat anything you can get, whenever you can get it, and as much as you crave. Keep hard at work. Eternal vigilance is the price of a good collection... After the noises of the day the hush which comes at night-fall causes even the hardened traveler at times to shudder. No man over the age of fifty years of age should attempt to enter this region. A hard heart and cold blood are useful to him who invades it.

—John D. Haseman (1912)

More than a century has passed since John Haseman wrote these hard words in a report to his patrons at the Carnegie Museum in Philadelphia. His expeditions to the Amazon in the early twentieth century occurred before the region had begun its historical transformation from uncharted wilderness to managed agricultural landscapes. Though much of tropical South America today remains thickly forested, great swaths have been converted to industrial farms in the Atlantic Rainforest of eastern Brazil, along the southern margin of the Amazon rainforest in Brazil and Bolivia, and in the Andean foothills and savannas of Colombia, Venezuela, and the Guianas (Albert et al. 2023; Guayasamin et al. 2024). Rivers are being damned across the Brazilian shield and Andean foothills, severing Andes-Amazon connections (Anderson et al. 2018). Anthropogenic climate change is driving increasingly severe droughts across the whole continent (Wunderling et al. 2022). For the first time in modern history, more Amazonian people live in cities than in villages or on farms (Erickson 2008).

Yet despite these many changes, the Amazon River Basin (hereafter Amazon *sensu* Albert et al. 2021) remains the global center of biodiversity for most groups of organisms. The numbers are breathtaking: 10% of all the world's described species of plants and animals are compressed into a tiny fraction (about 1%) of the Earth's total surface area (Rull and Carnaval 2020; Guayasamin et al. 2024). The enormous dimensions of this biota have been both an inspiration and a challenge to generations of ichthyologists. The seemingly endless forms of Neotropical organisms first attracted scientific attention in the nineteenth century (Albert and Reis 2011; Dagosta and de Pinna 2018). Famous early accounts were provided by von Humboldt and Bonpland (1811), Darwin (1839), Wallace (1854), and Bates (1863). Early expeditions by Spix and Martius (1819–1820), Natterer (1820–1838), Castelnau (1843–1847), Schomburgk (1835–1839), and Agassiz (1865–1866) produced the collections used in the pioneering taxonomic works on fishes by Franz Steindachner (Steindachner 1879) and Carl Eigenmann (Eigenmann and Eigenmann 1891; Eigenmann 1924, 1927). Yet even after generations of intensive research, the evolutionary processes and ecological conditions underlying the formation and maintenance of the megadiverse Amazonian aquatic ecosystem remain poorly understood. Even in the early twenty-first century, many basic facts concerning Amazonian fishes remain elusive, like the numbers of species, their geographic distributions, ecological requirements, behaviors, and other functional traits. Despite the ecological and economic importance of Amazonian fishes, information on this fauna has until now been scattered among hundreds of primary journal articles published in several different languages.

Discovering and describing tropical fishes are laborious and challenging, often involving fieldwork in remote wilderness settings with little or no scientific infrastructure, followed up by extensive studies in the lab, compiling and analyzing morphological and molecular datasets gathered across multiple locations and many years. Information from all these sources is used to elucidate the individual species' geographic distributions and preferred habitats and dietary requirements and to reconstruct their evolutionary interrelationships. Yet despite decades of efforts by fish taxonomists and aquatic ecologists, many of the phenotypes remain poorly documented. These phenotypes include the myriad physiological, morphological, and behavioral specializations and tolerances that allow members of a species to grow, reproduce, and complete their life cycle. We still have much to learn about the factors that propelled Amazonian species to exhibit such a broad range of ecological and physiological specializations and tolerances.

The Biodiversity Enterprise

Biodiversity may be defined broadly as the diversity of genes, traits, species, habitats, and landscapes in the biosphere (Seddon et al. 2016), and the many facets of biodiversity have been partitioned into a variety of classification schemes (see, e.g., Lyashevska and Farnsworth 2012; Magurran et al. 2018). Facing the daunting task of documenting biodiversity, a general principle has emerged, which is, to paraphrase Dobzhansky (1973): "*Nothing in evolution or ecology makes sense except in light of a good taxonomy.*"

Identifying and naming each *species* is critical to the biodiversity enterprise, as species are the fundamental units of ecology and evolution (Van Valen 1976; Magurran 2021). The species boundary defines the limits of gene flow and natural selection, and characteristic species phenotypes represent a relatively stable part of the environmental fabric in which other species evolve and adapt over millions of years (Lenormand 2002; Losos 2008; Tokeshi 2009; Agrawal and Zhang 2021). Further, the characteristic phenotypes of species and higher taxa (i.e., clades) are thought to constrain and channel lineages along distinct evolutionary pathways (Losos 2011; Wiens 2011; Harmon et al. 2019).

Teasing apart closely related and phenotypically similar species requires rigorous training in the science and practice of biological *taxonomy*. Ichthyologists must invest years learning the many tools of the trade, how to conduct effective and ethical scientific collecting, how to acquire and analyze complex biodiversity datasets, how to curate specimens and data, and how to disseminate scientific results in the form of presentations and peer-reviewed publications. For the biodiversity community as a whole, the investment involves permanent archiving and care for voucher specimens, tissue samples, digital images, and associated environmental data for millions of individual specimens representing thousands of species (Fig. P1).

Although counting the numbers of species in a clade or region is a central task in biodiversity assessment, *species richness* represents only one of many biodiversity dimensions. The Neotropics is the global center of freshwater fish diversity, as assessed by either genetic (Manel et al. 2020), morphological (Su et al. 2019), or ecological (Winemiller 1991; Matthews 2012) traits.

Neotropical Freshwater Fish (NFF) species possess a broad range of anatomical and physiological adaptations for utilizing and partitioning microhabitats and food resources and life history specializations associated with seasonal, climatic, and elevational gradients. NFF species also exhibit functionally important variation during growth, among habitats, across geographic space, and between the sexes.

While fieldwork is essential for asking interesting research questions, lab work is equally important in generating data to help answer these questions. New technological and conceptual advances are rapidly expanding opportunities to study Neotropical biodiversity (Antonelli et al. 2018; Albert et al. 2020). From the study of genetics and genomics to the mechanistic understanding of behavioral and ecological traits, it is an exciting time to be a Neotropical biologist. Knowledge of biological systematics is acquired from synergistic activities of field and laboratory studies, the results of each informing the priorities of the other (Fig. P2). Fieldwork

Fig. P1 Images of ichthyological fieldwork in the Amazon. (**A**) Fernando Jerep and Emmanuel Maxime deploying a trawl net in the Madre de Dios River, Peru. (**B**) Roberto Reis demonstrating the use of a cast net in the Alto Purús River, Peru. (**C**) A pair of freshly caught *Sternarchella patriciae* using a trawl net in the Rio Madre de Dios near Puerto Maldonado. (**D**) Nathan Lovejoy excising a sample of muscle tissue from a pacu (*Colossoma macropomum*) in the Rio Pacaya. Photos by JSA

Fig. P2 Knowledge of NFF systematics and evolution is acquired by work in the field and in museum and university laboratories. (**A**) Roberto Reis and James Albert sorting small tetras in the Rio Pacaya, Peru. (**B**) Tiago Carvalho, Jessica Espino, Fernando Jerep, and Emmanuel Maxime conducting field photography on a beach of the Rio Las Piedras, Peru. (**C**) Fish collection of the Museo Universidade de San Marcos (MUSM), Lima, Peru: Giannina Trevejo, Jack Craig, Hernán Ortega, and Lesley Kim. (**D**) Museu de Ciências e Tecnologia da PUCRS (MCP), Porto Alegre, Brazil: Rafael Lugo, Bárbara Calegari, Vanessa Meza, Roberto Reis, Maria Laura S. Delapieve, Margarete Lucena, Carlos Lucena, Álvaro Brum Neto, and Alejandro Londoño-Burbano. (**E**) University of Louisiana at Lafayette: Kassie Ford, Kory Evans, Brandon Waltz, Max Bernt, and James Albert involves scientific collecting and preserving specimens and multiple associated data streams. Data are carefully tracked to coordinate information on voucher specimens with field and museum catalog numbers, photographs showing live coloration, environmental data on chemistry and GPS locality, tissue samples preserved for genetic or stable isotope analyses, and specialized data from certain taxa like sound or electric signal recordings to document social and sexual communication signals or karyotypes to document chromosomal diversity. Everything must be carefully cross-referenced and tabulated, as unlabeled specimens and tissue vials just muddle an already overwhelmingly confusing system.

The taxonomic and biogeographic foundations of the NFF fauna, built over many decades, may be likened to the bones of the system, providing a framework on which to organize information on the immense diversity of organismal forms and functions. Now the community of Neotropical ichthyologists can be said to be adding flesh and blood to this understanding, with detailed species-level data on diet, habitat, growth, sexual dimorphism, and reproduction and life history data on traits ranging from courtship displays to seasonal long-distance migrations.

The time has arrived to synthesize data and ideas from all these many fields and perspectives into a single volume. The time has arrived for: *Ecology and Evolution of Amazonian Fishes*.

Plan and Goals of this Volume

This book addresses the evolutionary origins and ecological maintenance of the megadiverse Amazonian fish fauna as part of a larger continental diversification of Neotropical freshwater fishes. The contributing authors are all experts on the ecology, evolution, and conservation of different groups of fishes. The goals are to summarize and synthesize current knowledge on the diversity of physiological and behavioral traits of fishes from Amazonia and throughout tropical America by presenting four novel and coordinated datasets of Neotropical freshwater fish diversity: a comprehensive time-calibrated phylogeny of all 6342 species valid as of December 2021, geographic distributions of all these species, an exhaustive compilation of adult modal values for 42 physiological, behavioral, and ecological traits (i.e., *ecotraits*) scored for all these species, and a synthetic chronology of the main paleogeographic events and conditions of South America over the past 100 million years (Ma). We then merge ecological and evolutionary patterns in these datasets to test several prominent theories on the ecology and evolution of the NFF fauna, the most diverse continental vertebrate fauna on Earth.

The chapters are presented in four sections. Section 1 presents the results of NFF-wide analyses. Chapter 1 introduces the major geographic conditions and environments of modern South America and the main paleogeographic events of the past 100 million years thought to have governed the accumulation of its megadiverse biotas. Chapter 2 presents a time-calibrated phylogeny for all valid NFF species as of this compilation, representing 875 genera in 96 families and 42 taxonomic orders. Chapter 3 introduces the Big Nine NFF clades, which together comprise 5670 species, or about 90% of all NFF species. Each of the Big Nine clades represents an endemic continental radiation, including three clades of the teleostean superorder Acanthopterygii (i.e., Cichlinae, Poecilioidea, and Rivulidae) and six clades of the superorder Ostariophysi (i.e., Characoidea, Curimatoidea, Gymnotiformes, Aspredinoidea, Pimelodoidea, and Loricarioidea). Chapter 4 employs historical biogeographic analyses of the Big Nine NFF clades to estimate geographic areas of origin and early diversification. Chapter 5 examines rates of net lineage diversification in the Big Nine NFF clades to evaluate the roles of ecophysiological traits and environmental conditions in generating prominent macroevolutionary and macroecological patterns. Chapter 6 presents a multivariate statistical analysis of the ecological data matrix (i.e., *ecomatrix*) of 42 ecophysiological traits pertaining to life history, habitat utilization, and diet for all known NFF species. Chapter 7 uses ancestral state estimation to trace the phylogenetic history of 14 ecotraits on the synthetic NFF phylogeny. Chapter 8 reviews the classification of Amazonian rivers by properties of soil and water chemistry and some of the main macroecological consequences for South American fishes.

The next two sections summarize the phylogenetic distributions and evolutionary transformations of fishes that occupy the major aquatic habitats (Sect. 2) and dietary (i.e., trophic; Sect. 3) phenotypes of South American fishes. Chapter 9 overviews the evolution of fishes specialized to inhabit small upland streams, Chap. 10 the fishes of blackwater floodplains (i.e., *igapós*), Chap. 11 the fishes of whitewater floodplains (i.e., *várzeas*), Chap. 12 the fishes of seasonally hypoxic waters, Chap. 13 the fishes of deep river channels, and Chap. 14 the fishes of caves and aquifers (i.e. hypogean habitats). Chapter 15 focuses on fishes that consume fruits and leaves (frugivores and folivores), Chap. 16 fishes that consume benthic algae and detritus, Chap. 17 fishes that consume other fishes (e.g., piscivores), and Chap. 18 fishes that consume ectoparasites and the fins and tails of other fishes.

Section 4 presents several synthetic analyses that draw information and ideas from all the preceding chapters. Chapter 19 examines the evolution of body-size reduction in NFF groups, presenting a new definition of miniaturization from the perspective of comparative physiology, and Chap. 20 reviews the ecology and evolution of continental-scale migrations of large river fishes. Chapter 21 provides a general synthesis of the many spatial and temporal patterns of diversification in NFF clades, combining insights from the ecospace and ecotrace analyses of the whole fauna to test the macroevolutionary hypotheses introduced in Chap. 1.

A Grand Transformation

As this volume was being prepared from 2018 to 2025, the study of Neotropical ichthyology was undergoing a grand transformation, from that of a descriptive pioneering stage of natural history and species discovery to a more mature stage of posing and testing ecological and evolutionary hypotheses. The descriptive stage continues at a breakneck pace, with hundreds of new species described during the writing of this book, establishing phenotypic and geographic limits, and elucidating ecological roles. Yet the community is also developing a more complete scientific understanding of the fauna and its many taxonomic, biogeographic, behavioral, and ecological dimensions.

Some species formerly subsumed under a single name are now being recognized as distinct (e.g., *Electrophorus* by de Santana et al. 2021; *Sternopygus* by Torgersen et al. 2023), while other species previously separated based on ontogenetic, geographic, or sexual variation are now being synonymized (e.g., *Rhamphichthys* by Carvalho and Albert 2023; *Sternarchogiton* by Bernt et al. 2023). The limits and composition of higher taxa like genera and families are being reorganized to reflect an improved understanding of phylogenetic relationships (e.g., Ochoa et al. 2020; Melo et al. 2022; Astudillo-Clavijo et al. 2023; Nogueira et al. 2023; Melo et al. 2024; Silva et al. 2024). Major patterns of biodiversity and biogeography have been documented for most clades, and we now have a PhD or other monographic taxonomic revision completed for almost all of the valid NFF genera.

This volume is part of a long-term effort by the community of Neotropical Ichthyologists to describe the Earth's greatest continental vertebrate fauna. The rapid growth of this field has been fueled by extraordinary investments of human effort, ingenuity, and perseverance by investigators at every level of professional development. The community is currently publishing more than 1000 primary research papers per year, and recent meetings attract hundreds of participants from as many as 17 countries, a majority of whom are graduate and undergraduate students (Birindelli and Sidlauskas 2018). This growth has been advanced by a series of collaborative (multi-authored) books (Fig. P3), the journal *Neotropical Ichthyology* (Fig. P4), and substantial financial investments from scientific funding agencies (e.g., CNPq and FAPs—FAPESP, FAPEMIG, FAPERGS in Brazil, NRC in Canada, and NSF in the United States).

We present this book, *Ecology and Evolution of Amazonian Fishes*, with the goal of helping advance scientific understanding of the greatest continental vertebrate fauna on Earth. The goals of this book are to summarize disparate empirical observations from across a vast and previously fragmented literature and to illuminate the ecological and evolutionary processes that generate megadiverse tropical aquatic biotas. We emphasize that the novel datasets and analyses presented are part of a long series of investigations based on the pioneering studies of Carl Eigenmann, Gerald Smith, and many others. This book therefore represents a snapshot of

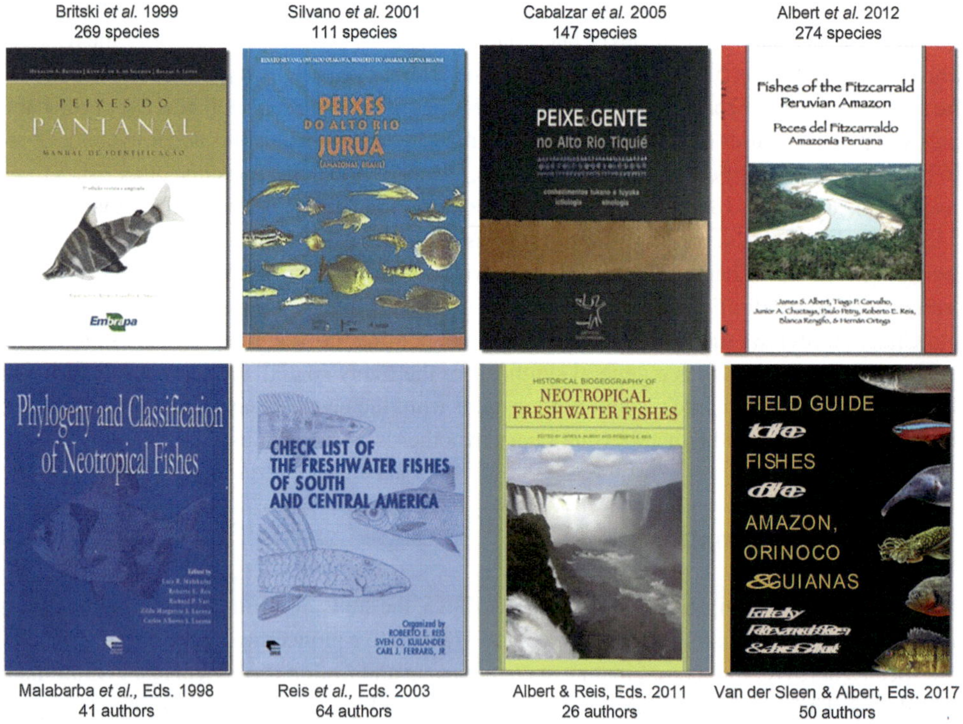

Fig. P3 Cornerstone volumes in the literature of Neotropical Freshwater Fishes. **Top row**: Regional faunas. Left to right: *Fishes of the Pantanal* reporting 269 species (Britski et al. 1999); *Fishes and People of the Upper rio Tiquié* reporting 147 species (Cabalzar et al. 2005); *Fishes of the upper rio Juruá* reporting 111 species (Silvano et al. 2001); *Fishes of the Fitzcarrald, Peruvian Amazon* reporting 274 species (Albert et al. 2012). **Bottom row: Synthetic compilations** requiring collective efforts of the whole community of Neotropical ichthyologists. Left to right: *Phylogeny and Classification of Neotropical Fishes* (Malabarba et al. Eds. 1998) with 41 contributing authors; *Check List of the Freshwater Fishes of South and Central America* (Reis et al. Eds. 2003) with 64 contributing authors; *Historical Biogeography of Neotropical Freshwater Fishes* (Albert and Reis Eds. 2011) with 26 contributing authors; *Field Guide to the Fishes of the Amazon, Orinoco, and Guianas* (Van der Sleen and Albert Eds. 2017) with 50 contributing authors

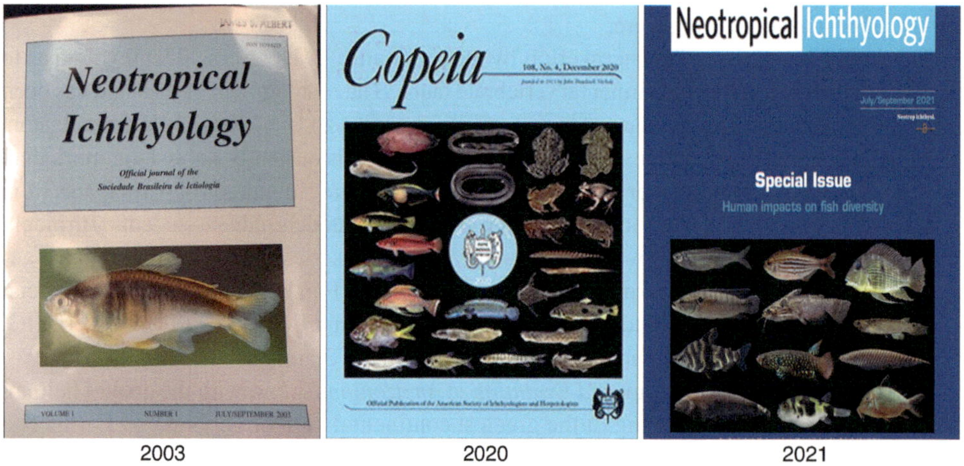

Fig. P4 Primary literature of NFF taxonomy and systematics. **Left**: Inaugural issue of *Neotropical Ichthyology*, July/Sept. 2003, illustrating *Lophiobrycon weitzmani* from the Rio Grande in Minas Gerais, Brazil. Photo by Ricardo Castro. **Center**: Final issue of *Copeia* (now *Ichthyology and Herpetology*) illustrating species published in the journal in 2020. **Right**: Special issue of *Neotropical Ichthyology* in 2021 illustrating species published in 2021. Alpha taxonomy or species delimitation is the process of identifying and formally naming species

Fig. P5 Neotropical ichthyology uses all the tools of integrative biology, merging morphological, molecular, and environmental data to study the evolution of ecological traits. (**A**) Victor Tagliacollo fishing at dusk near the mouth of the Tapajós River, Brazil. (**B**) Emmanuel Maxime (left) and Tiago Carvalho (right) collecting electric fishes from a stream in the Alto Purús River, Peru

our current understanding and is by no means to be considered the final word. There are still many gaps and errors of omission, as many new species, phylogenies, and ecotrait data have been published since the datasets for this book were compiled. Indeed, the facts and concepts presented in this volume are offered to stimulate more research. The father of one of the editors (JSA) sometimes says: "*never do something for the first time, since you do it so much better the second time*." It is in this spirit that we offer the data and analyses of this book as a contribution to ongoing and future studies on the ecology and evolution of Amazonian fishes.

Building on this rich research history and the ongoing efforts and investments of the whole research community, it is easy to state with confidence that the future looks bright indeed for the field of Neotropical Ichthyology (Fig. P5).

Lafayette, LA, USA — James S. Albert
Porto Alegre, RS, Brazil — Roberto E. Reis

References

Agrawal AA, Zhang X (2021) The evolution of coevolution in the study of species interactions. Evolution 75(7):1594–1606

Albert JS, Reis RE (2011) Introduction to Neotropical freshwaters. In: Historical biogeography of neotropical freshwater fishes, University of California Press

Albert JS, Carnaval AC, Flantua SG, Lohmann LG, Ribas CC, Riff Det al (2023) Human impacts outpace natural processes in the Amazon. Science 379(6630):eabo5003

Albert JS, Tagliacollo VA, Dagosta F (2020) Diversification of neotropical freshwater fishes. Annu Rev Ecol Evol Syst 51(1):27–53

Albert J, Hoorn C, Malhi Y, Phillips O, Encalada AC, Steege H (2021) The multiple viewpoints for the Amazon: geographic limits and meanings. In: Amazon assessment report, Science Panel For the Amazon, United Nations Sustainable Development Solutions Network, New York

Anderson EP, Jenkins CN, Heilpern S, Maldonado-Ocampo JA, Carvajal-Vallejos FM, Encalada AC et al (2018) Fragmentation of Andes-to-Amazon connectivity by hydropower dams. Sci Adv 4(1):eaao1642

Antonelli A, Ariza M, Albert JS, Andermann T, Azevedo J, Bacon C et al (2018) Conceptual and empirical advances in neotropical biodiversity research. PeerJ 6:e5644

Astudillo-Clavijo V, Stiassny ML, Ilves KL, Musilova Z, Salzburger W, López-Fernández H (2023) Exon-based phylogenomics and the relationships of African cichlid fishes: tackling the challenges of reconstructing phylogenies with repeated rapid radiations. Syst Biol 72(1):134–149

Bates HW (1863) The naturalist on the River Amazons, 2 volumes. Murray, London

Bernt MJ, Fronk AH, Evans KM, Albert JS (2023) Sexual dimorphism in the electric knifefish *Sternarchogiton labiatus* (Gymnotiformes: Apteronotidae) with a hypermorphic snout and oral jaws. Ichthyol Herpetol 111(4):621–633

Boyens P, del Toro G, Jackson P, Walsh F (2015) The hobbit: an unexpected journey. The Scripted Edit

Carvalho TP, Albert JS (2023) A taxonomic review of the Neotropical electric fish *Rhamphichthys* (Gymnotiformes: Rhamphichthyidae). Neotrop Ichthyol 21:e230012

Dagosta FC, de Pinna MC (2018) A history of the biogeography of Amazonian fishes. Neotrop Ichthyol 16
Darwin CR (1839) Voyages of the adventure and beagle. Henry Colburn, London
de Santana CD, Crampton WG, Dillman CB, Frederico RG, Sabaj MH, Covain R et al (2019) Unexpected species diversity in electric eels with a description of the strongest living bioelectricity generator. Nat Commun 10(1):1–10
Dobzhansky T (1973) Nothing in biology makes sense except in the light of evolution. Am Biol Teacher 35(3):125–129
Eigenmann CH (1924) The fishes of Western South America: part I: the fresh-water fishes of Northwestern South America, including Colombia, Panama, and the Pacific Slopes of Ecuador and Peru, together with an appendix upon the fishes of the Rio Meta in Colombia. (No. 1-4). Carnegie Museum
Eigenmann CH (1927) The American Characidae, vol 43. Carnegie Museum
Eigenmann CH, Eigenmann RS (1891) A Catalogue of the fresh-water fishes of South America. U.S. Government Printing Office
Erickson CL (2008) Amazonia: the historical ecology of a domesticated landscape. In: The handbook of South American archaeology. Springer New York, New York, NY, pp 157–183
Guayasamin JM, Ribas CC, Carnaval AC, Carrillo JD, Hoorn C, Lohmann LG et al (2024) Evolution of Amazonian biodiversity: a review. Acta Amazonica 54(spe1):e54bc21360
Harmon LJ, Andreazzi CS, Débarre F, Drury J, Goldberg EE, Martins AB et al (2019) Detecting the macroevolutionary signal of species interactions. J Evol Biol 32(8):769-782
Haseman JD (1912) Some factors of geographical distribution in South America. Ann N Y Acad Sci 22:9–112
Lenormand T (2002) Gene flow and the limits to natural selection. Trends Ecol Evol 17(4):183-189
Losos JB (2008) Phylogenetic niche conservatism, phylogenetic signal and the relationship between phylogenetic relatedness and ecological similarity among species. Ecol Lett 11(10):995-1003
Losos JB (2011) Convergence, adaptation, and constraint. Evolution 65(7):1827–1840
Lyashevska O, Farnsworth KD (2012) How many dimensions of biodiversity do we need? Ecol Indic 18:485–492
Magurran AE (2021) Measuring biological diversity. Curr Biol 31(19):R1174–R1177
Magurran AE, Deacon AE, Moyes F, Shimadzu H, Dornelas M, Phillip DA, Ramnarine IW (2018) Divergent biodiversity change within ecosystems. Proc Natl Acad Sci, 115(8):1843–1847
Matthews WJ (2012) *Patterns in freshwater fish ecology*. Springer Science & Business
Melo BF, Ota RP, Benine RC, Carvalho FR, Lima FC, Mattox GM et al (2024) Phylogenomics of characidae, a hyper-diverse neotropical freshwater fish lineage, with a phylogenetic classification including four families (Teleostei: Characiformes). Zool J Linn Soc 202(1):zlae101
Melo BF, Sidlauskas BL, Near TJ, Roxo FF, Ghezelayagh A, Ochoa LE et al (2022). Accelerated diversification explains the exceptional species richness of tropical characoid fishes. Syst Biol 71(1):78–92
Nogueira AF, Oliveira C, Langeani F, Netto-Ferreira AL (2023) Phylogenomics, evolution of trophic traits and divergence times of hemiodontid fishes (Ostariophysi: Characiformes). Mol Phylogenet Evol 186:107864
Ochoa LE, Datovo A, DoNascimiento C, Roxo FF, Sabaj MH, Chang J et al (2020) Phylogenomic analysis of trichomycterid catfishes (Teleostei: Siluriformes) inferred from ultraconserved elements. Sci Rep 10(1):2697
Rull V, Carnaval AC (eds) (2020) Neotropical diversification: patterns and processes. Springer, Berlin, pp 13–49
Seddon N, Mace GM, Naeem S, Tobias JA, Pigot AL, Cavanagh R et al (2016). Biodiversity in the anthropocene: prospects and policy. Proc R Soc B Biol Sci 283(1844):20162094
Silva GS, Rocha MS, Melo BF, Reia L, Roxo FF, Sabaj MH, Oliveira C (2024) Phylogenomics of the catfish family Pimelodidae with focus on the genus Pimelodus support the recognition of Sorubiminae and Pimelodinae (Teleostei, Siluriformes). Zool Scr 53(5):541-554
Steindachner F (1879) Beiträge zur Kenntniss der Flussfische Südamerika's, vol 1. Gerold
Su G, Villéger S, Brosse S (2019). Morphological diversity of freshwater fishes differs between realms, but morphologically extreme species are widespread. Glob Ecol Biogeogr 28(2):211–221
Tokeshi M (2009) Species coexistence: ecological and evolutionary perspectives. John Wiley & Sons
Torgersen KT, Galindo-Cuervo AM, Reis RE, Albert JS (2023) A new species of barred *Sternopygus* (Gymnotiformes: Sternopygidae) from the Orinoco River. Neotrop Ichthyol 21(1):e220088
Tyler JC, Johnson GD, Nakamura I, Collette BB (1989) Morphology of Luvarus imperialis (Luvaridae) with a phylogenetic analysis of the Acanthuroidei (Pisces). Smithson Contrib Zool 48(78):1-78
van der Sleen P, Albert JS (2022) Patterns in freshwater fish diversity. In: Reference Module in Earth Systems and Environmental Sciences 26(3), pp 894–907
Van Valen L (1976) Ecological species, multispecies, and oaks. Taxon 233–239.
von Humboldt A (1811) Voyage de Humboldt et Bonpland, vol 1. F. Schoell
Wallace AR (1854) XXIX. On the habits of the butterflies of the Amazon Valley. Trans R Ent Soc Lond 7(8):253–264
Wiens JJ (2011) The niche, biogeography and species interactions. Philos Trans R Soc B Biol Sci 366(1576):2336–2350
Winemiller KO (1991) Ecomorphological diversification in lowland freshwater fish assemblages from five biotic regions. Ecol Monogr 61(4) 343–365
Wunderling N, Staal A, Sakschewski B, Hirota M, Tuinenburg OA, Donges JF et al (2022) Recurrent droughts increase risk of cascading tipping events by outpacing adaptive capacities in the Amazon rainforest. Proc Natl Acad Sci 119(32):e2120777119

Acknowledgments

A book of this scope and detail cannot be prepared without the contributions of a large community of dedicated and collegial experts, working together over many years and decades. First and foremost, the authors are indebted to the many hundreds of local and Indigenous people from across much of tropical America with whom we have worked over many years for sharing their deep knowledge of fish habits and habitats.

For many useful conversations regarding ideas developed in this book, we are obliged to Alexandre Antonelli, Joel Cracraft, Scott Edwards, Carina Hoorn, Lucia Lohman, Camila Ribas, Victor Sacek, Hans Ter Steege, Hanna Tuomisto, and Pedro Val. For their many contributions to the data and ideas presented in this volume, we acknowledge the following friends and colleagues: João Abreu, Alberto Akama, Carolina Arantes, Maria Angeles Arce Hernández, Jairo Arroyave, Valter Azevedo-Santos, Gustavo Ballen, Ricardo Benine, Vinicius Bertaco, Luiz Bertollo, José Birindelli, Devin Bloom, Flávio Bockmann, Juan David Bogota, Marcelo Britto, Paulo Buckup, Cristina Bührnheim, Bárbara Calegari, Priscila Camelier, Fernando Carvalho, Lilian Casatti, Jorge Casciotta, Otto Castillo, Ricardo Castro, Prosanta Chakrabarty, Junior Chuctaya, Raphael Covain, Christina Cox-Fernandes, Jack Craig, Christian Cramer, Weferson da Graça, Fernando Dagosta, Alessio Datovo, Marco de Azevedo, Mario de Pinna, Luiz de Queiroz, Fabio Di Dario, Murilo Dias, Cleber Duarte, Jessica Espino, Izeni Farias, Carl Ferraris, Zé Lima Figueiredo, João Pedro Fontenelle, John Friel, Carmen Garcia-Davila, Michael Ghedotti, Julia Gioria, Max Hidalgo, Eric Hilton, Tomas Hrbek, Jean Huber, Fernando Jerep, Abreu João, Axel Katz, Philippe Keith, Lesley Kim, Matthew Kolmann, Francisco Langeani, Carlos Lasso, Oscar Lasso, Pierre-Yves Lebail, Pablo Lehmann, Flávio Lima, José Lima, Alejandro Londoño-Burbano, Hugo Lopez, Hernán López-Fernández, Nathan Lovejoy, Carlos Lucena, Margarete Lucena, Paulo Lucinda, Rafael Lugo, Claudia Malabarba, Javier Maldonado-Ocampo, Wilfredo Matamoros, Luiz Malabarba, Caleb McMahan, Vanessa Meza-Vargas, Juan Marcos Mirande, Ivan Mojica, Jan Mol, Carmen Montaña, Juan Montoya-Burgos, Cristiano Moreira, Sonia Muller, Alexandre Neto-Ferreira, Leo Nico, Luz Ochoa, Willian Ohara, Cláudio Oliveira, Rafaela Ota, Renata Ota, Osvaldo Oyakawa, Murilo Pastana, Carla Pavanelli, Luiz Peixoto, Fernando Pelicice, Edson Pereira, Paulo Petry, Nivaldo Piorski, Marc Pouilly, Lucia Py-Daniel, Robson Ramos, Vinícius Reis, Alexandre Ribeiro, Marcelo Rocha, Rocío Rodiles-Hernández, Douglas Rodríguez-Olarte, Ricardo Rosa, Fábio Roxo, Ramiro Royero, Mark Sabaj, José Sabino, Homero Sanchez, Scott Schaefer, Oscar Shibatta, Brian Sidlauskas, Gabriel Silva, Leandro Souza, Andrea Thomaz, Mônica Toledo-Piza, Giannina Trevejo, Mauro Triques, Vera Val-Almeida, Francisco Villa Navarro, Brandon Waltz, David Werneke, Phillip Willink, Stuart Willis, Wolmar Wosiacky, Angela Zanata, and Jansen Zuanon.

Finally, we express our enduring gratitude to the following colleagues and mentors for their lifelong devotion to Neotropical Ichthyology: Angelo Agostinho, Gloria Arratia, Mercedes Azpelicueta, Ramiro Barriga, Reinaldo Barthem, Jonathan Baskin, Heraldo Britski, Reeve Baily, Labbish Chao, Efrem Ferreira, Fausto Foresti, William Fink, Michael Goulding, William Gosline, Isaäc Isbrücker, Michel Jegu, Sven Kullander, Rosemary Lowe-McConnell, John Lundberg, Antonio Machado-Allison, Naércio Menezes, Robert Miller, Hernán Ortega, Francisco Provenzano, Tyson Roberts, Gerald Smith, Donald Stewart, Richard Vari, and Stanley Weitzman.

Epigraph

The world is not in your book and maps, it's out there.

— Boyens et al. (2015), inspired by J. R. R. Tolkien

Contents

Contributors

Daniel R. Akin University of Louisiana at Lafayette, Lafayette, LA, USA

James S. Albert University of Louisiana at Lafayette, Lafayette, LA, USA

Jonathan Allen University of Louisiana at Lafayette, Lafayette, LA, USA

Vera Maria F. Almeida-Val Instituto Nacional de Pesquisas da Amazônia, Manaus, AM, Brazil

Jonathan W. Armbruster Auburn University, Auburn, AL, USA

Naraiana L. Benone Universidade do Estado de Minas Gerais, Belo Horizonte, MG, Brazil

Maxwell J. Bernt Susquehanna State University, Selinsgrove, PA, USA

Maria Elina Bichuette Universidade Federal de São Carlos, São Carlos, SP, Brazil

Juan D. Bogotá-Gregory Instituto Amazónico de Investigaciones Científicas, Bogotá, Colombia

Karine O. Bonato Universidade Federal do Rio Grande do Sul, Porto Alegre, RS, Brazil

Pedro H. N. Bragança American Museum of Natural History, New York, NY, USA

Matthew R. Brander University of Louisiana at Lafayette, Lafayette, LA, USA

Susana Braz-Mota Instituto Nacional de Pesquisas da Amazônia, Manaus, AM, Brazil

Gabriel L. Brejão Universidade Estadual Paulista Rio Claro, Rio Claro, SP, Brazil

Michael D. Burns Oregon State University, Corvallis, OR, USA

Derek Campos College of Agricultural and Veterinary Sciences, São Paulo, SP, Brazil

Lorrane G. Cantanhêde Instituto Chico Mendes De Conservação, São Luiz, MA, Brazil

Tiago P. Carvalho Universidad Nacional de Colombia, Bogotá, Colombia

Lilian Casatti Universidade Estadual Paulista Rio Claro, Rio Claro, SP, Brazil

Jaqueline Custódio da Costa Instituto Nacional de Pesquisas da Amazônia, Manaus, AM, Brazil

Jack M. Craig Drexel University, Philadelphia, PA, USA

William G. R. Crampton University of Central Florida, Orlando, FL, USA

Joshua P. Egan University of Idaho, Moscow, USA

Kory M. Evans Rice University, Houston, TX, USA

Aaron H. Fronk University of Louisiana at Lafayette, Lafayette, LA, USA

Mark A. Genung University of Louisiana at Lafayette, Lafayette, LA, USA

Eric B. Haddad University of Louisiana at Lafayette, Lafayette, LA, USA

Waldir Heinrichs-Caldas Instituto Nacional de Pesquisas da Amazônia, Manaus, AM, Brazil

Jonathan M. Huie George Washington University, Washington, USA

Daiani Kochhann Acaraú Valley State University, Sobral, CE, Brazil

Matthew A. Kolmann University of Louisville, Louisville, KY, USA

Arnold Lugo-Carvajal Instituto Nacional de Pesquisas da Amazônia, Manaus, AM, Brazil

Nathan K. Lujan Royal Ontario Museum, Toronto, ON, Canada

Luiz R. Malabarba Universidade Federal do Rio Grande do Sul, Porto Alegre, RS, Brazil

Bruno F. Melo American Museum of Natural History, New York, NY, USA
UNICAMP Universidade Estadual de Campinas, Campinas, Brazil

Luciano F. A. Montag Federal University of Pará, Belém, PA, Brazil

Xavier Plasencia II University of Louisiana at Lafayette, Lafayette, LA, USA

Roberto E. Reis Pontifícia Universidade Católica Rio Grande do Sul, Porto Alegre, RS, Brazil

Peter van der Sleen Wageningen University, Wageningen, Netherlands

Bruno E. Soares University of Regina, Regina, SK, Canada

Victor A. Tagliacollo Universidade Federal do Uberlândia, Uberlândia, MG, Brazil

Milton Tan Illinois Natural History Survey, Urbana, IL, USA

Kevin T. Torgersen University of Louisiana at Lafayette, Lafayette, LA, USA

Adalberto Luis Val Instituto Nacional de Pesquisas da Amazônia, Manaus, AM, Brazil

Kirk O. Winemiller Texas A&M University, College Station, TX, USA

Jansen Zuanon Instituto Nacional de Pesquisas da Amazônia, Manaus, AM, Brazil

1 Geographic Template for the Diversification of Amazonian Fishes

James S. Albert and Roberto E. Reis

Abstract

This introductory chapter to the book summarizes the major geographic features of *Greater Amazonia* and South America, and the major paleogeographic conditions and Earth-history events, which together form the geographic template for the evolutionary origins and *diversification* of Neotropical freshwater fish (NFF) *clades*. Amazonia is a vast and geographically heterogeneous region of more than 8.4 million square kilometers of northern South America east of the Andes mountain range, and home to the greatest concentration of *biodiversity* on Earth. More than 3000 fish species inhabit Amazonian waters, exhibiting a bewildering display of phenotypic, physiological, and behavioral specializations. Amazonia also stands as the biodiversity core for the larger Neotropical Realm, with more than 6700 fish species distributed across South America, Central America, and the Greater Antilles, allotted to 875 genera, 96 families, and 42 taxonomic orders. This larger fauna represents about one-sixth (18%) of all fish biodiversity globally, and about 9% of all the vertebrate species on the Earth. *Any complete understanding of vertebrate diversity must, therefore, account for the extraordinary diversity of Amazonian fishes*. By the same token, the conditions and events that gave rise to Amazonian fishes are also implicated in the diversification of other megadiverse clades of Amazonian plants and animals.

J. S. Albert (✉)
University of Louisiana at Lafayette, Lafayette, LA, USA
e-mail: jsalbe@gmail.com

R. E. Reis
Pontifícia Universidade Católica Rio Grande do Sul, Porto Alegre, RS, Brazil
e-mail: reis@pucrs.br

1.1 Introduction

The chapter first reviews the major features of South American *geodiversity*, including the regional-scale landscape and riverscape formations, and the major geographic processes affecting fish diversity. These processes including the multiple effects of river dynamics on the ecology and evolution of aquatic species, with a focus on longitudinal stream order and river capture. The chapter, then, reviews the major features of South American fish biodiversity, including the distinctive core-periphery pattern, concluding these patterns make little or no sense without knowledge of events and conditions as they were in the past. Toward this goal, the chapter presents a syncretic chronology of the major paleogeographic, climatic, and geological conditions and events affecting the biotas of South America over the past 100 million years, compiled from numerous literature sources. Throughout this lengthy time interval, global climate changes and sea-level fluctuations strongly impacted aquatic paleoenvironments, driving marine transgressions that flooded, and marine regressions that exposed, lowland coastal and interior river basins. During this time, Amazonian rivers experienced dramatic changes in connectivity, precipitation, and water chemistry, driven largely by global climate changes and tectonic uplift, primarily on the geologically active northern and western continental margins, and erosional stream dynamics like *watershed* migration on the passive eastern continental margin. The fossil record indicates a strong macroevolutionary turnover between the Mesozoic and Cenozoic fish faunas, overall low extinction rates during the Cenozoic with few documented extinct genera, rapid phenotypic evolution in atleast some clades in the early Cenozoic with several highly-derived *extant* forms present by the Eocene, and a mostly modern fauna at the genus level by the Miocene. South American fish faunas were noticeably affected by the Eocene-Oligocene cooling event *c.* 34 Ma and associated marine regression that converted much of the sub-Andean foreland basin from shallow marine to

J. S. Albert and R. E. Reis (eds.), *Ecology and Evolution of Amazonian Fishes*, Fascinating Life Sciences,
https://doi.org/10.1007/978-3-032-09328-8_1

freshwater habitats, the formation and geological persistence of the Pebas megawetlands in the area of the western Amazon *c.* 22–10 Ma, and the dramatic drainage reorganization of northern South America *c.* 10–4.5 Ma creating the modern Amazon River and Orinoco River basins.

The last sections of the chapter introduce several interrelated macroevolutionary hypotheses based on events and conditions through geological time. Eigenmann's *Uplands Origins hypothesis*, positing ancestors during the Upper Cretaceous or early Cenozoic in upland clearwater rivers and streams of the Brazilian Shield and Guiana Shield, is contrasted with Smith's *Lowland Origins hypothesis* proposing an ancestral condition with large adult body sizes and broad geographic ranges in large lowland rivers, and a derived condition with small adult body sizes and narrow geographic ranges in small headwater rivers and streams. These sections also propose several hypotheses on the effects of the Cretaceous-Paleogene (K-Pg) asteroid impact and global mass extinction event on South American fishes. The chapter concludes by describing how the several macroevolutionary hypotheses presented will be tested and evaluated using data from three novel and taxonomically coordinated datasets presented in subsequent chapters.

1.1.1 Dimensions of Amazonian Diversity

Agassiz secured more species from a small lake in the valley of the Amazon than there are in all the fresh waters of Europe.

— Carl Eigenmann (1906, p. 17)

Greater Amazonia (hereafter, Amazonia) is a vast and heterogeneous region of tropical rainforests, wetlands, and savannas that extends over more than 8.4 million square kilometers of northern South America east of the Andes (Fig. 1.1; Albert et al. 2021b; Jézéquel et al. 2020). This is a region of geographic superlatives. The Amazon Rainforest is the largest tropical rainforest on Earth, either as measured by area, with 5.5 million km^2 or about 40% of the total rainforest area, or by volume, with about 20% of all forest biomass, sequestering about 100 billion tons (Gt) of carbon in aboveground biomass (i.e., trees and other vegetation) and another 100 Gt of carbon in its soils (Borrelli et al. 2020). South America also includes two other large tropical rainforest regions: the Chocó-Caribbean *ecozone* in the northwest covering about 70,000 km^2 separated from the Amazon Rainforest ecozone by the Northern Andes, and the Atlantic Rainforest (i.e., Mata Atlântica) covering about 1.3 million km^2 in the southeast, separated from the Amazon Rainforest by the Dry Diagonal region covering about 1.5 million km^2 of seasonally arid landscapes; i.e., Chaco, Cerrado, and Caatinga ecozones. The Amazon Rainforest constitutes the largest carbon sink and oxygen source of all terrestrial ecosystems, conducting about 20% of all global photosynthesis (Green et al. 2020). The Amazon River Basin (hereafter, Amazon Basin) is the largest river basin on Earth by area (7.1 million km^2 or about 40% of South America), with an annual freshwater discharge of about 200,000 cubic meters per second, or about 16%–20% of the Earth's total river discharge depending on the year (Richey et al. 1986). The margins of Amazon Rainforest and Amazon Basin do not precisely coincide, with the Amazon Rainforest margins extending up to 400 km north of those of the Amazon Basin margins.

Amazonian rivers, streams, lakes, and wetlands teem with fishes, in both abundance and variety. More than 3000 species inhabit these waterways, exhibiting a bewildering display of phenotypic, physiological, and behavioral specializations (Fig. 1.2). The Amazon basin is drained by tens of thousands of small rivers and streams that run under a closed forest canopy. In some lazy-flowing streams, you may find freshwater hatchet fishes (Gasteropelecidae) with long pectoral fins and an enlarged ventral keel, which are the only fishes on Earth capable of genuine powered flight (Burns 2017). The riffles, pools, and undercut banks of small rainforest streams are also home to a myriad of colorful fishes with complex courtship and mating behaviors, and many with small adult body sizes. Hundreds of Amazonian fish species grow to less than 2.6 cm (*c.* 1.0 in.) standard length and are classified as miniatures (Chap. 19). Prominent among these tiny fishes are tetras and piábas (Characidae), killifishes (Rivulidae), guppies (Poeciliidae), and armored catfishes (Loricariidae), each with a distinct set of morphological, behavioral, and ecological traits and tolerances (Weitzman and Vari 1988; Carvalho et al. 2009; Crampton 2011; Borghezan et al. 2021).

The large channels of lowland Amazonian rivers traverse broad sunlit floodplains and serve as the primary habitat for hundreds of riverine-adapted fish species (Chap. 11). Prominent among these are piranhas and pacus (Serrasalmidae), famous for biting off pieces of larger animals, but many of which also consume and disperse fruits and seeds of floodplain trees during the high-water period (Gottsberger 1978; Goulding 1980). The largest river channels flow across thousands of kilometers of the Amazon Basin serving as dispersal corridors for many large-bodied and commercially important migratory fishes. Among the most abundant migratory fishes are detritivorous characins (Curimatidae, Prochilodontidae) who feed on decaying organic matter from the river bottom, and predatory (i.e., piscivorous) Dogtooth characins (Cynodontidae) and Long-whiskered catfishes (Pimelodidae) who feed on other large river fishes. The channel fauna also includes small-bodied pencil catfishes (Trichomycteridae), some of which feed by sucking blood from gills and body surfaces of larger fishes, and perhaps occasionally from humans as well (Ohara et al. 2025).

The largest Amazonian river channels run deep, to more than 100 m (*c.* 300 feet) in some places, where they harbor a diverse and specialized fauna of benthic fishes (Chap. 13). Many deep-channel Amazonian fishes exhibit cave-like *adaptations*, like the reduction or loss of the eyes and body pigments; e.g., *Cetopsis oliveirai* (Cetopsidae), *Xyliphius sofiae* (Aspredinidae), *Orthosternarchus tamandua* (Apteronotidae).

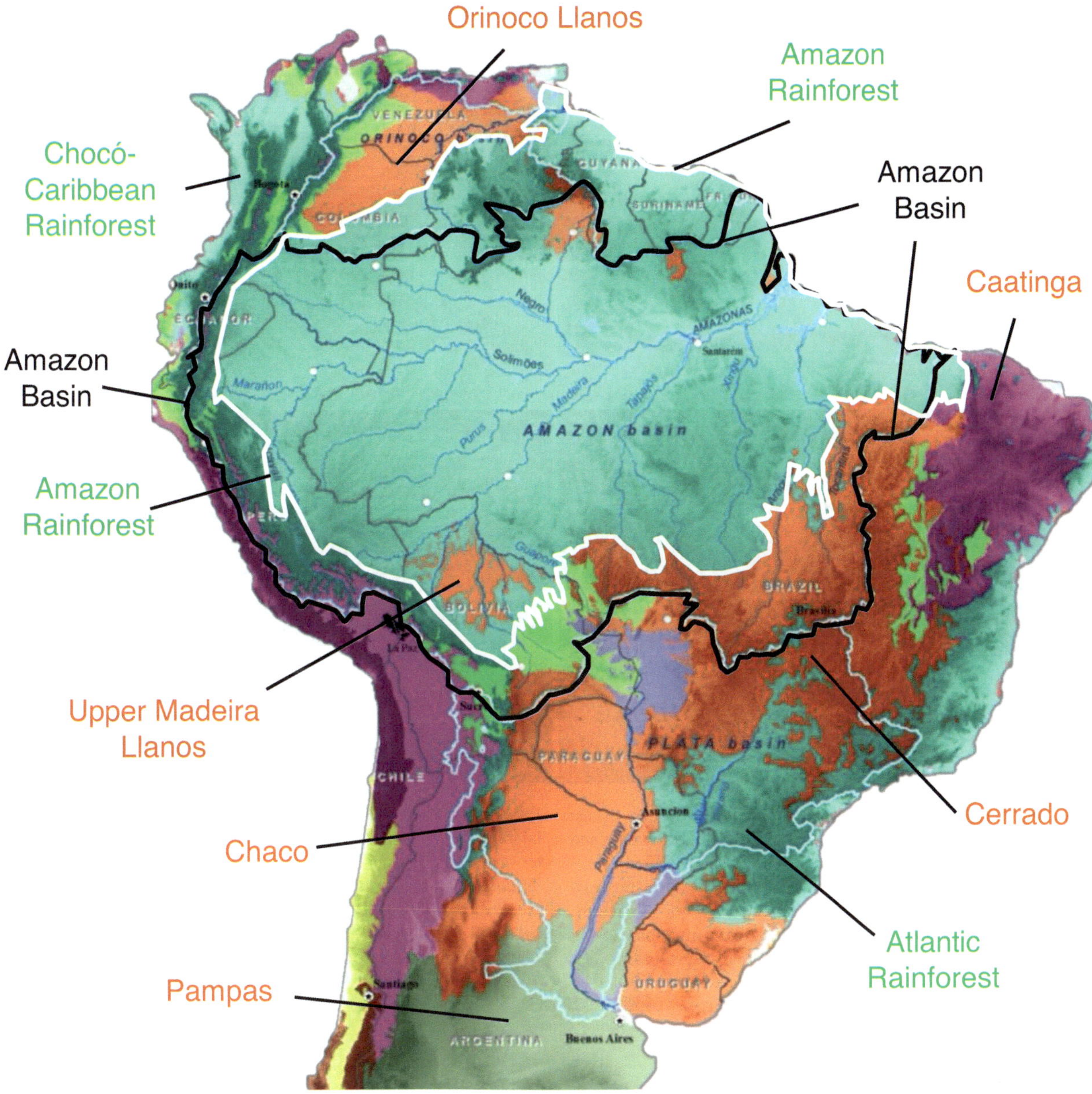

Fig. 1.1 Major ecozones of modern tropical and subtropical South America. Note three rainforest ecozones (Chocó-Caribbean, Amazon, Atlantic) in green, Dry Diagonal and other xeric ecozones (in orange), and offset margins of Amazon Rainforest (white border) and Amazon Basin (black border). Modified from Guayasamin et al. (2021)

These deep river channels also support a large biomass of Ghost electric fishes (Apteronotidae), a trophically diverse assemblage that includes midwater planktivores such as *Adontosternarchus balaenops* and *Sternarchella calhamazon*, benthic invertivores such as *Pariosternarchus amazonensis* and *Sternarchorhamphus muelleri*, and ectoparasites like *Sternarchella raptor* who nip off the tails of other deep-channel electric fishes (Evans et al. 2019; Chap. 18). The large biomass of electric fishes represents an important component of Amazonian river food webs, supporting populations of large-bodied predatory catfishes like *Brachyplatystoma filamentosum* and other commercially important food species consumed by tens of millions of people throughout the region (Lundberg et al. 1987; Fernandes et al. 2004; Goulding et al. 2019).

Amazonian fish diversity is further enriched by its geographical connections to other river basins of tropical South America, and from there to the even larger Neotropical Realm that extends from Central America to Patagonia (Albert et al. 2011a; Casemiro et al. 2023). Across this immense realm, from Mexico to Argentina, and from the high Andes to deep river channels and subterranean aquifers, Neotropical freshwaters

Fig. 1.2 Distinctive phenotypes of Neotropical freshwater fishes, illustrating species with extreme morphological, physiological, and behavioral specializations. (**A**) *Carnegiella strigata* (Marbled freshwater hatchetfish, Gasteropelecidae) with elongate pectoral fins and deep pectoral keel used for powered flight in small rainforest streams. (**B**) *Paracheirodon axelrodi* (Cardinal tetra, Acestrorhamphidae) in acidic (pH <4.0) waters of the upper Orinoco and Negro basins. (**C**) *Phreatobius cisternarum* (Heptapteridae) in artesian wells of Marajó Island. (**D**) *Tarumania walkerae* (Tarumaniidae) in rainforest leaf litter of the Rio Tarumã-Mirim, lower Rio Negro basin. (**E**) *Steatogenys duidae* (Rhamphichthyidae) from blackwater rainforest streams. (**F**) *Steatogenys elegans* (Rhamphichthyidae) from large river channels. (**G**) *Cetopsis coecutiens* (Cetopsidae), and (**H**) *Xyliphius sofiae* (Aspredinidae), both from benthos of deep (to 100 m) river channels

are home to an astonishing 6700 fish species, allotted to 875 genera, 96 families, and 42 taxonomic orders (as of this writing in July 2024). The Neotropical Freshwater Fish (NFF) fauna as a whole represents about one-sixth (18%) of all fish diversity globally, and about 9% of all the known vertebrate species that live on the Earth today (Lundberg et al. 2000; van der Sleen and Albert 2022). All these complex dietary, habitat, and life-history specializations of Amazonian fishes illustrate the fundamentally multidimensional nature of *biodiversity* (Magurran 2021; Díaz and Malhi 2022).

1.1.2 The Several Meanings of Amazonia

In this book, the term *Greater Amazonia* refers to the area of Amazon and Orinoco river basins and Guiana Shield, extending over an area of about 8.4 million km^2 (Denevan 1992; van der Sleen and Albert 2017). The term *Amazonia* has several distinct although overlapping meanings in different areas of research (Albert et al. 2021). Amazonia may refer to: (1) the Amazon River drainage basin as a hydrological unit, (2) the Amazonas sedimentary basin as a geological unit, (3) the Amazon Rainforest as a biodiversity province, and (4) Amazonas state or department, as one of several administrative or political units. The Amazon Basin is the largest drainage basin on Earth, encompassing about 7.1 million km^2, or about 40% of the South American continent. The Amazonas sedimentary basin is an intracratonic geological depression of about 500,000 km^2 that lies between the Brazilian and Guiana shields, and between the Purús Arch and Gurupá Arch, to the west and east, respectively. The Amazon Rainforest is a biogeographic province extends over about 5.5 million km^2 of northern South America constituting the largest contiguous region of moist tropical forest on Earth. The Amazon biogeographic province encompasses extensive areas of forest in the Guiana Shield, including much of the Brazilian state of Amapá, the countries French Guiana, Guyana, and Suriname with rivers that drain directly into the Atlantic Ocean, and much of forested areas of southern Venezuela and eastern Colombia that drain into the Orinoco River. The Amazon biogeographic province also includes the forested eastern flanks of the northern and central Andes from Colombia to Bolivia, seasonally-inundated floodplains and wetlands, and seasonally-burned savannahs of the Brazilian and Guiana shields.

1.1.3 Models of Biodiversity

Essentially, all models are wrong, but some are useful.

— George Box (1976)

How can we study ecology and evolution of megadiverse tropical biotas, with so many interconnected and overlapping biodiversity dimensions? As it is not possible to directly comprehend the full scope and scale of biodiversity, we must rely on simplified representations, referred to as *models*. A useful scientific model aids investigators in explaining or predicting natural systems, processes, or phenomena (Box 1976). Models in ecology and evolution help scientists understand complex and multidimensional biodiversity systems by simplifying and abstracting key conceptual components, and reducing dimensionality to highlight critical patterns (Cracraft 1974; Mooers and Heard 1997; Lawton 1999; Dodds 2009; Pietsch 2013). The most useful models provide both quantitative and qualitative insights, reveal emergent properties, predict system behaviors, validate theories, and integrate knowledge across disciplines. A good model is a powerful tool to analyze, interpret, and make decisions about complex systems that are otherwise difficult to grasp (Wiens and Morrill 2011; Adams and Collyer 2018).

Two widely used representational models in ecology and evolution are geographic *maps* and evolutionary *phylogenies*. A map is a diagram of spatially arranged data (Rigaux et al. 2002; Edler et al. 2017), while a phylogeny is a branching or tree-shaped diagram representing genealogical relationships among groups of organisms through time (Darwin 1859; Haas and Simpson 1946; Wiley and Lieberman 2011). Scientific modeling using maps and phylogenetic trees allows investigators to manage and understand the deep complexity and interrelationships present in all biodiversity datasets (von Neumann 1955; Elith et al. 2006; Elith and Leathwick 2009; Pollock et al. 2020). From earlier descriptive studies (e.g., Brown and Nicoletto 1991; Cracraft 1994) to more recent quantitative simulation analyses (e.g., Rangel et al. 2018; Hagen et al. 2021), maps allow investigators to merge data and ideas across a range of perspectives and disciplines. Some of the most important biodiversity data types summarized by maps and phylogenies come from biogeography, comparative physiology, conservation biology, ecology, ethology (i.e., animal behavior), paleontology, and population genetics.

Spatial models of the changing ecological conditions of South America, across geographic space and through geological time, serve as a dynamic *geographic template* on which to study the ecological and evolutionary processes that generate biodiversity (Lomolino 2020). The use of maps and phylogenies to model spatially and temporally structured biodiversity data has substantially advanced our understanding of Neotropical biodiversity by illuminating the principal ecological and evolutionary processes underlying the formation of continental biotas through time (Antonelli et al. 2018b; Rangel et al. 2018; McGill et al. 2019; Boschman et al. 2023; Guayasamin et al. 2024). Of course, no single map or set of maps can fully represent the multidimensional surface of the Earth, nor can one phylogeny ever tell a complete ecological or evolutionary story. To paraphrase Ecclesiastes 12:12 (Longman 1997): *There is no end to the making of maps and phylogenies, for each one describes a distinct facet of the Earth's multidimensional surface, or feature of its complex evolutionary history.*

1.1.4 Goals of This Chapter

This chapter summarizes the complex literature regarding paleogeographic conditions and events that have been hypothesized to explain the origins and diversification of the NFF fauna. This literature includes the main geographic models (i.e., *maps*) of modern and ancient landscapes used to understand the origins and diversification of other South American biotas (Lundberg et al. 1998; Albert et al. 2011a; Antonelli et al. 2009; Antonelli et al. 2018a; Hoorn et al. 2010; Rull and Carnaval 2020; Albert et al. 2020; Hoorn et al. 2022; Cassemiro et al. 2023). Subsequent chapters of this book will document the distinct evolutionary histories of different fish groups, each with a distinct phylogeny (Chaps. 2 and 3), biogeographic history (Chap. 4), rates of lineage diversification (Chap. 5), unique set of ecophysiological traits (i.e., *ecotraits*, Chap. 6), and each exhibiting distinct patterns of evolutionary diversification (Chap. 7). The central goals of this chapter for the book as a whole are to explain: (1) How ecotraits have facilitated and constrained diversification in different fish clades evolving on the same geographic template; and (2) How *evolutionary radiations* of NFF clades diversified into forming the most diverse continental vertebrate fauna on Earth.

1.2 South American Geodiversity

Non-living and living nature are inherently connected.

— Helena Tukiainen et al. (2023)

The landscapes and riverscapes of South America represent a global reservoir of geological diversity or *geodiversity*, defined as the variety of geological and physical elements of nature (Gray 2008; Tukiainen et al. 2023). These Earth system resources include geophysical features of the Earth's surface and subsurface (e.g., mountains, canyons, plateaus, rivers, wetlands, aquifers), the materials of which they are composed (e.g., minerals, sediments, fossils, soil, and water chemistry), and the geophysical processes that generate these materials (e.g., tectonics, weathering, erosion, sedimentation, pedogenesis; Quesada et al. 2011; Val et al. 2022). Furthermore, as on all the restless continents, South American geography, land, and riverscapes are forever changing (e.g., de Gamero 1996; Hoorn et al. 1995; Figueiredo et al. 2009; Montes et al. 2021; Albert et al. 2018).

1.2.1 Landscapes and Landforms

From a geological perspective, South America is readily divided into two broad regions of unequal area, age, *geodiversity*, and biodiversity: the South American Platform and the Andean Cordilleras (Fig. 1.3A). About 13 million km^2 or 75% of continental South America consists of the geologically stable continental platform, the low relief (i.e., relatively flat) and low elevation (below 1 km) portion of the South American plate composed of Proterozoic (2500–541 *Mega anum* or Ma) and Paleozoic (541–252 Ma) rocks of the cratons, with multifarious and composite origins, and their intervening intracratonic sedimentary basins (de Almeida et al. 2000; Albert et al. 2018a, 2018b). By contrast, the Andean mountain ranges (i.e., cordilleras) extend along the northwestern and western continental margins from Colombia and Venezuela in the north to southern Argentina, are composed of geologically younger rocks, mostly of Cretaceous (140–66 Ma) and Cenozoic (66–0 Ma) ages, and reaching to 4–6 km elevation. Both the continental platform and the Andean cordilleras have been sculpted by geological and climatic processes over tens to hundreds of millions of years. Several low-elevation (<500 m) watershed divides between adjacent basins are referred to as biogeographic *portals*, allowing partial or episodic hydrological and biotic connections among adjacent river basins (Fig. 1.3B).

As on all continents except Asia, South American exhibits a strong core-periphery pattern of drainage network geometry (Albert et al. 2011b). The hydrological core of South America is composed of three large river basins, the Orinoco, Amazon, La Plata basins, which drain the interior lowlands, a vast area of *c*. 17.8 million km^2 that represents a majority (63%) of the total continental surface area. By contrast, the hydrological periphery of South American lying near the continental margins is composed of numerous smaller rivers, each with a separate mouth to the sea (Fig. 1.3C, D). South America has the most pronounced core-periphery geodiversity pattern on Earth, in which the centripetal (i.e., moving toward the center) plus mainstem drainage networks of the Amazon, La Plata (i.e., Paraguay-Paraná), and Orinoco River basins occupy a larger proportion of the total continental surface area than do the centripetal networks of other continents or the centrifugal (i.e., radial or moving away from the center) networks of continental islands (Lehner et al. 2021; Ali 2018; Fernandes et al. 2019).

Some major geomorphological features affecting the biodiversity of Amazonian taxa are the Vaupés and Fitzcarrald structural arches, which separate the Western Amazon from the Orinoco and Upper Madeira sedimentary basins, respectively, and the Casiquiare, Rupununi, and Izozog portals that form semipermeable *dispersal filters* or corridors between the Negro (Amazon), Essequibo River, and Upper Paraguay River basins, respectively (Yamazaki et al. 2019; Fig. 1.4A). These geophysical surface features are based on differential uplift and erosion of landforms with widely different geological ages, ranging from Proterozoic basalts of the Brazilian and Guiana shields to Quaternary and Holocene sediments of Amazonian lowlands (Peters et al. 2018; Jones et al. 2024; Fig. 1.4B). As measured by average annual precipitation (Maeda et al. 2017; Fig. 1.4C) and precipitation runoff (Carmona et al. 2021; Fig. 1.4D), the wettest areas of the Amazon are along Andean foothills and in northwestern Amazonian of Colombia and the Negro River basin.

The principal landforms controlling patterns of river work connectivity through time in northern South America are depicted in Fig. 1.5 (Albert et al. 2018). The main features of this landscape are lithospheric swells (i.e., anticlines or subbasement rises) called *structural arches* (e.g., Purús and Vaupes Arches) of heterogenous geological origin, and lithospheric depressions (i.e., synclines) called *sedimentary basins* (e.g., Amazonas, Solimoes; Potter, 1997; Goudie, 2005).

These geomorphological features support an extraordinary heterogeneity of habitats, as measured by environmental variables related to flooding, soils, and vegetation. *Height Above Nearest Drainage* (HAND) is a good metric for local flooding because it quantifies how much higher a location is relative to the nearest stream, directly indicating its potential to flood during rising water levels (Yamazaki et al. 2019). The HAND metric allows investigators to discriminate active floodplains from nonfloodplain *terra firme* landscapes using satellite-derived data in a Digital Elevation Model (DEM; Fig. 1.6A). Soil chemistry affects, and is affected by, hydrology and vegetation cover in both terrestrial and aquatic Amazonian ecosystems (Chap. 9; Zuquim et al. 2023; Fig. 1.6B, C). Soil phosphorus P, which is a limiting nutrient in plant growth, exhibits a strong geographical gradient with the highest values in the west near the Andes to the lowest values in the most heavily weathered soils of the Brazilian and Guiana shields (Darela-Filho et al. 2024; Fig. 1.6C). The forest canopy is the highest in the far northeast and southwest of the Greater Amazonian, in the eastern Guiana Shield of Suriname and on Fitzcarrald Arch in southern Peru, where the average trees reach over 50 m (Lang et al. 2023; Fig. 1.6C, D).

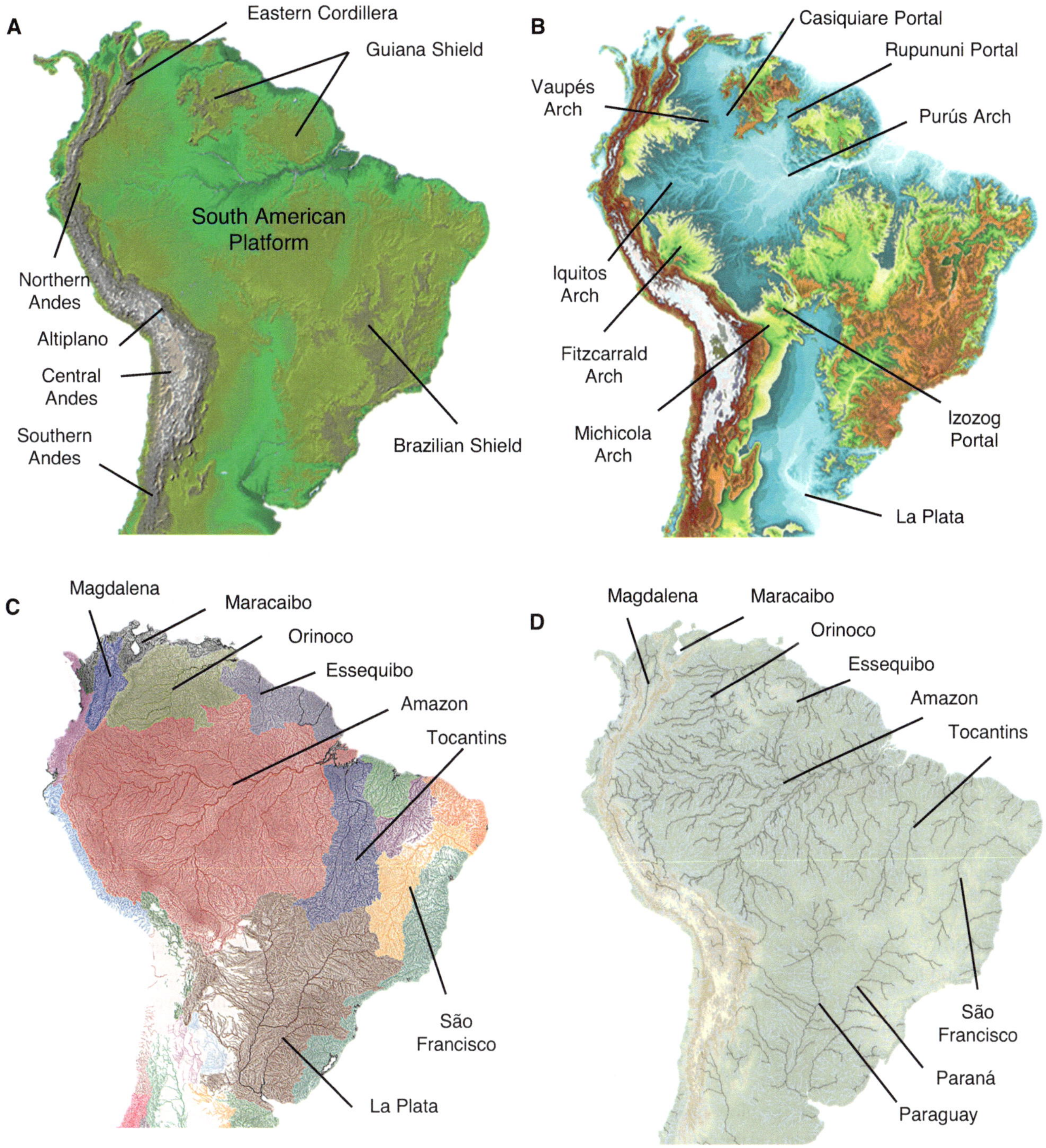

Fig. 1.3 Landforms and river basins of modern South America. (**A**) Topography colored on a linear scale, illustrating the relatively flat (i.e., low relief) continental platform and Andean Cordilleras. (**B**) False-colored elevation map emphasizing differences at lower elevations. Low-elevation (<500 m) watersheds (i.e., portals and arches) divide major river basins. (**C**) Hydrological drainage networks from PythonMaps. (**D**) Major rivers by Strahler stream order (SO): SO 4–5 blue, SO 6–10 black. Note the core-periphery pattern of river basins, with a few large basins draining most of the continental surface area, and many small river with separate mouths to the sea draining a small total area

1.2.2 Riverscapes and Stream Order

The waterways of hierarchically-branching river drainage networks exhibit high regular patterns and processes of both hydrology and ecology, with statistical regularities known as *Horton's Law* (Horton 1945; Shreve 1966) and the *River Continuum* (Vannote et al. 1980), respectively. Stream order is a system for classifying river segments that summarizes many important physical and biological attributes of continental freshwater environments (Leopold 1994; Hansen 2001;

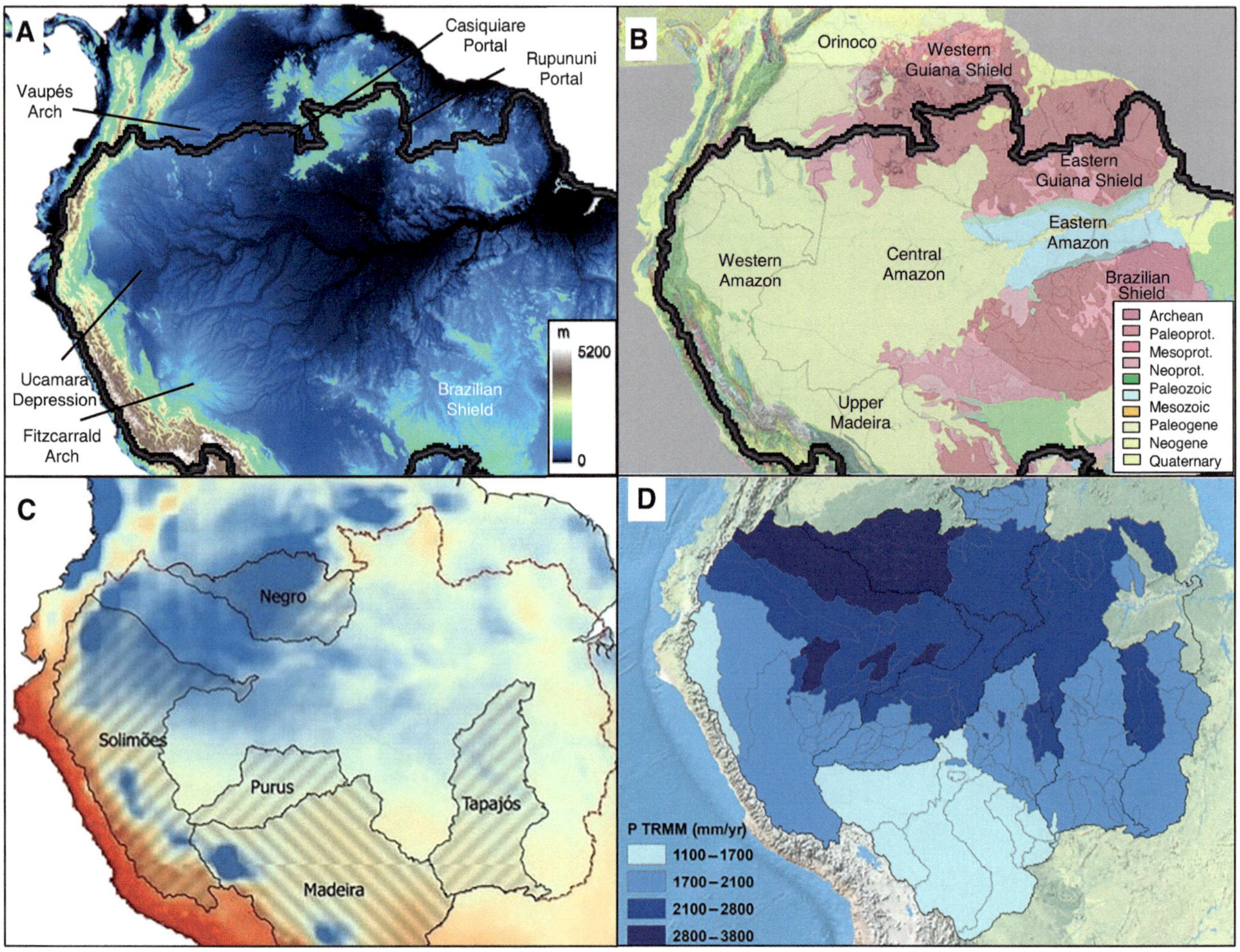

Fig. 1.4 Major geographical features of Greater Amazonia. (**A**) Elevation (Yamazaki et al. 2019). (**B**) Geological age of substrates from Macrostrat (Peters et al. 2018; Jones et al. 2024). Amazon Basin watershed in black. (**C**) Average annual precipitation (Maeda et al. 2017). (**D**) Precipitation runoff by subbasin in mm/yr (Carmona et al. 2021). *TRMM* The Tropical Rainfall Measuring Mission

Altermatt 2013). Stream order has been shown empirically to reflect the morphology of portions of a river catchment and is an important hydrographical indicator of river network structure, such as drainage density and frequency (Horton 1945; Hughes et al. 2011; McManamay and DeRolph 2019).

The Strahler classification of *stream order* treats the outermost tributaries as first-order (i.e., low-order) streams, with stream order increasing by one unit at the confluence of streams that have the same order, and the river mouth having the highest stream order (Strahler 1957; Fig. 1.7A). Like river systems worldwide, stream order in South America is highly correlated with measures of total aquatic habitat volume, local habitat heterogeneity, distance, and geographic isolation (Gorman and Karr 1978; Tarboton et al. 1988; Rodriguez-Iturbe and Rinaldo 1997; Fig. 1.7B). While headwater tributaries with Strahler orders 1–3 constitute more than 80% of the stream segments and occupy the greatest proportion of the continental landscape, large rivers with Strahler orders 6–10 represent a trivial 3% of total river-network length but account for more than half of the total river surface area, and a large majority (90%) of the total habitat volume (Fig. 1.7C). According to the Strahler stream-order system, the Lower Mississippi River below its confluence with the Missouri River is an eighth-order river, and the Lower Amazon below its confluence with the Madeira River is a tenth-order river.

The differing properties of habitats at either end of the stream-order spectrum are associated with contrasting predictions regarding life history and evolutionary ecology (Larsen et al. 2021). Headwater localities are substantially more fragmented from one another in the dendritic geometry of a river network, as compared with the same localities on the Euclidean geometry of a terrestrial habitat (Fagan 2002; Fagan et al. 2009). Riverine biodiversity scales with both ecosystem size (e.g., local habitat area or volume) and complexity (e.g., habitat heterogeneity). These regularities are referred to as the dual scaling law, in which larger and more branched (i.e., more complex) river networks support higher species richness due to increased space and environmental heterogeneity (Terui

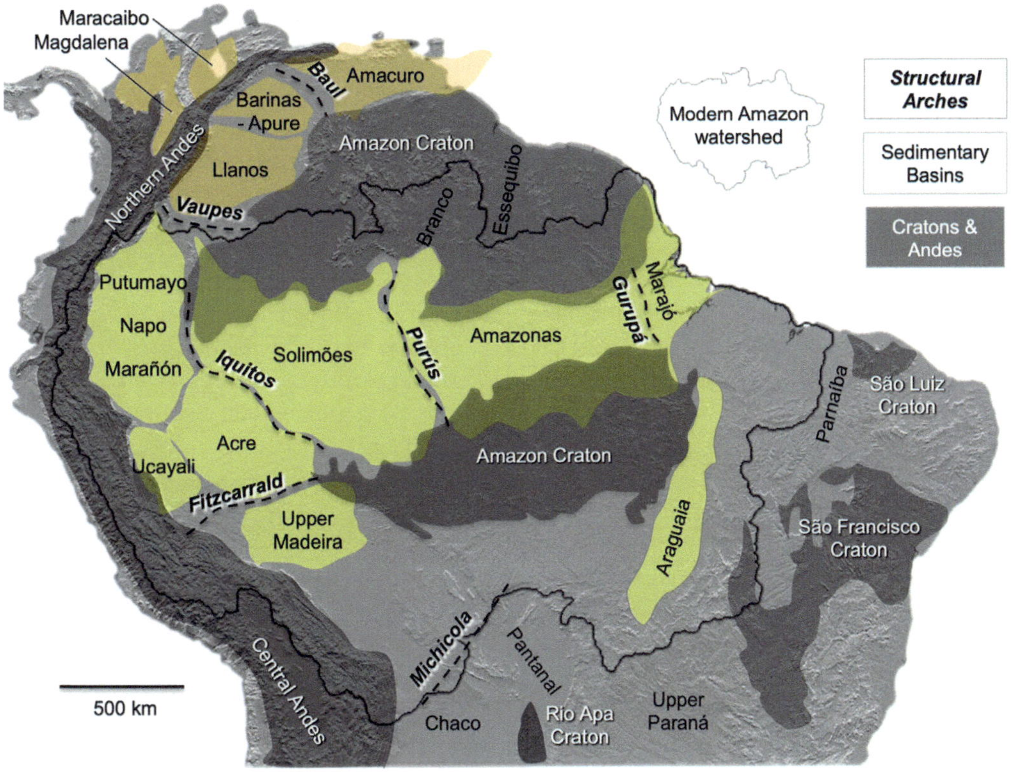

Fig. 1.5 Principal landforms controlling basin connectivity in northern South America. Structural arches (dashed lines) of diverse geological origins partially bounding sedimentary basins. Sedimentary basins draining to the Caribbean in orange, and to the Atlantic in yellow. Note that the Guiana Shield is composed mostly of the northern Amazon Craton, the Brazilian Shield by the southern Amazon Craton, and the Atlantic Shield by the São Luís, São Francisco, Luís Alves, and Río de la Plata cratons, all overlaid in parts with Paleozoic sedimentary formations. From Albert et al. (2018)

et al. 2021). Fishes adapted to small headwater tributaries with low habitat volume, separated by larger distances, and exposed to more habitat heterogeneity, are predicted to be more genetically isolated and have higher probabilities of *allopatric* speciation and stochastic extinction (Mari et al. 2014; Thomaz et al. 2016). The strategy of evolving more stenotopic (i.e., narrowly adapted) ecological requirements or life-history strategies is the evolutionary equivalent of putting all the eggs in one basket (White et al. 2013). By the same token, for fishes with more eurytopic (i.e., broadly adapted) traits, allowing them to inhabit large lowland rivers with larger population sizes and greater genetic connectivity, the prediction is for broader geographic ranges, less genetic isolation, and lower rates of speciation and extinction (Vrba 1974; 1980; Albert et al. 2011c). Fishes that inhabit small upland streams across a broad geographic range may better survive large scale but spatially localized environmental disturbances, like episodic marine incursions into the Amazonian lowlands (Lovejoy et al. 2006; Hoorn et al. 2022), or episodic savannization of the upland Brazilian Shield (Azevedo et al. 2020) (Figs. 1.8, 1.9, and 1.10).

1.2.3 River Capture

On the same continent freshwater fish often range widely, and as if capriciously; for in two adjoining river systems some of the species may be the same and some wholly different… Their dispersal may, however, be mainly attributed to changes in the level of the land within the recent period, causing rivers to flow into each other… Some freshwater fish belong to very ancient forms, and in such cases there will have been ample time for great geographical changes, and consequently time and means for much migration.

— Charles Darwin (1859: 301)

If the static geometry of the fragmented river network is an excellent substrate promoting evolutionary diversification, how much more effective is the dynamic landscape evolution of river network rearrangements (Albert et al. 2017; Val et al. 2022). Certain landscape evolution processes predictably alter the hydrological connections of adjacent river basins, separating and merging of portions of adjacent drainage networks, and the genetic connections of their resident aquatic biotas (Dias et al. 2014; Albert et al. 2017, 2018). Here, we describe the pronounced effects of two such landscape evolution processes on diversification of South American fishes: erosion-driven

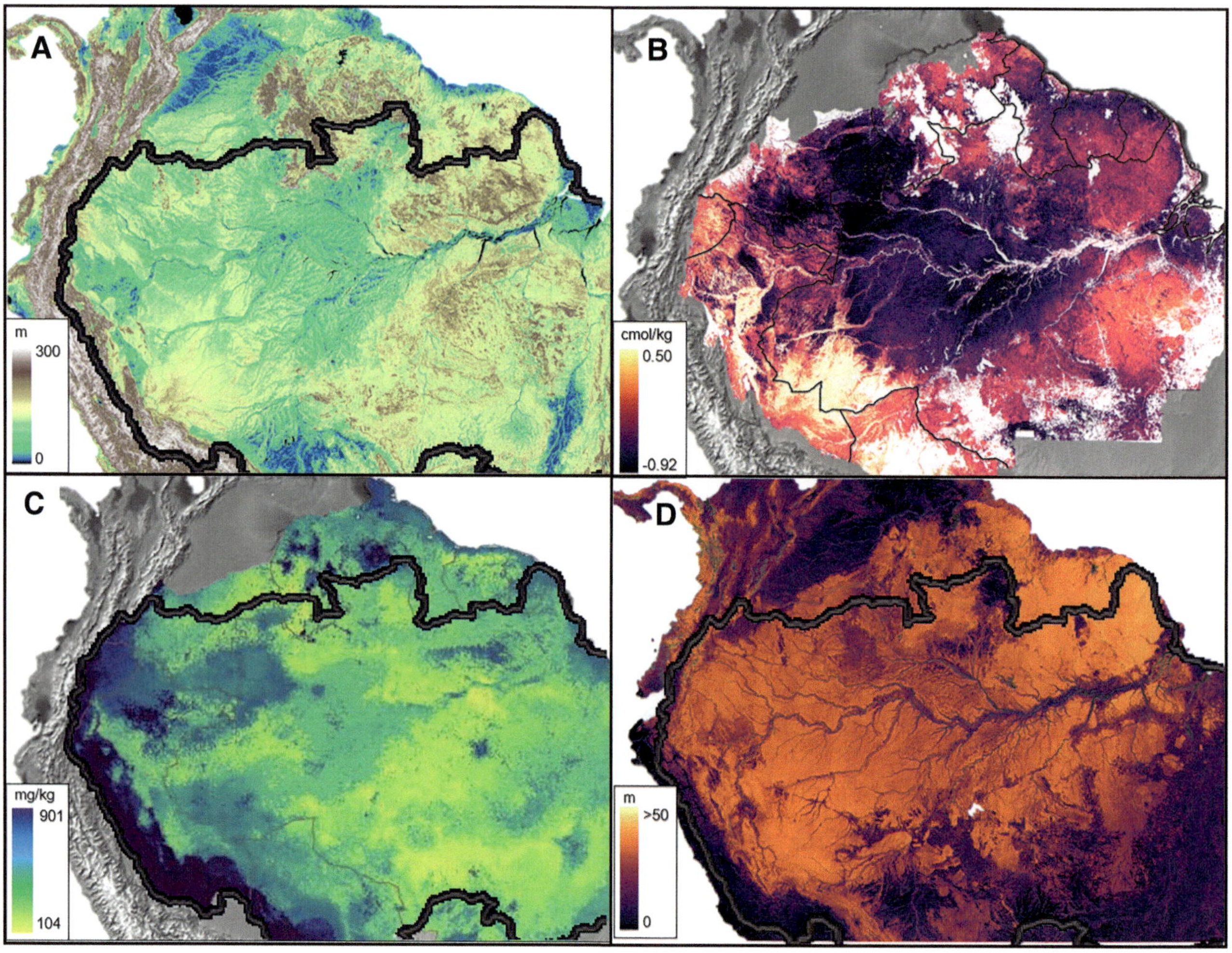

Fig. 1.6 Aspects of habitat heterogeneity across Greater Amazonia. (**A**) Local flood index. Height Above Nearest drainage (HAND; Yamazaki et al. 2019). (**B**) Forest soil basic cation concentrations (i.e., calcium $Ca^{2}+$, magnesium $Mg^{2}+$, potassium K+, and sodium Na+) are important for soil fertility and plant nutrition. Distinguished from acid cations (i.e., hydrogen H+, aluminum $Al^{3}+$; Zuquim et al. 2023). Note national borders within colored region. (**C**) Soil phosphorus P, a limiting nutrient in plant growth (Darela-Filho et al. 2024). (**D**) Forest canopy height (Lang et al. 2023). Note the black line around Amazon Basin watershed

river capture and climate-driven eustatic (global) sea-level oscillations.

River capture, sometimes referred to as stream capture or stream piracy, is a perennial geomorphological process on all ice-free continental surfaces that moves the geographic location of watershed divides (Howard 1971; Potter 1978; Bishop 1995). Understanding how river capture affects the evolutionary diversification of freshwaters fishes emerged gradually in the 20th century (Woodworth 1894; Eigenmann 1906; 1912; Ortmann 1913; Snyder 1917; Eigenmann 1921; Hubbs 1931; Hora 1937; Wheeler and Cook 1954; Tsai and Raney 1974; Hocutt 1979; Bailey and Smith 1981; Smith 1981; Mayden 1988; Skelton et al. 1994; Waters and Wallis 2000).

River capture effectively promotes diversification in freshwater taxa because it simultaneously separates and merges different portions of the same drainage network (Albert et al. 2011; Albert et al. 2017, 2018a, 2018b). These changing connections promote biotic dispersal and gene flow among some river reaches and inhibit dispersal and gene flow among others, thus inhibiting extinction and promoting speciation, respectively (Lyons et al. 2020; Stokes and Perron 2020; Val et al. 2022). River capture has been implicated in the diversification of freshwater taxa clades on all the ice-free continents (e.g., Albert and Carvalho 2011; Bloom et al. 2013; Tagliacollo et al. 2015b; Abreu et al. 2019; Stokes et al. 2023; Maia et al. 2024; He et al. 2024; Conde-Saldaña et al. 2025; Hoagstrom et al. 2025).

Three nonexclusive hypotheses have been proposed to explain the generation of high fish species richness under different conditions: the Megacapture, Intermediate, and Topographic controls hypotheses (Val et al. 2022). The *Megacapture hypothesis* predicts high species richness in

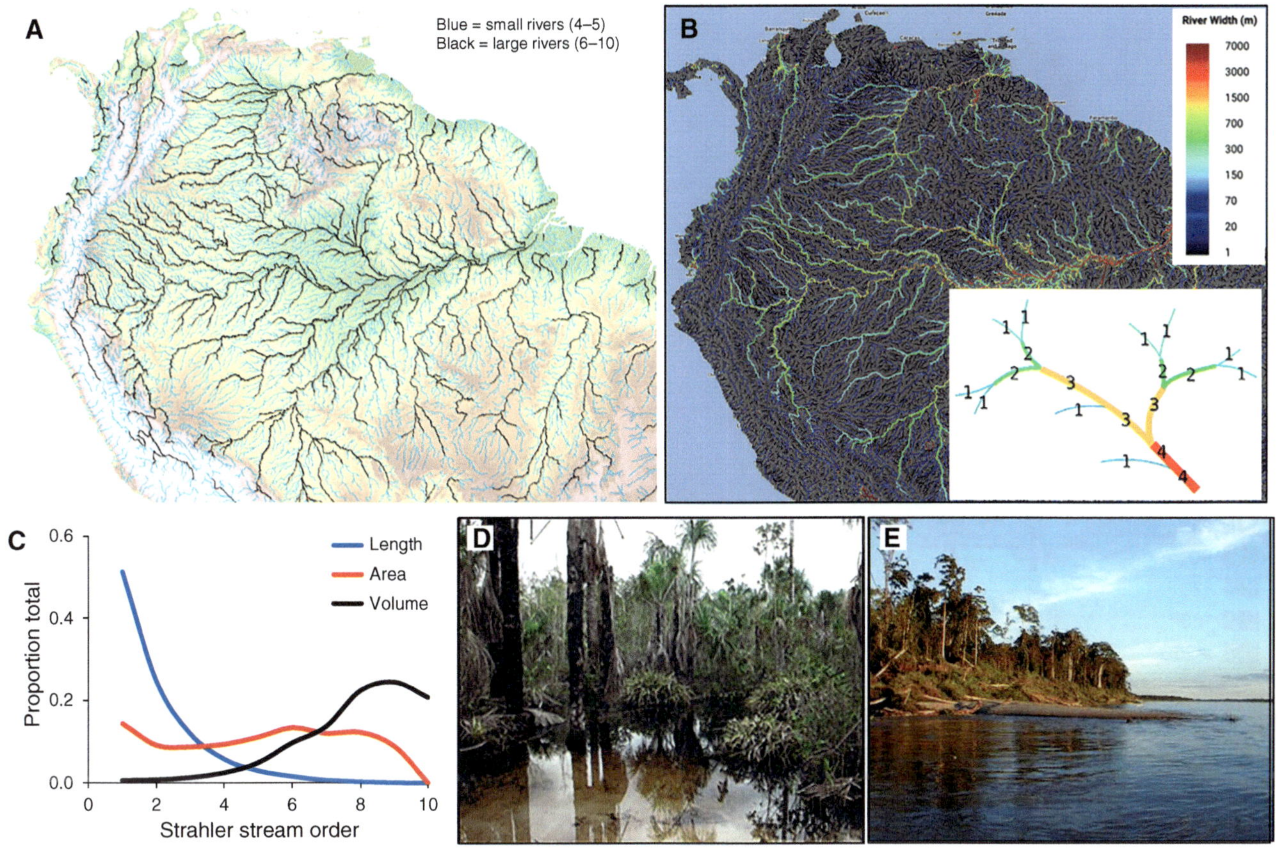

Fig. 1.7 River continuum across northern South America. (**A**) Strahler stream order: Medium-sized rivers (4–5) in blue, large rivers (6–10) in black. (**B**) River width in meters. From MERIT Hydro (Yamazaki et al. 2019). SO6 Amazonian rivers are usually *c.* 200 meters wide at low water. Inset: Schematic illustration of Strahler stream order system. (**C**) Fraction of South American riverine habitat by stream order, assessed as length, area, and volume. Calculations based on a total stream length of about 5.3 million km, a total stream surface area of about 764,000 km^2, and a total habitat volume of about 410 million km^3. Note that larger rivers (6–10) representing 3% of total river-network length account for a high proportion 53% of the total river surface area, and 90% of the total habitat volume. Data from HydroSHEDS analyzed by Jon Allen. (**D**) Small blackwater stream in the Arapiuns basin, Pará State, Brazil (SO1). (**E**) Margin of Solimões River near Alvarães, Amazonas State, Brazil (SO10; photos by Will Crampton)

riverine and riparian taxa exposed to the largest river capture events (>10,000 km^2; Tagliacollo et al. 2015; Albert et al. 2018; 2021). Among ecoregions of South America, the highest species richness is observed in lowlands (less than 250–300 m elevation) in the Western Amazon, Orinoco, and Paraguay basins that were exposed to megariver captures along the tectonically active Sub-Andean foreland within the past 20 Ma (e.g., Musher et al. 2019; Bragança and Costa 2019; Albert et al. 2021; Říčan et al. 2023b). Due to their large spatial scales, megacaptures can synchronize the vicariance and dispersal of whole faunal assemblages (e.g., Albert et al. 2022), resulting in the *coordinated dispersal* of multiple species under common geographic and environmental conditions (Ali and Huber 2010; Tagliacollo et al. 2016). The capture of the Western Amazon basin by the Eastern Amazon basin, and concomitant separation of the Western Amazon and Orinoco basins *c.* 10.0–4.5 Ma, represents the largest known megariver capture event in Earth history, transferring over 1 million km^2 from discharging north into the Caribbean to East into the Atlantic (Albert et al. 2018a, 2018b). This Earth history episode resulted in the coordinated dispersal of many taxa between the Western and Eastern Amazon basins, referred to as the *Great Amazonian Biotic Interchange*, involving numerous terrestrial and aquatic plant and animal taxa (Albert et al. 2021; Frable et al. 2022; Musher et al. 2022; Casemiro et al. 2023).

The effects of river capture on fish diversification depend strongly on geological setting (Fig. 11A). On the tectonically active western margin of South America, pulsed vicariance barriers, such as orogenic uplifts and megariver capture events predominate, while gradual watershed migration and geodispersal predominate in basins draining to the tectonically passive eastern (Atlantic) margin of South America (Griffiths 2018; Boschman et al. 2023). The opposing directions of watershed migration and geodispersal act as a perennial mechanism driving fish dispersal among river basins (Fig. 1.11B; Crelier et al. 2025). Escarpment retreat along the tectonically passive

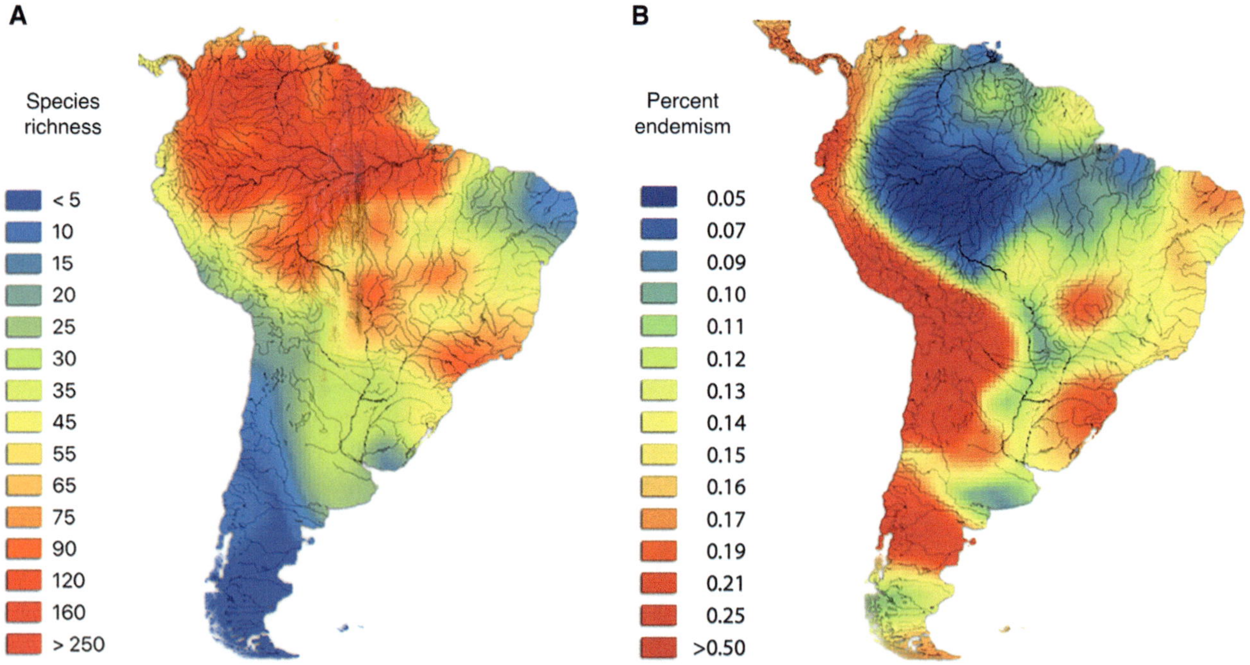

Fig. 1.8 Major biogeographic patterns of the megadiverse South American fish fauna. (**A**) Species richness: Note the highest values in the Amazon-Orinoco-Guiana (AOG) core, with secondary centers of diversity in the Pacific coast of Colombia (Choco) and southeastern Brazil. (**B**) Percent endemism: Note the highest values in the peripheral portions of the continent, on the Brazilian Shields, the high Andes, trans-Andean Pacific coast basins, and the Southern Cone. From Albert et al. (2020)

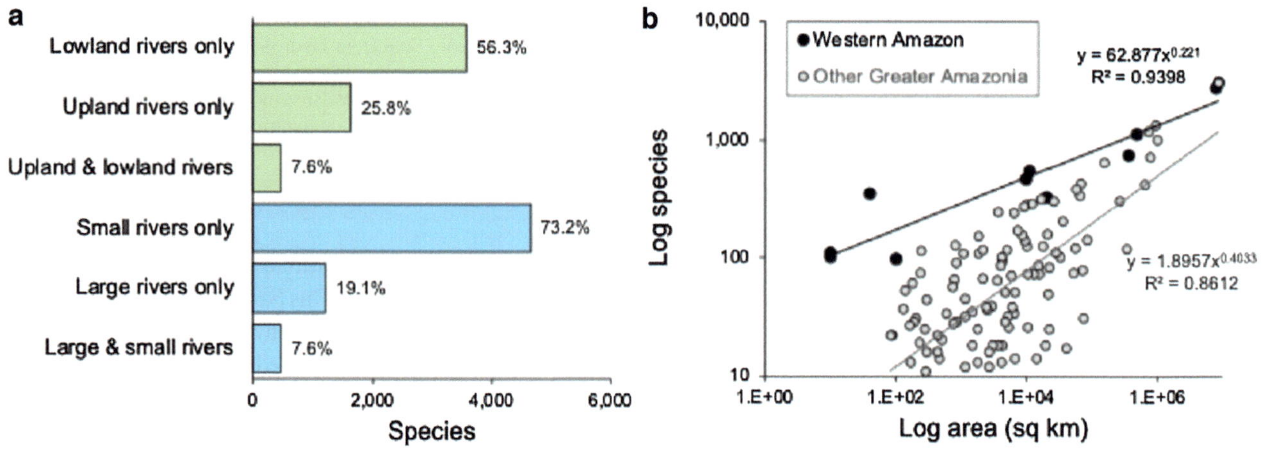

Fig. 1.9 Patterns of NFF fish species richness. (**A**) Numbers of NFF species by elevation zone (green), and river size (blue). Upland and lowland zones are divided at 250 m; small rivers and streams with Strahler stream order (SO) 1–5, large rivers SO 6–10. (**B**) Species area relationship illustrating species density (y intercept) for areas in the Western Amazon ($n = 11$) than other regions of Greater Amazonia. Data for 127 rivers and study sites. Note that most species inhabit small, lowland rivers of the Western Amazon

Atlantic margin systematically moves fishes from interior basins (e.g., La Plata, São Francisco Rivers) to coastal basins (Lima et al. 2017; Barreto et al. 2022; Mendes et al. 2022; Ramirez et al. 2023; Oliveira-Silva et al. 2024).

The *Intermediate capture hypothesis* predicts the highest species richness on landscape surfaces with "just right" rates of river captures (Albert et al. 2018). River capture occurs the fastest across the watersheds of lowland river basins with mobile sediment platforms (Albert et al. 2018a; 2018b; Ruokolainen et al. 2018; Musher et al. 2020; Val 2022; Fig. 1.12A). The diversity of NFF species is indeed the highest in lowland basins, where landscape surfaces are composed of easily-eroded

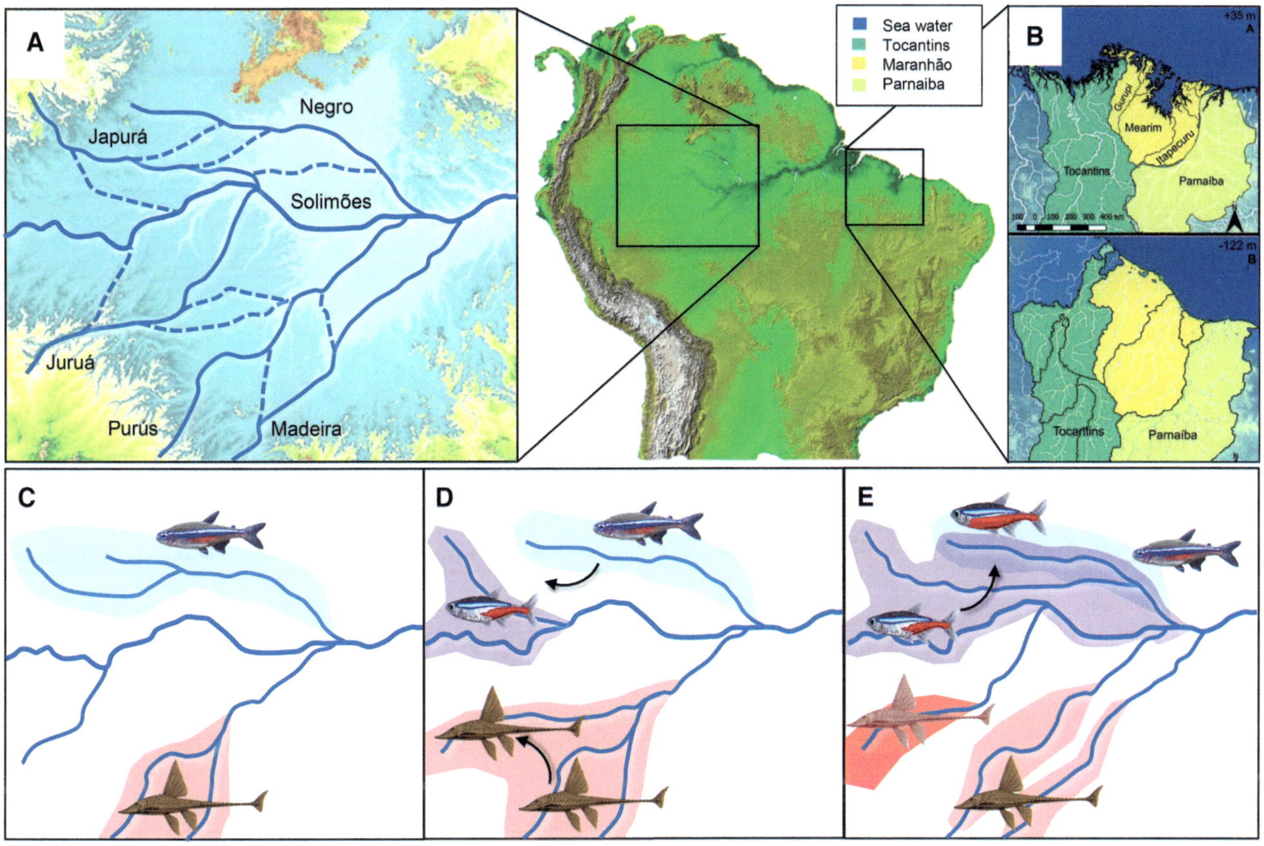

Fig. 1.10 River capture helps drive diversification in South American freshwaters. (**A**) Reticulated network of large river (stream orders 7–10) *thalwegs* (i.e., "valley ways" excluding meanders or braided channels). Note modern (solid blue) and paleo (dashed blue) thalwegs that have been reused over multiple rounds of river network rearrangements during the Plio-Pleistocene in Central Amazonia. (**B**) Plio-Pleistocene shoreline fluctuations and coastal paleodrainages of Maranhão at 35 m above modern at *c.* 5 Ma (top) and 122 m below modern at the LGM *c.* 20 Ka (Abreu et al. 2019). (**C–E**) Hypothetical species range evolution for *Paracheirodon* (Acestrorhamphidae) for north bank tributaries (ranges in blues), and *Sturisoma* (Loricariidae) for south bank tributaries (ranges in reds). Note that river capture both separates (vicariance) and connects (geodispersal) portions of river basins, potentially leading to speciation and dispersal

sediments such as mud and sand, allowing rapid river captures at the spatial scale of large rivers (i.e., stream orders 6–9; Fig. 1.12C; Val et al. 2022). By contrast, species richness is lower in basins of seismically active regions (e.g., Magdalena and Maracaibo basins) exposed to high extinction rates (Albert et al. 2006), and on seismically stable upland basins with hard substrates and low rates of river capture. The *Topographic controls hypothesis* predicts that faster rates of river capture in lowland basins than on the upland shields will result in higher species richness on lowlands and higher species endemism on uplands (Lyons et al. 2020; Val et al. 2022).

The means by which river capture affects the processes of biodiversification are similar to the actions of other earth system processes that exhibit stochastic or episodic fluctuations. These include shoreline fluctuations associated with marine transgressions and regressions (Jablonski 1980; Bloom and Lovejoy 2017) and other geographic changes associated with Pleistocene glacial–interglacial cycles; e.g., *Irion cycles* of *várzea* formation and erosion (Irion et al. 1995, 2009), and flickering connectivity of elevation zones in Andean *páramos* vegetation (Flantua et al. 2019). Pleistocene climate oscillations affect the historical connectivity of river mouths, and therefore, the genetic connections obligate riverine taxa (Dias et al. 2014; Albert et al. 2018a, 2018b). The highest sea-level in the past ten million years was +35 m above modern at *c.* 5.5 Ma, and the lowest level was –122 m at the Last Glacial Maximum (*c.* 21–25 Ka) (Spratt and Lisiecki 2016). These sea-levels correspond to 46 (max) and 16 (min) independent river mouths discharging to the sea, as compared with 35 on the modern landscape. Responses of subbasins to sea-level fluctuations depend on elevation and topography, with stronger responses in Eastern Amazon lowlands and the Maranhão coastal plain, and stronger responses when sea-levels rise than when they fall. Global climate oscillations and sea-level fluctuations over the past ten million years increased extinction rates in the Eastern Amazon Lowlands, affecting the biodiversity and biogeography of fishes in northern South America.

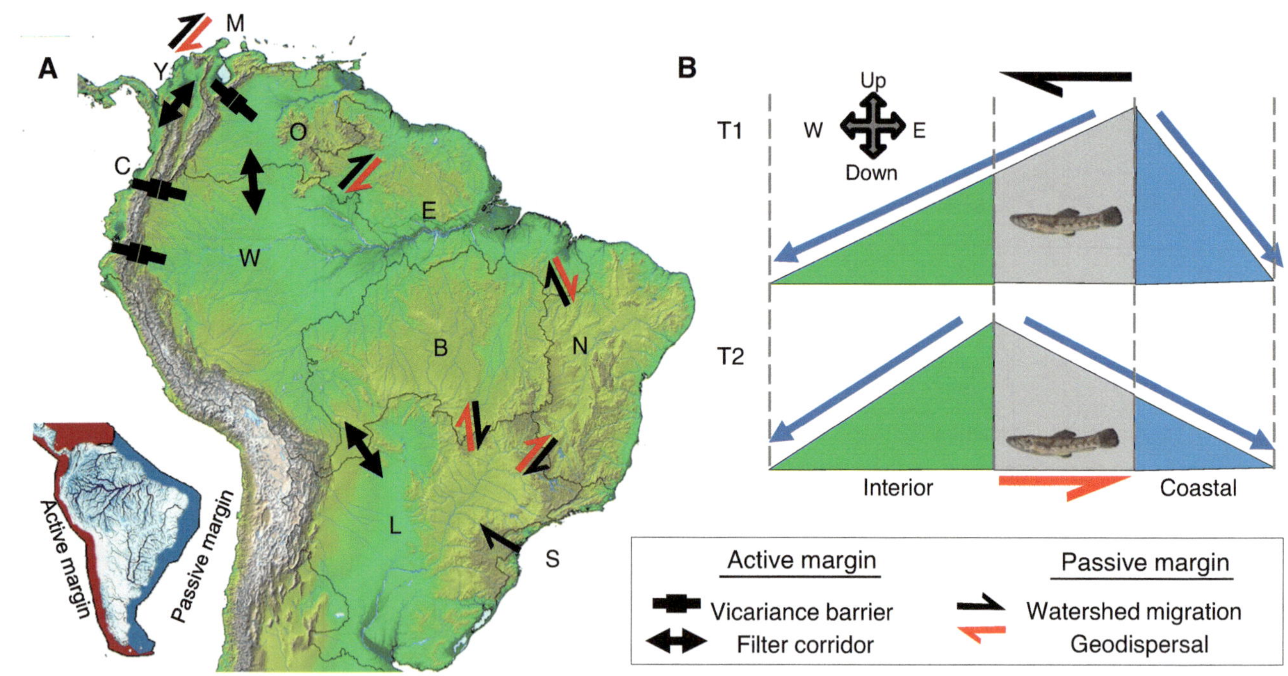

Fig. 1.11 Freshwater biogeography on tectonically active and passive continental margins. (**A**) Vicariance barriers, filter corridors, and punctuated megariver capture events predominate on the active Western margin, while gradual watershed migration and geodispersal predominate on the passive Eastern margin. Area abbreviations: C, Choco; Y, Magdalena-Sinú; M, Maracaibo; O, Orinoco; W, Western Amazon; E, Eastern Amazon-Guiana Shield; B, Amazon Brazilian Shield; N, Northeast; S, Southeast; L, La Plata. (**B**) Predicted effects of river capture on the geodispersal of obligate freshwater taxa on a passive continental margin, e.g., the L-S watershed. Long-profile schematic of rivers with blue arrows indicating direction of river flow. At T1 a headwater basin (gray) drains to an interior basin (green), and at T2 to a coastal basin (blue). Note eastward direction of watershed migration (black half-arrow a top) results in westward movement (red half-arrow) of headwater fishes from an interior to a coastal basin. Modified from Conde-Saldaña et al. (2025)

Fluctuating sea levels and shorelines directly affect the quality and quantity of habitat for both plant and animal species on coastal plains and interior continental lowlands (López-Fernández and Albert 2011; Roxo et al. 2014). Shoreline fluctuations may also affect habitats indirectly by changing regional precipitation regimes and vegetation cover (Hoorn et al. 2010; Svenning et al. 2015). For example, Pleistocene sea-level fluctuations of up to 100 m generated the periodic formation and erosion of lowland riverine floodplains; i.e., Irion cycles; see Fig. 1.10B. During interglacial periods, high sea levels backed-up and slowed the flow of lowland sediment-rich (i.e., whitewater) rivers, allowing sediments to settle out and build up to form floodplains. When sea levels fell during glacial periods, floodplains eroded as the river channel cut down to the lower base level (Irion et al. 1995; 2009). Irion cycles strongly affect the number and distribution of NFF species of Amazonian white-water (*várzea*) floodplains, one of the most productive and species-dense continental ecosystems on Earth (Junk 1984).

1.3 South American Biodiversity

A common goal is to explain and understand the origin and maintenance of these patterns, thus linking patterns and processes. Nevertheless, to achieve this goal, it is necessary to discuss first what we mean by "explain" and "understand" and what we mean by "processes" and "mechanisms".

— Alexandre Diniz-Filho (2023: 1)

1.3.1 Major Macroecological Patterns

Many prominent macroecological patterns observed in NFF taxa are similar to those of other Neotropical taxa (Albert et al. 2011b; Rull and Carnaval 2020, Diniz-Filho 2023). These including both terrestrial (e.g., birds, mammals, flowering plants, lepidopteran, and coleopteran insects) and aquatic (decapod crustaceans, opisthobranch mollusks, odonate, and ephemeropteran insects) clades. Some of the most prominent of these general biodiversity patterns are: *Tobler's first law of geography* (i.e., spatial autocorrelation) stating that near things are more related or more similar to one another than they are to more distant things (Tobler 1959); *Latitudinal diversity gradient* with higher species richness and other biodiversity metrics near the equator (Rohde 1992); *Mid-domain effect* with more species near the center of a domain like the continental platform (Colwell and Lees 2000); and *Rapaport's rule* of greater species richness in areas where species have smaller

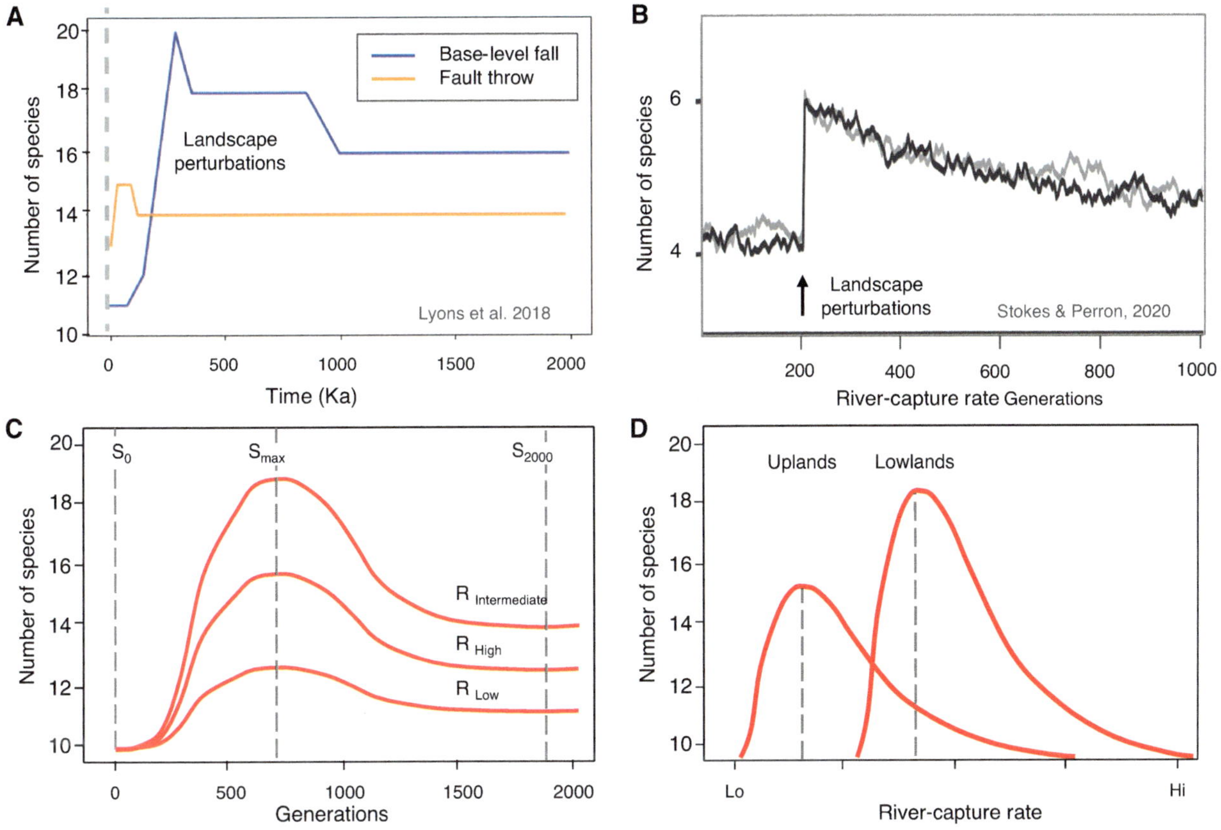

Fig. 1.12 Quantitative models of river capture driving species richness. Top panels: Numerical simulation studies showing transient increase in species richness after river capture events due to speciation and dispersal, followed by a decay in species richness due to local extirpation to an elevated species baseline higher than the number before the landscape perturbations (vertical arrows). (**A**) Evolutionary responses under two landscape perturbations initiating propagating waves of upstream knickpoint migrations and river captures; base-level fall and fault throw (Lyons et al. 2018). (**B**) Similar results obtained by Stokes and Perron, 2020. Bottom panels: Quantitative predictions of (**C**) *Intermediate capture hypothesis* (Val et al. 2022) and (**D**) *Topographic controls hypothesis* (Albert et al. 2018; Lyons et al. 2018). These results also conform to expectations of the *Ephemeral speciation hypothesis* (Rosenblum et al. 2012; Anderson and Lopez-Fernandez 2022) that most new species rapidly go extinct or are merged together via hybridization

geographic ranges (Rapoport 1975; Lomolino et al. 2017). These near-universal biodiversity patterns help reveal general mechanisms and conditions of organismal diversification; i.e., the processes that are not sensitive to taxon-specific dispersal modes or other life-history attributes, genetic background, or historically contingent Earth history events. Other biodiversity patterns vary substantially among taxa with different ecotraits or other biological properties. Such taxon-specific macroecological patterns arise from the different ways that taxa, each exposed to the same history of landscape dynamics, responds with speciation, extinction, dispersal, and adaptation.

1.3.2 Core-Periphery Biodiversity Patterns

A *core-periphery* biodiversity pattern is characterized by the contrasting patterns of local species richness (i.e., alpha diversity) and percent endemism (i.e., beta diversity) between core lowland and peripheral upland river basins (Albert et al. 2011b, 2016; Dagosta et al. 2024). Core lowland basins exhibit higher alpha and lower beta NFF diversity than those of upland basins at the continental periphery, which by contrast, have fewer species and species with smaller geographic ranges (greater percent endemism). This core-periphery biodiversity pattern roughly corresponds to the core-periphery geodiversity pattern described above (Albert et al. 2011b; Reis et al. 2016; Griffiths 2018; Albert et al. 2020; Casemiro et al. 2023; Fig. 1.8A). Similar contrasting core-periphery patterns are widely observed in Neotropically-distributed taxa, including birds, mammals, snakes, frogs, and plants (Antonelli et al. 2108; Azevedo et al. 2019; Rahbek et al. 2019a, 2019b; Vasconcelos et al. 2019; Thomson et al. 2021).

The region of South America with the highest NFF species richness is the low-relief (i.e., flat), lowland (<250 m elevation) river basins of the Amazon and Orinoco basins. There are also secondary NFF species richness hotspots in the Choco (Pacific slope of Colombia), the upper Paraguay basin (Pantanal wetlands) of central Brazil and Paraguay, and the Atlantic rainforest of southeastern Brazil. By contrast, the regions with the highest species endemism are in the high-relief uplands basins of the continental periphery, especially the Northern Andes and Pacific slope, the Atlantic rainforest of southeastern Brazil, and the Southern Cone. By contrast, the strongest predictors of NFF species richness and functional composition at the mesoscale (100–1000 m) and local microscale (less than 100 m) are environmental factors related to water volume and flow, temperature, sediment load, as proxies for autochthonous productivity (Winemiller et al. 2008; Albert et al. 2011b; Bogotá-Gregory et al. 2022).

1.3.3 Peripheral Relicts

South American fishes exhibit a curious and repeated phenomenon of taxa that can be called *peripheral relicts* (*sensu* Antonelli et al. 2018a). These taxa are low-diversity clades, often with just one or few species, that have small and isolated geographic ranges somewhere along the continental periphery. Remarkably, these clades are often an early-branching sister taxon to a highly diverse clade with broad geographic distributions (Ribeiro 2006; Albert et al. 2011b; 2020). This distinctive combination of biogeographic and phylogenetic attributes has the hallmarks of an ancient and relictual geographic distribution, indicating a long evolutionary history of isolation at the continental margin. Many peripheral relicts also often exhibit a generally plesiomorphic phenotype, thus resembling so-called "*living fossils*" (*sensu* Darwin 1859). Peripheral relicts are observed in macroevolutionary studies of many living and fossil taxa (Hennig 1966; Eldredge and Stanley 1984; Goldberg et al. 2005; Rieppel 2023), and are predicted from evolutionary theory (Brown 1957; Bellwood and Wainwright 2002; Crisp and Cook 2005; Albert et al. 2017). However, it must be noted that all *extant* (i.e., living) species are derived in some phenotypes (i.e., *mosaic evolution sensu* Stebbins 1983; Grandcolas et al. 2014). Furthermore, some NFF taxa resembling living fossils have broad geographic distributions across the continental core; e.g., *Arapaima*, *Lepidosiren*.

Regardless of their origins, many clades characterized here as peripheral relicts inhabit the upland Brazilian and Guiana shields and Atlantic coastal basins draining these uplands. These taxa include: *Lignobrycon myersi* (Triportheidae) from southeastern Brazil (Malabarba 1998), *Bivibranchia* within Hemiodontidae from the upland shields (Nogueira et al. 2023), *Conorhynchos conirostris* (Conorhynchidae) within Pimelodoidea (Sullivan et al. 2006), *Franciscodoras marmoratus* and *Wertheimeria maculata* (Doradidae) from Atlantic coastal rivers of Brazil (Arce et al. 2013; Birindelli 2014; Sabaj and Arce 2021), *Lithogenes* and *Delturus* (Loricariidae) from the Guiana Shield and Atlantic coastal rivers of Brasil (Armbruster 2004; Reis et al. 2006; Londoño-Burbano and Reis 2021), *Trichogenes* + *Copiodontinae* (de Pinna 1992; Ochoa et al. 2017) from Atlantic coastal rivers of Brasil, *Gymnotus* (*Pantherinus*) from southeastern Brazil (Craig et al. 2019), and *Retroculus* (Cichlidae) from rapids on the Brazilian and Guiana Shields (López-Fernández 2021). Other early-branching, low-diversity NFF clades are also endemic to the Southern Cone (e.g., *Diplomystes*, Diplomystidae), and trans-Andean basins of Chile (e.g., *Nematogenys inermis*, Nematogenyidae), and Colombia and Panamá (e.g., *Saccodon*, Parodontidae).

The preservation of peripheral relicts in peripheral upland river basins may reflect a systematic effect of lower river capture rates, as compared with the mobile sediments and basin watersheds of the interior lowlands (Albert et al. 2018a; 2018b; Ruokolainen et al. 2018; Musher et al. 2020; Val 2022). Most localities containing the earliest known NFF fossils are located on the continental periphery, especially on the Brazilian Shield. Some of the most famous early fossil NFF taxa are from the Upper Cretaceous Santana Formation (†*Santanichthys diasi*) of Ceará state in northeastern Brazil, the late Paleocene-early Eocene Maiz Gordo Formation of Salta, Argentina (†*Hoplisoma revelatus*), and the Eocene-Oligoene Tremembé Formation of São Paulo State in southeastern Brazil (†*Gymnogeophagus eocenicus*). All these fossil-bearing formations are located in the geologically and hydrologically stable sedimentary basins of the upland Brazilian Shield, and the location of the fossils mat, therefore, be taphonomic artefact reflecting better preservation and recovery than sediments of more mobile sedimentary basins of the interior lowlands (Lundberg et al. 1998; López-Fernández and Albert 2011).

1.3.4 Distinctive Fish Patterns

Other major diversity patterns differ qualitatively between terrestrial and aquatic taxa. These patterns illuminate how phenotypes associated with feeding, habitat utilization, dispersal and sexual signaling affect gene flow and population structure, and ultimately the macroevolutionary processes of speciation and extinction (Albert et al. 2018a, 2018b; Fig. 3). Whereas many terrestrial taxa exhibit a midelevation peak in species richness (birds, mammals, flowering plants), species richness for freshwater fishes is greatest at low elevations (see Hoorn et al. 2018). Other prominent differences between terrestrial and aquatic taxa include the roles of large rivers as semipermeable barriers or corridors for dispersal and gene flow (Musher et al. 2019), the effects of Euclidean vs. dendritic geometry on landscape connectivity and population genetic isolation (Smith et al. 2010; Dias et al. 2014), and the position of diversity peaks along *elevational gradients* (Antonelli et al. 2018a; Hoorn et al. 2018).

The larger NFF fauna is composed of several regionally defined faunas, each with somewhat distinct although overlapping taxonomic compositions. Three of these regional faunas have ancient origins during the Paleogene or Upper Cretaceous (Tagliacollo et al. 2015a; 2015b): (1) Amazonian Lowlands and Pacific slope of northern South America, including the Amazon, Orinoco, and Essequibo basins and Pacific basins of Colombia and Ecuador; (2) Amazon Craton, including upland basins (>250 m) on the Brazilian and Guiana shields and their Amazonian and Atlantic coastal drainages; and (3) Southern Cone, including Pacific basins of Chile and Atlantic basins of Argentina in which the diversity was severely reduced by late Cenozoic climatic cooling. The NFF fauna also includes at least two distinct Neogene faunas; (1) an upland fauna of the Northern Andes (Colombia to Bolivia above 500 m); and (2) a mixed upland and lowland fauna of the Caribbean Plate, including nuclear and southern Middle America and the Greater Antilles. Examples of taxa exhibiting these distinct biogeographic patterns in northern South America are provided by Dagosta and de Pinna (2019).

1.3.5 Limits of Macroecology

Macroecology is the study of relationships among organisms and modern environments at large spatial scales, focusing, for example, on correlations among climatic and landscape variables with metabolic scaling, extinction risk and diversity gradients (Diniz-Filho 2023). Despite many successes, macroecological studies have fallen short in explaining some of the most important biodiversity patterns at continental scales (Rangel et al. 2018). Several prominent NFF biodiversity patterns are poorly predicted by modern environmental conditions (Burns and Sidlauskas 2019; López-Fernández 2021; Soares et al. 2023) and can only be understood from studies of lineage diversification and adaptation in an explicit phylogenetic context (Burns et al. 2024; Kolmann et al. 2024; Valente-Neto et al. 2024).

Among the most fascinating of these mismatches is the unexpected location of the NFF center of species richness in the lowlands (below *c.* 250 m elevation) of Western Amazonia (Albert et al. 2011b; Oberdorff et al. 2019; Casemiro et al. 2023; Fig. 1.8A). At the subbasin-scale, NFF species richness is the highest in the Amazon mainstem and lowland reaches of its major whitewater tributaries, from the confluence of the Marañón and Ucayali rivers in Peru, to the confluence of Solimões (Amazon) and Japurá rivers in Brazil, and including the lowland portions of the Caquetá-Japurá, Putumayo, Napo, Amazonas, Javari, and Japurá rivers (Oberdorff et al. 2019; Cassemiro et al. 2023). This area was recognized as the Western Amazonian Endemic Area for cichlid fishes (Kullander 1986).

At the local scale, fish species richness exceeds 100 species on whitewater river floodplains of the Pacaya river in Peru (Correa et al. 2008) with a total of 320 species in whitewater floodplains of the Pacaya-Samiria National Park (*c.* 20,000 sq km; Crampton et al. 2008), and 734 species in the whole of the Ucayali basin (Chuctaya et al. 2022). Among the hottest of the hotspots for freshwater fish globally are all in the Western and Central Amazon, including 344 fish species in just a 40 sq. km in the vicinity of Letícia, Colombia (Bertaco 2006), 541 species in *c.* 11,000 sq km in the Mamirauá Reserve, Brazil (Hercos et al. 2021), and 458 fish species in the Yasuní National Park (*c.* 9800 sq km) of Ecuador (Escobar-Camacho et al. 2025).

Yet based on macroecological principles, the western Amazon is not predicted to be the center of NFF species richness by any of the prominent models of biodiversity (Oberdorff et al. 2019). In the context of macroevolution, a neutral model assumes that individual species and geographic regions do not differ functionally with respect to rates of speciation, extinction or dispersal (Pennell et al. 2014; Worm and Tittensor 2018). The species-area relationship (SAR) predicts greatest species richness in the Amazonas sedimentary basin, the largest in South America (Albert et al. 2011b), and the species-habitat-volume relationship that predicts higher species richness in larger downstream rivers (Griffiths 2018). Similarly, the unexpectedly high fish species density observed in the Orinoco basin is also poorly matched by macroecological factors like basin area or aquatic habitat volume (Albert et al. 2011b; 2020). The *mid-domain effect* is another neutral model, which predicts the center of diversity where species geographic ranges have the greatest overlap (Colwell and Lees 2000). Yet, the geographic center of South America, lies in the Upper Paraguay (La Plata) basin near the city of Cuiabá in Mato Grosso state. This subbasin has a moderately high, but not highest, fish species richness, despite being located near headwaters of other large lowlands rivers; Aripuana (Madeira basin), and Xingu (Amazon basin).

NFF species richness patterns also do not match predictions of some prominent ecological theories. For example, the *river continuum concept* predicts fish species richness to peak in downstream portions of large river systems, based on physical factors along the river axis, such as altitude, temperature, stream order, and channel width (Vannote et al. 1980). This pattern is observed in most river systems worldwide (Poff and Allan 1995; Matthews 1998). Other important environmental parameters, like precipitation, temperature, and hydrological connectivity, all of which help predict fish species diversity in North America (Smith et al. 2010) have less predictive power in South America. For example, the freshwater ecoregions of the Central Amazon have more hydrological (pertaining to rivers) connections and pedological (pertaining to soils) similarities to the ecoregions of adjacent biogeographic provinces than they do to the ecoregions of the Western or Eastern Amazon (Abell et al. 2008; Tuomisto et al. 2019). Furthermore, precipitation and temperature do not vary sufficiently across the lowland Amazon and Orinoco basins to predict the observed patterns of fish species richness (Almazroui et al. 2021).

Although many NFF functional traits are strongly correlated with environmental variables at the microscale (1–10 m), there is little to no correlation of these traits at macroscales above 1000 m (Bracciali et al. 2016; Craig et al. 2020). This is because the turnover in species composition among habitats (i.e., beta diversity) is greater than the spatial grain of the habitat heterogeneity (Oberdorff et al. 2019; Doretto et al. 2020; Gomes et al. 2023). Among fishes across biomes and continents, trophic level and food web patterns are not predicted by body size, due to the diverse ecological strategies employed by different fish groups with similar body sizes (Keppeler et al. 2020).

1.4 South American Paleogeography

The Americas were torn away from Europe and Africa… by earthquakes and floods… The vestiges of the rupture reveal themselves, if someone brings forward a map of the world and considers carefully the coasts of the three continents (Translated from Latin).

— Abraham Ortelius (1596)

1.4.1 The Past Is the Key

Geologists have long understood that conditions of the Earth observed around us today accumulated from landscape evolution processes operating over immense time intervals (Lyell 1830; Spalding 2002). By the same token, many dimensions of biodiversity only fully make sense in the light of events and conditions through deep time, an idea aptly summarized with the phrase: "*The past is the key to the present*" (Jablonski and Sepkoski 1996; Blundell and Scott 1998). Just as the Earth's surface was sculpted by tectonic and erosional processes through a lengthy history, so too continental biotas became assembled through the cumulative actions of numerous ecological and evolutionary processes, all operating in concert over many millions of years, and often over a continental arena encompassing millions of square kilometers (Tagliacollo et al. 2017; Albert et al. 2020; Busch et al. 2023). Many patterns of NFF biodiversity make little or no sense without knowledge of these events and conditions as they were in the deep past (Fig. 1.13). A synthetic chronology of major climatic and geological events affecting South American for the past 120 million years was compiled from numerous sources (Jaramillo et al. 2019; Azevedo et al. 2020; Kirschner and Hoorn 2020; Martínez et al. 2020; Antoine et al. 2021; Scotese 2021; Scotese et al. 2021; de la Parra et al. 2022; Table 1.1).

Global climatic and sea-level fluctuations strongly impacted aquatic paleoenvironments, driving *marine transgressions* (MT in pink) and *marine regressions* (MR in blue; Fig. 1.14) Other important factors were drainage shifts of Amazonian rivers

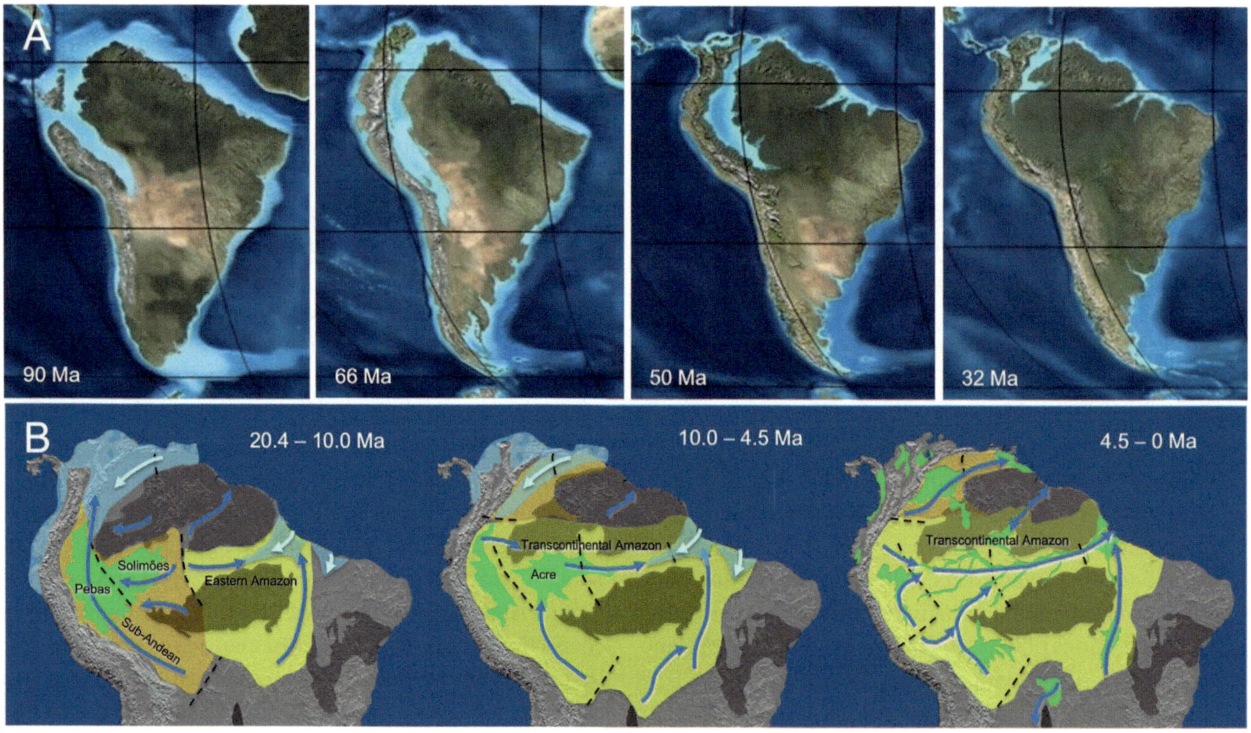

Fig. 1.13 Paleogeographic reconstructions of South America. (**A**) Upper Cretaceous and Paleogene (90–32 Ma). Images from Blakey (2016). (**B**) Neogene (20.4–2.6 Ma). Images from Albert et al. (2018). Left: Early and Middle Miocene (20.4–9.0 Ma). North-trending Sub-Andean river basin orange; east-trending Eastern Amazon river basin yellow. Middle: Late Miocene (9.0–5.3 Ma). East-trending transcontinental Amazon River basin. Right: Middle Pliocene–*Recent* (*c.* 4.5–0 Ma). Landforms and notations as in Fig. 2.2. Note marine seaway in Sub-Andean Foreland from 9–20 Ma, and freshwater megawetlands from 20–0 Ma, including modern seasonally flooded *Llanos* of the Orinoco and Upper Madeira, and floodplains of the Amazon, Negro, and Tocantins river basins

Table 1.1 Syncretic chronology of major paleogeographic and climatic events ($n = 20$) affecting the evolutionary histories of South American fishes

Epoch	Ma 1	Events 2	Drivers	References
Upper Cretaceous	110–100	Final separation South America & Africa	Tectonic breakup of Western Gondwana	Müller et al. 2018
Upper Cretaceous	100–66	Regression of epicontinental seas; UK-Pg river drainage networks (3)	Global cooling and eustatic sea level fall	Graham 2011; Hurtado et al. 2018
K-Pg boundary	66	K-Pg mass extinction	Chicxulub asteroid impact, Deccan Traps	Longrich et al. 2011; 2012; Vajda and Bercovici 2014; Sibert and Norris 2015
Paleocene	64–56	Modern rainforest floras & fish faunas (4)	Multiple biotic & abiotic factors	López-Fernández and Albert 2011
Paleocene	64–56	Megathermal forests to southern Patagonia	Multiple biotic & abiotic factors	Burnham and Johnson 2004; Wing et al. 2009
Early Oligocene	34	Separation Amazon and Atlantic biotas, formation Dry Diagonal (5)	Global cooling, contraction of rainforests to tropical latitudes	Strömberg 2011; Werneck et al. 2012; Thode et al. 2019
Oligocene	32–23	Shoreline fluctuations in SAF and coastal basins	Global climate and sea-level oscillations	Liebrand et al. 2017; De Lira Mota et al. 2023; Buffan et al. 2025
Oligocene	32–23	Mega-river captures in Sub-Andean foreland	Uplift of Central Andes, formation of Central Andean (Bolivian) Orocline	Garzione et al. 2008; 2017; Tagliacollo et al. 2015
Miocene	22–12	Pebas mega-wetland system	Global warming, high eustatic sea-levels, Andean tectonics	Hoorn 2010
Middle Miocene	17–14	Expansion of C4 grasses & mammalian grazers	Middle Miocene climatic optimum	Carrapa et al. 2019
Late Miocene	12–8	Separation cis- and trans-Andean lowland biotas	Uplift northern Andes; Eastern Cordillera of Colombia, Merida Andes	Albert et al. 2006; Horton et al. 2020
Late Miocene	12–8	Mega-river captures and formation transcontinental Amazon River	Uplift Vaupez Arch, overfilling Western Amazon sedimentary basins	Figueiredo et al. 2009; Hoorn et al. 2017; Jaramillo et al. 2017; Albert et al. 2018; 2021
Late Miocene– Early Pliocene	10– 4.5	Desertification at continental periphery (6)	Major uplifts Central and Northern Andes, Neogene cooling	da Silva et al. 2018
Late Miocene	10	Onset South American Monsoon	Major uplifts Central and Northern Andes, Neogene cooling	Poulsen et al. 2010
Pliocene	4.5	Increase discharge sediment & water at Amazon mouth	Uplifts Central & Northern Andes, Mega-river captures in Sub-Andean foreland	Hoorn et al. 2017; van Soelen et al. 2017
Pliocene	5–3	Separation Western Amazon & Upper Madeira	Uplift Fitzcarrald Arch, Subduction Nazca Ridge	Espurt et al. 2007
Late Miocene–Pliocene	12–3	Great American Biotic Interchange	Uplift Panama Isthmus	Bacon et al. 2015; O'Dea et al. 2016; Carrillo et al. 2020
Late Miocene– Pliocene	7–4	Alpine tundra (Páramo)	Major uplifts Northern Andes	Madriñán et al. 2013
Pliocene-Quaternary	5.3–0	Megafaunal extinctions	Global climate oscillations, i.e., ice age cycles	Barnosky et al. 2016; Carrillo et al. 2020
Pliocene-Quaternary	5.3–0	Fluctuating landscape & habitat connections (7)	Climate fluctuations, river capture	Albert et al. 2017; Flantua et al. 2019; Musher et al. 2020
Holocene	<0.01	Anthropogenic habitat alteration, deforestation & defaunation	Global climate change, regional development/ urbanization	Malhi et al. 2008; Roosevelt 2013; Sandom et al. 2014; Boivin et al. 2016; Bar-On 2018; Lovejoy and Nobre 2019

Notes: 1. Origins and notable tectonic, climatic, paleogeographic and biotic formations and events. 2. Dates approximate & rounded to Ma except in Quaternary. 3. Paleogene drainage networks: Northern sub-Andean (proto-Orinoco-Amazon) to Caribbean/Pacific, Southern sub-Andean (La Plata) to North Atlantic, Eastern (intracratonic) to South Atlantic. 4. Composed of taxonomically modern genera & physiognomically recognizable lowland rainforest Neotropical taxa. Taxonomically modern aquatic angiosperms known from Middle to Late Cretaceous. 5. Tropical savanna and temperate grasslands, including Caatinga, Cerrado, and Chaco. 6. Expansion Dry Diagonal, desertification in Patagonia, High Andes, Pacific slope, Maracaibo. 7. Driving dispersal, speciation and extinction among forest-savannah, rivers-interfluves, and elevation zones.

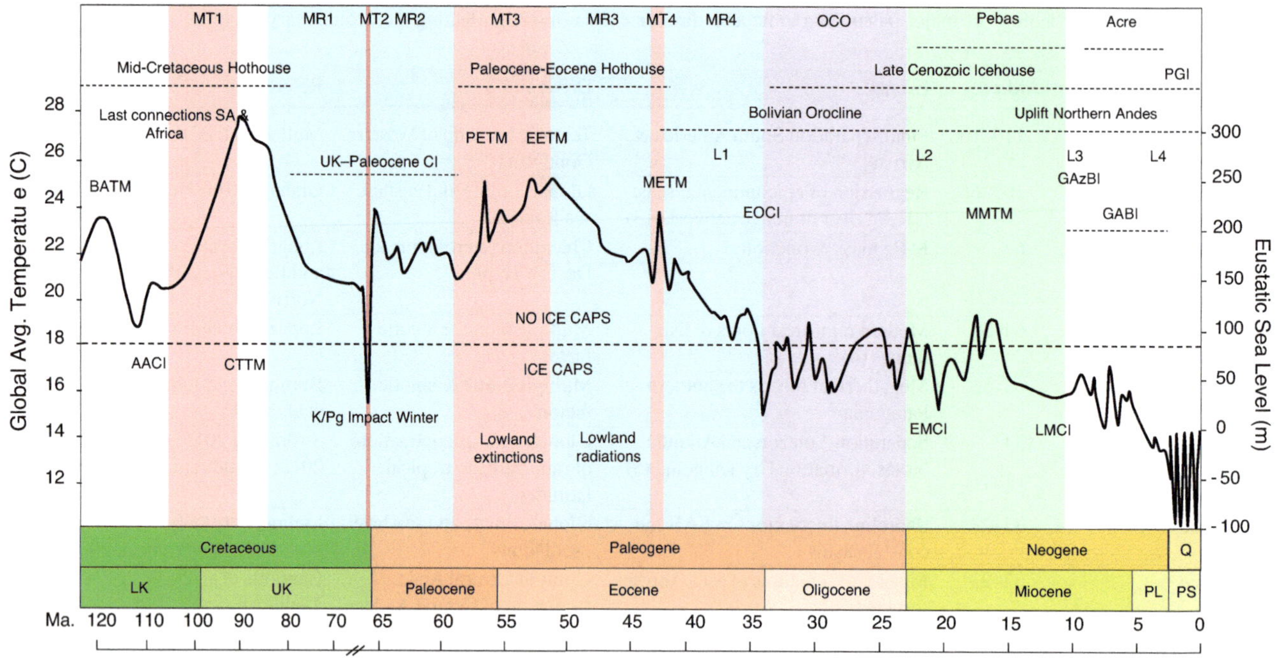

Fig. 1.14 Timeline of major paleogeographic events in South America. Modified from Scotese (2021) and Scotese et al. (2021). Acronyms: CI Cooling Interval; GABI Great American Biotic Interchange; GAzBI Great Amazonian Biotic Interchange; L1-L4 La Plata biogeographic events (Chap. 4); MR Marine Regression; MT Marine Transgression; PGI Pleistocene Global Ice age; TM Thermal Maximum

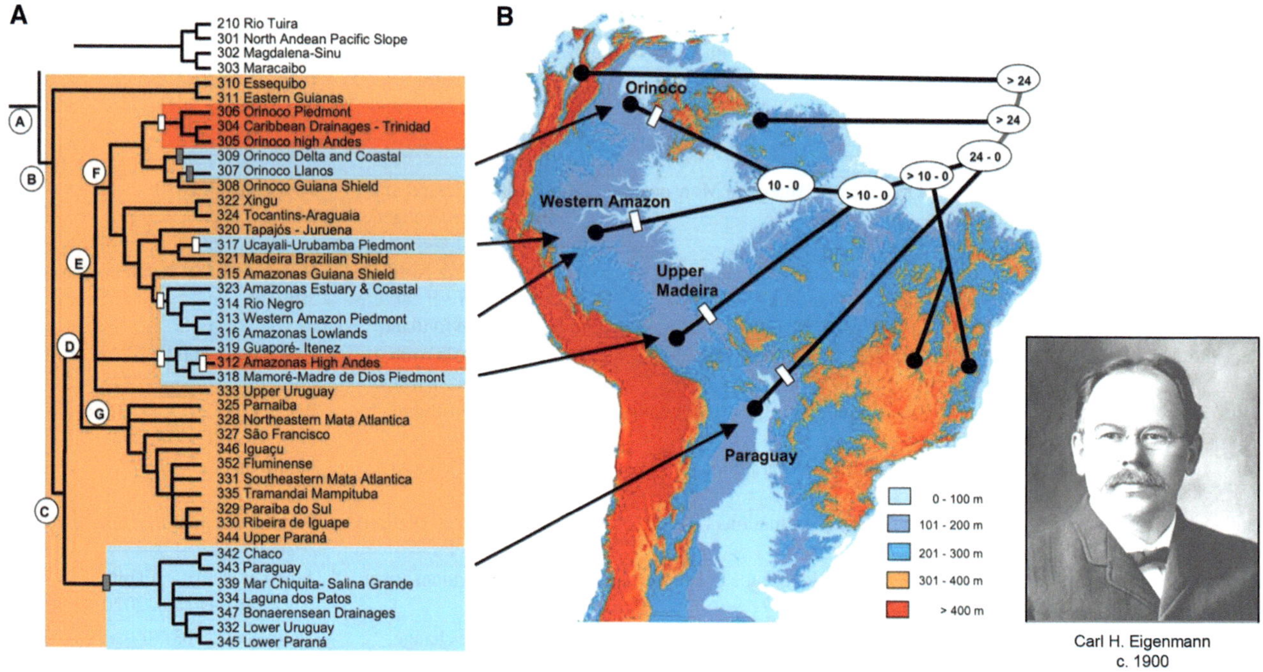

Fig. 1.15 Eigenmann's *Uplands Origins* hypothesis. Early (Upper Cretaceous–Early Paleogene; *c.* 100–34 Ma) origins of NFF clades in upland clearwater rivers and streams, and later (Oligocene–Neogene; *c.* 34–1.8 Ma) radiations in lowland whitewater rivers (Eigenmann 1906; 1912). (**A**) General area cladogram of freshwater ecoregions of South America colored by elevation. Root optimized to uplands with polyphyletic dispersal and radiations of clades in lowland basins (Albert and Carvalho 2011: fig. 7.3). Data for 333 species in 33 genera. Under this model, the primitive condition is smaller adult body sizes inhabiting smaller rivers with smaller geographic ranges. The derived condition is to have a larger body size, inhabit larger rivers, and have fewer species with larger geographic ranges. (**B**) Backbone of the phylogeny in A superimposed on a map of part of South America, with elevation bands colored in blues for lowlands and orange & red for uplands

during the Cretaceous and early Paleocene driven by plate tectonics that opened the equatorial Atlantic Ocean and uplifted the Andean cordilleras (Hurtado et al. 2018). There was a strong macroevolutionary turnover between the Mesozoic and Cenozoic fish faunas, in part from the K-Pg mass extinction (Freidman 2022), whereas NFF faunas were already mostly modern at the genus level by the Miocene (Lundberg et al. 1998; Ballen et al. 2019). The fossil record of NFF fishes across the Upper Cretaceous-Paleogene boundary indicates rapid replacement of Mesozoic taxa after the catastrophic K-Pg event (Brito et al. 2008; Martin-Abad and Poyato-Ariza 2013; Malabarba and Di Dario, 2017; Parmera et al. 2017). Low extinction rates are indicated by the few documented extinct NFF genera (e.g., †*Megapiranha* Cione et al. 2009; †*Acregoliath* Ballen et al. 2019), and presence of several highly-derived *extant* forms by the Eocene (e.g., †*Hoplisoma revelatus*, †*Gymnogeophagus eocenicus*).

The formation and persistence of the *Pebas megawetland* system is another unique paleogeographic event with prominent influence on the formation of NFF faunas (green area at left in Fig. 1.13). This megawetland was an immense lacustro-fluvial complex extending up to one million km^2 across much of what is now the western Amazon and southern Orinoco basins, and which persisted in different configurations for about 12 million years, from 22 to 10 Ma (Wesselingh et al. 2001; Wesselingh 2006; Hoorn et al. 2010; Jaramillo et al. 2017; Hoorn et al. 2022). The Pebas megawetlands deposits, up to about one km thick, gray-colored muddy sediments, are known as the Pebas Formation in Peru and the Solimões Formation in Brazil (Zapata et al. 2023).

The timing of the Pebas megawetlands coincides with the origins of several *marine-derived clades* (MDCs), *or peripheral freshwater fishes*, that invaded South American freshwaters during the Middle Miocene Thermal Optimum (MMTO), when marine waters extended deep into the lower Sub-Andean foreland basin. These clades include South American freshwater stingrays (Potamotrygonidae), anchovies (Clupeidae), needlefishes (Belonidae), drums (Sciaenidae), pufferfishes (Tetraodontidae), and flatfishes (Achiridae; Bloom et al. 2011; 2017; Bitencourt et al. 2023), as well as other freshwater clades of late Cenozoic marine origins; e.g., mangroves (Hoorn 2006; Boonstra et al. 2015; Hoorn et al. 2022), bivalve mollusks (Wesselingh, 2007), ostracod crustaceans (Wesselingh and Ramos 2010, Alvim et al. 2023), and sirenid mammals (de Souza et al. 2021).

1.4.2 Uplands Origins Hypothesis

Based on available taxonomic and geographic data, two alternative macroevolutionary hypotheses have been proposed for the evolution of freshwater fishes. Eigenmann's *Uplands Origins hypothesis* posits NFF ancestors during the early Tertiary (i.e., Paleogene) in upland clearwater rivers and streams of the Guiana and Brazilian Shields (Eigenmann, 1906; 1912; Fig. 1.15). Most modern fishes of these rivers have small adult body sizes, a generalized benthic invertebrate diet, and relatively small geographic ranges (Box 1.1).

Box 1.1: Eigenmann's Uplands Origins hypothesis

Inspired by the then-new ideas of former geological connections among the southern continents (von Ihering 1900; Suess 1901), Carl H. Eigenmann, invoked the *Archelenis-Archiguiana theory* for the origins and diversification of South American freshwater fishes. This theory was a product of evolutionary thinking during the Victorian era, based on concepts of "dispersal from centers of origin" and "adaptive radiations" (Osborn 1902; Matthew 1915). An abstract of this theory based on paleogeographic events is presented below from quotes in Eigenmann (1912), lightly edited for clarity:

1. Origins in Western Gondwana. "There was a land mass between Africa and South America—possibly in contact with Guiana in South America and some point in tropical Africa... inhabited, among others, by Lepidosirenidae, Poeciliidae, Characidae, Cichlidae, and Siluridae. This land mass sank beneath the surface of the ocean, forcing the fauna in two directions, toward Africa and towards South America, exterminating all types not moved to the east or the west. From these two rudiments have developed the present diverse faunas of Africa and South America, each reinforced by intrusives from the ocean and by autochthonous development... The connection between Africa and South America existed before the origin of present genera and even before the origin of some of the present subfamilies and families, some time before the earlier Tertiary." Note Archelenis refers to the "land mass between Africa and South America."

2. Dispersal to Brazilian Shield. "The East Brazilian land mass south of the Amazon (Archamazona) must have become stocked from the western end of Helenis (Archiguiana) very early, for it contains many genera peculiar to the region, indicating a long separation, and Tertiary freshwater deposits in this area contain existing genera of freshwater fishes." Note Archamazona refers to the Brazilian Shield and Archiguiana to the Guiana Shield.

3. Marine-derived clades. "The Pacific slope fauna is a more recent acquisition… the Cordilleras arose out of the ocean… their developing streams and arms of the sea, connected with brackish, and later freshwater lakes, all became populated with marine types from the surrounding sea." Note these freshwater fish groups of predominantly marine families with late Cenozoic origins in continental freshwaters are also known as Peripheral freshwater fishes (Myers 1966) or marine-derived clades (MDLs; Lovejoy et al. 2006; Albert et al. 2011b).

4. Amazon adaptive radiations. "With the further elevation of the Cordilleras into a continuous barrier and the formation of the Orinoco, Amazon, and La Plata valleys through elevation and the debris brought from the land masses, and the development of the enormous freshwater system occupying these valleys, this system, particularly the Amazon, became colonized from the older land areas and became the center of unparalleled *adaptive radiation* and a new center for distribution, which it has remained to the present time.

5. Dispersal to the periphery. "From the Amazon species moved in all directions till they met barriers of one sort or another. The Pacific slope fauna is derived to a very large extent from this latter divergent migration over the isthmus of Panama and through the Atrato valley between the western and coast Cordilleras of Colombia. Others possibly crossed over the Andes east of Guayaquil before the Andes reached their present height… Amazonian types moved south till climate and barriers checked them south of Buenos Aires. They migrated northward till they came in competition with emigrants from the north in the lowlands of Mexico… This fauna is in part of marine origin and antedates, in a large measure, the mongrel fauna of the Mexican lowlands."

Under Eigenmann's *Archelenis-Archiguiana theory*, here, referred to as the *Uplands Origins* hypothesis, the diverse faunas and derived ecophysiological traits of fishes in large lowland whitewater rivers (e.g., Amazon, Orinoco) evolved later, in the Sub-Andean Foreland basin during the Oligocene to early Neogene (34–10 Ma). Traits associated with large rivers include larger body sizes, specialized piscivorous, planktivorous and detritivorous diets, and large-scale (> 100 km) migratory behaviors.◄

1.4.3 Lowland Origins Hypothesis

In a seminal study on the evolution of North American Suckers (Family Catostomidae), Gerald Smith (1992) proposed a macroevolutionary model for freshwater fishes in which the ancestral condition has a large adult body size and a broad geographic range in large lowland rivers, and the derived condition is to have small adult body size, and a narrow geographic range in small headwater rivers and streams (Smith 1992; Box 1.2; Fig. 1.16). Smith's *Lowland Origins* hypothesis has subsequently been proposed for some Neotropical fishes, including riverine-adapted Ghost electric fishes (Apteronotidae; Albert et al. 2021) and Toothless characins (Curimatidae; Melo et al. 2021), and in the paleotropics in cyprinin minnows (Cyprininae; Yang et al. 2021). Generally, most freshwater fishes on earth are thought to have originated in large lowland river basins or coastal estuarine habitats (Burr and Mayden 1992; Chen et al. 2013; Ross and Matthews 2014; Tao et al. 2019; Lavoué 2020; Cheng et al. 2022; Chen et al. 2023; van der Sleen and Albert 2022; Melo and Stiassny 2024).

Box 1.2: Smith's Lowland Origins hypothesis

Merging data from phylogeny, paleontology, biogeography, morphology, ecology, and population genetics, Gerald R. Smith advanced an integrated evolutionary theory for the diversification of North American suckers (Catostomidae; Smith 1992). This theory posited origins and early diversification in large lowland rivers, followed by radiations of some subclades in headwater tributaries.

"As noted for other North American fluvial fish groups (Wiley and Mayden, 1985) vicariance by drainage allopatry is the dominant mode of speciation in catostomids. Most sympatric species pairs represent more ancient clades and probably came into being via dispersal following speciation in allopatry. Many examples of parallel evolution of small size, small fins, scales, and vertebrae, ventral scraping jaws, and associated opercular and hyoid changes, increased taste organs on the lips (and associated changes in the nervous system), and reduced swim bladders, provide clues to the selective forces in sucker evolution. These features are responses to conditions in cool, intermediate gradient headwater habitats. The opposite characteristics in large-river species are apparently retained from ancestors and seem to have evolved slowly as a result of the size and connectedness of the large river habitat (pp 812–813)."

Smith's hypothesis is framed using modern evolutionary concepts and principles such as geographic range fragmentation, genetic isolation, and habitat specialization, in contrast to the somewhat archaic language used by Eigenmann. Smith's hypothesis included the effects of prominent paleogeographic events and conditions (e.g., "mid-Cenozoic marine embayment, which isolated the rivers flowing into the Mississippi valley as far north as southern Illinois" (p 807)), and data from fossil taxa which improved the dating and resolution of the tree and also helped polarize trait state; "The fossil record is consistent with the conclusion that the large deep-bodied species are basal because they are found earlier in time."◄

1.4.4 K-Pg Hypotheses

One of the most important *stochastic* (i.e., statistically random) events to occur in the Neotropical region was the Chicxulub asteroid impact in the northern Yucatán peninsula of Mexico (Schulte et al. 2010; Chiarenza et al. 2020; Fig. 1.17). The impact event itself, and the climatic and geological consequences, produced a geologically-instantaneous global mass extinction marking the end of the Cretaceous (K), and start of the Paleogene (Pg), geological periods (Richards et al. 2015).

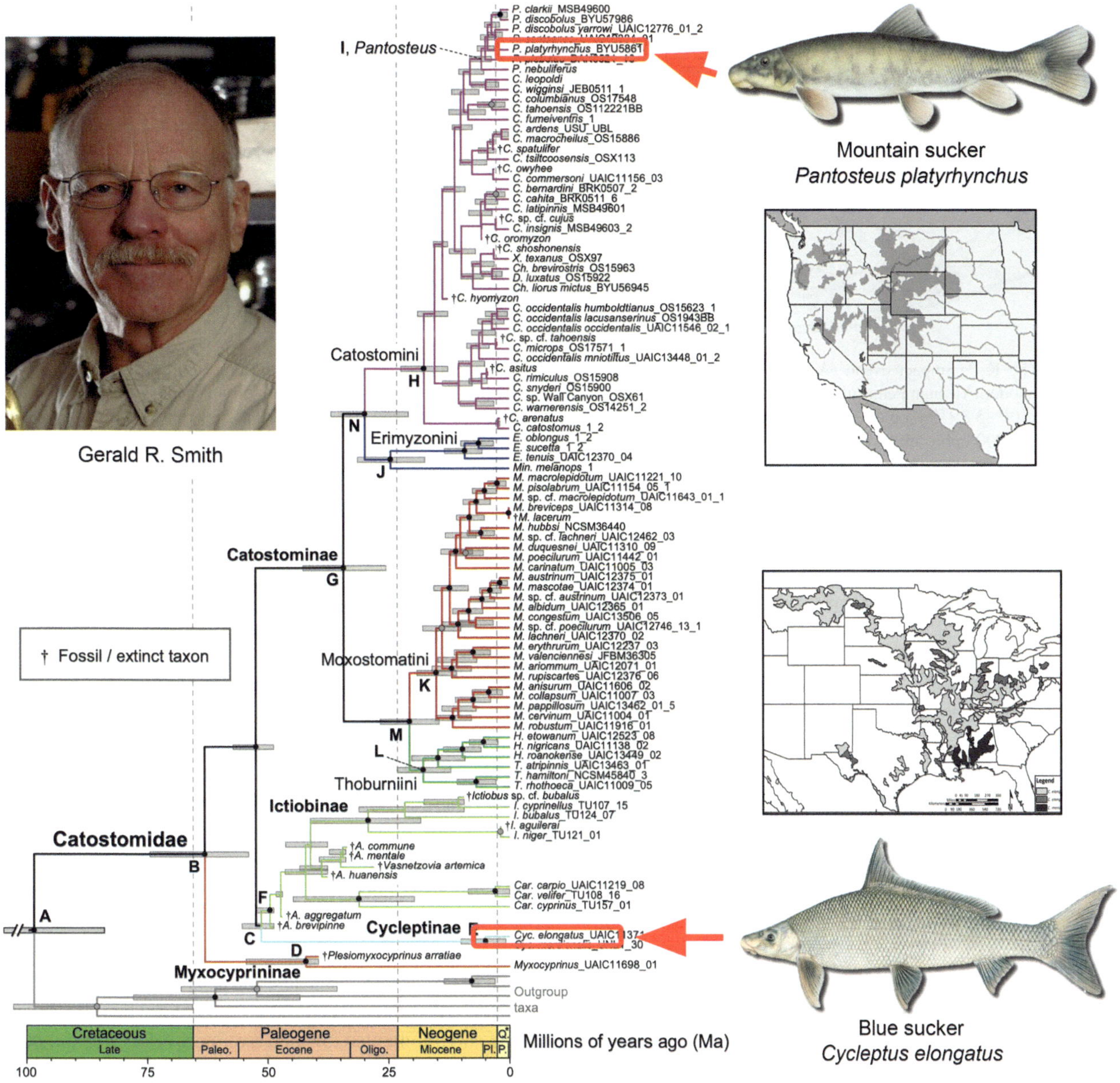

Fig. 1.16 Smith's *Lowlands Origins* hypothesis. Historical ecology and niche evolution in North American Suckers (Catostomidae). Concept from G. R. Smith (1991). Phylogeny from Bagley et al. (2018). Under this hypothesis, the primitive condition is to have a larger adult body size, to inhabit larger lowland rivers, and to have fewer species with broad geographic ranges, e.g., *Cycleptus elongatus* from the Mississippi-Missouri basin. By contrast, the derived condition is to have a smaller adult body size, to inhabit small headwater tributaries, and to have more species with smaller geographic ranges; e.g., *Catostomus platyrhynchus* from the upper Missouri, Columbia, Sacramento, and Colorado basins

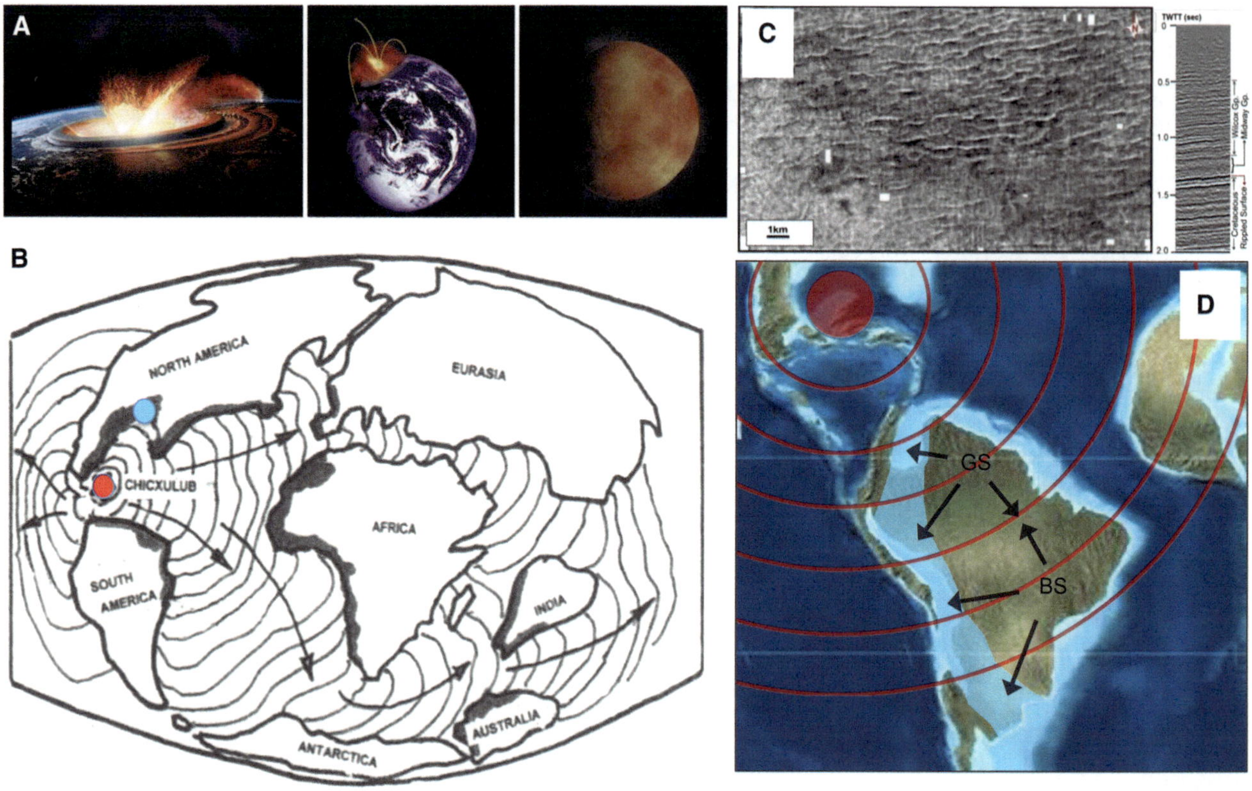

Fig. 1.17 The K-Pg impact event. (**A**) Chicxulub impact 66.043 ± 0.011 Ma in modern Yucatán Peninsula (red circle in **B** and **D**). Estimated energy: 1.3×10^{24} –5.8×10^{25} joules (21–921 billion Hiroshima A-bombs). Mobilized ~48,000 cubic miles of sediment. Global shock (pressure) waves emanating from the impact site triggered earthquakes and volcanic eruptions; e.g., the Deccan Traps in India. Impact fallout and muddy rains fill lowlands and coastal plains with acidic debris (Bailey 2002). Global "impact mega-winter" of months to years, when dust covered the atmosphere and blocked the sun triggering ecosystem collapse in both marine and terrestrial systems. (**B**) Mega-tsunami spreading waves 4.6–1.5 km high (Shonting and Ezrailson 2017; Range et al. 2022). Shaded areas depict severely eroded shorelines. Southern arm of the Chicxulub tsunami with wave heights below 100 m to India and Australia. (**C**) Seismic imaging showing mega-ripples in the Mississippi Embayment in modern Louisiana (blue circle in **B**) from Kinsland et al. (2025). Left: Planiform view showing the largest sediment ripples documented on Earth, average wavelengths 600 m, average ripple heights 16 m. Right: vertical view showing K-Pg horizon at about 1550 m depth. (**D**) Paleogeographic model for mass extinction of lowland taxa following K-Pg bolide impact. Sub-Andean foreland in blue indicating transient inundation. Black arrows indicate dispersal and recolonization of lowlands from survivors located in the upland Guiana Shield (GS) and Brazilian Shield (BS)

During the Mesozoic, Neotropical forests are thought to have had an open canopy that was dominated by conifers and ferns, until after the Chicxulub impact approximately 66 Ma when modern closed-canopy rainforests became dominated by angiosperms (Carvalho et al. 2021).

Here, we evaluate three hypotheses proposed to explain the differential extinction rates in lowland than upland regions of tropical South America in the early evolution of the South American fish fauna. These hypotheses must be assessed against a null hypothesis that extinctions rates were not significantly different between the lowland and upland shield basins of South America. The *Epicontinental sea hypothesis* posits higher extinction rates in lowland than upland taxa due to marine transgressions into the continental interior during the global super-greenhouse climate of the Upper Cretaceous and early Paleogene (*c.* 120–35 Ma; Ma et al. 2021). Such extinctions could have been protracted over the whole of this 70 Ma time interval, or focused on time periods of rising seas during the Mid-Cretaceous Hothouse (MT1 at 105–90 Ma), Paleocene-Eocene Hothouse (MT3 at 59–51 Ma), and Mid-Eocene Thermal Maximum (MT4 at 43 Ma) global warming events (Scotese 2021; Fig. 1.18A). Marine seaways intermittently covered much of the continental interior, from the north, east, and south, through mouths of the Proto-Orinoco-Amazon, Proto-Eastern Amazon, and Proto La Plata rivers basins, respectively, and seaways also covered the lowland coastal plains across the whole of the Atlantic margin. These seaways both reduced the total volume of lowland freshwater habitat and fragmented lowland river networks, contributing to elevated extinction risk (Albert et al. 2006; López-Fernández and Albert 2011).

The *Uplands tolerance hypothesis* posits that species that inhabited upland shield rivers preferentially survived the global cataclysm of the K-Pg impact event because upland

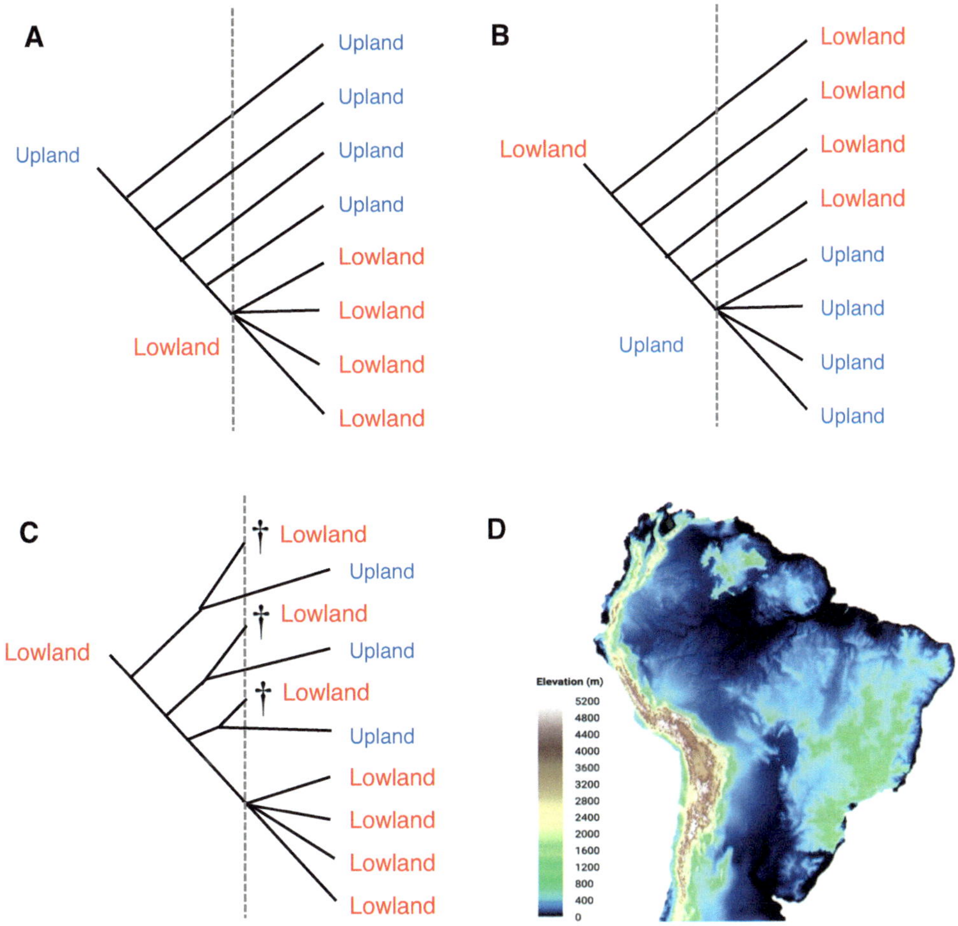

Fig. 1.18 Alternative models of evolutionary and ecological diversification among taxa inhabiting lowland and upland river basins of tropical South America. Terminal taxa may represent single species or monophyletic higher taxa. Each model makes a distinct prediction regarding tree topology, relative branch lengths, and ecological traits of species inhabiting lowland and upland rivers and streams. (**A**) Eigenmann's Upland-Origins hypothesis, with higher net diversification in lowlands. (**B**) Smith's Lowlands-Origins hypothesis, with higher net diversification in uplands, and early branches retaining original location on uplands as peripheral relicts. (**C**) Smith's Lowlands-Origins hypothesis, with diversity-dependent speciation and extinction (i.e., high species-turnover in species-rich lowlands), resulting in relictual accumulation of upland species. (**D**) False-color elevation map of part of South America. Lowlands in blues, upland shields in greens, Andes in yellows to browns

climates are more seasonally variable, and upland taxa are, therefore, expected to have greater ecophysiological tolerances to habitat disturbances (Janzen 1967; Brown 2014). Among living fishes, upland species are often better adapted to more seasonally variable habitats (Bloom et al. 2011; Crampton 2011; Lennox et al. 2019), and are, therefore, more likely to have survived a protracted impact mega-winter lasting years to decades (Robertson et al. 2013; Morgan et al. 2022). However, counterpoints to this hypothesis are that upland fish taxa are less buffered from climate change because they often have smaller geographic ranges and inhabit smaller water bodies with lower population sizes, all factors associated with greater extinction risk for freshwater fishes (Tagliacollo et al. 2021). Further, climate modelling for the K-Pg event shows more cooling in uplands than lowlands (Chiarenza et al. 2020).

The *Mega-tsunami hypothesis* posits greater survivorship in upland habitats from lower exposure to a large (1.5–4.6 km tall) tsunami thought to have swept across the continental lowlands in just a few hours, decimating lowland habitats and ecosystem functions (Shonting and Ezrailson 2017; Kinsland et al. 2025; Range et al. 2022). The mega-tsunami hypothesis predicts that relictual upland distributions should be observed across taxa, in both freshwater and terrestrial, as well as plant and animal groups. To date, such patterns have been observed in frogs (Vasconcelos et al. 2019), but most studies of taxa with relictual distributions focus on higher latitudes in the Southern Cone. The mega-tsunami hypothesis also predicts that upland taxa should exhibit distinct functional traits associated with habitat variability, which has been observed in squamate reptiles (Longrich et al. 2012), birds (Field et al. 2018), and some plants (Blonder et al. 2014). Alternative models of evolutionary and ecological diversification assuming different polarity of transitions between lowland and upland basins of tropical South America is provided in Fig. 1.18.

1.5 Causes of Evolution

The reductionist view that evolution can ultimately be understood in terms of genetics and molecular biology is clearly in error. We must turn not to population genetic studies of established species, but to studies of speciation and extinction in order to decipher the higher-level process that governs the general course of evolution.

— Steven Stanley (1975)

1.5.1 Eigenmann's Hypothesis in the Twenty-First Century

Modern research supports all five of Eigenmann's (1912) hypothesized paleogeographic events as having been instrumental in the origins and diversification of South American fishes (Lundberg et al. 1998; Albert et al. 2011b, 2020). We now have a qualitatively improved understanding of the taxonomic, biogeographic, and ecological scope of the NFF fauna, time-calibrated and species-dense phylogenies of most clades, and paleogeographic information regarding critical geomorphological and climatic events and conditions.

Modern evolutionary theory has also improved on Eigenmann's view, with at least four important new insights (Albert et al. 2011b, 2020). (1) Interactions matter; evolutionary transformations generally involve a complex, *many-to-many mapping* of causes and effects in the integrated hierarchy of biology organization, with single strong drivers being unusual (Stern and Orgogozo 2008; Svensson 2018). (2) Paleogeographic events usually have disparate influences on taxa with different *ecotraits* (Benton 2009). (3) Absolute dating of geophysical events provides chronological sequences that constrain hypotheses on drivers of biotic events (Benton and Donoghue 2007; Diniz-Filho 2023). (4) Evolutionary sequences usually result from probabilistic or *stochastic* regularities, rather than from *deterministic* drivers, with paleogeographic events and conditions serving as boundary conditions with statistical tendencies that influence and guide phylogeny and the production of biodiversity (Losos 2017). Each of these insights has been incorporated into the modern understanding of evolution of the South American fish fauna.

The first of these insights advances the idea that evolutionary diversification involves interactions among multiple environmental and organismal drivers. The paleogeographic events depicted in Fig. 1.14 and Table 1.1 are of variable tectonic, climatic, cosmic, and biotic origins, occur at different rates, and often have complex nonlinear synergistic and antagonistic effects on the other events and conditions. Some of these events are rare or even unique, confounding efforts at systematic study. For example, the tectonically driven separation of South America and Africa along with their resident biotas, was a singular Earth history event in the Upper Cretaceous, with broad consequence for the formation of Neotropical and Afrotropical biotas (Sanmartín and Ronquist 2004). Similarly, the unique tectonic uplifts of the northern Andean cordilleras during the Neogene isolated Pacific and Atlantic lowland basins and contributed strongly the divergences of their fish faunas (Albert et al. 2006). Rare tectonic events also drove most of the large-scale (>1 million Ha) mega-river capture events that have so strongly increased diversification rates in lowland fish clades (Albert et al. 2018a, 2018b).

Other events are irregular or episodic, like the climate fluctuations that resulted in marine transgressions into, and regressions out of, the continental interior lowlands and coastal plains (López-Fernández and Albert 2011). The formation of the seasonally The *Dry Diagonal* zone of arid ecosystems separating the Amazonian and Atlantic rainforests, also developed from a protracted series of regional aridification events associated with global cooling during the Neogene (Werneck et al. 2012; Antonelli et al. 2018a; Thode et al. 2019; Collevatti et al. 2020). Variability in Milankovitch orbital cycles drove the Pleistocene glacial-interglacial climate cycles (Bennett 1990), underling the Irion cycle of *várzea* formation and erosion in the eastern and central Amazon valley (Albert and Reis 2011), and the flickering connectivity system of *páramos* vegetation at high elevations in the northern Andes (Flantua et al. 2019).

Many paleogeographic events arose from the idiosyncratic interactions of multiple drivers working together. For example, the Pebas megawetland system, a central event/structure in the origins of lowland Amazonian fish faunas, formed by tectonic subduction that deformed the continental platform, in combination with eustatic sea-level rise during the MMTO (Shephard et al. 2010; Sacek 2014). The modern transcontinental Amazon River formed from the synergistic effects of several events (Figueiredo et al. 2009; Hoorn et al. 2017; Albert et al. 2018a; 2018b; 2021). These included tectonic uplifts of the Northern Andean cordilleras that increased water and sediments flows that buried the Purus Arch, and rise of the Vaupes arch that (almost) isolated the modern Amazon and Orinoco basins (Mora et al. 2010; 2020; Horton et al. 2020; Sacek et al. 2023). The increased stream power of the enlarged Amazon drainage basin after 10 Ma resulted in multiple additional river captures (Val et al. 2014; Fadul et al. 2022; Stokes et al. 2018). Uplift of the Eastern Cordillera of Colombia also isolated the trans-Andean Magdalena and Cauca valleys from the cis-Andes Orinoco and Amazon valleys (Albert et al. 2006; Montes et al. 2021). The last riverine connections between the cis-Andean and trans-Andean basins bay have been located in the area of the Colombian Massif in southern Colombia, in the time frame from the Late Miocene to Pliocene *c.* 10–3 Ma (Zapata et al. 2023; Sanín et al. 2024).

All the drivers discussed so far are episodic or relatively rare, exerting strong influences over a circumscribed time interval. Other environmental processes are more common, often exerting smaller but cumulative and protracted effects on the processes of biodiversification. These drivers include the perennial river captures and shoreline fluctuations discussed above that separate and merge portions of adjacent river networks and their resident aquatic biotas, driving rates of dispersal, speciation, and extinction (Albert et al. 2018, 2021). Perennial climate change and landscape evolution processes contribute to habitat heterogeneity across space and through time, which drives the beta diversity of species turnover among adjacent habitats and along environmental gradients (Arantes et al. 2019; Benone et al. 2020). Estimates of the rates of hypothesized drivers can be constrained by absolute dating techniques on the ages of rocks and sedimentary layers (Chesley et al. 1991; Albert et al. 2018a, 2018b). Absolute dating also narrows the ages of fossil priors used in phylogenetic time calibration, thereby clarifying the timing and sequence of events used to infer historical causal contentions (Chap. 2).

The last conceptual advance of modern evolutionary theory recognizes the fundamental roles of both "chance and necessity" in evolutionary history. As the molecular geneticist and Nobel laureate Jacques Monod famously observed, biological phenomena arise from the combined effects of *chance* (i.e., coincidence or fortuitous events) and *necessity* (i.e., which is impossible to avoid; Monod 1971). The indelible signatures of chance and necessity are also widely observed in the fields of ecology and evolution, where the responses of individual organisms, species, and clades to changing landscapes and climates cab partly be predicted by their ecotraits, but with a lot of ontological noise in the system (Losos 2017). From a modern perspective, chance and necessity are not discrete drivers but rather ends of a continuous spectrum of probability estimates, which apply throughout the natural sciences (Prigogine and Stengers 1997), both of which are widely observed in evolutionary biology (De Queiroz 2014; Losos 2018). Examples from the NFF fauna abound, including the probabilistic extinctions of *Lepidosiren* and *Arapaima* from the Orinoco basin (Albert et al. 2011b), the overseas dispersal of cichlids from Africa (Matschiner 2019; Matschiner et al. 2020; López-Fernández 2021), and the overseas dispersal of cichlids and poecilids to Central America (Matamoros et al. 2015; Tagliacollo et al. 2017).

1.5.2 Marine and Freshwater Transitions

The alternative hypotheses of Carl Eigenmann and Gerald Smith advance contrasting predictions about the physiological, morphological, and ecological phenotypes, and the phylogenetic positions, of upland and lowland fishes. Between the time that Eigenmann and Smith proposed their hypotheses, most discussion of ecological evolution in freshwater fishes focused on the biogeographic effects of saltwater tolerances, primarily under lowland conditions (Myers 1949). Eigenmann's hypothesis is based mainly on habitats associated with topography (e.g., elevation, relief), predicting plesiomorphic (ancestral) trait values of smaller body sizes and other adaptations to inhabit small upland rivers and streams. Smith's hypothesis is based on habitats associated with hydrology and geography, predicting a broader range of ancestral trait values, including a wider range of body sizes and other adaptations for large river channels and floodplains; e.g., high tolerance for hypoxia and predation pressures. Smith's hypothesis also predicts fewer species with larger geographic ranges in large lowland rivers, and more species with smaller geographic ranges in headwater tributaries. Freshwater fishes like all animals ultimately have marine origins; thus, a transitional sequence from marine to lowland to uplands seems intuitive. Thus, from the larger perspective of vertebrate evolution, Eigenmann's hypothesis implies a history of extinctions of the intermediate lowland ancestors, with subsequent recolonizations of lowland basins from relictual upland survivors.

The circumstances of marine to freshwater transitions among NFF clades vary considerably by clade (Lovejoy et al. 2006; Bloom and Lovejoy 2017; de Brito et al. 2022). Most NFF species are members of the ostariophysan clades Characiformes, Gymnotiformes, and Siluriformes, considered primary freshwater fishes with Mesozoic origins before or associated with the breakup of Gondwana (Myers 1927; 1938). Primary freshwater fishes have little or no ecophysiological tolerance to salty seawater and a negligible capacity to cross marine barriers. Other NFF clades are regarded as secondary or peripheral freshwater fishes, meaning they have higher salt tolerances, greater capacity to disperse overseas, and more recent origins in continental freshwaters (Miller 1966; Myers 1966; Tagliacollo et al. 2017). The most diverse secondary NFF taxa are Cichlinae and Cyprinodontiformes (including Rivulidae and Poeciliidae), and the most diverse peripheral NFF taxa is Potamotrygonidae.

1.5.3 Limits of Paleogeographic Maps

Said the Red Queen to Alice: "We now use the country itself as its own map, and I assure you, it does nearly as well" Lewis Carroll (1889)

The multifarious geophysical processes and paleogeographic events described in this chapter span a vast continental arena, encompassing ecological and evolutionary processes operating among thousands of species over tens of million years, many of which possess distinct ecological and physiological traits. Given this exceptionally high level of complexity, it is fair to ask: *How accurate are the paleogeographic maps and evolutionary models evaluated in this chapter?*

There are clearly many limits on what it is possible to infer about evolutionary processes underlying macroevolutionary diversification. Geological erosion removes paleoenvironmental and taxonomic information from the rock record, and biological extinction degrades information about the processes of biodiversification from the phylogeny. Entropy is always generating, as well as degrading biological information, and there are no guarantees that a particular paleogeographic events or condition can ever be known with certainty (Brooks et al. 1988).

Yet, hard-won experience teaches us hope—evolutionary biology has resolved many puzzles that seemed all but impenetrable to previous generations of investigators. Consider Charles Darwin's "abominable mystery" on the origin of flowering plants (Darwin 1879), Alfred Wegener's mechanism-free theory of continental drift (Wegener 1922), or the evolution of mammalian middle ear bones from the lower-jaw bones of vertebrate precursors (Westoll 1945; Hopson 1966), to take a few notable examples.

Although Eigenmann identified many important biogeographic and taxonomic patterns over 100 years ago, some critical paleogeographic features and events described in this chapter were unknown until just the past few decades. Paramount among these are the K-Pg mass extinction at *c.* 66 Ma (Alvarez et al. 1980), the Eocene-Oligocene global cooling event and marine regression at *c.* 34 Ma (Zachos et al. 2001), the Pebas megawetlands at *c.* 22–10 Ma (Hoorn 1993), and the formation of the modern transcontinental Amazon river at *c.* 10 Ma (Lundberg et al. 1998; Hoorn 2006).

As illustrated by the quotes above, a map has long been understood to represent a useful simplification of Nature, designed to illuminate selected landscape features of interest by filtering out other confounding factors. We argue that the geographic models of modern and ancient landforms, riverscapes, and impactful paleogeographic events described in this chapter provide critical information to understand and explain formation of NFF diversity. Yet, models alone are clearly not sufficient for this task. Information on the biological attributes of organisms is equally important, as will be explored in subsequent chapters of this book.

1.6 Conclusions

The largest landforms constraining NFF evolution are the geologically older, stable, and lower-relief Brazilian and Guiana shields of the South American platform, the younger, more dynamic, and higher-relief Andean Cordilleras along the western and northern margins, and the structural arches and sedimentary depressions that channel the changing courses of large river drainages through time. South American riverscapes assume a strong *core-periphery pattern* of drainage network geometry, with a biogeographic core composed of (from North to South) the Orinoco, Amazon, and La Plata basins) that drain the interior lowlands, and a biogeographic periphery along the continental margins composed of numerous smaller rivers. The ecological distributions of Amazonian and other Neotropical freshwater fishes are strongly constrained by the amount (i.e., volume) and connections of aquatic habitats and waterways through time, and by water chemistry profiles forged in soils that reflect the chemical composition of the source bedrock and paleoclimatic conditions through deep time.

The NFF fauna exhibits several prominent biodiversity patterns widely observed in biotas worldwide: *Core-periphery* biodiversity pattern with high-diversity and phylogenetically under-dispersed (more-closely related) species assemblages in lowland (i.e., Amazon, Orinoco, La Plata) basins, and low-diversity and phylogenetically overdispersed assemblages in basins of the upland periphery (e.g., Guiana and Brazilian shields, Andean Cordilleras and trans-Andean catchments). *Hollow-curve* (i.e., power function) frequency distribution of species richness among clades in which a few clades include most of the species, and where most clades have only a few species.

Other major patterns of NFF diversity differ from those of terrestrial taxa (i.e., amniotes, angiosperms), illuminating the role of phenotypes associated with aquatic habitat utilization and dispersal in evolutionary diversification. Most NFF clades exhibit the highest diversity at low elevations (<100 m) in contrast to the *mid-elevation peak* in species richness observed in birds, mammals, and flowering plants. The Western Amazon is the center of diversity for NFF species, as for most Neotropical taxa, but patterns of NFF diversity do not support some of the most prominent models of diversity (e.g., *mid-domain effect*; *river continuum concept*). NFF diversity does match expectations of the species-area relationship and the species-habitat volume relationship. Environmental parameters, such as precipitation, temperature, and hydrological connectivity, which predict fish species diversity in temperate fish faunas, have less predictive power in South America.

Major patterns of NFF biodiversity only make sense in the light of events and conditions occurring over tens of million years. These include global climatic and sea-level fluctuations driving *marine transgressions* and *marine regressions*, changes in the hydrological connection of large lowland sedimentary basins driven by plate tectonics and erosion, and changing rates of speciation and extinction under the influences of macroevolutionary abiotic (e.g., climatic) and biotic drivers (biotic interchanges, evolutionary novelties).

Data from the biodiversity and evolution of NFF clades support aspect of both the *Uplands* and *Lowlands origins hypotheses*. The *Uplands origins hypothesis* posits ancestors with small adult body size, a generalized benthic invertebrate diet, and small geographic ranges in upland clearwater rivers and streams of the Guiana and Brazilian Shields during the Upper Cretaceous and Early Paleogene (*c.* 100–4 Ma). The *Lowlands*

Origins hypothesis posits an ancestral condition with large adult body size and a broad geographic range inhabiting large lowland rivers in the Middle to Late Cenozoic (after *c.* 34 Ma). These data also support aspects of all three K-Pg hypotheses regarding differential rates of net diversification in lowland and upland regions: (1) The *Epicontinental sea hypothesis* posits higher extinction rates in lowland than upland taxa due to marine transgressions into the continental interior during the K-Pg global super-greenhouse climate (*c.* 120–35 Ma). (2) The *Uplands tolerance hypothesis* posits that species of upland shield rivers preferentially survived the global cataclysm of the K-Pg impact event, because upland climates are more seasonally variable, and upland taxa are, therefore, expected to have greater ecophysiological tolerances to habitat disturbances. (3) The *Mega-tsunami hypothesis* posits greater survivorship in upland habitats from lower exposure to a large (1.5–4.6 km tall) tsunami thought to have swept across the continental lowlands in just a few hours, decimating lowland habitats and ecosystem functions.

More generally, data on NFF taxa support several axioms of modern evolutionary theory. (1) *Interactions matter*; evolutionary transformations involve a complex, many-to-many mapping of causes and effects among levels in the integrated hierarchy of biology organization, with single strong drivers being unusual. Many pronounced paleogeographic events and conditions (e.g., K-Pg event, Pebas megawetland system, modern transcontinental Amazon river) arose from the idiosyncratic interactions of multiple drivers working together. (2) *Biological traits matter*; paleogeographic events and conditions have disparate influences on taxa with different biological traits related to life-history strategy, habitat utilization, and trophic position. (3) *Timing matters*: new geophysical tools for dating paleogeographic events have greatly improved the time-resolution for drivers of biotic events. (4) *Rare but impactful (Black Swan events) matter*, with a spectrum of biodiversification drivers, including factors of "chance" (i.e., stochastically idiosyncratic) and "necessity" (i.e., stochastically deterministic). These drivers range from rare or unique events that exert strong influences over restricted time intervals, to the many common processes and perennial conditions that exert small but cumulative effects over protracted time frames. Such drivers include perennial river captures and shoreline fluctuations that separate and merge portions of adjacent river networks and their resident aquatic biotas, driving rates of dispersal, speciation, and extinction (Albert et al. 2025; Arantes et al. 2019; Bar-On et al. 2018; Barbosa et al. 2024; Bertaco 2006; Bogotá-Gregory et al. 2022; Carmona et al. 2021; Rapoport 1975; Crelier et al. 2025; Jablonski 2022; Jézéquel et al. 2020; Kinsland et al. 2025; Lasso et al. 2016; Lehner et al. 2021; Montaña and Winemiller (2009); Müller et al. 2018; Ohara et al. 2025; Scotland and Sanderson 2004; Symington 2024; van der Meer et al. 2022; Willis 1922).

Acknowledgments We are grateful to José Birindelli, Carina Hoorn, Ian Solon Bortoloci Ito, Humberto Lima Filho, Rafael Marques, and Brian Sidlauskas for insightful comments.

References

Abreu JMS, Craig JM, Albert JS, Piorski NM (2019) Historical biogeography of fishes from coastal basins of Maranhão State, northeastern Brazil. Neotrop Ichthyol 17(2):e180156

Adams DC, Collyer ML (2018) Multivariate phylogenetic comparative methods: evaluations, comparisons, and recommendations. Syst Biol 67(1):14–31

Albert JS, Reis RE (2011) Introduction to Neotropical freshwaters. In: Historical biogeography of neotropical freshwater fishes, University of California Press, pp 3–20

Albert JS, Bart Jr HL, Reis RE (2011a) Species richness and cladal diversity. In: Albert JS, Reis RE (eds) Historical biogeography of neotropical freshwater fishes. University of California Press, Berkeley, pp 89–104

Albert JS, Bernt MJ, Fronk AH, Fontenelle JP, Kuznar SL, Lovejoy NR (2021) Late Neogene megariver captures and the Great Amazonian biotic interchange. Glob Planet Change 205:103554

Albert JS, Carvalho TP, Petry P, Holder MA, Maxime EL, Espino J et al (2011c) Aquatic biodiversity in the Amazon: habitat specialization and geographic isolation promote species richness. Animals 1(2):205–241

Albert JS, Craig JM, Tagliacollo VA, Petry P (2018b). Upland and lowland fishes: a test of the river capture hypothesis. Mt Clim Biodiversity 273–294

Albert JS et al (2025) An ecological trait matrix of Neotropical freshwater fishes, the most diverse continental vertebrate fauna. Nat Sci Data (in press)

Albert JS, Lovejoy NR, Crampton WG (2006) Miocene tectonism and the separation of cis- and trans-Andean river basins: evidence from Neotropical fishes. J South Am Earth Sci 21(1–2):14–27

Albert JS, Petry P, Reis RE (2011b) Major biogeographic and phylogenetic patterns. In: Albert JS, Reis RE (eds) Historical biogeography of neotropical freshwater fishes. University of California Press, Berkeley, pp 3–20

Albert JS, Schoolmaster Jr DR, Tagliacollo V, Duke-Sylvester SM (2017) Barrier displacement on a neutral landscape: toward a theory of continental biogeography. Syst Biol 66(2):167–182

Albert JS, Val P, Hoorn C (2018a) The changing course of the Amazon River in the Neogene: center stage for Neotropical diversification. Neotrop Ichthyol 16, e180033. https://doi.org/10.1590/1982-0224-201901

Albert J, Hoorn C, Malhi Y, Phillips O, Encalada AC, Steege H (2021) The multiple viewpoints for the Amazon: geographic limits and meanings. Amazon Assess Rep, pp 1-5

Altermatt F (2013) Diversity in riverine metacommunities: a network perspective. Aquat Ecol 47:365–377

Alvarez LW, Alvarez W, Asaro F, Michel HV (1980) Extraterrestrial cause for the Cretaceous-Tertiary extinction. Science 208(4448):1095–1108

Antonelli A, Ariza M, Albert J, Andermann T, Azevedo J, Bacon C et al (2018a) Conceptual and empirical advances in Neotropical biodiversity research. PeerJ 6:e5644

Antonelli A, Nylander JA, Persson C, Sanmartín I (2009) Tracing the impact of the Andean uplift on Neotropical plant evolution. Proc Natl Acad Sci 106(24):9749–9754

Antonelli A, Zizka A, Carvalho FA, Scharn R, Bacon CD, Silvestro D, Condamine FL (2018b) Amazonia is the primary source of Neotropical biodiversity. Proc Natl Acad Sci 115(23):6034–6039

Arantes CC, Winemiller KO, Petrere M, Freitas CE (2019) Spatial variation in aquatic food webs in the Amazon River floodplain. Freshwater Sci 38(1):213–228

Azevedo JA, Collevatti RG, Jaramillo CA, Strömberg CA, Guedes TB, Matos-Maraví P et al (2020) On the young savannas in the land of ancient forests. InNeotropical diversification: Patterns and Processes. Springer International Publishing, Cham, pp 271-298

Bacon CD, Silvestro D, Jaramillo C, Smith BT, Chakrabarty P, Antonelli A (2015) Biological evidence supports an early and complex emergence of the Isthmus of Panama. Proc Natl Acad Sci 112(19):6110–6115

Bailey RM, Smith GR (1981) Origin and geography of the fish fauna of the Laurentian Great Lakes basin. Can J Fish Aquat Sci 38(12):1539–1561

Bar-On YM, Phillips R, Milo R (2018) The biomass distribution on Earth. Proc Natl Acad Sci 115(25):6506–6511

Barbosa TAP, Prudente BDS, Mendonça MB, Salvador GN, Montag LFDA, Akama A (2024) The ichthyofauna of the Potiritá River basin: assessing the conservation status in a mining area in the Eastern Amazon. Acta Limnol Bras 36:e31

Barnosky AD, Lindsey EL, Villavicencio NA, Bostelmann E, Hadly EA, Wanket J, Marshall CR (2016) Variable impact of late-Quaternary megafaunal extinction in causing ecological state shifts in North and South America. Proc Natl Acad Sci 113(4):856–861

Barreto SB, Knowles LL, Mascarenhas R, Affonso PRADM, Batalha-Filho H (2022) Drainage rearrangements and in situ diversification of an endemic freshwater fish genus from north-eastern Brazilian rivers. Freshwater Biol 67(5):759–773

Bellwood DR, Wainwright PC (2002) The history and biogeography of fishes on coral reefs. Coral Reef Fishes: Dyn Divers Complex Ecosyst 5:32

Bennett KD (1990) Milankovitch cycles and their effects on species in ecological and evolutionary time. Paleobiology 16(1):11–21

Benone NL, Leal CG, dos Santos LL, Mendes TP, Heino J, de Assis Montag LF (2020) Unravelling patterns of taxonomic and functional diversity of Amazon stream fish. Aquat Sci 82:1–11

Benton MJ (2009) The Red Queen and the Court Jester: species diversity and the role of biotic and abiotic factors through time. Science 323(5915):728–732

Benton MJ, Donoghue PC (2007) Paleontological evidence to date the tree of life. Mol Biol Evol 24(1):26–53

Bertaco VA (2006) Peces del medio Amazonas: Región de Leticia. Neotrop Ichthyol 4

Bishop P (1995) Drainage rearrangement by river capture, beheading and diversion. Prog Phys Geogr 19(4):449–473

Blonder B, Royer DL, Johnson KR, Miller I, Enquist BJ (2014) Plant ecological strategies shift across the Cretaceous–Paleogene boundary. PLoS Biol 12(9):e1001949

Bloom DD, Lovejoy NR (2017) On the origins of marine-derived freshwater fishes in South America. J Biogeogr 44(9):1927–1938

Bloom DD, Lovejoy NR, Albert JS, Reis RE (2011) The biogeography of marine incursions in South America. In: Albert JS, Reis RE (eds) Historical biogeography of neotropical freshwater fishes. University of California Press, Berkeley, pp 137–144

Bloom DD, Weir JT, Piller KR, Lovejoy NR (2013) Do freshwater fishes diversify faster than marine fishes? A test using state-dependent diversification analyses and molecular phylogenetics of new world silversides (Atherinopsidae). Evolution 67(7):2040–2057

Blundell DJ, Scott AC (1998) Lyell: The past is the key to the present. Geological Society, London, p 376

Bogotá-Gregory JD, DoNascimiento C, Lima FC, Acosta-Santos A, Villa-Navarro FA, Urbano-Bonilla A et al (2022) Fishes from the Colombian Amazonia region: Species composition from the river systems within the rainforest biome. Biota Neotropica 22(4):e20221392

Boivin NL, Zeder MA, Fuller DQ, Crowther A, Larson G, Erlandson JM et al (2016) Ecological consequences of human niche construction: examining long-term anthropogenic shaping of global species distributions. Proc Natl Acad Sci 113(23):6388–6396

Borrelli P, Robinson DA, Panagos P, Lugato E, Yang JE, Alewell C et al (2020) Land use and climate change impacts on global soil erosion by water (2015-2070). Proc Natl Acad Sci 117(36):21994–22001

Boschman LM, Carraro L, Cassemiro FA, de Vries J, Altermatt F, Hagen O et al (2023) Freshwater fish diversity in the western Amazon basin shaped by Andean uplift since the Late Cretaceous. Nat Ecol Evol 7(12):2037–2044

Brooks DR, Wiley EO, Brooks DR (1988) Evolution as entropy. University of Chicago Press, Chicago

Brown Jr WL (1957) Centrifugal speciation. Q Rev Biol 32(3):247–277

Brown JH (2014) Why are there so many species in the tropics? J Biogeogr 41(1):8–22

Buffan L, Condamine FL, Stutz NS, Pujos F, Antoine PO, Marivaux L (2025) The fate of South America's endemic mammalian fauna in response to the most dramatic Cenozoic climate disruption. Proc Natl Acad Sci 122(20):e2419520122

Brown JH, Nicoletto PF (1991) Spatial scaling of species composition: body masses of North American land mammals. Am Nat 138(6):1478–1512

Burnham RJ, Johnson KR (2004) South American palaeobotany and the origins of neotropical rainforests. Philos Trans R Soc London [Biol] 359(1450):1595–1610

Burns MD, Knouft JH, Dillman CB (2024) The role of abiotic and biotic factors in the unequal body shape diversification of a Gondwanan fish radiation (Otophysi: Characiformes). Evolution 78(2):253–266

Burr BM, Mayden RL (1992) Phylogenetics and North American freshwater fishes. In: Mayden RL (ed) Systematics, historical ecology, and north american freshwater fishes. Stanford University Press, Stanford, pp 18–75

Carmona AM, Renner M, Kleidon A, Poveda G (2021) Uncertainty of runoff sensitivity to climate change in the Amazon River basin. Ann NY Acad Sci 1504(1):76–94

Carrapa B, Clementz M, Feng R (2019) Ecological and hydroclimate responses to strengthening of the Hadley circulation in South America during the Late Miocene cooling. Proc Natl Acad Sci 116(20):9747–9752

Carrillo JD, Faurby S, Silvestro D, Zizka A, Jaramillo C, Bacon CD, Antonelli A (2020) Disproportionate extinction of South American mammals drove the asymmetry of the Great American Biotic Interchange. Proc Natl Acad Sci 117(42):26281–26287

Carroll L (1889) Sylvie and Bruno: By Lewis Carroll. With 46 Illus. by Harry Furniss. Macmillan, London

Cassemiro FA, Albert JS, Antonelli A, Menegotto A, Wüest RO, Cerezer F et al (2023) Landscape dynamics and diversification of the megadiverse South American freshwater fish fauna. Proc Natl Acad Sci 120(2):e2211974120

Chen F, Xue G, Wang Y, Zhang H, Clift PD, Xing Y et al (2023). Evolution of the Yangtze River and its biodiversity. The Innov 4(3):100417

Chen WJ, Lavoué S, Mayden RL (2013) Evolutionary origin and early biogeography of otophysan fishes (Ostariophysi: Teleostei). Evolution 67(8):2218–2239

Cheng P, Yu D, Tang Q, Yang J, Chen Y, Liu H (2022) Macro-evolutionary patterns of East Asian opsariichthyin-xenocyprinin-cultrin fishes related to the formation of river and river-lake environments under monsoon climate. Water Biol Secur 1(2):100036

Chesley JT, Halliday AN, Scrivener RC (1991) Samarium-neodymium direct dating of fluorite mineralization. Science 252(5008):949–951

Chiarenza AA, Farnsworth A, Mannion PD, Lunt DJ, Valdes PJ, Morgan JV, Allison PA (2020) Asteroid impact, not volcanism, caused the end-Cretaceous dinosaur extinction. Proc Natl Acad Sci 117(29):17084–17093

Chuctaya J, Meza-Vargas V, Faustino-Fuster DR, Hidalgo M, Ortega H (2022) Checklist of the ichthyofauna of Rio Ucayali basin, Peru. 29:e20049

Collevatti RG, Lima NE, Vitorino LC (2020) The diversification of extant angiosperms in the South America dry diagonal. In: Rull V, Carnaval AC (Eds.) Neotropical diversification: patterns and processes. Springer, Berlin, pp 547–568

Colwell RK, Lees DC (2000) The mid-domain effect: geometric constraints on the geography of species richness. Trends Ecol Evol 15(2):70–76

Conde-Saldaña CC, Melo BF, Roxo FF, Toledo-Piza M, Tagliacollo VA, Oliveira C, Albert JS (2025) Landscape evolution drives continental diversification in Neotropical freshwater fishes (Teleostei, Erythrinoidea). J Biogeography 52(5):e15129

Correa SB, Crampton WGR, Chapman LJ, Albert JS (2008) A comparison of flooded forest and floating meadow fish assemblages in an upper Amazon floodplain. J Fish Biol 72(3):629–644

Cracraft J (1974) Phylogenetic models and classification. Syst Biol 23(1):71–90

Cracraft J (1994) Species diversity, biogeography, and the evolution of biotas. Amer Zool 34(1):33–47

Crampton WGR (2011) An ecological perspective on diversity and distributions. In: Albert JS, Reis RE (Eds.) Historical biogeography of neotropical freshwater fishes. University of California Press, Berkeley, pp 165–192

Crampton WGR, Ortega H, Lovejoy NR, Reis RE, Albert JS (2008) Fishes of the Pacaya-Samiria National Reserve, Perú. University of Louisiana, Lafayette, p 182

Crelier C, Zumba A, Peifer D, Val P (2025) Mobility of South America's transcontinental drainage divide and shrinkage of the Paraná river basin linked to lithologic and geodynamic controls. Sci Rep 15(1):3134

Crisp MD, Cook LG (2005) Do early branching lineages signify ancestral traits? Trends Ecol Evolut 20(3):122–128

da Silva JMC, Leal IR, Tabarelli M (eds) (2018) Caatinga: The largest tropical dry forest region in South America. Springer, Switzerland, p 503

Dagosta FCP, Monção MS, Nagamatsu BA, Pavanelli CS, Carvalho FR, Lima FCet al (2024) Fishes of the upper rio Paraná basin: diversity, biogeography and conservation. Neotrop Ichthyol 22(1):e230066

Darela-Filho JP, Rammig A, Fleischer K, Reichert T, Lugli LF, Quesada CA et al (2024) Reference maps of soil phosphorus for the pan-Amazon region. Earth Syst Sci Data 16(1):715–729

Darwin C (1879) More letters of Charles Darwin: a record of his work in a series of Hitherto Unpublished Letters, vol 2. 1903. D. Appleton

Darwin CR (1859) On the origin of species by means of natural selection, or the preservation of favoured races in the struggle for life. John Murray, London

De Lira Mota MA, Dunkley Jones T, Sulaiman N, Edgar KM, Yamaguchi T, Leng MJ et al (2023) Multi-proxy evidence for sea level fall at the onset of the Eocene-Oligocene transition. Nat Commun 14(1):4748

DeQueiroz A (2014) The Monkey's voyage: how improbable journeys shaped the history of life. Basic Books

Denevan WM (eds) (1992) The native population of the Americas in 1492. University of Wisconsin Press, Madison

Diniz-Filho JAF (2023) The macroecological perspective. In: The macroecological perspective: theories, models and methods. Springer International Publishing, Cham, pp 1–13

Dodds W (2009) Laws, theories, and patterns in ecology. University of California Press, Berkeley

Edler D, Guedes T, Zizka A, Rosvall M, Antonelli A (2017) Infomap bioregions: Interactive mapping of biogeographical regions from species distributions. Syst Biol 66(2):197–204

Eigenmann CH (1906) The fresh-water fishes of South and Middle America. Pop Sci Mon 515–530

Eigenmann CH (1912) The freshwater fishes of British Guiana, including a study of the ecological grouping of species and the relation of the fauna of the plateau to that of the lowlands, vol 5. Carnegie Institute, Pittburgh

Eigenmann CH (1921) The nature and origin of the fishes of the Pacific slope of Ecuador, Peru and Chili. Proc Am Philos Soc 60(4):503–523

Escobar-Camacho D, Valdiviezo-Rivera J, Carrillo-Moreno C, Argüello P, Swing K (2025) Exceptional concentration of fish diversity in Yasuní National Park, Ecuador (Napo River Basin). Biodivers Data J 13:e136476

Espurt N, Baby P, Brusset S, Roddaz M, Hermoza W, Regard V et al (2007) How does the Nazca Ridge subduction influence the modern Amazonian foreland basin? Geology 35(6):515–518

Evans KM, Kim LY, Schubert BA, Albert JS (2019) Ecomorphology of neotropical electric fishes: an integrative approach to testing the relationships between form, function, and trophic ecology. Integr Org Biol 1(1):obz015

Fadul CM, Oliveira P, Val P (2022) Ongoing landscape transience in the eastern Amazon Craton consistent with lithologic control of base level. Earth Surf Processes Landforms 47(13):3117–3132

Fagan WF (2002) Connectivity, fragmentation, and extinction risk in dendritic metapopulations. Ecology 83(12):3243–3249

Fagan WF, Grant EHC, Lynch HJ, Unmack PJ (2009) Riverine landscapes: ecology for an alternative geometry. In: Spatial ecology. Chapman and Hall/CRC, Boca Raton, pp 107–122

Fernandes CC, Podos J, Lundberg JG (2004) Amazonian ecology: tributaries enhance the diversity of electric fishes. Science 305(5692):1960–1962

Field DJ, Bercovici A, Berv JS, Dunn R, Fastovsky DE, Lyson TR et al (2018) Early evolution of modern birds structured by global forest collapse at the end-Cretaceous mass extinction. Curr Biol 28(11):1825–1831

Figueiredo JJJP, Hoorn C, Van der Ven P, Soares E (2009) Late Miocene onset of the Amazon River and the Amazon deep-sea fan: evidence from the Foz do Amazonas Basin. Geology 37(7):619–622

Flantua SG, O'Dea A, Onstein RE, Giraldo C, Hooghiemstra H (2019) The flickering connectivity system of the north Andean páramos. J Biogeography 46(8):1808–1825

Garzione CN, Hoke GD, Libarkin JC, Withers S, MacFadden B, Eiler J et al (2008) Rise of the Andes. Science 320(5881):1304–1307

Garzione CN, McQuarrie N, Perez ND, Ehlers TA, Beck SL, Kar N et al (2017) Tectonic evolution of the Central Andean plateau and implications for the growth of plateaus. Annu Rev Earth Planet Sci 45(1):529–559

Goldberg EE, Roy K, Lande R, Jablonski D (2005) Diversity, endemism, and age distributions in macroevolutionary sources and sinks. Am Nat 165(6):623–633

Gorman OT, Karr JR (1978) Habitat structure and stream fish communities. Ecology 59(3):507–515

Goudie AS (2005) The drainage of Africa since the Cretaceous. Geomorphology 67:437–456

Graham A (2011) The age and diversification of terrestrial New World ecosystems through Cretaceous and Cenozoic time. Am J Bot 98(3):336–351

Grandcolas P, Nattier R, Trewick S (2014) Relict species: a relict concept? Trends Ecol Evolut 29(12):655–663

Gray M (2008) Geodiversity: the origin and evolution of a paradigm. In: Burek CV, Prosser CD (eds) The history of geoconservation, vol 300. Geological Society of London, London. https://doi.org/10.1144/sp300.4

Green JK, Berry J, Ciais P, Zhang Y, Gentine PJSA (2020) Amazon rainforest photosynthesis increases in response to atmospheric dryness. Sci Adv 6(47):eabb7232

Guayasamin JM, Ribas CC, Carnaval AC, Carrillo JD, Hoorn C, Lohmann LG et al (2024) Evolution of Amazonian biodiversity: a review. Acta Amazonica 54(spe1):e54bc21360

Haas O, Simpson GG (1946) Analysis of some phylogenetic terms, with attempts at redefinition. Proc Am Philos Soc 90:319–349
Hagen O, Skeels A, Onstein RE, Jetz W, Pellissier L (2021) Earth history events shaped the evolution of uneven biodiversity across tropical moist forests. Proc Natl Acad Sci 118(40):e2026347118
Hansen WF (2001) Identifying stream types and management implications. Forest Ecol Manag 143(1-3):39–46
He C, Braun J, Tang H, Yuan X, Acevedo-Trejos E, Ott RF, Stucky de Quay G (2024) Drainage divide migration and implications for climate and biodiversity. Nat Rev Earth Environ 5(3):177–192
Hennig W (1966) Phylogenetic systematics. University of Illinois Press, Champaign-Urbana, p 280
Hercos AP, Oliveira JAD, Oliveira JCD, Rodrigues EKDQ, Barbosa RL, Queiroz HLD (2021) Checklist of the ichthyofauna of Mamirauá Sustainable Development Reserve, Middle Solimões, Amazonas, Brazil: high richness in a large protected area of Western-Central Amazonia. Biota Neotropica 21(4):e20211207
Hoagstrom CW, Davenport SR, Osborne MJ (2025) Assembling the Pecos River fish fauna: barrier displacement on the Southern Great Plains. North Am Biol Rev 100(4):1534-1556
Hocutt CH (1979) Drainage evolution and fish dispersal in the central Appalachians. Geol Soc Am Bull 90(2_Part_II):197–234
Hoorn C (1993) Marine incursions and the influence of Andean tectonics on the miocene depositional history of northwestern Amazonia: results of a palynostratigraphic study. Palaeogeog Palaeoclim Palaeoecol 105(3–4):267–309
Hoorn C (2006) The birth of the mighty Amazon. Sci Am 294(5):52–59
Hoorn C, Bogotá-A GR, Romero-Baez M, Lammertsma EI, Flantua SG, Dantas EL et al (2017) The Amazon at sea: onset and stages of the Amazon River from a marine record, with special reference to Neogene plant turnover in the drainage basin. Glob Planet Change 153:51–65
Hoorn C, Boschman LM, Kukla T, Sciumbata M, Val P (2022) The Miocene wetland of western Amazonia and its role in Neotropical biogeography. Bot J Linean Soc 199(1):25–35
Hoorn C, Wesselingh FP, Ter Steege H, Bermudez MA, Mora A, Sevink J et al (2010) Amazonia through time: Andean uplift, climate change, landscape evolution, and biodiversity. Science 330(6006):927–931
Hopson JA (1966) The origin of the mammalian middle ear. Amer Zool 6(3):437–450
Hora SL (1937) Geographical distribution of Indian freshwater fishes and its bearing on the probable land connections between India and the adjacent countries. Curr Sci 5(7):351–356
Horton BK, Parra M, Mora A (2020) Construction of the Eastern Cordillera of Colombia: insights from the sedimentary record. Geol Colombia 3:ca-3
Horton RE (1945) Erosional development of streams and their drainage basins; hydrophysical approach to quantitative morphology. Geol Soc Am Bull 56(3):275–370
Howard AD (1971) Simulation model of stream capture. Geological Society of
Hubbs CL (1931) *Parexoglossum laurae*, a new cyprinid fish from the Upper Kanawha River system. Occ Pap Mus Zool, Univ Mich 234:1–12
Hughes RM, Kaufmann PR, Weber MH (2011) National and regional comparisons between Strahler order and stream size. JNABS 30(1):103–121
Hurtado C, Roddaz M, Santos RV, Baby P, Antoine PO, Dantas EL (2018) Cretaceous-early Paleocene drainage shift of Amazonian rivers driven by Equatorial Atlantic Ocean opening and Andean uplift as deduced from the provenance of northern Peruvian sedimentary rocks (Huallaga basin). Gondwana Res 63:152–168
Ihering HV (1900) The history of the Neotropical region. Science 12(310):857–864
Irion G, Müller J, Morais JO, Keim G, de Mello JN, Junk WJ (2009) The impact of Quaternary sea level changes on the evolution of the Amazonian lowland. Hydrol Processes: Int J 23(22):3168–3172
Irion G, Müller J, Nunes de Mello J, Junk WJ (1995) Quaternary geology of the Amazonian lowland. Geo-Mar Lett 15:172–178
Jablonski D (1980) Apparent versus real biotic effects of transgressions and regressions. Paleobiology 6(4):397–407
Jablonski D (2022) Evolvability and macroevolution: overview and synthesis. Evol Biol 49(3):265–291
Jablonski D, Sepkoski Jr JJ (1996) Paleobiology, community ecology, and scales of ecological pattern. Ecology 77(5):1367–1378
Janzen DH (1967) Why mountain passes are higher in the tropics. Am Nat 101(919):233–249
Jaramillo C, Romero I, D'Apolito C, Bayona G, Duarte E, Louwye S et al (2017) Miocene flooding events of western Amazonia. Sci Adv 3(5):e1601693
Jézéquel C, Tedesco PA, Bigorne R, Maldonado-Ocampo JA, Ortega H, Hidalgo M et al (2020) A database of freshwater fish species of the Amazon Basin. Sci Data 7(1), 96
Jones LA, Dean CD, Gearty W, Allen BJ (2024) rmacrostrat: an R package for accessing and retrieving data from the Macrostrat geological database. Geosphere 20(6):1456–1467
Kinsland GL, Zhang R, Burr R, Klug S (2025) Chicxulub impact tsunami megaripples, imaged in 3D seismic data: distribution and characteristics on the northern Gulf of Mexico shelf and slope. Mar Geol 481:107466
Kolmann MA, Poulin E, Rosen J, Hemraj-Naraine D, Burns MD (2024) Phenotypic convergence is more frequent and stronger in herbivorous freshwater fishes. Integr Comp Biol icae037
Lang N, Jetz W, Schindler K, Wegner JD (2023) A high-resolution canopy height model of the Earth. Nat Ecol Evol 7(11):1778-1789
Larsen S, Majone B, Zulian P, Stella E, Bellin A, Bruno MC, Zolezzi G (2021) Combining hydrologic simulations and stream-network models to reveal flow-ecology relationships in a large Alpine catchment. Water Resour Res 57(4):e2020WR028496
Lasso CA, Machado-Allison A, Taphorn DC (2016) Fishes and aquatic habitats of the Orinoco River Basin: diversity and conservation. J Fish Biol 89(1):174–191
Lavoué S (2020) Origins of Afrotropical freshwater fishes. Zool J Linnean Soc 188(2):345–411
Lawton JH (1999) Are there general laws in ecology? Oikos 84:177–192
Lehner B, Roth A, Huber M, Anand M, Grill G, Osterkamp N et al (2021) HydroSHEDS v2.0–refined global river network and catchment delineations from TanDEM-X elevation data. In: EGU General Assembly Conference Abstracts, pp EGU21-9277
Lennox RJ, Crook DA, Moyle PB, Struthers DP, Cooke SJ (2019) Toward a better understanding of freshwater fish responses to an increasingly drought-stricken world. Rev Fish Biol Fish 29:71–92
Liebrand D, de Bakker AT, Beddow HM, Wilson PA, Bohaty SM, Ruessink G et al (2017) Evolution of the early Antarctic ice ages. Proc Natl Acad Sci 114(15):3867–3872
Lima SM, Berbel-Filho WM, Araújo TF, Lazzarotto H, Tatarenkov A, Avise JC (2017) Headwater capture evidenced by paleo-rivers reconstruction and population genetic structure of the armored catfish (*Pareiorhaphis garbei*) in the Serra do Mar mountains of southeastern Brazil. Front Genet 8:199
Lomolino MV, Riddle BR, Whittaker RJ (2017) Biogeography. Oxford University Press, Oxford, p 784
Longman T (1997) The Book of Ecclesiastes. Wm. B. Eerdmans Publishing
Longrich NR, Bhullar BAS, Gauthier JA (2012) Mass extinction of lizards and snakes at the Cretaceous–Paleogene boundary. Proc Natl Acad Sci 109(52):21396–21401
Longrich NR, Tokaryk T, Field DJ (2011) Mass extinction of birds at the Cretaceous–Paleogene (K–Pg) boundary. Proc Natl Acad Sci 108(37):15253–15257

López-Fernández H (2021) Neotropical riverine cichlids: adaptive radiation and macroevolution at continental scales. In: The behavior, Ecology and Evolution of Cichlid Fishes. Springer Netherlands, Dordrecht, pp 135-173

López-Fernández H, Albert JS (2011) Paleogene radiations. Hist Biogeogr Neotrop Freshwater Fishes 1:105–118

Losos JB (2017) Improbable destinies: fate, chance, and the future of evolution. Riverhead Books, New York

Lovejoy NR, Albert JS, Crampton WG (2006) Miocene marine incursions and marine/freshwater transitions: evidence from Neotropical fishes. J South Am Earth Sci 21(1–2):5–13

Lovejoy TE, Nobre C (2019) Amazon tipping point: last chance for action. Sci Adv 5(12):eaba2949

Lundberg JG, Kottelat M, Smith GR, Stiassny ML, Gill AC (2000) So many fishes, so little time: an overview of recent ichthyological discovery in continental waters. Ann Mo Bot Gard 87:26–62

Lundberg JG, Lewis Jr WM, Saunders III JF, Mago-Leccia F (1987) A major food web component in the Orinoco River channel: evidence from planktivorous electric fishes. Science 237(4810):81–83

Lundberg JG, Marshall LG, Guerrero J, Horton B, Malabarba MCSL, Wesselingh F (1998) The stage for Neotropical fish diversification: a history of tropical South American rivers. Phylog Classif Neotrop Fishes 27:13–48

Lyons NJ, Val P, Albert JS, Willenbring JK, Gasparini NM (2020) Topographic controls on divide migration, stream capture, and diversification in riverine life. Earth Surf Dyn 8(4):893–912

Ma M, He M, Zhao M, Peng C, Liu X (2021) Evolution of atmospheric circulation across the Cretaceous–Paleogene (K–Pg) boundary interval in low-latitude East Asia. Glob Planet Change 199:103435

Madriñán S, Cortés AJ, Richardson JE (2013) Páramo is the world's fastest evolving and coolest biodiversity hotspot. Front Genet 4:192

Maeda EE, Ma X, Wagner FH, Kim H, Oki T, Eamus D, Huete A (2017) Evapotranspiration seasonality across the Amazon Basin. Earth Syst Dyn 8(2):439–454

Maia GF, Ferrão M, Moraes LJ, Kaefer IL, Lima AP (2024) Cryptic diversity and historical biogeography of a nurse-frog species complex (*Allobates* tinae): Insights on biotic diversification in Brazilian southwestern Amazonia. Zool Scr 54(3):317-331. https://doi.org/10.1111/zsc.12711

Malhi Y, Roberts JT, Betts RA, Killeen TJ, Li W, Nobre CA (2008) Climate change, deforestation, and the fate of the Amazon. Science 319(5860):169–172

Mari L, Casagrandi R, Bertuzzo E, Rinaldo A, Gatto M (2014) Metapopulation persistence and species spread in river networks. Ecol Lett 17(4):426–434

Matamoros WA, McMahan CD, Chakrabarty P, Albert JS, Schaefer JF (2015) Derivation of the freshwater fish fauna of Central America revisited: Myers's hypothesis in the twenty-first century. Cladistics 31(2):177–188

Matschiner M (2019) Gondwanan vicariance or trans-Atlantic dispersal of cichlid fishes: a review of the molecular evidence. Hydrobiologia 832:9–37

Matschiner M, Böhne A, Ronco F, Salzburger W (2020) The genomic timeline of cichlid fish diversification across continents. Nat Commun 11(1):5895

Matthew WD (1915) Climate and evolution, vol 24. New York Academy of Sciences, New York, p 223

Mayden RL (1988) Vicariance biogeography, parsimony, and evolution in North American freshwater fishes. Syst Biol 37(4):329–355

McManamay RA, DeRolph CR (2019) A stream classification system for the conterminous United States. Sci Data 6(1):1–18

Melo BF, Stiassny ML (2024) Phylogenomic and anatomical evidence for the Late Cretaceous diversification of African characiform fishes, including a new family, under the influence of the Trans-Saharan Seaway. Evol J Linn Soc 3(1):kzae030

Mendes IS, Melo BF, Damasceno JS, Teixeira DF, Carvalho DC (2022) Phylogeography of *Hypomasticus copelandii* (Teleostei, Anostomidae) reveals distinct genetic lineages along Atlantic coastal drainages of Eastern Brazil. Diversity 14(1):29

Monod JL (1971) Chance and necessity: an essay on the Natural Philosophy of Modern Biology. Knopf, New York, pp 199

Montaña CG, Winemiller KO (2009) Comparative feeding ecology and habitats use of Crenicichla species (Perciformes: Cichlidae) in a Venezuelan floodplain river. Neotrop Ichthyol 7:267–274

Montes C, Silva CA, Bayona GA, Villamil R, Stiles E, Rodriguez-Corcho AF et al (2021) A middle to late Miocene trans-Andean portal: geologic record in the Tatacoa Desert. Front Earth Sci 8:587022

Mooers AO, Heard SB (1997) Inferring evolutionary process from phylogenetic tree shape. Q Rev Biol 72(1):31–54

Mora A, Baby P, Roddaz M, Parra M, Brusset S, Hermoza W, Espurt N (2010) Tectonic history of the Andes and sub-Andean zones: implications for the development of the Amazon drainage basin. In: Hoorn C, Wesselingh F (eds) Amazonia, landscape and species evolution: a look into the past. Wiley-Blackwell, Chichester, pp 38–60

Mora A, Villagómez D, Parra M, Caballero VM, Spikings R, Horton BK et al (2020) Late Cretaceous to Cenozoic uplift of the northern Andes: paleogeographic implications. Geol Colomb 3:ca-4

Morgan JV, Bralower TJ, Brugger J, Wünnemann K (2022) The Chicxulub impact and its environmental consequences. Nat Rev Earth Environ 3(5):338–354

Müller RD, Hassan R, Gurnis M, Flament N, Williams SE (2018) Dynamic topography of passive continental margins and their hinterlands since the Cretaceous. Gondwana Res 53:225–251

Musher LJ, Giakoumis M, Albert J, Del-Rio G, Rego M, Thom G et al (2022) River network rearrangements promote speciation in lowland Amazonian birds. Sci Adv 8(14):eabn1099

Myers GS (1949) Salt-tolerance of fresh-water fish groups in relation to zoogeographical problems. Bijdr Dierkd 28(1):315–322

O'Dea A, Lessios HA, Coates AG, Eytan RI, Restrepo-Moreno SA, Cione AL et al (2016) Formation of the Isthmus of Panama. Sci Adv 2(8):e1600883

Oberdorff T, Dias MS, Jézéquel C, Albert JS, Arantes CC, Bigorne R et al (2019) Unexpected fish diversity gradients in the Amazon basin. Sci Adv 5(9):eaav8681

Ohara WM, Zdradek DL, Tomaz FS (2025) The stream fish fauna of a pristine area in the South Amazon, rio Machado basin, Rondônia. Papéis Avulsos de Zoologia 65:e202565002

Oliveira-Silva L, Batalha-Filho H, Camelier P, Zanata AM (2024) Underestimated diversity in *Characidium* (Characiformes: Crenuchidae) from Neotropical rivers revealed by an integrative approach. Syst Biodiversity 22(1):2346510

Ortelius A (1596) Thesaurus geographicus: a new body of geography: or, a compleat description of the Earth. Christopher Plantin, Antwerp, pp 380

Ortmann AE (1913) The Alleghenian divide, and its influence upon the freshwater fauna. Proc Am Philos Soc 52(210):287–390

Osborn HF (1902) The law of adaptive radiation. Am Nat 36(425):353–363

Peters SE, Husson JM, Czaplewski J (2018) Macrostrat: a platform for geological data integration and deep-time earth crust research. Geochem Geophys Geosyst 19(4):1393–1409

Pietsch TW (2013) Trees of life: a visual history of evolution. JHU Press, Baltimore, p 376

Pollock LJ, O'connor LM, Mokany K, Rosauer DF, Talluto MV, Thuiller W (2020) Protecting biodiversity (in all its complexity): new models and methods. Trends Ecol Evolut 35(12):1119–1128

Potter PE (1978) Significance and origin of big rivers. J Geol 86(1):13–33

Potter PE (1997) The Mesozoic and Cenozoic paleodrainage of South America: a natural history. J South Am Earth Sci 10(5–6):331–344

Poulsen CJ, Ehlers TA, Insel N (2010) Onset of convective rainfall during gradual late Miocene rise of the central Andes. Science 328(5977):490–493

Ramirez JL, Machado CB, de Mello Affonso PRA, Galetti Jr PM (2023) Speciation in coastal basins driven by staggered headwater captures: dispersal of a species complex, *Leporinus bahiensis*, as revealed by genome-wide SNP data. Syst Biol 72(5):973–983

Range MM, Arbic BK, Johnson BC, Moore TC, Titov V, Adcroft AJ et al (2022) The Chicxulub impact produced a powerful global tsunami. AGU Adv 3(5):e2021AV000627

Rangel TF, Edwards NR, Holden PB, Diniz-Filho JAF, Gosling WD, Coelho MTP et al (2018) Modeling the ecology and evolution of biodiversity: biogeographical cradles, museums, and graves. Science 361(6399):eaar5452

Rapoport EH (1975) Areografía. Estrategias Geográficas de las Especies. Fondo de Cultura Económica, México

Ribeiro AC (2006) Tectonic history and the biogeography of the freshwater fishes from the coastal drainages of eastern Brazil: an example of faunal evolution associated with a divergent continental margin. Neotrop Ichthyol 4:225–246

Richey JE, Meade RH, Salati E, Devol AH, Nordin Jr CF, Santos UD (1986) Water discharge and suspended sediment concentrations in the Amazon River: 1982–1984. Water Resour Res 22(5):756–764

Rigaux P, Scholl M, Voisard A (2002) Spatial databases: with application to GIS. Morgan Kaufmann, San Francisco, p 410

Robertson DS, Lewis WM, Sheehan PM, Toon OB (2013) K-Pg extinction patterns in marine and freshwater environments: the impact winter model. J Geophys Res: Biogeosci 118(3):1006–1014

Rodriguez-Iturbe I, Rinaldo A (1997) Fractal river basins: chance and self-organization. Cambridge University Press, Cambridge, p 547

Rohde K (1992) Latitudinal gradients in species diversity: the search for the primary cause. Oikos 65:514–527

Roosevelt AC (2013) The Amazon and the Anthropocene: 13,000 years of human influence in a tropical rainforest. Anthropocene 4:69–87

Ross ST, Matthews WJ (2014) Evolution and ecology of North American freshwater fish assemblages. Freshwater fishes North Am 1:1–49

Rull V, Carnaval AC (2020) Neotropical diversification: patterns and processes. Springer, Berlin

Sacek V (2014) Drainage reversal of the Amazon River due to the coupling of surface and lithospheric processes. Earth Planet Sci Lett 401:301–312

Sacek V, Mutz SG, Bicudo TC, de Almeida RP, Ehlers TA (2023) The Amazon paleoenvironment resulted from geodynamic, climate, and sea-level interactions. Earth Planet Sci Lett 605:118033

Sandom C, Faurby S, Sandel B, Svenning JC (2014) Global late quaternary megafauna extinctions linked to humans, not climate change. Proc R Soc B 281(1787):20133254

Sanín MJ, Cardona A, Céspedes Arias LN, González-Arango C, Pardo N, Cadena CD (2024) Volcanoes, evolving landscapes, and biodiversity in Neotropical mountains. Front Biogeogr 16(1):e61882

Sanmartín I, Ronquist F (2004) Southern hemisphere biogeography inferred by event-based models: plant versus animal patterns. Syst Biol 53:216–243

Scotland RW, Sanderson MJ (2004) The significance of few versus many in the tree of life. Science 303(5658):643–643

Shephard GE, Müller RD, Liu L, Gurnis M (2010) Miocene drainage reversal of the Amazon River driven by plate–mantle interaction. Nat Geosci 3(12):870–875

Shonting D, Ezrailson C (2017) Chicxulub: the impact and tsunami: the story of the largest known asteroid to hit the Earth. Springer International Publishing, p 148

Shreve RL (1966) Statistical law of stream numbers. J Geol 74(1):17–37

Sibert EC, Norris RD (2015) New age of fishes initiated by the Cretaceous– Paleogene mass extinction. Proc Natl Acad Sci 112(28):8537–8542

Skelton PH, Teugels GG, Guegan JF, Albaret JJ (1994) Diversity and distribution of freshwater fishes in East and Southern Africa. Am Bull 82(5):1355–1376

Smith GR (1981) Late Cenozoic freshwater fishes of North America. Annu Rev Ecol Syst 12:163–193

Smith GR (1992) Phylogeny and biogeography of the Catostomidae, freshwater fishes of North America and Asia. In: Mayden R (ed) Systematics, Historical Ecology, and North American Freshwater Fishes, Stanford University Press, Stanford, pp 778–826

Smith GR, Badgley C, Eiting TP, Larson PS (2010) Species diversity gradients in relation to geological history in North American freshwater fishes. Evol Ecol Res 12(6):693–726

Snyder JO (1917) An account of some fishes from Owens River, California. Proc U S Nat Mus 54:201–205

Stanley SM (1975) A theory of evolution above the species level. Proc Natl Acad Sci 72(2):646–650

Stebbins GL (1983) Mosaic evolution: an integrating principle for the modern synthesis. Experientia 39(8):823–834

Stokes MF, Perron JT (2020) Modeling the evolution of aquatic organisms in dynamic river basins. J Geophys Res: Earth Surf 125(9):e2020JF005652

Stokes MF, Goldberg SL, Perron JT (2018) Ongoing river capture in the Amazon. Geophys Res Lett 45(11):5545–5552

Stokes MF, Kim D, Gallen SF, Benavides E, Keck BP, Wood J et al (2023) Erosion of heterogeneous rock drives diversification of Appalachian fishes. Science 380(6647):855–859

Strahler AN (1957) Quantitative analysis of watershed geomorphology. Eos, Trans Am Geophys Union 38(6):913–920

Strömberg CA (2011) Evolution of grasses and grassland ecosystems. Annu Rev Earth Planet Sci 39:517–544

Suess E (1901) Das Antlitz der Erde (The Face of the Earth), vol 1. G. Freytag, Leipzig, Germany

Symington A (2024). https://www.pythonmaps.com/

Tagliacollo VA, Dagosta FCP, Pinna MD, Reis RE, Albert JS (2021) Assessing extinction risk from geographic distribution data in Neotropical freshwater fishes. Neotrop Ichthyol 19(03):e210079

Tagliacollo VA, Duke-Sylvester SM, Matamoros WA, Chakrabarty P, Albert JS (2017) Coordinated dispersal and pre-isthmian assembly of the Central American ichthyofauna. Syst Biol 66(2):183–196

Tagliacollo VA, Roxo FF, Duke-Sylvester SM, Oliveira C, Albert JS (2015) Biogeographical signature of river capture events in Amazonian lowlands. J Biogeography 42(12):2349–2362

Tao W, Yang L, Mayden RL, He S (2019) Phylogenetic relationships of Cypriniformes and plasticity of pharyngeal teeth in the adaptive radiation of cyprinids. Sci China Life Sci 62:553–565

Tarboton DG, Bras RL, Rodriguez-Iturbe I (1988) The fractal nature of river networks. Water Resour Res 24(8):1317–1322

Terui A, Kim S, Pregler KC, Kanno Y (2021) Non-random dispersal in sympatric stream fishes: influences of natural disturbance and body size. Freshwater Biol 66(10):1865–1875

Thode VA, Sanmartín I, Lohmann LG (2019) Contrasting patterns of diversification between Amazonian and Atlantic forest clades of Neotropical lianas (*Amphilophium*, Bignonieae) inferred from plastid genomic data. Mol Phylogenet Evol 133:92–106

Thomaz AT, Christie MR, Knowles LL (2016) The architecture of river networks can drive the evolutionary dynamics of aquatic populations. Evolution 70(3):731–739

Tobler WR (1959) Automation and cartography. Geogr Rev 49(4):526–534

Tsai C, Raney E (1974) Systematics of the banded darter, *Etheostoma zonale* (Pisces: Percidae). Copeia 1974:1–24

Tukiainen H, Toivanen M, Maliniemi T (2023) Geodiversity and biodiversity. In: Kubalíková L, Coratza P, Pál M, Zwoliński Z, Irapta PN, van Wyk de Vries B (eds) Visages of Geodiversity and Geoheritage, vol 530. Geological Society London Special Publications, London, pp 31–47

Vajda V, Bercovici A (2014) The global vegetation pattern across the Cretaceous–Paleogene mass extinction interval: a template for other extinction events. Glob Planet Change 122:29–49

Val P, Lyons NJ, Gasparini N, Willenbring JK, Albert JS (2022) Landscape evolution as a diversification driver in freshwater fishes. Front Ecol Evol 9:788328

Val P, Silva C, Harbor D, Morales N, Amaral F, Maia T (2014) Erosion of an active fault scarp leads to drainage capture in the Amazon region, Brazil. Earth Surf Processes Landforms 39(8):1062–1074

Valente-Neto F, Mello JL, Pestana GC, Shimabukuro EM, de Siqueira AS, Covich AP, Saito VS (2024) Ecological perspectives on the organization of biodiversity in Neotropical streams. Hydrobiologia 852(12):3025-3047

van der Meer DG, Scotese CR, Mills BJ, Sluijs A, van de Weg RM (2022) Long-term Phanerozoic global mean sea level: insights from strontium isotope variations and estimates of continental glaciation. Gondwana Res 111:103–121

van der Sleen P, Albert JS (2022) Patterns in freshwater fish diversity. Ref Mod Earth Syst Environ Sci 26(3):894–907

van der Sleen P, Albert JS (eds) (2017) Field guide to the fishes of the Amazon, Orinoco, and Guianas. Princeton University Press, Princeton, p 464

van Soelen EE, Kim JH, Santos RV, Dantas EL, de Almeida FV, Pires JP et al (2017) A 30 Ma history of the Amazon River inferred from terrigenous sediments and organic matter on the Ceará Rise. Earth Planet Sci Lett 474:40–48

Vannote RL, Minshall GW, Cummins KW, Sedell JR, Cushing CE (1980) The river continuum concept. Can J Fish Aquat Sci 37(1):130–137

Vasconcelos, TS, Da Silva, FR, Dos Santos, TG, Prado, VH, Provete, DB, 2019. Biogeographic Patterns of South American Anurans, vol. 2019. Springer International Publishing, Switzerland, p 149

Vrba ES (1974) Chronological and ecological implications of the fossil Bovidae at the Sterkfontein australopithecine site. Nature 250:19–23

Vrba ES (1980) Evolution, species, and fossils: How does life evolve? S Afr J Sci 76:61–84

Waters JM, Wallis GP (2000) Across the southern Alps by river capture? Freshwater fish phylogeography in South Island, New Zealand. Mol Ecol 9(10):1577–1582

Wegener A (1922) Die Entstehung der Kontinente und Ozeane [The Origin of Continents and Oceans] (in German). Borntraeger, F. Vieweg & Sohn, p 144

Werneck FP, Gamble T, Colli GR, Rodrigues MT, Sites JrJW (2012) Deep diversification and long-term persistence in the South American 'dry diagonal': integrating continent-wide phylogeography and distribution modeling of geckos. Evolution 66(10):3014–3034

Westoll TS (1945) The mammalian middle ear. Nature 155(3926):114–115

Wheeler HE, Cook EF (1954) Structural and stratigraphic significance of the Snake River capture, Idaho-Oregon. J Geol 62(6):525–536

White AE, Li YJ, Griskevicius V, Neuberg SL, Kenrick DT (2013) Putting all your eggs in one basket: life-history strategies, bet hedging, and diversification. Psychol Sci 24(5):715–722

Wiens JJ, Morrill MC (2011) Missing data in phylogenetic analysis: reconciling results from simulations and empirical data. Syst Biol 60(5):719–731

Wiley EO, Lieberman BS (2011) Phylogenetics: theory and practice of phylogenetic systematics. John Wiley & Sons, p 432

Willis JC (1922) Age and area. Cambridge University Press, Cambridge, p 259

Wing SL, Herrera F, Jaramillo CA, Gómez-Navarro C, Wilf P, Labandeira CC (2009) Late Paleocene fossils from the Cerrejón Formation, Colombia, are the earliest record of Neotropical rainforest. Proc Natl Acad Sci 106(44):18627–18632

Woodworth JB (1894) The relation between base-leveling and organic evolution. Am Geol 14:209–235

Yamazaki D, Ikeshima D, Sosa J, Bates PD, Allen GH, Pavelsky TM (2019) MERIT Hydro: a high-resolution global hydrography map based on latest topography datasets. Water Resour Res 55:5053–5073

Zachos J, Pagani M, Sloan L, Thomas E, Billups K (2001) Trends, rhythms, and aberrations in global climate 65 Ma to present. Science 292(5517):686–693

Zapata S, Calderon-Diaz L, Jaramillo C, Oboh-Ikuenobe F, Piedrahita JC, Rodríguez-Cuevas M et al (2023) Drainage and sedimentary response of the Northern Andes and the Pebas system to Miocene strike-slip tectonics: a source to sink study of the Magdalena Basin. Basin Res 35(5):1674–1717

Zuquim G, Chaves PP, Quesada CA, Ruokolainen K, Tuomisto H (2023) Introducing a map of soil base cation concentration, an ecologically relevant GIS-layer for Amazonian forests. Geoderma Reg 33:e00645

2 Phylogeny and Divergence Time Estimates of Neotropical Freshwater Fishes

Victor A. Tagliacollo, Milton Tan, Roberto E. Reis and James S. Albert

Abstract

The Neotropics harbor the most diverse ichthyofauna on Earth, a vast mosaic of more than 6342 fish species, thriving in the tropical waters of Central and South America. This assemblage is a testament to sheer species richness and a living archive of evolutionary experimentation, one of the most striking examples of phenotypic diversity among continental vertebrates. From the slender-bodied characins darting through blackwater streams to the armor-plated catfishes scouring the benthos of Amazonian rivers, Neotropical fishes embody a spectrum of ecological strategies shaped by deep evolutionary histories. The past few decades have witnessed an unprecedented surge in molecular phylogenetic studies, many resolving relationships at the species level, offering a clearer, though still evolving, portrait of fish diversification in the Neotropics. This chapter endeavors to discuss a comprehensive, genome-wide, and time-calibrated phylogeny of NFF, utilizing DNA sequences generated over three decades through the collaborative efforts of a multinational ichthyological research community. These sequences, generated through the concerted efforts of an international community of ichthyologists, seek to reconcile molecular evidence with the intricate evolutionary pathways that define NFF diversity. By integrating thousands of publicly available DNA sequences from dozens of published studies, we have constructed a phylogenetic tree comprising nearly six thousand terminals, representing approximately 51% of known NFF species. It stands as one of the most comprehensive evolutionary frameworks for Neotropical fishes ever assembled. Yet, as with all scientific endeavors, its relevance is necessarily transient; species boundaries are constantly refined, as new taxa emerge from unexplored habitats, and analytical methodologies continue to evolve. Our aim is not to present a static, definitive tree, but rather to provide a foundational framework, one that serves as a reference point for future investigations, inviting revisions, refinements, and expansions as new data emerge. We acknowledge that the story of NFF is far from complete. Instead, it unfolds with each new discovery, each unexpected clade, and each evolutionary insight that reshapes our understanding of one of the richest fish assemblages on the planet. This time-calibrated tree of NFF encapsulates the collective efforts of a research community over decades and served as a framework for some of the evolutionary studies presented in this book focused on deciphering the macroevolutionary and macroecological patterns of the planet's most species-rich ichthyofauna.

2.1 Introduction

2.1.1 Phylogenetic Trees

Phylogenetic trees play a crucial role in understanding the evolutionary processes underlying the formation of biodiversity and the generation of large-scale ecological patterns (Pagel 1997; Smith et al. 2020). Phylogenetic trees are often used in macroecological studies, extending beyond their primary role of depicting evolutionary relationships for taxonomic classification. Indeed, phylogenies serve as frameworks for understanding historical patterns and processes, illuminating how lineages have expanded, contracted, and diversified across

V. A. Tagliacollo
Universidade Federal do Uberlândia, Uberlândia, MG, Brazil
e-mail: vatagliacollo@gmail.com

M. Tan
Illinois Natural History Survey, Urbana, IL, USA
e-mail: miltont@illinois.edu

R. E. Reis
Pontifícia Universidade Católica Rio Grande do Sul, Porto Alegre, RS, Brazil
e-mail: reis@pucrs.br

J. S. Albert (✉)
University of Louisiana at Lafayette, Lafayette, LA, USA
e-mail: jsalbe@gmail.com

J. S. Albert and R. E. Reis (eds.), *Ecology and Evolution of Amazonian Fishes*, Fascinating Life Sciences,
https://doi.org/10.1007/978-3-032-09328-8_2

vast landscapes. These studies reveal patterns of trait evolution, highlighting how species adapt to environmental shifts over time, while also offering a lens for examining community assembly processes, helping to discern whether species co-occurrence is shaped by environmental filtering, competitive exclusion, or historical contingency. Phylogenies also provide insights into macroecological diversity metrics, accounting for species diversity and evolutionary distinctiveness; how deeply species are embedded within the tree of life, how much history their lineages contain, and how that history shapes ecological interactions today. By integrating molecular data, morphological traits, and ecological characteristics, phylogenies contribute to a more holistic understanding of how Neotropical fish diversity has emerged and continues to evolve as one of Earth's most extraordinary groups. The comparative power of large-scale phylogenies enables researchers to uncover the intricate evolutionary narratives that have unfolded across South American waterways for tens of millions of years, revealing a history still being written with each new discovery.

Estimating the phylogenetic relationships among Neotropical freshwater fishes (NFF) is an ongoing challenge, due to the exceptional diversity of this group and the complexity of its evolutionary history. The dynamic nature of riverine environments, driven by river captures, marine incursions, and shifting drainage patterns, has repeatedly disrupted fish spatial distributions, creating both barriers to gene flow and opportunities for speciation. Freshwater fishes, more than many other vertebrate groups, have experienced lineage diversification at the mercy of these ever-changing hydrological networks. Beyond the complexities of biogeographic history, morphological diversity adds another layer of difficulty. The evolution of specialized body forms, feeding structures, and ecological adaptations has led to instances of convergent evolution that obscure phylogenetic signals, making it challenging to resolve relationships among certain groups. Rapid evolutionary radiations and hybridization events further complicate efforts to construct robust phylogenies. To navigate these challenges, ichthyologists increasingly rely on molecular data, integrating multilocus datasets and genome-wide approaches to improve phylogenetic resolution. Advances in sequencing technologies and analytical methodologies continue to refine our understanding of fish evolution, allowing researchers to untangle the intricate web of relationships that define the Neotropical ichthyofauna.

Large-scale phylogenetic reconstructions are essential for studying the macroecology of Neotropical fishes, offering a framework for exploring patterns of diversification across deep evolutionary time. These trees enable researchers to investigate morphological evolution, shedding light on how ecological pressures have shaped fish anatomy and function. They also provide a foundation for biogeographic analyses, tracing the movements of lineages across ancient river systems and floodplains. The statistical power afforded by large phylogenetic datasets ensures that evolutionary hypotheses are tested with greater robustness, minimizing uncertainties and allowing for more precise inferences on a large scale. Beyond their role in academic research, these trees have practical applications in conservation, helping identify evolutionarily significant units, prioritize regions for protection, and safeguard the most phylogenetically unique lineages. As biodiversity faces mounting threats from habitat destruction, climate change, and invasive species, an evolutionary perspective is essential for guiding conservation strategies that preserve not only species but also the deep evolutionary heritage they represent.

2.1.2 Goals of This Chapter

This chapter presents a comprehensive, genomic-wide, and time-calibrated tree of NFF species using DNA sequences from GenBank, accumulated over three decades by an international ichthyological research community. These data represent a collective effort resulting in the compilation of a supermatrix and estimation of relationships and lineage divergence times that summarize our understanding of the evolutionary relationships among the most diverse ichthyofauna on Earth. By elucidating NFF relationships on a broad scale, this chapter establishes an initial framework for further investigations and opens new avenues for research in the realms of macroecology and macroevolution within the Neotropical ichthyofauna. This phylogeny is the product of a specific set of methods, which resulted in a summary of our current understanding of the timing and relationships of NFF species. There are however some known discrepancies with the published literature, and future work will need to continually update this phylogeny in light of new discoveries.

2.2 Methods

2.2.1 Taxon Sampling

A list of 6342 valid species was compiled using the information from the Check List of the Freshwater Fishes of South and Central America (CLOFFSCA; Reis et al. 2003), subsequently updated with data from the Catalog of Fishes (Fricke 2021) up to 2021. Each species name was reviewed to verify its current taxonomic status. New species described since the publication of CLOFSCA were identified through searches conducted at the family level within the Catalog of Fishes. All available names from these searches were examined, and only those recognized as valid were included. Major changes to the taxonomy of NFF species since the data for this analysis were assembled in 2021 include: cichlid genus *Crenicichla* raised to subtribe Crenicichlina with six five new genera (*Hemeraia*, *Lugubria*, *Saxatilia*, *Wallaciia*; Varella et al. 2023); new genus of Characidae (*Makunaima*; Terán et al. 2020); Characidae partitioned into four families (Acestrorhamphidae, Characidae, Spintherobolidae, Stevardiidae; Melo et al. 2024); three newly recognized subfamilies of Anostomindae (Anostominae, Leporellinae, Leporininae; Sidlauskas et al. 2025); four newly recognized subfamilies of Pimelodidae (Phractocephalinae, Pimelodinae, Sorubininae, Steindachneridioninae; Silva et al. 2024); two newly recognized subfamilies of Pseudopimelodidae (Batrochoglaninae, Pseudopimelodidae, Silva et al. 2021); two newly recognized subfamilies of Heptapteridae (Heptapterinae, Rhamdiinae; Silva et al. 2021); three newly recognized subfamilies of Doradidae (Acanthodoradinae, Agamyxinae, Rhinodoradinae; Sabaj and Arce 2021); Callichthyid genus *Corydoras* raised to subfamily Corydoradinae with seven newly recognized genera: *Aspidoras*, *Brochis*, *Gastrodermus*, *Hoplisoma*, *Osteogaster*, *Scleromystax* (Dias et al. 2024); new genus of Loricariidae (*Rhinotocinclus*, Reis and Lehmann 2022); two newly recognized genera (*Colossimystax*, Stellantia), and three newly recognized tribes of Hypostominae (Peckoltini, Pseudancistrini, Stellantini; Armbruster and Lujan 2025).

2.2.2 Acquiring DNA Sequences

To assemble a comprehensive tree of NFF, we established a chronological threshold, systematically seeking published DNA sequences for valid species described up to 2021. While this list aims to be comprehensive, it is not definitive, as new species continue to be discovered. We employed customized scripts and the R package PhylotaR (Bennett et al. 2018) to retrieve published DNA sequences from public repositories. We retrieved orthologous sequences from GenBank, using the GenBank Taxonomy Database as a reference (Schoch et al. 2020), and downloaded sequences at the family level. DNA sequences from different markers were retrieved independently for each family and subsequently merged into a single supermatrix encompassing all sequences. To minimize errors from misidentification or low-quality sequencing, we retrieved up to three sequences per species and manually inspected alignments to remove problematic sequences with poor homology or uncertain identification (e.g., cf., sp., aff.). The outcome is a curated and expandable dataset, including GenBank accession numbers, that can be continuously updated with newly available sequences in public repositories. The list of DNA markers and their corresponding availability for each species is available in the Zenodo repository (10.5281/zenodo.13221492).

The retrieved DNA markers exhibit broad genomic diversity, including both mitochondrial and nuclear, as well as coding and noncoding genes commonly employed in multilocus phylogenies of NFF clades (e.g., Oliveira et al. 2011; Tagliacollo et al. 2012; Ochoa et al. 2017; Melo et al. 2018; Fontenelle et al. 2021). Because orthologous sequences were retrieved independently at the family level, some markers were available for certain families but not for others. Nevertheless, there was sufficient overlap of markers across families to ensure reliable empirical estimates of evolutionary relationships. To resolve relationships at finer taxonomic scales, we partitioned certain highly variable markers, such as the D-LOOP region, into separate alignments for three major clades: Siluriformes, Poeciliidae, and Cichlidae. This strategy ensured a more robust hypothesis of homologies for rapidly evolving molecular regions. Sequences were derived from the same identified species, and whenever possible, from the same individual specimens. A similar approach combining sequences from different individuals of the same species has been successfully used in other phylogenetic studies (Arce-H et al. 2017; Mirande 2019). The majority of markers retrieved are commonly sequenced for inclusion in multilocus datasets (Li et al. 2007).

2.2.3 Alignment, Trimming, and Partitioning Scheme

After retrieving DNA sequences from GenBank, we performed independent alignments for each genetic marker using MAFFT v.7 (Katoh and Toh 2008) with default settings. To enhance alignment quality and avoid subjectivity in manual editing, we applied an automated trimming procedure with GBlocks (Castresana 2000; Talavera and Castresana 2007), which removed gap-rich and poorly aligned regions, particularly at the flanks of each alignment, based on conserved sequence blocks. The resulting cleaned alignments were then concatenated into a supermatrix, where each marker was treated as a separate partition. For each partition, we specified the GTR + GAMMA model of nucleotide substitution. To identify the most appropriate partitioning scheme, we used PartitionFinder2 (Lanfear et al. 2017), applying a greedy search algorithm under the assumption of unlinked branch lengths across partitions. We fixed the substitution model to GTR + GAMMA across all partitions to focus the search on the optimal partition configuration while maintaining model consistency.

2.2.4 Tree Inferences

Phylogenetic inference was conducted using RAxML-HPC v.8 (Stamatakis 2014), employing a maximum likelihood (ML) approach to identify the best-scoring tree topology. The analysis was based primarily on genetic markers commonly utilized in multilocus phylogenetic studies (e.g., mitochondrial and nuclear gene regions), and we incorporated a backbone tree constrained at key nodes, especially at the family level, following well-supported hypotheses from prior studies (e.g., Armbruster et al. 2015; Tagliacollo et al. 2016; Lujan et al. 2018; Melo et al. 2022; Reis and Lehmann

2022). These topological constraints were necessary due to the limited phylogenetic resolution of the selected markers at deeper nodes. The backbone tree and its respective node constraints can be seen in the Zenodo repository (10.5281/zenodo.13221492). We did not include ultraconserved elements (UCEs), which have been shown to better resolve deep phylogenetic relationships (Faircloth et al. 2012), due to the lack of available sequences for many taxa within the NFF clade. The tree was rooted using the two lamprey species *Geotria australis* and *Mordacia lapicida* as outgroup taxa, placed at their most recent common ancestor. No additional outgroups were required because time-reversible substitution models used in ML methods do not require explicit character polarization (Felsenstein 2004). RAxML estimated all free parameters and optimized the likelihood under the GAMMA model using its rapid hill-climbing algorithm. To assess tree support, we conducted 100 bootstrap replicates on the resulting likelihood tree topology. All analyses were carried out on the CIPRES Science Gateway (Miller et al. 2010).

2.2.5 Time Calibration

Given the size of the supermatrix and the number of taxa included, estimating divergence times using Bayesian methods was computationally prohibitive. As an alternative, we employed the congruification method available in the R package geiger v2 (Pennell et al. 2014), which allows the estimation of divergence times by aligning nodes between an uncalibrated phylogeny and a time-calibrated reference tree. This approach relies on identifying shared clades between the focal tree (with branch lengths in substitutions per site) and a dated reference tree, using the overlapping divergence times as secondary calibrations (Eastman et al. 2013; Pennell et al. 2014). To reduce sampling redundancy, we randomly selected a single representative per species. For the reference framework, we used a comprehensive, time-calibrated phylogeny of ray-finned fishes derived from 1105 phylogenomic markers across 303 extant species, encompassing 66–72 recognized orders and incorporating 31 fossil constraints (Hughes et al. 2018). Divergence times between cartilaginous fish genera were retrieved from the shark tree of life, which is based on ten fossil calibrations (Stein et al. 2018). Clade matches between our uncalibrated tree and the reference trees were determined using the congruify.phylo function (Pennell et al. 2014) in the R environment (R Core Team 2023). The estimated clade ages were then used to calibrate node times in our phylogeny using treePL (Smith and O'Meara 2012), a penalized likelihood method suited for large trees. We initialized the treePL analysis with thorough = TRUE and prime = TRUE options, followed by the default parameter priming to fine-tune the optimization. Based on the suggested configuration, we applied the following settings for time estimation: opt = 1, optad = 1, and optcvad = 4.

2.3 Results

2.3.1 Assembled Matrix

The resulting supermatrix constitutes one of the most comprehensive DNA sequence datasets assembled to date for NFF taxa, encompassing approximately half of all recognized species. Despite considerable efforts to include all valid species, nearly 50% remain unrepresented due to the absence of sequences in GenBank, unverifiable sample identifiers, or insufficient sequence quality for accurate alignment. Using PhylotaR, we retrieved sequences via BLAST queries to GenBank, identifying 51 orthologous clusters. The final dataset comprises 43,850 base pairs, 21,223 sequences, 5984 terminals, and 3167 species distributed across 741 genera, representing about 86% of all NFF genera. The data consist of approximately 61% mitochondrial and 39% nuclear markers, primarily coding regions, with a few non-coding sequences such as tRNAs and rRNAs. To optimize phylogenetic inference, we applied PartitionFinder2 (Lanfear et al. 2017), which simultaneously selected the best partitioning scheme and molecular evolution models. The best scheme identified 30 distinct partitions, all modeled under GTR + GAMMA, but with independent parameter estimates for each. Information on GenBank accession numbers, alignments, and species-level phylogenetic relationships is archived in the Zenodo repository (10.5281/zenodo.13221492). Additional metrics regarding taxon sampling completeness are presented in Fig. 2.1 and Table 2.1.

2.3.2 Estimated Phylogeny

The phylogeny depicted in Fig. 2.1 is generally consistent with information in the existing literature, as expected given the compilation of original datasets from GenBank into a supermatrix. To enhance interpretation and visualization of the evolutionary relationships, we divided the time-calibrated tree into five sections (Figs. 2.2–2.6) including relationships among major NFF clades (Fig. 2.2), Characiformes (Fig. 2.3), Siluriformes (Fig. 2.4), Cichlinae (Fig. 2.5), Cyprinodontiformes (Fig. 2.6). More information, including species-level relationships, is available in the Zenodo repository (10.5281/zenodo.13221492).

2.3.3 Clade Ages Estimates

The time-calibrated phylogeny indicates that the earliest lineages of Neotropical freshwater fishes (NFF), represented by two lamprey species, diverged from all other NFF during the Ordovician period of the Paleozoic Era, approximately 480 million years ago (Ma) (Fig. 2.2). Among gnathostome vertebrates with extant freshwater representatives

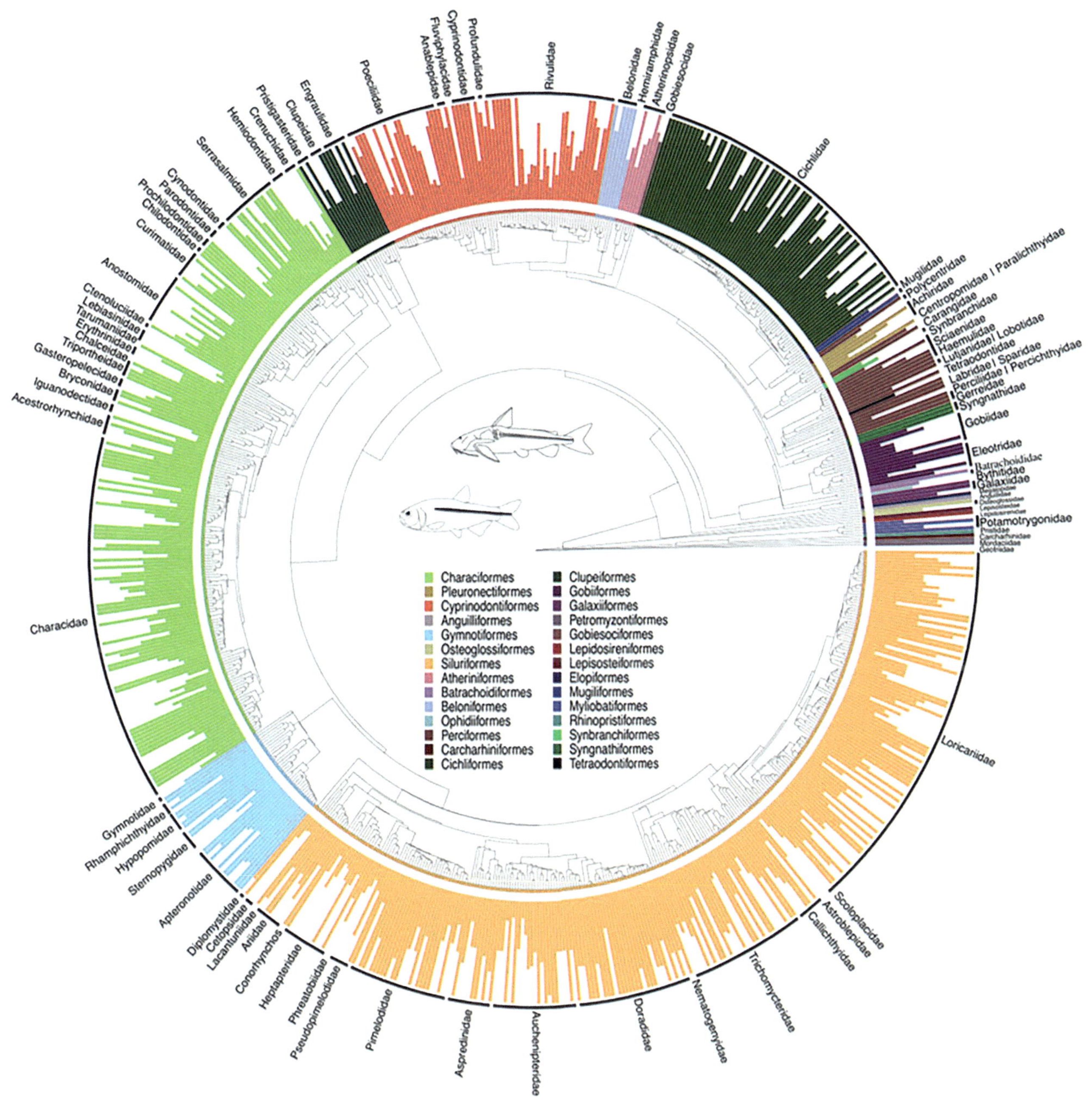

Fig. 2.1 Time-calibrated phylogeny of 3166 neotropical freshwater fish (NFF) species, representing 51% of all described NFF (Table 2.1). The figure illustrates evolutionary relationships among 741 genera (86% of NFF genera) across 38 orders. The phylogeny was estimated using RAxML-HPC from alignments of 51 orthologous markers, totaling 43,850 base pairs, derived from 21,223 sequences and comprising 5984 terminals. Bars at the tips are proportional to the fraction of valid species from each genus included in the phylogeny. Figure from Tagliacollo et al. (2024)

in the Neotropics—Chondrichthyes, Sarcopterygii, and Actinopterygii—the most recent common ancestor dates to the Silurian, with subsequent divergences occurring in the Permian (299–252 Ma; Fig. 2.2). Within Osteichthyes, the Actinopterygii, the most species-rich clade in the Neotropics, encompassing ostariophysians, cichlids, and cyprinodontiforms, diverged from their sister group Sarcopterygii in the Upper Cretaceous (Fig. 2.2). Most actinopterygian families diversified during the Paleogene (66–23 Ma). In Characiformes, a clade that includes piranhas, tetras, and their relatives, the initial divergence also dates to the Upper Cretaceous, with most families originating during the Eocene (56–34 Ma; Fig. 2.3). The Siluriformes, comprising whiskered catfishes, armored catfishes, and allied taxa, similarly

Table 2.1 Summary of the dataset used to estimate the time-calibrated phylogeny of neotropical freshwater fish (NFF) species

Order	Family	Species		%	MRCA (Ma)	Group
		Tree	NFF			
Petromyzontiformes	Mordaciidae	1	1	100.0	480.7	Stem
	Geotriidae	1	1	100.0	480.7	Stem
Carcharhiniformes	Carcharhinidae	1	1	100.0	276.8	Stem
Rhinopristiformes	Pristidae	1	1	100.0	200.9	Stem
Myliobatiformes	Potamotrygonidae	13	35	37.1	39.0	Crown
Ceratodontiformes	Lepidosirenidae	1	1	100.0	440.8	Stem
Lepisosteiformes	Lepisosteidae	1	1	100.0	298.8	Stem
Osteoglossiformes	Arapaimidae	1	2	50.0	117.2	Stem
	Osteoglossidae	2	2	100.0	117.2	Stem
Anguilliformes	Anguillidae	1	1	100.0	182.5	Stem
Elopiformes	Megalopidae	1	1	100.0	182.5	Stem
Clupeiformes	Engraulidae	13	20	65.0	27.5	Crown
	Clupeidae	4	6	66.7	80.4	Crown
	Pristigasteridae	5	7	71.4	27.4	Crown
Characiformes	Acestrorhynchidae	17	25	68.0	56.1	Crown
	Anostomidae	70	148	47.3	42.2	Crown
	Bryconidae	31	44	70.5	44.2	Crown
	Chalceidae	3	6	50.0	24.5	Crown
	Characidae	438	1088	40.3	51.7	Crown
	Chilodontidae	8	8	100.0	52.3	Crown
	Crenuchidae	24	76	31.6	63.0	Crown
	Ctenoluciidae	5	7	71.4	67.5	Crown
	Curimatidae	82	115	0.7	49.7	Crown
	Cynodontidae	4	8	50.0	31.5	Crown
	Erythrinidae	11	16	68.8	57.7	Crown
	Gasteropelecidae	6	9	66.7	33.8	Crown
	Hemiodontidae	7	31	22.6	44.1	Crown
	Iguanodectidae	5	36	13.9	51.8	Crown
	Lebiasinidae	11	71	15.5	71.1	Crown
	Parodontidae	9	32	28.1	38.6	Crown
	Prochilodontidae	21	21	100.0	43.6	Crown
	Serrasalmidae	76	99	76.8	33.7	Crown
	Tarumaniidae	1	1	100.0	67.6	Stem
	Triportheidae	19	21	90.5	50.2	Crown
Siluriformes	Ariidae	22	32	0.7	25.0	Crown
	Aspredinidae	30	46	65.2	55.1	Crown
	Astroblepidae	4	80	5.0	19.0	Crown
	Auchenipteridae	71	123	57.7	38.3	Crown
	Callichthyidae	146	221	66.1	57.6	Crown
	Cetopsidae	5	34	14.7	57.1	Crown
	Diplomystidae	6	7	85.7	3.4	Crown
	Doradidae	78	91	85.7	41.2	Crown
	Heptapteridae	46	184	25.0	38.4	Crown
	Lacantuniidae	1	1	100.0	54.0	Stem
	Loricariidae	518	971	53.3	52.6	Crown
	Nematogenyidae	1	1	100.0	84.5	Stem
	Phreatobiidae	2	3	66.7	18.2	Crown
	Pimelodidae	67	109	61.5	46.4	Crown
	Pseudopimelodidae	15	49	30.6	32.1	Crown
	Scoloplacidae	1	6	16.7	58.7	Stem
	Trichomycteridae	119	299	39.8	68.0	Crown
Gymnotiformes	Apteronotidae	60	95	63.2	32.2	Crown
	Gymnotidae	35	46	76.1	62.6	Crown
	Hypopomidae	31	34	91.2	45.6	Crown
	Rhamphichthyidae	19	26	73.1	47.5	Crown
	Sternopygidae	26	54	48.1	53.4	Crown

(continued)

Table 2.1
(continued)

Order	Family	Species		%	MRCA (Ma)	Group
		Tree	NFF			
Galaxiiformes	Galaxiidae	7	8	87.5	61.9	Crown
Ophidiiformes	Bythitidae	3	7	42.9	15.2	Crown
Batrachoidiformes	Batrachoididae	2	3	66.7	35.3	Crown
Gobiiformes	Eleotridae	15	19	78.9	54.6	Crown
	Gobiidae	10	24	41.7	69.6	Crown
Syngnathiformes	Syngnathidae	5	5	100.0	86.1	Crown
Acanthuriformes	Lobotidae	1	1	100.0	85.0	Stem
	Gerreidae	7	7	100.0	67.0	Crown
	Sciaenidae	17	26	65.4	20.3	Crown
	Haemulidae	3	3	100.0	31.0	Crown
	Lutjanidae	1	2	50.0	67.0	Stem
Perciformes	Labridae	1	1	100.0	89.3	Stem
	Sparidae	1	1	100.0	88.7	Stem
Tetraodontiformes	Tetraodontidae	1	2	50.0	88.7	Stem
Centrarchiformes	Percichthyidae	1	5	20.0	18.0	Stem
	Perciliidae	2	2	100.0	2.0	Crown
Synbranchiformes	Synbranchidae	2	5	40.0	97.0	Crown
Carangiformes	Carangidae	2	2	100.0	69.4	Crown
	Achiridae	11	21	52.4	27.0	Crown
	Paralichthyidae	1	1	100.0	55.0	Stem
	Centropomidae	8	8	100.0	25.2	Crown
Cichliformes	Polycentridae	2	3	66.7	29.5	Crown
	Cichlidae	371	557	66.6	35.2	Crown
Mugiliformes	Mugilidae	3	4	75.0	37.3	Crown
	Gobiesocidae	3	4	75.0	23.5	Crown
Cyprinodontiformes	Anablepidae	16	19	84.2	29.6	Crown
	Cyprinodontidae	27	53	50.9	52.6	Crown
	Fluviphylacidae	2	7	28.6	14.0	Crown
	Poeciliidae	139	211	65.9	28.0	Crown
	Profundulidae	11	15	73.3	25.0	Crown
	Rivulidae	221	450	49.1	42.5	Crown
Atheriniformes	Atherinopsidae	57	88	64.8	53.0	Crown
Beloniformes	Belonidae	15	16	93.8	48.9	Crown
	Hemiramphidae	1	2	50.0	48.0	Stem
Total		3166	6036	52.0		

"Species TREE" indicates the number of species per family included in the phylogeny. "Species NFF" refers to the total number of described species per family (likely underestimated due to the high rate of ongoing species descriptions). "MRCA" denotes the estimated node age of the most recent common ancestor for each family. Node ages correspond to either crown or stem groups, as specified in the "Group" column. For additional details, see the Zenodo repository: 10.5281/zenodo.13221492

originated in the Upper Cretaceous, with the bulk of their diversification occurring in the Paleogene, particularly during the Eocene (Fig. 2.4). Neotropical cichlids of the subfamily Cichlinae (Cichlidae), including heroines, cichlasomatines, geophagines, among others, represent the youngest of the major NFF clades, with origins around 40 Ma in the Cenozoic (Fig. 2.5). All South American cichlid genera emerged during the Miocene, while a subset of Central American genera appeared more recently, between the Pliocene and Pleistocene (Fig. 2.5). The Cyprinodontiformes, which include killifishes, livebearers, and their allies, also originated in the Upper Cretaceous (Fig. 2.6). Within this group, the most recent common ancestors of Rivulidae, Cyprinodontidae, and Fluviphylacidae diversified during the Eocene, while those of Profundulidae, Poeciliidae, and Anablepidae originated in the Oligocene (Fig. 2.6). Divergence time estimates for NFF families are presented in Table 2.1.

A time-calibrated phylogeny of all 6342 NFF species used in the ecospace and ecotrace analyses of subsequent chapters is provided in Fig. 2.7. This "All NFF species tree" was assembled

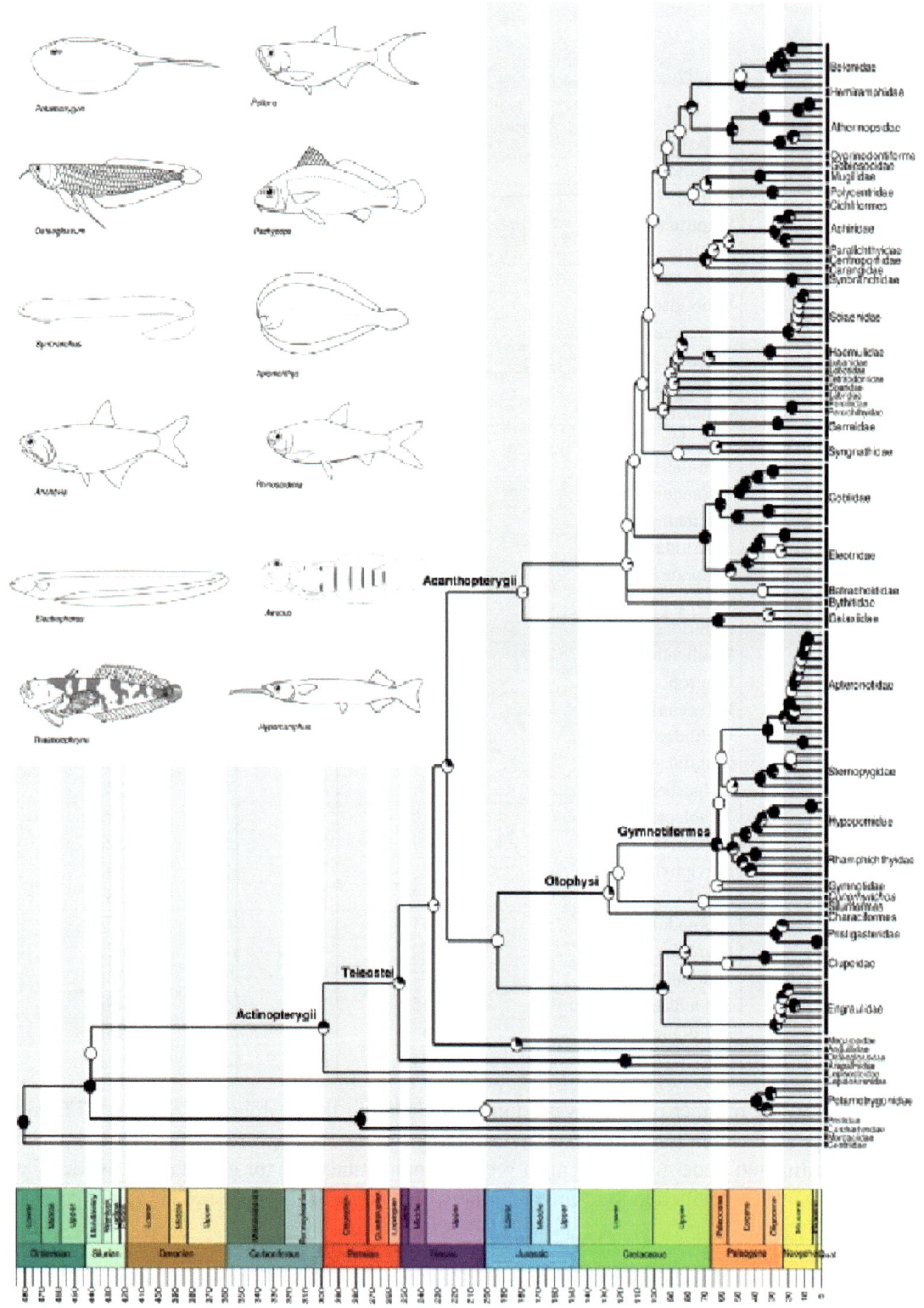

Fig. 2.2 Time-calibrated phylogeny of neotropical freshwater fishes (NFF) among genera (unlabeled tips) and families (labelled). See details for Characiformes in Fig. 2.3, Siluriformes in Fig. 2.4, Cichliformes in Fig. 2.5, and Cyprinodontiformes in Fig. 2.6. Pie charts at nodes indicate bootstrap support values, with proportions in black corresponding to bootstrap support values, and solid black circles representing 100% support. All GenBank accession numbers, alignments, species-level relationships, divergence clade estimates, and species representativeness percentages provided in Tagliacollo et al. (2025). Line-art images from van der Sleen and Albert (2017)

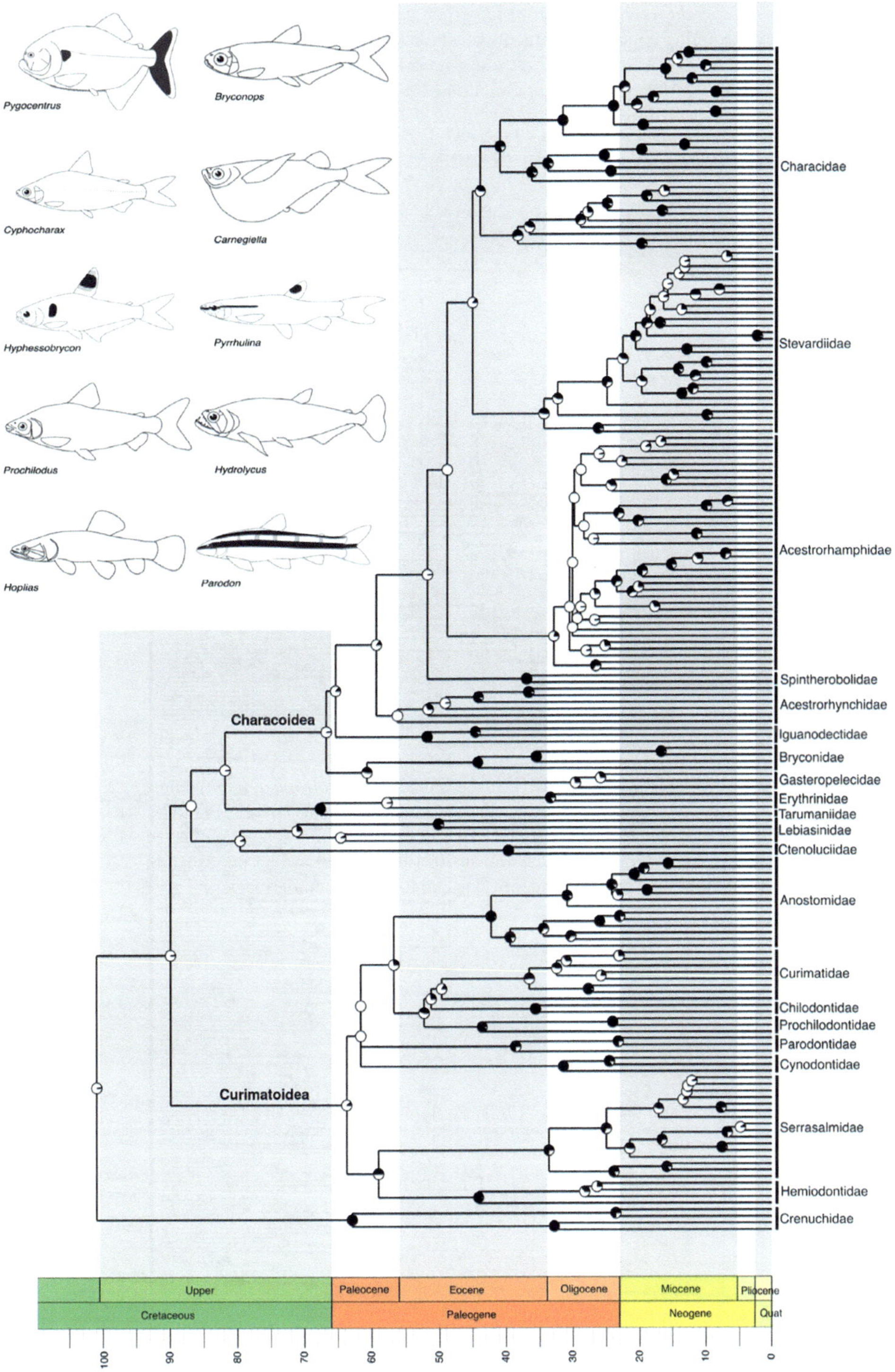

Fig. 2.3 Time-calibrated phylogeny of neotropical characiformes illustrating evolutionary relationships among all valid genera. Note origins at c. 100 Ma approximately coeval with the last continental connection between South America and Africa. Data and graphical conventions as in Fig. 2.2

using the phylogeny of 3166 species in 741 genera in Fig. 2.2 as a backbone phylogeny and manually adding the additional 3176 species, including 139 genera, by 44 taxonomic experts, using a combination of information sources; taxonomic identity (i.e., genus, species-complex), published phylogenies, and expert knowledge of each genus (see Albert et al. 2025).

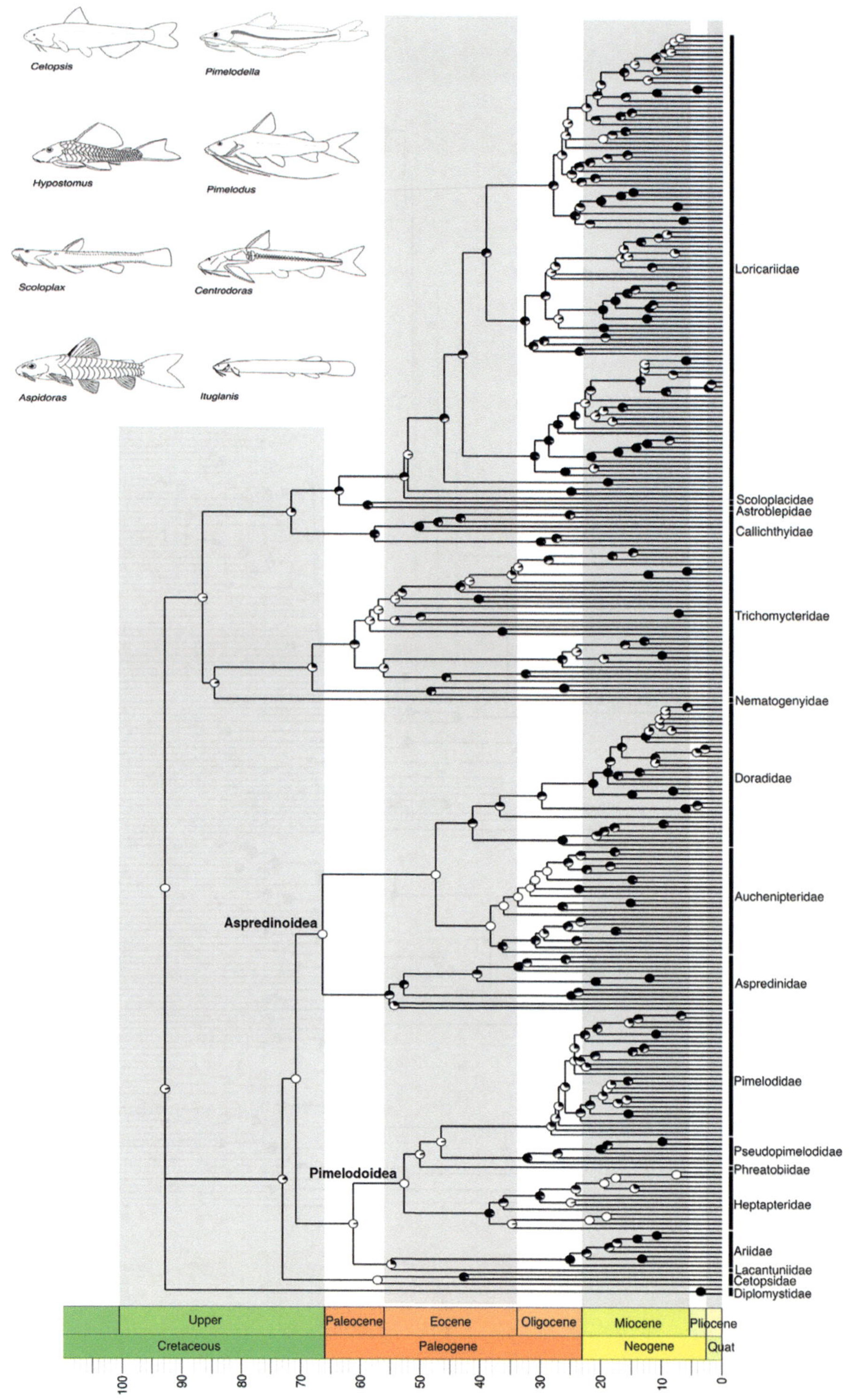

Fig. 2.4 Time-calibrated phylogeny of neotropical siluriformes illustrating evolutionary relationships among all valid genera. Note origins at c. 94 Ma soon after the last continental connections between South America and Africa. Data and graphical conventions as in Fig. 2.2

In order to calculate diversification rates and to estimate species ages, species were added at the middle of branches, and all polytomies were arbitrarily resolved. This "All NFF species tree" is therefore presented as a working hypothesis to be improved with results of future studies, and inferences are not made about Ancestral State Estimation or Ancestral Area Estimation among species with divergence times less than 5 Ma.

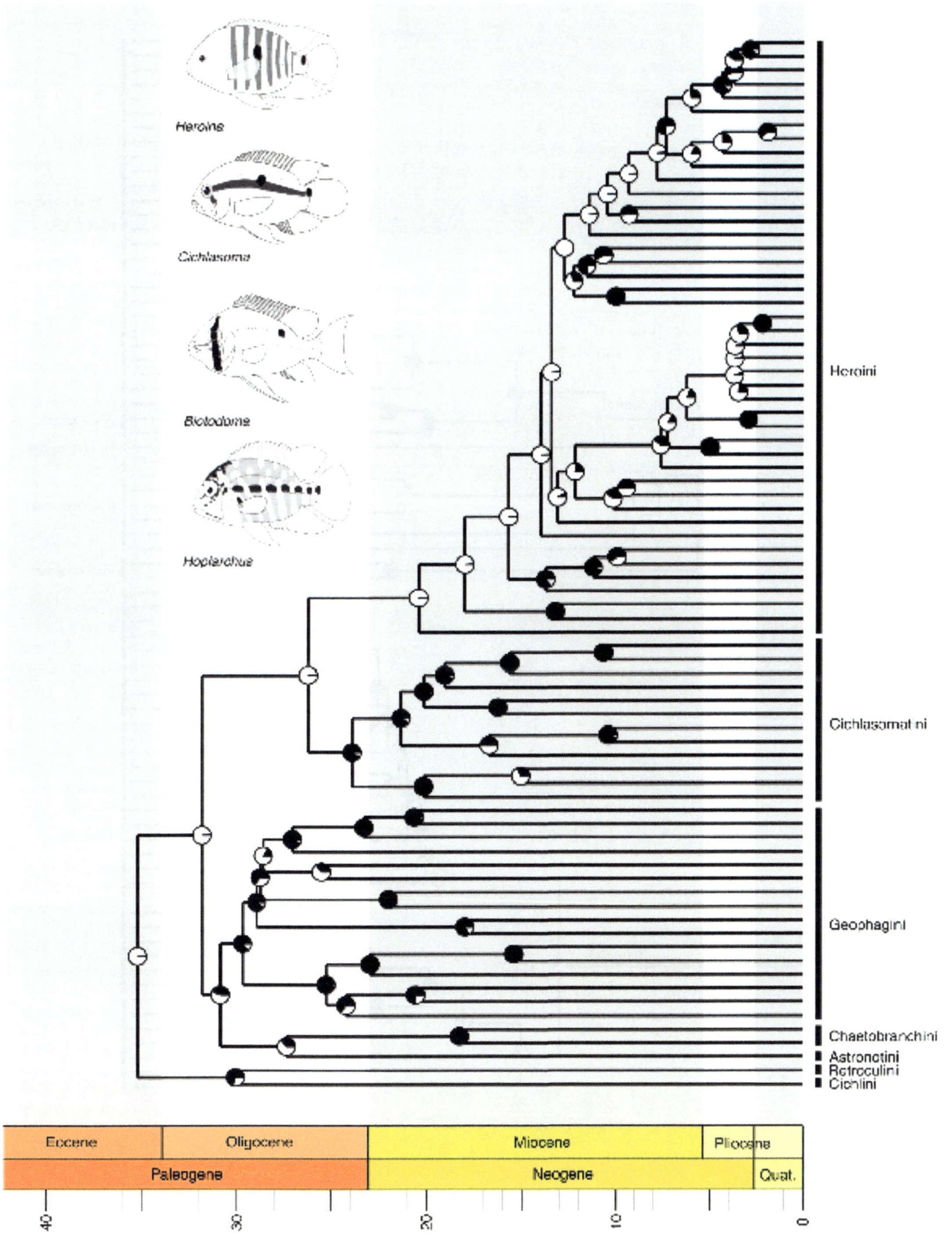

Fig. 2.5 Time-calibrated phylogeny of neotropical cichliformes (Cichlinae) illustrating evolutionary relationships among all valid genera. Note late Eocene origins (c. 36 Ma) in this calibration, long after the last continental connections between South America and Africa. Data and graphical conventions as in Fig. 2.2

2.4 Advances and Limitations of the All NFF Phylogeny

The construction of a comprehensive, time-calibrated phylogeny for NFF stands as a scientific endeavor of both intellectual rigor and deep scientific consequence. It is, at its core, an attempt to piece together an evolutionary mosaic, one that captures the branching patterns of shared ancestry and the nuanced historical processes that have shaped one of the world's richest freshwater faunas. And in doing so, it illuminates far more than the tree itself; it reveals the rhythms of speciation, the echoes of ancient

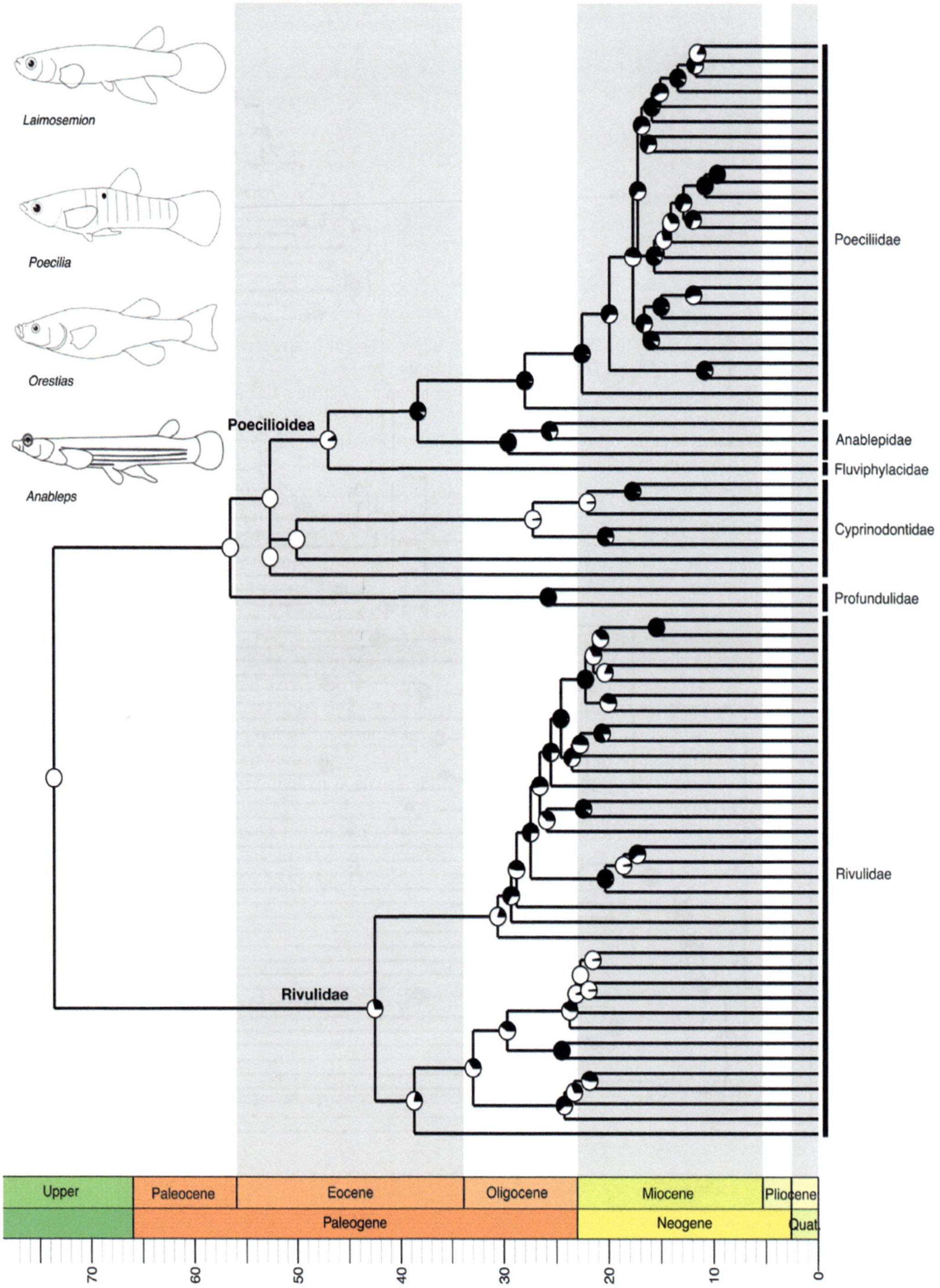

Fig. 2.6 Time-calibrated phylogeny of neotropical cyprinodontiformes illustrating evolutionary relationships among all valid genera. Note two clades in South America, Poecilioidea and Rivulidae, each with Paleogene origins, c. 47 and 43 Ma, respectively, long after the last continental connection between South America and Africa. Data and graphical conventions as in Fig. 2.2

drainages, and the evolutionary signatures etched across a continental waters. This chapter presents the most species-rich NFF phylogeny assembled to date, an effort borne of multinational collaboration, cumulative knowledge, and the curiosity that defines the Neotropical ichthyological community. It is, as all scientific constructs are, provisional. Its structure is constrained by current data availability, by the unresolved boundaries of certain taxa, and by the inherent dynamism of evolutionary discovery. Yet within those constraints, it offers a robust framework upon which future studies, including those in this book, can be built.

Building a comprehensive phylogenetic tree encompassing nearly half of all valid NFF species offers profound value across scientific, conservation, and educational spheres. At its core, such a tree provides a holistic narrative of lineage relationships, evolutionary pathways, and biodiversity patterns shaped by deep time. Scientifically, it acts as both a lens and a scaffold; clarifying taxonomic uncertainties while illuminating macroevolutionary processes that govern diversification across river basins and ecological gradients. From a conservation standpoint, its relevance is no less compelling; by highlighting phylogenetic distinctiveness and evolutionary isolation, the tree informs efforts to safeguard not just species, but evolutionary heritage. Educationally, it becomes a living archive, accessible, interpretable, and engaging, encouraging dialogue, curiosity, and collaboration within and beyond the scientific community. Methodologically, its assembly represents an ambitious step forward in scaling up phylogenetic inference, pushing the boundaries of computational biology and data integration. Globally, it serves as a reference point for comparative biogeography and systematics, embedding Neotropical ichthyofauna within broader frameworks of biodiversity science. And, crucially, because the tree is not static but adaptive, amenable to updates, corrections, and the addition of new data, it stands as a dynamic resource, ever evolving alongside our understanding, and thus indispensable for ongoing research and stewardship.

The construction of a comprehensive phylogenetic tree for NFF is also an indispensable endeavor for unraveling the intricate web of biodiversity, evolutionary history, and ecological complexity that defines this group. By capturing phylogenetic relationships across hundreds, even thousands, of lineages, this large-scale tree reveals who is related to whom and how diversification has unfolded through pulses of speciation, bursts of adaptive radiation, and the quiet persistence of ancient lineages surviving in refugia. At a continental scale, it enables researchers to track macroevolutionary patterns shaped by geological upheaval, hydrological connectivity, and climatic oscillations, patterns that would otherwise remain obscured in fragmented or regional

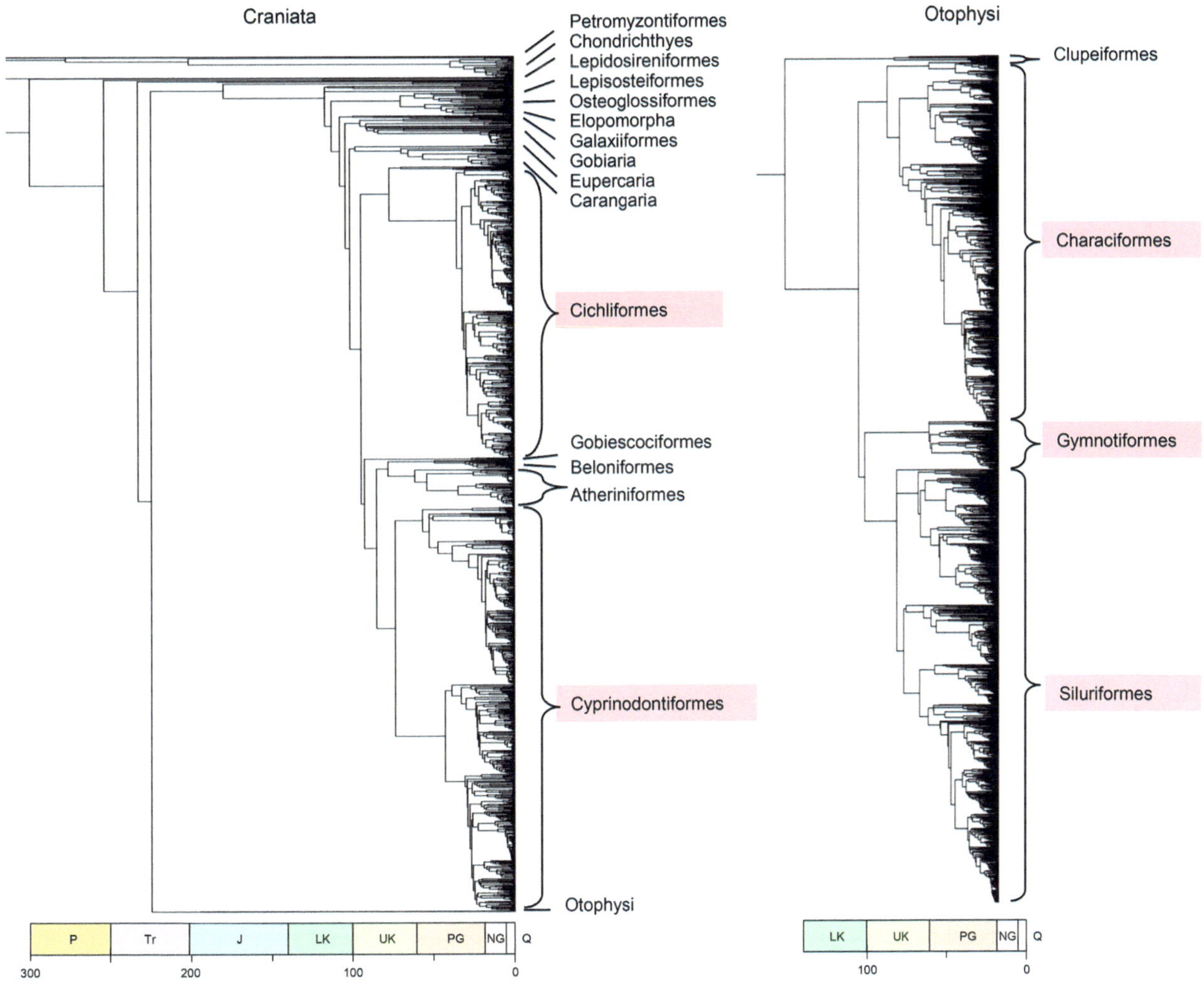

Fig. 2.7 Time calibrated phylogeny of all valid NFF species used in ecospace and ecotrace analyses of subsequent chapters. Data for 6342 species grouped by orders and super orders. Otophysi at bottom left expanded at right. Note five most species-rich orders colored

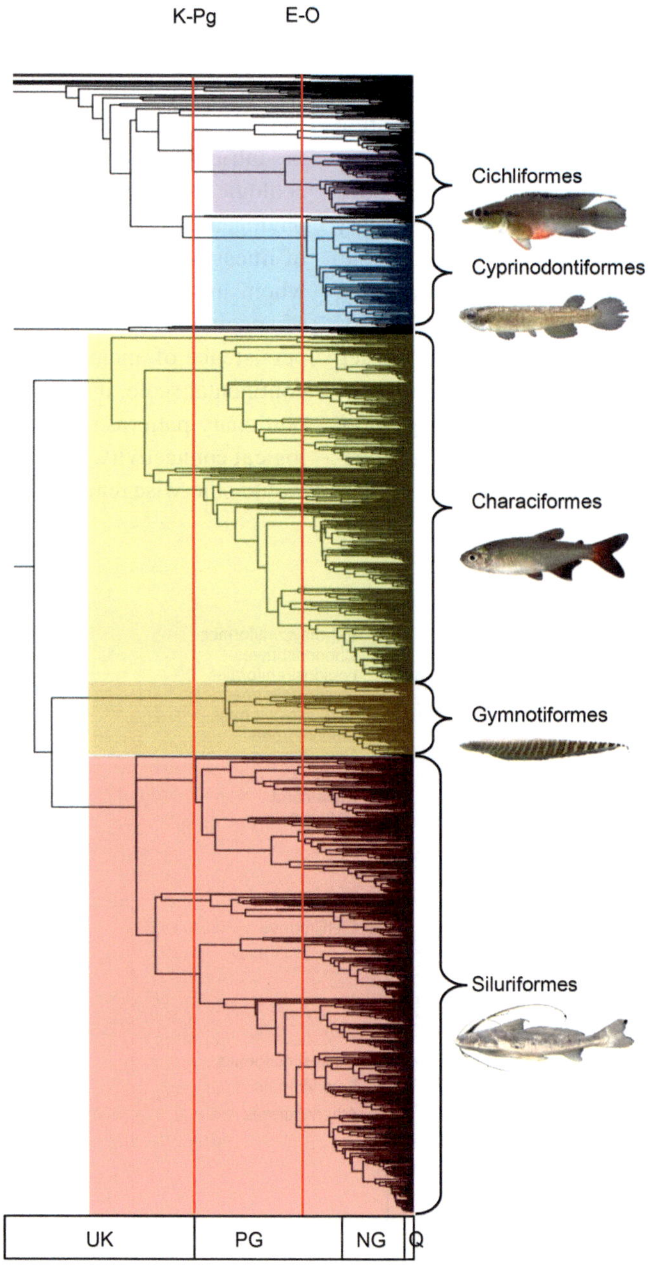

Fig. 2.7 (continued)

datasets. Conservation, too, finds a vital ally in this framework; evolutionarily distinct lineages, often hidden in taxonomic uncertainties, can now be identified and prioritized for protection. The tree can enrich our understanding of community assembly: how species co-occur, interact, and coexist within diverse freshwater systems shaped by both history and habitat. As a foundational tool, this phylogeny supports the emergence of more precise, hypothesis-driven research, guiding new inquiries into the tempo and mode of NFF evolution. It brings coherence to the complexity of Neotropical ichthyofauna, offering a scaffold for science, and a lens through which we might more clearly see, and more responsibly conserve, the biological richness of tropical freshwater ecosystems.

The elaboration of a comprehensive phylogenetic tree for NFF species is a pivotal act, one that illuminates the intricate tapestry of biodiversity woven through the evolutionary fabric of freshwater ecosystems. More than a tree showing species relationships, this expansive tree functions as a conceptual roadmap, tracing shared ancestries, reconstructing historical biogeography, and revealing the deep-time patterns that have shaped the emergence of diversity. Such a tree becomes a linchpin for conservation biology, providing a rigorous basis for identifying evolutionarily distinct and irreplaceable clades, and thus enabling more strategic and grounded approaches to biodiversity preservation. But its relevance stretches further. Embedded within its branches are insights into ecosystem dynamics, species assemblages, and potential responses to environmental perturbations—be they gradual or abrupt, natural or anthropogenic. As a scientific framework, it informs current understanding and also the questions we ask next, guiding future research into the evolutionary history of Neotropical freshwater systems.

Enhancing this large-scale phylogenetic tree is not a singular act, but a complex, iterative endeavor, one that reflects both the expanding frontiers of ichthyological research and the evolving demands of evolutionary science. This process begins with the essential task of broadening taxonomic coverage, ensuring that underrepresented lineages and newly described species are brought into the fold. It continues with the integration of additional molecular markers, high-throughput genomic data, and curated sequences, which collectively deepen the resolution of phylogenetic relationships. Equally important are the technical refinements, accurate sequence alignment, judicious trimming, and the careful optimization of partitioning schemes, all of which enhance the reliability of the resulting tree. Fossil calibrations, when anchored in robust paleontological evidence, introduce a temporal dimension that contextualizes divergence events within geological history, while the thoughtful inclusion of morphological characters bridges the gap between extant and extinct taxa. Improved computational frameworks, ever more powerful and accessible, allow for the analysis of increasingly large datasets with greater precision. Yet, perhaps most critically, this is a communal effort driven by collaboration across laboratories, institutions, and borders. Ongoing dialogue, data sharing infrastructures, and the incorporation of environmental metadata all contribute to the continuous updating of the tree. As knowledge advances, so too must the tree evolve, flexible enough to accommodate new discoveries, yet stable enough to serve as a reference point. Our tree becomes a dynamic, living resource, continually refined to mirror our deepening understanding of NFF biodiversity, evolutionary history, and ecological complexity.

The multinational ichthyological community now stands at a promising threshold, prepared to engage in a truly collaborative work to refine and expand the large phylogenetic tree of NFF. This collective effort, grounded in a shared commitment to open data, sustained taxonomic discovery, and methodological innovation, reflects more than just technical progress; it embodies a vision of science as a communal enterprise, responsive to intellectual curiosity. Through interdisciplinary collaboration, dialogue across academics, and the inclusion of diverse data sources, the tree becomes a research product and a platform for shared understanding of species relationships. Educational initiatives and public engagement efforts will play a vital role in democratizing access to phylogenetic insights, translating complex patterns into narratives that resonate across sectors and societies. At the heart of this endeavor lies conservation: the tree will help pinpoint evolutionary distinctiveness, highlight vulnerable lineages, and guide policies that prioritize species and their evolutionary histories. The formal validation of methods, the establishment of publicly accessible repositories, and the alignment with policy advocacy efforts ensure that this framework remains both rigorous and actionable. Integration with complementary datasets, from hydrological models to climate projections, and a proactive stance toward identifying and addressing data gaps reinforce the community's resolve.

2.5 Conclusions

The comprehensive, time-calibrated phylogeny presented in this chapter stands as the culmination of years of collaborative effort within the ichthyological community, aimed at untangling the evolutionary intricacies of the Neotropical realm. Representing the most species-rich NFF phylogeny ever assembled, it captures a moment in our scientific trajectory, a snapshot of collective knowledge, while acknowledging the limitations imposed by unresolved lineages, ongoing species discoveries, and the ever-evolving understanding of species boundaries. This tree is not an endpoint, nor should it be. Rather, it serves as a foundational scaffold upon which future investigations will be constructed, and from which new questions will emerge. Recognizing the dynamic nature of this field, this chapter embraces the principle of continual refinement, committing to regular updates that align with new data, new insights, and the expanding reach of ichthyological research. In doing so, it underscores the enduring importance of collaboration. At the interface of macroecology and macroevolution, this phylogeny opens new avenues for exploring the patterns and processes that structure freshwater biodiversity across time and space. And perhaps most importantly, by fostering open inquiry and collective stewardship, we hope this work contributes meaningfully to the preservation of NFF diversity, marking a decisive step toward a future in which Neotropical aquatic ecosystems are both understood and protected.

References

Albert JS, Abrahão V, Akin D R, et al. (2025) An ecological trait matrix for neotropical freshwater fishes, the most diverse continental vertebrate fauna. Nat Sci Data, in press 12(1):1127

Arce-H M, Lundberg JG, O'Leary MA (2017) Phylogeny of the North American catfish family Ictaluridae (Teleostei: Siluriformes) combining morphology, genes and fossils. Cladistics 33:406–428

Armbruster JW, Lujan NK (2025) New tribe-level classification of Hypostominae (Loricariidae) based on optimization of morphological states on DNA-based relationships, with descriptions of three new tribes and two new genera. Neotrop Ichthyol 22:e240108

Armbruster JW, Werneke DC, Tan M (2015) Three new species of saddled loricariid catfishes, and a review of *Hemiancistrus*, Peckoltia, and allied genera (Siluriformes). Zookeys 480:97

Bennett DJ, Hettling H, Silvestro D, Zizka A, Bacon CD, Faurby S et al. (2018) phylotaR: an automated pipeline for retrieving orthologous DNA sequences from GenBank in R. Life 8:20. https://doi.org/10.3390/life8020020

Castresana J (2000) Selection of conserved blocks from multiple alignments for their use in phylogenetic analysis. Mol Biol Evol 17:540–552

Dias AC, Tencatt LF, Roxo FF, Silva GDSDC, Santos SA, Britto MR, et al. (2024) Phylogenomic analyses in the complex neotropical subfamily Corydoradinae (Siluriformes: Callichthyidae) with a new classification based on morphological and molecular data. Zool J Linn Soc 203, zlae053. https://doi.org/10.1093/zoolinnean/zlae053

Eastman JM, Harmon LJ, Tank DC (2013) Congruification: support for time scaling large phylogenetic trees. Methods Ecol Evol 4:688–691

Faircloth BC, McCormack JE, Crawford NG, Harvey MG, Brumfield RT, Glenn TC (2012) Ultraconserved elements anchor thousands of genetic markers spanning multiple evolutionary timescales. Syst Biol 61:717–726

Felsenstein J (2004) Inferring phylogenies. In: Inferring phylogenies. Sinauer Associates, Sunderland, MA, pp 664–664

Fontenelle JP, Lovejoy NR, Kolmann MA, Marques FP (2021) Molecular phylogeny for the Neotropical freshwater stingrays (Myliobatiformes: Potamotrygoninae) reveals limitations of traditional taxonomy. Biol J Linn Soc 134:381–401

Fricke, R, (2021). Genera/species by family/subfamily in Eschmeyer's catalog of fishes. California Academy of Sciences, California

Hughes LC, Ortí G, Huang Y, Sun Y, Baldwin CC, Thompson AW, Shi Q (2018) Comprehensive phylogeny of ray-finned fishes (Actinopterygii) based on transcriptomic and genomic data. Proc Natl Acad Sci U S A 115:6249–6254

Katoh K, Toh H (2008) Recent developments in the MAFFT multiple sequence alignment program. Brief Bioinf 9:286–298

Lanfear R, Frandsen PB, Wright AM, Senfeld T, Calcott B (2017) PartitionFinder 2: new methods for selecting partitioned models of evolution for molecular and morphological phylogenetic analyses. Mol Biol Evol 34:772–773

Li C, Orti G, Zhang G, Lu G (2007) A practical approach to phylogenomics: the phylogeny of ray-finned fish (Actinopterygii) as a case study. BMC Evol Biol 7(1):44.

Lujan NK, Armbruster JW, Lovejoy NR (2018) Multilocus phylogeny, diagnosis and generic revision of the Guiana Shield endemic suckermouth armoured catfish tribe Lithoxini (Loricariidae: Hypostominae). Zool J Linn Soc 184:1169–1186

Melo BF, Sidlauskas BL, Hoekzema K, Vari RP, Dillman CB, Oliveira C (2018) Molecular phylogenetics of neotropical detritivorous fishes of the family Curimatidae (Teleostei: Characiformes). Mol Phylogenet Evol 127:800–812

Melo BF, Sidlauskas BL, Near TJ, Roxo FF, Ghezelayagh A, Ochoa LE et al. (2022) Accelerated diversification explains the exceptional species richness of tropical characoid fishes. Syst Biol 71:78–92

Melo BF, Ota RP, Benine RC, Carvalho FR, Lima FC, Mattox GM et al. (2024) Phylogenomics of characidae, a hyper-diverse neotropical freshwater fish lineage, with a phylogenetic classification including four families (Teleostei: Characiformes). Zool J Linn Soc 202(1):zlae101

Miller MA, Pfeiffer W, Schwartz T (2010) Creating the CIPRES science gateway for inference of large phylogenetic trees. In: Gateway computing environments workshop (GCE) IEEE, pp 1–8

Mirande JM (2019) Morphology, molecules and the phylogeny of Characidae (Teleostei, Characiformes). Cladistics 35:282–300

Ochoa LE, Roxo FF, Do Nascimiento C, Sabaj MH, Datovo A, Alfaro M et al. (2017) Multilocus analysis of the catfish family Trichomycteridae (Teleostei: Ostariophysi: Siluriformes) supporting a monophyletic Trichomycterinae. Mol Phylogenet Evol 115:71–81

Oliveira C, Avelino GS, Abe KT, Mariguela TC, Benine RC, Ortí G et al. (2011) Phylogenetic relationships within the speciose family Characidae (Teleostei: Ostariophysi: Characiformes) based on multilocus analysis and extensive ingroup sampling. BMC Evol Biol 11:275

Pagel M (1997) Inferring evolutionary processes from phylogenies. Zoologica Scr 26(4):331–348

Pennell MW, Eastman JM, Slater GJ, Brown JW, Uyeda JC, FitzJohn RG et al. (2014) geiger v2. 0: an expanded suite of methods for fitting macroevolutionary models to phylogenetic trees. Bioinformatics 30:2216–2218

R Core Team R (2023) _R: A Language and Environment for Statistical. Computing_. R Foundation for Statistical Computing, Vienna, Austria. https://www.R-project.org/

Reis RE, Lehmann A P (2022) A new genus of armored catfish (Siluriformes: Loricariidae) from the Greater Amazon, with a review of the species and description of five new species. Neotrop Ichthyol 20(02):e220002

Reis RE, Kullander SO, Ferraris CJ Jr (eds) (2003) Check list of the freshwater fishes of South and Central America. Edipucrs.

Sabaj MH, Arce HM (2021) Towards a complete classification of the neotropical thorny catfishes (Siluriformes: Doradidae). Neotrop Ichthyol 19(04):e210064

Schoch CL,Ciufo S,Domrachev M, Hotton CL,Kannan SK,hovanskaya R, (2020) NCBI taxonomy: a comprehensive update on curation, resources and tools. Database 2020:baaa062. https://doi.org/10.1093/database/baaa062

Sidlauskas BL, Melo BF, Birindelli JL, Burns MD, Frable BW, Hoekzema K et al. (2025) Molecular phylogenetics, a new classification, and a new genus of the Neotropical fish family Anostomidae (Teleostei: Characiformes). Neotrop Ichthyol 23(1):e240076

Silva GS, Roxo FF, Melo BF, Ochoa LE, Bockmann FA, Sabaj MH et al. (2021) Evolutionary history of Heptapteridae catfishes using ultraconserved elements (Teleostei, Siluriformes). Zool Scr 50(5):543–554

Silva GS, Melo BF, Roxo FF, Ochoa LE, Shibatta OA, Sabaj MH, Oliveira C (2021) Phylogenomics of the bumblebee catfishes (Siluriformes: Pseudopimelodidae) using ultraconserved elements. J Zool Syst Evolut Res 59(8):1662–1672

Silva GS, Rocha MS, Melo BF, Reia L, Roxo FF, Sabaj MH, Oliveira c (2024) Phylogenomics of the catfish family Pimelodidae with focus on the genus Pimelodus support the recognition of Sorubiminae and Pimelodinae (Teleostei, Siluriformes). Zool Scr 53(5):541–554.

Smith SA, O'Meara BC (2012) treePL: divergence time estimation using penalized likelihood for large phylogenies. Bioinformatics 28:2689–2690

Smith SD, Pennell MW, Dunn CW, Edwards SV (2020) Phylogenetics is the new genetics (for most of biodiversity). Trends Ecol Evol 35(5):415–425

Stamatakis A (2014) RAxML version 8: a tool for phylogenetic analysis and post-analysis of large phylogenies. Bioinformatics 30:1312–1313

Stein RW, Mull CG, Kuhn TS, Aschliman NC, Davidson LN, Joy JB et al. (2018) Global priorities for conserving the evolutionary history of sharks, rays and chimaeras. Nat Ecol Evol 2:288–298. https://doi.org/10.1038/s41559-017-0448-4

Stein RW, Mull CG, Kuhn TS, Aschliman NC, Davidson LN, Joy JB, Mooers AO (2018) Global priorities for conserving the evolutionary history of sharks, rays and chimaeras. Nat Ecol Evol 2:288–298

Tagliacollo VA, Souza-Lima R, Benine RC, Oliveira C (2012) Molecular phylogeny of Aphyocharacinae (Characiformes, Characidae) with morphological diagnoses for the subfamily and recognized genera. Mol Phylogenet Evol 64:297–307

Tagliacollo VA, Bernt MJ, Craig JM, Oliveira C, Albert JS (2016) Model-based total evidence phylogeny of neotropical electric knifefishes (Teleostei, Gymnotiformes). Mol Phylogenet Evol 95:20–33

Tagliacollo VA, Tan M, Reis RE, Gaia R, Carrijo V, Ranuzi M, et al. (2024) Time-calibrated phylogeny of neotropical freshwater fishes. Frontiers in Bioinformatics 4, 1433995

Talavera G, Castresana J (2007) Improvement of phylogenies after removing divergent and ambiguously aligned blocks from protein sequence alignments. Syst Biol 56:564–577

Terán GE, Benitez MF, Mirande JM (2020) Opening the Trojan horse: phylogeny of *Astyanax*, two new genera and resurrection of *Psalidodon* (Teleostei: Characidae). Zool J Linn Soc 190(4):1217–1234

Van der Sleen, P, Albert, JS (2017) Field guide to the fishes of the Amazon, Orinoco, and Guianas. Princeton University Press, Princeton, p. 464

Varella HR, Kullander SO, Menezes NA, Oliveira C, López-Fernández H (2023) Revision of the generic classification of pike cichlids using an integrative phylogenetic approach (Cichlidae: tribe Geophagini: subtribe Crenicichlina). Zool J Linn Soc 198(4):982–1034

3 The Big Nine Clades of Amazonian Fishes

James S. Albert, Jack M. Craig, Kevin T. Torgersen, Milton Tan, Victor A. Tagliacollo and Roberto E. Reis

Abstract

Biodiversity measures are highly unevenly distributed among the 145 fish *clades* with phylogenetically distinct (i.e., evolutionarily independent) origins in Neotropical freshwaters. These fishes exhibit a *hollow curve species* frequency distribution, with most species members of a few dominant clades, and most clades having few species. Most NFF species (5670 or 89.3%) are members of just nine highly diverse clades, herein referred to as the *Big Nine NFF clades*. The remaining NFF species constitute a heterogeneous assemblage of 675 species (10.6% of all NFF species) in 136 distinct clades representing 31 different taxonomic orders. Most NFF clades (101 clades or 70% of the total) are *Marine Derived Lineages* (MDLs; i.e., *peripheral freshwater fish* species that are a member of an otherwise marine family) represented by a single species (i.e., singleton). The Big Nine NFF clades are phylogenetically clustered among fishes worldwide, all with origins in the super-greenhouse world of the Upper Cretaceous (c. 85–66 Ma) or early Paleogene (c. 66–44 Ma), crown-group *clade ages* ranging from c. 85 Ma in Loricarioidei to c. 43 Ma in Rivulidae, and Big Nine NFF clades are members of the two most diverse teleostean super-orders; Acanthopterygii and Ostariophysi. Three of the Big Nine NFF clades are *secondary freshwater fishes*, with greater salinity tolerances and *dispersal* capacities over marine barriers than *primary freshwater fishes*. The three acanthopterygian NFF clades are the cichliform Cichlinae (669 species), and the cyprinodontiforms Poecilioidea (364 species) and Rivulidae (482 species), with a combined total of 1515 valid species, or 23.9% of all NFF species. The other six of the Big Nine NFF clades are ostariophysan fishes, with 4798 valid species or 75.6% of all NFF species, of which 4371 species (91%) are included in the *macroevolutionary* analyses of this book. The six ostariophysan NFF clades are primary freshwater fishes with lesser salinity tolerances, including the characiforms Curimatoidea (471 species) and Characoidea (1571 species), Gymnotiformes (271 species), and the siluriforms Aspredinoidea (273 species), Pimelodoidea (409 species), and Loricarioidei (1803 species). Body size is strongly structured phylogenetically among NFF clades, with some clades overrepresented among species with extreme (small and large) body sizes. NFF species with the largest body sizes are either MDLs (e.g., chondrichthyans like the Largetooth sawfish *Pristis pristis* and Bull shark *Carcharhinus leucas*, holosteans like the Alligator gars (*Atractosteus* spp.), or teleosts like the Atlantic tarpon (*Megalops atlanticus*), or primary freshwater ostariophysans (e.g., Goliath catfishes, Electric eels). NFF species with the smallest adult body sizes are secondary freshwater fishes in the gobiiform family Eleotridae and poecillioid family Fluviphylacidae, and primary freshwater fishes in the siluriform families Scoloplacidae and Trichomycteridae and the characiform families Crenuchidae and Acestrorhamphidae. As estimated from branch lengths on the time-calibrated

J. S. Albert (✉)
University of Louisiana at Lafayette, Lafayette, LA, USA
e-mail: jsalbe@gmail.com

J. M. Craig
Drexel University, Philadelphia, PA, USA
e-mail: jack.m.craig@gmail.com

K. T. Torgersen
University of Louisiana at Lafayette, Lafayette, LA, USA
e-mail: kevintorgersen@gmail.com

M. Tan
Illinois Natural History Survey, Urbana, IL, USA
e-mail: miltont@illinois.edu

V. A. Tagliacollo
Universidade Federal do Uberlândia, Uberlândia, MG, Brazil
e-mail: vatagliacollo@gmail.com

R. E. Reis
Pontifícia Universidade Católica Rio Grande do Sul, Porto Alegre, RS, Brazil
e-mail: reis@pucrs.br

J. S. Albert and R. E. Reis (eds.), *Ecology and Evolution of Amazonian Fishes*, Fascinating Life Sciences,
https://doi.org/10.1007/978-3-032-09328-8_3

phylogeny of 6143 NFF species, the median species age (MSA) of all NFF taxa is 2.3 Ma, ranging only about two-fold in values, from 1.6 Ma in Poecilioidea to 4.4 Ma in Rivulidae. Among the 44 families of the Big Nine, species richness is negatively correlated with MSA, and there is a positive correlation between species-richness and skewness values of species ages among families. In other words, more diverse clades tend to have younger species and a more asymmetric tree topology with longer earlier branches. These age estimates are about half that of literature estimates for other fish groups, perhaps due to the high taxonomic sampling and concomitant lower average branch lengths of the phylogeny used to estimates these MSAs. *Ecospace* occupancy was assessed as the proportion of the multivariate hypervolume calculated from a statistical analysis of 42 ecotrait variables. Among members of the Big Nine, Characoidea has the largest ecospace volume, which was used to assess the proportional ecospace occupancy of the other clades. Ecospace occupancy ranges from 91% in Loricariodei to 0.3% in Rivulidae. When assessed as deviations from their species richness values, Aspredinoidea has the largest ecospace, and the two cyprinodontiform clades (Poecilioidea and Rivulidae) the smallest ecospaces.

3.1 A Bewildering Array of Phenotypes

The Neotropical Freshwater Fish (NFF) fauna confronts human observers with a bewildering array of physiological, morphological, behavioral and ecological *traits* or *phenotypes* (Fig. 3.1). The members of this fauna represent thousands of different forms distributed across tens of millions of square kilometers and constitute some of the most diverse ecosystems on Earth (Albert et al. 2020; Rull and Carnaval 2021; Guayasamin et al. 2024). In the face of such vast complexity and variety it is a fair question to ask: *How it is possible to study the ecology and evolution of megadiverse tropical biotas*?

Comparative biologists have developed numerous strategies to compile and organize the many disparate sources of information about biological diversity (i.e., *biodiversity*), to compare these data with those from the environmental sciences, e.g., climatology, geography, geology; Magurran, 2021; Diniz-Filho, 2023). The introduction to this chapter reviews three classifications used in the analyses presented in this book that help us partition the immense biodiversity of NFF taxa by species, *functional traits*, and higher *taxa* (Wiens 2023). The chapter then introduces the major fish *clades* (i.e., monophyletic branches of a phylogenetic tree) that have evolutionarily independent origins in Neotropical freshwaters, and which are used in the comparative analyses presented in subsequent chapters. Finally, this chapter ends with a brief overview of the major macroevolutionary patterns of diversity in taxonomic, body size, species age, and ecospace occupancy observed in NFF clades. Detailed analyses of trait evolution among NFF taxa are provided in subsequent chapters.

3.1.1 Species as Natural Kinds

A core observation of ecology and evolution is that, with few exceptions, most groups of plants and animals are readily divided into natural kinds called *species* (Lowe-McConnell 1969; Mayden 2002; Sexton et al. 2009). In particular, each species exhibits a characteristic set of morphological, ecological, and genetic traits, however cryptic or subtle these features may seem to the human eye (Bickford et al. 2007; Craig et al. 2017; Struck et al. 2018). Although the differences may be invisible to all but the most experienced taxonomist, a tremendous volume of careful and detailed studies conducted over many decades and across broad range of geographic scopes, reveal there is almost always a distinct genetic and phenotypic gap between most-closely related species (Smith and Carstens 2022). The causal mechanistic reasons underlying the species phenomenon are incompletely known, and many competing hypotheses have been proposed. The reader is directed to the following references for an entry into this large and complex literature (Mishler and Donoghue 1982; Agapow et al. 2004; Stankowski and Ravinet 2021; Cazzolla Gatti et al. 2022). These multiple alternative *species concepts* are based on or more genetic, reproductive, ecological, or evolutionary perspectives. In this context, it is useful to consider that individuals of a given NFF species recognize and mate with one another using a species-specific set of cues from the visual (e.g., color patterns, locomotor displays), auditory (e.g., buzzes, chirps, songs), chemical (e.g., pheromones), or electric (Borghezan et al. 2021) sensory cues. That is to say, the members of each fish species recognize themselves (Crampton 2011). For these reasons, and many others, species are widely thought to represent fundamental units of ecology, evolution, and conservation (De Queiroz 2007).

Biodiversity is discovered and documented one species at a time, with each species representing a distinct and indivisible "atom" in the fabric of ecosystems or higher taxa (e.g., genera, families; Claridge et al. 1997; Colwell 2009). Like all natural entities, species are composed of parts (e.g., populations, organisms, cells), and also have boundaries which define the limits of gene flow, *natural selection*, and *adaptation* (Lenormand 2002; Petit and Excoffier 2009). The genome and ontogeny of each species lineage are designed to produce a relatively stable phenotype, which may persist for long periods of time, often millions of years (Wake et al. 1983; Eldredge et al. 2005). Species and their dependable phenotypes are therefore reliable features of the environment that affect the adaptations and ecological interactions of other

Fig. 3.1 Distinctive phenotypic specializations of South American freshwater fishes. Species by Latin and common names, characteristic habitat or diet, and taxonomic family. **A** *Retroculus lapidifer*, Rapids eartheater, rapids of large clearwater shield rivers (Cichlidae); **B** *Symphysodon aequifasciatus*, Blue discus, Crowns of drowned trees in oxbow lakes (Cichlidae); **C** *Melanorivulus atlanticus*, Atlantic forest killifish, clearwater rainforest streams (Rivulidae); **D** *Phalloceros caudimaculatus*, Dusky millions livebearer, clearwater rainforest streams and forest pools (Poeciliidae); **E** *Anableps anableps*, Largescale four-eyes, Brackish to freshwater coastal streams and mangroves (Anablepidae); **F** *Paracheirodon innesi*, Neon tetra, small blackwater streams (Acestrorhamphidae); **G** *Thoracocharax stellatus*, Spotfin hatchetfish, lowland river beaches and small rivers (Gasteropelecidae); **H** *Pygocentrus nattereri*, Red-bellied piranha, lowland rivers and floodplains (Serrasalmidae); **I** *Semaprochilodus insignis*, Flagtail *Prochilodus*, migratory in large lowland rivers (Prochilodontidae); **J** *Tarumania walkerae*, No common name, ephemeral forest pools and moil leaf litter (Tarumaniidae); **K** *Sternarchella raptor*, Tail-biting bulldog knifefish, large lowland river channels (Apteronotidae); **L** *Gymnotus chaviro*, Fitzcarrald banded knifefish, floodplain forests and oxbow lakes (Gymnotidae); **M** *Pterodoras granulosus*, Granulated catfish, widespread in lowland rivers and floodplains (Doradidae); **N** *Phreatobius cisternarum*, No common name, aquifers, ground water and artesian wells (Heptapteridae); **O** *Phractocephalus hemioliopterus*, Red-tail catfish, widespread in lowland rivers and floodplains (Pimelodidae); **P** *Peckoltia vittata*, Candy-striped pleco, river rapids (Loricariidae).

species, and which constrain ecosystem processes (Fine 2015; Zamudio et al. 2016; Turcotte and Levine 2016). *Species richness* is the number of species that constitute different habitats, geographic regions, or higher taxa (Wiens 2011; Chao and Chiu 2016). Determining the identity, life history traits, and interrelationships of all the myriad NFF species is therefore of highest priority to ecologists and conservation biologists interested in preserving tropical aquatic biodiversity (Mace 2004).

3.1.2 Functional Diversity and Ecotraits

The diversity of functional traits (i.e., *functional diversity*) includes physiological, behavioral, and ecological differences among species, higher taxa (i.e., clades), habitats, and biogeographic regions. Functional diversity has important consequences for ecology, evolution, and conservation (Winemiller, 1991; Tilman, 2001; Carmona et al., 2016; Freudenstein et al., 2017). The ecological traits (*ecotraits*) used in the analyses of this book are introduced and defined in Chap. 6, as heritable (or phenotypically plastic) physiological or behavioral aspects of the integrated *phenotype*, which constrain resource acquisition and utilization; i.e., habitat, diet, and life history traits (Chap. 7, Fig. 7.1). The integrated phenotype refers to the full suite of organismal traits across the hierarchy of organismal design, from the molecular genetic to ecological levels of organization. Although the organismal phenotype always functions as a single integrated entity, the ontogenetic development and evolutionary changes of individual traits are often somewhat disintegrated or modular, with leading to the phenomenon referred to as *mosaic evolution* (Hall and Hallgrímsson, 2007; Evans et al., 2019; Ford et al., 2022).

Organisms exhibit important genetic and phenotypic differences in many dimensions of the integrated phenotype. Differences are observed within species (i.e., ontogenetic and geographic *variation*) and among species (i.e., phylogenetic *diversity*), and in traits spanning the full hierarchy of biological organization, from molecular to behavioral and even ecological attributes (O'Leary et al. 2024). Functional diversity in some physiological, behavioral, and ecological traits has been used to create predictive special-purpose classifications of biodiversity, based, for example, on stages of life cycles (e.g., haploid vs. diploid; larvae vs. adults), physiological tolerances (e.g., thermoconformers, osmoregulators), habitat preferences (rheophilic vs. psammophilic), trophic status (e.g., algivore vs. piscivore), trophic position (e.g., primary producers, top carnivores, parasites), and locomotion (e.g., anguilliform vs. carangiform). Fishes have also grouped based on different life history traits, such as reproductive modes (e.g., oviparous vs. viviparous), diel activity patterns (e.g., diurnal vs. nocturnal), and migratory behaviors (e.g., anadromous vs. catadromous).

3.1.3 Phylogeny as the Natural System

Despite all these special-purpose eco-physiological classifications, centuries of study have taught us that the most predictive classification of organisms is based on genealogical or phylogenetic relationships (Wiley and Lieberman 2011; Hall 2013). By this, we mean that the *phylogeny* of species, represented as a branching "Tree of Life," has the greatest predictive power for storing and retrieving information of the many different dimensions of the integrated phenotype (Murren 2012). The phylogenetic relationships discovered from comparative studies of one organ system (e.g., bones and cartilages) or one set of genes reliably predict patterns observed in other phenotypes (e.g., nervous system) and other genes. This is because the phylogenetic diversity observed among species emerged from a unitary evolutionary history of speciation and extinction events descending from a common ancestor (Mayr and Bock 2002). Taxonomy refers to the names applied to the parts of a classification, and in the case of phylogenetic systematics, the taxonomy refers to the species and higher taxa (i.e., clades). Thus, as noted in the Preface to this volume, and to paraphrase Dobzhansky (1973), *Nothing in ecology or evolution makes sense except in light of a good taxonomy.*

3.2 NFF Clades

Following the methodology of Albert et al. (2011), we used the comprehensive phylogeny of species in Chap. 2 (Fig. 2.1) to assign all NFF species to one of 145 clades, each with an independent phylogenetic origin in Neotropical freshwaters. NFF clades were derived from one of three biogeographic sources: (1) persistence in South America from an ancestral distribution in Western Gondwana after the final separation from Africa c. 100 Ma (e.g., Characoidea); (2) dispersal into Neotropical freshwaters from another continent (e.g., Cichlinae from Africa); (3) dispersal into Neotropical freshwaters from the sea (e.g., Potamotrygonidae).

3.3 The Big Nine NFF Clades

Most NFF species are members of just nine highly diverse clades, hereafter referred to as the *Big Nine NFF clades* (Fig. 3.2). Summaries of the phylogenetic composition and biodiversity dimensions of the Big Nine are presented in the following sections. At the other end of the biodiversity spectrum, most NFF clades are represented by a single species in Neotropical freshwaters. Among these singletons are species of otherwise marine groups that enter large rivers (e.g., Bull shark *Carcharhinus leucas;* Atlantic tarpon *Megalops atlanticus*), obligate freshwater species of an otherwise marine family (e.g., Bobo mullet *Joturus pichardi;* Mugilidae), and freshwater species that have been restricted to continental waters for more than 100 million years (e.g., South American lungfish *Lepidosiren paradoxa*).

3.3.1 Phylogenetic Composition

The Big Nine are strongly clustered phylogenetically among higher teleostean clades, all with relatively ancient origins in the super-greenhouse world of the Late Cretaceous (c. 85–66 Ma) or the early Paleogene (c. 66–44 Ma). All of the Big Nine NFF clades are teleostean fishes, and six of these clades are members of a megadiverse superorder of teleost fishes called Ostariophysi. Ostariophysi includes four meadiverse clades ranked as taxonomic orders (i.e., Cypriniformes, Characiformes, Siluriformes, and Gymnotiformes) and almost 11,000 living species. Ostariophysi fishes therefore encompass about 30% of all known fish species, or about 60% of

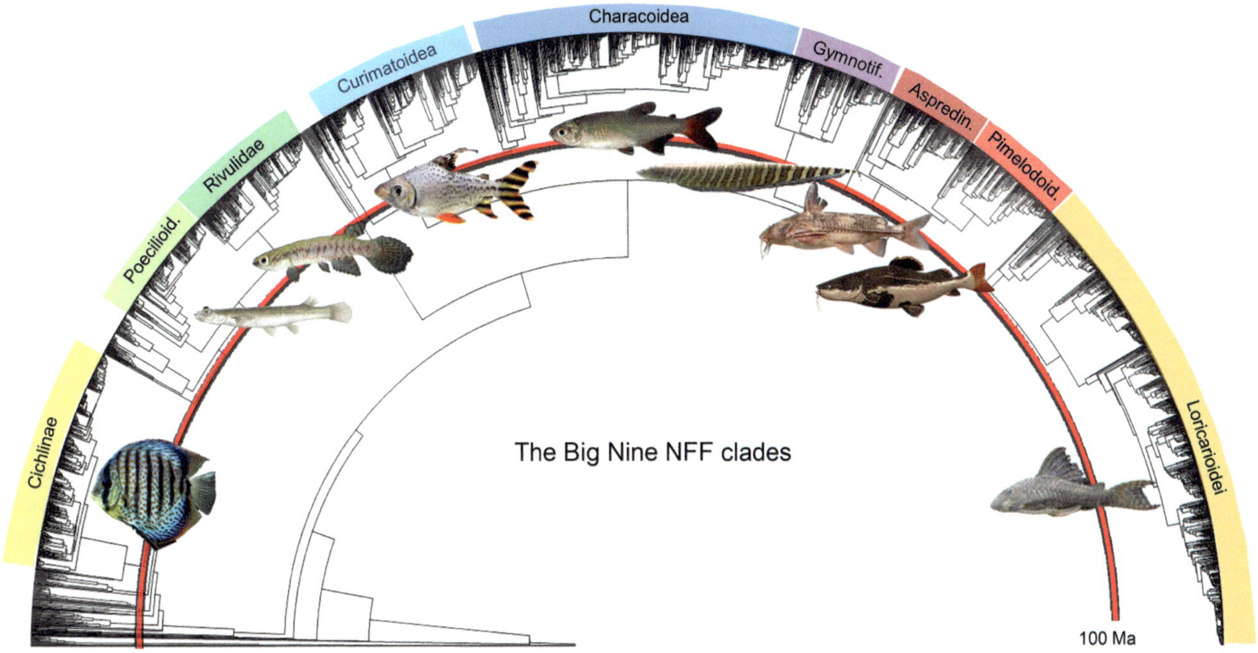

Fig. 3.2 Time-calibrated phylogeny of Neotropical freshwater fishes highlighting the Big Nine NFF clades (colored regions). Terminals represent 3167 species (49.6% of NFF total) in 865 genera, 96 families, and 40 orders (Chap. 2). Astroblepidae and Profundulidae were excluded due to poor phylogenetic resolution and lack of relevant ecotrait diversity for this study. The Big Nine NFF includes 5740 species or 89% of the NFF fauna. Red semicircle at c. 100 Ma marks the time of the final separation of South America and Africa and the latest time of origin for phylogenetically independent NFF clades (Albert et al. 2011)

all known freshwater fish species (Fricke et al. 2025). These ostariophysan clades all originated during the Early Cretaceous (c. 140–100 Ma) in continental freshwaters of the supercontinent Pangaea (Chen et al. 2013; Hughes et al. 2018). With the notable exceptions of two clades of predominantly marine catfishes (i.e., Hardhead catfishes, Ariidae; Eel-tail catfishes, Plotosidae), these clades are *primary freshwater fishes* (*sensu* Myers 1949; see Cano-Barbacil et al. 2024), with little to no tolerance for saltwater or ability to disperse across marine barriers. Neotropical ostariophysans include 34 family-level clades in the Big Nine (Table 3.1), including 17 families of Curimatoidea and Characoidea in the order Characiformes (i.e., tetras, piranhas, and allies), five families in the order Gymnotiformes (i.e., Neotropical electric fishes), and 12 families of Aspredinoidea, Pimelodoidea and Loricarioidei in the order Siluriformes (i.e., catfishes).

The other three Big Nine NFF clades are members of another megadiverse superorder of teleost fishes called Acanthopterygii, the Spiny-ray fishes (Hughes et al. 2018), the most diverse superorder of fishes worldwide. Acanthopterygians of the Big Nine include 29 orders, 337 families, and about 18,700 valid species, thereby constituting about 25% of all living vertebrate species (Fricke et al. 2025). The Big Nine acanthopterygian clades include Neotropical cichlids (Cichlinae; e.g., Freshwater Angel fishes, Pike Cichlids and allies), and two clades of Neotropical tooth-carps in the order Cyprinodontiforms; Poecilioidea including guppies (Poeciliidae) and foureye fishes (Anablepidae), and Rivulidae commonly known as the South American killifishes. These three clades (i.e., Cichlinae, Poecilioidea, and Rivulidae) are members of taxonomic orders with somewhat more recent origins than otophysans in continental freshwaters during the Late Cretaceous (c. 100–66 Ma), and members of these clades exhibit higher tolerances for saltwater and better abilities to cross marine barriers (secondary freshwater fishes; Myers 1949; Matamoros et al. 2015; Tagliacollo et al. 2017).

3.3.2 Biodiversity Dimensions

This section summarizes some of the major features of biodiversity in the Big Nine NFF clades, in terms of taxonomic diversity (species richness of major taxonomic subgroups), body size, habitats, diets, and geographic distributions. Ecological trait data are from Albert et al. (2025) and summarized in Chap. 6. Geographic data are summarized in Chap. 4. Clades are presented below in the order they are displayed from left to right on the phylogeny depicted in Fig. 3.2. Species richness values are from Eschmeyer's catalog as of May 1, 2025.

Cichlinae

- *Diversity:* 669 valid species in 81 genera and six tribes; Retroculini with four species in the genus *Retroculus*, Cichlini with approximately 15 species in the genus

Table 3.1 Salient features of the Big Nine NFF clades

Clade	Species[a]	Families	Genera	Size	Age	S/Ma	MSA	Ecosp	Salt
Cichlinae	569	1	81	14.7	54	10.5	2.0	38%	Sec
Poecilioidea	341	5	61	4.6	47	7.3	1.6	7%	Sec
Rivulidae	431	1	35	5.4	43	10.0	4.4	0%	Sec
Curimatoidea	463	8	53	18.2	70	6.6	3.3	23%	Pri
Characoidea	1388	12	176	9.4	85	16.3	1.7	100%	Pri
Gymnotiformes	259	5	36	30.7	60	4.3	3.1	22%	Pri
Pimelodoidea	390	4	60	23.5	55	7.1	1.7	39%	Pri
Aspredinoidea	267	3	69	15.2	65	4.1	3.6	74%	Pri
Loricarioidei	1628	5	146	12.2	85	19.2	2.4	91%	Pri
Other Primary[b]	331	15	52	17.3	NA	NA	6.4	60%	Pri
MDL[b]	344	41	111	33.8	NA	NA	4.0	58%	Per
Sum or Avg	**6411**	**100**	**880**	**14.3**	**63**	**9.5**	**2.3**	**47%**	

Size as average maximum body size (cm SL); Age (crown clade age) and median species age (MSA) in Ma. Data summarized from Table 3.2. Taxa arranged in the phylogenetic sequence of Fig. 3.2. Body size as average of maximum recorded standard length (cm); Age (crown clade age) MSA (median species age) in Ma. Ecospace given as percentage of Characoidea, the clade with the largest ecospace volume; see Chap. 6. Salt as salinity tolerance categories of Myers (1938): Per = peripheral, Pri = Primary, Sec = Secondary. Note Big Nine include 5736 species or 90.4% of NFF total, and others include 675 species or 10.5% of NFF total [a]Species in ecomatrix [b]Aggregate data for nonmonophyletic groups

Cichla, Astronotini with three species in the genus *Astronotus*, Chaetobranchini with four species in two genera, Geophagini with more than 200 species in 16 genera, Cichlasomatini with 260 species in 12 extant genera and two extinct species of †*Tremembichthys*, and Heroini with 184 species in 44 genera.
- *Body size*: Mostly small to medium adult sizes, average body size of adults 14.7 cm SL, ranging from 2.1 cm SL in mature *Apistogramma staecki* from the Rio Mamoré basin in Bolivia (Koslowski 1985) to 74 cm SL and 6.8 kg in a male *Cichla ocellaris* introduced to Florida (IGFA 2001).
- *Habitats:* Many cichline species occupy a wide range of habitats and substrates, and are often generalists with respect to water chemistry. Some exceptions to these generalities are the strong preference of *Retroculus* for river rapids, and of *Cichla* and *Astronotus* for the margins and floodplain lakes of lowland rivers.
- *Diets*: All major trophic categories, including some specialized herbivores (e.g., *Symphysodon*), invertivores (e.g., *Mesonauta*), omnivores (e.g., *Hypsophrys*), and facultative piscivores (e.g., *Cichla*).
- *Geographic distribution*: Widespread throughout the humid Neotropics, from Patagonia to Texas. Heroini is most diverse in Central America with about 120 species or 65% of the tribe total (Tagliacollo et al., 2017).

Poecilioidea

- *Diversity:* Five families and 61 genera and 341 species in Neotropical freshwaters, including: Anablepidae with 18 species in three genera (Aguilera et al. 2019; Amorim and Costa 2019); Cyprinodontidae with 55 species in seven genera, not monophyletic in Neotropical freshwaters (Hernández-Ávila et al. 2024); Fluviphylacidae with seven species in the genus *Fluviphylax*; Poeciliidae with 276 species in 25 genera (Matamoros et al. 2012, 2015; Tagliacollo et al. 2017; Rodríguez-Machado et al. 2024); and Profundulidae with 16 species in three genera (Morcillo et al. 2016).
- *Body size*: Mostly small adult body sizes, average size 4.5 mm SL, ranging from 1.4 cm SL in mature male *Fluviphylax wallacei* (Fluviphylacidae) from the Negro River (Bragança 2018) to 32 cm SL in *Anableps microlepis* (Anablepidae) from coastal streams of northeastern Brazil and the Guianas (Schwassmann and Kruger 1965).
- *Habitats:* Fresh and brackish waters of coastal drainages (Poeciliidae, Cyprinodontidae), tidal flats and mangroves (Anablepidae), clearwater and blackwater terra firme streams (Poeciliidae, Profundulidae), high-elevation lakes and streams (*Orestias*, Cyprinodontidae).
- *Diets*: Mostly benthic microvores, consuming a variety of small-sized (0.01–1.0 mm or 10–1000 μm) microalgae (e.g., chlorophytes and diatoms), animal larvae (e.g., cladocerans, copepods, and insects), and other microfauna and meiofauna (e.g., rotifers, ciliated protozoa, nematodes, tardigrades).
- *Geographic distribution*: Full extent of tropical American freshwaters and adjacent temperate areas of North and South America, being most diverse in Central America with 261 recorded species (Gómez-Martínez et al. 2024).

Rivulidae

- *Diversity:* 488 species in 35 genera and one family.
- *Habitats:* Most rivulids inhabit slowly moving streams or standing pools of clearwater river basins.

- *Body size*: Mostly small adult sizes, average maximum body size 5.4 cm SL, ranging from 1.8 cm SL in *Laimosemion ubim* from streams affluent to Lago Amanã in the Central Amazon, to 22 cm SL in *Austrolebias elongatus* from the La Plata Basin.
- *Diets*: Mostly benthic microvores.
- *Geographic distribution*: Full extent of tropical American freshwaters and adjacent temperate areas of North and South America, most diverse in upland streams and small river of the Brazilian Shield. Rivulids are sometimes referred to as *peixes de nuvem* (fish from the clouds) or *peixes do céu* (sky fishes) in Brazilian folklore, as they often appear in temporary water bodies after dry periods (Santelli and Fonseca 2023). The center of rivulid diversity is the Dry Diagonal (*Caatinga* and *Cerrado* ecoregions) of the Brazilian Shield, from more than 200 species are described. There are about 150 species in just four rivulid genera largely confined to this region: *Melanorivulus* (61 species), *Hypsolebias* (52 species), *Cynolebias* (22 species), and *Atlantirivulus* (14 species).

Curimatoidea

- *Diversity:* 463 species in 53 genera and eight families (Anostomidae, Chilodontidae, Curimatidae, Cynodontidae, Hemiodontidae, Parodontidae, Prochilodontidae, and Serrasalmidae), all with small to medium body sizes (5–50 cm SL).
- *Habitats:* River channels and floodplains, with species in pelagic, benthic, and littoral habitats.
- *Body size*: Mostly moderate adult sizes, average maximum body size 18.2 cm SL, ranging from 3.3 cm SL in *Cyphocharax signatus* (Curimatidae) from the Vermelho River tributary of the Araguaia River in the Tocantins Basin, to *Piaractus mesopotamicus* (Serrasalmidae) at a maximum of 88 cm SL, but more commonly 62 cm SL and 20 kg, in the Paraguay and Paraná rivers in the La Plata Basin (Agostinho et al. 2003).
- *Diets*: Many curimatids, chilodontids, and prochilodontids are detritivores, consuming a wide range of living and non-living organic matter (decomposing leaf litter, shed insect casings, fecal matter, etc.), as well as the microbial communities that grow on the detritus. Detritus in aquatic systems can be divided into two broad categories: particulate organic matter (POM) and dissolved organic matter (DOM; also called dissolved organic carbon or DOC; Chap. 16). Anostomids exhibit a range of dietary specializations from benthic invertivores to omnivores (many *Leporinus* and *Synaptolaemus* species), consuming both plant and animal tissues. Most hemiodontids (e.g., *Hemiodus*) are omnivores and all 14 species of Cynodontinae are piscivores (e.g., *Cynodon, Roestes*).
- *Geographic distribution*: All major South American river basins, north into the Pacific drainages of the Puntarenas Province of Costa Rica and south to central Argentina (Vari 1983).
- Two families (Curimatidae and Prochilodontidae) undertake long-distance seasonal migrations (Melo et al. 2021).

Characoidea

- *Diversity:* 1388 species in 176 genera and 12 families of tetras and allies: Acestrorhamphidae, Acestrorhynchidae, Bryconidae, Chalceidae, Characidae, Ctenoluciidae, Gasteropelecidae, Iguanodectidae, Lebiasinidae, Spintherobolidae, Stevardiidae, and Triportheidae. Taxonomic notes: The traditional family Characidae was divided into four families based on a robust phylogenomic analysis of 1348 UCE loci with 538,472 bp for 494 species in 123 genera (Melo et al. 2024). These families are: Spintherobolidae (six species in three genera), Stevardiidae (324 species in 40 genera), Characidae (207 species in 39 genera), and Acestrorhamphidae (688 species in 56 genera).
- *Habitats:* Taxonomically diverse and ecologically abundant in all major habitats in Neotropical freshwaters.
- *Body size*: Mostly small to moderate sizes (5–20 cm SL), average maximum body size 9.4 cm SL, ranging from 1.4 cm SL in *Xenurobrycon polyancistrus* (Stevardiidae) from forest pools adjacent to rivers in savannahs of the Mamore and Isiboro basins of Bolivia, and *X. varii* from forest pools in the Tapajós and Jamanxim rivers in Pará, Brazil, to the Golden Dorado *Salminus brasiliensis* (Bryconidae) at 130 cm SL from the Paraguay, Paraná, Uruguay, Lagoa dos Patos-Jacui, and Upper Madeira basins (Vitule et al. 2014).
- *Diets*: All major trophic categories, including invertivores (e.g., *Lebiasina*), omnivores (e.g., *Bryconops*), and piscivores (e.g., *Boulengerella*).
- *Geographic distribution*: From the southern USA to northern Patagonia, and from the trans-Andean Pacific drainages to the Atlantic and Caribbean drainages, including all major river basins.

Gymnotiformes

- *Diversity:* 275 species in 36 genera and five families: Apteronotidae, Gymnotidae, Hypopomidae, Rhamphichthyidae, and Sternopygidae.
- *Habitats:* Most major freshwater habitats below c. 500 m elevation. Apteronotidae and Sternopygidae are often amongst the most diverse and abundant fish groups in deep channels (5–100 m depth) of large lowland rivers (Lundberg et al. 1987; Duarte et al. 2022). Some species of *Gymnotus* (Gymnotidae) and *Sternarchorhynchus* (Apteronotidae)

reach c. 1000 m elevation in the northern Andes of Colombia and Ecuador (Albert et al. 2005; de Santana and Vari 2010). *Eigenmannia vicentespelaea* (Sternopygidae) facultatively inhabits caves in the São Domingos karst area of the Brazilian Shield in the state of Goiás (Bichuette and Trajano 2017). A population of *Sternopygus cf. macrurus* (Sternopygidae) inhabits mangroves on the northeastern coast of Pará, Brazil, where they forage for shrimps (decapod crustaceans) in salty marine waters c. 2 m below the low-salinity surface water (Jonathan Ready, pers. comm.).

- *Body size*: Highly variable by clade, but mostly moderate adult sizes, average maximum body size 30.7 cm TL, ranging from 5.3 cm TL in *Procerusternarchus pixuna* (Hypopomidae) from the Negro River of Brazil (Fernandes et al. 2014) and 6.2 cm TL in *Hypopygus minissimus* (Hypopomidae) from the Cunucunuma and Sipapo river basins of Venezuela, to 200 cm TL and 20 kg in the Electric Eel *Electrophorus electricus* (Gymnotidae) from the Guianas (Albert 2001).
- *Diets*: Most species are invertivores, including members of each family; e.g., *Apteronotus leptorhynchus* (Apteronotidae), *Brachyhypopomus brevirostris* (Hypopomidae), *Gymnotus coropinae* (Gymnotidae), *Rhamphichthys rostratus* (Rhamphichthyidae), and *Eigenmannia trilineata* (Sternopygidae). Some riverine species are planktivores feeding on individual food particles (e.g., *Adontosternarchus baleanops*, Apteronotidae; *Steatogenys elegans*, Rhamphichthyidae), or filtering food particles with gill rakers (e.g., *Rhabdolichops stewarti*, Sternopygidae). Some larger-bodied species are omnivores, consuming fruits or fishes adventitiously (e.g., *Electrophorus electricus*), piscivores mainly consuming other fishes as adults (e.g., *Gymnotus tigre*, Gymnotidae), or ectoparasites consuming the tails of other electric fishes (e.g., *Sternarchella raptor*, Apteronotidae).
- *Geographic distribution*: Widespread throughout the humid Neotropics, from Costa Rica to the pampas of northern Argentina and Uruguay (Albert and van der Sleen 2017).

Pimelodoidea

- *Diversity:* 419 species in 60 genera and five families; Conorhynchidae, Heptapteridae, Phreatobiidae, Pimelodidae, and Pseudopimelodidae.
- *Habitats:* Species of Heptapteridae and Pseudopimelodidae inhabit small *terra firme* rivers and streams, Phreatobiidae lives in hypogean waters, and Conorhynchidae and Pimelodidae inhabit in large river channels and floodplains.
- *Body size*: Highly variable among families, average maximum body size 23.5 mm SL, ranging from 2.0 cm SL in *Microglanis zonatus* (Pseudopimelodidae) from small rivers in the Upper Amazon basin of Peru (Ruiz and Shibatta 2010), to a maximum recorded 280 cm SL, but more commonly to 58.1 cm and 32.7 kg, in the Dark-caped Goliath catfish or False Piraiba *Brachyplatystoma capapretum* (Pimelodidae) from the Amazon river (Lundberg and Akama 2005) and a maximum of 170 cm, but more commonly 117 cm, in the Slobbering catfish *Brachyplatystoma platynemum* (Lundberg and Akama 2005).
- *Diets*: Mainly benthic invertivores (e.g., *Imparfinis*), omnivores (e.g., *Pimelodella*), and piscivores (e.g., *Brachyplatystoma*).
- *Geographic distribution*: Widespread through the Neotropics, from southern Mexico to the pampas of Argentina.
- *Note*: One family, Pimelodidae, has species with medium to large body sizes (200–2000 mm SL) and seasonal long-distance migratory behaviors.

Aspredinoidea

- *Diversity:* 273 species, 69 genera, and three families; Aspredinidae, Auchenipteridae, and Doradidae.
- *Habitats:* Habitats associated with deep river channels, as well as tidal estuaries and shallow backwaters. Some aspredinids are semi-fossorial and live in leaf litter. Species of Doradidae often have high biomasses in deep channel habitats (Duarte et al. 2022).
- *Body size*: Mostly small to moderate adult sizes, average maximum body size 15.2 mm SL, ranging from 1.5 cm SL in *Micromyzon orinoco* from the Orinoco River to 76 cm SL in the Ripsaw catfish or Cuiu-Cuiu *Oxydoras niger* (Doradidae) from large rivers of the Amazon, Essequibo, and São Francisco river basins (Birindelli 2014), and the Sierra Copora *Oxydoras sifontesi* from large rivers in the Orinoco basin (Cardillo and Guerrero 2008).
- *Diets*: Mainly benthic invertivores (e.g., *Trachelyopterus*), but also include some piscivores (e.g., *Ageneiosus*), detritivores (e.g., *Doras*), omnivores (e.g., *Acanthodoras*), and frugivores (e.g., *Lithodoras*).
- *Geographic distribution*: All major tropical rivers in South America, a few rivers west of the Andes, Atlantic drainages and coastal marine waters from the Orinoco to Amazon deltas, Panama to Argentina.

Loricarioidei

- *Diversity:* 1834 species in 146 genera and six families; Astroblepidae, Callichthyidae, Loricariidae, Nematogenyidae, Scoloplacidae, and Trichomycteridae. Loricariidae is the second most species-rich fish family, with 997 valid species in five subfamilies and 100 genera (1065 species in 115 genera listed on Eschmeyer's Catalog; Gobiidae with 1420 species in 167 genera).

- *Habitats:* All Amazonian habitats, with many whitewater, blackwater, and clearwater endemics, as well as high-altitude river endemics (Eigenmann 1922).
- *Body size*: Mostly small, average maximum body size 12.2 cm SL, ranging from 1.2 cm SL in *Scoloplax dolicholophia* (Scoloplacidae) from the Negro river basin in Brazil and *Miuroglanis platycephalus* (Trichomycteridae) from the Solimões River of Brazil, to 70 cm SL in the Vermiculated Sailfin catfish *Pterygoplichthys disjunctivus* (Loricariidae) from the Madeira river (Santos and Velez-Gavilan 2024) and *Hypostomus topavae* (Loricariidae) from the Paraná Basin Dias and Zawadzki, 2018).
- *Diets*: Primarily benthic feeders (e.g., *Loricaria*) and algivores (e.g., *Peckoltia*), and the hematophagous (blood-eating) Vandeliinae.
- *Geographic distribution*: Widespread throughout the humid Neotropics (Eigenmann 1922).

3.4 Major Macroevolutionary Patterns

Distinct patterns of ecology and evolution and observed among the many dimensions of biodiversity, with pronounced differences among branches of the phylogeny, among organisms that possess different kinds of traits (e.g., body sizes, habitats, diets, other life history parameters), and in the spatial distributions of species and higher taxa across regions of the Earth's surface (biogeography; Wiens 2023). Here, we discuss major patterns in the distributions of species among prominent phylogenetic and trait categories. Patterns in the biogeographic distributions of the Big Nine are the subject of Chap. 4.

3.4.1 Taxonomic Diversity

Species richness is distributed highly unevenly among the 145 phylogenetically independent clades of Neotropical freshwater fishes (Fig. 3.3). As of the date of the data compilation for this study, the Big Nine NFF clades represented 44 families constituting 46% of all NFF families, 717 genera constituting 82% of the NFF genera, and 5736 species constituting 90% of all NFF species (Table 3.1; Fig. 3.4A). The Big Nine NFF clades exhibit species richness values ranging from 1834 species in Loricarioidei to 275 species in Gymnotiformes (Table 3.2).

The other 136 NFF clades have a combined total of 675 species or 9.5% of the NFF total. At the other end of the diversity spectrum, about 70% (101 of the 145 NFF clades) are represented by a single species. These singleton clades encompass a very broad taxonomic composition, being assigned to 66 genera and 35 families. All 101 of these singleton clades are Marine Derived Lineages (MDLs), meaning each of these freshwater fish species is a member of an otherwise marine family, with relatively recent (Neogene or Quaternary) origins in continental freshwaters. Gobiidae is the most diverse MDL family in Neotropical freshwaters, with 20 species in nine genera. Snooks of the genus *Centropomus* (Centropomidae) have the greatest number of independent evolutionary transitions into Neotropical freshwaters, with six species in Atlantic and four species in Pacific draining basins, and with all species inhabiting both freshwater and marine waters; i.e., no snook species is endemic to freshwaters.

A frequency distribution in which most species are members of a few dominant clades, and most clades have few species, is known as a *hollow curve* (McGill 2010; Bennett et al. 2017). Such hollow-curve frequency distributions are widely observed in biodiversity datasets across the Tree of Life (Scotland and Sanderson 2004). Hollow-curve frequency distributions may be assessed either as variance in the numbers of species per genus or other supraspecific Linnean rank, or in the highly unbalanced shape of most phylogenetic tree topologies (Mooers and Heard 1997; Blum and François 2006; Gascuel et al. 2015).

During the 10 years from the previous compilation of NFF taxa by Albert et al. (2011, table 5.1), the number of species in the Big Nine expanded from 5164 to 6342 species, an increase of 1179 species or about 23%. During this same time period, the number of known singleton NFF clades (represented by one species each) expanded from 16 to 101 species, based primarily on new records of MDLs collected in coastal rivers of the Pacific coast of Colombia and Ecuador (i.e., Choco) and northeastern Brazil.

3.4.2 Body Size

As with most fishes and animals generally (Albert and Johnson 2012), body size is strongly structured phylogenetically among NFF clades, with some clades overrepresented among species with extreme body size values (Chap. 19). Among the top 10 largest species by standard body length (measured from tip of snout to end of caudal-fin base), three of the four largest species are MDLs; two chondrichthyans, the Largetooth sawfish *Pristis pristis* at 610 cm, and *Carcharhinus leucas* at 340 cm, and an elopiform teleost fish, the Atlantic tarpon *Megalops atlanticus* at 250 cm. All the other largest species are primary freshwater species; the osteoglossiform *Arapaima gigas* at 300 cm, the pimelodid catfishes *Brachyplatystoma capapretum* at 280 cm and *B. platynemum* at 192 cm, the gymnotid electric eels *Electrophorus electricus* at 200 cm and *E. multivalvulus* (the senior synonym of *E. varii*) at 148 cm, and the Cuban gar *Atractosteus tristoechus* at 180 cm.

Among the 12 NFF species with the smallest standard body lengths, three of these species are MDLs in the gobiiform family Eleotridae (*Leptophilypnion fittkaui* at 0.9 cm, *Hemieleotris levis* at 1.0 cm, and *Microphilypnus acangaquara* at 1.4 cm), one species is a secondary freshwater poecillioid in the family

Table 3.2 Salient features of 44 families constituting the Big Nine NFF clades

Clade	Family	Genera	Spp. in tree	Total spp.[a]	Min size	Max size	MSA	Adult modal habitat
Cichlinae	Cichlinae	81	569	669	2.0	100	2.0	Rivers, streams, and rapids
Poecilioidea	Anablepidae	3	18	19	22.0	32.0	1.6	Coastal drainages, tidal flats, mangroves, and brackish
	Cyprinodontidae[b]	7	55	55	3.5	27.0	16.1	High-elevation lakes and streams
	Fluviphylacidae	1	7	7	1.4	2.4	1.6	Blackwater streams
	Poeciliidae	47	250	276	1.7	6.0	1.6	Fresh and brackish waters
	Profundulidae	3	11	16	5.5	11.5	4.2	Clearwater upland streams
Rivulidae	Rivulidae	35	431	488	2.5	15.0	0.0	Fresh and brackish waters
Curimatoidea	Anostomidae	15	148	150	7.7	60.0	2.7	Most habitats
	Chilodontidae	2	8	8	6.4	16.3	17.3	Rivers, floodplains, oxbow lakes, and streams
	Curimatidae	8	115	121	3.9	32.0	4.8	Most freshwater ecosystems
	Cynodontidae	3	8	8	28.0	66.0	1.9	River channels, margins, and vegetated margins
	Hemiodontidae	4	32	34	7.0	31.0	3.0	Lakes, large rivers, and forest streams
	Parodontidae	3	32	32	5.6	15.0	2.0	Small streams and forest pools
	Prochilodontidae	3	21	21	24.0	80.0	2.6	Rivers, floodplains, oxbow lakes, and streams
	Serrasalmidae	15	99	102	6.0	100	2.5	Rivers, streams, floodplain lakes, and flooded forests
Characoidea	Acestrorhamphidae	56	632	688	1.5	32.0	2.9	Streams and rivers
	Acestrorhynchidae	6	26	26	10.0	40.0	14.8	Lakes, lagoons, river margins, and small streams
	Anostomidae	14	74	147	1.6	16.0	2.7	Small streams and forest pools
	Bryconidae	4	44	49	15.0	100	5.0	Streams and rivers
	Chalceidae	1	5	5	14.0	25.0	2.6	Flooded forests, river channels, and forest streams
	Characidae	39	207	207	1.8	40.0	1.9	Rivers, lakes, flooded forests, and mountain streams
	Ctenoluciidae	2	7	7	25.0	70.0	27.0	Streams, rivers, and lentic water bodies
	Gasteropelecidae	3	9	9	2.2	9.0	13.0	Open rivers, streams, and lakes
	Iguanodectidae	3	34	39	3.1	13.0	1.7	Small streams and others
	Spintherobolidae	3	6	6	2.0	6.1	4.9	Small streams and others
	Stevardiidae	40	324	376	1.4	40.0	1.5	Small streams and others
	Triportheidae	5	20	21	4.2	24.0	11.1	Surface of lakes, flooded forests, and streams
Gymnotiformes	Apteronotidae	16	96	98	15.0	129	2.2	Deep channels, small streams, and rapids
	Gymnotidae	2	49	51	10.0	216	2.1	Broadly eurytopic except river channels
	Hypopomidae	6	34	38	5.3	46.0	4.6	Streams, savannas, oxbow lakes, and floating meadows
	Sternopygidae	7	54	62	14.0	70.0	3.0	Deep channels, floodplains, rapids, and streams
	Rhamphichthyidae	5	26	26	5.6	100	6.1	River channels and small streams
Pimelodoidea	Heptapteridae	23	223	240	2.4	47.0	1.5	Small forest streams, benthic, and subterranean
	Phreatobiidae	1	3	3	4.0	6.0	18.2	Phreatic environments
	Pimelodidae	30	114	117	20.0	200	4.4	Large rivers, high-gradient streams, and lakes
	Pseudopimelodidae	6	50	59	2.0	80.0	1.6	Benthic habitats

Clade	Family	Genera	Spp. in tree	Total spp.[a]	Min size	Max size	MSA	Adult modal habitat
Aspredinoidea	Aspredinidae	13	47	49	1.6	38.0	11.1	Shallow backwaters, deep channels, and tidal estuaries
	Auchenipteridae	25	125	128	2.5	59.0	2.0	Main channels, flooded forests, and marginal lakes
	Doradidae	31	95	96	2.2	100	3.2	Main channels, flooded forests, and marginal lakes
Loricarioidei	Astroblepidae	1	80	82	2.5	30.0	4.2	High-elevation streams
	Callichthyidae	8	221	229	2.5	24.0	3.0	Streams, floodplains, and margins
	Loricariidae	104	997	1070	2.2	60.0	2.1	Nearly all freshwater habitats <1500 m elevation
	Scoloplacidae	1	6	6	1.0	2.0	7.5	Terra firme streams, oxbow lakes, and backwater pools
	Trichomycteridae	32	324	447	1.3	38.0	2.0	Wide variety of habitats, including extremes
Sum (Avg)	**44**	**717**	**5736**	**6387**	**–7.3**	**–52.3**	**–5.4**	
% NFF	**45.4%**	**81.5%**	**90.4%**	**90.4%**				

Species in phylogeny of Chapter 2 with 6342 NFF species. Body size as maximum recorded standard length (cm) for smallest and largest species per family; MSA, median species age (Ma) from Chap. 5; Habitat data summarized from Albert et al., in press [a]Species totals as of May 1, 2025 [b]Neotropical species only

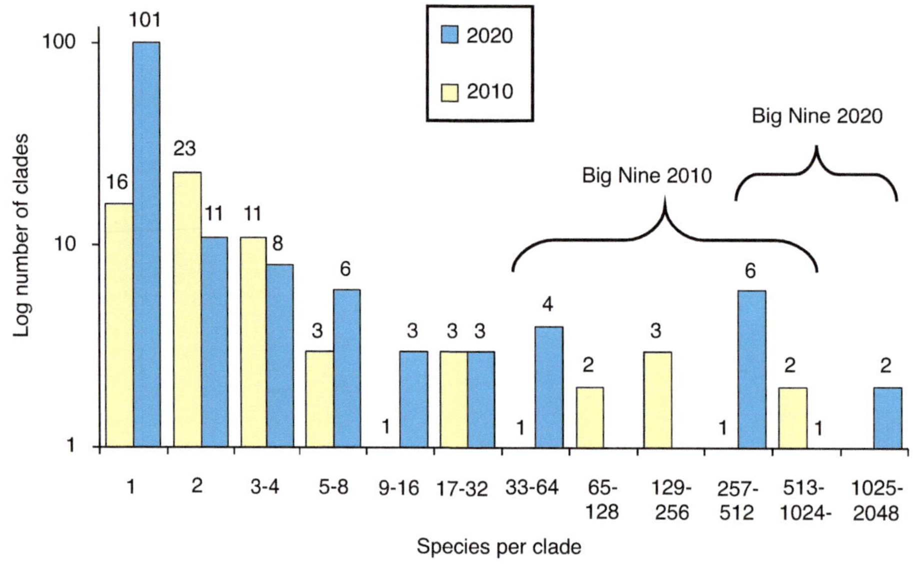

Fig. 3.3 Frequency distribution of species among NFF clades. Data for 2020 include 6342 species in 145 clades; data for 2010 include 5762 species in 66 clades. Note the Big Nine NFF clades include 5736 species representing 93% of the whole NFF fauna. Note also that a large majority (101 of 145, or 70%) of NFF clades are represented by a single species, allotted to 66 genera and 35 families. Note further from 2010 to 2020 the singleton clades expanded from 16 to 101 species, all from newly recorded MDLs in coastal rivers of northwestern South America (Colombia and Ecuador), and northeastern Brazil. 2010 data from Albert et al. (2011)

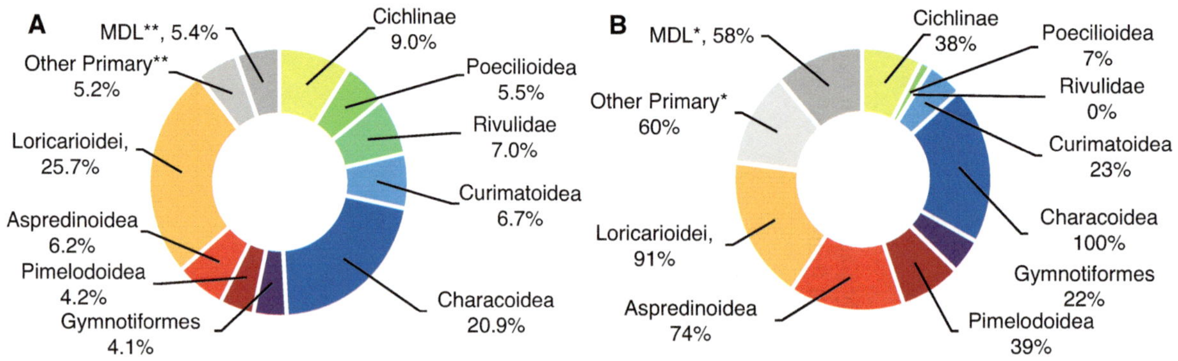

Fig. 3.4 Biodiversity metrics for Big Nine NFF clades. **A** Relative proportions of species richness. Values sum to 100%. **B** Relative proportions of ecospace volume. Values as a proportion of Characoidea. Species richness from phylogeny in Chap. 2 and ecospace in Chap. 6. Ecospace volume from Chap. 6. *, aggregate data for multiple clades. MDL marine derived lineages ($n = 101$); Other Primary clades ($n = 15$). Note the predominance of ostariophysan fishes in both species richness and ecospace volume

Fluviphylacidae (*Fluviphylax gouldingi* at 1.4 cm), and 10 species are otophysan fishes, including six catfishes in the siluriform families Scoloplacidae (*Scoloplax dolicholophia* at 1.2 cm and *S. dicra* at 1.4 cm) and Trichomycteridae (*Miuroglanis platycephalus* at 1.2 cm, and *Potamoglanis anhanga* at 1.3 cm), and four in the characiform families Crenuchidae (*Microcharacidium weitzmani* at 1.2 cm and *Elachocharax mitopterus* at 1.4 cm) and Stevardiidae (*Xenurobrycon polyancistrus* at 1.4 cm and *X. varii* at 1.4 cm).

The Big Nine NFF clades with the largest average adult body sizes are Gymnotiformes (30.7 cm), Pimelodoidea (23.5 cm), and Curimatoidea (18.2 cm). These clades include the ecologically dominant riverine families Apteronotidae, Pimelodidae, and Prochilodontidae, respectively (Fig. 3.5A). Big Nine NFF clades with the smallest average adult body sizes are the two cyprinodontiform clades, Poecilioidea (4.6 cm) and Rivulidae (5.4 cm), both clades of which are almost entirely absent from large river habitats.

3.4.3 Species Ages

A species-lineage age is the divergence time since the most recent common ancestor of sister lineages; i.e., pendant edges (Robinson and Foulds 1981; Stadler et al. 2014; Craig et al. 2022). Divergence times for this study were estimated from

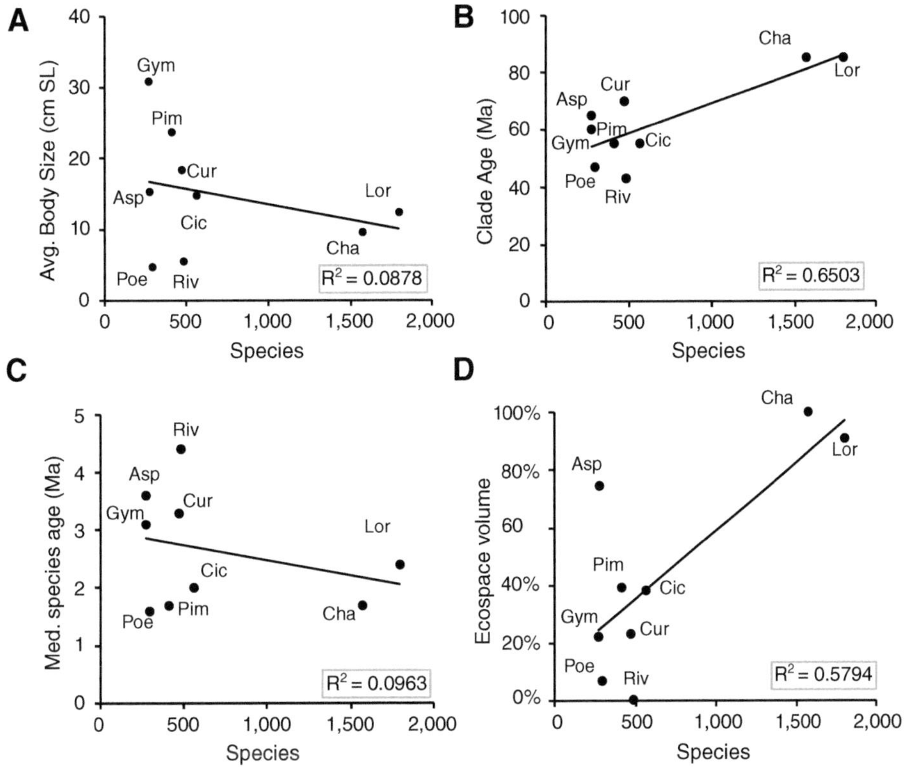

Fig. 3.5 Correlations between species richness and three other biodiversity metrics for the Big Nine NFF clades. **A** Average adult body size (cm SL). **B** Crown-group clade age (Ma). **C** Median species age (Ma). **D** Ecospace volume (see Chap. 6). Note positive correlations between species richness and clade age and ecospace volume, but not with body size or median species age. Abbreviations as first three letters of clade name. Data in Table 3.1

branch lengths on a time-calibrated phylogeny of 6143 NFF species, referred to here as the 6 K tree. The 6 K tree was built using the phylogeny presented in Chap. 2 with 3167 species, referred to here as the 3 K tree, by manually attaching 2976 species (representing 48.4% of the NFF total) for which no genetic sequence data are yet available, based on the expert opinion of contributing taxonomic authorities.

The median species age (MSA) of all NFF taxa is 2.3 Ma, ranging from 1.6 Ma in Poecilioidea to 4.4 Ma in Rivulidae. Due to the significant positive skew (3.8) in the distribution of NFF species ages, the average species age of all NFF species is notably older at 4.3 Ma. MSA is therefore a more useful estimate of the central tendency of species age values for each clade (Hedges et al., 2015). Among the Big Nine NFF clades, Rivulidae has the oldest MSA, with 34 species being older than 10.0 Ma. This pattern of ancient rivulid lineages is well exemplified by the Mexican rivulus *Millerichthys robustus*, an annual hermaphroditic species with small adult body size (to 4 cm SL) and an estimated branch length of 38.7 Ma. This species inhabits ephemeral pools in Vera Cruz, Mexico, and is geographically separated from its closest living relatives that live in the Greater Antilles (Cuba and Hispaniola; Domínguez-Castanedo et al. 2024; Fig. 3.5C). Other ancient rivulid species with apparently relictual distributions are *Micromoema xiphophora* (MSA = 20.4 Ma) from the Orinoco basin and *Hypsolebias adornatus* (MSA = 19.2 Ma) from the São Francisco River basin.

The MSA values recovered here are similar among the Big Nine clades, ranging only about two-fold from 1.6 Ma in Poeciliodea to 4.4 Ma in Rivulidae. Species richness is negatively correlated with MSA among 55 family- and tribe-level clades of the Big Nine (Fig. 3.6C), meaning that, on average, more diverse clades have younger species. There is also a strong positive correlation between species-richness and the skewness value of species ages among these clades (Fig. 3.6D). A positive skew in the distribution of MSA values results from the presence of more old species in a clade, and therefore a more asymmetric tree topology (Mooers and Heard 1997). Therefore, the positive correlation observed between the skew of MSA and species richness among NFF families describes the same phenomenon as the hollow-curve distribution of species among higher taxa referred to in the text above. These observations that are consistent with the *Effect Hypothesis* of Vrba (1984; Vrba and Eldredge 1984; Albert et al. 2011).

The age estimates for NFF species reported here are slightly younger than literature estimates for other fish groups generated

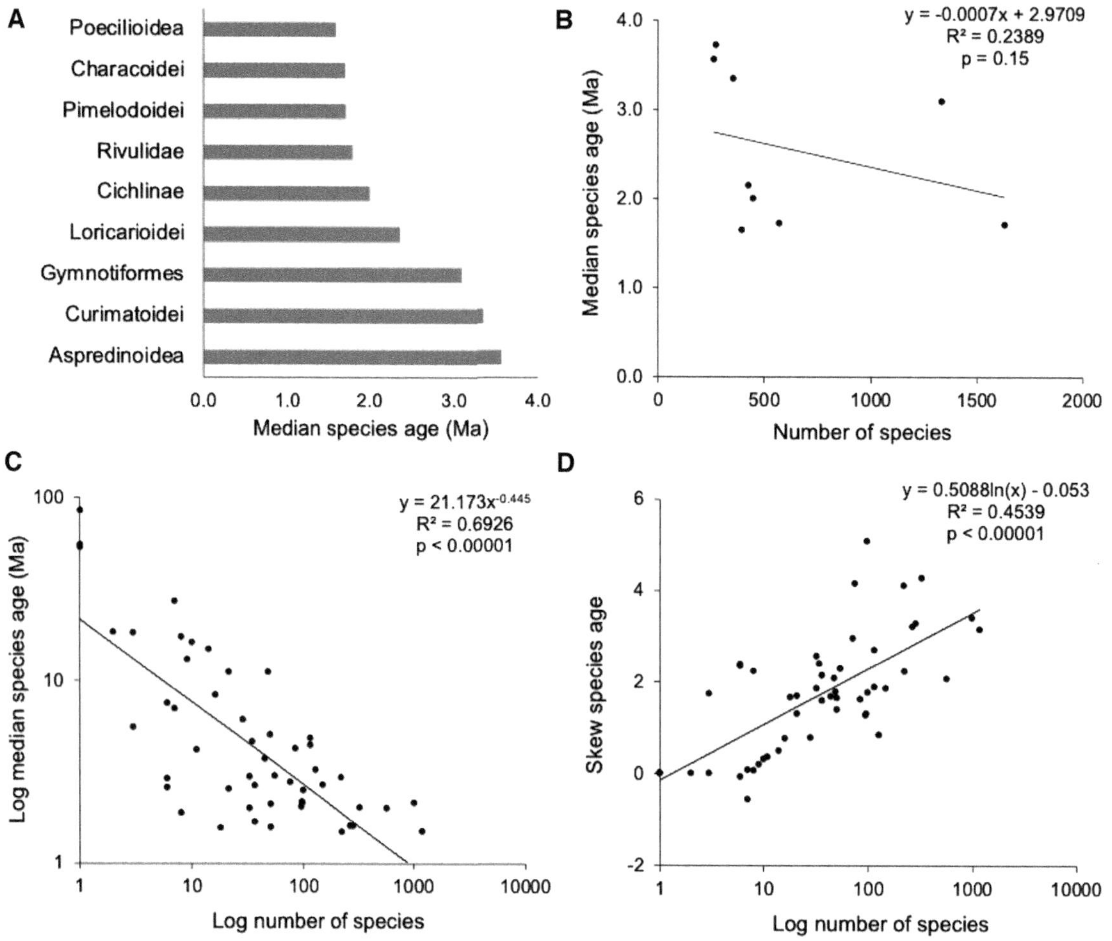

Fig. 3.6 Median species age (MSA) of NFF clades. **A** Similar MSA values among the Big Nine clades, ranging about two-fold from 1.6 Ma in Poecilioidea to 3.6 Ma in Aspredinoidea. **B** Species richness is not significantly correlated with MSA among the Big Nine. **C** Strong negative correlation between species richness and MSA among 55 family- and tribe-level clades of the Big Nine. **D** Strong positive correlation between species-richness and the skewness value of species ages among family- and tribe-level clades of the Big Nine. High skew values indicate a preponderance of older early-branching species in the species-rich clades

from molecular phylogenetic and paleontological datasets (Hedges et al. 2015; Pie and Caron 2023). Poikilothermic species of fishes, amphibians, and reptiles are generally thought to be older on average than homeothermic species of birds and mammals; 4–10 vs. 2–3 Ma, respectively. Tropical species generally are thought to exhibit a wider range of species age values due to phylogenetic age and niche conservatism (Wiens and Donoghue 2004), and tropical fishes are reported to be younger than their polar relatives due to Pleistocene extinctions and *evolutionary radiations* (Rabosky 2020). However, estimated ages for vertebrate species vary over many orders of magnitude, from 10 Ka to more than 100 Ma, and the age distribution is log-normal, meaning there are many more ancient species than would be expected from a statistically normal distribution (Guo et al. 2024). Further, MSAs estimated from branch lengths of the 6 K tree are expected to be about half that estimated from the 3 K tree, highlighting the importance of taxon sampling in estimates of MSA (Stadler et al. 2014; Craig et al. 2022).

3.4.4 Ecospace Occupancy

Ecospace occupancy was assessed as the proportion of the multivariate hypervolume calculated from a statistical analysis of 42 ecotrait variables; see details in Chap. 6. Briefly, we used principal components analysis (PCA) to extract linear combinations of the 42 ecotrait variables that maximally explain the variance. We call this multivariate hypervolume of ecological trait diversity the NFF *ecospace*. Ecospace hypervolumes for NFF clades were calculated based on 41 principal component (PC) dimensions derived from a PCA of the 42 ecological traits across 6342 species. We used a custom R script to calculate the size of the ecospace as a convex hull around all the points in the 41-dimensional hypervolume.

Among members of the Big Nine, Characoidea has the largest ecospace volume, which was used assess the proportional ecospace occupancy of the other clades (Fig. 3.4B). Ecospace occupancy ranges from 91% in Loricariodei to 0.3% in Rivulidae. This means that Loricarioidea occupies the greatest range of ecological modalities due to its enormous species richness and morphological disparity, while Rivulidae is much more ecologically specialized, both due to their lower species richness and conserved morphologies. When assessed as deviations from their species richness values, Aspredinoidea has the largest ecospace, and the two cyprinodontiform clades (Poecilioidea and Rivulidae) have the smallest ecospaces (Fig. 3.5D).

References

Agapow PM, Bininda-Emonds OR, Crandall KA, Gittleman JL, Mace GM, Marshall JC, Purvis A (2004) The impact of species concept on biodiversity studies. Q Rev Biol 79(2):161–179

Agostinho AA, Gomes LC, Suzuki HI, Júlio JrHF (2003) Migratory fishes of the Upper Paraná River Basin, Brazil. In: Carolsfeld J, Harvey B, Ross C, Baer A (Eds.)Eds. Chapter 19 in Migratory fishes of South America. World Fisheries Trust, Victoria, BC

Aguilera G, Terán GE, Mirande JM, Alonso F, Rometsch S, Meyer A, Torres-Dowdall J (2019) Molecular and morphological convergence to sulfide-tolerant fishes in a new species of *Jenynsia* (Cyprinodontiformes: Anablepidae), the first extremophile member of the family. PLoS One 14(7):e0218810

Albert JS (2001) Species diversity and phylogenetic systematics of American knifefishes (Gymnotiformes, Teleostei). Miscellaneous Publications Museum of Zoology, University of Michigan (190):1–127

Albert JS, Johnson DM (2012) Diversity and evolution of body size in fishes. Evol Biol 39:324–340

Albert JS, Crampton WGR, Thorsen DH, Lovejoy NR (2005) Phylogenetic systematics and historical biogeography of the Neotropical electric fish *Gymnotus* (Teleostei: Gymnotidae). Syst Biodiversity 2(4):375–417

Albert JS, Petry P, Reis RE (2011) Major biogeographic and phylogenetic patterns. In: Albert JS, Reis RE (Eds.)Eds. Historical biogeography of neotropical freshwater fishes. University of California Press, Berkeley

Albert JS, Tagliacollo VA, Dagosta F (2020) Diversification of Neotropical freshwater fishes. Annu Rev Ecol Evol Syst 51(1):27–53

Amorim PF, Costa WJ (2019) Reconstructing biogeographic temporal events in the evolution of the livebearer fish genus *Jenynsia* based on total evidence analysis (Cyprinodontiformes: Anablepidae). Syst Biodiversity 17(2):124–133

Bennett DJ, Sutton MD, Turvey ST (2017) Evolutionarily distinct "living fossils" require both lower speciation and lower extinction rates. Paleobiology 43(1):34–48

Bichuette ME, Trajano E (2017) Biology and behavior of *Eigenmannia vicentespelaea*, a troglobitic electric fish from Brazil (Teleostei: Gymnotiformes: Sternopygidae): a comparison to the epigean species, *E. trilineata*, and the consequences of cave life. Trop Zool 30(2):68–82

Bickford D, Lohman DJ, Sodhi NS, Ng PK, Meier R, Winker K et al. (2007) Cryptic species as a window on diversity and conservation. Trends Ecol Evolut 22(3):148–155

Birindelli JL (2014) Phylogenetic relationships of the South American Aspredinoidea (Ostariophysi: Siluriformes). Neotrop Ichthyol 12(03):451–564

Blum MG, François O (2006) Which random processes describe the tree of life? A large-scale study of phylogenetic tree imbalance. Syst Biol 55(4):685–691

Borghezan EDA, Pires THDS, Ikeda T, Zuanon J, Kohshima S (2021) A review on fish sensory systems and Amazon water types with implications to biodiversity. Front Ecol Evol 8:589760

Bragança PHN (2018) *Fluviphylax gouldingi* and *F. wallacei*, two new miniature killifishes from the middle and upper Rio Negro drainage, Brazilian Amazon. Spixiana 41(1):133–146

Cano-Barbacil C, Olden JD, García-Berthou E (2024) Myers' divisions revisited: contemporary evidence for distinct trait differences among global inland fishes. Fish Fish 25(4):672–685. https://doi.org/10.1111/faf.12832

Cardillo E, Guerrero HY (2008) Reproductive cycle of the Venezuelan catfish "Sierra Negra" (*Oxydoras sifontesi*). Cybium 32(2):172–173

Carmona CP, De Bello F, Mason NW, Lepš J (2016) Traits without borders: integrating functional diversity across scales. Trends Ecol Evolut 31(5):382–394

Cazzolla Gatti R, Reich PB, Gamarra JG, Crowther T, Hui C, Morera A et al. (2022) The number of tree species on Earth. Proc Natl Acad Sci 119(6):e2115329119

Chao A, Chiu CH (2016) Species richness: estimation and comparison. Wiley StatsRef: Statistics Reference Online 1:26

Chen WJ, Lavoué S, Mayden RL (2013) Evolutionary origin and early biogeography of otophysan fishes (Ostariophysi: Teleostei). Evolution 67(8):2218–2239

Claridge MF, Dawah HA, Wilson MR (eds) (1997) Species: the units of biodiversity, pp xvi+–439

Colwell RK (2009) Biodiversity: concepts, patterns, and measurement. Princeton Guide Ecol 663:257–263

Craig JM, Crampton WG, Albert JS (2017) Revision of the polytypic electric fish *Gymnotus carapo* (Gymnotiformes, Teleostei), with descriptions of seven subspecies. Zootaxa 4318(3):401–438

Craig JM, Kumar S, Hedges SB (2022) Limitations of phylogenomic data can drive inferred speciation rate shifts. Mol Biol Evol 39(3):msac038

Crampton WG (2011) An ecological perspective on diversity and distributions. In: Albert J, Reis RE (eds) Historical biogeography of Neotropical freshwater fishes. University of California Press, Berkeley, Los Angeles and London, pp 165–189.

De Queiroz K (2007) Species concepts and species delimitation. Syst Biol 56(6):879–886

de Santana CD, Vari RP (2010) Electric fishes of the genus *Sternarchorhynchus* (Teleostei, Ostariophysi, Gymnotiformes); phylogenetic and revisionary studies. Zool J Linnean Soc 159(1):223–371

Dias AC, Zawadzki CH (2018) Identification key and pictures of the *Hypostomus* Lacépède, 1803 (Siluriformes, Loricariidae) from the rio Ivaí, upper rio Paraná basin. Check List 14(2):393–414

Diniz-Filho, JAF, 2023. The Macroecological Perspective: Theories, Models and Methods. Springer Nature, Cham, Switzerland, 449

Dobzhansky T (1973) Nothing in biology makes sense except in the light of evolution. Am Biol Teach 75(2):87–91

Domínguez-Castanedo O, Thompson AW, Meredith RW, Furness AI (2024) An independent origin of an annual life cycle in a North American killifish species. Biol J Linn Soc 144(2):blae029

Duarte C., Farago TLB, dos Anjos CS, dos Santos NR, do Nascimento LM, Cella-Ribeiro A, de Deus CP (2022) Spatial and seasonal variation of benthic fish assemblages in whitewater rivers of Central Amazon. Biota Neotrop 22(4), e20211312

Eigenmann CH (1922) The Fishes of Western South America, Part 1. The fresh-water fishes of northwestern South America, including Colombia, Panama, and the Pacific slopes of Ecuador and Peru, together with an appendix upon the fishes of the Rio Meta in Colombia. Mem Carnegie Mus 9(9):1–346

Eldredge N, Thompson JN, Brakefield PM, Gavrilets S, Jablonski D, Jackson JB et al. (2005) The dynamics of evolutionary stasis. Paleobiology 31(S2):133–145

Evans KM, Vidal-García M, Tagliacollo VA, Taylor SJ, Fenolio DB (2019) Bony patchwork: mosaic patterns of evolution in the skull of electric fishes (Apteronotidae: Gymnotiformes). Integr Comp Biol 59(2):420–431

Fernandes CC, Nogueira A, Alves-Gomes JA (2014) *Procerusternarchus pixuna*, a new genus and species of electric knifefish (Gymnotiformes: Hypopomidae, Microsternarchini) from the Negro River, South America. Proc Acad Nat Sci Philadelphia 163(1):95–118

Fine PV (2015) Ecological and evolutionary drivers of geographic variation in species diversity. Annu Rev Ecol Evol Syst 46(1):369–392

Ford KL, Bernt MJ, Summers AP, Albert JS (2022) Mosaic evolution of craniofacial morphologies in ghost electric fishes (Gymnotiformes: Apteronotidae). Ichthyol Herpetol 110(2):315–326

Freudenstein JV, Broe MB, Folk RA, Sinn BT (2017) Biodiversity and the species concept—lineages are not enough. Syst Biol 66(4):644–656

Fricke, R, Eschmeyer, WN, Van derLaan, R, 2025. Catalog of fishes: genera, species, references. California Academy of Sciences, San Francisco, CA, USA, http://researcharchive.calacademy. org/research/ichthyology/catalog/fishcatmain.asp

Gascuel F, Ferrière R, Aguilée R, Lambert A (2015) How ecology and landscape dynamics shape phylogenetic trees. Syst Biol 64(4):590–607

Gómez-Martínez RF, Matamoros WA, Hoagstrom CW, López-Vila JM, Bonilla EPDD (2024) Phylogenetic diversity of live-bearing fishes (Poeciliidae) peaks on peninsulas, isthmuses, and in deserts. J Biogeogr 52(3):641–656. https://doi.org/10.1111/jbi.15066.

Guayasamin JM, Ribas CC, Carnaval AC, Carrillo JD, Hoorn C, Lohmann LG et al. (2024) Evolution of Amazonian biodiversity: a review. Acta Amazonica 54(spe1):e54bc21360

Guo Q, Qian H, Zhang J, Liu P (2024) On the relationship between species diversity and range size. J Biogeogr 49:1911–1919

Hall BK (2013) Homology: The Hierarchical Basis of Comparative Biology. Elsevier, San Diego, CA, 504

Hall, BK, Hallgrímsson, B, 2007. Strickberger's Evolution. Jones & Bartlett Publishers, Sudbury, MA, p. 762

Hedges SB, Marin J, Suleski M, Paymer M, Kumar S (2015) Tree of life reveals clock-like speciation and diversification. Mol Biol Evol 32(4):835–845

Hernández-Ávila SG, Hoagstrom CW, Matamoros WA (2024) Historical biogeography of North American killifishes (Cyprinodontiformes) recapitulates geographical history in the Gulf of México watershed. Zool J Linnean Soc 202(2):zlae105

Hughes LC, Ortí G, Huang Y, Sun Y, Baldwin CC, Thompson AW et al. (2018) Comprehensive phylogeny of ray-finned fishes (Actinopterygii) based on transcriptomic and genomic data. Proc Natl Acad Sci 115(24):6249–6254

IGFA (2001) Database of IGFA angling records until 2001. International Game and Fishing Association, Fort Lauderdale, Florida.

Koslowski I (1985) Description of new species of *Apistogramma* (Teleostei: Cichlidae) from the Rio Mamoré system in Bolivia. Bonner zoologische Beiträge 36:145–162

Lenormand T (2002) Gene flow and the limits to natural selection. Trends Ecol Evolut 17(4):183–189

Lowe-McConnell RH (1969) Speciation in tropical freshwater fishes. Biol J Linnean Soc 1(1-2):51–75

Lundberg JG, Akama A (2005) *Brachyplatystoma capapretum*: a new species of Goliath catfish from the Amazon basin, with a reclassification of allied catfishes (Siluriformes: Pimelodidae). Copeia 2005(3):492–516

Lundberg JG, Lewis WM Jr, Saunders JF III, Mago-Leccia F (1987) A major food web component in the Orinoco River channel: evidence from planktivorous electric fishes. Science 237(4810):81–83

Mace GM (2004) The role of taxonomy in species conservation. Philos Trans R Soc London [Biol] 359(1444):711–719

Magurran AE (2021) Measuring biological diversity. Curr Biol 31(19):R1174–R1177

Matamoros WA, Kreiser BR, Schaefer JF (2012) A delineation of Nuclear Middle America biogeographical provinces based on river basin faunistic similarities. Rev Fish Biol Fish 22:351–365

Matamoros WA, McMahan CD, Chakrabarty P, Albert JS, Schaefer JF (2015) Derivation of the freshwater fish fauna of Central America revisited: Myers's hypothesis in the twenty-first century. Cladistics 31(2):177–188

Mayden RL (2002) On biological species, species concepts and individuation in the natural world. Fish Fish 3(3):171–196

Mayr E, Bock WJ (2002) Classifications and other ordering systems. J Zool Sys Evol Res 40(4):169–194

McGill BJ (2010) Towards a unification of unified theories of biodiversity. Ecol Lett 13(5):627–642

Melo BF, Albert JS, Dagosta FC, Tagliacollo VA (2021) Biogeography of curimatid fishes reveals multiple lowland–upland river transitions and differential diversification in the Neotropics (Teleostei, Curimatidae). Ecol Evol 11(22):15815–15832

Melo BF, Ota RP, Benine RC, Carvalho FR, Lima FC, Mattox GM et al. (2024) Phylogenomics of Characidae, a hyper-diverse Neotropical freshwater fish lineage, with a phylogenetic classification including four families (Teleostei: Characiformes). Zool J Linnean Soc 202(1):zlae101

Mishler BD, Donoghue MJ (1982) Species concepts: a case for pluralism. Syst Zool 31(4):491–503

Mooers AO, Heard SB (1997) Macroevolution and the shapes of phylogenetic trees. Q Rev Biol 72:31–54

Morcillo F, Ornelas-García CP, Alcaraz L, Matamoros WA, Doadrio I (2016) Phylogenetic relationships and evolutionary history of the Mesoamerican endemic freshwater fish family Profundulidae (Cyprinodontiformes: Actinopterygii). Mol Phylogenet Evol 94:242–251

Murren CJ (2012) The integrated phenotype. Integr Comp Biol 52(1):64–76

Myers GS (1949) Salt-tolerance of fresh-water fish groups in relation to zoogeographical problems. Bijdr Dierkd 28(1949):315–322

O'Leary NA, Cox E, Holmes JB, Anderson WR, Falk R, Hem V et al. (2024) Exploring and retrieving sequence and metadata for species across the tree of life with NCBI Datasets. Sci Data 11(1):732

Petit RJ, Excoffier L (2009) Gene flow and species delimitation. Trends Ecol Evolut 24(7):386–393

Pie MR, Caron FS (2023) Substantial variation in species ages among vertebrate clades. Evol J Linn Soc 2(1):kzad006

Rabosky DL (2020) Speciation rate and the diversity of fishes in freshwaters and the oceans. J Biogeogr 47(6):1207–1217

Robinson DF, Foulds LR (1981) Comparison of phylogenetic trees. Math Biosci 53(1-2):131–147

Rodríguez-Machado S, Elías DJ, McMahan CD, Gruszkiewicz-Tolli A, Piller KR, Chakrabarty P (2024) Disentangling historical relationships within Poeciliidae (Teleostei: Cyprinodontiformes) using ultraconserved elements. Mol Phylogenet Evol 190:107965

Ruiz WBG, Shibatta OA (2010) A new species of *Microglanis* (Siluriformes, Pseudopimelodidae) from lower Rio Tocantins basin, Pará, Brazil, with description of superficial neuromasts and pores of lateral line system. Zootaxa 2632(1):53–66

Santelli A, Fonseca G (2023) Amazonian expedition searches out rare 'fish from the clouds'. Mongabay. https://news.mongabay.com/2023/12/

Santos C, Velez-Gavilan J (2024) *Pterygoplichthys disjunctivus* (Vermiculated Sailfin catfish). CABI Compendium. https://www.cabi.org/ISC/Datasheet/11659

Schwassmann HO, Kruger L (1965) Experimental analysis of the visual system of the four-eyed fish *Anableps microlepis*. Vision Res 5(6–7):269-IN1

Scotland RW, Sanderson MJ (2004) The significance of few versus many in the tree of life. Science 303(5658):643–643

Sexton JP, McIntyre PJ, Angert AL, Rice KJ (2009) Evolution and ecology of species range limits. Annu Rev Ecol Evol Syst 40:415–436

Smith, ML, Carstens, BC, 2022. Species delimitation using molecular data, in: Species Problems and Beyond. CRC Press, Boca Raton, pp. 145–160

Stadler T, Rabosky DL, Ricklefs RE, Bokma F (2014) On age and species richness of higher taxa. Am Nat 184(4):447–455

Stankowski S, Ravinet M (2021) Defining the speciation continuum. Evolution 75(6):1256–1273

Struck TH, Feder JL, Bendiksby M, Birkeland S, Cerca J, Gusarov VI et al. (2018) Finding evolutionary processes hidden in cryptic species. Trends Ecol Evolut 33(3):153–163

Tagliacollo VA, Duke-Sylvester SM, Matamoros WA, Chakrabarty P, Albert JS (2017) Coordinated dispersal and pre-isthmian assembly of the Central American ichthyofauna. Syst Biol 66(2):183–196

Tilman, D, 2001. Functional diversity, Pp. 587-596 in Encyclopedia of Biodiversity, 2nd edn. Elsevier Inc., San Diego, p. 596

Turcotte MM, Levine JM (2016) Phenotypic plasticity and species coexistence. Trends Ecol Evolut 31(10):803–813

Vari RP (1983) Phylogenetic relationships of the families Curimatidae, Prochilodontidae, Anostomidae, and Chilodontidae (Pisces: Characiformes). Smithson Contrib Zool 378:iii+60

Vitule JRS, Bornatowski H, Freire CA, Abilhoa V (2014) Extralimital introductions of *Salminus brasiliensis* (Cuvier, 1816) (Teleostei, Characidae) for sport fishing purposes: a growing challenge for the conservation of biodiversity in neotropical aquatic ecosystems. BioInvasions Records 3(4):291–296

Vrba ES (1984) What is species selection?. Syst Zool 33(3):318–328

Vrba ES, Eldredge N (1984) Individuals, hierarchies and processes: towards a more complete evolutionary theory. Paleobiology 10(2):146–171

Wake DB, Roth G, Wake MH (1983) On the problem of stasis in organismal evolution. J Theor Biol 101(2):211–224

Wiens JJ (2011) The causes of species richness patterns across space, time, and clades and the role of "ecological limits". Q Rev Biol 86(2):75–96

Wiens JJ (2023) Trait-based species richness: ecology and macroevolution. Biol Rev 98(4):1365–1387

Wiens JJ, Donoghue MJ (2004) Historical biogeography, ecology and species richness. Trends Ecol Evolut 19(12):639–644

Wiley, EO, Lieberman, BS, 2011. Phylogenetics: Theory and Practice of Phylogenetic Systematics. John Wiley & Sons, Hoboken, NJ, p. 432

Winemiller KO (1991) Ecomorphological diversification in lowland freshwater fish assemblages from five biotic regions. Ecol Monograph 61(4):343–365

Zamudio KR, Bell RC, Mason NA (2016) Phenotypes in phylogeography: species' traits, environmental variation, and vertebrate diversification. Proc Natl Acad Sci 113(29):8041–8048

4 Geographic Origins of South American Freshwater Fishes

Milton Tan, Victor A. Tagliacollo, Roberto E. Reis and James S. Albert

Abstract

This chapter uses spatial and temporal patterns of diversification to test the two hypotheses outlined in Chap. 1 on the geographic origins of NFF clades. Geographic origins were estimated using ancestral area estimation (AAE) with on the time-calibrated phylogeny of Chap. 2, and geographic distributions of 5933 species of South American fishes among seven modern bioregions four paleo-bioregion for tree nodes older than 10 Ma. Uplands origins were inferred from AAE in bioregions of the Continental Periphery or Eastern Highlands, and lowland origins from ancestry in the Sub-Andean Foreland (SAF) basin. AAE analyses found support for both the *Uplands Origins* and *Lowlands Origins* hypotheses. Among the Big Nine NFF clades, upland origins are estimated for Neotropical Cyprinodontiformes, Loricarioidei, Gymnotiformes, and Cichlinae, and lowland origins for Curimatoidea, Characoidea, Aspredinoidea, and Pimelodoidea. The 44 subclades representing families or tribes of the Big Nine clades were split approximately evenly between those with upland and lowland origins, with 21 clades (48%) having lowland origins, and 23 (52%) clades upland origins. Upland origins on the Continental Periphery were estimated for all cyprinodontiform families and all loricarioid families except Callichthyidae. Eastern Highlands origins were estimated for Anostomidae, all cichline tribes except Heroini and all gymnotiform families except Apteronotidae. Atlantic Coast origins were estimated for *Conorhynchos*, La Plata origins for Heptapteridae, and Guiana Shield origins for Hemiodontoidae (Hemiodontidae+Serrasalmidae). SAF basin (lowland) origins were estimated for all Characoidea and Curimatoidea families except Anostomidae, Apteronotidae, all Aspredinoidea families, all Pimelodoidea subfamilies, and Callichthyidae. All three of the secondary freshwater clades (i.e., Cichlinae, Poecilioidea, Rivulidae) Paleogene origins in uplands, and three cichline subclades originated in SAF lowlands during the Paleogene; Crenicaratina, Geophagini, and Heroini. Among 75 South American freshwater fish clades (ranked as families or cichlid tribes) represented by two or more species, a plurality (31% or 41%) have origins in the Eastern Highlands, followed by SAF lowlands (15% or 20%) and Coastal (25% or 33%). The proportion of species belonging to family-level clades with Eastern Highland ancestry is very large (83%). At least 34 of 44 (77%) Big Nine clades have estimated origins during the super greenhouse world of the Upper Cretaceous to Eocene (*c.* 100–35 Ma). These early clades included 19 clades with upland origins and 15 clades with lowland origins, and included members of all the Big Nine clades. Clades with early (*c.* 100–35 Ma) upland origins include representatives from six of the Big Nine NFF clades (i.e., Cichlinae, Poecilioidea, Rivulidae, Curimatoidea, Gymnotiformes, Loricarioidei), and include both primary and secondary freshwater fish groups. All but one of these clades with early upland origins exhibit greatest diversity among living forms in lowland basins, with only Cyprinodontiformes exhibiting highest extant diversity in upland basins. All clades with early upland origins originated in the geographic periphery of South America. The locations of early diversification were either in the Eastern Highlands of the Guiana and/or Brazilian

M. Tan (✉)
Illinois Natural History Survey, Urbana, IL, USA
e-mail: miltont@illinois.edu

V. A. Tagliacollo
Universidade Federal do Uberlândia, Uberlândia, MG, Brazil
e-mail: vatagliacollo@gmail.com

R. E. Reis
Pontifícia Universidade Católica Rio Grande do Sul, Porto Alegre, RS, Brazil
e-mail: reis@pucrs.br

J. S. Albert
University of Louisiana at Lafayette, Lafayette, LA, USA
e-mail: jsalbe@gmail.com

J. S. Albert and R. E. Reis (eds.), *Ecology and Evolution of Amazonian Fishes*, Fascinating Life Sciences,
https://doi.org/10.1007/978-3-032-09328-8_4

Shield (i.e., Cichlinae, Hemiodontoidae, Anostomidae, Gymnotiformes, *Conorhynchos*, Scoloplacidae, Lithogeninae, Copiodontinae+Trichogeninae, and Delturini), or along the western margin in the Northern Andes (*Saccodon*, Parodontidae), Central Andes (*Orestias*, Cyprinodontidae), or Southern Andes (Nematogenyidae; Diplomystidae). The three oldest NFF clades with upland origins were estimated to have evolved within rainforest refugia of the Guiana Shield during the Upper Cretaceous. These clades are Neotropical Cyprinodontiformes and Loricarioidei during the Santonian (*c.* 86–84 Ma), and *Nematogenys* in the Southern Cone during the Campanian (83–72 Ma). The Santonian was characterized by extreme hothouse global climates with a low latitudinal temperature gradient, and high levels of storm action (Burgener et al. 2023). The Campanian was characterized by progressive global cooling, with pronounced cooling during the transition to the Maastrichtian *c.* 71–69 Ma. All the oldest lowland NFF clades originated in the northern SAF basin during times of global cooling events and eustatic sea-level falls. The early lowland NFF clades represent seven of the Big Nine clades (excluding the two cyprinodontiform clades), and include both primary and secondary freshwater fish taxa. The earliest estimated NFF clades with lowland origins are Characoidea and Curimatoidei, both coinciding in time and place a major marine regression at *c.* 83–72 Ma in the SAF basin. An additional 16 NFF clades are estimated to have diversified in the northern SAF basin during the time of other marine regressions at 65–58 Ma and 42–37 Ma. Body size is correlated with clade age among clades with lowland but not upland origins. Species richness is not correlated with crown-group clade age or body size. Early NFF diversification corresponded with numerous major paleogeographic events and paleoenvironmental conditions.

4.1 Introduction

The origins and early evolution of South American fishes involved innumerable phenotypic changes among many clades over tens of millions of years, and across the vast evolutionary arena of continental South America (Chap. 1). Over these immense geological and evolutionary scales, each of Big Nine continental radiations described in Chap. 3 evolved a distinctive set of physiological, behavioral, and ecological traits and tolerances facilitating life histories in different aquatic habitats and ecological regions (i.e., *ecoregions*; Albert et al. 2011; 2020). Diversification of the Big Nine was guided in part from the ways in which clade-specific ecological traits (i.e., *ecotraits*) associated with habitat utilization, diet, and life history strategies, interacted with prevailing and changing environmental conditions (Albert et al. 2018).

The study of historical biogeography among NFF clades is a large literature with many threads (Hubert and Renno 2006; Albert and Reis 2011; Matamoros et al. 2015; Abreu et al. 2019; Albert et al. 2020). The most important environmental drivers of evolution in South American fishes are the paleogeographic events and climatic conditions described in Chap. 1. Paramount among these events were the K-Pg extinction at *c.* 66 Ma, the Eocene–Oligocene Cooling Event and Bolivian Orocline at *c.* 34 Ma, the Pebas Megawetlands in the Western Amazon from 22–10 Ma, the formation of the modern Amazon and Orinoco basins at *c.* 10 Ma, and the Pleistocene climate oscillations starting *c.* 2.6 Ma. These events established riverine and riparian connections among paleo-geographic regions and influenced whether these regions represent uplands or lowlands. For example, the uplift of the Northern Andean cordilleras that resulted in the Great Amazonian Biotic Interchange (GAzBI) also caused the Fitzcarrald and Vaupes arches of the lowland Sub-Andean Foreland (SAF) to transition from depositional to erosional surfaces (Albert et al. 2017). These tectonic events affecting elevation and topography also affected the prevailing climatic conditions governing precipitation and erosion, and the lithological and biotic conditions underlying the chemistry of soils and river waters. Topography and erosion also affected the locations and movements of *watershed* (i.e., drainage basin) boundaries, and the ever-changing dendritic geometry of river drainage networks with stream-order effects on habitat quality, quantity, and connectivity. River capture and shore-line fluctuations have operated continuously on all ice-free continental surfaces, serving to merge and separate segments of adjacent river basins and their resident aquatic biotas (Albert et al. 2018).

The paleogeographic hypotheses introduced in Chap. 1 make somewhat different predictions regarding associations of fluvial and hydrological conditions with fish species richness and ecotrait values. To briefly recapitulate, the *Uplands Origins* hypothesis posits evolutionary origins in upland clearwater rivers and streams of the Guiana and Brazilian Shields during the Upper Cretaceous and Early Paleogene (*c.* 100–34 Ma) in fishes with small adult body sizes, a generalized benthic invertebrate diet, and which inhabit small and fragmented geographic ranges in headwater rivers and streams (Eigenmann 1912; Eigenmann and Allen 1942). In contrast, the *Lowland Origins* hypothesis proposes contrasting ancestral conditions, with origins among species with large adult body sizes inhabiting large lowland rivers with broad and contiguous geographic ranges (Smith 1992).

The contrasting predictions of these macroecological and macroevolutionary hypotheses can be evaluated using statistical analyses to assess correlations among environmental (e.g., elevation, connectivity) and organismal (e.g., body size, vagility) variables (Diniz-Filho et al. 2023). This chapter presents results of historical biogeographic analyses to estimate the ecological and geographic origins of fish clades in a probabilistic framework.

4.2 Methods

4.2.1 Historical Biogeography

Historical biogeography integrates phylogenetics and geographic data to reconstruct the evolutionary history of species distributions (Sanmartín and Ronquist 2004; Matzke 2013; Albert and Antonelli 2017). The process begins with phylogeny estimation, where a time-calibrated tree is inferred to establish evolutionary relationships. Next, bioregionalization divides the study area into discrete regions, scoring species as present or absent in each to analyze distribution patterns. Finally, *ancestral area estimation* (AAE) employs models like the Dispersal-Extinction-Cladogenesis (DEC) model to infer historical biogeographic processes, including dispersal between regions, local extinctions, and range evolution during speciation. By combining these steps, researchers can uncover how geological events, dispersal routes, and evolutionary dynamics shaped the current geographic distribution of biodiversity.

4.2.2 Ancestral Area Estimation

We used AAE to study the evolution of species geographic ranges (Ree and Sanmartín 2009; Albert et al. 2021). Biogeographical histories of the Big Nine NFF clades (Fig. 4.1) were estimated using a probabilistic model of geographical range evolution performed in the R package *BioGeoBEARS* (Matzke, 2013). This package uses Maximum Likelihood to estimate ancestral range probabilities for the nodes of the dated phylogeny in Chap. 2. South America was partitioned into seven modern *bioregions* and four paleo-bioregions (Fig. 4.2) based on combinations of freshwater ecoregions (Abell et al. 2008) delimited by hypothesized modern and paleogeographic barriers (Albert et al. 2011; Dagosta and de Pinna 2017; Albert et al. 2018). The modern bioregions corresponded to modern river basins and ecographic regions; Trans-Andean (T), Orinoco (O), Guianas (G), Western Amazon (W), Brazilian Shield (B), Atlantic Coastal Basins (A), and La Plata (L). Paleo-bioregion estimates for tree nodes older than 10 Ma include: SAF (Trans-Andean, Orinoco, and Western Amazon), Continental Periphery (Trans-Andean, Guiana, and Atlantic), Eastern Highlands (Eastern Amazon and Guianas), and a region representing all of South America. Geographic origins in uplands were inferred from ancestry in Continental Periphery

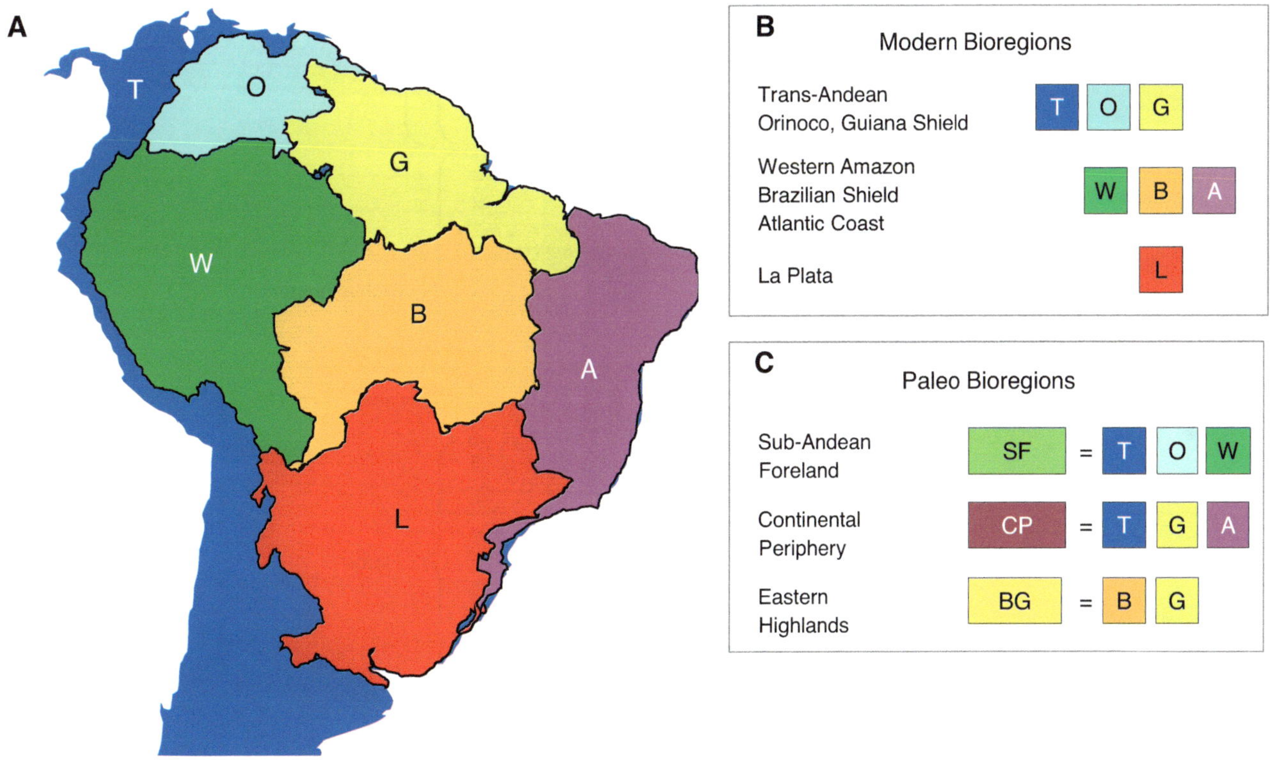

Fig. 4.1 Paleogeographic model used in Ancestral Area Estimation (AAE) using BioGeoBEARS (Matzke 2013) and the Dispersal-Extinction Cladogenesis (DEC) model of geographic range evolution (Ree and Smith 2008; Matzke 2013). (**A** and **B**) Modern bioregions based on Freshwater Ecoregions of the World (Abell et al. 2008): A = Atlantic, L = La Plata, B = Brazilian Shield, G = Guiana Shield, O = Orinoco, T = Trans-Andean, W = Western Amazon. Paleo Bioregions: CP = Continental Periphery, BG = Eastern Highlands (Brazilian Shield + Guiana Shield), SAF = Sub-Andean Foreland. (**C**) Paleo Bioregions represent undivided Bioregions before *c.* 10 Ma

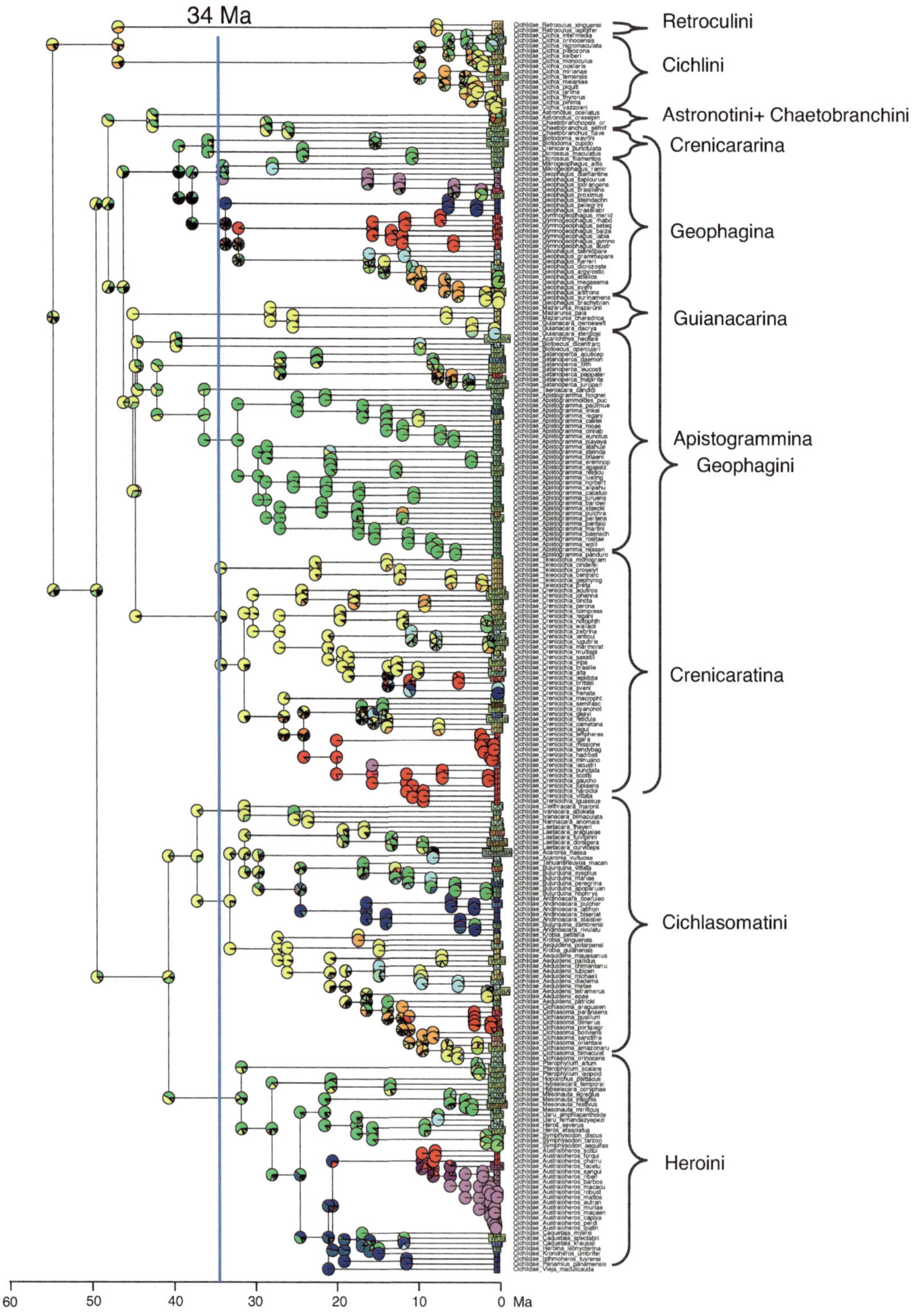

Fig. 4.2 AAE of Cichlinae. Data for 569 species in six tribe-level clades. Root age: 54 Ma. Number of estimated biogeographic events: 59

or Eastern Highlands, and lowland origins from ancestry in the SAF lowlands. Biogeographical inference was performed using the likelihood-based Dispersal-Extinction Cladogenesis (DEC—Ree and Smith 2008) model of geographic range evolution, which allows for multiple biogeographic events along tree branches, including dispersal (range expansion to a new bioregion), local (within bioregion) extinction, range switching (dispersal plus local extinction), sympatric speciation within a bioregion, and vicariant speciation among adjacent bioregions (Ree and Smith 2008).

We performed biogeographic analyses on three datasets: 1) 43 taxa of the Big Nine clades ranked as families (or tribes for cichlids) including 5933 species (Table 4.1), 2) 75 clades of all NFF clades ranked as families (or tribes for cichlids) that have two or more species, and 3) 558 taxa of the Big Nine clades representing 285 unique taxonomic names, including 221 generic and subgeneric taxa with unique names, and 64 supergeneric taxa with unique names (Table 4.2). Associations between species richness, crown clade age, and body size were assessed distinguishing among taxa with upland vs. lowland origins. Associations were also assessed between species richness, crown clade age, and body size among clades of upland vs. lowland origins. Analyses of Cichlinae were performed using two alternative root ages; 34 Ma from Tagliacollo et al. (2024) and 54 Ma from a synthesis of nine studies available the *TimeTree of Life* (Kumar et al. 2022).

Body sizes were compiled from the literature and summarized in the ecomatrix described in Chap. 6. Body size is a fundamental feature of the ecological phenotype in fishes and other animals, affecting metabolic rate and other physiological processes, behavior, resource utilization, locomotion and life history (Weitzman and Vari 1988; McNab 2002). NFF body size was assessed as the maximum recorded standard length (SL), or total length (TL) for taxa without a caudal fin. Maximum known standard or total lengths are highly correlated with body mass in most fishes and this a widely used metric of overall body size in fishes (Albert and Johnson 2012). Body size data are available for 5400 NFF species representing 85.1% of all 6342 valid NFF species and all 881 valid NFF genera.

4.3 The Big Nine NFF clades

4.3.1 Cichlinae

For the Neotropical cichlids, we rescaled the subtree for a root age of 54 Ma, corresponding to a roughly Early Eocene origin in the Eastern Highlands (Brazilian Shield+Guiana Shield; Fig. 4.3). Nearly all major tribes originated between 44 and 37 Ma during a time of global cooling when marine regressions (MR4) converted large lowland areas from epicontinental seas to freshwater ecosystems (López-Fernández and Albert 2011). This ancestral region was retained in the Guianacarina, Crenicaratina, and Cichlasomatini, while the remaining clades dispersed to the SAF (primarily western Amazon) including the Astronotini+Chaetobranchini, Crenicaratina+Geophagina, Apistogrammina, and Heroini. We estimated 59 later dispersal events between the Western and Eastern Amazon (Brazilian Shield) occurred within numerous tribes around the 10 Ma range when the Andes reversed the direction of the Amazon. We also identified numerous deep transitions and recent transitions (associated with tip branches) into the Atlantic Coast, Trans-Andean, and La Plata regions.

4.3.2 Poecilioidea

We estimated a deep divergence among cyprinodontiform families ~85 Ma, with an estimated ancestral range of the Continental Periphery (Atlantic Coast, Trans-Andean, Guiana Shield), likely explained by tolerance to saltwater (Fig. 4.4). This timeframe is consistent with that of other recent phylogenetic studies of cyprinodontiform taxa outside the Neotropics (Hernandez et al. 2024). During the Eocene–Oligocene cooling, we estimated early transitions into the Guiana Shield of *Anableps* (Anablepidae) and a transition back to the Atlantic Coast (or potentially retention if deeper ancestors are uncertain) in *Jenynsia* (Anablepidae). During the Oligocene T1 event that drove dispersal into the Trans-Andean region, we found dispersal events resulting in the origin of *Oxyzygonectes* (Anablepidae). Poeciliidae has highest regional species density in Central America, with local hotspots on the Isthmus of Panamá, the Grijalva–Usumacinta drainage, the Isthmus of Tehuantepec, and western Hispaniola (Gómez-Martínez et al. 2024).

4.3.3 Rivulidae

Rivulids originated potentially prior to the 34 Ma Eocene–Oligocene cooling event in the Atlantic Coast and Shield Regions, and Cynolebiasinae also originated from this ancestral Atlantic Coast region during this time. Anablepids and poeciliids derived a little later, with ancestral ranges also corresponding to the Continental Periphery. During the Eocene–Oligocene cooling, we estimated a transition back to the Atlantic Coast (or potentially retention if deeper ancestors are uncertain) in *Kryptolebias* (Rivulidae). During the Oligocene T1 event that drove dispersal into the Trans-Andean region, we found dispersal events resulting in the origin of *Cynodontichthys* (Rivulidae). Marine Regressions in the Oligocene (MR5) resulted in Guiana Shield region transitioning to the SAF basin, corresponding to the origins of the *Laimosemion* and the clade formed by *Anablepsoides*+*Moema*+*Gnatholebias*+*Renova*. Subsequent dispersal events include deep and recent transitions to the Trans-Andean, La Plata, and transitions back to the Atlantic Coast.

Table 4.1 Summary of data for 43 taxa (families or cichlid tribes) in the Big Nine NFF clades

Taxon (family or tribe)	Order	Big nine	Avg SL (cm)	Age (Ma)	Species	Origin: bioregion	Origin: elevation	Origin: water	Salinity tolerance
Astronotini + Chaetobranchini	Cichliformes	Cichlinae	46	28	6	EG	Uplands	C	Secondary
Cichlasomatini	Cichliformes	Cichlinae	20	22	80	EG	Uplands	C	Secondary
Geophagini	Cichliformes	Cichlinae	26	33	284	EG	Uplands	C	Secondary
Heroini	Cichliformes	Cichlinae	15	22	183	SAF	Lowlands	C	Secondary
Retroculini + Cichlini	Cichliformes	Cichlinae	100	30	19	EG	Uplands	C	Secondary
Anablepidae	Cyprinodontiformes	Poecilioidea	32	33	19	CP	Uplands	E	Secondary
Cyprinodontidae	Cyprinodontiformes	Poecilioidea	22	10	45	CP	Uplands	C	Secondary
Poeciliidae	Cyprinodontiformes	Poecilioidea	6	22	273	CP	Uplands	E	Secondary
Rivulidae	Cyprinodontiformes	Rivulidae	8	42	470	CP	Uplands	CB	Secondary
Acestrorhynchidae	Characiformes	Characoidea	19	58	31	SAF	Lowlands	CBW	Primary
Bryconidae	Characiformes	Characoidea	80	60	49	SAF	Lowlands	CW	Primary
Chalceidae	Characiformes	Characoidea	25	52	5	SAF	Lowlands	C	Primary
Characidae	Characiformes	Characoidea	10	42	1251	SAF	Lowlands	C	Primary
Ctenoleuciidae	Characiformes	Characoidea	100	65	7	SAF	Lowlands	CBW	Primary
Gasteropelecidae	Characiformes	Characoidea	4	42	9	SAF	Lowlands	W	Primary
Iguanodectidae	Characiformes	Characoidea	13	52	38	SAF	Lowlands	W	Primary
Triportheidae	Characiformes	Characoidea	36	38	21	SAF	Lowlands	CW	Primary
Anostomidae	Characiformes	Curimatoidea	21	46	150	EG	Uplands	C	Primary
Chilodontidae	Characiformes	Curimatoidea	10	39	8	SAF	Lowlands	C	Primary
Curimatidae	Characiformes	Curimatoidea	13	36	119	SAF	Lowlands	W	Primary
Cynodontidae	Characiformes	Curimatoidea	36	32	33	SAF	Lowlands	CBW	Primary
Hemiodontidae	Characiformes	Curimatoidea	17	46	34	G	Uplands	C	Primary
Parodontidae	Characiformes	Curimatoidea	10	40	32	SAF	Lowlands	C	Primary
Prochilodontidae	Characiformes	Curimatoidea	33	45	21	SAF	Lowlands	W	Primary
Serrasalmidae	Characiformes	Curimatoidea	100	34	100	G	Uplands	C	Primary
Apteronotidae	Gymnotiformes	Gymnotiformes	34	32	99	SAF	Lowlands	W	Primary
Gymnotidae	Gymnotiformes	Gymnotiformes	34	60	50	EG	Uplands	CW	Primary
Hypopomidae	Gymnotiformes	Gymnotiformes	19	46	46	EG	Uplands	C	Primary
Rhamphichthyidae	Gymnotiformes	Gymnotiformes	51	48	28	EG	Uplands	B	Primary
Sternopygidae	Gymnotiformes	Gymnotiformes	41	55	59	EG	Uplands	W	Primary
Heptapteridae	Siluriformes	Pimelodoidea	12	38	233	L	Uplands	C	Primary
Pimelodidae	Siluriformes	Pimelodoidea	61	43	116	SAF	Lowlands	C	Primary
Pseudopimelodidae	Siluriformes	Pimelodoidea	19	33	54	SAF	Lowlands	C	Primary
Aspredinidae	Siluriformes	Aspredinoidea	10	55	49	SAF	Lowlands	C	Primary
Auchenipteridae	Siluriformes	Aspredinoidea	14	40	128	SAF	Lowlands	C	Primary
Doradidae	Siluriformes	Aspredinoidea	23	42	96	SAF	Lowlands	C	Primary
Astroblepidae	Siluriformes	Loricarioidei	9	62	82	CP	Uplands	C	Primary
Callichthyidae	Siluriformes	Loricarioidei	5	58	222	SAF	Lowlands	C	Primary
Loricariidae	Siluriformes	Loricarioidei	16	48	1041	CP	Uplands	C	Primary
Nematogenyidae	Siluriformes	Loricarioidei	41	80	1	CP	Uplands	C	Primary
Phreatobiidae	Siluriformes	Loricarioidei	6	19	3	SAF	Lowlands	C	Primary
Scoloplacidae	Siluriformes	Loricarioidei	2	62	6	CP	Uplands	C	Primary
Trichomycteridae	Siluriformes	Loricarioidei	8	64	393	CP	Uplands	C	Primary
Min			2	10	1				
Max			100	80	1251				
Avg			28	43	139				

Taxa arranged phylogenetically as in Fig. 3.2. Estimated crown clade ages (Ma) from Chap. 2. Estimated bioregion and elevation origins, and abbreviations, in Fig. 4.2. Estimated water chemistry origins in Chap. 8, abbreviations: *C*, clearwater; *B*, blackwater; *W*, whitewater; *E*, estuarine. Data for 5933 species representing *c*. 95% of the NFF fauna

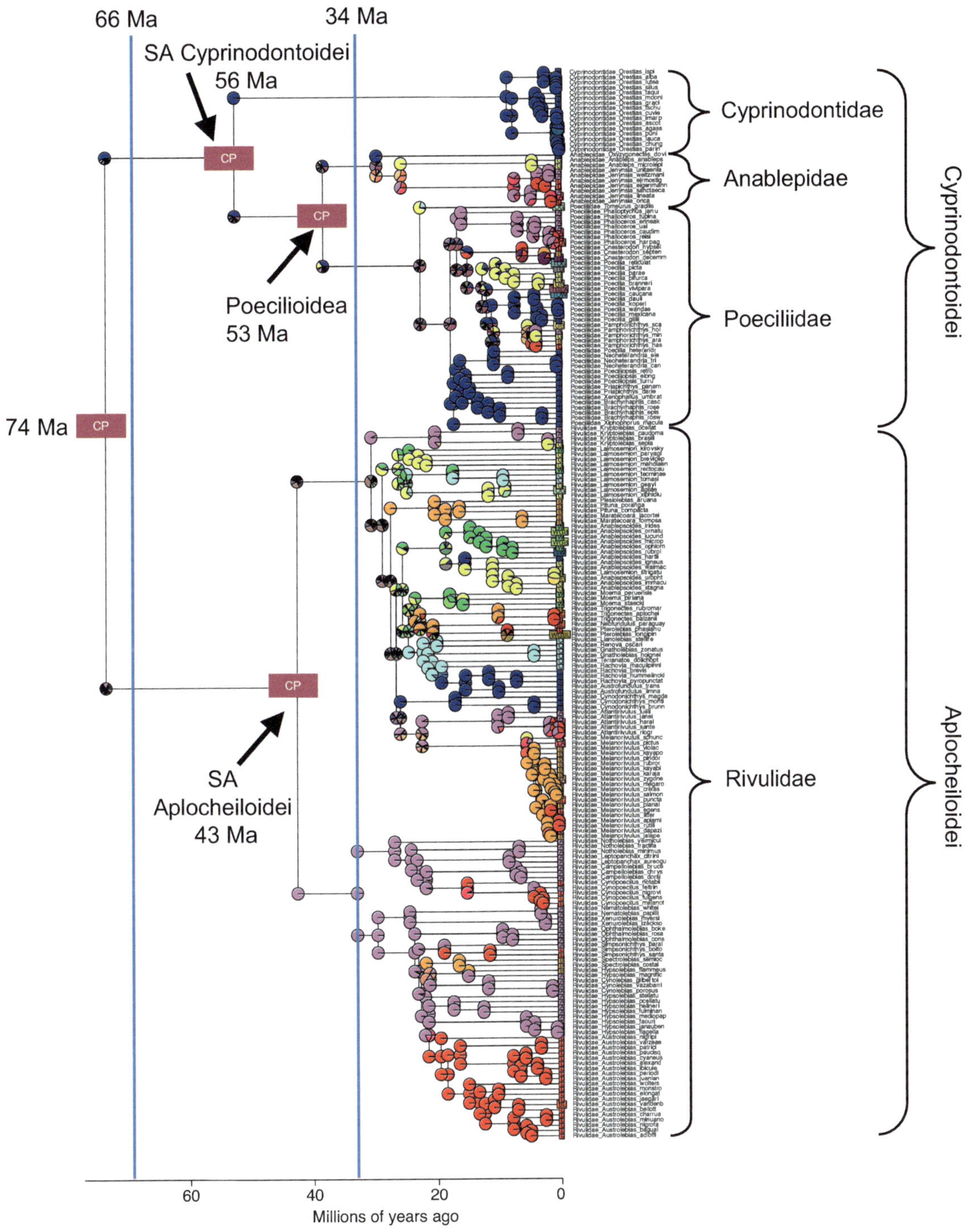

Fig. 4.3 AAE of South American Cyprinodontiformes. Data for 803 species in four family-level clades. Root age estimate 74 Ma. Number of estimated biogeographic events: 36. Note the clade ages estimated for Poecilioidea and Cyprinodontidae are younger than published estimates, as this tree excludes five Cyprinodontoidei families from North and Central America (Profundulidae, Goodeidae, Fundulidae) and Europe (Aphaniidae, Valenciidae), and two Aplocheiloidei families from Africa (Aplocheilidae, Nothobranchiidae)

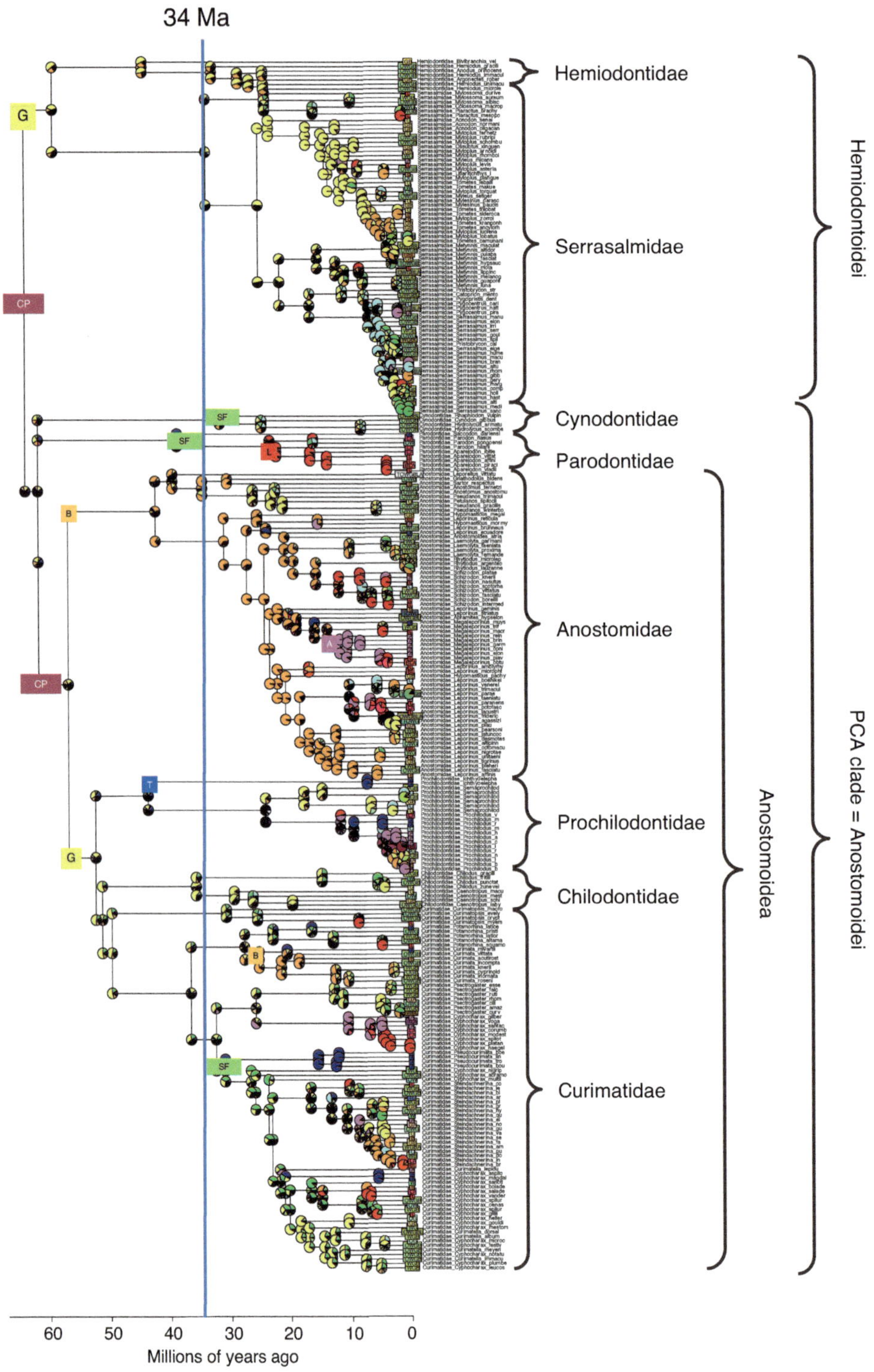

Fig. 4.4 AAE of Curimatoidea. Data for 467 species in eight family-level clades. Root age estimate 70 Ma. Number of estimated biogeographic events: 111

4.3.4 Curimatoidea

We estimated an ancient age of origin for the Curimatoidea prior to the K-Pg extinction event. Within the Curimatoidea, the epifamily Curimatoidae (Parodontidae+Cynodontidae+traditional "anostomoids") originated first in the SAF prior to 80 Ma (Fig. 4.5). The epifamily Hemiodontoidae (Hemiodontidae+Serrasalmidae) and Anostomidae remained relict lineages in the Continental Periphery in the Guiana Shield and Brazilian Shield, respectively. The Trans-Andean Scrapetooth genus *Saccodon* (Parodontidae) from the Tumbes-Chocó-Magdalena region from northwestern Peru to Panama, originated in the Early Eocene, likely due to an Early Eocene uplift of the northern Andes (Gregory-Wodzicki 2000; Mora et al. 2020). During the late Eocene, dispersal events to the Guiana Shield occurred in the MRCA of the *Anostomus+Pseudanos+Petulanos* clade (Anostomidae) and *Semaprochilodus* (Prochilodontidae) (Table 4.2), while vicariance of the La Plata region from the SAF corresponded to the origin of the Parodontinae. The Eocene-Oligocene cooling event corresponded to the origin of *Curimatella lepidura* in the Atlantic Coast (São Francisco River) and *Cyphocharax* (Curimatidae) to the Guiana Shield. The Oligocene T1 event split the Trans-Andean region, leading to origins within *Leporinus* and the origins of *Megaleporinus* (Anostomidae), and the origins of *Potamorhina* and *Curimata* (Curimatidae).

The Hemiodontidae and Serrasalmidae were both estimated to retain an ancestral Guiana Shield origin, with several more recent transitions between 10 and 15 Ma to the La Plata, Brazilian Shield, and between the Western and Eastern Amazon. Throughout the PCA clade, the numerous families diversified, most also were estimated to originate in the ancestral SAF region; however, in the

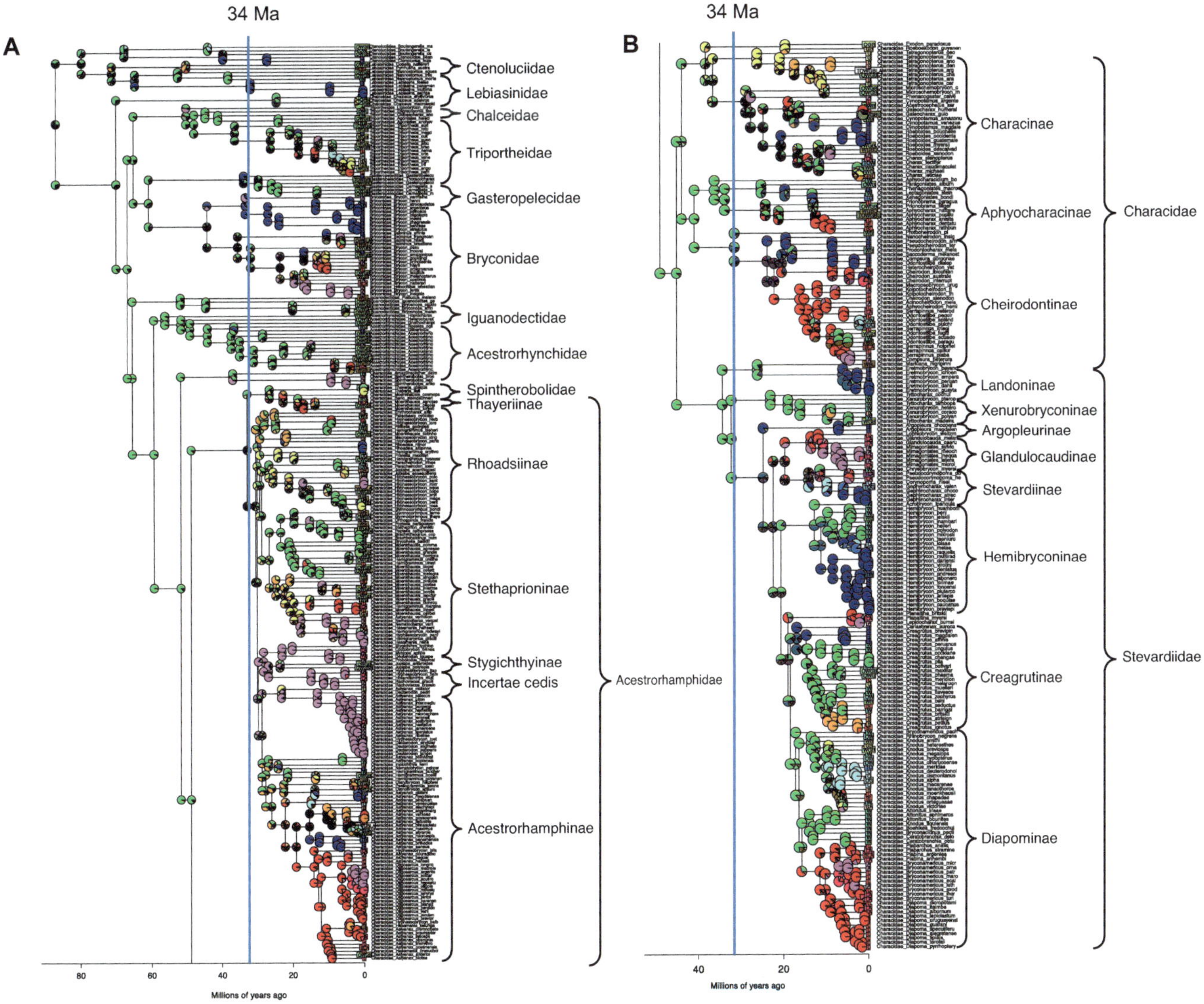

Fig. 4.5 AAE of Characoidea (*sensu* Melo et al. 2022). Data for 1357 species in nine family-level clades. Root age estimate 85 Ma. Number of estimated biogeographic events: 110

Table 4.2 Full list of biotic responses of NFF clades. Data for 558 clades. Events as in Fig. 1.10 and Table 1.1. Bioregions and color scheme as in Fig. 4.2. AAEs for crown groups. A. Events from 86 to 25 Ma. B. Events from 25 to 11 Ma. C. Events from 10 to 5 Ma

Period	Epoch	Ma	Events	Biotic reponses	Bioregions	Cichlinae		Cyprinodontoidei		Rivulidae		Curimatoidea	
UK	Santon.	86-83	Global greenhouse	Radiations	CP								
	Campan.	83-72	Southern Andes	Vicariance	CP: T								
	Maastric.	72-66		Radiations	CP							Curimatoidea	1
	Campan.	83-72	MR1	Radiations	SF								1
PG	P	66-55	Relicts	Vicariance	CP: T								
	P	65-58	MR2	Radiations	SF								
	EE	55-48	Relicts	Vicariance	B, G, A	Cichlinae	1	SA Cyprinodontoidei	1			Anostomoidei, Hemiodontoidei, Parodontidae *Saccodon*	3
	EE	52-43	MR3							Rivulidae	1		
	LE	42-37	MR4	Radiations	SF	Astronotini+ Chaetobranchini, Crenicararina+ Geophagina, Apistogrammina, Heroini	4						
	LE	42-38	G1		G	Guianacarina, Crenicaratina, Cichlasomatini	3					*Anostomus+ Pseudanos+ Petulanos, Semaprochilodus*	2
	LE	38	L1	Vicariance	SF-->L							Parodontinae	1
	E-O	34	EO cooling	Vicariance	G-->A, SF-->G			*Anableps*	1			*Curimatella, Cyphocharax*	2
	E-O	34	A1	Vicariance	SF x B, SF-->A			*Jenynsia*	1	*Kryptolebias*	1		
	O	34-30	T1	Vicariance	CP: T; SF-->T			*Oxyzygonectes*	1	*Cynodonichthys*	1	*Leporinus, Megaloporinus, Potamorhina, Curimata*	4
	O	34-30	OCO	Dispersal	G--> SF			*Anablepsoides+ Moema+ Gnatholebias+ Renova*	1	*Laimosemion*	1		
NG	EM-MM	23-15	L2	Vicariance	A-->L, SF-->L; W-->L	*Gymnogeophagus, Crenicichla*	2			*Cynolebias, Simpsonichthys, Austrolebias*	3	*Mylosoma, Myloplus, Schizodon, Leporinus, Curimatopsis, Potamorhina, Cyphocharax x3*	9
	EM-MM	23-15	A2	Vicariance	B, G, A, L-->A; B-->A	*Geophagus, Crenicichla*	2			*Atlantirivulus*	1	*Hypomasticus, Schizodon, Megaloporinus, Leporinus, Prochilodus, Cyphocharax*	6
		23-15	Dry Diagonal	Vic-Dis-Rad	CP: B; G-->B			*Cnesterodon*	1	*Plesiolepis+ Pituna+ Maratecoara, Hypsolebias, Spectrolebias*	3	*Tometes, Steindachnerina*	2
	EM	20-15	T2	Vic-Dis-Rad	CP: T	*Andinoacara*, CA Heroini	2	*Orestias*, CA Poeciliini, *Poecilia*	3	*Rachovia+ Austrofundulus, Anablepsoides*	2	*Ichthyoelephas, Pseudocurimata*	2

								Big Nine Clades				Radiations
Characoidea		**Gymnotiformes**		**Pimelodoidea**		**Aspredinoidea**		**Loricarioidei**		**Other**		**All clades**
								Loricarioidei	1	Diplomystidae	1	2
								Nematogenys	1			1
												1
Characoidea	1											1
		Gymnotiformes	1									1
				Pimelodoidea		Doradoidea	1	TSVSG clade, Callichthyidae	2			4
				Conorhynchos	1			Scoloplacidae, Lithogeninae, Copiodontinae+ Trichogeninae, Delturini, Astroblepidae	5			11
										Cetopsidae		2
		Rhamphichthyidae	1					Loricariinae	1			6
												5
				Heptapteridae	1							2
Stethaprioninae, Characini	2					Hoplomyzontinae	1	Hartiini, Hypostominae	2			8
Lignobrycon, Brycon, Salminus	*3*			*Gladioglanis*	1	*Wertheimeria+ Kalyptodoras+ Franciscodoras*	1					7
Ctenolucius, Lebiasina, Gasteropelecus, Brycon x2, Gilbertolus, Rhoadsia, Argopleura	*8*											14
		Brachyhypopomus, Apteronotidae	2									4
Bario, Stethaprionini *x4, Astyanax* B x2, Aphyocharacini, Compsurini, Glandulocaudini, *Planaltina*, Creagrutini	*12*	*G. Tigre, Gymnorhamphichthys, Apteronotus*	3	*Pimelodus+ Parapimelodus, Luiciopimelodus, Megalonema*	3	*Pseudobunocephalus, Bunocephalus, Tatia, Amblydoras*	*4*	Trichomycterinae	1			37
Acestrorhynchus, Glandulocaudini	*2*	*G. Pantherinus*	1	Pseudopimelodidae, *Steindachneridion*	2			*Aspidoras*, Hartiini, *Lampiella+ O. affinis*, Neoplecostominae	4			18
												6
Triportheus, Astyanax x2, Compsurini, Creagrutini	*5*	*A. cuchillo*	1	*Heptapterus, Rhamdia, Pimelodella, Leiarius*	4							19

(continued)

Table 4.2 (continued)

Period	Epoch	Ma	Events	Biotic reponses	Bioregions	Cichlinae		Cyprinodontoidei		Rivulidae		Curimatoidea	
	EM-LM	22-10	Pebas	Vic-Dis-Rad	W								
		18	G2	Vicariance	CP: G			*Tomeurus, Poecilia*	2				
	LM	11-8	T3	Vicariance	T, O, W	*Geophagus*	1					*Prochilodus, Cyphocharax*	2
				Vicariance	T, L								
	LM	10	GAzBI	Vic-Dis-Rad	W, O, B. G&B-->W	*Cichla x2, Astronotusx2, Chaetobranchus x2, Biotodoma, Mikrogeophagus, Geophagus, Acarichthys, Satanoperca, Apistogramma x2, Crenicichla x4, Laetecara, Acaronia, Aequidens x2, Cichlasoma, Pterophylum, Hypselecara, Mesonauta, Heros, Caquetaia*	27			*Anablepsoides, Pterolebias, Laiosemion*	3	*Hemiodus, Anodus, Mylossoma, Colossoma, Piaractus, Myloplus x4, Myleus, Metynnis x3, Pristobrycon, Catoprion, Pygopristis, Pygocentrus, Serrasalmus, Rhaphiodon, Cynodon, Hydrolycus, Leporellus, Anostomus x2, Pseudanos x2, Hypomasticus, Leporinus x4, Anostomoides, Laemolyta, Rhytiodus, Schizodon, Abramites, Megaloporinus, Semaprochilodus x2, Prochilodus, Chilodus x2, Caenotropus, Curimatopsis x3, Potamorhina x3, Curimata x4, Psectrogaster x2, Cyphocharax x5, Steindachnerina x7, Curimatella x3*	70
	LM	10	L3	Dispersal	A-->L	*Australoheros*	1	*Jenynsia*	1				
	LM	10-6	A3	Vic-Dis-Rad	L-->A, B-->A, AxB	*Australoheros*	1	*Phallocerus, Pamphorichthys*	2	*Melanorivulus*	1		

								Big Nine Clades				Radiations
Characoidea		**Gymnotiformes**		**Pimelodoidea**		**Aspredinoidea**		**Loricarioidei**		**Other**		**All clades**
		G. Lamontianus, G. Tigrinus, G. Tigre, G. carapo, Steatogenys, Hypopygus, Brachyhypopomus, Gymnorhamphichthys, Sternopygus, Rhabdolichops, Eigenmannia, Navajini	12					*Corydoras*	1			13
												2
Roeboides, Phenagoniates, Xenagoniates	3	*Gymnotus, Brachyhypopmus, Sternopygus, Eigenmannia, Apteronotus x2*	6	*Pimelodus, Cheirocerus, Platysilurus, Sorubim, Cheirocerus, Megalonema*	6	*Xyliphius, Bunocephalus, Trachelyopterus, Rhinodoras, Doraops, Centrochir*	6	*Corydoras, Scleromystax, Hoplosternum,* Loricariini x6, *Eremophilus clade x4, Loricariini x2*	15			39
								Hatcheria+ Bullockia	1			1
Boulengerella x3, Copella, Nannostomus x2, Chalceus x2, Triportheus x6, Thoracocharax, Carnegiella x2, Brycon x3, Iguanodectes, Bryconops x2, Hoplocharax, Heterocharax, Gnathocharax, Acestrorhynchus x5, Rhoadsiini *x5,* Stethaprionini *x7, Astyanax B x5, Tetragonopterus, Microschemobrycon, Acestrocephalus, Galeocharax, Roeboides, Charax,* Xenurobryconini, *Hemibrycon+Acrobrycon, Creagrutus, Knodus, Attonitus, Bryconacidinus, Ceratobranchia*	61	*Electrophorus, Gymnotus x2, Steatogenys x2, Hypopygus x2, Gymnorhamphichthys x3, Rhamphichthys, Microsternarchus, Brachyhypopomus x6, Sternopygus, Rhabdolichops, Distocyclus, Eigenmannia x2,* Sternarchorhamphinae, *Adontosternarchus, Platyurosternarchus, Sternarchorhynchus x3, Porotergus, Sternarchella, Sternarchogiton+ Compsaraia*	32	***Imparfinis, Rhamdia,*** *Pimelodella x2, Batrochoglanis, Lophiosilurus, Phractocephalus, Leiarius, Pimelodina, Pinirampus, Calophysus, Cheirocerus, Pimelodus x2, Hypopthalmus, Platysilurus, Sorubimichthys, Zungaro, Pseudoplatystoma, Hemidorubim, Sorubim, Brachyplatystoma*	22	*Tatia, Auchenipterus, Trachelyopterus, Rhinodoras, Doras, Trachydoras, Leptodoras, Hassar, Hemidoras, Ossancora, Tenellus, Nemadoras*	12	*TSVSG clade x14, Eremophilus clade,* Vandeliinae *x3; Dianema, Callichthys, Corydoras* x15, *Pseudorinelepis,* Loricariini *x10,* Otocinclini x7, Neoplecostomini x6	59			286
Brycon, Hyphessobrycon, Diapomini	3	*Rhamphichthys, Eigenmannia, Apteronotus*	3	*Microglannis*	1							9
Spintherobolus, Bryconamericus x2	3	*Brachyhypopomus, Eigenmannia, Apteronotus*	3	*Heptapterus, Pimelodella, Acentronichthys, Pseudoplatystoma, Pimelodus*	5							15

(continued)

Table 4.2 (continued)

Period	Epoch	Ma	Events	Biotic reponses	Bioregions	Cichlinae		Cyprinodontoidei		Rivulidae		Curimatoidea	
	PL	<5	L4	Dispersal	B-->L; W-->L	*Cichlasoma*	1			*Atlantorivulus* x2, *Melanorivulus* x2, *Trigonectes*	5	*Piaractus*	1
	PI	<5	T4	Dispersal		*Satanoperca, Apistogramma, Crenicichla x2, Bujurquina, Australoheros, Acaronia*	7					1+K29	
	PL	<5	A4	Dispersal	A-->L; L--A	*Gymnogeophagus x2, Acarichthys, Crenicichla, Acaronia, Aequidens, Cichlasoma*	7	*Poecilia*	1			*Metynnis, Serrasalmus, Pygocentrus, Parodon, Apareiodon x2*	6
Events	110	29					59		15		22		112

Anostomidae, a transition to the Brazilian Shield region was estimated, such that they ancestrally originated in the Eastern Highlands. Throughout the PCA clade, numerous transitions occurred near the Eocene–Oligocene cooling event and Andean uplift event to the Guiana Shield, Trans-Andean, Atlantic Coast, and La Plata regions.

4.3.5 Characoidea

The characoids include characins and numerous related families. We estimated an ancestry for Characoidea in the SAF (specifically Western Amazon) during the Campanian marine regression, and all constituent families also originated within this ancestral region (Fig. 4.6). During the Eocene-Oligocene cooling event that drove vicariance that split the Guiana Shield and the Atlantic region, we estimated ancestral areas in the Guiana Shield early in the Stethaprioninae and Characini and the Atlantic Coast for *Lignobrycon* (Triportheidae) and *Brycon vermelha* and *Salminus* (Bryconidae). From the predominantly SAF ancestral characoid families, the Oligocene T1 vicariance event that split the Trans-Andean region from the SAF corresponded to origins of numerous clades with ancestral areas in the Trans-Andean ancestry, including *Ctenolucius* (Ctenoluciidae), *Lebiasina* (Lebiasinidae), *Gasteropelecus* (Gasteropelecidae), two clades within *Brycon* (Bryconidae), *Gilbertolus* (Acestrorhynchidae), *Rhoadsia* (Rhoadsiini), *Eretmobrycon* (Eretmobryconini), and *Argopleura* (Argopleurini). Many of the main remaining transitions after 34 Ma were to the La Plata, Trans-Andean, and Atlantic Coast, and 61 transitions occurred through the GAzBI.

4.3.6 Gymnotiformes

For the electric knifefishes, we reconstructed an origin of the Guiana Shields that is consistent with an origin within the Continental Periphery for the clade, and consistent as representing relictual lineages originating during the Paleocene (Fig. 4.7). Nearly all constituent families inherited this ancestral range as the region transitioned into the Eastern Highlands. The Rhamphichthyidae originated in the Late Eocene during a marine regression (MR4), which was associated with dispersal into the SAF basin. During the Oligocene marine regression (MR5), the Apteronotidae and the radiation of *Brachypopomus* (Hypopomidae) both originated in the SAF basin. Between 15 and 34 Ma, we estimated numerous deep transitions to the Trans-Andean and La Plata. Around 10 Ma, we estimated 32 transitions through the GAzBI and a few transitions to the Atlantic Coast.

4.3.7 Pimelodoidea

The Pimelodoidea is the superfamily including Pimelodidae and related families including Pseudopimelodidae, Heptapteridae, and the enigmatic genera *Phreatobius* (Phreatobiidae) and *Conorhynchos* (family *incertae sedis*) (Sullivan et al. 2013; Lundberg et al. 2014). The presence of *Conorhynchos* in the Atlantic Coast was consistent with a relictual vicariance during the Early Eocene that separated that Atlantic Coast from the Guiana Shield (Fig. 4.8). The Late Eocene marine regression (MR4) drove the formation of the SAF where the remaining Pimelodoidea originated

								Big Nine Clades				Radiations
Characoidea		Gymnotiformes		Pimelodoidea		Aspredinoidea		Loricarioidei		Other		All clades
Triportheus, Acestrorhynchus, Creagrutini	3	*Brachyhypopomus x2, Sternopygus, Porotergus*	4									14
Charax, Knodus, Eretmobrycon	3	*Gymnotus*	1									11
Planaltina	1	*Gymnotus x3, Sternopygus*	4									19
	110		74		46		25		93			558

and radiated, resulting in most of the remaining families. The Late Eocene vicariance event that split the La Plata (L1) coincided with the origin of Heptapteridae through dispersal from the SAF into the La Plata region. Within the Heptapteridae, we reconstructed an early transition to the Atlantic Coast in *Gladioglanis*, which may be consistent with the A1 event during the Eocene-Oligocene cooling. Between 15 and 34 Ma, we inferred a SAF ancestral range for Pseudopimelodidae, followed by an early divergence during the Early Miocene–Mid Miocene associated with the vicariance of the Atlantic Coast (A2). As the SAF changed into the Western Amazon during the Eocene-Oligocene cooling, the SAF ancestry of the Pimelodoidea was retained in ancestors of Pimelodidae (Late Eocene for MRCA of all Pimelodidae, Oligocene origin of all Pimelodidae except *Aguarunichthys*) and Phreatobiidae (Miocene origin of crown). Among Pimelodoidea, we estimated additional transitions after 34 Ma to the Trans-Andean, Atlantic Coast, and Brazilian Shield, with 22 shifts between the Western and Eastern Amazon throughout the GAzBI.

4.3.8 Aspredinoidea

We estimated an origin *c.* 65 Ma of Aspredinoidea in the SAF basin, consistent with the Paleocene marine regression (MR2; Fig. 4.9). The three constituent families remained in this region, and all originated in the SAF less than *c.* 34 Ma. We estimated dispersal of the Hoplomyzontinae into the Guiana Shield corresponding to the Eocene-Oligocene cooling event. The Eocene-Oligocene separation of the Atlantic Coast also corresponded closely to the origin of the clade including *Wertheimeria*, *Franciscodoras*, and *Kalyptodoras*. Numerous subsequent dispersals included transitions to the La Plata during the Early Miocene–Middle Miocene (L2), and vicariance between the Trans-Andean, Orinoco, and Western Amazon during the Late Miocene, as well as 12 events during the GAzBI.

4.3.9 Loricarioidei

We estimated an ancient origin for the Loricarioidei in the Santonian in the Continental Periphery (Fig. 4.10). We estimated a Continental Periphery ancestral region for *Nematogenys* within the Trans-Andean Region associated with vicariance through the formation of the Southern Andes in the Campanian. We also estimated Continental Periphery origins that were retained from across Loricarioidei for the Trichomycteridae and the Loricariidae prior to 66 Ma. The TSVSG clade and the Callichthyidae radiated in the SAF basin associated with the Paleocene Marine Regression (MR2). During the Early Eocene, we estimated numerous clade origins associated with vicariant events and the origins of relictual lineages including *Scoloplax* (Scoloplacidae) and *Lithogenes* (Loricariidae: Lithogeninae) in the Eastern Highlands and the Copiodontinae+Trichogeninae clade (Trichomycteridae) and Delturini (Loricariidae) in the Atlantic Coast. During the Late Eocene marine regression (MR4), we estimated an origin of Loricariinae in the SAF basin. During the Eocene–Oligocene cooling event, we estimated dispersal events for the Harttiini and Hypostominae. Numerous later transitions within the Loricarioidei occurred after 34 Ma occurred to the La Plata, Atlantic Coast, Western Amazon, and Trans-Andean. Across Loricarioidei, we identified 59 events during GAzBI.

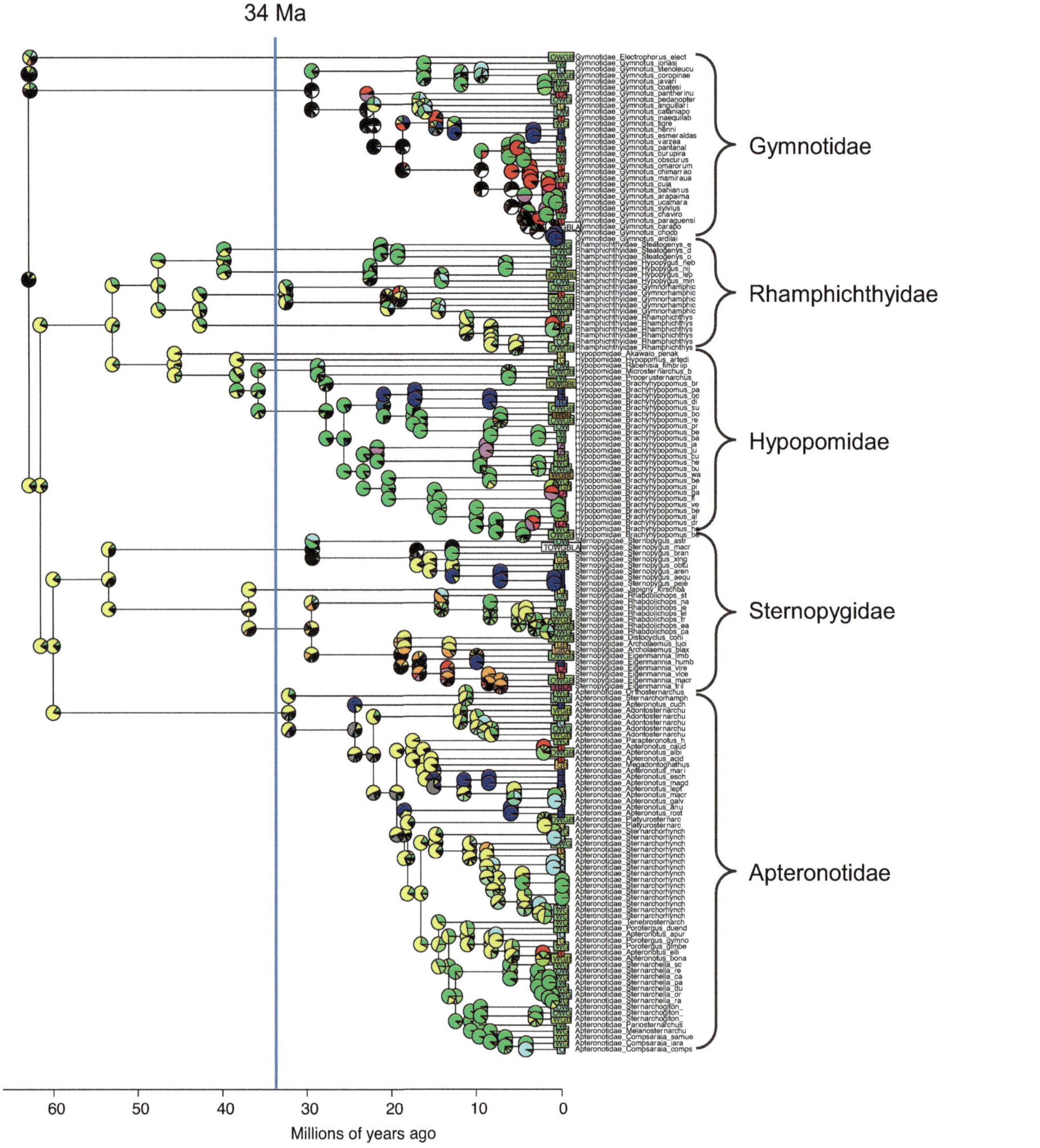

Fig. 4.6 AAE of Gymnotiformes. Data for 271 species in five family-level clades. Root age estimate 65 Ma. Number of estimated biogeographic events: 74

4.4 Discussion

4.4.1 Upland vs. Lowland Origins

Understanding the biogeographic history of NFF and its relationship to ecology is a fundamental aspect of understanding the stage that potentiated NFF ecological diversification. Although the River Continuum Concept is a powerful framework for understanding the distribution of present-day taxa, this macroecological phenomenon played out over evolutionary timescales, with shifting uplands and lowlands, affords some macroevolutionary predictions for the role of ecology in biogeographic evolution and subsequent diversification. We tested two alternative hypotheses here, where Eigenmann proposed upland origins may be subsequently followed by diversification in the lowlands, whereas Smith proposed lowland origins that may be subsequently followed by diversification in the uplands. We found support for both of these hypotheses across and within the Big Nine clades of NFF. We summarized the

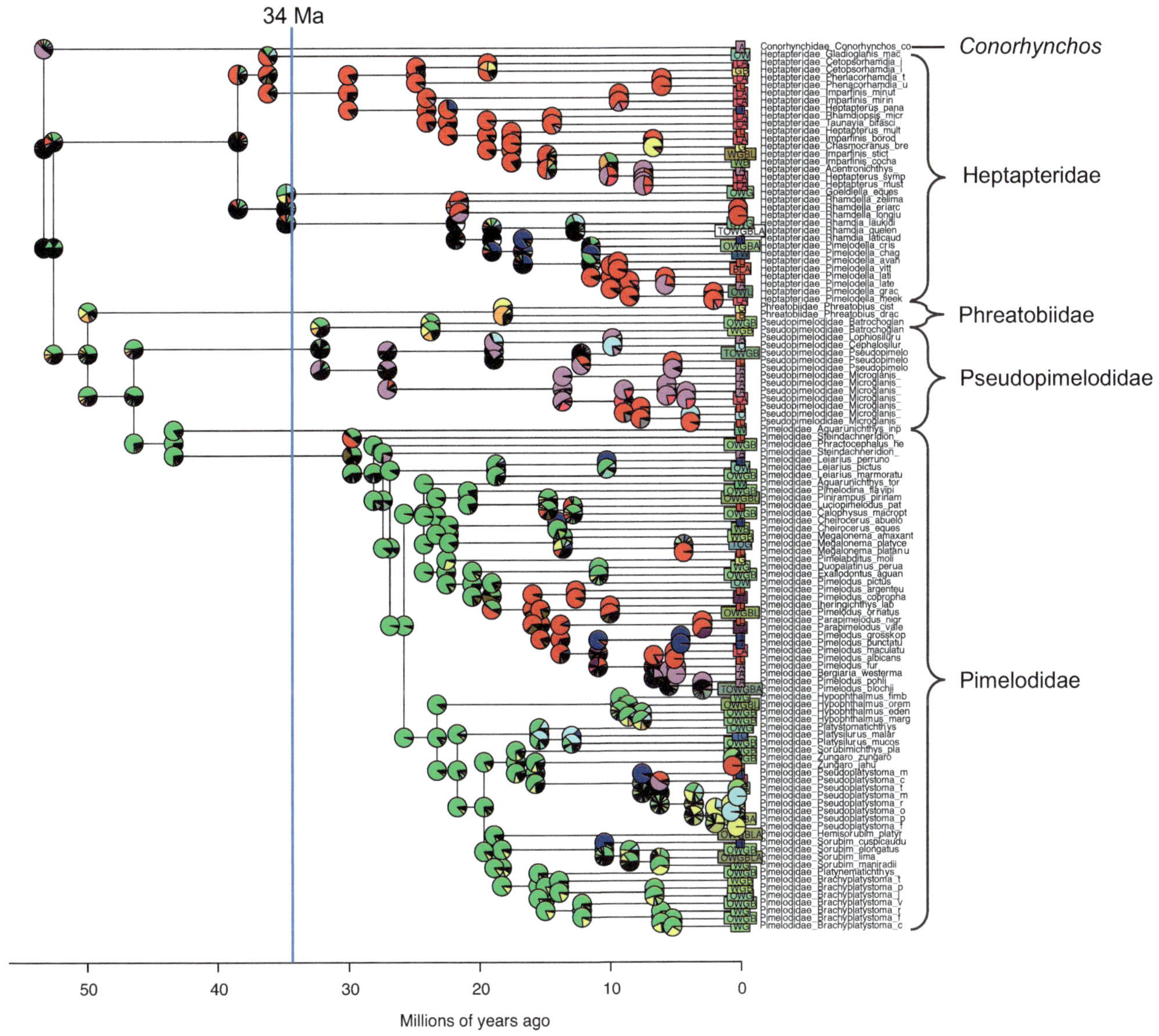

Fig. 4.7 AAE of Pimelodoidea. Data for 395 species in five family-level clades. Root age estimate 55 Ma. Number of estimated biogeographic events: 47

results in terms of clades, but here we shall discuss how these correspond to sensible paleobiogeographic scenarios.

To test whether we see evidence for Eigenmann's *Uplands origins* hypothesis with subsequent lowland diversification, or Smith's *Lowlands origins* hypothesis with subsequent upland diversification, we summarize reconstructions for ancestral origins of major clades, first for the Big Nine NFF clades. To summarize, we estimated ancestral upland origins for the more ancient Continental Periphery for the Cyprinodontiformes and Loricarioidei and the Eastern Highlands for the Gymnotiformes and Cichlinae. By contrast, we estimated SAF basin (lowland) origins for the Curimatoidea, Characoidea, Aspredinoidea, and Pimelodoidea. We then considered the ancestral ranges of 44 clades representing families within the Big Nine, except for Cichlinae where we considered five clades comprising the Retroculini+Cichlini, Astronotini+Chaetobranchini, Geophagini, Cichlasomatini, and Heroini. Again, clades were split roughly half with upland ancestral regions (21/44 clades = 48%) versus lowland ancestral regions (23/44 = 52%) (Table 4.1). A variety of upland origins for these clades were estimated including the Continental Periphery (all Cyprinodontiformes families, all Loricarioidei except Callichthyidae), Eastern Highlands (Anostomidae, all Cichlinae groups except Heroini, all gymnotiform families except Apteronotidae), Atlantic Coast (*Conorhynchus*), La Plata (Heptapteridae), and Guiana Shield (Hemiodontidae, Serrasalmidae).

Notably all of the *secondary freshwater clades* (i.e., Cichlinae, Poecilioidea, Rivulidae) originated in the uplands,

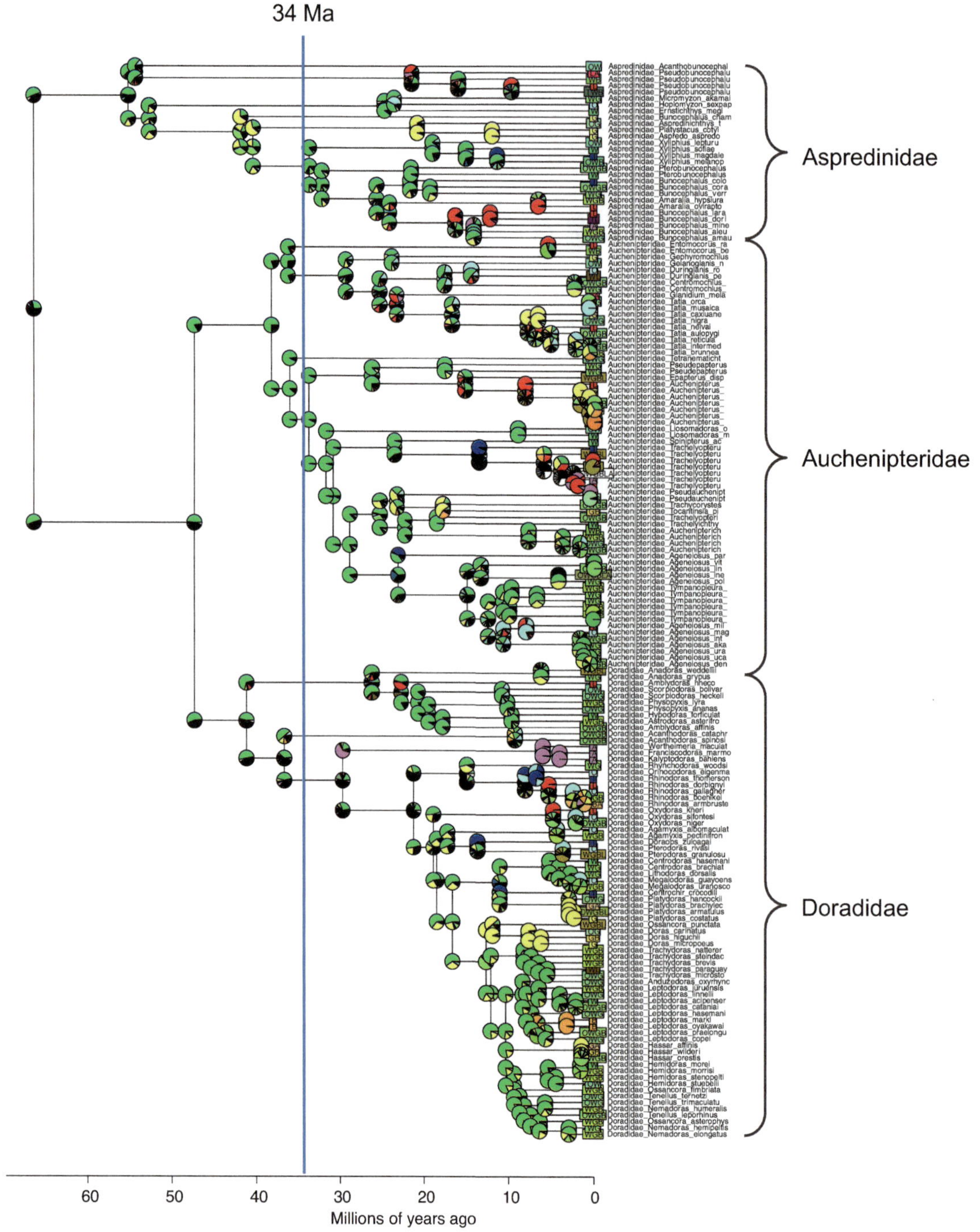

Fig. 4.8 AAE of Aspredinoidea. Data for 269 species in three family-level clades. Root age estimate 65 Ma. Number of estimated biogeographic events: 25

except for Heroini. We estimated SAF basin (lowland) origins for the Heroini, all Characoidea families, Apteronotidae, all Curimatoidea families except Anostomidae, all Aspredinoidea families, all Pimelodoidea subfamilies, and Callichthyidae. We then considered families that did not belong to Big Nine clades based on the region these species are presently found in 75 clades (including 31 additional clades; Fig. 4.11). The majority of the additional clades are coastal, representing secondarily marine taxa (25 clades, e.g., Atherinopsidae, Ariidae). We also assigned origins in the Eastern Highlands (Crenuchidae, Cetopsidae, Erythrinidae), SAF basin (Potamotrygonidae, Arapaimidae, Osteoglossidae, Tetraodontidae, Atherinidae, Hemiramphidae), Southern Cone (Galaxiidae, Diplomystidae, and Perciliidae), and

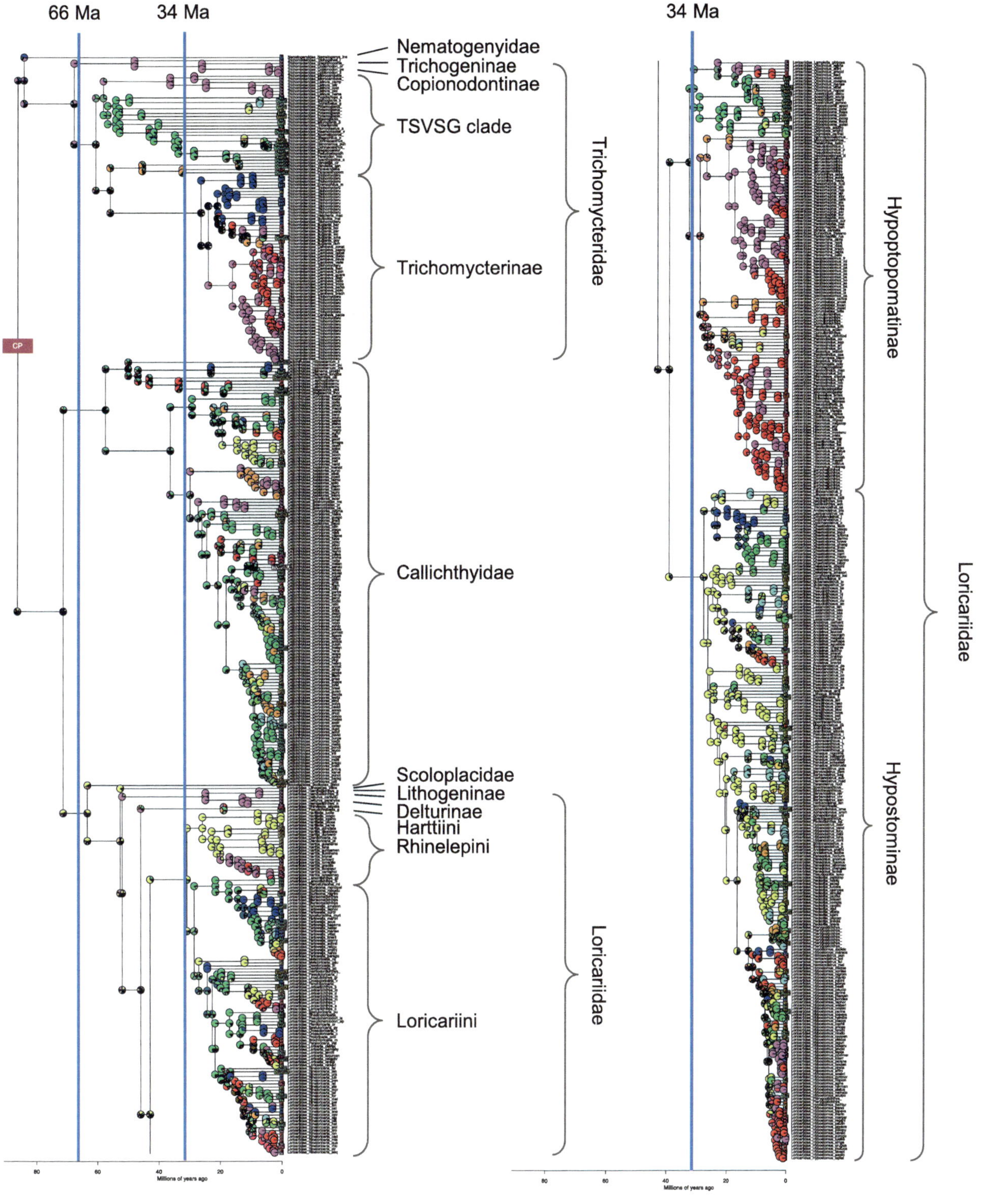

Fig. 4.9 AAE of Loricarioidei. Data for 1636 species in five family-level clades. Root age estimate 85 Ma. Number of estimated biogeographic events: 93

Central America (Profundulidae). The Eastern Highlands are the most dominant ancestral area of these 75 clades (31 clades total), although lowland SAF basin (15) and Coastal (25) families are also higher in number (40 total clades). The proportion of species belonging to clades with Eastern Highland ancestry is very large (~83%).

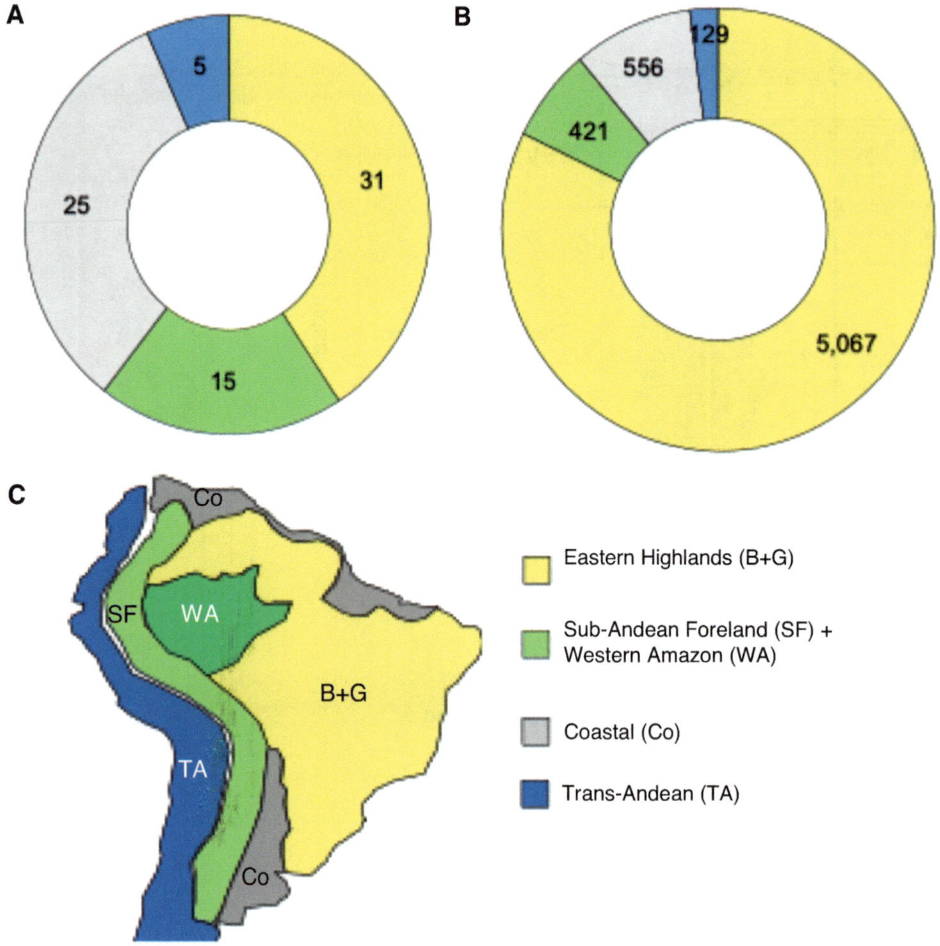

Fig. 4.10 Estimated geographic origins of 75 South American freshwater fish clades (families and cichlid tribes) among paleobioregions. (**A**) Numbers of clades represented by two or more species. Note three clades are estimated with widespread geographic origins in lowlands+uplands. (**B**) Total numbers of species (in both uplands and lowlands) that are members of clades with upland or lowland origins. (**C**) Paleo Bioregions. EH=A, B, G, and parts of O and L; SAF=parts of O, W, and L; CO=parts of O, G, and L. Note slightly more NFF clades with lowland (SAF+W=40) than upland (EH=31) origins, but that most NFF species (5067=*c.* 83%) are members of clades with upland origins. Data from Table 4.1

Evolution and diversification may be influenced by their origins within either upland or lowland regions. Body size is correlated with clade age among with lowland origins (black) but not those with upland origins (Fig. 4.12A). NFF clades of lowland origins with largest average body sizes are Ctenoluciidae (Pike characins) at 100 cm SL, Bryconidae (Dorados) at 80 cm SL, and Pimelodidae (Long-whiskered catfishes) at 61 cm SL, all of which are older clades that originated in the Paleogene. NFF clades with upland origins that grow to the largest sizes are Retroculini+Cichlini (retroculine cichlids and Peacock basses) at 100 cm SL, Serrasalmidae (piranhas and allies) at 100 cm SL, both with younger origins in the Oligocene, and both with largest-bodied species in the lowlands. Species richness is not correlated with crown-group clade age (Fig. 4.12B) or body size (Fig. 4.12C).

4.4.2 Caveats: Interpretations and Uncertainties

We acknowledge difficulties when estimating or inferring ancestral ranges from extant taxa data (Wisniewski et al. 2022). Some of the main caveats are with how to interpret ancestral areas given potentially unobserved extinctions, phylogenetic uncertainty in divergence time estimates, and the effect of missing taxa. There are two main potential alternative hypotheses for ancestral area reconstructions. For example, a reconstruction of ancestral area in the upland shields is consistent with two different explanations: 1) Early branches may retain the ancestral location on the upland shields (a typical assumption when interpreting biogeographic reconstructions), or 2) alternatively, ancestral clades may have been widespread, but early branches may preferentially survive on shields as peripheral relicts while lowland species went extinct, being replaced by

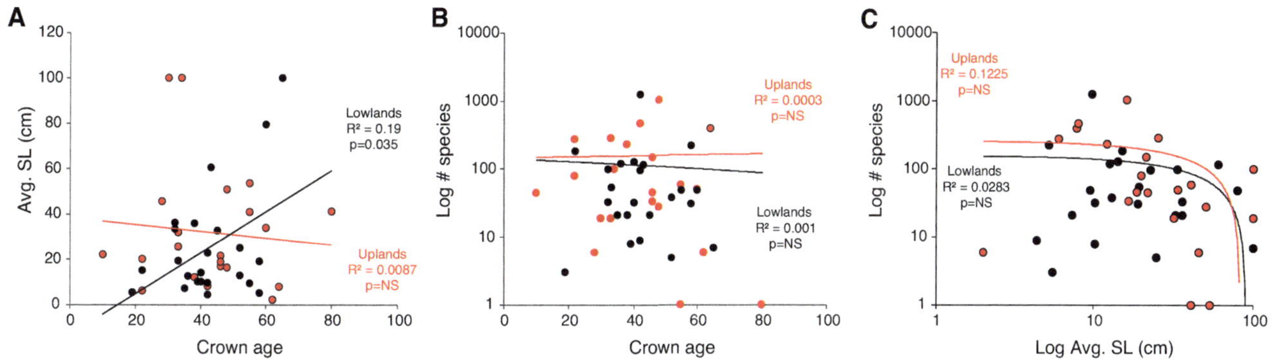

Fig. 4.11 Correlations among body size, clade age, and species richness among NFF clades. Data for 5933 species representing 44 clades (families or cichlid tribes; Table 4.1). (**A**) Body size is correlated with clade age among with lowland origins (black) but not upland origins (red). This means the oldest taxa are riverine fishes with large body sizes in lowlands (e.g., Pimelodidae, Curimatidae) but small stream fishes in uplands (e.g., Loricariidae, Trichomycteridae). Note similar numbers of clades with upland ($n = 21$) and lowland ($n = 23$) origins meaning the polyphyletic NFF fauna had broad geographic and ecological origins. Note also no significant correlations among crown clade age and species richness (**B**) or body size (**C**), for either lowland or upland faunas

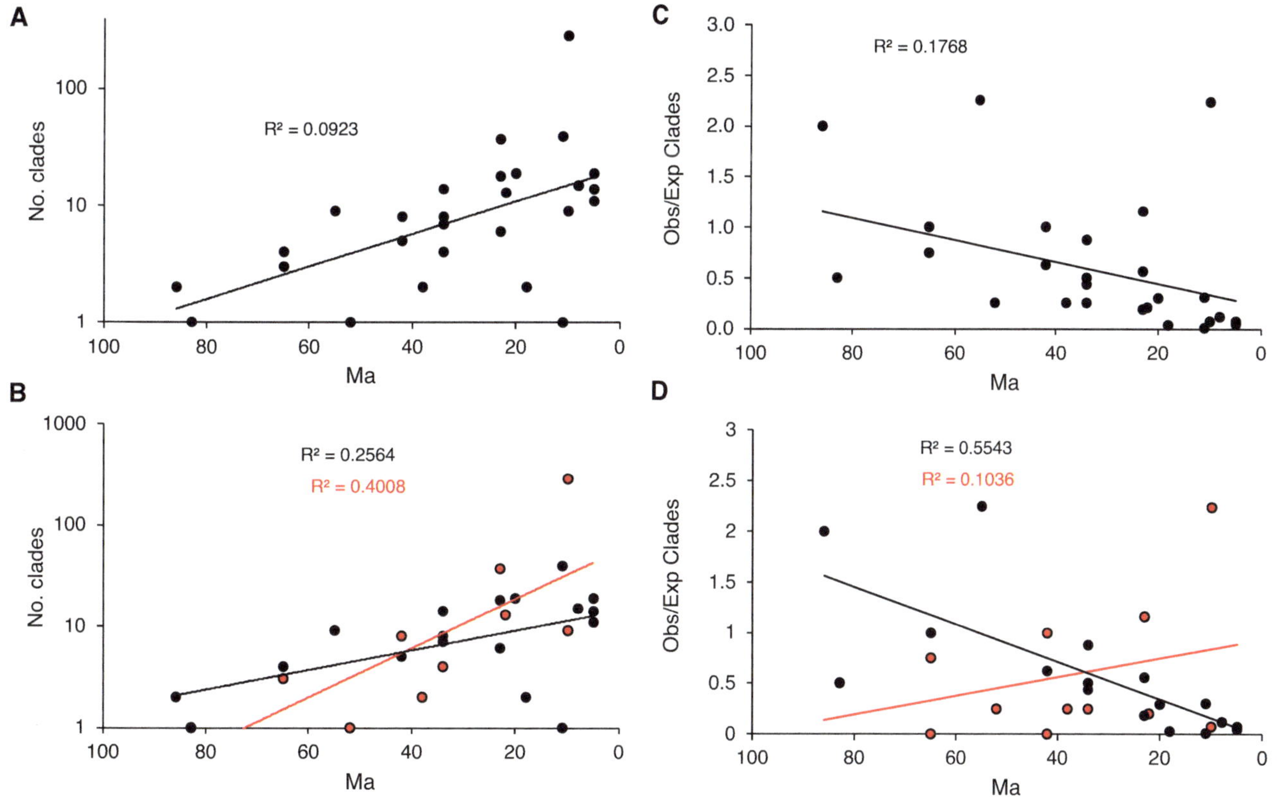

Fig. 4.12 Temporal trends in 558 NFF radiations across 27 paleogeographic events in 10 time horizons (Table 4.2).(**A**) Data for 364 lowland radiations (65.2%) and 194 upland radiations (34.8%), data combined; (**B**) Data separated (upland clades in red). Note lowlands contribute more than uplands to the temporal increase in number of clades. (**C**) Ratio of observed/expected clades from exponential expansion over 10 time intervals, data combined. (**D**) Data separated (upland clades in red). Note the opposite temporal trends, highlighting the stronger influences of earlier K and PG upland radiations and later NG lowland evolutionary radiations

younger radiations (e.g., Fig 1.15C). Incorporating ancestral ranges using fossil data can help to distinguish between these alternative hypotheses. For example, the recent discovery of a putative loricariid fossil from Africa, would, if confirmed, add additional context for the Continental Periphery ancestral state inferred for loricariids and other loricarioid families (Brito et al. 2024). However, the taxonomic identity of this fossil is disputed, and it may represent a juvenile gar (Britz et al. 2024).

Evolutionary interpretations of this study were based on the time-calibrated phylogeny of Chap. 2, with a range of errors on lineage divergence times. We preformed ancestral area estimates on the incomplete (sequence-only) phylogenies of 3 K species for the Big Nine clades. We suspect our results will still be robust despite at least two main concerns. First, there may be some potential biases on taxa being nonrandomly sampled with missing taxa resulting in an underestimation of ancestral area transitions (e.g., by dispersal or vicariance). Given that taxa are quite reasonably sampled especially for deeper lineages, we believe that most of the missing transitions are <5 Ma, which we therefore exclude from further analysis. Therefore, our conclusions are based primarily on parts of the tree unlikely to change with more complete sampling. While we acknowledge, we reconstructed a taxon-complete phylogeny for NFF using taxonomic assignment and polytomy resolution for the Big Nine, such trees are best used for diversification rate analyses (Chap. 5) (Chang et al. 2020). Polytomy-resolution for taxon placement can result in increased rates of events and greater uncertainty of ancestral area estimates due to the added taxa (Rabosky 2015). That said, just like missing taxa, incorrectly placed taxa would mostly affect reconstructions near the tips including the large number of reconstructions <5 Ma, we did not focus on in the present work. Second, focusing on the Big Nine clades means that biogeographic drivers affecting other NFF clades were not assessed. Nearly all of these remaining clades are extremely small and young marine derived clades that did not experience the extensive biogeographic histories of the Big Nine. Furthermore, given that they comprise <10% of all NFF species, we assume that biogeographic patterns within these clades would not greatly affect conclusions on the drivers of major patterns of the dominant >90% of species.

4.4.3 Open Questions

Some open questions remain from our various results. Why do almost all secondary freshwater clades among the Big Nine originate in uplands? This finding is difficult to explain as upland origins would be unexpected for marine or brackish clades that are secondary freshwater taxa, as these would be primarily expected to enter the Neotropics at sea level. One particular example of this are the cyprinodontid pupfishes of the genus *Orestias*, which are inferred as a relatively recent migrant into the Neotropics among the Big Nine clades, and for which species are currently isolated in upland lakes in the Andes (Parenti 1984). However, numerous other cyprinodontiform and cichlid groups show ancestral upland origins in the Continental Periphery. Of course, this finding ignores the origins of secondary clades not represented among the Big Nine (e.g., clupeids, potamotrygonids, etc.), for which we infer Coastal origins (Fig. 4.11), but these origins were not directly estimated. These may be best explained by the inundation of the Pebas Wetland within the SAF basin (Chap. 1).

We found crown age and species richness were not correlated. Perhaps this is not surprising. Such a correlation might be expected only if diversification rates are constant within clades (Rabosky and Adams 2012). However, diversification rates are highly variable among NFF (Cassemiro et al. 2023; Chap. 5). The relationship between age and richness, in a continental radiation, may be best explained by the timing of paleogeographic events. Diversification might also be driven by intrinsic properties such as sex chromosome evolution (Chap. 3); if these intrinsic properties vary across taxa they may also introduce variation in diversification rates across clades that would not be explained well by crown age.

Why is there a correlation between age of origin and body size in lowlands, but not uplands? We found older lowland taxa generally had larger body sizes, but younger lowland taxa had smaller body sizes. This is not explained purely by the variation in body sizes, as both uplands and lowlands can have large-bodied taxa (i.e., mean ≥80 cm SL), but only in lowlands are these larger bodied taxa also representing older clades. This may imply some consequence of either upland or lowland ecology affecting the tempo or mode of body size evolution. Under neutral body size evolution from a small ancestor (typical of many clades), an expectation of increasing mean body size over time could be expected through neutral processes, known as Deperet's Rule (Bokma 2016). It is possible the patterns of body size evolution might be explained by such a scenario in clades that originated in the lowlands, for which larger body sizes might be less constrained. However, clades that originated in the uplands may have evolved under a different evolutionary mode, which may explain the apparent differences in the patterns and overall rates of body size evolution such that crown age no longer explains mean body size. That said, the relationship between crown age and mean body size in lowland taxa is fairly weak, so it may be driven as an artefact of the few outliers.

4.5 Conclusions: NFF Clade Origins

Early NFF diversification corresponded to numerous major paleogeographic events (Table 4.2). At least 34 of the 44 Big Nine clades have estimated origins during the super greenhouse world of the Upper Cretaceous and Eocene (*c.* 100–34 Ma). These early clades included 19 clades with upland origins and 15 clades with lowland origins, and included members of all the Big Nine clades (Table 4.2).

Clades with early (*c.* 100–35 Ma) upland origins include representatives from five of the Big Nine NFF clades (i.e., Cichlinae, Poecilioidea, Rivulidae, Curimatoidea, Gymnotiformes, Loricarioidei), and include both primary and secondary freshwater fish groups (Table 4.1). All but one of these clades with early upland origins exhibit greatest diversity among living forms in lowlands, with only Cyprinodontiformes exhibiting highest extant diversity in upland basins. All clades with early upland origins originated in the geographic periphery of South America. The locations of early diversification were either in the "Eastern Highlands" of the Guiana and/or Brazilian Shield (i.e., Cichlinae, Hemiodontoidae, Anostomidae, Gymnotiformes, *Conorhynchus*, Scoloplacidae, Lithogeninae, Copiodontinae+Trichogeninae, Delturini), or along the western margin in the Northern Andes (*Saccodon*, Parodontidae), Central Andes (*Orestias*, Cyprinodontidae), or Southern Andes (Nematogenyidae; Diplomystidae) (Table 4.1). The three oldest NFF clades with upland origins were estimated to have evolved within rainforest refugia of the Guiana Shield during the Upper Cretaceous. These clades are Neotropical Cyprinodontiformes and Loricarioidei during the Santonian (*c.* 86–84 Ma), and *Nematogenys* in the Southern Cone during the Campanian (83–72 Ma). The Santonian was characterized by extreme hothouse global climates with a low latitudinal temperature gradient, and high levels of storm action (Burgener et al. 2023). The Campanian was characterized by progressive global cooling, with pronounced cooling during the transition to the Maastrichtian *c.* 71–69 Ma (Scotese et al. 2021).

All the oldest lowland NFF clades originated in the northern SAF basin during times of global cooling events and eustatic sea-level falls. The early lowland NFF clades represent seven of the Big Nine clades (excluding the two cyprinodontiform clades) and include both primary and secondary freshwater fish taxa. The earliest estimated NFF clades with lowland origins are Characoidea and Curimatoidea, both coinciding in time and place with Marine Regression 1 (MR1; *c.* 83–72 Ma) in the SAF basin of northern South America. An additional 16 NFF clades are estimated to have diversified in the northern SAF basin during the time of MR2 (65–58 Ma) and MR3-4 (42–37 Ma). The earliest lowland cyprinodontiform clades are *Oxyzygonectes* (Anablepidae) and *Cynodonichthys* (Rivulidae) in trans-Andean basins and Central America, both dated to *c.* 34 Ma, and two clades of Rivulidae (*Laimosemion*, *Anablepsoides*+*Moema*+*Gnatholebias*+*Renova*), which colonized the SAF basin *c.* 34 Ma from ancestors on the Guiana Shield.

References

Abell R, Thieme ML, Revenga C, Bryer M, Kottelat M, Bogutskaya N et al (2008) Freshwater ecoregions of the world: a new map of biogeographic units for freshwater biodiversity conservation. BioScience 58(5):403–414

Abreu JMS, Craig JM, Albert JS, Piorski NM (2019) Historical biogeography of fishes from coastal basins of Maranhão State, northeastern Brazil. Neotrop Ichthyol 17(2):e180156

Albert JS, Johnson DM (2012) Diversity and evolution of body size in fishes. Evol Biol 39(3):324–340

Albert JS, Bart HL Jr, Reis RE (2011) Species richness and cladal diversity. In: Albert JS, Reis RE (eds) Historical biogeography of neotropical freshwater fishes. University of California Press, Berkeley, pp 89–104

Albert JS, Antonelli A (2017) Society for the study of systematic biology symposium: frontiers in parametric biogeography. Syst Biol 66(2):125–127

Albert JS, Val P, Hoorn C (2018) The changing course of the Amazon River in the Neogene: center stage for Neotropical diversification. Neotrop Ichthyol 16(03):e180033. https://doi.org/10.1590/1982-0224-201901

Albert JS, Schoolmaster Jr DR, Tagliacollo V, Duke-Sylvester SM (2017) Barrier displacement on a neutral landscape: toward a theory of continental biogeography. Syst Biol 66(2):167–182

Albert JS, Tagliacollo VA, Dagosta F (2020) Diversification of Neotropical freshwater fishes. Annu Rev Ecol Evol Syst 51:27–53

Albert JS, Bernt MJ, Fronk AH, Fontenelle JP, Kuznar SL, Lovejoy NR (2021) Late Neogene megariver captures and the Great Amazonian biotic interchange. Glob Planet Change 205:103554

Bokma F, Godinot M, Maridet O, Ladevèze S, Costeur L, Solé F, Gheerbrant E, Peigné S, Jacques F, Laurin M (2016) Testing for Depéret's rule (body size increase) in mammals using combined extinct and extant data. Syst Biol 65(1):98–108

Brito PM, Dutheil DB, Gueriau P, Keith P, Carnevale G, Britto M, Meunier FJ, Khalloufi B, King A, de Amorim PF, Costa WJEM (2024) A Saharan fossil and the dawn of Neotropical armoured catfishes in Gondwana. Gondwana Res 132:103–112

Britz R, Pinion AK, Kubicek KM, Conway KW (2024) Comment on "A Saharan fossil and the dawn of Neotropical armoured catfishes in Gondwana" by Brito et al Gondwana Res 133:267–269

Burgener L, Hyland E, Reich BJ, Scotese C (2023) Cretaceous climates: mapping paleo-Köppen climatic zones using a Bayesian statistical analysis of lithologic, paleontologic, and geochemical proxies. Palaeogeogr Palaeoclimatol Palaeoecol 613:111373

Cassemiro FAS, Albert JS, Antonelli A, Menegotto A, Wüest RO, Cerezer F, Coelho MTP, Reis RE, Tan M, Tagliacollo V, Bailly D, da Silva VFB, Frota A, da Graça WJ, Ré R, Ramos T, Oliveira AG, Dias MS, Colwell RK, Graham CH (2023) Landscape dynamics and diversification of the megadiverse South American freshwater fish fauna. Proc Natl Acad Sci 120(2):e2211974120

Chang J, Rabosky DL, Alfaro ME (2020) Estimating diversification rates on incompletely sampled phylogenies: theoretical concerns and practical solutions. Syst Biol 69(3):602–611

Dagosta FCP, de Pinna M (2017) Biogeography of Amazonian fishes: deconstructing river basins as biogeographic units. Neotrop Ichthyol 15(3):e170034. https://doi.org/10.1590/1982-0224-20170034

Diniz-Filho JAF, Coelho MTP, Terribile LC (2023) Unpacking underdetermination: theoretical challenges in understanding macroecological and biogeographic patterns. J Biogeogr 50(11):1890–1898

Eigenmann CH (1912) The freshwater fishes of British Guiana, including a study of the ecological grouping of species and the relation of the fauna of the plateau to that of the lowlands. Mem Carnegie Mus 5:1–578

Eigenmann CH, Allen WR (1942) Fishes of western South America. I. The Intercordilleran and Amazonian lowlands of Peru. II. The high pampas of Peru, Bolivia, and northern Chile. The University of Kentucky, pp 1–524

Gómez-Martínez RF, Matamoros WA, Hoagstrom CW, López-Vila JM, Bonilla EPDD (2024) Phylogenetic diversity of live-bearing fishes (Poeciliidae) peaks on peninsulas, isthmuses, and in deserts. J Biogeogr 52:641–656. https://doi.org/10.1111/jbi.15066

Gregory-Wodzicki KM (2000) Uplift history of the Central and Northern Andes: a review. Geol Soc Am Bull 112(7):1091–1105

Hernandez SG, Hoagstrom CW, Matamoros WA (2024) Historical biogeography of North American Killifishes (Cyprinodontiformes) recapitulates geographical history in the Gulf of México watershed. Zool J Linn Soc 202(2):zlae105. https://doi.org/10.1093/zoolinnean/zlae105

Hubert N, Renno JF (2006) Historical biogeography of South American freshwater fishes. J Biogeogr 33(8):1414–1436

Kumar S, Suleski M, Craig JM, Kasprowicz AE, Sanderford M, Li M, Stecher G, Hedges SB (2022) TimeTree 5: an expanded resource for species divergence times. Mol Biol Evol 39(8):msac174

Lewis Jr WM, Hamilton SK, Saunders III JF (2006) Rivers of northern South America. River and Stream Ecosystems of the World. University of California Press, pp 219–256

López-Fernández H, Albert JS (2011) Paleogene radiations. In: Albert JS, Reis R (eds) Historical biogeography of neotropical freshwater fishes. University of California Press, Berkeley, pp 105–118

Lundberg JG, Luckenbill KR, Babu KS, Ng HH (2014) A tomographic osteology of the taxonomically puzzling catfish *Kryptoglanis shajii* (Siluriformes, Siluroidei, incertae sedis): description and a first phylogenetic interpretation. Proc Acad Nat Sci Phila 163(1):1–41

Matzke NJ (2013) Probabilistic historical biogeography: new models for founder-event speciation, imperfect detection, and fossils allow improved accuracy and model-testing. Front Biogeogr 5(4):242–248

Matamoros WA, McMahan CD, Chakrabarty P, Albert JS, Schaefer JF (2015) Derivation of the freshwater fish fauna of Central America revisited: Myers's hypothesis in the twenty-first century. Cladistics 31(2):177–188

Melo BF, Sidlauskas BL, Near TJ, Roxo FF, Ghezelayagh A, Ochoa LE et al (2022) Accelerated diversification explains the exceptional species richness of tropical characoid fishes. Syst Biol 77(1):78–92

McNab BK (2002) The physiological ecology of vertebrates: a view from energetics. Cornell University Press

Mora A, Villagómez D, Parra M, Caballero VM, Spikings R, Horton BK et al (2020) Late cretaceous to cenozoic uplift of the northern Andes: paleogeographic implications. Geol Colombia 3:ca–4

Parenti LR (1984) A taxonomic revision of the Andean killifish genus *Orestias* (Cyprinodontiformes, Cyprinodontidae). Bull Am Mus Nat Hist 178(2):107–214

Rabosky DL (2015) No substitute for real data: a cautionary note on the use of phylogenies from birth-death polytomy resolvers for downstream comparative analyses. Evolution 69(12):3207–3216

Rabosky DL, Adams DC (2012) Rates of morphological evolution are correlated with species richness in salamanders. Evolution 66(6):1807–1818

Ree RH, Smith SA (2008) Maximum likelihood inference of geographic range evolution by dispersal, local extinction, and cladogenesis. Syst Biol 57(1):4–14

Ree RH, Sanmartín I (2009) Prospects and challenges for parametric models in historical biogeographical inference. J Biogeogr 36(7):1211–1220

Scotese CR, Song H, Mills BJW, Van Der Meer DG (2021) Phanerozoic paleotemperatures: the earth's changing climate during the last 540 million years. Earth-Sci Rev 215:103503

Sanmartín I, Ronquist F (2004) Southern hemisphere biogeography inferred by event-based models: Plant versus animal patterns. Syst Biol 53:216–243

Smith GR (1992) Phylogeny and biogeography of the Catostomidae, freshwater fishes of North America and Asia. Systematics, Historical Ecology, and North American Freshwater Fishes. Stanford University Press, pp 778–826

Sullivan JP, Muriel-Cunha J, Lundberg JG (2013) Phylogenetic relationships and molecular dating of the major groups of catfishes of the Neotropical superfamily Pimelodoidea (Teleostei, Siluriformes). Proc Acad Nat Sci Phila 162(1):89–110

Tagliacollo VA, Tan M, Reis RE, Gaia R, Carrijo V, Ranuzi M, Albert JS (2024) Time-calibrated phylogeny of neotropical freshwater fishes. Front Bioinform 4:1433995

Weitzman SH, Vari RP (1988) Miniaturization in South American freshwater fishes; and overview and discussion. Proc Biol Soc Wash 101(2):444–465

Wisniewski AL, Lloyd GT, Slater GJ (2022) Extant species fail to estimate ancestral geographical ranges at older nodes in primate phylogeny. Proc R Soc B 289(1975):20212535

5 Neotropical Freshwater Fishes: Diversification Rates Through Time

Jack M. Craig and James S. Albert

Abstract

Different diversification patterns of Neotropical freshwater fish clades through the Cenozoic indicate distinct evolutionary responses to common environmental factors, indicating the influences of distinct functional or ecotrait organismal properties of perhaps stochastic factors. Most clades of fishes adapted to streams and small rivers with small adult body sizes exhibit close to a regular clock-like speciation, and therefore clade age is a good predictor of species richness for these fishes. Clades that inhabit the channels and floodplains of larger rivers do not follow this pattern, and therefore species richness in these clades is not correlated with clade age. Conversely, the estimated rate of diversification is a good predictor of species richness in clades that inhabit larger rivers, but not those of smaller nonfloodplain (*terra firme*) rivers and streams. These contrasting patterns may be explained by the greater frequency of vicariance and geodispersal events that impact fishes in small rivers and streams from river capture, as compared to the relatively lower frequency of these events affecting larger-bodied and more vagile fishes of large rivers.

5.1 Introduction

The humid Neotropics is home to the greatest freshwater biodiversity on the planet. More than 6342 described fish species inhabit South America, Central America, and the Greater Antilles, representing about 20% of all freshwater fishes, and about 10% of all vertebrate species on Earth (Reis et al. 2016; van der Sleen and Albert 2022). This remarkable assemblage represents an excellent system in which to investigate the patterns and processes of biodiversification at a continental scale (Albert et al. 2020; Val et al. 2022). While the *Neotropical Fish Fauna* (NFF) has been the subject of taxonomic work for centuries, including several in the *Systema Naturae* (Linnaeus 1758), there have been to date only a few studies quantifying the macroevolutionary processes giving rise to this remarkable biodiversity (Cassemiro et al. 2023; Cerezer et al. 2023).

While we have increasingly robust accounts of NFF species richness and evolutionary relationships among taxa (Chap. 2), we have not yet been able to meaningfully answer some of the most important questions about the origins of this fauna. Prominent among these is the question of relative species richness. Why do some clades have so many more species and ecological functions than others? Two testable explanations have been proposed. First, it may simply be the case that more species-rich clades with more functional disparity are older, allowing more time for the diversification to occur, without the need to invoke any processes particular to a clade or a region (Coyne and Orr 1998; Hedges et al. 2015; Henao Diaz et al. 2019; Miller and Román-Palacios 2021). Second, species-rich and functionally diverse clades may be characterized by more rapid rates of speciation, producing greater species richness and functional disparity in the same amount of time (Rabosky et al. 2013; Seehausen and Wagner 2014; Rabosky 2020).

Differential rates of lineage diversification arise from the balance between rates of speciation and extinction, hereafter called diversification. These differences could arise from intrinsic characteristics of a species, like smaller body size, or anatomical and physiological specializations that affect habitat utilization, diet, or dispersal ability (Boucher et al. 2017; Fontanillas et al. 2007; Han et al. 2020; Sanderson and Donoghue 1994; Sayol et al. 2019; Wilson 2003). Alternatively, differential diversification could arise from

J. M. Craig
Drexel University, Philadelphia, PA, USA
e-mail: jack.m.craig@gmail.com

J. S. Albert (✉)
University of Louisiana at Lafayette, Lafayette, LA, USA
e-mail: jsalbe@gmail.com

J. S. Albert and R. E. Reis (eds.), *Ecology and Evolution of Amazonian Fishes*, Fascinating Life Sciences,
https://doi.org/10.1007/978-3-032-09328-8_5

external environmental factors such as climatic or landscape changes (Arbour and Santana 2017; Hou et al. 2011; Koenen et al. 2013; Mitter et al. 1988; Murray et al. 2018; Phillimore et al. 2006; Price et al. 2012; Upham et al. 2019; Wang et al. 2018). Furthermore, it may be the case that biogeographic history has resulted in a random distribution of clade sizes among NFF taxa, necessitating a thorough understanding of historical geological events to fully grasp the patterns of modern species richness (Albert and Carvalho 2011; Baker and Couvreur 2013; Claramunt and Cracraft 2015; Hughes and Eastwood 2006; Jansson et al. 2013; Liu et al. 2018; Pontarp and Wiens 2017; Quintero and Jetz 2018; Ricklefs 2006; Roelants et al. 2007).

Here, we test these hypotheses by quantifying the rates of diversification across 11 NFF clades comprising 5771 species, or about 90% of the entire described NFF fauna. We take the robust, taxonomically rich phylogeny presented in Chap. 2 as a reference for the phylogenies of each of these clades. We first perform a simple regression between the total described species richness of a clade and its phylogenetic age. We then employ two more approaches to infer net diversification rates across the phylogenies of these clades. First, we use the ClaDS model (Maliet et al. 2019; Maliet and Morlon 2021) to infer a time-varying or episodic speciation rate for each branch in each phylogeny independently, providing the maximum phylogenetic resolution possible. Second, we will use a Bayesian approach (Höhna et al. 2016) to quantify diversification rates across the entire tree of each clade, providing a summary of the *phylodynamic* pattern of branching events within a shared evolutionary lineage. Phylodynamics refers to the study of phylogenetic tree parameters (e.g., branch-length frequencies, branching symmetry) to estimate evolutionary processes (Grenfell et al. 2004). We apply a robust and rapidly developing set of macroevolutionary methods to understand one of the greatest vertebrate assemblages on Earth. Through the synthesis of classical taxonomy with modern phylogenetics and phylodynamics, we can begin to answer some of the foundational questions in evolutionary biology.

5.2 Methods

5.2.1 Taxa

All phylogenetic analyses in this chapter are based on the time-calibrated phylogeny presented in Chap. 2 (Fig. 5.1). For all analyses of the whole NFF fauna, we examined a subset of these taxa capturing teleost fishes only ($N = 6306$), as many basal taxa represent marine-derived lineages which are highly incompletely represented in Neotropical freshwaters. Importantly, as a regional fauna, the NFF is not monophyletic. For example, significant portions of the Characiformes and Siluriformes are found outside the humid Neotropics in Africa and North America. Similarly, while the NFF fauna includes representatives of many marine-derived clades (e.g., Engraulidae, Potamotrygonidae, Tetraodontidae), the majority of species in these clades are marine which were excluded from the NFF phylogeny. Performing a phylodynamic analysis on such a broadly nonmonophyletic assemblage violates the underlying assumptions of the models being used, so it is necessary to examine only monophyletic clades. We identified 11 species-rich monophyletic NFF clades, including the Big Nine NFF clades of Chap. 3 (i.e., Characoidea, Cichlinae, Curimatoidea, Aspredinoidea, Gymnotiformes, Loricarioidei, Pimelodoidea, Poeciliinae, and Rivulidae), with the addition of two additional clades (i.e., Ctenolucoidea and Cetopsidae; Table 5.1). We then portioned these clades into those which predominantly inhabit streams and small rivers with Strahler stream order (SO) 1–5 and large rivers with SO 6–10 (Chap. 1). We estimated crown-group clade ages (i.e., *clade ages*) from the time-calibrated phylogeny as the most recent common ancestor of all currently existing (i.e., extant) species within a particular clade (Rabosky 2010; Title and Rabosky 2017).

5.2.2 Net Diversification Rates

We employed three methods to estimate net diversification rates. First, we calculated a simple ratio of species richness and clade age as a simple diversification metric: species per million years (Ma). A positive correlation of these metrics is interpreted as evidence that time (T) is an underlying driver of species richness. Next, we used the ClaDS model (Maliet et al. 2019; Maliet and Morlon 2021), which infers the rate of lineage branching for each daughter lineage at a phylogenetic divergence based on its number of descendants. Using this approach gives us a species-specific rotation of speciation for every tip on the tree, as well as a median rate of speciation across the clade, and estimates of the heterogeneity in rates and their general trend, either accelerating or decelerating. We ran the ClaDS model in Julia 1.6.3 (Bezanson et al. 2017), allowing the program to stop upon reaching a convergence among three concurrent Markov chains when the Gelman statistic fell below 1.05 (Maliet and Morlon 2021). Because performing a phylodynamic analysis in a field where taxonomic knowledge is changing rapidly risks not correctly accounting for the true species richness, the sampling fraction in this analysis was treated as 0.66. This means that we can assume our phylogeny of 6306 species represents about two-thirds of the actual total species richness of the NFF fauna, including presently undescribed species, of about 9000 species (Reis et al. 2016). Ignoring a full third of the NFF species richness for this model would result in a serious underestimation of the speciation rate. This also helps to account for biases due to cryptic diversity (Utami et al. 2022).

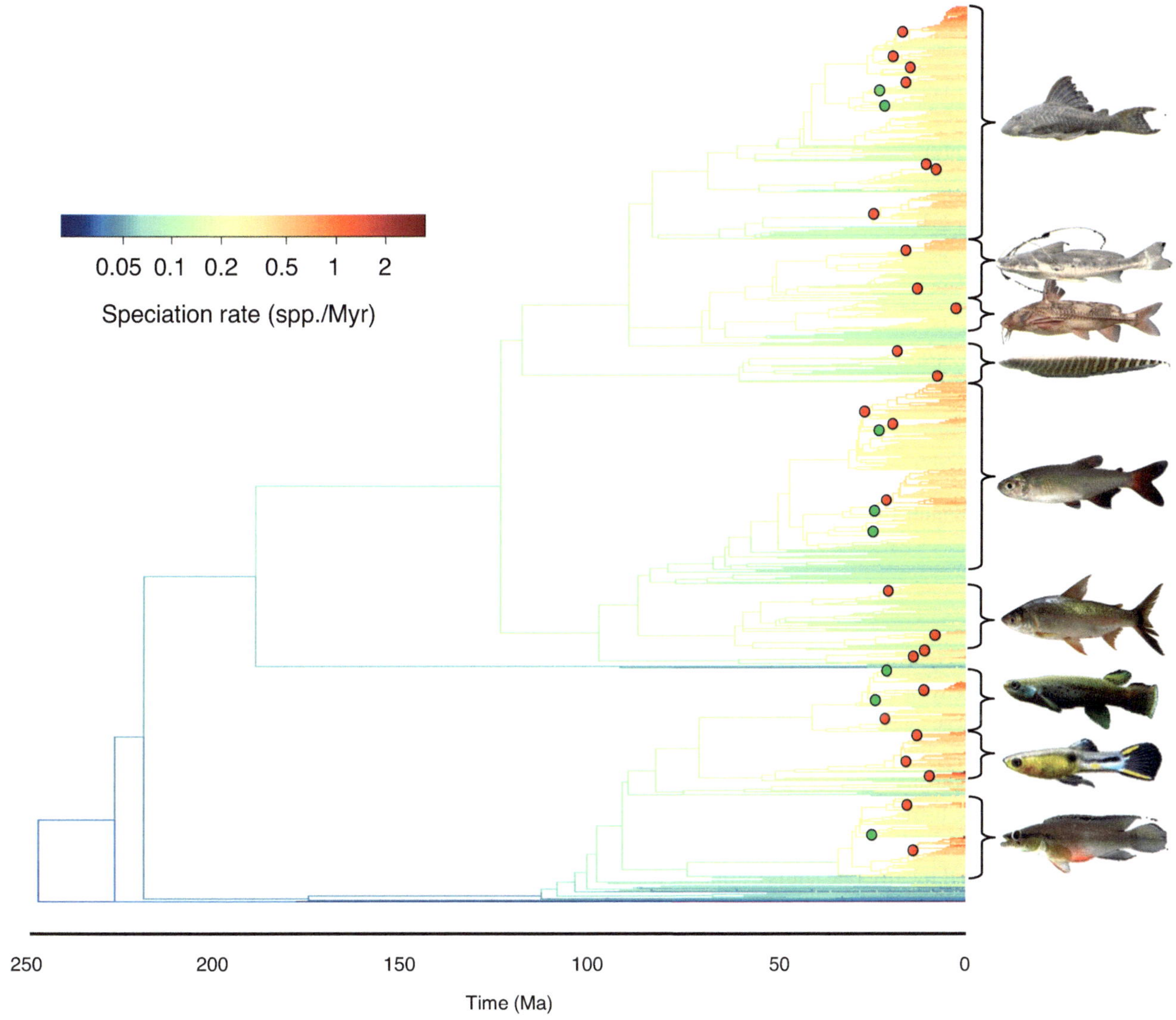

Fig. 5.1 Net lineage diversification rates through time among NFF clades using the CLaDS birth–death model. Phylogeny of 5679 extant teleost species from Chap. 2. The Big Nine NFF clades (*sensu* Chap. 3) depicted at right from top to bottom, with shifts to highest (>0.30, orange and red) and lowest rates (<2.0, green and blue) in parentheses. Big Nine combined (27, 8), Loricarioidei (3, 2), Pimelodoidea (2, 0), Aspredinoidea (2, 0), Gymnotiformes (2, 0), Characoidea (4, 3), Curimatoidea (4, 0), Rivulidae (2, 2), Poeciliodea (3, 0), and Cichlinae (2, 0)

Finally, we used a Bayesian approach (Fabreti and Höhna 2022; Höhna 2015; Höhna et al. 2016; May et al. 2016) to estimate the clade-wide rate of diversification through time generating both an initial hyperparameter of speciation for the whole clade, which may be compared to the median rate from ClaDS, and a plot of the changes in rate through time. This hyperparameter may correlate to species richness, making it potentially unsuitable for direct comparisons between clades without validation by comparison to another model like ClaDS (Craig et al. 2022). The Bayesian model was implemented in the R (Core Development Team 2020) package TESS (Höhna et al. 2016). We allowed TESS to infer hyperparameters directly from each given phylogeny prior to each run, then ran the chain for 200,000 iterations, removing the first 10,000 as burn-in. We chose not to parameterize any mass extinctions. The sampling fraction in this analysis was treated as 0.66. If either model recovers rapid increases in rates over time in clades characterized by high species richness, we can conclude that the rate diversification, not necessarily clade age, is the key driver in species richness.

Some caution must be exercised in the biological interpretation of rates estimated from phylogenies of extant taxa. Known biases may arise from incomplete taxon sampling and insufficient genetic variation in the molecular alignment underlying phylogeny (Craig et al. 2022). A lack of adequate genetic divergence due to the use of only a few genetic markers used

Table 5.1 Summary of data used in speciation rate analyses

Clade	Spp. in tree	Valid spp.	%spp in tree	Crown age	River size[a]	Spp./Ma	TESS	ClaDS
Cetopsidae	44	44	100%	57	Large	0.77	0.13	0.11
Ctenolucoidea	82	82	100%	80	Large	1.03	0.16	0.14
Gymnotiformes	261	271	96%	63	Large	4.15	0.20	0.16
Pimelodoidea	394	409	96%	53	Large	7.38	0.29	0.23
Doradoidea	269	273	99%	66	Large	4.06	0.18	0.17
Curimatoidea	463	471	98%	64	Large	7.27	0.19	0.14
Rivulidae	347	486	71%	43	Small	8.15	0.35	0.22
Poecillioidei	351	364	96%	56	Small	6.27	0.50	0.27
Cichlinae	569	669	85%	35	Small	16.16	0.33	0.25
Characoidea	1371	1571	87%	87	Small	15.79	0.38	0.34
Loricarioidei	1620	1803	90%	86	Small	18.74	0.31	0.23
Sum (Avg)	5771	6443	90%	60		8.43	0.27	0.22

Data from 11 clades spanning two orders of magnitude in species richness, and comprising c. 90% of the whole NFF fauna by species count [a]Streams and small rivers = SO 1–5; Large rivers = SO 6–10. Speciation rates assessed by TESS and ClaDS (see text)

to generate the phylogeny of Chap. 2 could potentially lead to apparent spikes in these rates, because the model identifies branch lengths indicating little divergence. Further, the taxonomic shortfall could potentially cause premature spikes and crashes in inferred rates. Increasing the number of species and genetic loci can reduce these biases. Nevertheless, the patterns of net diversification and relative rate shifts estimated by the model may be informative. There is no obvious reason why the NFF tree would have many short branches at an internal time point due and the large number of species (~6 K) and multiple ($n = 11$) monophyletic groups should buffer the results against these stochastic effects.

5.3 Macroevolutionary Patterns

Lineage diversification rates of the Big Nine NFF clades fluctuated unevenly through the Cenozoic, with each clade exhibiting a distinct pattern of accelerations and decelerations (Fig. 5.2A). The ensemble diversification rates among clades, estimated using the CLaDS birth–death model, vary over a relatively narrow range (i.e., 2.4-fold) from a low of 0.14 in Curimatoidea to a high of 0.34 in Characoidea (Table 5.1; Fig. 5.2B). However, diversification rates assessed as species per Ma vary considerably through time, ranging over 60-fold from 0.03 in Loricaroidei at c. 34 Ma to 1.90 in Rivulidae at c. 20 Ma.

Each clade also exhibits a distinct pattern of accelerated and decelerated diversification regimes across three Cenozoic time intervals of unequal duration: an early phase from c. 60 to 34 Ma, a middle phase from c. 34 to 10 Ma, and a late phase from 10 to 2.6 Ma (Fig. 5.2C). Cichlinae and Rivulidae exhibit the highest absolute diversification rates of all Big Nine NFF clades, in the time period immediately after colonization from Africa c. 40 to 35 Ma. These were followed by the rapid deceleration of Cichlinae after c. 25 Ma and of Rivulidae after c. 20 Ma. Poeciliodea exhibited sustained acceleration from c. 55 to 18 Ma with a slight deceleration from c. 18 to 0 Ma. Curiously, the decelerated diversification of Cichlinae and Poeciliodea took place during the Neogene time frame in which these clades radiated in Central America (Matamoros et al. 2015). Cichlinae and Characoidei exhibit a similar pattern with a generally monotonic declaration over their whole history.

Characoidei is unique among the Big Nine NFF clades in undergoing approximately sustained deceleration throughout the entire time frame of the Upper Cretaceous and Cenozoic (c. 90–0 Ma), interrupted only by a slight acceleration in the early Miocene (c. 22–18 Ma). By contrast, three of the Big Nine NFF clades with many species that inhabit large rivers experienced different diversification regimes across each the three Cenozoic time intervals defined above; Curimatoidei with deceleration from c. 60 to 40 Ma followed by acceleration from c. 40 to 20 Ma and then approximately even diversification from c. 20 to 0 Ma; Gymnotiformes with deceleration from c. 60 to 34 Ma followed by acceleration from c. 22 to 15 Ma and approximately even diversification from c. 15 to 0 Ma; and Pimelodoidea with deceleration from c. 60 to 34 Ma followed by acceleration from c. 34 to 20 Ma and then deceleration from c. 20 to 2.6 Ma.

The other two siluriform clades exhibit distinct diversification patterns across these time intervals; Aspredinoidea with acceleration through most of Cenozoic from c. 60 to 5 Ma followed by rapid deceleration from c. 5 to 2.6 Ma;

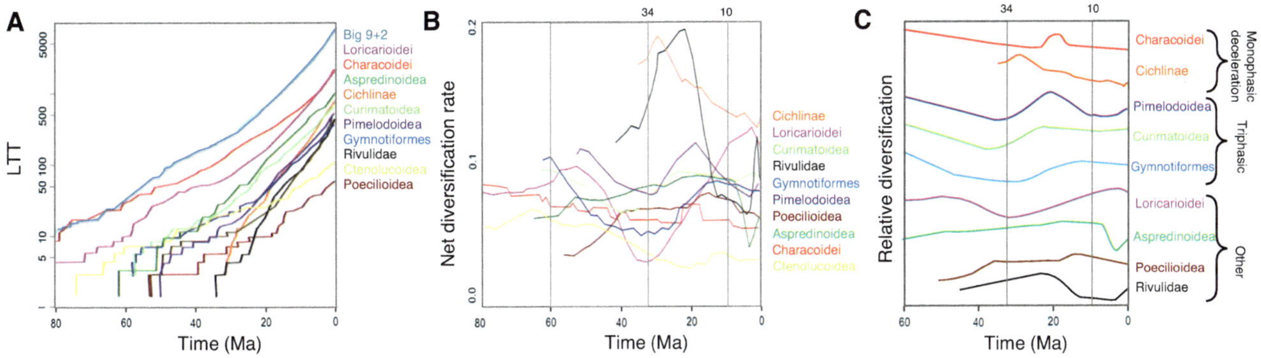

Fig. 5.2 Lineage diversification profiles for Big Nine NFF clades plus two. **A** Log lineages through time (LTT) plots. Cetopsidae not shown due. **B** Ensemble net diversification rates (using CLaDS birth–death model) among clades ranging from a low of 0.14 in Curimatoidea to a high of 0.34 in Characoidea. **C** Smoothed diversification rate plots for the Big Nine NFF clades with amplitudes normalized to compare major phases of acceleration and deceleration. No single pattern fits all clades, with clades accelerating and deceleration at different times. Note a slight general trend toward higher variability in diversification rates toward the recent, possibly due to greater sampling of tip taxa

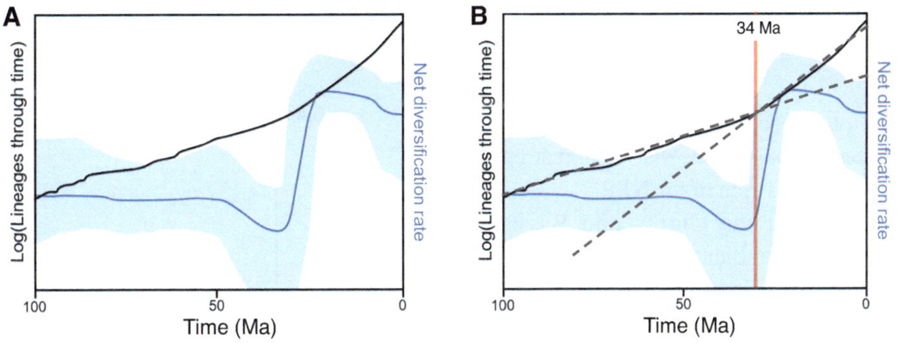

Fig. 5.3 Accelerated diversification of NFF lineage after the Eocene-Oligocene cooling event c. 34 Ma. **A** Superimposed LTT (black) and net diversification (blue) curves for Big Nine NFF clades plus two. **B** Dashed lines illustrating the statistically different slope values (n > 1000, p = 10–[14]) of the LLT curve before (b = 0.30) and after (b = 0.76) 34 Ma, with an angle of c. 20 degrees. The slope values are represented by exponent coefficients (b) of the power law LTT curve on this semi-log plot

Loricarioidei with deceleration from c. 60 to 35 Ma followed by acceleration from c. 34 to 5 Ma. Only Aspreinoidea exhibited accelerated diversification in the time frame c. 60–34 Ma, and the other five clades decelerated in this interval. The rapid rate changes observed in several clades during the Quaternary (c. 2.6–0 Ma) may be artefacts arising from incomplete or inaccurate taxon sampling or poor species delimitation (Craig et al. 2022).

Analysis of slope values for the LTT curve of all NFF lineages shows a dramatic acceleration in net diversification rates of NFF lineage after the Eocene-Oligocene cooling event (c. 34 Ma, Fig. 5.3). On these LTT plots the dashed lines have statistically different slope values ($n > 1000$, $p = 10–$[14]) before ($b = 0.30$) and after ($b = 0.76$) 34 Ma, with an angle of c. 20 degrees, where the exponent coefficients (b) of the LTT curve represented the slope values on this semi-log plot.

Among members of 11 NFF clades examined (i.e., Big Nine Plus Two), representing about 90% of all NFF species, species richness is only weakly correlated with clade age ($R^2 = 0.29$, $P = 0.09$; Fig. 5.4A), failing to support the theory that time is the primary driver of species richness; i.e., contra interpretation of some previous studies (Wiens 2011; Li and Wiens 2019; Miller and Román-Palacios 2021). However, when partitioned into taxa that primarily inhabit smaller rivers with SO 1–5 (i.e., Crenuchidae, Rivulidae, Poeciliidae, Cichlinae, Characoidea, Loricarioidei) and larger rivers with SO 6–10 (Cetopsidae, Ctenolucoidea, Curimatoidea, Aspredinoidea, Gymnotiformes, and Pimelodidae), fishes of small rivers provide weak support for the hypothesis that species richness is influenced by clade age ($R^2 = 0.80$, $P = 0.04$), while larger river fishes are uniformly species-poor and show little to no trend at all ($R^2 = 0.13$, $P = 0.48$; Fig. 5.4B). Thus, we may conclude that

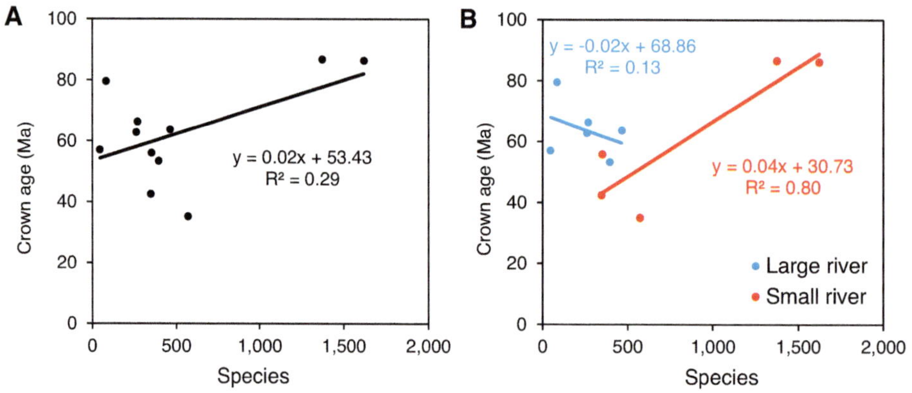

Fig. 5.4 NFF clades adapted to small rivers and streams (Strahler stream orders SO 1–5) diversify faster than those adapted to large rivers (SO 6–10). Linear regressions of species richness against crown clade age for the Big Nine NFF clades plus two additional clades: Cetopsidae and Ctenolucoidea (Ctenolucidae + Lebiasinidae). Note in panel **A** the weak relationship between clade age and species richness for all clades combined ($R^2 = 0.29$, $P = 0.09$), but in panel **B** a stronger correlation between clade age and species richness for clades adapted to small rivers (red circles; $R^2 = 0.80$, $P = 0.04$) than large rivers (blue circles; $R^2 = 0.13$, $P = 0.48$). Note that there are about three times more species in small rivers (4330 species or 73.9% of NFF total) than in large rivers (1532 species or 26.1% NFF total) Data in Table 5.1

the regular accumulation of species through time is a major driver of richness at least among clades that inhabit streams, which are also the most species-rich of the NFF.

To validate our usage of the ClaDS and Bayesian approaches, we compared diversification rates estimated by each of the Big Nine Plus Two clades. Across all clades, we found a strong correlation between the two models. The correlation remained strong among river clades ($R^2 = 0.89$, $P <$ 0.003) and stream clades ($R^2 = 0.14$, $P = 0.56$) independently (Fig. 5.5). Thus, we can conclude that these two models are estimating fundamentally similar things, supporting their usage in inferring macroevolutionary patterns across the major clades of the NFF.

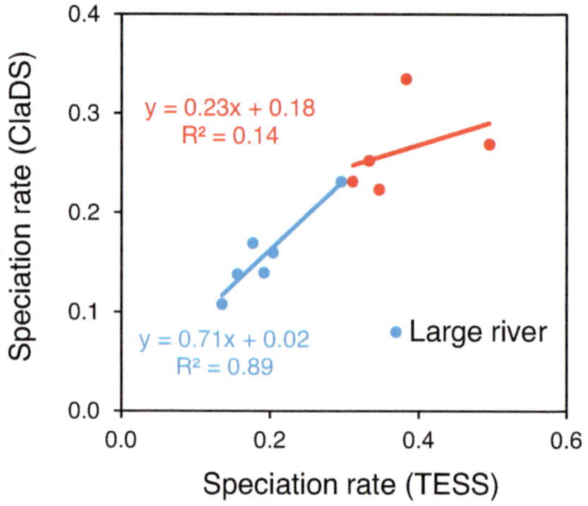

Fig. 5.5 NFF clades adapted to small rivers and streams diversify faster than those adapted to large rivers. Small rivers have Strahler stream orders (SO) 1–5, large rivers SO 6–10. Linear regression of speciation rate inferred by TESS Bayesian model implemented in R against the median per-species rate of speciation inferred by ClaDS likelihood model implemented in Julia

We found that diversification rates estimated by both methods were weakly correlated with the observed species richness values of each clade, following a similar pattern as crown age. This result suggests that the species richness observed in each clade resulted from a regular accumulation of species over time, in contrast to accelerating rates within each clade. We first considered the Bayesian TESS method across all NFF taxa, and found only a weak correlation ($R^2 = 0.19$, $P = 0.19$) (Fig. 5.6A). However, when considering river and fishes of small rivers independently (Fig. 5.6B), it is clear that a correlation exists among river fishes ($R^2 = 0.49$, $P = 0.007$) but not among fishes of small rivers ($R^2 = 0.21$, $P = 0.12$). Similarly, when considering the likelihood-based ClaDS model for all clades, we found a negligible correlation ($R^2 = 0.42$, $P = 0.03$) (Fig. 5.6C), with a trend for large river fishes ($R^2 = 0.35$, $P = 0.60$) than small river fishes ($R^2 = 0.09$, $P = 0.22$; Fig. 5.6D).

Taken together, these results suggest that while species richness in stream clades is not influenced by the diversification rate of that clade in either model, the richness of river clades seems to be influenced by its estimated diversification rate. Compellingly, the opposite pattern is observed in the influence of clade age, where stream clade richness was primarily driven by age, but river clade richness was not. Thus, we conclude that in the major rivers of the humid Neotropics,

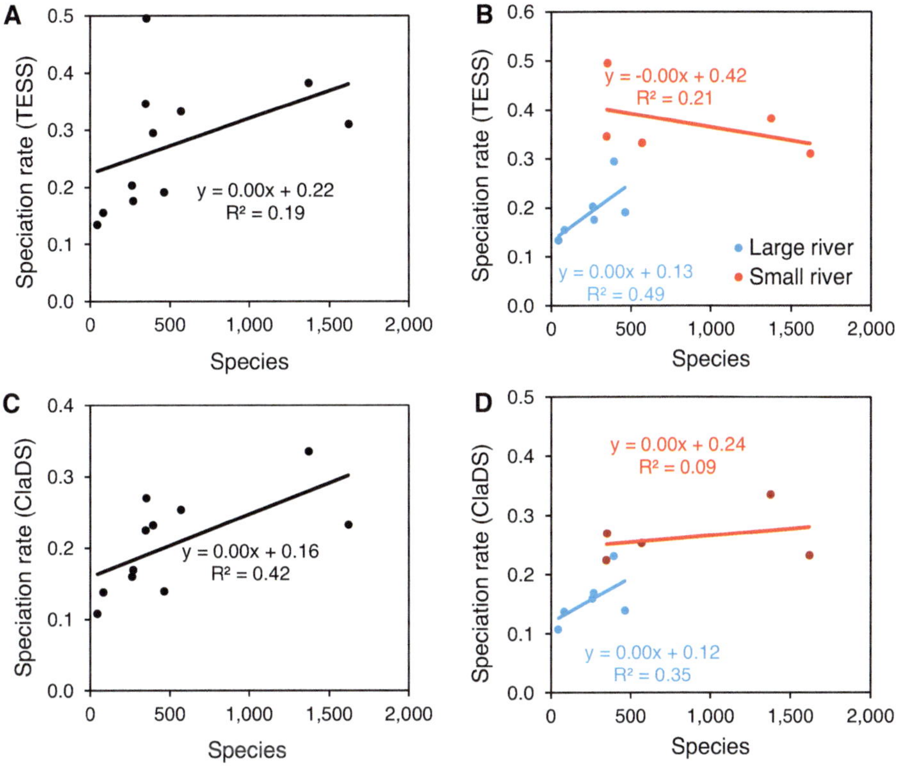

Fig. 5.6 Linear regressions of speciation rates against species richness for the Big Nine NFF clades plus two. Note weak correlations for analyses of all clades combined (left column) as estimated by TESS (**A**) or CLaDS (**B**). Note however stronger correlations for large river clades but not small river clades (right column) as estimated by TESS (**B**) or CLaDS (**D**) Data from Table 5.1

species richness is driven more by diversification rate than clade age, whereas the opposite is true of clades in small rivers. Finally, plots of diversification rates through time for each of these clades on a set of common axes reveal that some clades experience increases in estimated diversification rate from the period of 30 to 20 Ma, while many others remain mostly constant through time.

5.4 Discussion

5.4.1 Stochastic Speciation Hypothesis

We have shown that clade age is a good predictor of species richness in stream fishes, consistent with the hypothesis of a *speciation clock* (Coyne and Orr 1998; McPeek and Brown 2007; Hedges et al. 2015; Henao Diaz et al. 2019). The *Stochastic speciation hypothesis* (SSH; Hedges et al. 2015) proposes that speciation results from the regular accumulation of genetic incompatibilities by drift in allopatry. Allopatry is created by stochastic vicariance events, like orogeny or river capture, which divide populations of a single species into multiple different lineages (Albert et al. 2017, 2020). After sufficient time in isolation, these lineages are no longer capable of interbreeding even if the barrier separating them is removed, after which they become distinct species. Because these earth history events are stochastic, at a large enough time and spatial scales, they generate a stochastically constant rate of speciation, with some variation among smaller subsets of the fauna (e.g., different species and genera).

This stochastic speciation hypothesis is similar to the *Intermediate-rate capture* hypothesis (ICH, Val et al. 2022). Under this hypothesis, river captures in lowland rivers occur at "just right" spatial and temporal scales to generate more divergences than convergences, and this drive up the total number of species. The effect of river capture on NFF diversification has been documented in many studies (e.g., Albert et al. 2011, 2018; Roxo et al. 2014; Tagliacollo et al. 2015; Camelier et al. 2018; Bagley et al. 2021; Leitão et al. 2021; Pio and Carvalho 2021; Val et al. 2022; Boschman et al. 2023).

5.4.2 Major Paleogeographic Events

By contrast, fishes that inhabit large river channels and floodplains have much greater vagility and geographic ranges, and are therefore less susceptible to smaller-scale events. River capture events are known to occur with a frequency inversely proportional to the area they affect (Albert et al. 2018). A fish clade with access to the entire Amazon basin may only have ever experienced the impact of a single large river capture event (Lundberg et al. 1998), while remaining free to traverse the entire continent for the rest of its history. In such riverine lineages, we hypothesize that the scale of speciation in time and space may simply be too great enough for the stochastic occurrence of vicariance events to have an averaging effect as it does at smaller scales, where many more events occur. As we see in our results, this minimizes the impact of time alone on species richness, leaving room for explanations based on other properties of the organism or environment.

In estimates of the pattern of diversification through time, we see some evidence of the impacts of major paleogeographic events on NFF clades (Chap. 1). The Cretaceous-Paleogene extinction (66 Ma) event and subsequent early Paleogene (c 60–50 Ma) radiations on the upland shields were likely too old to have impacted these groups, as most of them had not yet emerged, or were represented by only a handful of species. The Eocene-Oligocene (34 Ma) cooling and associated extinctions and concurrent separation of Amazon and Atlantic forests do coincide with an acceleration in diversification among the Loricarioidei and Crenuchidae, but the latter in particular was still a very young clade at this time (crown age 63 Ma).

The Miocene-Pliocene fragmentation of the Sub-Andean Foreland basin (23–3 Ma), which drove many vicariance events across the continent, may have impacted the accelerated rates of diversification observed in many fishes of small rivers such as the Cichlinae, Loricarioidei, and Rivulidae, but it is important to note here that terminal increases in rate in such phylodynamic analyses are commonly associated with incomplete taxonomic sampling, which may have been an issue despite our attempts to account for it (Craig et al. 2022). The Early-Middle Miocene (22–10 Ma) megathermal and formation of the Pebas mega-wetland is commonly presented as a driver of diversification in the NFF (Albert and Reis 2011), and like the fragmentation of the Sub-Andean Foreland basin (23–3 Ma), it coincides with diversification rate shifts in several clades (Roxo et al. 2014; Tagliacollo et al. 2015; Albert et al. 2018). The uplift of Northern Andean cordilleras and final separation of cis- and trans-Andean river basins in the Late Miocene (c 12–10; Albert et al. 2006), and the Late Miocene (10–4.5) formation of modern transcontinental Amazon River and Great Amazonian Biotic Interchange (Albert et al. 2021), also coincide with terminal rate shifts in these clades.

5.5 Conclusions

A comprehensive analysis of the diversification rates among major clades of the NFF fauna has not been possible until now. As our taxonomic and phylogenetic knowledge grows, and with inputs from novel methods in phylodynamics and macroevolution, we are reaching the point where it is possible to make testable hypotheses and draw supportable conclusions about the causes of the remarkable species richness of the NFF fauna. We find that while the theory of regular clock-like speciation is well supported for fishes of streams and small rivers, this is not the case for clades that inhabit the channels and floodplains of large rivers. We propose that the greater frequency of smaller-scale stochastic vicariance events affecting *terra firme* rivers may have resulted in more diversification overall, whereas in the larger, more interconnected habitats of large rivers, species experience vicariance events less often, creating a more irregular pattern at this larger scale (Barreto et al. 2022; Rabosky et al. 2012).

References

Albert JS, Carvalho TP (2011) Neogene assembly of modern faunas. In: Albert JS, Reis RE (eds) Historical biogeography of neotropical freshwater fishes. University of California Press, Berkeley, CA, pp 119–136

Albert JS, Reis RE (2011) Historical biogeography of neotropical freshwater fishes. Univ of California Press, Berkeley, CA

Albert JS, Petry P, Reis RE (2011) Major biogeographic and phylogenetic patterns. In: Historical biogeography of neotropical freshwater fishes. University of California Press, pp 3-20

Albert JS, Bernt MJ, Fronk AH, Fontenelle JP, Kuznar SL, Lovejoy NR (2021) Late neogene megariver captures and the Great Amazonian biotic interchange. Glob Planet Change 205:1–6

Albert JS, Craig JM, Tagliacollo VA, Petry P (2018) Upland and lowland fishes: a test of the river capture hypothesis. In: Hoorn C, Perrigo A, Antonelli A (eds) Mountains, climate and biodiversity, pp 273-294

Albert JS, Destouni G, Duke-Sylvester SM, Magurran AE, Oberdorff T, Reis RE, Winemiller KO, Ripple WJ (2021) Scientists' warning to humanity on the freshwater biodiversity crisis. Ambio 50(1):85–94

Albert JS, Lovejoy NR, Crampton WG (2006) Miocene tectonism and the separation of cis-and trans-Andean river basins: evidence from neotropical fishes. J South AmEarth Sci 21(1–2):14–27

Albert JS, Schoolmaster JrDR, Tagliacollo V, Duke-Sylvester SM (2017) Barrier displacement on a neutral landscape: toward a theory of continental biogeography. Syst Biol 66(2):167–182

Albert JS, Tagliacollo VA, Dagosta F (2020) Diversification of neotropical freshwater fishes. Annu Rev Ecol Evol Syst 51:27–53

Arbour JH, Santana SE (2017) A major shift in diversification rate helps explain macroevolutionary patterns in primate species diversity. Evolution 71(6):1600–1613

Bagley JC, de Aquino PDPU, Hrbek T, Hernandez-Rangel S, Langeani F, Colli GR (2021) Using ddRAD-seq phylogeography to test for genetic effects of headwater river capture in suckermouth armored catfish (Loricariidae: *Hypostomus*) from the central Brazilian shield. bioRxiv 2021-04

Baker WJ, Couvreur TLP (2013) Global biogeography and diversification of palms sheds light on the evolution of tropical lineages. II. Diversification history and origin of regional assemblages. J Biogeogr 40(2):286–298

Barreto SB, Knowles LL, Mascarenhas R, Affonso PRADM, Batalha-Filho H (2022) Drainage rearrangements and in situ diversification of an endemic freshwater fish genus from north-eastern Brazilian rivers. Freshwater Biol 67(5):759–773

Bezanson J, Edelman A, Karpinski S, Shah VB (2017) Julia: A fresh approach to numerical computing. SIAM Rev 59(1):65–98

Boschman LM, Carraro L, Cassemiro FA, de Vries J, Altermatt F, Hagen O et al (2023) Freshwater fish diversity in the western Amazon basin shaped by Andean uplift since the Late Cretaceous. Nat Ecol Evol 7(12):2037–2044

Boucher FC, Verboom GA, Musker S, Ellis AG (2017) Plant size: a key determinant of diversification?. New Phytol 216(1):24–31

Camelier P, Menezes NA, Costa-Silva GJ, Oliveira C (2018) Molecular and morphological data of the freshwater fish *Glandulocauda melanopleura* (Characiformes: Characidae) provide evidences of river captures and local differentiation in the Brazilian Atlantic Forest. PLoS One 13(3):e0194247

Cassemiro FA, Albert JS, Antonelli A, Menegotto A, Wüest RO, Cerezer F et al (2023) Landscape dynamics and diversification of the megadiverse South American freshwater fish fauna. Proc Natl Acad Sci U S A 120(2):e2211974120

Cerezer FO, Dambros CS, Coelho MT, Cassemiro FA, Barreto E, Albert JS et al (2023) Accelerated body size evolution in upland environments is correlated with recent speciation in South American freshwater fishes. Nat Commun 14(1):6070

Claramunt S, Cracraft J (2015) A new time tree reveals Earth history's imprint on the evolution of modern birds. Sci Adv 1(e1501005 11):1–13

Core Development Team R (2020) A language and environment for statistical computing (Version 3). R Foundation for Statistical Computing, Vienna, Austria. http://www.r-project.org

Coyne JA, Orr HA (1998) The evolutionary genetics of speciation. Philos Trans R Soc London Ser B 353(1366):287–305

Craig JM, Kumar S, Hedges SB (2022) Limitations of phylogenomic data can drive estimated speciation rate shifts. Mol Biol Evol 39(3):msac038. https://doi.org/10.1093/molbev/msac038

Fabreti LG, Höhna S (2022) Bayesian inference of phylogeny is robust to substitution model over-parameterization. bioRxiv 2022.02.17.480861

Fontanillas E, Welch JJ, Thomas JA, Bromham L (2007) The influence of body size and net diversification rate on molecular evolution during the radiation of animal phyla. BMC Evol Biol 7(95):1–12

Grenfell BT, Pybus OG, Gog JR, Wood JL, Daly JM, Mumford JA, Holmes EC (2004) Unifying the epidemiological and evolutionary dynamics of pathogens. Science 303(5656):327–332

Han TS, Zheng QJ, Onstein RE, Rojas-Andrés BM, Hauenschild F, Muellner-Riehl AN, Xing YW (2020) Polyploidy promotes species diversification of Allium through ecological shifts. New Phytol 225(1):571–583

Hedges SB, Marin J, Suleski M, Paymer M, Kumar S (2015) Tree of life reveals clock-like speciation and diversification. Mol Biol Evol 32(4):835–845

Henao Diaz LF, Harmon LJ, Sugawara MTC, Miller ET, Pennell MW (2019) Macroevolutionary diversification rates show time dependency. Proc Natl Acad Sci U S A 116(15):7403–7408

Höhna S (2015) The time-dependent reconstructed evolutionary process with a key-role for mass-extinction events. J Theor Biol 380:321–331

Höhna S, May MR, Moore BR (2016) TESS: an R package for efficiently simulating phylogenetic trees and performing Bayesian inference of lineage diversification rates. Bioinformatics 32(5):789–791

Hou Z, Sket B, Fišer C, Lia S (2011) Eocene habitat shift from saline to freshwater promoted Tethyan amphipod diversification. Proc Natl Acad Sci U S A 108(35):14533–14538

Hughes C, Eastwood R (2006) Island radiation on a continental scale: exceptional rates of plant diversification after uplift of the Andes. Proc Natl Acad Sci U S A 103(27):10334–10339

Jansson R, Rodríguez-Castañeda G, Harding LE (2013) What can multiple phylogenies say about the latitudinal diversity gradient? A new look at the tropical conservatism, out of the tropics, and diversification rate hypotheses. Evolution 67(6):1741–1755

Koenen EJM, de Vos JM, Atchison GW, Simon MF, Schrire BD, de Souza ER, de Queiroz LP, Hughes CE (2013) Exploring the tempo of species diversification in legumes. S Afr J Bot 89:19–30

Leitão CSDS, Souza ÉM, Santos CH, Val P, Val AL, Almeida-Val VM (2021) River reorganization affects populations of dwarf cichlid species (*Apistogramma* genus) in the lower Negro River, Brazil. Front Ecol Evol 9:760287

Li H, Wiens JJ (2019) Time explains regional richness patterns within clades more often than diversification rates or area. Am Nat 193(4):514–529

Linnaeus C (1758) Systema Naturae per Regna Tria Naturae secundum Classes, Ordinus: vol. Editio dec (ED Tomus I (ed), p 824

Liu B, Le CT, Barrett RL, Nickrent DL, Chen Z, Lu L, Vidal-Russell R (2018) Historical biogeography of Loranthaceae (Santalales): diversification agrees with emergence of tropical forests and radiation of songbirds. Mol Phylogenet Evol 124:199–212

Lundberg JG, Marshall LG, Guerrero J, Horton B, Malabarba MCSL, Wesselingh F (1998) The stage for neotropical fish diversification a history of tropical South American rivers. In: Malabarba LR, Reis RE, Vari RP, Lucena ZMS, Lucena CAS (eds) Phylogeny and classification of neotropical fishes, 27, 13-48. EDIPUCRS (Editora da Pontifícia Universidade Católica do Rio Grande do Sul)

Maliet O, Morlon H (2021) Fast and accurate estimation of species-specific diversification rates using data augmentation. Syst Biol 71(2):353–366

Maliet O, Hartig F, Morlon H (2019) A model with many small shifts for estimating species-specific diversification rates. Nat Ecol Evol 3(7):1086–1092

Matamoros WA, McMahan CD, Chakrabarty P, Albert JS, Schaefer JF (2015) Derivation of the freshwater fish fauna of Central America revisited: Myers's hypothesis in the twenty-first century. Cladistics 31(2):177–188

May MR, Höhna S, Moore BR (2016) A Bayesian approach for detecting the impact of mass-extinction events on molecular phylogenies when rates of lineage diversification may vary. Methods Ecol Evol 7(8):947–959

McPeek MA, Brown JM (2007) Clade age and not diversification rate explains species richness among animal taxa. Am Nat 169(4):1–10

Miller E, Román-Palacios C (2021) Evolutionary time explains the global distribution of freshwater fish diversity. Glob Ecol Biogeogr 30:749–763

Mitter C, Farrell B, Wiegmann B (1988) The phylogenetic study of adaptive zones: has phytophagy promoted insect diversification?. Am Nat 132(1):107–128

Murray EA, Bossert S, Danforth BN (2018) Pollinivory and the diversification dynamics of bees. Biol Lett 14(11):0–3

Phillimore FC, Orme DL, Owens IPF (2006) Ecology predicts large-scale patterns of phylogenetic diversification in birds. Am Nat 168(2):220

Pio NL, Carvalho TP (2021) Evidence on the paleodrainage connectivity during Pleistocene: phylogeography of a hypoptopomatine endemic to southeastern Brazilian coastal drainages. Neotrop Ichthyol 19:e200128

Pontarp M, Wiens JJ (2017) The origin of species richness patterns along environmental gradients: uniting explanations based on time, diversification rate and carrying capacity. J Biogeography 44(4):722–735

Price SA, Hopkins SSB, Smith KK, Roth VL (2012) Tempo of trophic evolution and its impact on mammalian diversification. Proc Natl Acad Sci U S A 109(18):7008–7012

Quintero I, Jetz W (2018) Global elevational diversity and diversification of birds. Nature 555(7695):246–250

Rabosky DL (2010) Primary controls on species richness in higher taxa. Syst Biol 59(6):634–645

Rabosky DL (2012) Testing the time-for-speciation effect in the assembly of regional biotas. Methods Ecol Evol 3(2):224–233

Rabosky DL (2020) Speciation rate and the diversity of fishes in freshwaters and the oceans. J Biogeogr 47(6):1207–1217

Rabosky DL, Santini F, Eastman J, Smith SA, Sidlauskas B, Chang J, Alfaro ME (2013) Rates of speciation and morphological evolution are correlated across the largest vertebrate radiation. Nat Commun 4(1):1958

Rabosky DL, Slater GJ, Alfaro ME (2012). Clade age and species richness are decoupled across the eukaryotic tree of life

Reis RE, Albert JS, Di Dario F, Mincarone MM, Petry P, Rocha LA (2016) Fish biodiversity and conservation in South America. J Fish Biol 89(1):12–47

Ricklefs RE (2006) Global variation in the diversification rate of passerine birds. Ecology 87(10):2468–2478

Roelants K, Gower DJ, Wilkinson M, Loader SP, Biju SD, Guillaume K, Moriau L, Bossuyt F (2007) Global patterns of diversification in the history of modern amphibians. Proc Natl Acad Sci U S A 104(3):887–892

Roxo FF, Albert JS, Silva GS, Zawadzki CH, Foresti F, Oliveira C (2014) Molecular phylogeny and biogeographic history of the armored neotropical catfish subfamilies Hypoptopomatinae, Neoplecostominae and Otothyrinae (Siluriformes: Loricariidae). PLoS One 9(8):e105564

Sanderson MJ, Donoghue MJ (1994) Shifts in diversification rate with the origin of angiosperms. Science 264(5165):1590–1593

Sayol F, Lapiedra O, Ducatez S, Sol D (2019) Larger brains spur species diversification in birds. Evolution 73(10):2085–2093

Seehausen O, Wagner CE (2014) Speciation in freshwater fishes. Annu Rev Ecol Evol Syst 45:621–651

Tagliacollo VA, Roxo FF, Duke-Sylvester SM, Oliveira C, Albert JS (2015) Biogeographical signature of river capture events in Amazonian lowlands. J Biogeogr 42(12):2349–2362

Title PO, Rabosky DL (2017) Do macrophylogenies yield stable macroevolutionary inferences? An example from squamate reptiles. Syst Biol 66(5):843–856

Upham NS, Esselstyn JA, Jetz W (2019) Ecological causes of uneven diversification and richness in the mammal tree of life. bioRxiv 504803. https://doi.org/10.1101/504803

Utami CY, Sholihah A, Condamine FL, Thébaud C, Hubert N (2022) Cryptic diversity impacts model selection and macroevolutionary inferences in diversification analyses. Proc Biol Sci 289(1987):20221335

Val P, Lyons NJ, Gasparini N, Willenbring JK, Albert JS (2022) Landscape evolution as a diversification driver in freshwater fishes. Front Ecol Evol 9:788328

van der Sleen P, Albert JS (2022) Patterns in freshwater fish diversity. Ref Mod Earth Syst Environ Sci 26(3):894–907

Wang Q, Wu S, Su X, Zhang L, Xu X, Lyu L, Cai H, Shrestha N, Liu Y, Wang W, Wang Z (2018) Niche conservatism and elevated diversification shape species diversity in drylands: Evidence from Zygophyllaceae. Proc R Soc B 285(1890):1–9

Wiens JJ (2011) The causes of species richness patterns across space, time, and clades and the role of "ecological limits". Q Rev Biol 86(2):75–96

Wilsone EO (2003) Pheidole in the new world: a dominant, hyperdiverse ant genus. Harvard University Press

The First Ecological Trait Space of Neotropical Freshwater Fishes

6

Kevin T. Torgersen, William G. R. Crampton, Mark A. Genung, Milton Tan, Kirk O. Winemiller and James S. Albert

Abstract

This chapter presents the first comprehensive analysis of ecological traits (*ecotraits*) among *Neotropical Freshwater Fishes* (NFF), using an extensive ecotrait matrix (*ecomatrix*) compiled from published data and expert input for 6345 valid species across 880 genera, 97 families, and 40 orders. The matrix includes data on 42 ecophysiological traits, habitat use across water chemistry and spatial scales, diet, and behavior. Principal component analysis (PCA) of this matrix revealed that stream order is a better predictor of species distribution in ecological space than elevation, and that body size correlates with habitat preference, with large-bodied species found in large rivers, and smaller species in smaller streams and floodplains. Only 17% of NFF species are adapted to large rivers, highlighting the specialization and potential vulnerability of most species. This ecological trait space (*ecospace*) offers a foundational tool for studying the adaptive diversity and conservation needs of Neotropical freshwater fishes amid environmental change.

K. T. Torgersen (✉) · M. A. Genung · J. S. Albert
University of Louisiana at Lafayette, Lafayette, LA, USA
e-mail: kevintorgersen@gmail.com

M. A. Genung
e-mail: mark.genung@louisiana.edu

J. S. Albert
e-mail: jsalbe@gmail.com

W. G. R. Crampton
University of Central Florida, Orlando, FL, USA
e-mail: crampton@ucf.edu

M. Tan
Illinois Natural History Survey, Urbana, IL, USA
e-mail: miltont@illinois.edu

K. O. Winemiller
Texas A&M University, College Station, TX, USA
e-mail: k-winemiller@tamu.edu

6.1 Introduction

6.1.1 Functional Traits and Biodiversity

> Evolutionary processes are usually multidimensional and complicated; the models we work with represent simple theories about what might be important, and it would be naive to expect them to capture all the relevant biology. Hence, even good models are not expected to explain all or even most of the variation. Most of the "error" in evolutionary model fitting is likely to be "biological" error and not "statistical' measurement error. (Hansen 2024).

Biodiversity may be defined broadly as the diversity of genes, traits, species, habitats, and landscapes in the biosphere (Seddon et al. 2016; Magurran et al. 2018). Estimating the species richness of a clade or region is a central task in biodiversity assessment, yet raw species counts represent only one of many biodiversity dimensions. *Functional traits* of organisms provide critical information for understanding community assembly, species coexistence, and biodiversity loss (Dawson et al. 2021). Trait-based approaches to the study of ecology and evolution illuminate how organisms interact with their environments and each other, which is essential for understanding community structure and ecosystem functioning, both now and in the future (Klausmeier et al. 2020). Functional traits include the ecological, life-history, morphological, physiological, and behavioral adaptations commonly used to categorize species into functional groups and guilds (Goldstein and Meador 2005; Frimpong and Angermeier 2009,2010).

6.1.2 Ecotraits

Among animals generally, ecophysiological traits (*ecotraits*) are linked to developmental stage and body size, habitat utilization, trophic utilization, and reproductive behaviors. Together, these traits inform conservation strategies aimed at preserving biodiversity and ecosystem functionality (Brejão et al.

J. S. Albert and R. E. Reis (eds.), *Ecology and Evolution of Amazonian Fishes*, Fascinating Life Sciences,
https://doi.org/10.1007/978-3-032-09328-8_6

2013; Costello et al. 2015; Galán-Acedo et al. 2019; Heilpern et al. 2022). With a few notable exceptions (e.g., Frimpong and Angermeier 2009; Winemiller et al. 2015; Torgersen et al. 2023a), most previous studies on functional trait diversity within freshwater fishes have used linear measurements to assess body proportions (e.g., body aspect ratio, relative eye size, and gape size) as proxies for functionally adaptive traits (Hubbs et al. 2004; Oliveira et al. 2010; Villéger et al. 2017; Su et al. 2019; Brosse et al. 2021; Friedman et al. 2022). Here, we define an *ecotrait* as any heritable (or phenotypically plastic) physiological or behavioral aspect of the integrated organismal phenotype (Pigliucci 2003; Felice et al. 2018) that constrains resource acquisition and utilization (i.e., habitat, diet, life history). There is a large and growing literature on the role of ecotraits in the evolutionary diversification of NFF clades (e.g., Burress 2015; Correa et al. 2015; Bloom and Lovejoy 2017; Evans et al. 2019; Melo et al. 2021; Kolmann et al. 2021,2022) and the formation of species-rich NFF assemblages (e.g., Dias et al. 2016; Gomes et al. 2023; Leitão et al. 2016; Lima et al. 2018; Soares et al. 2023; Baldasso et al. 2024).

6.1.3 Goals of This Chapter

This chapter presents the first dataset of ecotraits scored for all valid NFF species. These ecotraits include aspects of the *phenotype* related to habitat and diet utilization, and behaviors related to life-history activities. This study aims to characterize the range of physiological and behavioral adaptations among NFF species for utilizing and partitioning environmental and food resources in time and space, and life history specializations associated with seasonal, climatic, and elevational gradients.

6.2 Methods

6.2.1 List of Valid NFF Species

The list of authenticated species names was compiled by consulting recent taxonomic reviews, and the list of valid species names available at Eschmeyer's online Catalog of Fishes (Eschmeyer et al. 2024). Valid species names exclude synonyms and were assigned to a genus name based on reliable sources, including original species descriptions (e.g., Vanegas-Ríos et al. 2018; Gamarra et al. 2019; Craig et al. 2019), taxonomic revisions (e.g., Crampton et al. 2016; Villa-Navarro et al. 2017; Ferrer and Malabarba 2020), checklists (e.g., Ferraris et al. 2017; Ota et al. 2018; Ribeiro et al. 2019), books (e.g., Reis et al. 2003; Queiroz et al. 2013; van der Sleen and Albert 2017), and online catalogs, such as FishBase, SpeciesLink, and GBIF, among others. In this process, the spelling of scientific names was verified, synonyms were identified and corrected, and nonnative species to the Neotropical region were excluded. Following this step, a preliminary list containing all compiled species was made available to collaborators for validation and refinement.

Dozens of ichthyologists specializing in different taxonomic groups contributed to the consolidation of this list with updated information, attesting to the taxonomic validity of the nominal species, the inclusion of missing species, and other relevant aspects. To the best of our knowledge, this list encompasses all valid species names from the Neotropical region as of December 31, 2021. Published literature references with ecotrait data were compiled using Google Scholar for taxa representing all valid NFF genus names and 6345 species names representing 32% of the NFF total.

6.2.2 Ecotrait Classification

The ecotrait categories used in this study for NFF species (Table 6.1) were defined using criteria developed in previous studies of Neotropical aquatic habitats, fish trophic guilds and fish behavior related to ecological functions (Saint-Paul and Soares 1987; Winemiller 1991; Zuanon and Ferreira 2008; Crampton 2011; Junk et al. 2011; Winemiller et al. 2015; Duarte et al. 2019; Val 2019). In some cases, it was necessary to modify the categories and states based on conditions observed under varied freshwater regions. For example, the ecotrait of "blackwater utilization" in this study is restricted to fishes that inhabit waters with both high organic content and low pH, which distinguishes the distinct fauna of acidophilic Amazonian blackwater rivers (e.g., Negro River, Amazonas, Brazil; Val et al. 1998) from that of the darkly colored waters with near-neutral pH of floodplain lakes along the Amazon River (Junk 1984), or of the Negro River of Uruguay (Beretta-Blanco and Carrasco-Letelier 2021). The ecotraits represent aspects of body size, four physiological attributes related to habitat utilization of water chemistry, 21 related to physical habitat structure, ten related to diet (i.e., trophic guilds), and seven behavioral traits related to daily or seasonal life-history activities. Quantitative data on life history traits of NFF species, including fecundity, survivorship, and generation time, are limited to commercially important species in the fisheries or ornamental fish trades, and these life history data are unknown for a majority of NFF species, which represents an important knowledge shortfall (Winemiller 2005; Hortal et al. 2015).

Body size is a fundamental feature of the ecological phenotype in fishes and other animals, affecting metabolic rate and other physiological processes, behavior, resource utilization, locomotion, and life history (Weitzman and Vari 1988; McNab 2002). NFF body size was assessed as the maximum recorded standard length (SL), or total length (TL) for taxa without a caudal fin, scored to four size categories: large size as 100–610 cm (2.00–2.79 $\log_{10}$ cm); medium size as 10–99 cm

Table 6.1 Abbreviated descriptions of 42 ecophysiological traits scored for all species of Neotropical freshwater fishes. ET#, ecotrait number; total species numbers (spp.) and proportions (% spp.) of species scored as possessing each ecotrait. Percentages do not add to 100% as some species are scored as modally present in more than one ecotrait state

Category	ET#	Ecotrait	Spp.	% Spp.
Adult body size	1	Log cm SL (or TL): large, 2.00–2.79; medium, 1.00–1.99; small, 0.43–0.99; miniature, −0.05 to 0.41	5338	84.1
Water chemistry	2	Clearwater: high transparency, low sediments, and dissolved ions, near neutral pH (5.6–8.0)	4690	73.9
	3	Whitewater: low transparency, high sediments, and dissolved ions, near neutral pH (5.6–8.0)	1742	27.5
	4	Blackwater: low transparency, low sediments, and dissolved ions, low pH (3.5–5.5)	1185	18.7
	5	Estuarine: variable salinity from fresh to brackish	263	4.1
Macrohabitat	6	Low-gradient (slope less than 2%) streams and small rivers (Strahler stream orders 1–5)	3998	63.0
(100–1000 m)	7	High-gradient (slope more than 2%) riffles in streams and rapids in small rivers	1257	19.8
	8	Ancient lakes, more than 10,000 ka; e.g., volcanic calderas, tectonic rifts	68	1.1
	9	Hypogean caves and aquifers	48	0.8
	10	Low-gradient channels and margins of large rivers (Strahler stream orders 6–10)	1561	24.6
	11	High-gradient rapids of large rivers	150	2.4
Mesohabitat	12	Lotic (flowing) streams (English), *quebradas* (Spanish), *igarapés* (Portuguese)	4422	69.7
(10–100 m)	13	Lentic (standing) waters of small rivers; pools and ponds	539	8.5
	14	Seasonally dry (e.g., ephemeral) aquatic habitats	302	4.8
	15	Benthic (bottom) of large river channels	730	11.5
	16	Pelagic (midwater) of large river channels	397	6.3
	17	Littoral (margins) of large river channels	892	14.1
	18	Seasonal floodplains (oxbow lakes and flooded forest) of large rivers	1051	16.6
Microhabitat	19	Benthic, unspecified	1578	24.9
(1–10 m)	20	Benthic, mud, and silt	374	5.9
	21	Benthic, sand	588	9.3
	22	Benthic, gravel/rocks	1262	19.9
	23	Benthic, submerged wood, and vegetation	710	11.2
	24	Benthic, leaf litter, and packs	239	3.8
	25	Epibenthic in water-column above bottom	2816	44.4
	26	Epibenthic at or near surface	418	6.6
Diet:	27	Benthic microvores, small animals 0.5–5 mm	4143	65.3
(Trophic guild)	28	Surface invertivores: allochthonous arthropods (insects, arachnids)	218	3.4
	29	Pelagic planktivores: autochthonous arthropod and fish larvae	418	6.6
	30	Benthic detritivores: nonliving decomposing organic (mostly plant) matter	205	3.2
	31	Benthic algivores: living algaes and bacterial biofilm	1003	15.8
	32	Folivores: submerged plant leaves, riverweeds	57	0.9
	33	Frugivores: fruits and seeds	34	0.5
	34	Piscivores: fish tissue	527	8.3
	35	Omnivores: both plant and animal tissues	1317	20.8
	36	Ectoparasites, scales, and mucous	114	1.8
Behavior	37	Potamodromous: longitudinal river migrations more than 100 km	844	13.3
	38	Viviparous: live birth	276	4.3
	39	Aestivation: prolonged infaunal torpor (i.e., lowered metabolic activity) and desiccation	4	0.1
	40	Venomous: phenotypes that deliver injected toxin(s) and associated aposematic behaviors	912	14.4
	41	Passive electroception: sensory use of weak (mV) ambient electric fields	2685	42.3
	42	Active electroreception and electrogeneration using electric organs	261	4.1

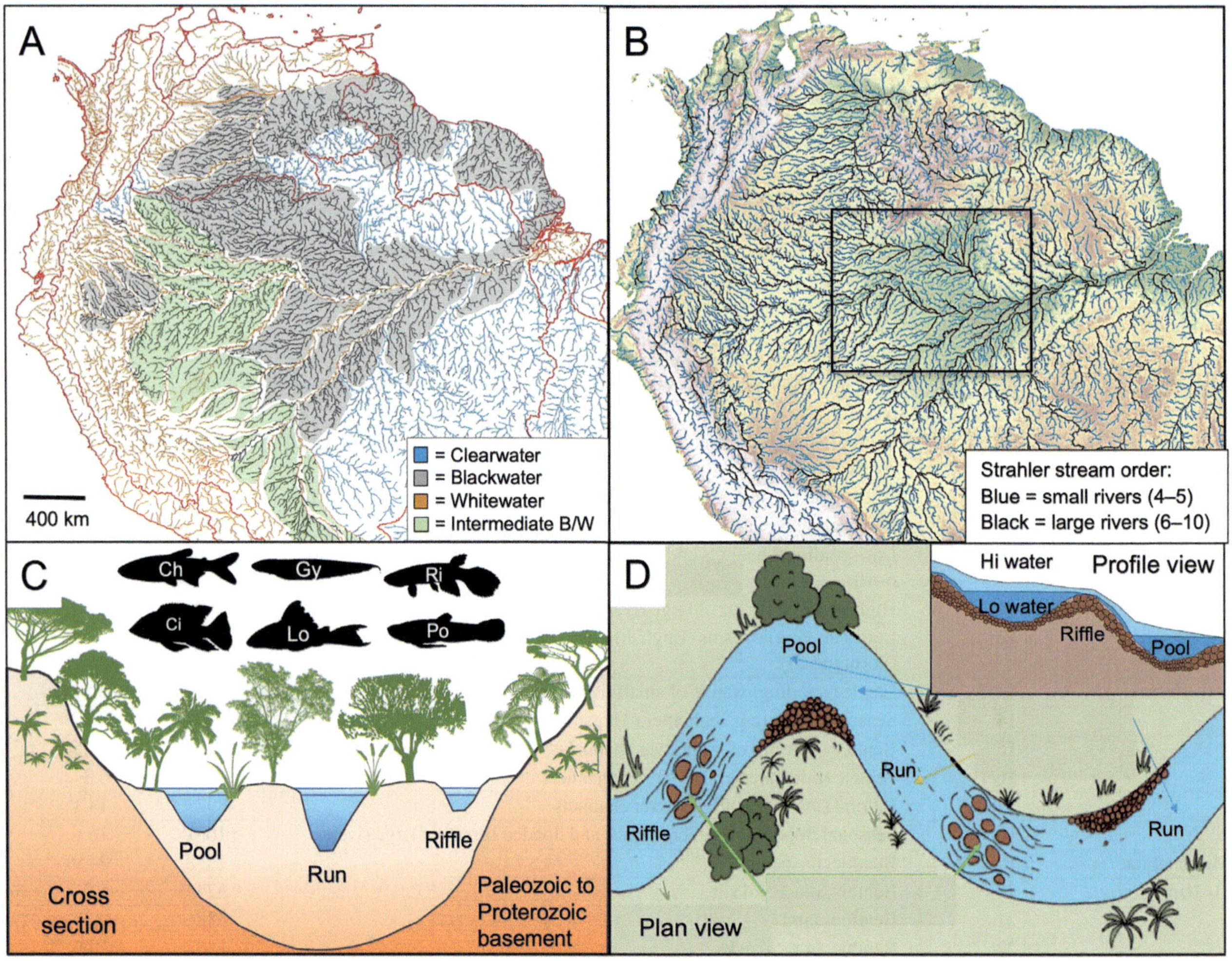

Fig. 6.1 Major riverscape features used to score habitats of NFF species. See Table 6.1 for numerical limits of ecotrait categories. (A) Water chemistry based on combination of color, sediment load, dissolved ions, and pH. (B) Strahler stream order (SO) within tributary hierarchy; blue = small rivers (SO 4–5; SO 1–3 omitted for clarity); black = large rivers (SO 6–10). Black box indicated location of map in Fig. 6.2A. (C) Schematic cross section of small rivers (SO 1–5) illustrating some of the main structural microhabitats (1–10 m) and mesohabitats (10–100 m), and ecologically dominant fish taxa (silhouettes). Ch = Characidae, Ci = Cichlidae, Lo = Loricariidae, Ri = Rivulidae, Po = Poeciliidae, Gy = Gymnotidae, Pi = Pimelodidae, Ap = Apteronotidae, Cu = Curimatidae, Do = Doradidae. (D) Plan and profile views of same river in panel C. Note longitudinal sequence of swift shallow flow in rifles, slow deep flow in pools, and smooth unbroken flow in runs

(1.00–1.99 $\log_{10}$ cm); small size as 2.7–9.9 cm (0.43–0.99 $\log_{10}$ cm); and miniature size as 0.9–2.6 cm (0.05–0.41 $\log_{10}$ cm). Maximum known standard or total lengths are highly correlated with body mass in most fishes and is a widely used metric of overall body size in fishes (Albert and Johnson 2012). Body size data are available for 5400 NFF species representing 85.1% of the fauna and all the valid NFF genera. Body size data for other NFF species were scored by the editors by phylogenetic imputation based on the body sizes of their closest relatives. In several studies of NFF faunas, body size has been shown to be correlated with trophic position in predator but not prey species (e.g., Layman et al. 2005).

The ecotraits related to water chemistry (Fig. 6.1A) are physiological tolerances to inhabit clearwaters with high transparency, low sediment load, low concentrations of dissolved ions, and near neutral pH (5.6–8.0), whitewaters with low transparency, high sediment load, high dissolved ions, near neutral pH (5.6–8.0), blackwaters with low transparency, low sediment load, low dissolved ions, low pH (3.5–5.5), and estuarine waters located at river mouths and coastal estuaries with variable salinity values from fresh to brackish. Ecotraits of physical habitat structure represent physiological tolerances to distinct aspects of environmental variation at three levels of spatial scale: macrohabitat (100–1000 m), mesohabitat (10–100 m), and microhabitat (1–10 m). Within each of these scales, habitats were characterized using environmental variables of Strahler stream order (1–5 streams and small rivers, 6–10 large rivers; Figs. 6.1B and 6.2) as a metric for total aquatic habitat volume (Strahler 1957). Additional mesohabitat attributes include hydrological setting (benthic channel,

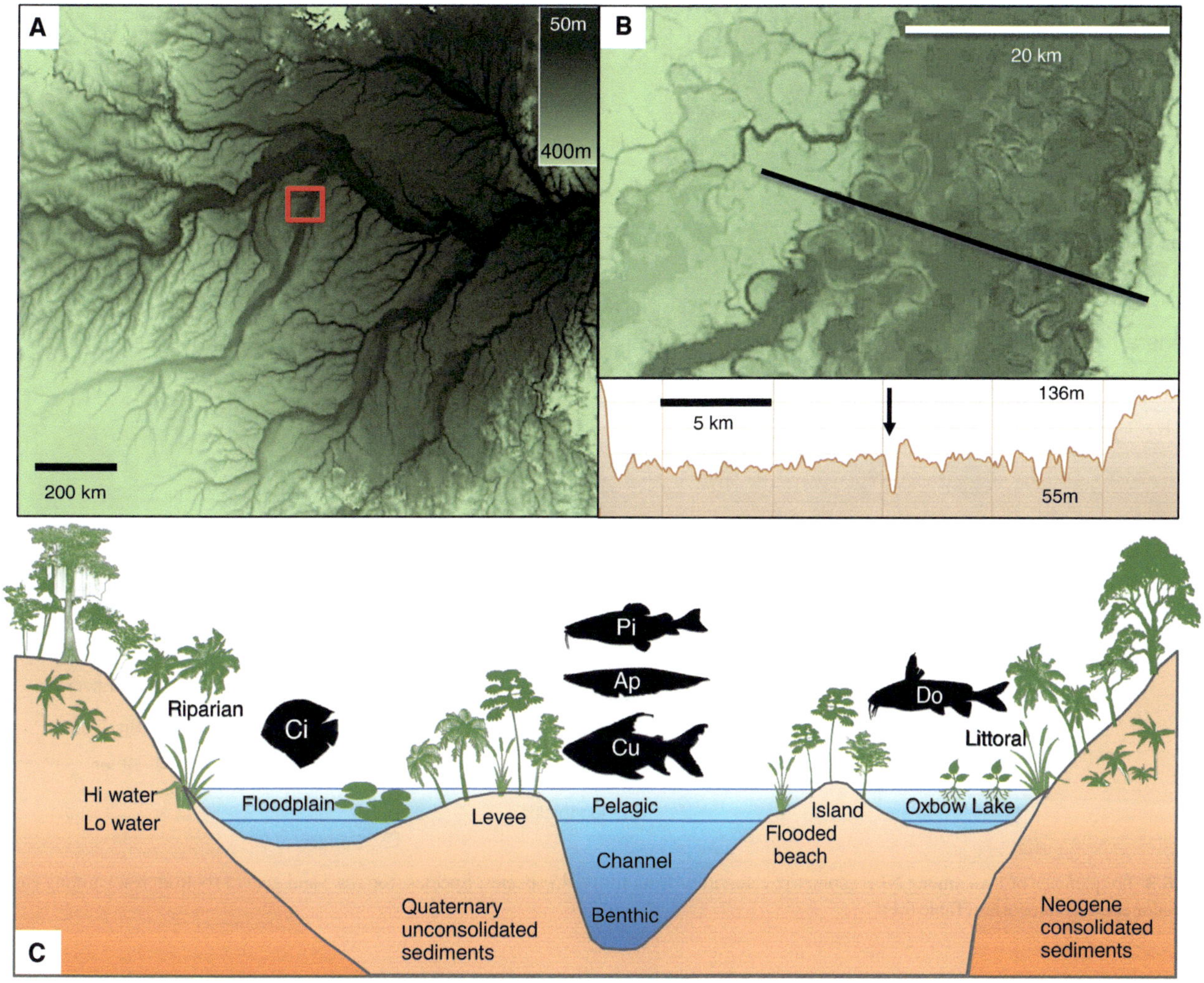

Fig. 6.2 Main aquatic habitats of large lowland Amazonian rivers. (A) Elevation map of Central Amazonia (black box in Fig. 6.1B). Inset indicates color scheme. (B) Elevation map of red box in panel A near mouth of Juruá river, Brazil, centered on lat −3.58, lon −66.20. Inset: elevation profile of transect across floodplain, from a high of 136 m on terra firme to a low of 55 m at Juruá river channel (arrow). Elevation maps from MERIT Hydro; elevation profile from Google Earth. (C) Schematic cross section of large lowland rivers (SO 6–10) and associated floodplain habitats illustrating main structural microhabitats (1–10 m) and mesohabitats (10–100 m), and ecologically dominant fish taxa (silhouettes). Ci = Cichlidae, Pi = Pimelodidae, Ap = Apteronotidae, Cu = Curimatidae, Do = Doradidae. Note most large river fishes use both floodplain and channel habitats during their life cycle (Chaps. 11 and 13)

pelagic channel, littoral channel, floodplain; Figs. 6.1C and 6.2), water flow (lotic vs. lentic), flow gradient (low gradient runs vs. high gradient riffles or rapids; Fig. 6.1D), geological setting (tectonic lakes, hypogean aquifers, and caves), seasonality (annual vs. perennial), water-column position (surface, pelagic, benthic), and substrate composition (unspecified, mud/silt, sand, gravel/rocks, submerged wood, vegetation, leaf packs; Fig. 6.3).

The trophic guilds are benthic invertivores (including meiovores/microvores), surface invertivores, benthic detritivores, benthic algivores, pelagic planktivores, folivores, frugivores (fruits and seeds), piscivores, omnivores (both plant and animal parts), and ectoparasites (eating scales, fins, or blood of other fishes). Many NFF species are trophic opportunists and dietary generalists, and it is not possible to identify adult modal diet more finely than that of a generalized microvore (Brazil-Sousa et al. 2023). Ecotraits of behavioral attributes include species with long-distance (>100 km) longitudinal seasonal migrations, viviparity (live-birth), venom injection with associated behaviors, passive electroreception with ampullary electroreceptor organs, active electroreception with electric organs and tuberous electroreceptor organs, and physiological aestivation or prolonged infaunal torpor (i.e., lowered metabolic activity) with desiccation during the low water season.

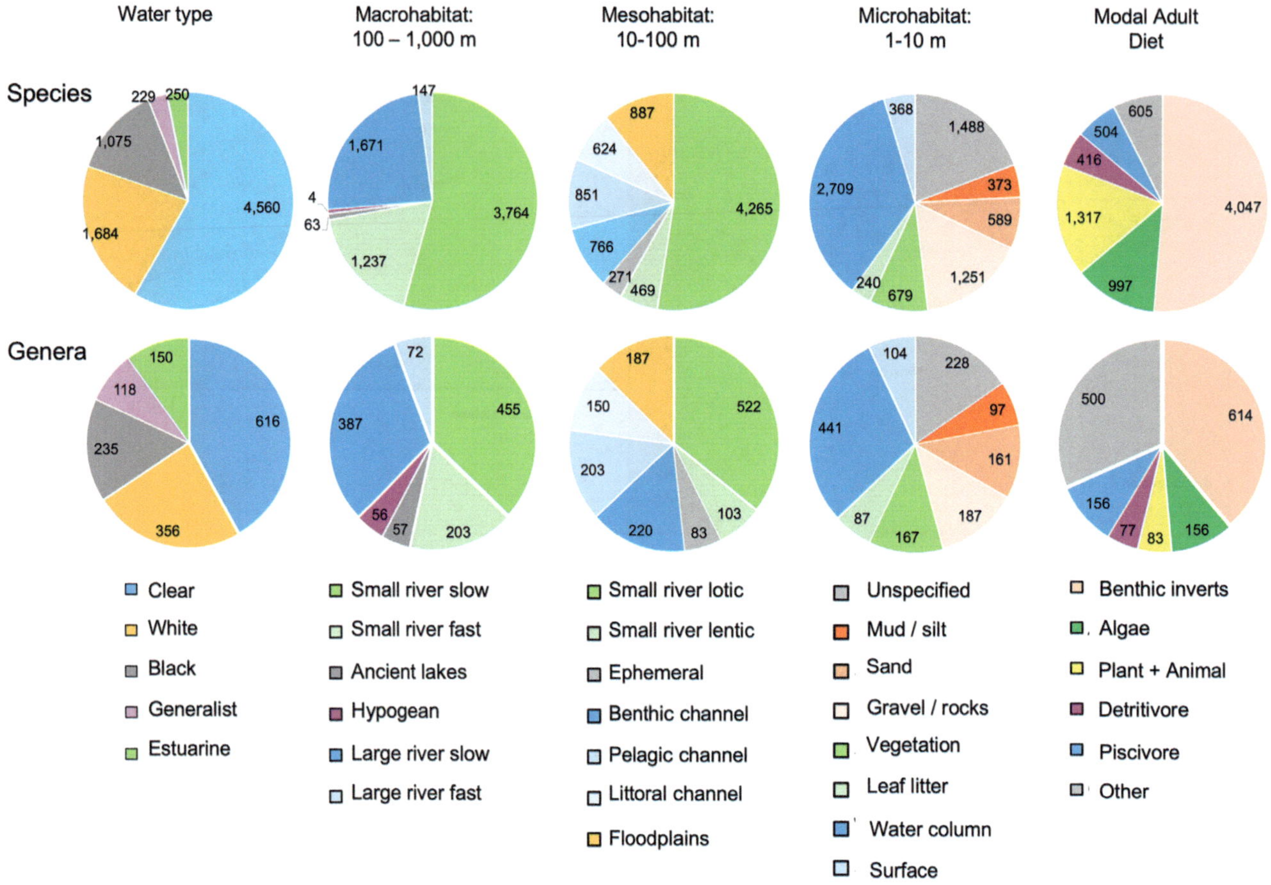

Fig. 6.3 Frequencies of taxa among NFF ecomatrix categories. Data for 6,345 species. Species (top row) and genera (bottom row) among each of the major ecotrait categories (Table 6.1)

6.2.3 NFF Ecomatrix

The NFF ecomatrix synthesizes taxonomic and ecological information for all freshwater fishes of South and Central America and the Greater Antilles considered valid as of Dec. 2019. The study design was for project editors (JSA and RER) to build a draft version of the whole NFF ecomatrix by scoring taxa for all ecotraits to the genus level, and then disseminating subsets of this draft dataset by family, or subfamily in the case of the most diverse families (i.e., Characidae and Loricariidae) to each of 55 contributing experts (Albert et al. 2025). The edited versions of these data subsets and their associated literature references were then reassembled into the final NFF ecomatrix, which was used by the editors to conduct data analyses and prepare the final manuscript for publication. The NFF ecomatrix is available at the Harvard Dataverse online at: https://doi.org/10.7910/DVN/WZJSNX.

The NFF ecomatrix includes observations of 6345 NFF species representing 880 genera, 97 families, and 40 taxonomic orders scored for 42 ecotraits of habitat utilization, diet, and other ecologically relevant behaviors described previously. Species in the ecomatrix are arranged in a conventional phylogenetic sequence by Class, Order, and Family and alphabetically within genus. Adult modal values for each ecotrait for each species were qualitatively assessed by 55 experts of NFF biology (see Acknowledgments), using data from 1294 literature references, reporting data for taxa representing all 880 valid NFF genera and 32% of all valid NFF species. The resulting ecomatrix includes 266,490 observations with 50,688 (19%) as positive observations.

The NFF ecomatrix has no empty cells or question marks, and estimated trait values of some poorly known species are provided based on information gleaned from collection records and knowledge of their closest relatives. At least three experts laid eyes on every cell in the ecomatrix, including the two project editors and one or more contributing experts. Cells were scored using all available sources of information, including literature references when available, observations of the contributing experts based on their personal knowledge from field collections and museum records, and in some cases unpublished data or consultations with other professional colleagues.

Data on the NFF fauna have several documented collection biases (Dagosta and De Pinna 2019; Tagliacollo et al. 2021), which may also influence knowledge of fish species' ecological traits. The more well-documented NFF species tend to have larger body sizes, are exploited for fisheries with commercial value as food items or in the aquarium trade, have other special interest to the scientific community, have broader geographic ranges, live near cities with major universities and museums of natural history, inhabit low-order streams along major roads, high stream-order tributaries along major rivers, inhabit places that will be affected by projects that need to obtain environmental licenses that require collections (Jézéquel et al. 2020), or belong to well-studied ecological model systems, such as hypopomid knifefish (e.g., Waddell et al. 2019; Waddell and Crampton 2022). Additionally, fine-scale details for some studies are available from single-species autecological studies (e.g., López-Rodriguez et al. 2021 for *Helogenes*, or Crampton 2008 for *Symphysodon*), but often conspicuously absent for other closely related species. Data compilation required more than 5 years, during which time many new publications became available, including about 300 new species published from 2020 to 2023, with many new ecological observations of NFF fishes. The final ecomatrix of this study is, therefore, presented as a starting point for continuing studies of the megadiverse NFF fauna.

6.2.4 Multivariate Statistical Analyses

We used *principal components analysis* (PCA) to extract linear combinations of the 42 ecotrait variables that maximally explain the variance (Jolliffe 2002). We call this multivariate hypervolume of ecological trait diversity the NFF *ecospace*. Ecospace hypervolumes for NFF clades were calculated based on 41 principal component (PC) dimensions derived from a PCA of 42 ecological traits across all 6345 NFF species. We used a custom R script to calculate the size of the ecospace,

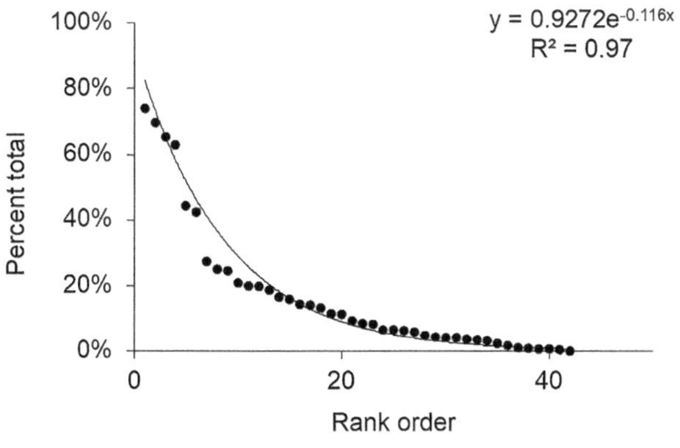

Fig. 6.4 Frequency distribution of the 366 unique ecotrait combinations among NFF species. Data for 6345 species scored for 42 ecotraits (Table 6.1). Hollow curve frequency distribution approximating a power function ($R^2 = 0.97$), illustrating the widely observed phenomenon that most ecotraits are rare and few are common

as a convex hull around all the points in the 41-dimensional hypervolume (Appendix 6.1). The hypervolume measure allows direct comparisons of the ecological disparity among and between groups. This analysis examines the distribution of NFF species within the multidimensional trait space, with comparisons made across major phylogenetic groups, body size categories, elevation, and salt tolerance [i.e., primary, secondary, and peripheral freshwater fishes (*sensu* Myers 1949; Matamoros et al. 2015)].

The concept of an ecospace as a multidimensional ecological hypervolume was first presented by Hutchinson 1957 (see also Blonder et al. 2014), and thereafter was frequently used in the paleontological literature (Valentine 1969; McKenna 1972; Bambach et al. 2007), and in community-wide studies of animal communication signals (for an example involving Neotropical fish see Crampton et al. 2011). It is now gaining traction in aquatic ecology (Haug et al. 2024). However, it is not to be confused with the Ecospace model that is commonly seen in the fish ecology literature which was developed from the Ecopath and Ecosim food web models to add a spatial dimension for studying marine ecosystems (Walters et al. 1999; Pauly et al. 2000).

6.3 Results

6.3.1 Unique Ecotrait Combinations

The 6345 NFF species in this ecomatrix exhibit a total of 366 unique combinations (Fig. 6.4). Despite representing 42 different dimensions of ecophysiological and behavioral diversity, a unique combination of these ecotraits does not necessarily imply a single ecological niche. Some of the ecotraits in this study may represent multiple functional traits, depending on the habitat or diet breadth of individual species. For example, "benthic invertivores" includes fishes that consume aquatic and aerial insects, crustaceans, and mollusks, often involving distinct physiological, morphological, and behavioral phenotypes (Winemiller 1990; López-Fernández et al. 2014; Kolmann et al. 2022; Lima et al. 2023). Similarly, the terms "algivore," "herbivore," and "detritivore" include species with potentially distinct modal diets as ascertained by stable isotope analysis (Lujan et al. 2012; Fitzgerald et al. 2017). A similar logic applies to food and habitat traits, where species may differ in the breadth of resource utilization seasonally or through life history (Winemiller et al. 2010; Correa and Winemiller 2014; Lima et al. 2023). We therefore present these unique combinations not as species niches but rather as ecologically conserved clade-level (mostly super-specific) attributes.

A majority of NFF species exhibit a distinctive combination of ecotraits, inhabiting slowly flowing clearwater streams and small rivers, and consuming benthic invertebrates (e.g., insects, crustaceans). Note also that ecotrait profiles differ by

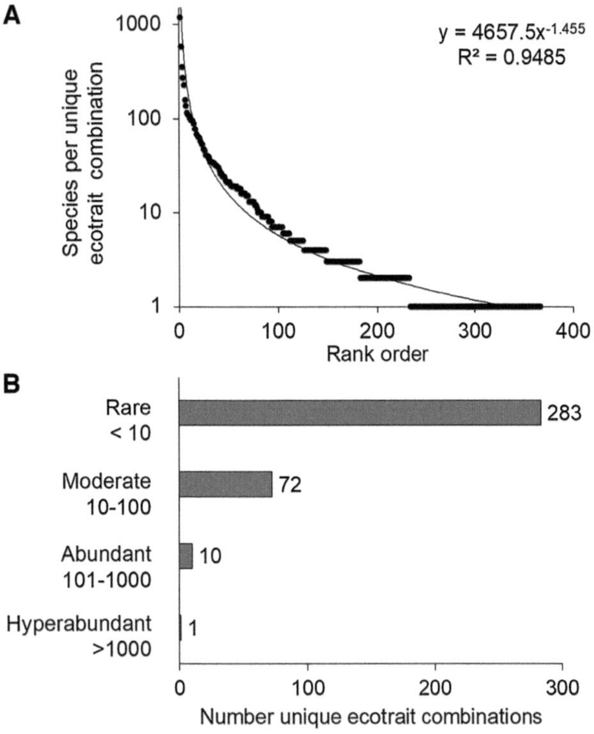

Fig. 6.5 Proportion of NFF species exhibiting each ecotrait. Data for 6345 species scored for 42 ecotraits in Table 6.1. (A) The rank–order frequency distribution closely approximates a power function or Yule distribution ($R^2 = 0.9485$) as expected from preferential attachment theory (Newman 2001). (B) Frequency distribution of unique ecotrait combinations. Note that most ecotrait combinations are rare and only one is hyperabundant, characterizing 1198 species

taxonomic level, with greater ecotrait diversity among genera, illustrating the effect of functional redundancy among closely related species that almost always exhibit allopatric geographic distributions (Albert et al. 2020). The proportion of NFF species exhibiting each ecotrait is presented in Fig. 6.5. Four ecotraits (i.e., 8, 18, 33, and 12, listed here in descending order of their frequencies) characterize more than 60% of NFF species, while six other ecotraits (e.g., 14, 38, 15, 3, 39, and 7) characterize less than 1% of the NFF fauna. The four most common ecotraits are benthic microvores inhabiting small, slowly moving, clearwater streams, and small rivers (stream order 1–5).

Almost all (5891 or 95.9%) of NFF species exhibit at least one of the four most common ecotraits, and almost one-third (1914 species or 31.4%) exhibit all four of these ecotraits. The 366 unique ecotrait combinations represent about 5.8% of the whole NFF fauna, setting a minimum estimate for the proportion of distinct functional groups or *guilds*. The actual number of unique functional groups is presumably larger than this because there are many ecological differences among NFF species that are not included in this study (e.g., Lujan and Armbruster 2012; Lujan et al. 2013). In this study, a single ecotrait combination is hyperabundant, characterizing 18.6% of all NFF species with values of "1" for four ecotraits; i.e., 110001000001000000000000000000000000000000. This ecotrait combination is composed of species with small adult body sizes (c. 3–10 cm SL) that inhabit clearwater, low-gradient streams, and small rivers (with stream orders 1–5). At the other end of the diversity spectrum, a large majority (283 or 77.3%) of the ecotrait combinations are rare, represented by 10 or fewer species.

6.3.2 The NFF Ecospace

The PCA of 42 ecotraits for 6345 NFF species produced an ecospace of the NFF fauna, in which each PC represents a mathematically independent axis of variance, with variable loadings for each of the ecotraits. PC1 of the ecospace is dominated by variance in ecotraits associated with stream order along the longitudinal river axis from headwaters to river main stem, including ecotraits associated with river size (i.e., stream order, volume of local mesoscale and macroscale aquatic habitats, pelagic, and migratory behaviors), river flow (i.e., stream gradient), and water chemistry (i.e., sediment load and dissolved ions/conductivity) (Fig. 6.6; Appendix 6.1).

In this regard, PC1 is similar to variance along the river continuum described for fluvial ecosystems (Vannote and Sweeney 1980), with additional mechanistic environmental drivers at the macroscale (100–1000 m) and mesoscale (10–100 m) (Clarke et al. 2008; Vander Vorste et al. 2017). PC2 of the ecospace is dominated by variance in ecotraits associated with environmental drivers at the microscale (1–10 m); e.g., diet and position in the water column (i.e., epibenthic and benthic invertivores vs. benthic algivores) and substrate (i.e., gravel, sand, mud).

The distribution of species across the first two principal components (PC1 and PC2) in the ecospace accounts for a combined 18.5% of the total variance (Fig. 6.6). The top 12 PCs sum to 56% of the total variance (Table 6.2). Analysis of the phylogenetic structure of the ecospace reveals substantial overlap among the Big Nine NFF clades, marine-derived lineages (MDLs), and other primary freshwater taxa, particularly at the plesiomorphic values concentrated near the top right of the ecospace in Fig. 6.7. This concentration may reflect evolutionary conservatism in traits related to habitat use and trophic function among many clades, particularly those occupying small stream habitats.

A notable pattern in this structure is that all clades of *secondary freshwater fishes*, including Cichlinae, Poecilioidea, and Rivulidae, are predominantly restricted to the upper-right portion of the ecospace comprised by PC1 and PC2 (Fig. 6.7). In contrast, other clades have expanded into unique regions of the ecospace. These include large river fishes, such as some MDLs and Gymnotiformes, which extend into lower PC1 values. Benthic algivorous loricariids show expansion into lower PC2 values, representing adaptations to a benthic, algae-consuming niche. The Pimelodoidea clade shows

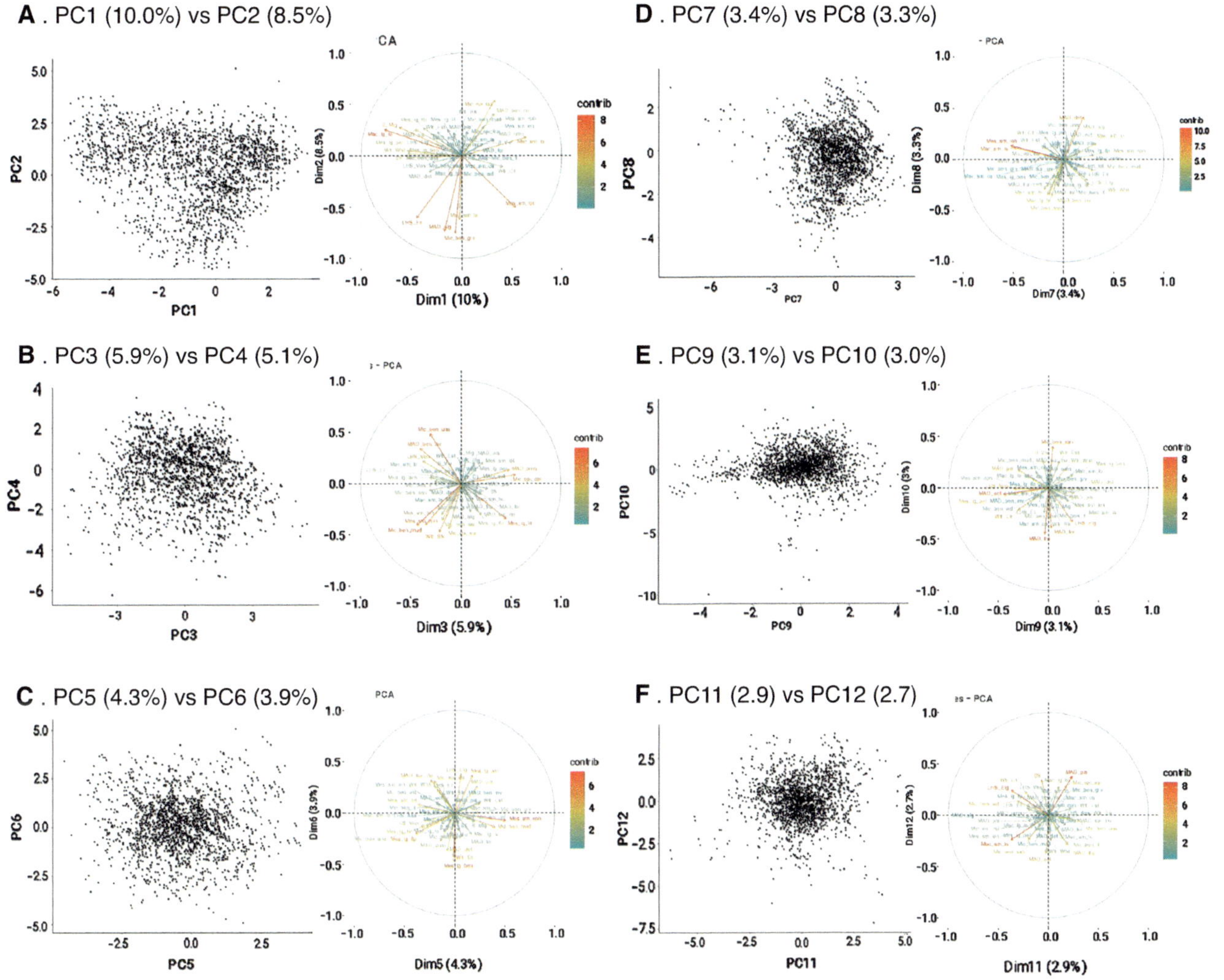

Fig. 6.6 Part of the NFF ecological trait space. (A) PC1 and PC2 represent 9.8% and 8.7%, respectively, of the total ecospace variance. (B) Vector loadings of 42 ecotraits on PC1 and PC2. PC1 trait loadings indicate an axis of ecotraits along the river continuum, from upland streams and small rivers to large lowland rivers. PC2 trait loadings indicate an axis of ecotraits from benthic invertivores in river runs to algivores in river riffles and rapids

Table 6.2 Qualitative descriptions of top 12 PCs accounting for 56% of the total variance in the ecospace analysis. Descriptions using terms from ecotraits that load most strongly on each PC

PC	%	Negative PC scores	Positive PC scores
1	10.0	Large, low gradient whitewater rivers with pelagic migratory fishes	Small, low-gradient clearwater rivers
2	8.5	Benthic gravel, benthic algivores, electeroreception	Epibenthic and benthic invertivores
3	5.9	Small river, benthic mud, ephemeral streams, benthic intertivores	Large river littoral, epibenthic omnivores
4	5.1	Blackwater and ephemeral streams, surface feeders	Benthic invertivores, substrate unspecified
5	4.3	Large river floodplains	Ephemeral streams, benthic mud
6	3.9	Large clearwater rivers, benthic detritivores	Large river pelagic planktivores and surface invertiviores, elecrogenic
7	3.4	Small, low gradient river, ancient lakes	No large ecotrait loadings
8	3.3	Benthic detritivores	Benthic invert sand
9	3.1	Larger river pelagic ectoparasitivores and pisciovres	Larger river benthic detrivores
10	3.0	Fruivores and folivores	Psammophlic
11	2.9	Small rivers, low gradient	Small river riffles, benthic leaf litter, and unspecified substrate
12	2.7	Piscivores, sandy bottom	Ectoparasitivores, frugivores, surface
Total	56.1		

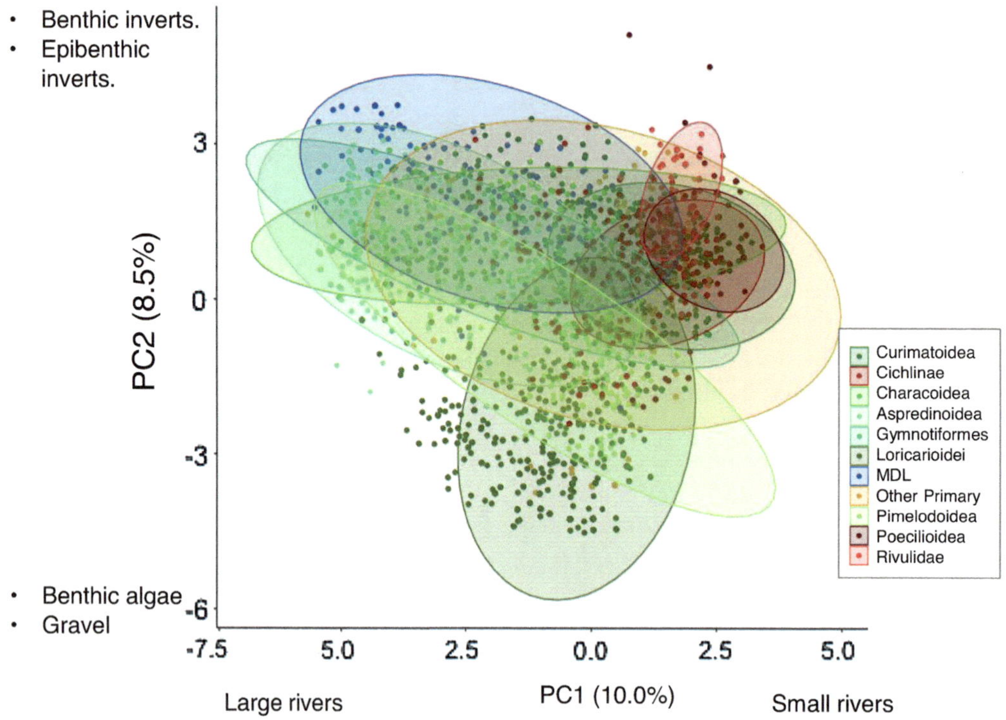

Fig. 6.7 Phylogenetic structure of part of the NFF ecospace. Partitioning of the first ecospace of the first two PCs accounting for a combined one-sixth (18.5% of the total variance). Data for 6345 species. Species partitioned by the Big Nine NFF clades + MDLs + other primary freshwater taxa. MDLs = marine-derived lineages

a unique expansion, with its ecospace centroid shifted away from the plesiomorphic region. Within Pimelodoidea, the Pimelodidae subclade aligns with habitats in larger rivers (top left), while Pseudopimelodidae and Heptapteridae occupy intermediate values on PC2, between the plesiomorphic invertivorous and derived algivorous regions. This pattern suggests a range of ecological adaptations within Pimelodoidea, including differentiation in trophic niches that contrasts with the limited divergence seen in some other clades.

Analyses of the functional composition of the ecospace partitioned by salt tolerance, stream order, elevation, and body size reveal several patterns of NFF ecological diversification (Fig. 6.8A). In general, species with higher salt tolerance (MDLs and secondary freshwater fishes) are concentrated in higher PC2 values, with MDLs with the highest salt tolerance largely found in large rivers and secondary freshwater fishes with moderate tolerance found in small streams. In contrast, primary freshwater fishes with no salt tolerance are distributed across a wider range of habitats in the ecospace of PC1 and PC2. A partition of the species by stream order reveals two distinct faunas for large and small rivers with no overlap, and eurytopic species overlapping at the center (Fig. 6.8B). Upland and lowland species exhibit heavy overlap within the ecospace (Fig. 6.8C), suggesting that elevation has less influence on trait divergence than other habitat characteristics contrasting with previous hypotheses that the two habitats compose distinct ecological faunas (Albert et al. 2018,2020). The ecospace occupancy of lowland species trends toward large rivers (PC1), which compose a major proportion of the available lowland freshwater habitats.

Body size is another important aspect of functional diversity (Chap. 19) that is structured in the ecospace (Fig. 6.8D). The largest species are primarily associated with large rivers, demonstrated by the relatively strong loading of the ecotrait on PC1. NFF species range in size over about 2.5 orders of magnitude, from *Leptophilypnion fittkaui* (Eleotridae) at 0.9 cm to *Pristis pristis* (Pristidae) at 610 cm (Fig. 6.9A). The largest species that is found only in SO 1–5 is *Lophiosilurus alexandri* (Pseudopimelodidae) at 72 cm SL. The smallest species found only in SO 6–10 is *Potamoglanis anhanga* (Trichomycteridae) at 1.3 cm SL. Most NFF species are confined to small rivers (SO 1–5; $n = 3381$ species, 71% NFF total), and only a small fraction are generalists (SO 1–10; $n = 444$ species, 8% NFF total; Fig. 6.9B). A bimodal size distribution of the generalists fauna (Fig. 6.9C) resulting from many eurytopic species of Characidae (e.g., *Gephyrocharax*, *Phenacogaster*) and Loricariidae (e.g., *Otocinclus*, *Parotocinclus*) with 4–6 cm SL, and many eurytopic species of Serrasalmidae (e.g., *Metynnis*, *Serrasalmus*) and Loricariidae (e.g., *Farlowella*, *Lamontichthys*) in the size range 16–20 cm SL. Thus, while the majority of species occupy the ancestral trait space characterized by small body size, low stream order, and benthic feeding habits, certain lineages exhibit significant divergence,

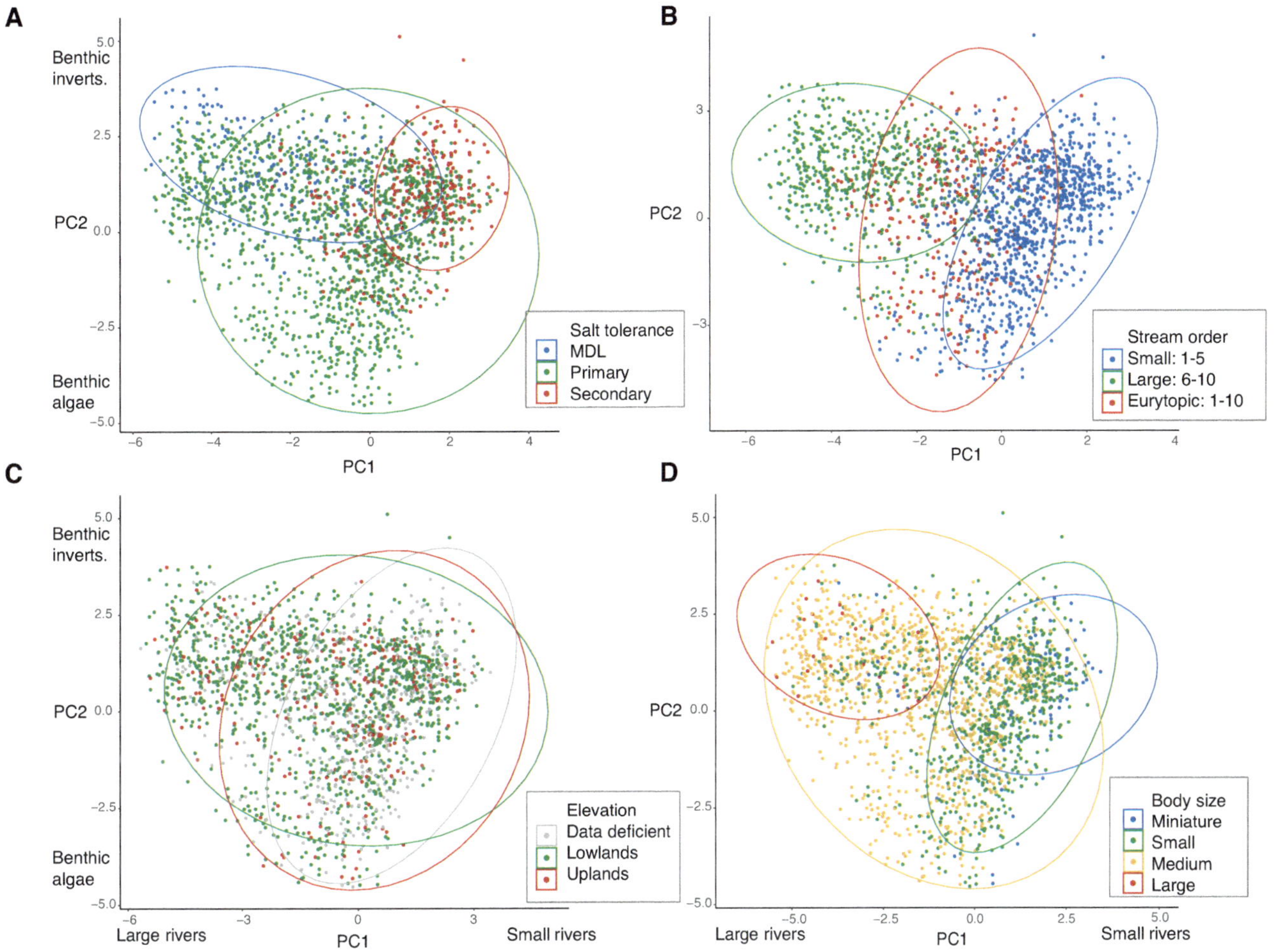

Fig. 6.8 Functional composition of the NFF ecological trait space of PC1 vs. PC2. Ellipses represent 95% CI. Data partitioned by four classifiers. (A) Species grouped by salt tolerance categories. Primary freshwater, no tolerance; secondary freshwater, some tolerance; MDL, marine-derived lineage, high tolerance. (B) Species grouped by characteristic Strahler stream order (SO). Small rivers (SO = 1–5), large rivers (SO = 6–10), eurytopic tolerances (SO = 1–10). (C) Species grouped by elevation, defined as median elevation range of georeferenced localities. DD = data deficient; lowland and upland = geographic range centroid below or above 250 m. (D) Species grouped by maximum adult body size. Miniature (0.9–2.5 cm SL), small (2.6–9.9 cm), medium (10.0–99 cm), large (100–610 cm)

particularly into large river habitats, highlighting the ecological diversity within the NFF ecospace.

6.3.3 Ecospace Hypervolumes

Calculations of ecospace hypervolumes allow the comparison of ecospace utilization of the major phylogenetic groups, including the Big Nine NFF clades, other primary freshwater fishes, and MDLs (Fig. 6.10A). While hypervolume size varied across clades, no significant correlation was found between hypervolume size and species richness for phylogenetic categories ($R = 0.066$, $N = 11$, $p = 0.27$). This may indicate that the ecological disparity among clades is not strongly tied to the number of species within each group, suggesting that factors other than species richness may play a larger role in shaping the trait diversity within each clade.

Hypervolumes were also calculated for functional categories based on body size, elevation, and salt tolerance (Fig. 6.10B). We found a significant positive correlation between ecospace hypervolume and species richness for the combined salt tolerance and elevation categories ($R = 0.82$, $N = 6$, $p = 0.046$). This correlation suggests that species occupying a broader range of salinities and elevations tend to have greater ecological disparity. However, no significant correlation was found with body size, indicating that trait space disparity in the NFF assemblage is not closely related to adult body size alone.

NFF species with medium adult body sizes (9.0–99 cm SL) exhibit a markedly smaller ecospace hypervolume compared to other body size categories, despite high species richness and representation across all major phylogenetic groups. This deviation from the expected relationship between body size and ecospace volume suggests a convergence in ecological traits among medium-sized species, potentially driven by shared

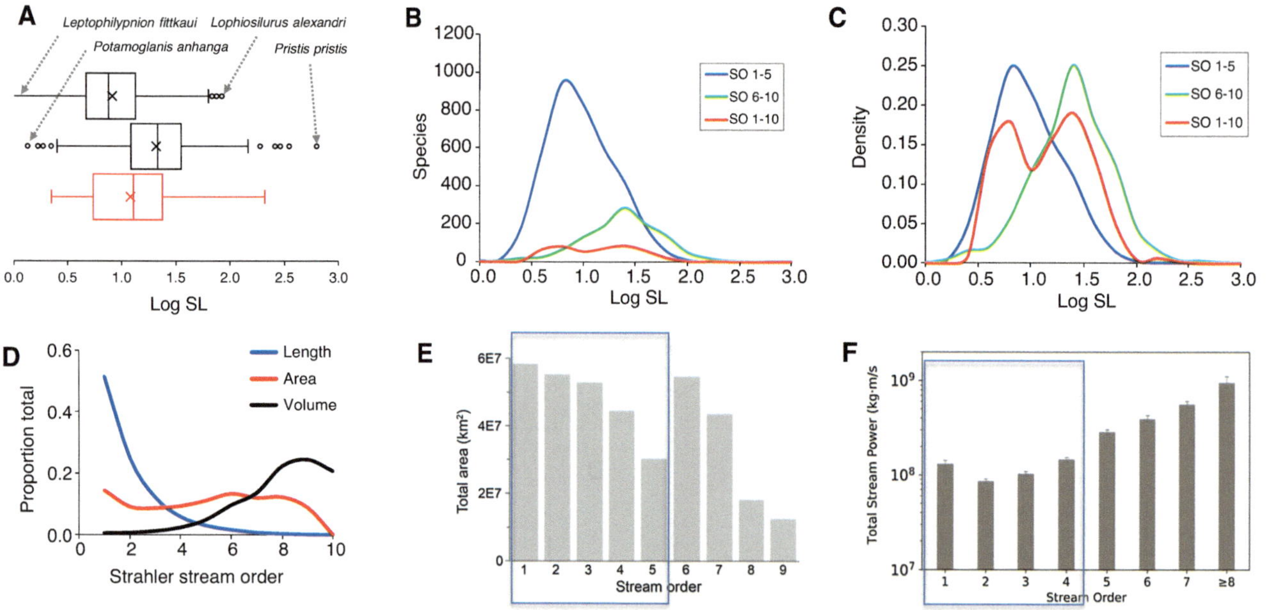

Fig. 6.9 Body size distributions of NFF species in small and large rivers. Fishes in small rivers in blue (SO 1-5; $n=3{,}381$ species, 71% NFF total); large rivers in green (SO 6-10; $n=1{,}131$ species, 21% NFF total), and generalists in red (SO 1-10; $n=444$ species, 8% NFF total). Size as log maximum standard length (SL); data for $n=5{,}430$ species. A. Box and whisker plots illustrating the larger body sizes of fishes in large rivers. B. Numbers of NFF species across body sizes by SO category. C. Body-size frequency distribution of NFF species by SO category. D. Proportions of river segments by SO for length, area and volume. E. Total area of global drainages by Stream Order (He et al., 2024). F. Total stream power by Stream Order (Feng and Gleason, 2024; fig. 4A).

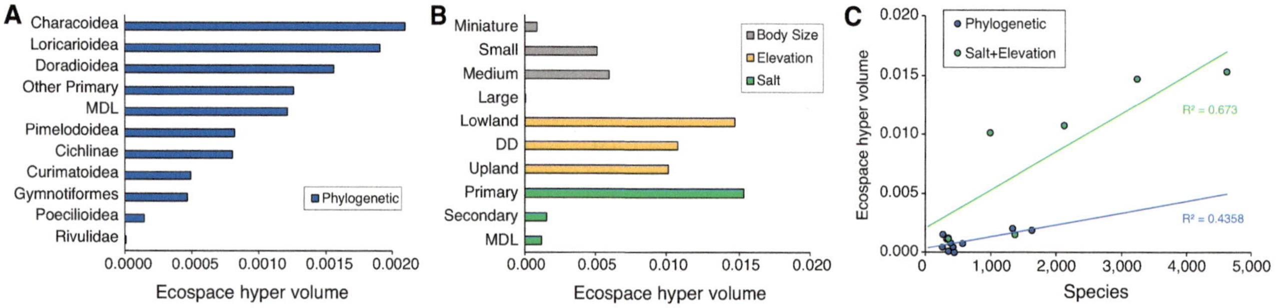

Fig. 6.10 Comparisons of ecospace hypervolumes among NFF species. Volumes calculated from the convex hull of the data in 41 PC dimensions, from PCA of 42 ecotraits and 6345 species. (A) Among major phylogenetic groups, including the Big Nine clades, other primary freshwater fishes, and MDLs (see Table 6.1). (B) Among adult body size, elevation, and salt tolerance categories. (C) Correlations between ecospace hypervolume and species richness

habitat or dietary preferences that limit ecological expansion within this size class. Consequently, the medium-sized category stands out as an exception in the general pattern, showing lower disparity despite species richness.

6.4 Discussion

6.4.1 Limitations of the NFF Ecomatrix

The 6345 NFF species in this ecomatrix exhibit a total of 366 unique combinations (Fig. 6.4). Despite representing 42 ecotrait dimensions, a unique combination of these ecotraits does not necessarily constitute a distinct functional role or ecological niche. Some ecotraits represent multiple functional traits depending on the habitat or diet breadth of individual species. For example, "benthic invertivores" includes fishes that consume aquatic and aerial insects, crustaceans, and mollusks, often involving distinct physiological, morphological, and behavioral phenotypes (Winemiller 1990; López-Fernández et al. 2014; Kolmann et al. 2022; Lima et al. 2023). Similarly, the terms "algivore," "herbivore," and "detrivore" include species with potentially distinct modal diets as ascertained by stable isotope analysis (Lujan et al. 2012; Fitzgerald et al. 2017). A similar logic applies to food and habitat traits, where species may differ in the breadth of resource utilization seasonally or through life history (Winemiller et al. 2010; Correa and Winemiller 2014; Lima et al. 2023). We, therefore, present these unique combinations not as species niches but rather as ecologically conserved clade-level (mostly super-specific) attributes.

Except for the Poeciliidae (e.g., Reznick et al. 1990; Johnson and Zúñiga-Vega 2009; Hulthén et al. 2021) and Hypopomidae (Waddell et al. 2019; Waddell and Crampton 2022), we currently lack data regarding ecological and life history trait data for a majority of NFF taxa at the species level. This lack of information includes important data on size at onset of sexual reproduction, number and size of eggs and larvae, and reproductive lifespan (Stearns 2000). Other important data types with poor representation across NFF taxa include the allocation of resources to the competing ends of growth, reproduction, survivorship, maturation, growth, and mortality (Matthews 2012).

6.4.2 Large River Channels as a Niche Fauna

Data from the ecomatrix record indicate that 1560 species, or about 24% of all NFF species, are modally present in large river channels, of which about 1128 (72%) are endemic to large rivers and rarely if ever occupy smaller rivers (Chap. 13). By contrast, a great majority (93%) of NFF fishes inhabit smaller rivers (SO 1–5), and 81% of the NFF fauna is only present in these smaller rivers. The number of total fish species in large river channels may be surprisingly low to some observers, given the well-known diversity of the deep channel fauna (Crampton 2007; Lasso and Morales-Betancourt 2022; Torgersen et al. 2023b). However, the large river channels constitute a small fraction of the total river length (~3%) of the South American rivers, and are very well connected along their length, facilitating high local or alpha diversity at any given location, but low species turnover or beta diversity across the riverscape (Fig. 1.4C). Further, the total riverine footprint occupies a meager 3500 km^2, or only about half the area of metropolitan Sao Paulo. Finally, an additional 760 NFF species (12%) are eurytopic in terms of stream order, inhabiting both large and small rivers (SO 1–10).

The fauna of NFF species that inhabit large river channels represents a broad range of the higher NFF taxa, including 368 genera and 68 families. Of NFF species endemic to large rivers, 1365 species (68%) are primary freshwater fishes, mostly ostariophysans, 191 species (3%) are marine-derived lineages (MDLs), and only four species are secondary freshwater fishes, all rivulids. However, as proportions of their respective totals, about 64% of the primary fishes inhabit large rivers as compared with 93% of the MDLs. In this sense large rivers can be considered a "niche fauna," as in vernacular English a "niche occupation" requires specialized knowledge or skills not common in the general job market, and a "niche market" is the subset of the market on which a specific product is focused.

6.4.3 NFF Ecological Trait Space

The ecospace of NFF species reveals a notable pattern of trophic diversity with small rivers and streams supporting a broader range of feeding strategies as compared with larger rivers (Fig. 6.8). This distribution might be viewed as indicating a greater complexity and heterogeneity of small river habitats offering more diverse ecological niches (Fig. 6.1C, D), thus supporting a wider variety of dietary specializations. However, large rivers also exhibit high habitat heterogeneity, with varied environments vertically (i.e., benthos to surface), and laterally (e.g., marginal meadows, log jams, eddies, and side bays), and large river floodplains have perhaps the most heterogeneous environments in the Amazon Basin (Fig. 6.2; Crampton 2011).

Additionally, the PCA of ecological traits indicates a relatively low total variance in the first two principal components, especially when compared to PCAs focused solely on body shape (e.g., Torgersen et al. 2023a). This difference could be explained by a hypothesis that traits under strong genetic control, like body shape, may be influenced by fewer dominant evolutionary or ecological drivers, resulting in more constrained trait variation. In contrast, life history and dietary traits may be shaped by a broader range of environmental and ecological factors, allowing for greater disparity. This hypothesis implies that body shape, being more constrained, occupies a narrower trait space, while life history and dietary traits, subject to complex selective pressures, contribute more substantially to the diversity observed within ecological trait spaces.

6.5 Conclusions

This chapter presents results of the first comprehensive analysis of ecotraits among NFF species using published data and expert knowledge from the community of Neotropical ichthyologists. The ecotrait matrix, or *ecomatrix*, includes adult modal values of 42 ecophysiological traits related to adult body size, four of habitat utilization related to water chemistry, 21 of physical habitat structure, ten of diet (i.e., trophic guilds), and six other behavioral traits, for all 6345 valid NFF species as of 2021, representing 880 genera, 97 families, and 40 orders. The ecomatrix is based on data from literature references for 1346 (21.4%) NFF species, with the other species codings provided by 55 experts in NFF diversity and ecology.

The habitat data are qualitative estimates of adult modal habitat utilization for 26 aspects of habitat diversity in four broad categories; water type (water chemistry) and three levels of spatial scale (macrohabitat 100–1000 m; mesohabitat

10–100 m; and microhabitat 1–10 m). The four water types are: clearwaters with high transparency due to low sediment and organic compounds, slightly acidic (pH = 6.1–6.7); whitewaters with high sediment load, low transparency, and almost neutral pH (pH= 6.8–7.0); blackwaters with high content organic compounds, low sediment load and low transparency, and more acidic (pH = 6.0–3.5); generalist species with about equal frequency in all three main water types (white, black, and clearwaters); and estuarine species in freshwater coastal river mouths and deltas. The 21 structural habitat types are based on aspects of spatial scale (macro, meso, micro), flow volume (Strahler stream orders 1–5 vs. 6–10), water flow (lotic vs. lentic), flow gradient (low vs. high, i.e., rheophilic), geological setting (tectonic lakes, hypogean), seasonality (annual vs. perennial), hydrological setting (benthic channel, pelagic channel, littoral channel, floodplain), water-column position (surface, pelagic, benthic), and substrate composition (unspecified, mud/silt, sand, gravel/rocks, submerged wood, vegetation, leaf packs). The trophic data are qualitative estimates of adult modal diet for ten categories: benthic invertivores (including meiovores/microvores), surface invertivores, benthic detritivores, benthic algivores, pelagic planktivores, folivores, frugivores (fruits and seeds), piscivores, omnivores (both plant and animal parts), and ectoparasites (eating scales and fins of other fishes).

We used PCA of the ecomatrix to produce and analyze the first ecospace of the whole NFF fauna. When mapped across principal components of the ecospace, NFF species exhibit overlap and ecological divergence among clades. This distribution illustrates that different clades often occupy unique regions in ecospace, based on habitat adaptations. For example, certain clades are strongly associated with specific habitat conditions, such as high-gradient, flowing (rheophilic) environments versus slower, lentic systems. Some clades (Loricarioidea, Pimelodoidea) show broad expansion and occupancy of the ecospace while other clades of primary freshwater fishes are predominantly restricted to the plesiomorphic region, indicating limited ecological divergence in those groups from the ancestral condition. We found that stream order, rather than elevation, latitude, or climate zone, bests predicts the distribution of species within the ecospace. Body size also structures NFF species with large-bodied species modally inhabiting large rivers, small and miniature species largely inhabiting small rivers, and medium-sized species as ecological generalists. We found only 17% of the analyzed NFF species are adapted to large river habitats, indicating that the majority of species are more specialized for small streams and floodplain environments. This low percentage underscores the specialized nature of most NFF species and may have implications for conservation strategies, particularly concerning the preservation of smaller, more vulnerable habitat types.

This first ecological trait space (ecospace) for Neotropical freshwater fishes provides an unprecedented new resource for understanding the ecological dynamics, adaptive strategies, and conservation needs of this biodiverse group. The ecomatrix will serve as a critical reference point for future studies, allowing researchers to track shifts in habitat preferences, trophic interactions, and ecological roles in response to environmental change and anthropogenic pressures (Carvalho et al. 2023; Escamilla-Pinilla et al. 2019; Fricke et al. 2020; Hansen 2024).

Acknowledgments The data compiled for this study required the coordinated activities of the entire community of Neotropical ichthyologists and their extended professional networks. For proofing the ecomatrix for particular taxa, we thank: Vitor Abrahão, Daniel Akin, Marcelo Ândrade, Mariangeles Arce, Jonathan Armbruster, Ricardo Benine, Maxwell Bernt, Maria Bichuette, José Birindelli, Pedro Bragança, Marcelo Britto, Paulo Buckup, Michael Burns, Bárbara Calegari, Tiago Carvalho, Lilian Casatti, Wilson Costa, Fernando Dagosta, Joshua Egan, Luis Fernández, Daniel Fitzgerald, João Fontenelle, Matthew Kolmann, Francisco Langeani, Pablo Lehmann, Rafael P Leitão, Marina Loeb, Carlos Lucena, Paulo Lucinda, Nathan Lujan, Luiz Malabarba, Wilfredo Matamoros, Bruno Melo, Andre Netto-Ferreira, Carla Pavanelli, Fernando Pelicice, Edson Pereira, Robson Ramos, Marcelo Rocha, Fábio Roxo, Vivianne Sant'Anna, Oscar Shibatta, Veronica Slobodian, and Madlen Stange. We also thank the following colleagues for the ideas and motivational support for this project over many years: Devin Bloom, Prosanta Chakrabarty, Cristhián Conde, Jack Craig, Murilo Dias, Michael Goulding, Sven Kullander, John Lundberg, Hernán López-Fernández, Nathan Lovejoy, Thierry Oberdorff, Cláudio Oliveira, Hernán Ortega, Francisco Provenzano, Mark Sabaj, Brian Sidlauskas, Melanie Stiassny, Pablo Tedesco, Brandon Waltz, Ângela Zanata, and Jansen Zuanon. We also thank the many museum scientists and collections staff who curate and share specimens and collection information of NFF fishes with the community, including Oliver Crimmen, Caleb McMahan, James Maclaine, Lúcia Py-Daniel, Rob Robins, Enrique Santoyo-Brito, Randy Singer, Chris Taylor, and David Werneke.

Appendix 6.1: A Customized R Script to Calculate a Hypervolume from PCA Output Data.

```
##########
a = data.matrix(NFF.matrix2[, -c(43:51)])
# summarize rda() function version of PCA
b = summary(rda(a))
# here are the point values we need
b$sites
#install.packages("geometry")
library(geometry)
# this will give us a volume, but across all
points
convhulln(b$sites, output.options=T)$vol
# so we need to split it up by group, in this
case, the Big_Nine
NFF.bn.list = unique(NFF.matrix2[,48])
NFF.nsp = length(NFF.bn.list)
# empty vector
out = c()
# one time through the loop for each group
# filling the empty vector
for(i in 1:NFF.nsp){
```

```
  out[i] =
    convhulln(b$sites[
    which(NFF.matrix2[,48] %in% NFF.bn.list[i]),
  ],
  output.options=T)$vol
}
# name the volumes
names(out) = NFF.bn.list
# look at results
out
##########
```

References

Albert JS, Johnson DM (2012) Diversity and evolution of body size in fishes. Evol Biol 39:324–340

Albert JS, Tagliacollo VA, Dagosta F (2020) Diversification of Neotropical freshwater fishes. Annu Rev Ecol Evol Syst 51:27–53

Albert JS, Craig JM, Tagliacollo VA, Petry P (2018). Upland and lowland fishes: a test of the river capture hypothesis. Mt Clim Biodivers, 273–294

Albert JS, Abrahão V, Akin DR, Allen JG, Ândrade M, Arce M, et al. (2025) An ecological trait matrix of Neotropical freshwater fishes. Sci data 12(1):1127

Baldasso MC, Oliveira AGD, Kliemann BCK, Delariva RL (2024) Habitat modification driven by land use as an environmental filter on the morphological traits of neotropical stream fish fauna. Neotrop Ichthyol 22:e230119

Bambach RK, Bush AM, Erwin DH (2007) Autecology and the filling of ecospace: key metazoan radiations. Palaeontology 50(1):1–22

Beretta-Blanco A, Carrasco-Letelier L (2021) Relevant factors in the eutrophication of the Uruguay River and the Río Negro. Sci Total Environ 761:143299

Blonder B, Lamanna C, Violle C, Enquist BJ (2014) The n-dimensional hypervolume. Global Ecol Biogeogr 23(5):595–609

Bloom DD, Lovejoy NR (2017) On the origins of marine-derived freshwater fishes in South America. J Biogeography 44(9):1927–1938

Brazil-Sousa C, Soares BE, Svanbäck R, Albrecht MP (2023) Individual specialization is the highest in generalist populations from intermediary to high trophic positions in tropical freshwater fishes. Austral Ecol 49(2). https://doi.org/10.1111/aec.13368

Brejão GL, Gerhard P, Zuanon J (2013) Functional trophic composition of the ichthyofauna of forest streams in eastern Brazilian Amazon. Neotrop Ichthyol 11:361–373

Brosse S, Charpin N, Su G, Toussaint A, Herrera-r GA, Tedesco PA, Villéger S (2021) FISHMORPH: a global database on morphological traits of freshwater fishes. Global Ecol Biogeogr 30(12):2330–2336

Burress ED (2015) Cichlid fishes as models of ecological diversification: patterns, mechanisms, and consequences. Hydrobiologia 748:7–27

Carvalho RL, Resende AF, Barlow J, França FM, Moura MR, Maciel R et al. (2023) Pervasive gaps in Amazonian ecological research. Curr Biol 33(16):3495–3504

Clarke A, Mac Nally R, Bond N, Lake PS (2008) Macroinvertebrate diversity in headwater streams: a review. Freshw Biol 53(9):1707–1721

Correa SB, Winemiller KO (2014) Niche partitioning among frugivorous fishes in response to fluctuating resources in the Amazonian floodplain forest. Ecology 95(1):210–224

Correa SB, Costa-Pereira R, Fleming T, Goulding M, Anderson JT (2015) Neotropical fish–fruit interactions: eco-evolutionary dynamics and conservation. Biol Rev 90(4):1263–1278

Costello MJ, Claus S, Dekeyzer S, Vandepitte L, Tuama ÉÓ, Lear D, Tyler-Walters H (2015) Biological and ecological traits of marine species. PeerJ 3:e1201

Craig JM, Kim LY, Tagliacollo VA, Albert JS (2019) Phylogenetic revision of Gymnotidae (Teleostei: Gymnotiformes), with descriptions of six subgenera. PLoS One 7(14(11)):e0224599

Crampton WG, Lovejoy NR, Waddell JC (2011) Reproductive character displacement and signal ontogeny in a sympatric assemblage of electric fish. Evolution 65(6):1650–1666

Crampton WGR (2007) Diversity and adaptation in deep channel Neotropical electric fishes. In: Sebert P, Onyango DW, Kapoor BG (Eds.)Eds. Fish life in special environments. Science Publishers, Enfield, NH, 283–339pp

Crampton WGR (2008) Ecology and life history of an Amazon floodplain cichlid: the discus fish *Symphysodon* (Perciformes: Cichlidae). Neotrop Ichthyol 6(4):599–612

Crampton WGR (2011) An ecological perspective on diversity and distributions. In: Albert JS, Reis RE (Eds.) Historical Biogeography of Neotropical Freshwater Fishes. University of California Press, Berkeley, 165–192

Crampton WGR, Santana CD, Waddell JC, Lovejoy NR (2016) A taxonomic revision of the Neotropical electric fish genus *Brachyhypopomus* (Ostariophysi: Gymnotiformes: Hypopomidae), with descriptions of 15 new species. Neotrop Ichthyol 14:e150146

Dagosta FC, De Pinna M (2019) The fishes of the Amazon: distribution and biogeographical patterns, with a comprehensive list of species. Bull Am Mus Nat Hist 2019(431):1–163

Dawson SK, Carmona CP, González-Suárez M, Jönsson M, Chichorro F, Mallen-Cooper M et al. (2021) The traits of "trait ecologists": an analysis of the use of trait and functional trait terminology. Ecol Evol 11(23):16434–16445

Dias MS, Zuanon J, Couto TB, Carvalho M, Carvalho LN, Espírito-Santo HM et al. (2016) Trends in studies of Brazilian stream fish assemblages. Natureza Conservação 14(2):106–111

Duarte C, Magurran AE, Zuanon J, Deus CP (2019) Trophic ecology of benthic fish assemblages in a lowland river in the Brazilian Amazon. Aquat Ecol 53(4):707–718

Escamilla-Pinilla C, Mojica JI, Molina J (2019) Spatial and temporal distribution of *Gymnorhamphichthys rondoni* (Gymnotiformes: Rhamphichthyidae) in a long-term study of an Amazonian terra firme stream, Leticia-Colombia. Neotrop Ichthyol 17(3):e190006

Eschmeyer WN, Fricke R, Van der Laan R 2024. Catalog of fishes: genera, species, references. (Accessed date: Oct 2020). https://researcharchive.calacademy.org/research/ichthyology/catalog/fishcatmain.asp.

Evans KM, Kim LY, Schubert BA, Albert JS (2019) Ecomorphology of neotropical electric fishes: an integrative approach to testing the relationships between form, function, and trophic ecology. Integr Org Biol 1(1):obz015

Felice RN, Randau M, Goswami A (2018) A fly in a tube: macroevolutionary expectations for integrated phenotypes. Evolution 72(12):2580–2594

Ferraris Jr.CJ, de Santana CD, Vari RP (2017) Checklist of Gymnotiformes (Osteichthyes: Ostariophysi) and catalogue of primary types. Neotrop Ichthyol 15:e160067

Ferrer J, Malabarba LR (2020) Systematic revision of the Neotropical catfish genus *Scleronema* (Siluriformes: Trichomycteridae), with descriptions of six new species from Pampa grasslands. Neotrop Ichthyol 18:e190081

Fitzgerald DB, Winemiller KO, Sabaj Pérez MH, Sousa LM (2017) Using trophic structure to reveal patterns of trait-based community assembly across niche dimensions. Funct Ecol 31(5):1135–1144

Fricke R, Eschmeyer WN, Van derLaan R (2020) Catalog of fishes: genera, species, references. California Academy of Sciences, San Francisco, CA, USA, Catalog of fishes: genera, species, references. California Academy of Sciences, San Francisco, CA, USA, http://researcharchive.calacademy.org/research/ichthyology/catalog/fishcatmain.asp2020Oct.

Friedman ST, Collyer ML, Price SA, Wainwright PC (2022) Divergent processes drive parallel evolution in marine and freshwater fishes. Syst Biol 71(6):1319–1330

Frimpong EA, Angermeier PL (2009) Fish traits: a database of ecological and life-history traits of freshwater fishes of the United States. Fisheries 34(10):487–495

Frimpong EA, Angermeier PL (2010) Trait-based approaches in the analysis of stream fish communities. In: American Fisheries Society Symposium, vol 73, No. 109–136. American Fisheries Society, Bethesda, MD, 2010

Galán-Acedo C, Arroyo-Rodríguez V, Andresen E, Arasa-Gisbert R (2019) Ecological traits of the world's primates. Sci Data 6(1):55

Gamarra SP, Calegari BB, Reis RE (2019) A new species of *Curculionichthys* (Siluriformes: Loricariidae) from the north edge of the Brazilian Shield, lower Amazon basin. Neotrop Ichthyol 17:e190001

Goldstein RM, Meador MR (2005) Multilevel assessment of fish species traits to evaluate habitat degradation in streams of the upper Midwest. North Am J Fish Manage 25(1):180–194

Gomes LC, Dias RM, Ruaro R, Benedito E (2023) Functional diversity: a review on freshwater fish research. Neotrop Ichthyol 21(02):e230022

Hansen TF (2024) Three modes of evolution? Remarks on rates of evolution and time scaling. J Evol Biol 37:1523–1537

Haug JT, Baranov V, Braig F, Haug C (2024) Morpho-ecospaces, or how to measure biodiversity in a different way. Acta Zool 105(4):496–513

Heilpern SA, Sethi SA, Barthem RB, Batista VDS, Doria CR, Duponchelle F et al. (2022) Biodiversity underpins fisheries resilience to exploitation in the Amazon river basin. Proc R Soc B 289(1976):20220726

Hortal J, de Bello F, Diniz-Filho JAF, Lewinsohn TM, Lobo JM, Ladle RJ (2015) Seven shortfalls that beset large-scale knowledge of biodiversity. Annu Rev Ecol Evol Syst 46:523–549

Hubbs, CL, Lagler, KF, Smith, GR, 2004. Fishes of the Great Lakes region, revised edition. University of Michigan Press, Ann Arbor, p. 332

Hulthén K, Hill JS, Jenkins MR, Langerhans RB (2021) Predation and resource availability interact to drive life-history evolution in an adaptive radiation of livebearing fish. Front Ecol Evol 9, 619277

Hutchinson G (1957) Concluding remarks. Cold Spring Harb Symp Quant Biol 22:415–427

Jézéquel C, Tedesco PA, Bigorne R, Maldonado-Ocampo JA, Ortega H, Hidalgo M et al. (2020) A database of freshwater fish species of the Amazon Basin. Sci Data 7(1):96

Johnson JB, Zúñiga-Vega JJ (2009) Differential mortality drives life-history evolution and population dynamics in the fish *Brachyrhaphis rhabdophora*. Ecology 90(8):2243–2252

Jolliffe, IT, 2002. Principal component analysis. Springer, NY

Junk WJ (1984) Ecology of the várzea, floodplain of Amazonian whitewater rivers. In the Amazon: limnology and landscape ecology of a mighty tropical river and its basin. Springer Netherlands, Dordrecht, 215–243pp

Junk WJ, Piedade MTF, Schöngart J, Cohn-Haft M, Adeney JM, Wittmann F (2011) A classification of major naturally-occurring Amazonian lowland wetlands. Wetlands 31(4):623–640

Klausmeier CA, Kremer CT, Koffel T (2020). Trait-based ecological and eco-evolutionary theory. Theor Ecol, 161–194

Kolmann MA, Hughes LC, Hernandez LP, Arcila D, Betancur-R R, Sabaj MH et al. (2021) Phylogenomics of piranhas and pacus (*Serrasalmidae*) uncovers how dietary convergence and parallelism obfuscate traditional morphological taxonomy. Syst Biol 70(3):576–592

Kolmann MA, Marques FP, Weaver JC, Dean MN, Fontenelle JP, Lovejoy NR (2022) Ecological and phenotypic diversification after a continental invasion in Neotropical freshwater stingrays. Integr Comp Biol 62(2):424–440

Lasso CA, Morales-Betancourt MA (2022) Peces del fondo del río Orinoco y afluentes principales (Colombia-Venezuela): diversidad y aspectos bioecológicos. XXI. Serie Editorial Recursos Hidrobiológicos y Pesqueros Continentales de Colombia. Instituto de Investigación de Recursos Biológicos Alexander von Humboldt, Bogotá, pp 361

Layman CA, Winemiller KO, Arrington DA, Jepsen DB (2005) Body size and trophic position in a diverse tropical food web. Ecology 86(9):2530–2535

Leitão RP, Zuanon J, Villéger S, Williams SE, Baraloto C, Fortunel C et al. (2016) Rare species contribute disproportionately to the functional structure of species assemblages. Proc R Soc B 283(1828):20160084

Lima AC, Sayanda D, Agostinho CS, Machado AL, Soares AM, Monaghan KA (2018) Using a trait-based approach to measure the impact of dam closure in fish communities of a Neotropical River. Ecol Freshwater Fish 27(1):408–420

Lima MS, Martins SS, Montag LFA, Freitas TMS (2023) Dietary shift of a pimelodid catfish in response to the flood pulse in the Xingu River. Neotrop Ichthyol 21(4):e230097

López-Fernández H, Arbour J, Willis S, Watkins C, Honeycutt RL, Winemiller KO (2014) Morphology and efficiency of a specialized foraging behavior, sediment sifting, in Neotropical cichlid fishes. PLoS One 9(3):e89832

López-Rodríguez NC, Leão AH, Rocha RM, Prudente BS, Montag LF (2021) Environmental influence on the reproductive strategy of *Helogenes marmoratus* (Siluriformes: Cetopsidae) in the Amazonian streams. Neotrop Ichthyol 19(4):e210092

Lujan NK, Armbruster JW (2012) Morphological and functional diversity of the mandible in suckermouth armored catfishes (Siluriformes: Loricariidae). J Morphol 273(1):24–39

Lujan NK, Winemiller KO, Armbruster JW (2012) Trophic diversity in the evolution and community assembly of loricariid catfishes. BMC Evol Biol 12(1):1–13.

Lujan NK, Roach KA, Jacobsen D, Winemiller KO, Vargas VM, Ching VR, Maestre JA (2013) Aquatic community structure across an Andes-to-Amazon fluvial gradient. J Biogeography 40(9):1715–1728

Magurran AE, Deacon AE, Moyes F, Shimadzu H, Dornelas M, Phillip DA, Ramnarine IW (2018) Divergent biodiversity change within ecosystems. Proc Natl Acad Sci 115(8):1843–1847

Matamoros WA, McMahan CD, Chakrabarty P, Albert JS, Schaefer JF (2015) Derivation of the freshwater fish fauna of Central America revisited: Myers's hypothesis in the twenty-first century. Cladistics 31(2):177–188

Matthews, WJ, 2012. Patterns in Freshwater Fish Ecology. Springer Science & Business Media, Amsterdam

McKenna MC (1972) Possible biological consequences of plate tectonics. BioScience 22(9):519–525

McNab , BK, 2002. The physiological ecology of vertebrates: a view from energetics. Cornell University

Melo BF, Albert JS, Dagosta FC, Tagliacollo VA (2021) Biogeography of curimatid fishes reveals multiple lowland–upland river transitions and differential diversification in the Neotropics (Teleostei, Curimatidae). Ecol Evol 11(22):15815–15832

Myers GS (1949) Salt-tolerance of fresh-water fish groups in relation to zoogeographical problems. Bijdr Dierkd 28(1949):315–322

Newman ME (2001) Clustering and preferential attachment in growing networks. Phys Rev E 64(2), 025102

Oliveira EF, Goulart E, Breda L, Minte-Vera CV, Paiva LRDS, Vismara MR (2010) Ecomorphological patterns of the fish assemblage in a tropical floodplain: effects of trophic, spatial and phylogenetic structures. Neotrop Ichthyol 8:569–586

Ota RR, Deprá GD, Graça WJ, Pavanelli CS (2018) Peixes da planície de inundação do alto rio Paraná e áreas adjacentes: revised, annotated and updated. Neotrop Ichthyol 11(16):e170094

Pauly D, Christensen V, Walters C (2000) Ecopath, Ecosim, and Ecospace as tools for evaluating ecosystem impact of fisheries. ICES J Mar Sci 57(3):697–706

Pigliucci M (2003) Phenotypic integration: studying the ecology and evolution of complex phenotypes. Ecol Lett 6(3):265–272

Queiroz, LJ, Torrente-Vilara, G, Ohara, WM, Pires, THS, Zuanon, J, Doria, CRC, 2013. Peixes do Rio Madeira, 1 edn. Dialeto Latin American Documentary, São Paulo

Reis RE, Kullander SO, Ferraris CJ (Eds.) (2003). Check List of the Freshwater Fishes of South and Central America. Edipucrs, Porto Alegre

Reznick DA, Bryga H, Endler JA (1990) Experimentally induced life-history evolution in a natural population. Nature 346(6282): 357–359

Ribeiro DC, Chagas JM, Thereza MR, Langeani F (2019) Checklist and key for the identification of fish fauna of the Uberaba River, Upper Paraná River system, Brazil. Zookeys 875:129

Saint-Paul U, Soares GM (1987) Diurnal distribution and behavioral responses of fishes to extreme hypoxia in an Amazon floodplain lake. Environ Biol Fishes 20:91–104

Seddon N, Mace GM, Naeem S, Tobias JA, Pigot AL, Cavanagh R et al. (2016) Biodiversity in the Anthropocene: prospects and policy. Proc R Soc B 283(1844):20162094

Soares BE, Benone NL, Leitão RP, Leal CG, Santos LL, de Assis Montag LF, Caramaschi ÉP (2023) The ecomorphological diversity of Amazonian stream fishes is constrained by phylogenetic relationships. EcoEvoRxiv, https://doi.org/10.32942/X2N59N

Stearns SC (2000) Life history evolution: successes, limitations, and prospects. Naturwissenschaften 87:476–486

Strahler AN (1957) Quantitative analysis of watershed geomorphology. Eos Trans Am Geophys Union 38(6):913–920

Tagliacollo VA, Dagosta FCP, Pinna MD, Reis RE, Albert JS (2021) Assessing extinction risk from geographic distribution data in Neotropical freshwater fishes. Neotrop Ichthyol 19:e210079

Torgersen KT, Bouton BJ, Hebert AR, Kleyla NJ, Plasencia X, Rolfe GI et al. (2023a) Phylogenetic structure of body shape in a diverse inland ichthyofauna. Sci Rep 13(1):20758

Torgersen KT, Bogotá-Gregory JD, Acosta-Santos A, Potosí-Chuña JJ, Agudelo-Cordoba E, Albert JS (2023b) Gymnotiform electric fishes of the Tres Fronteras region of the western Amazon. Check List 19(5):767–790

Val AL (2019) Fishes of the Amazon: diversity and beyond. An Acad Bras Cienc 91:e20190260

Val AL, Gonzalez RJ, Wood CM, Wilson RW, Patrick ML, Bergman HL, Narahara A (1998) Effects of water pH and calcium concentration on ion balance in fish of the Rio Negro, Amazon. Physiol Zool 71(1):15–22

Valentine JW (1969) Patterns of taxonomic and ecological structure of the shelf benthos during Phanerozoic time. Palaeontology 12(4):684–709

van der Sleen P, Albert JS (Eds.) (2017). Field Guide to the Fishes of the Amazon, Orinoco, and Guianas. Princeton University Press, Princeton, Vol. 115

Vander Vorste R, McElmurray P, Bell S, Eliason KM, Brown BL (2017) Does stream size really explain biodiversity patterns in lotic systems? A call for mechanistic explanations. Diversity 9(3):26

Vanegas-Ríos JA, Azpelicueta MD, Malabarba LR (2018) A new species of *Diapoma* (Characiformes, Characidae, Stevardiinae) from the Rio Paraná basin, with an identification key to the species of the genus. J Fish Biol 93(5):830–841

Vannote RL, Sweeney BW (1980) Geographic analysis of thermal equilibria: a conceptual model for evaluating the effect of natural and modified thermal regimes on aquatic insect communities. Am Nat 115(5):667–695

Villa-Navarro FA, Acero AR, Cala P (2017) Taxonomic review of trans-Andean species of *Pimelodus* (Siluriformes: Pimelodidae), with the descriptions of two new species. Zootaxa 4299(3):337–360

Villéger S, Brosse S, Mouchet M, Mouillot D, Vanni MJ (2017) Functional ecology of fish: current approaches and future challenges. Aquat Sci 79:783–801

Waddell JC, Crampton WGR (2022) Reproductive effort and terminal investment in a multi-species assemblage of Amazon electric fish. Ecol Monograph 92:e1499

Waddell JC, Njeru SM, Akhiyat YM, Schachner BI, Correa-Roldán EV, Crampton WG (2019) Reproductive life-history strategies in a species-rich assemblage of Amazonian electric fishes. PLoS One 14(12):e0226095

Walters C, Pauly D, Christensen V (1999) Ecospace: prediction of mesoscale spatial patterns in trophic relationships of exploited ecosystems, with emphasis on the impacts of marine protected areas. Ecosystems 2:539–554

Weitzman SH, Vari RP (1988) Miniaturization in South American freshwater fishes; an overview and discussion. Proc Biol Soc Wash 101:444–465

Winemiller KO (1990) Spatial and temporal variation in tropical fish trophic networks. Ecol Monograph 60(3):331–367

Winemiller KO (1991) Ecomorphological diversification in lowland freshwater fish assemblages from five biotic regions. Ecol Monograph 61(4):343–365

Winemiller KO (2005) Life history strategies, population regulation, and implications for fisheries management. Can J Fish Aquat Sci 62(4):872–885

Winemiller KO, Flecker AS, Hoeinghaus DJ (2010) Patch dynamics and environmental heterogeneity in lotic ecosystems. J North Am Benthol Soc 29(1):84–99

Winemiller KO, Fitzgerald DB, Bower LM, Pianka ER (2015) Functional traits, convergent evolution, and periodic tables of niches. Ecol Lett 18(8):737–751

Zuanon, J, Ferreira, E, 2008. Feeding ecology of fishes in the Brazilian Amazon-a naturalistic approach, Feeding and Digestive Functions of Fishes. Science Publishers Inc., USA, New Yoro, 589pp. 1–34

Ecotrace Analyses: Evolution of Ecological Traits in Neotropical Freshwater Fishes

7

James S. Albert, Matthew R. Brander, Milton Tan and Kevin T. Torgersen

Abstract

This chapter summarizes the main patterns of ecotrait macroevolution of the Big Nine NFF clades. All clades originated in small upland rivers as invertivores, with small to moderate body sizes. Only the otophysan taxa evolved to larger body sizes and to dominate the fish faunas of large lowland rivers, while the secondary freshwater taxa (cichlids and cyprinodontiforms) tend to retain smaller body sizes and dominate the fish faunas of smaller upland rivers and streams. The two cyprinodontiform clades show little or no trophic evolution, with Rivulidae entirely lacking piscivores, and only one species of poecillioid being piscivorous; *Belonesox belizanus* from Central America. Body-size evolution is strongly correlated with evolution along into large rivers, although many large river species retain small body sizes. There are multiple transitions to both large and small body sizes from ancestral intermediate conditions in Cichlinae and Otophysi, while cyprinodontiforms always stay small, with the largest species (*Titanolebias*) in the temperate zone of Argentina. The highest rates of habitat evolution are out of caves and lakes into surface waters, and from clearwaters to white or blackwaters. The fastest evolving dietary traits, and the most asymmetric, are the evolution to and from folivory and frugivory. There is greater variance in trophic position at low transition rates, or higher variance in transition rates at lower trophic levels. The life history traits examined are highly conserved. The Big Nine NFF clades representing 95% of all NFF species originated during the Upper Cretaceous and Paleogene in small upland rivers and streams as invertivores and microvores with small to intermediate body sizes. This suite of ecotraits differs from the more general condition of freshwater fish faunas observed on the other ice-free continents, where most freshwater fishes originated as piscivores in large lowland river basins and coastal estuarine habitats.

7.1 Evolution of Ecological Traits

It might have been thought (and was in ancient times thought) that those parts of the structure which determined the habits of life, and the general place of each being in the economy of nature, would be of very high importance in classification. Nothing can be more false. No one regards the external similarity of a mouse to a shrew, of a dugong to a whale, of a whale to a fish, as of any importance (Darwin 1859, p. 414).

7.1.1 Form and Function in Fish Names

Early classifications of fishes were based on superficial similarities of external body shape and fin positions closely associated with behavior and habitat utilization. For example, Linnaeus (1758) allocated all fishes then known into four orders: Apodes (eels), Jugulares (fish with pelvic fins anterior to pectoral fins, e.g., cod), Thoracici (fish with pelvic fins below or near the pectoral fins, e.g., perch), and Abdominales (fish with pelvic fins well posterior to pectoral fins, e.g., salmon). Fish systematists have labored for almost 300 years to overcome groupings based on superficial similarities, often associated with ecological factors

J. S. Albert (✉) · M. R. Brander · K. T. Torgersen
University of Louisiana at Lafayette, Lafayette, LA, USA
e-mail: jsalbe@gmail.com

M. R. Brander
e-mail: matthew.brander1@louisiana.edu

K. T. Torgersen
e-mail: kevintorgersen@gmail.com

M. Tan
Illinois Natural History Survey, Urbana, IL, USA
e-mail: miltont@illinois.edu

J. S. Albert and R. E. Reis (eds.), *Ecology and Evolution of Amazonian Fishes*, Fascinating Life Sciences,
https://doi.org/10.1007/978-3-032-09328-8_7

like diet and locomotion (see Regan 1904; Greenwood et al. 1966). Through the centuries, the goal has always been to discover the "Natural System" (*sensu* Linnaeus 1758), later recognized as the "Phylogenetic System" (*sensu* Hennig 1966) based on evolutionary relationships (Betancur-R et al. 2017; Hughes et al. 2018).

Most morphology-based systematic studies construct trees, and therefore the classification, using hard physical structures of mineralized tissues, e.g., scales, bones, and cartilages. Characters of soft tissues (e.g., muscles, digestive tract, nervous systems, gonads) are sometimes employed (e.g., Winterbottom 1973; Albert et al. 1998; Birindelli and Shibatta 2011; Thacker 2011; Birindelli et al. 2012; Fontenelle and de Carvalho 2016; Pereira Arantes et al. 2017; Girard et al. 2020), but these are widely regarded to be phenotypically variable and phylogenetically labile (Strauss and Bond 1990; Blomberg et al. 2003). Most systematists avoid using these supposedly more labile traits. In other words, there has long been a preference for the study of form, and not function, in discovering the Natural System (see a history of this ancient debate in Russell 1916). This primacy of form over function has transcended the molecular revolution in systematics, with great efforts taken to exclude genes and gene sequences thought to have evolved under strong selection (Li et al. 2007; Alda et al. 2021).

Yet naturalists have long recognized the similarities of behavioral and ecological traits among fish groups. Generations of ichthyologists have observed the clustering of functional and ecological traits among species in higher taxonomic groups (Eigenmann 1906; Myers 1949 Lowe-McConnell 1975; Page and Swofford 1984; Albert 2001; Albert and Crampton 2005; Crampton 2011; Floeter et al. 2018; Borstein et al. 2019; Torgersen et al. 2023). Many of the traits that allow fishes to feed, survive, and reproduce in distinct environments are shared among close relatives, and are reasonably interpreted to represent similarities inherited from a common ancestor (Hall 2013; Hodos 2019). These traits include morphological aspects of body shape, mouth position, dentition, and coloration, physiological traits affecting thermal and water chemistry tolerances (Chaps. 8–15), and behavioral traits associated with foraging, reproduction, and migration (Chaps. 16–20). Some examples of phylogenetic conservatism in fish behavior and ecology include migratory behaviors (Melo et al. 2021; Herrera-R et al. 2024), climate-induced range shifts (Comte et al. 2014), and thermal tolerances (Comte and Olden 2017). The idea that closely related species share similar behavioral traits and ecological functions due to shared common ancestry is easy to understand and well documented in freshwater fishes (Burr and Mayden 1992; Chen et al. 2013, 2023; Ross and Matthews 2014; Tao et al. 2019; Lavoué 2020; Cheng et al. 2022; van der Sleen and Albert 2022; Melo and Stiassny 2024).

7.1.2 Fishes Named for Behaviors and Habitats

The common or vernacular names of many fish groups are based on characteristic behaviors and activities. Some well-known examples in English include Climbing perches (Anabantidae) which crawl short distances on land using paired fins and axial muscles, Flying fishes (Exocoetidae) which glide above the water using enlarged pectoral fins, Drums and Croakers (Sciaenidae), Grunts (Haemulidae), and Toadfishes (Batrachoididae) which produce sounds using swim bladder muscles, Cleaner wrasses (five species of *Labroides*, Labridae) which eat ectoparasites from larger fishes, Jawfishes (Opistognathidae) which brood eggs in their mouth until they hatch, Triggerfishes (Balistidae) which lock their dorsal-fin spine when threatened, and Flashlight fish (Anomalopidae) which emits light from bioluminescent organs under their eyes. Many other fish groups are named for their characteristic habitats. Some examples, again in English, include Reef Sharks (Carcharhinidae) which inhabit the waters around coral reefs, Mudskippers (23 species in ten genera of Oxudercinae) on intertidal mudflats, Deep-Sea Anglerfishes (Ceratioidei) in the bathypelagic realm, and Hillstream loaches (Balitoridae) in fast-flowing rivers and streams of south and southeast Asia.

The common names of many Amazonian fishes are derived from Indigenous words for their characteristic behaviors and habitats (López and Hernández 1986; Cabalzar et al. 2005; Tobes et al. 2022; Bogotá-Gregory et al. 2024; Castro e Silva and Hünemeier 2025). In particular, the etymological roots of many Amazonian fish names are derived from the Tupi-Guarani language family, which served as the *Língua Geral* (i.e., common language) across much of tropical South America for many centuries before and after the European conquest (Bueno 1974; Epps 2020). The meanings of Tupi words presented below are taken from the *Dicionário de Tupi Antigo* (Navarro 2013). The name *Arraia* or *Raia* for freshwater stingrays (Potamotrygonidae) is formed from the old Tupi word *ará* meaning "flat and wide" and the suffix *ia* used to indicate agency or motion. For example, the English word jaguar comes from the Tupi *îagûara* combined from *îagûa* meaning "to hunt" and *ia* for agency, indicating "the one who hunts." The name *Tucunaré* for the Peacock Bass *Cichla* (Cichlidae) is derived from *tukũ* or *tukuna* meaning "tree trunk" or "wood," and *aré* meaning "belonging to," referring to the habitat of hiding near submerged logs and tree roots.

The common names of many commercially important Amazonian fishes are also derived from old Tupi words describing their behaviors. The name *Curimatã* for *Prochilodus* (Prochilodontidae) is formed from the Tupi words *curi* meaning "to scrape" or "to nibble" and *matã* meaning "tough, durable, father," referring to the characteristic behavior of feeding on muddy riverbeds. The name *Jaraqui* for *Semaprochilodus* (Prochilodontidae) is formed from *iara* a mythical water spirit and *qui*, in this case meaning "small" or "sharp," describing the behavior as a "swift swimmer." Some well-known examples in the family Serrasalmidae include the names *Piranha* for *Pygocentrus* and *Serrasalmus* referring to their aggressive feeding behaviors (*pirãîa* means "tooth fish"), and *Tambaqui* for *Colossoma macropomum* from *tamba* meaning "fruit" and in this case *qui* meaning "to eat," referring to its fruit-eating behavior. The name *Tararira* or *Trahira* for the Wolfish *Hoplias malabaricus* (Erythrinidae) is derived from *tará* meaning "luminous" and *rirá* meaning "to laugh" or "to trick," referring to the aggressive, jerky movements of this fish when striking prey. The name *Piraíba* for the giant catfish *Brachyplatystoma filamentosum* (Pimelodidae) comes from the words *pirá* for "fish" and *îba* for "big" or "fierce," and this fish is called *Valentón* in Spanish meaning "brave" or "bold." The name *Piramutaba*, the giant catfish *B. vaillantii*, is derived from *pirá* for "fish" and *mutaba* meaning "soft" or "gentle," and the Portuguese common name for this species Laulau refers to the sound of the water movements it makes while swimming.

Using behavioral and ecological attributes to name Amazonian fishes is also common in European languages. Some well-known examples include the Portuguese and Spanish names *Candirú* or *Cañero* for catfishes in the families Cetopsidae and Trichomycteridae with parasitic or scavenging behaviors, and *Carachama* or *Giri-giri* for armoured suckermouth catfishes (Loricariidae) referencing the clicking and grinding sounds they produce when handled. The name *Pavón* for *Cichla* (Cichlidae) means "peacock" in Spanish referring in part to its territorial strikes and ambush hunting, and the phrase *Peces de las Nubes* meaning "fish from the clouds" is used for small killifishes (Rivulidae) referencing their appearance in seasonal ponds after rains (van der Sleen and Albert 2017). The English common names for several Amazonian fishes are similarly based on characteristic behaviors or habitats; e.g., River stingrays (Potamotrygoninae), Splashing tetras (*Copella arnoldi*), Climbing catfishes (Astroblepidae), Driftwood catfishes (Auchenipteridae), Sand knifefishes (*Gymnorhamphichthys*), and Swamp eels (Synbranchidae).

7.1.3 Historical Ecology

The characteristic behavioral or ecological attributes of these fish groups illustrates the general tendency, observed across the Tree of Life, for closely related organisms to share ecophenotypic traits, a phenomenon referred to as *phylogenetic niche conservatism* (Prinzing et al. 2001; Wiens 2004; Wiens and Graham 2005; Buckley et al. 2010; Wiens et al. 2010; Crisp and Cook 2012). This strong tendency forms the basis for much of modern comparative studies of animal ecology and evolution (Wake 1991; Cooper et al. 2010; Wiens et al. 2010; Wake et al. 2011; Allgeier et al. 2021; Ford et al. 2022; Saban et al. 2023; Ng et al. 2024).

The term *Historical Ecology* was originally proposed to describe the study of the evolutionary history of ecological associations; i.e., mutualisms and parasitism (Brooks 1985; Brooks and McLennan 1991). The scope of this field was later expanded to include the effects of speciation and adaptation on clades and community structure (Mayden 1987, 1992; Vitt et al. 1999; Wares and Cunningham 2001). This deep-time evolutionary perspective of studying ecological phenomena may be contrasted with non-historical ecological approaches seeking to explain associations between traits and environmental variables based on current conditions alone (Boucot 1975). For example, the field of *Evolutionary Ecology* investigates the evolution of ecological traits within species based on environmental conditions as they exist in the modern world today (Pianka 2011). As an aside, the term Historical Ecology also has a different meaning in the literature, where it can refer to the history of the science of ecology (see Szabó 2015).

Historical Ecology takes an explicitly hierarchical view of the organismal *phenotype*, including organismal traits from the molecular and genetic levels to the behavioral and ecological levels of organization (Striedter and Northcutt 1991; Post and Palkovacs 2009; Hallgrímsson and Hall 2011; Fig. 7.1). The integrated phenotype includes all heritable aspects of organismal form and function, including the behaviors and eco-physiological traits associated with habitat utilization, diet, and life history (Egan et al. 2018). This broad view allows researchers to trace ecophysiological traits on a phylogeny and apply the full panoply of models used in the study of morphological and genetic traits. Such an "*integrationist*" (i.e., eco-evo-devo) view of organismal design considers the interactions among external environmental and internal organismal factors as constraints and drivers in the evolution of the integrated phenotype (Abouheif et al. 2014). Such a view contrasts with the predominantly "*externalist*" (i.e. "top-down" or "outside-in") approach to studying evolutionary causation of the Neo-Darwinian synthesis, in which organismal traits were viewed as solutions to environmental problems or selection pressures (Gould 1977; Arthur 2000). The integrationist approach also incorporates the "*internalist*" (i.e., "bottom up" or "inside out") approach of the evo-devo synthesis, which views traits as emerging from genomic and developmental dynamics under environmental constraints (Müller 2007).

7.1.4 Goals of this Chapter

This chapter describes the computational methods used to study the evolution of ecological traits (i.e., *ecotraits*) in the context of a phylogenetic tree and then traces the evolution of several key ecotraits on the phylogeny of NFF species presented in Chap. 2. These *ecotrace* analyses employ ancestral trait estimation (*sensu* Revell 2025) for all 42 ecotraits; e.g., macrohabitat by stream order (Fig. 7.2), water chemistry (Fig. 10.5), whitewater floodplain (Fig. 11.2), deep river channels (Fig. 13.5), trophic position (Figs. 7.3, 16.2, 17.1, and 18.6), and body size (Fig. 7.4).

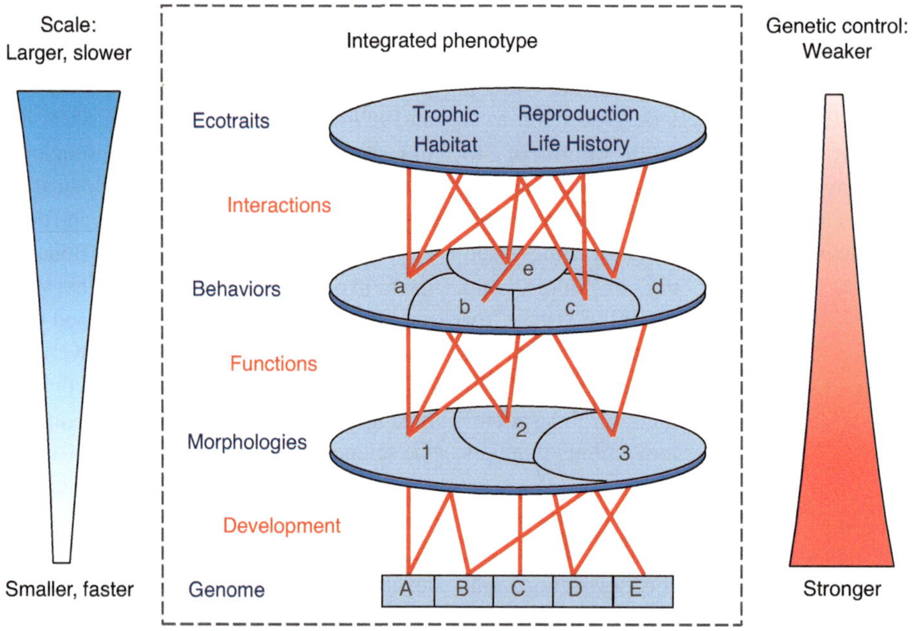

Fig. 7.1 Hierarchical organization of the extended phenotype in integrated biology. Note the many-to-many mapping of organismal traits (blue) and reciprocal information flows of biological processes (red) among hierarchical levels. Scheme of the extended phenotype modified from Striedter and Northcutt (1991). Characteristic scales and rates at left, relative influences of genetic vs. environmental factors at right. Note simplified depictions of discrete hierarchical levels and smoothly-graded scales and controls

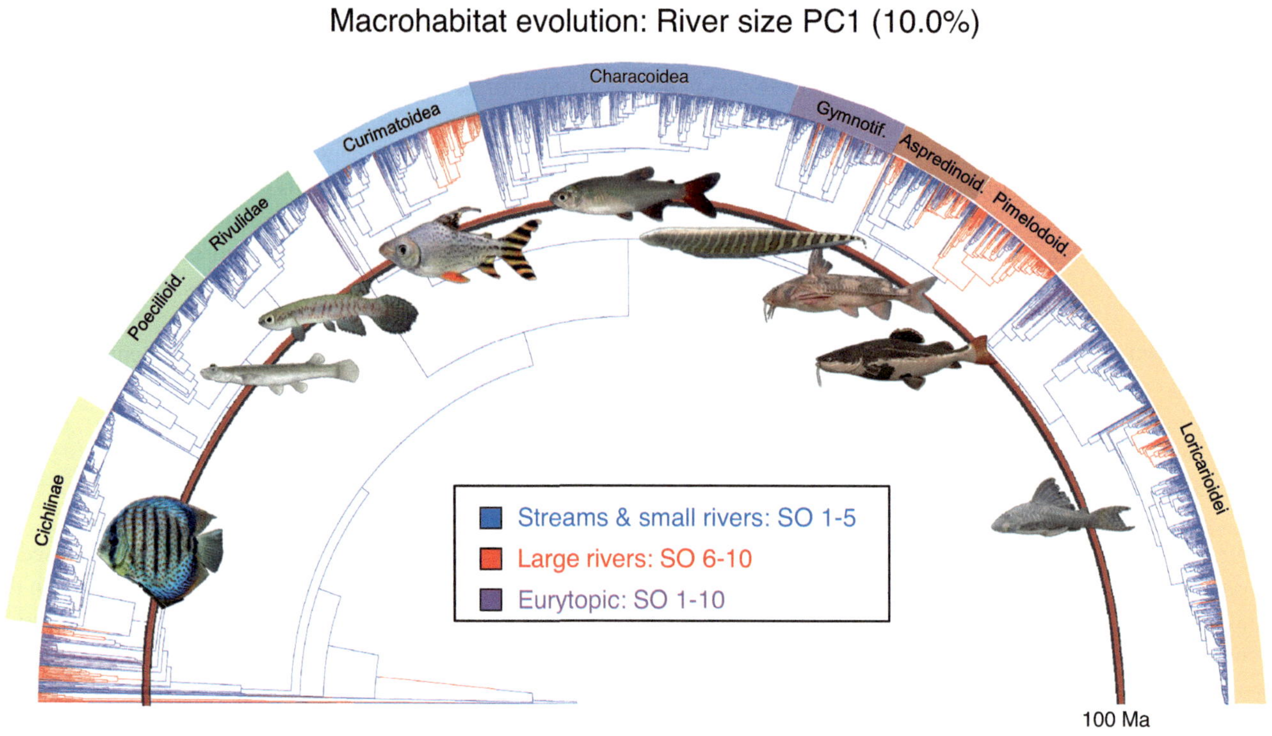

Fig. 7.2 Evolution of PC1 values from the ecospace analysis of Chap. 6, representing 10.0% of the total variance. Ecotraits loading most heavily on PC1 are related to longitudinal position along the river continuum, from streams and small rivers (SO 1–5) to large rivers (SO 6–10). Phylogeny of 3167 species (49.6% of NFF total) in 865 genera, 96 families, and 40 orders from Chap. 2; Big Nine NFF clades (colored regions) from Chap. 3. Red semicircle at c. 100 Ma marks the time of the final separation of South America and Africa and the latest origin of phylogenetically independent NFF clades (Albert et al. 2011)

7.2 Methods: Evolution of Ecological Traits

7.2.1 Multivariate Statistical Analyses

We used *principal components analysis* (PCA) to extract linear combinations of the 42 ecotrait variables that maximally explain the variance (Chap. 6). We call this multivariate hypervolume of ecological trait diversity the NFF *ecospace*. Ecospace hypervolumes for NFF clades were calculated based on 41 principal component (PC) dimensions derived from a PCA of 42 ecological traits across all 6342 NFF species. We used a custom R script to calculate the size of the ecospace, as a convex hull around all the points in the 41-dimensional hypervolume (Table 6.2).

7.2.2 Ancestral State Estimation

The fossil record represents a unique source of information in evolutionary inference, providing direct observations of ancestral phenotypes. Yet lowland tropical rivers are a poor environment for fossil preservation and recovery, and we lack fossils in most NFF clades (Lundberg et al. 1998; Albert et al. 2011). Fortunately, it is possible to study the evolution of trait values from observations of living taxa (i.e., tips of the tree) using *ancestral state estimation* (ASE), also sometimes called *ancestral state reconstruction*. ASE is a phylogenetic comparative method that involves estimating the unknown trait values of hypothetical ancestral taxa at internal nodes of a phylogenetic tree (Revell 2025). ASE infers phenotypic trait values of ancestral branches of a phylogenetic tree in the absence of a fossil record. ASE is widely used in comparative biology to estimate ancestral trait values and trace trait values from along the branches of a phylogenetic tree using one of several model-fitting methods (Cunningham et al. 1998; Martins 1999; Pagel 1999; Huelsenbeck and Bollback 2001).

This chapter presents ASE for ecotraits using maximum likelihood and the Brownian Motion model of trait evolution (O'Meara et al. 2006; Revell 2012, 2025). A critical assumption of ASE is that the phylogenetic diversity of taxa included in the study sufficiently retains information of ancient evolutionary transformations (Salisbury and Kim 2001; Slater et al. 2012). This is because extinction removes taxa and their trait values, increasing the branch lengths among the living (i.e., tip) species, which can produce positively misleading interpretations of trait evolution. This bias is readily observed in cases where ancestral trait values estimated from extant taxa alone were proved incorrect by considering the fossil record; e.g., large body size in ancestral vertebrates (Albert et al. 2009), or freshwater habitats in ancestral ray-finned fishes (Actinopterygii) (Betancur-R et al. 2015).

ASE can provide important insights into the conditions of character trait evolution when a sufficiently large representation of phylogenetic diversity is available for study (Revell

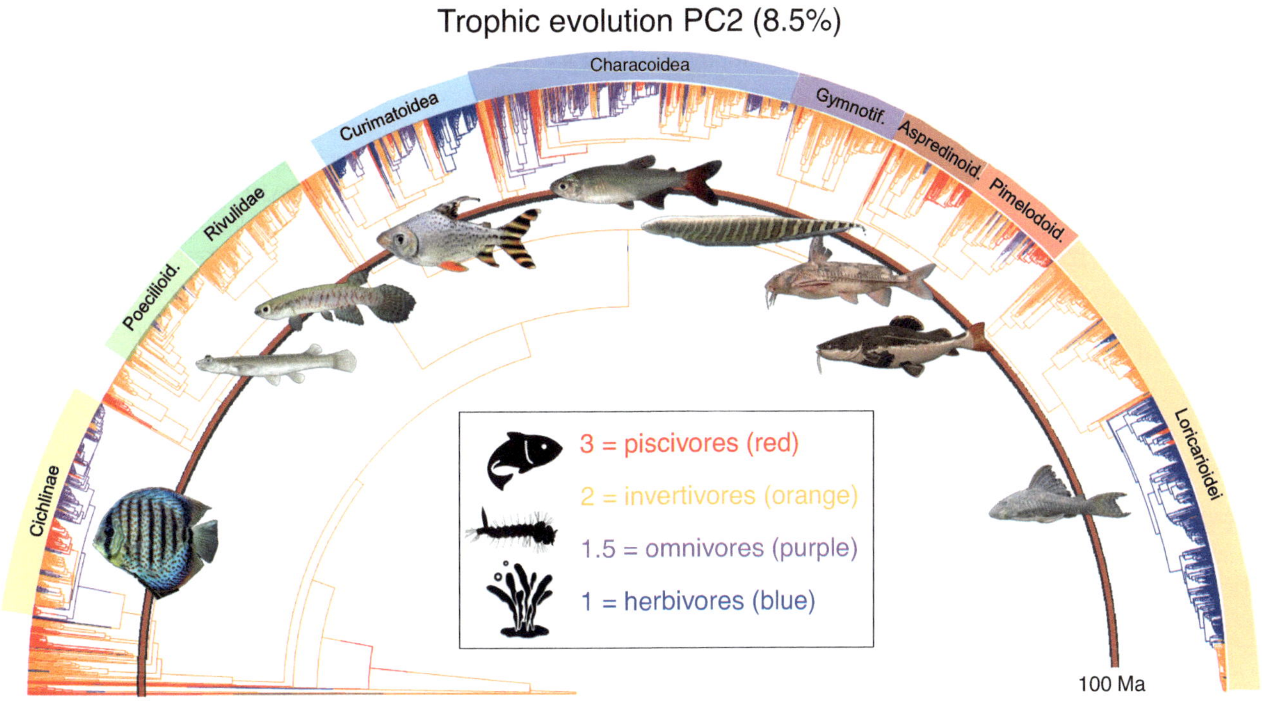

Fig. 7.3 Evolution of PC2 values from the ecospace analysis of Chap. 6, representing 8.5% of the total variance. Ecotraits loading most heavily on PC2 are related to diet. Graphical conventions as in Fig. 7.2

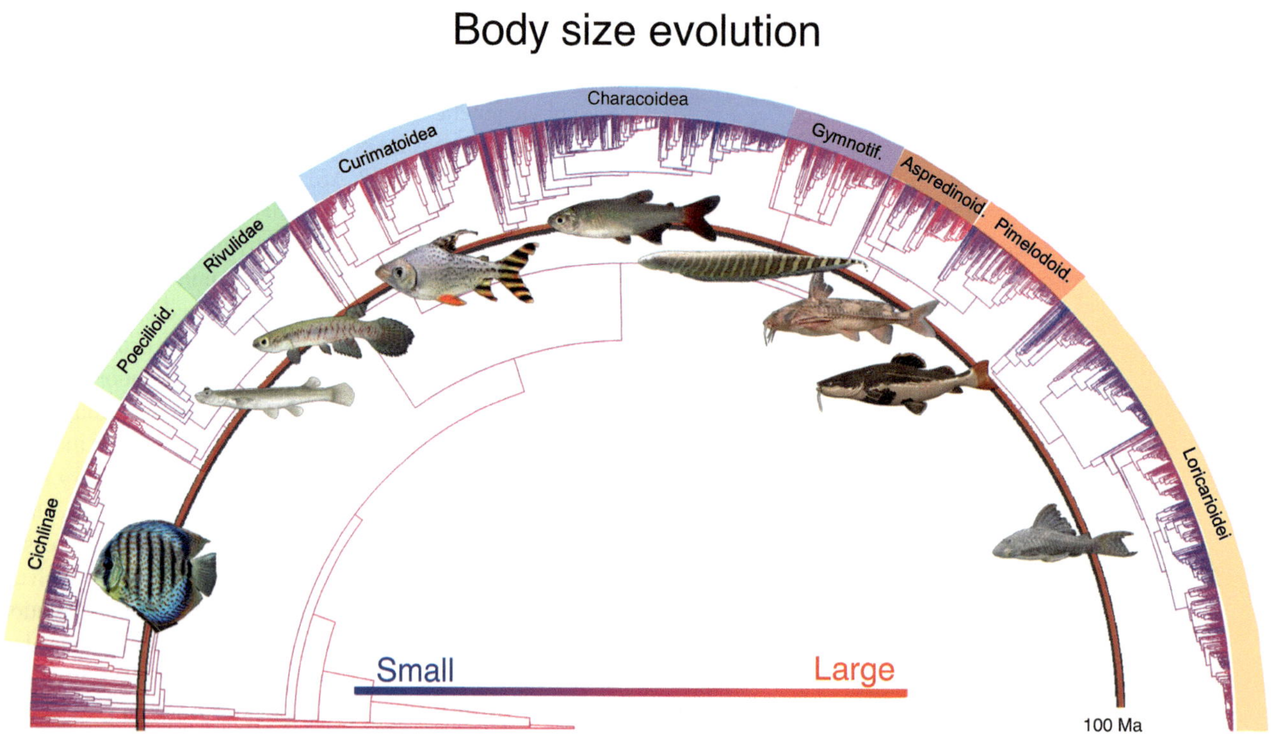

Fig. 7.4 Evolution of body size (cm SL) among NFF species. Graphical conventions as in Fig. 7.2. Note intermediate body sizes (10–99 cm SL) are estimated as the ancestral state for seven of the Big Nine NFF clades, with Poeciliodea and Rivulidae estimated to have had a small ancestral body size (2.6–9.9 cm SL)

2025). Studies using ASE have correlated the evolution of ecophenotypic traits of organisms, and with environmental and geographic conditions in which these phenotypes evolved, and evaluated the contributions of internal developmental constraints and external functional fits to environmental challenges. ASE has been successfully applied to the evolution of morphological and molecular characters (Finarelli and Flynn 2006; Ekman et al. 2008), as well as behavioral characters such as feeding modes and reproductive strategies (Hwang and Weirauch 2012; Burress and Wainwright 2019). ASE has been used to constrain estimates of ancestral habitats (Hwang and Weirauch 2012; Hundt et al. 2014) and behaviors (Ord and Cooke 2016; Rice et al. 2022) in fishes.

7.2.3 Evolutionary Rates Analyses

The phylogeny of 3169 species used in this analysis was constructed using DNA samples from GenBank and time-calibrated using the "congruification" approach (See Chap. 2 for a further discussion of the methodology used in the construction of this tree). An *ecomatrix* for the species included in this tree was constructed following the methods described in Chap. 6. As previously discussed in this chapter, mapping the traits from the ecomatrix onto the previously constructed phylogeny allows for the ancestral state values of the NFF fauna to be estimated.

For each trait, we used the fitDiscrete function in geiger (Pennell et al. 2014) to fit the Equal-Rates (ER) and All-Rates-Different (ARD) models of discrete trait evolution, which for a binary discrete trait either constraints forward transitions (q01, or $0 \rightarrow 1$) and reverse transitions (q10, or $1 \rightarrow 0$) to be of equal rates, or for these two rates to differ. We used the likelihood ratio test to determine using a two-tailed chi-squared distribution to determine if the fit of the ARD model was significantly better than the ER. If the ARD model fit significantly better, we reported these rates and used these rates for ASR; if not, we reported the ER rates and used these for ASR. Ancestral State Reconstruction for each trait was performed using the q-matrix of the best-fit model using stochastic mapping as implemented in phytools (Bollback 2006; Revell 2011), using 100 simulations.

7.3 Ecotrait Evolution

7.3.1 Macrohabitat Utilization

Ecotrait variance along PC1 reflects changes along the longitudinal river axis over regional scales (100–1000 km) from headwaters (SO 1) to river mainstems (SO 6–10) (Fig. 6.6; Table 6.2). The evolution of PC1 values on the phylogeny of NFF fishes shows that all Big Nine NFF clades originated in small rivers and streams with multiple independent evolutionary transitions into large rivers (Fig. 7.2). Ancestral state estimation of nodes older than c. 100 Ma is obscured by the massive polyphyly of the 145 clades which constitute the NFF fauna, with low and highly biased taxonomic sampling of the clades that represent deeper nodes of the phylogeny. Species with ecotraits that load most positively on PC1 (represented in red) have an adult modal utilization of large, low gradient flow rivers with pelagic and often migratory fishes. Species with ecotraits that load most negatively on PC1 (represented in blue), inhabit low gradient clearwater rivers. Species that possess ecotraits with intermediate PC1 values (represented in purple) are eurytopic, with adult modal utilization across a broad range of stream orders from smaller to larger rivers. *Phylogenetic lability*, or the propensity for evolutionary change, of habitat utilization with regard to river size is variable, with relatively high lability within the clades that comprise the primary freshwater fishes (Otophysi) and relatively low lability in the clades that comprise the secondary freshwater fishes (Cichliformes and Cyprinodontiformes) and peripheral freshwater fishes (i.e., MDLs).

Species that comprise the extant small river and stream fauna are found throughout the tree, with species from all of the Big Nine clades retaining this ancestral state. The extant large river fauna is polyphyletic in origin, being comprised of members of all of the Big Nine clades with the exceptions of the two cyprinodontiform clades. The adaptations required to inhabit large rivers evolved independently numerous times throughout the NFF fauna. Eurytopic river utilization evolved within Cichlinae, Gymnotiformes, Aspredinoidea, and Pimelodoidea, where it is particularly prominent. As noted in Chap. 6, *secondary freshwater fishes*, including Cichlinae, Poecilioidea, and Rivulidae, are predominantly restricted to the upper-right portion of the ecospace comprised by PC1 and PC2. This is the plesiomorphic region of the ecospace of Fig. 7.2, suggesting limited ecological diversification from ancestral conditions.

7.3.2 Trophic Position

Ecotrait variance along PC2 reflects changes associated with environmental drivers at the microscale (1–10 m). The ecotraits loading most strongly on this PC are related to aspects of diet and position in the water column (i.e., epibenthic and benthic invertivores vs. benthic algivores), and river substrate (i.e., gravel, sand, mud). The evolution of PC2 values on the phylogeny of NFF fishes shows that all Big Nine NFF clades originated as benthic invertivores with multiple independent evolutionary transitions into different trophic levels (Fig. 7.3). In this analysis, herbivores and detritivores (blue) consuming algae, leaves, fruits, or nuts are assigned a trophic level of 1, omnivores (purple) consuming both plant and animal tissues a trophic level of 1.5, invertivores and microvores (orange) a trophic level 2, piscivorous species (red) a trophic level of 3. See Table 6.1 for a further breakdown of the trophic categories.

Invertivory and microvory are the estimated ancestral state for each of the Big Nine NFF clades, and are the dominant condition observed in a majority of species of Poecilioidea, Rivulidae, Gymnotiformes, and Pimelodoidea. Piscivory has evolved at least once in all of the Big Nine clades with the exception of Rivulidae, including the remarkable poecillioid *Belonesox belizanus* from parts of Central America. Omnivory is prominent within the Cichlinae, Curimatoidea, Characoidea, and Aspredinoidea, whereas herbivory is widely present within Loricarioidei. Detritivory (listed under herbivory) has evolved several times within Curimatoidea.

7.3.3 Body Size

Species classified as "miniature" (darkest blue) have an adult body size of 0.9–2.5 cm SL cm, those classified as "small" (blue) 2.6–9.9 cm SL, those classified as "intermediate" (purple) 10–99 cm SL, and those classified as "large" (red) are >100 cm SL. According to this analysis, the ancestral state for the NFF fauna as a whole is to possess an intermediate body size (Fig. 7.4). This ancestral state is estimated for seven of the Big Nine clades, with the exception of Poecilioidea and Rivulidae which are estimated to have had a small body ancestral size. Within Cichlinae and the ostariophysan clades, there have been multiple evolutionary transitions to both larger and smaller body size categories from the ancestral condition of intermediate body size. Small body size is common for members of Pimelodoidea in the families Heptapteridae and Pseudopimelodidae.

There is no strict correlation between trophic level and body size. Many NFF species are small-bodied piscivores like the Leaf fish *Polycentrus jundia* (Polycentridae) at 2.8 cm, the Glass headstander *Charax hemigrammus* (Characidae) at 3.3 cm, and the Pike cichlid *Crenicichla alta* (Cichlidae) at 4.8 cm. Other NFF species are large-bodied omnivores like the electric eel *Electrophorus electricus* (Gymnotidae) at c. 200 cm, the Golden dorado *Salminus brasiliensis* (Bryconidae) at 100 cm, the Magdalena knifefish *Apteronotus magdalenensis* (Apteronotidae) at 129 cm, the lungfish *Lepidosiren paradoxa* (Lepidosirenidae) at 125 cm, and the freshwater stingray *Potamotrygon adamastor* (Potamotrygonidae) at 94 cm.

7.3.4 Rates of Ecotrait Evolution

Analyses of the rates of NFF ecotrait evolution show that traits related to water chemistry and macrohabitat are most labile, and life history traits are least labile, among the traits assessed. Macrohabitat is highest from evolution out of caves and lakes, while water chemistry is second from evolution

Table 7.1 Rates of evolution for 41 NFF ecotraits, excluding body size

Category	ET#	Ecotrait	q01	q10	qtot	q avg	q stdev
Water chemistry	2	Clearwater: high transparency, low sediments and dissolved ions, near neutral pH (5.6–8.0)	0.025	0.016	0.042		
	3	Whitewater: low transparency, high sediments and dissolved ions, near neutral pH (5.6–8.0)	0.015	0.025	0.040		
	4	Blackwater: low transparency, low sediments and dissolved ions, low pH (3.5–5.5)	0.010	0.030	0.040		
	5	Estuarine: Variable salinity from fresh to brackish	0.000	0.007	0.007	0.032	0.017
Macrohabitat	6	Low-gradient (slope less than 2%) streams and small rivers (Strahler stream orders 1–5)	0.007	0.007	0.015		
	7	High-gradient (slope more than 2%) riffles in streams and rapids in small rivers	0.000	0.062	0.062		
	8	Ancient lakes, more than 10,000 ka; e.g., volcanic calderas, tectonic rifts	0.002	0.024	0.027		
	9	Hypogean caves and aquifers	0.001	0.081	0.082		
	10	Low-gradient channels and margins of large rivers (Strahler stream orders 6–10)	0.006	0.006	0.012		
	11	High-gradient rapids of large rivers	0.001	0.045	0.047	0.041	0.028
Mesohabitat	12	Lotic (flowing) streams (English), *quebradas* (Spanish), *igarapés* (Portuguese)	0.005	0.005	0.010		
	13	Lentic (standing) waters of small rivers; pools and ponds	0.003	0.026	0.029		
	14	Seasonally dry (e.g., ephemeral) aquatic habitats	0.000	0.000	0.000		
	15	Benthic (bottom) of large river channels	0.002	0.010	0.012		
	16	Pelagic (midwater) of large river channels	0.000	0.013	0.013		
	17	Littoral (margins) of large river channels	0.002	0.009	0.011		
	18	Seasonal floodplains (oxbow lakes and flooded forest) of large rivers	0.006	0.012	0.018	0.013	0.009
Microhabitat	19	Benthic, unspecified	0.003	0.010	0.013		
	20	Benthic, mud, and silt	0.001	0.009	0.010		
	21	Benthic, sand	0.003	0.018	0.020		
	22	Benthic, gravel/rocks	0.002	0.015	0.018		
	23	Benthic, submerged wood, and vegetation	0.003	0.018	0.021		
	24	Benthic, leaf litter, and packs	0.001	0.006	0.007		
	25	Epibenthic in the water column above the bottom	0.001	0.004	0.006		
	26	Epibenthic at or near the surface	0.001	0.008	0.009	0.013	0.006
Trophic	27	Benthic microvores, small animals 0.5–5 mm	0.003	0.003	0.005		
	28	Surface invertivores: Allochthonous arthropods (insects, arachnids)	0.001	0.001	0.001		
	29	Pelagic planktivores: Autochthonous arthropod and fish larvae	0.001	0.001	0.001		
	30	Benthic detritivores: Nonliving decomposing organic (mostly plant) matter	0.000	0.004	0.004		
	31	Benthic algivores: Living algae and bacterial biofilm	0.000	0.001	0.001		
	32	Folivores: Submerged plant leaves, riverweeds	0.000	0.042	0.042		
	33	Frugivores: Fruits and seeds	0.000	0.039	0.039		
	34	Piscivores: Fish tissue	0.001	0.001	0.003		
	35	Omnivores: Both plant and animal tissues	0.001	0.018	0.019		
	36	Ectoparasites, scales, and mucous	0.000	0.013	0.013	0.013	0.016
Behavior	37	Potamodromous: longitudinal river migrations of more than 100 km	0.001	0.004	0.005		
	38	Viviparous: live birth	0.000	0.002	0.002		
	39	Aestivation: prolonged infaunal torpor (i.e., lowered metabolic activity) and desiccation	0.000	0.004	0.004		
	40	Venomous: phenotypes that deliver injected toxin(s) and associated aposematic behaviors	0.000	0.001	0.001		
	41	Passive electroception: sensory use of weak (mV) ambient electric fields	0.000	0.000	0.000		
	42	Active electroreception and electrogeneration using electric organs	0.000	0.000	0.000	0.002	0.002

Ecotrace analysis using ** from phylogeny in Fig. 2.1. q01, rates from state 0 to 1; q10, rates from state 1 to 0; qtot, total rates between states 0 and 1. Avg and stdev assessed for each ecotrait category

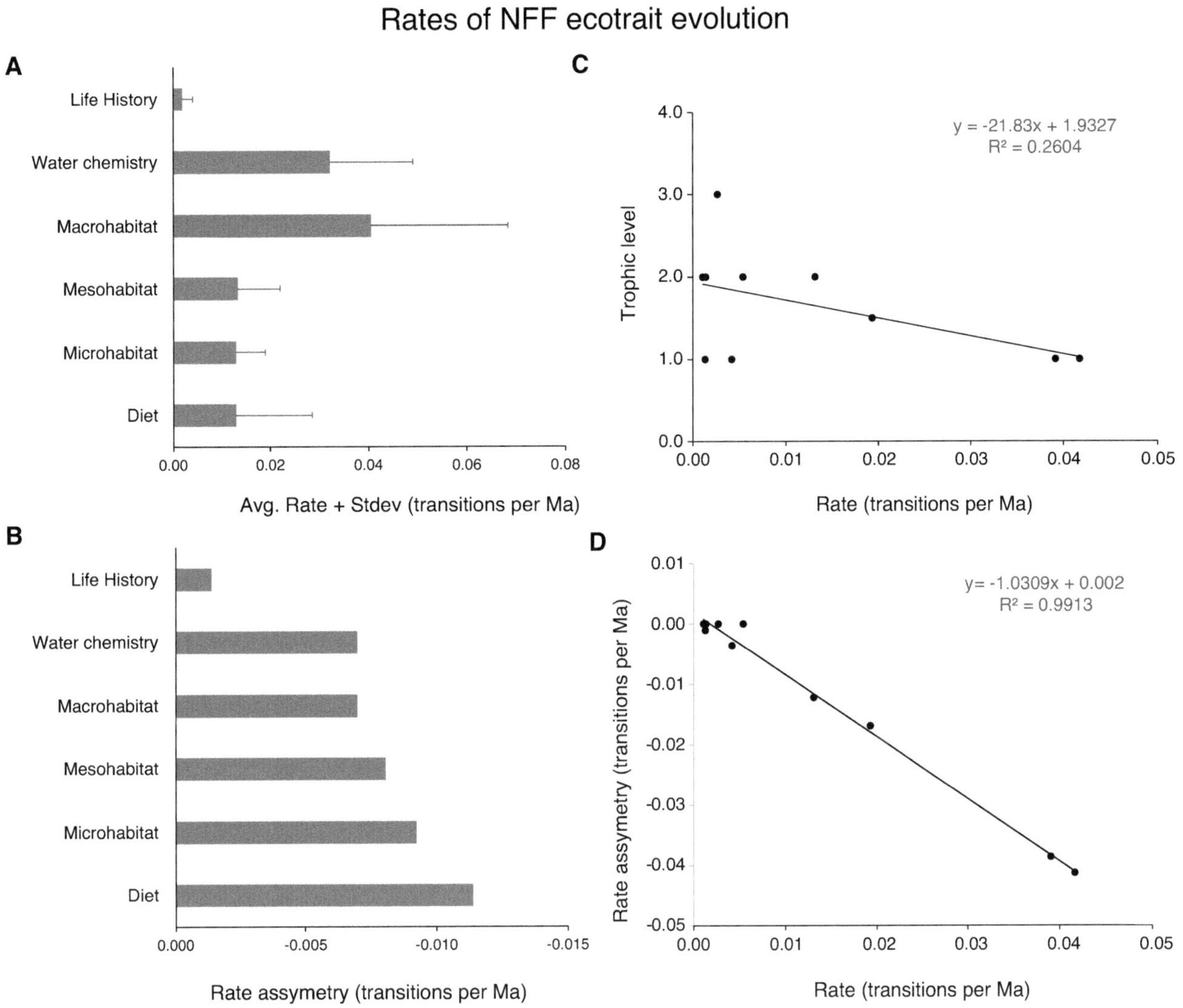

Fig. 7.5 Rates of NFF ecotrait evolution. **A** Rates by major category. Note that water chemistry and macrohabitat are most labile, and life history traits are least labile, among the traits assessed. Results summarize about 3451 trait changes on the time-calibrated phylogeny of NFF species. **B** Asymmetry of transitions (0 → 1, 1 → 0) among the Big Nine plus Cetopsidae. **C** Correlation between trophic position and evolutionary rates of the ten dietary ecotraits among NFF species. Note fastest evolving dietary traits, and the most asymmetric, are folivory and frugivory. Note also more variance in trophic position at low transition rates or more variance in transition rate at lower trophic levels. **D** Ecotraits with the highest rates also exhibit the greatest rate asymmetry; i.e., folivores and frugivores

out of clearwaters (Table 7.1). Life history traits examined are most conserved Fig. 7.5A. Asymmetry of transitions (0→ 1, 1→0) among the Big Nine plus Cetopsidae (Fig. 7.5B). Correlation between trophic position and evolutionary rates of the 10 dietary ecotraits among NFF species (Fig. 7.5C). There is strong heteroscedasticity in the relationship between variance in tropic and rate, indicating low trophic level is necessary but not sufficient to explain high rates of trophic evolution (Fig. 7.5C). Ecotraits with the highest rates also exhibit the greatest rate asymmetry; i.e., folivores and frugivores (Fig. 7.5D). Note that the fastest evolving dietary traits, and the most asymmetric, are folivory and frugivory. Note also more variance in trophic position at low transition rates or more variance in transition rate at lower trophic levels. Numbers of NFF species and body size by trophic level (Fig. 7.6). Most NFF species are invertivores and microvores, and there is no correlation between trophic level and body size (Fig. 7.6B). Trophic level 1 is the plesiomorphic condition for all Big Nine NFF clades.

7.4 Conclusions

The most important axis of ecotrait variance in NFF species relates to environmental differences along the longitudinal river axis, over regional scales (100–1000 km) from headwaters to

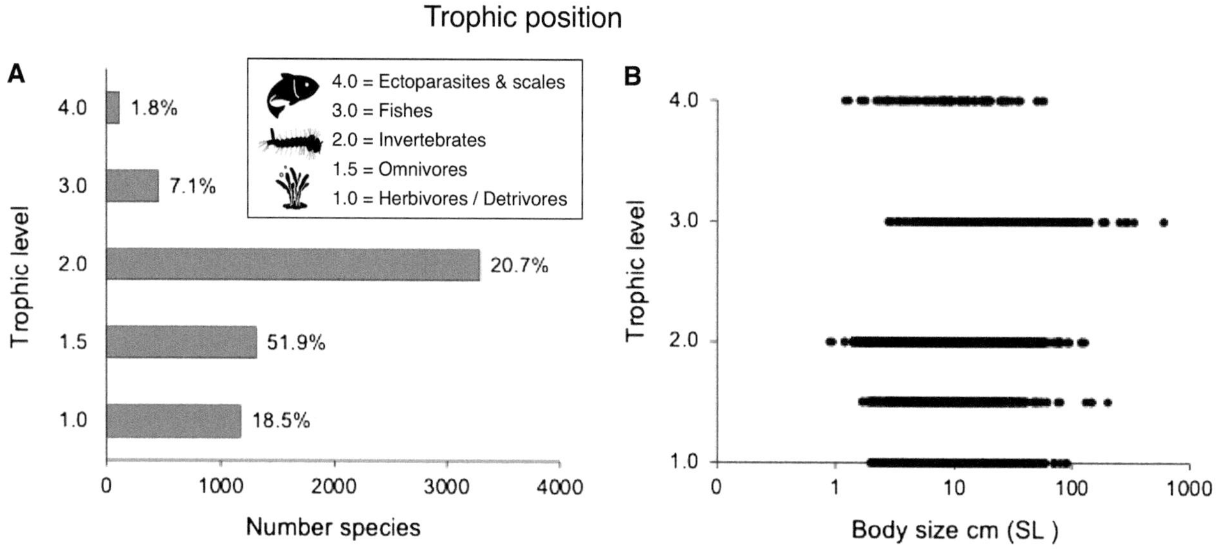

Fig. 7.6 Numbers of NFF species and body size by trophic level. **A** 6342 NFF species scored by adult modal trophic level: 1.0, algivores, detritivores, and folivores. 1.5, omnivores. 2.0, invertivores and microvores. 3.0, piscivores. 4.0, consuming ectoparasites, mucous, and scales from other fishes. **B** Maximum body size data for 5406 NFF species. Note most NFF species are invertivores and microvores, and there is no correlation between trophic level and body size. Note also that trophic level 1 is the plesiomorphic condition for all Big Nine NFF clades

river mainstems. All of the Big Nine NFF clades originated in small clearwater rivers and streams of the upland Brazilian and Guiana shields, with stream orders 1–5 (<60–70 m across), and there have been multiple independent evolutionary transitions of these clades into larger rivers with stream orders 6–10 (>60–70 m across) and to blackwater or whitewater rivers. The phylogenetic lability, or propensity for evolutionary change, among small and larger rivers is variable, with relatively high lability in Otophysan primary freshwater fishes, and relatively low lability in secondary (Cichliformes and Cyprinodontiformes) and peripheral freshwater fishes (i.e., clades of predominantly marine species). Species that comprise both small and large river faunas are polyphyletic and found throughout the phylogeny, with multiple transitions between the faunas, with species from all Big Nine clades inhabiting small rivers, and with species in seven of these clades inhabiting large rivers. The second most important axis of ecotrait variance relates to aspects of diet and position in the water column (i.e., epibenthic and benthic invertivores vs. benthic algivores), and river substrate (i.e., gravel, sand, mud). All Big Nine NFF clades originated as benthic invertivores and microvores (trophic level 2), and all these clades except Rivulidae experienced multiple independent evolutionary transitions to other trophic levels. Piscivory has evolved at least once in all of these clades except Rivulidae, omnivory is prominent in Cichlinae, Curimatoidea, Characoidea, and Aspredinoidea, and herbivory evolved several times in Loricarioidei, and detritivory (a form of herbivory) several times within Curimatoidea. The estimated ancestral body size is classified as intermediate (10–99 cm SL) in seven of the Big Nine clades, with Poecilioidea and Rivulidae having small estimated ancestral body sizes (2.6–9.9 cm SL). Miniature NFF fishes (0.9–2.5 cm SL) evolved at least 94 times in genera representing 22 families and all but two of the Big Nine clades (no miniature Curimatoidea or Gymnotiformes by the standard applied). NFF fishes with large body sizes (>100 cm SL) evolved at least 24 times, in genera representing 32 species in 18 families, including three of the Big Nine clades (Characoidei, Gymnotiformes, Pimelodoidea), nine marine derived clades in eight orders (Beloniformes, Carangiformes, Carcharhiniformes, Centropomiformes, Elopiformes, Rhinopristiformes, Lobotiformes, and Myliobatiformes), and three other primary freshwater clades (Lepidosireniformes, Lepisosteiformes, and Osteoglossiformes). Within Cichlinae and the six ostariophysan clades, there were multiple evolutionary transitions to both larger and smaller body size categories, from the ancestral intermediate body size. Small body size is common in the Pimelodoidea families Heptapteridae and Pseudopimelodidae. There is no strict correlation between trophic level and body size, with many small-bodied piscivores species and large-bodied omnivorous species in several families. Among the ecotraits examined, those related to water chemistry and macrohabitat exhibit the highest phylogenetic lability, and life history traits exhibit the lowest phylogenetic lability. Rates of macrohabitat evolution are highest out of caves and lakes, and out of clearwaters among ecotraits related to water chemistry. The ecotraits with the highest evolutionary transition rates are folivory and frugivory. There is more variance in transition rates in clades at lower trophic levels, such that low trophic position is necessary but not sufficient to explain high rates of trophic evolution.

References

Abouheif E, Favé MJ, Ibarrarán-Viniegra AS, Lesoway MP, Rafiqi AM, Rajakumar R (2014) Eco-evo-devo: the time has come. In: Ecol Genomics, pp 107–125. In: Landry C, Aubin-Horth N (eds) Ecological genomics. Advances in experimental medicine and biology, vol 781. Springer, Dordrecht

Albert JS (2001) Species diversity and phylogenetic systematics of American knifefishes (Gymnotiformes, Teleostei). Miscellaneous Publications Museum of Zoology, University of Michigan, Ann Arbor (190), pp 1–127

Albert JS, Crampton WG (2005) Diversity and phylogeny of neotropical electric fishes (Gymnotiformes). In: Electroreception. Springer New York, New York, NY, pp 360–409

Albert JS, Johnson DM, Knouft JH (2009) Fossils provide better estimates of ancestral body size than do extant taxa in fishes. Acta Zool 90:357–384

Albert JS, Lannoo MJ, Yuri T (1998) Testing hypotheses of neural evolution in gymnotiform electric fishes using phylogenetic character data. Evolution 52(6):1760–1780

Albert JS, Petry P, Reis RE (2011) Major biogeographic and phylogenetic patterns. In: Albert JS, Reis RE (eds) Historical biogeography of neotropical freshwater fishes. University of California Press, Berkeley, pp 21–57

Alda F, Ludt WB, Elías DJ, McMahan CD, Chakrabarty P (2021) Comparing ultraconserved elements and exons for phylogenomic analyses of Middle American cichlids: when data agree to disagree. Genome Biol Evol 13(8):evab161

Allgeier JE, Weeks BC, Munsterman KS, Wale N, Wenger SJ, Parravicini V et al (2021) Phylogenetic conservatism drives nutrient dynamics of coral reef fishes. Nat Commun 12(1):5432

Arthur W (2000) The concept of developmental reprogramming and the quest for an inclusive theory of evolutionary mechanisms. Evol Dev 2(1):49–57

Betancur-R R, Wiley EO, Arratia G, Acero A, Bailly N, Miya M et al (2017) Phylogenetic classification of bony fishes. BMC Evol Biol 17:1–40

Betancur-R R, Ortí G, Pyron RA (2015) Fossil-based comparative analyses reveal ancient marine ancestry erased by extinction in ray-finned fishes. Ecol Lett 18(5):441–450

Birindelli JL, Shibatta OA (2011) Morphology of the gas bladder in bumblebee catfishes (Siluriformes, Pseudopimelodidae). J Morphol 272(7):890–896

Birindelli JL, Akama A, Britski HA (2012) Comparative morphology of the gas bladder in driftwood catfishes (Siluriformes: Auchenipteridae). J Morphol 273(6):651–660

Blomberg SP, Garland T Jr, Ives AR (2003) Testing for phylogenetic signal in comparative data: behavioral traits are more labile. Evolution 57(4):717–745

Bogotá-Gregory JD, Hurtado LFJ, Samudio JFG, Córdoba EA (2024) Ethnoecology and use of fishes by the Cubeo people from the Cuduyarí River, Colombian Amazonia. J Ethnobiol Ethnomed 20(101). https://doi.org/10.1186/s13002-024-00737-1

Bollback JP (2006) SIMMAP: stochastic character mapping of discrete traits on phylogenies. BMC Bioinformatics 7:1–7

Borstein SR, Fordyce JA, O'Meara BC, Wainwright PC, McGee MD (2019) Reef fish functional traits evolve fastest at trophic extremes. Nat Ecol Evol 3(2):191–199

Boucot AJ (1975) Standing diversity of fossil groups in successive intervals of geologic time viewed in the light of changing levels of provincialism. J Paleontol 1105–1111

Brooks DR (1985) Historical ecology: a new approach to studying the evolution of ecological associations. Ann Missouri Bot Gard 72:660–680. St. Louis

Brooks DR, McLennan DA (1991) Phylogeny, ecology, and behavior: a research program in comparative biology. University of Chicago Press

Buckley LB, Davies TJ, Ackerly DD, Kraft NJ, Harrison SP, Anacker BL et al (2010) Phylogeny, niche conservatism and the latitudinal diversity gradient in mammals. Proc R Soc B 277(1691):2131–2138

Bueno FDS (1974) Grande dicionário etimológico-prosódico da língua portuguesa: vocábulos, expressões da língua geral e científica-sinônimos contribuições do tupi-guarani. Ed. Brasilia

Burr BM, Mayden RL (1992) Phylogenetics and North American freshwater fishes. In: Mayden RL (ed) Systematics, historical ecology, and North American freshwater fishes. Stanford University Press, Stanford, pp 18–75

Burress ED, Wainwright PC (2019) Adaptive radiation in labrid fishes: a central role for functional innovations during 65 My of relentless diversification. Evolution 73(2):346–359

Cabalzar et al (2005) Fishes and people of the Upper rio Tiquié. Instituto Socioambiental, Sao Paulo

Castro e Silva MA, Hünemeier T (2025) A multidisciplinary overview on the Tupi-speaking people expansion. Am J Biol Anthropol 186(1):e24876

Chen F, Xue G, Wang Y, Zhang H, Clift PD, Xing Y et al (2023) Evolution of the Yangtze River and its biodiversity. Innovation 4(3):3100417

Chen WJ, Lavoué S, Mayden RL (2013) Evolutionary origin and early biogeography of otophysan fishes (Ostariophysi: Teleostei). Evolution 67(8):2218–2239

Cheng P, Yu D, Tang Q, Yang J, Chen Y, Liu H (2022) Macro-evolutionary patterns of East Asian opsariichthyin-xenocyprinin-cultrin fishes related to the formation of river and river-lake environments under monsoon climate. Water Biol Secur 1(2):100036

Comte L, Olden JD (2017) Evolutionary and environmental determinants of freshwater fish thermal tolerance and plasticity. Global Change Biol 23(2):728–736

Comte L, Murienne J, Grenouillet G (2014) Species traits and phylogenetic conservatism of climate-induced range shifts in stream fishes. Nat Commun 5(1):5053

Cooper N, Jetz W, Freckleton RP (2010) Phylogenetic comparative approaches for studying niche conservatism. J Evol Biol 23(12):2529–2539

Crampton WGR (2011) An ecological perspective on diversity and distributions. In: Albert JS, Reis RE (eds) Historical biogeography of neotropical freshwater fishes. University of California Press, Berkeley, pp 165–192

Crisp MD, Cook LG (2012) Phylogenetic niche conservatism: what are the underlying evolutionary and ecological causes?. New Phytol 196(3):681–694

Cunningham CW, Omland KE, Oakley TH (1998) Reconstructing ancestral character states: a critical reappraisal. Trends Ecol Evolut 13(9):361–366

Darwin CR (1859) On the origin of species by means of natural selection: or, the preservation of favored races in the struggle for life. John Murray, London

Egan JP, Bloom DD, Kuo CH, Hammer MP, Tongnunui P, Iglésias SP, Sheaves M, Grudpan C, Simons AM (2018) Phylogenetic analysis of trophic niche evolution reveals a latitudinal herbivory gradient in Clupeoidei (herrings, anchovies, and allies). Mol Phylo Evol 124:151-161

Eigenmann CH (1906) The freshwater fishes of South and Middle America. Pop Sci 68(34):515–530

Ekman S, Andersen HL, Wedin M (2008) The limitations of ancestral state reconstruction and the evolution of the ascus in the Lecanorales (lichenized Ascomycota). Syst Biol 57(1):141-156

Epps P (2020) Amazonian linguistic diversity and its sociocultural correlates. In: Crevels M, Muysken P (eds) Language dispersal, diversification, and contact. Oxford University Press, Oxford, pp 275–290

Finarelli JA, Flynn JJ (2006) Ancestral state reconstruction of body size in the Caniformia (Carnivora, Mammalia): the effects of incorporating data from the fossil record. Syst Biol 55(2):301–313

Floeter SR, Bender MG, Siqueira AC, Cowman PF (2018) Phylogenetic perspectives on reef fish functional traits. Biol Rev 93(1):131–151

Fontenelle JP, de Carvalho MR (2016) Systematic implications of brain morphology in Potamotrygonidae (Chondrichthyes: Myliobatiformes). J Morphol 277(2):252–263

Ford KL, Bernt MJ, Summers AP, Albert JS (2022) Mosaic evolution of craniofacial morphologies in ghost electric fishes (Gymnotiformes: Apteronotidae). Ichthyol Herpetol 110(2):315–326

Girard MG, Davis MP, Smith WL (2020) The phylogeny of carangiform fishes: morphological and genomic investigations of a new fish clade. Copeia 108(2):265–298

Gould SJ (1977) Ontogeny and phylogeny. Belknap Press, Cambridge, MA

Greenwood PH, Rosen DE, Weitzman SH, Myers GS (1966) Phyletic studies of teleostean fishes, with a provisional classification of living forms. Bull AMNH v. 131:article 4

Hall BK (ed) (2013) Homology: the hierarchical basis of comparative biology. Elsevier, Amsterdam

Hallgrímsson B, Hall BK (eds) (2011) Epigenetics: linking genotype and phenotype in development and evolution. Univ of California Press, Berkely

Hennig W (1966) Phylogenetic systematics. University of Illinois Press, Champaign

Herrera-R GA, Heilpern SA, Couto TB, Victoria-Lacy L, Duponchelle F, Correa SB et al (2024) A synthesis of the diversity of freshwater fish migrations in the Amazon basin. Fish Fish 25(1):114–133

Hodos W (2019) The concept of homology and the evolution of behavior. In: Masterton B, Hodos W, Jeriso H (eds) Evolution, brain, and behavior, pp 153–167. Psychology Press, London/New York, 286 pp

Huelsenbeck JP, Bollback JP (2001) Empirical and hierarchical Bayesian estimation of ancestral states. Syst Biol 50(3):351–366

Hughes LC, Ortí G, Huang Y, Sun Y, Baldwin CC, Thompson AW et al (2018) Comprehensive phylogeny of ray-finned fishes (Actinopterygii) based on transcriptomic and genomic data. Proc Natl Acad Sci U S A 115(24):6249–6254

Hundt PJ, Iglésias SP, Hoey AS, Simons AM (2014) A multilocus molecular phylogeny of combtooth blennies (Percomorpha: Blennioidei: Blenniidae): multiple invasions of intertidal habitats. Mol Phylogenet Evol 70:47–56

Hwang WS, Weirauch C (2012) Evolutionary history of assassin bugs (Insecta: Hemiptera: Reduviidae): insights from divergence dating and ancestral state reconstruction. PLoS One 7(9):e45523

Ord TJ, Cooke GM (2016) Repeated evolution of amphibious behavior in fish and its implications for the colonization of novel environments. Evolution 70(8):1747–1759

Lavoué S (2020) Origins of Afrotropical freshwater fishes. Zool J Linnean Soc 188(2):345–411

Li C, Ortí G, Zhang G, Lu G (2007) A practical approach to phylogenomics: the phylogeny of ray-finned fish (Actinopterygii) as a case study. BMC Evol Biol 7:1–11

Linnaeus C (1758) Systema Naturae per Regna Tria Naturae secundum Classes, Ordinus: vol. Editio dec (E. D. Tomus I ed), p 824

López M, Hernández R (1986) Moa. Descriptiones en Lengua Cubea de los Habitos de Algunos Peces. Editorial Townsend, Lomalina, Meta, Colombia, p 50

Lowe-McConnell RH (1975) Fish communities in tropical freshwaters. Their distribution, ecology and evolution. Longman, London and New York

Lundberg JG, Marshall LG, Guerrero J, Horton B, Malabarba MCSL, Wesselingh F (1998) The stage for Neotropical fish diversification: a history of tropical South American rivers. In: Malabarba LR, Reis RE, Vari RP, Lucena ZMS, Lucena CAS (eds) Phylogeny and classification of neotropical fishes. Edipurcs, Porto Alegre, pp 13–48

Martins EP (1999) Estimation of ancestral states of continuous characters: a computer simulation study. Syst Biol 48(3):642–650

Mayden RL (1987) Historical ecology and North American highland fishes: a research program in community ecology. In: Community and evolutionary ecology of North American stream fishes. University of Oklahoma Press, Norman, pp 210–222

Mayden RL (1992) Explorations of the past. In: Mayden RL (ed) Systematics, historical ecology, and North American freshwater fishes. Stanford University Press, Stanford, pp 3–16

Melo BF, Stiassny ML (2024) Phylogenomic and anatomical evidence for the late cretaceous diversification of African characiform fishes, including a new family, under the influence of the Trans-Saharan Seaway. Evol J Linn Soc 3(1):kzae030

Melo BF, Albert JS, Dagosta FC, Tagliacollo VA (2021) Biogeography of curimatid fishes reveals multiple lowland–upland river transitions and differential diversification in the Neotropics (Teleostei, Curimatidae). Ecol Evol 11(22):15815–15832

Müller GB (2007) Evo–devo: extending the evolutionary synthesis. Nat Rev Genet 8(12):943–949

Myers GS (1949) Salt-tolerance of fresh-water fish groups in relation to zoogeographical problems. Bijdr Dierkd 28(1):315–322

Navarro, E deAlmeida (2013) Dicionário de Tupi Antigo: A Língua Indígena Clássica do Brasil (in Portuguese), 1st edn. Global, São Paulo

Ng I, Bellwood DR, Strugnell JM, Parravicini V, Siqueira AC (2024) The rise of dietary diversity in coral reef fishes. Proc R Soc B 291(2029):20241004

O'Meara BC, Ané C, Sanderson MJ, Wainwright PC (2006) Testing for different rates of continuous trait evolution using likelihood. Evolution 60(5):922–933

Page LM, Swofford DL (1984) Morphological correlates of ecological specialization in darters. In: Environmental biology of darters. Springer, Netherlands, pp 103–123

Pagel M (1999) The maximum likelihood approach to reconstructing ancestral character states of discrete characters on phylogenies. Syst Biol 48(3):612–622

Pennell MW, Eastman JM, Slater GJ, Brown JW, Uyeda JC, FitzJohn RG et al (2014) geiger v2. 0: an expanded suite of methods for fitting macroevolutionary models to phylogenetic trees. Bioinformatics 30(15):2216–2218

Pereira Arantes F, Andrade Silva F, dos Santos JE, Rizzo E, Sato Y, Bazzoli N (2017) Comparative morphology of gonads from six species of fish belonging to the family Anostomidae (Characiformes: Anostomidae). Rev Biol Trop 65(2):713–723

Pianka ER (2011) Evolutionary ecology, 7th edn. Harper and Row, New York, 512 pp

Post DM, Palkovacs EP (2009) Eco-evolutionary feedbacks in community and ecosystem ecology: interactions between the ecological theatre and the evolutionary play. Philos Trans R Soc B 364(1523):1629–1640

Prinzing A, Durka W, Klotz S, Brandl R (2001) The niche of higher plants: evidence for phylogenetic conservatism. Proc R Soc Lond [Biol] 268(1483):2383–2389

Regan CT (1904) The phylogeny of the Teleostomi. J Nat Hist 13(77):329–349

Revell LJ (2012) phytools: an R package for phylogenetic comparative biology (and other things). Methods Ecol Evol 2:217–223

Revell LJ (2025) Ancestral state reconstruction of phenotypic characters. Evol Biol 52(1):1–25

Rice AN, Farina SC, Makowski AJ, Kaatz IM, Lobel PS, Bemis WE, Bass AH (2022) Evolutionary patterns in sound production across fishes. Ichthyol Herpetol 110(1):1–12

Ross ST, Matthews WJ (2014) Evolution and ecology of North American freshwater fish assemblages. Freshw Fishes North Am 1:1–49

Russell ES (1916) Form and function: a contribution to the history of animal morphology. J. Murray, London

Saban KE, Qu YF, Wiens JJ (2023) Niche conservatism over deep timescales and diverse niche axes in land vertebrates. Global Ecol Biogeogr 32(9):1522–1534

Salisbury BA, Kim J (2001) Ancestral state estimation and taxon sampling density. Syst Biol 50(4):557–564
Slater GJ, Harmon LJ, Alfaro ME (2012) Integrating fossils with molecular phylogenies improves inference of trait evolution. Evolution 66(12):3931–3944
Strauss RE, Bond CE (1990) Taxonomic methods: morphology. In: Moyle P, Schreck C (eds) Methods for fish biology. American Fisheries Society, Bethesda, 684 pp.
Striedter GF, Northcutt RG (1991) Biological hierarchies and the concept of homology. Brain Behav Evol 38(4-5):177–189
Szabó P (2015) Historical ecology: past, present and future. Biol Rev 90(4):997–1014
Tao W, Yang L, Mayden RL, He S (2019) Phylogenetic relationships of Cypriniformes and plasticity of pharyngeal teeth in the adaptive radiation of cyprinids. Sci China Life Sci 62:553–565
Thacker CE (2011) Systematics of Gobiidae. In: The biology of gobies, pp 129–136. CRC Press, Boca Raton, 702 pp
Tobes I, Carrillo-Moreno C, Guarderas-Flores L, Jácome-Negrete I, Velázquez-Cárdenas Y (2022) Ethnoichthyology and ethnotaxonomy of the Kichwa Indigenous people of Arawanu (Arajuno), in the Ecuadorian Amazon. Front Ecol Evol 10:826781
Torgersen KT, Bouton BJ, Hebert AR, Kleyla NJ, Plasencia X, Rolfe GL et al (2023) Phylogenetic structure of body shape in a diverse inland ichthyofauna. Sci Rep 13(1), 20758
van der Sleen, Albert (eds) (2017) Field guide to the fishes of the Amazon, Orinoco, and Guianas. Princeton University Press, Princeton, NJ
van der Sleen P, Albert JS (2022) Patterns in freshwater fish diversity. Reference Module Earth Syst Environ Sci 26(3):894–907
Vitt LJ, Zani PA, Espósito MC (1999) Historical ecology of Amazonian lizards: implications for community ecology. Oikos 87:286–294
Wake DB (1991) Homoplasy: the result of natural selection, or evidence of design limitations?. Am Nat 138(3):543–567
Wake DB, Wake MH, Specht CD (2011) Homoplasy: from detecting pattern to determining process and mechanism of evolution. Science 331(6020):1032–1035
Wares JP, Cunningham CW (2001) Phylogeography and historical ecology of the North Atlantic intertidal. Evolution 55(12):2455–2469
Wiens JJ (2004) Speciation and ecology revisited: phylogenetic niche conservatism and the origin of species. Evolution 58(1):193–197
Wiens JJ, Graham CH (2005) Niche conservatism: integrating evolution, ecology, and conservation biology. Annu Rev Ecol Evol Syst 36:519–539
Wiens JJ, Ackerly DD, Allen AP, Anacker BL, Buckley LB, Cornell HV et al (2010) Niche conservatism as an emerging principle in ecology and conservation biology. Ecol Lett 13(10):1310–1324
Winterbottom R (1973) A descriptive synonymy of the striated muscles of the Teleostei. Proc Acad Nat Sci Philadelphia 125:225–317

8 Amazonian Fishes and Water Chemistry

Eric B. Haddad, Jonathan Allen, William G. R. Crampton, Juan D. Bogotá-Gregory, Xavier Plasencia II, Kevin T. Torgersen and James S. Albert

Abstract

Amazonian rivers exhibit a wide range of water chemistry profiles that affect fish physiology, ecology, and evolution. Water transparency affects the visual cues used in foraging and reproduction; water transparency, electrical conductivity, and pH affect autochthonous primary productivity (i.e., nutrients); and water transparency and color together affect temperature and dissolved oxygen. Amazonian rivers could be artificially divided into three broad categories, based on a combination of their water color, sediment load, dissolved organic carbon (DOC) content, pH, and electrical conductivity (i.e., dissolved ions). Clearwaters have high transparency due to low sediment load and DOC, a near-neutral but somewhat variable pH, and low dissolved inorganic ions, reflected by their low conductivity (typically <40 μS cm^{-1}), which is a key limiting factor on nutrient abundances in most Amazonian waters. Blackwaters have high DOC and low pH, and are low in nutrients, sediments, and dissolved inorganic ions. Whitewaters have high sediment loads and dissolved inorganic ions, with a low transparency, high nutrients, a near neutral pH, and often a high DOC. Many rivers with intermediate chemical profiles between these main categories have a greenish color. The water chemistry and physical profiles of Amazonian rivers are a direct consequence of underlying geology, soils, and vegetation. Clearwaters originate from basic granitic or calcium-rich limestone parent rocks, blackwaters from aluminum-rich soils under primary rainforests, and whitewaters from geologically young soils of Andean origin. Among rivers with these three water chemistry profiles, whitewaters have the highest fish abundance and diversity. The water-chemistry origins of South American fish clades can be divided approximately evenly into three major categories, with about one-third originating in clearwaters, another third in estuarine or marine waters, and the last third in waters with other chemistry profiles; i.e., blackwaters, clearwaters, or eurytopic combinations of waters. However, most (73.5%) South American fish species are members of clades with origins in the clearwaters of the Brazilian and Guiana Shields, while marine-derived species constitute only about 10% of the fauna. A general evolutionary trend observed in all Big Nine clades is a narrowing of habitat utilization from eurytopic water chemistry tolerances to specialized utilization of rivers with one of the three main water chemistry profiles.

Keywords

Blackwaters · clearwaters · ecology · evolution · whitewaters

E. B. Haddad (✉) · J. Allen · X. Plasencia II · K. T. Torgersen · J. S. Albert
University of Louisiana at Lafayette, Lafayette, LA, USA
e-mail: eric.bruschi@hotmail.com

J. Allen
e-mail: C00495995@louisiana.edu

X. Plasencia II
e-mail: xavier.plasencia1@louisiana.edu

K. T. Torgersen
e-mail: kevintorgersen@gmail.com

J. S. Albert
e-mail: jsalbe@gmail.com

W. G. R. Crampton
University of Central Florida, Orlando, FL, USA
e-mail: crampton@ucf.edu

J. D. Bogotá-Gregory
Instituto Amazónico de Investigaciones Científicas, Bogotá, Colombia
e-mail: juandbogota@gmail.com

8.1 Amazonian Waters

8.1.1 A Brief History of Discovery

The observation that different waters have distinct kinds of fishes is widely known among Indigenous and local peoples of Amazonia (e.g., Arbeláez and Orozco 1986; Lima et al. 2005;

J. S. Albert and R. E. Reis (eds.), *Ecology and Evolution of Amazonian Fishes*, Fascinating Life Sciences,
https://doi.org/10.1007/978-3-032-09328-8_8

Silvano et al. 2022; Bogotá-Gregory et al. 2024). However, Alexander von Humboldt was the first to report such phenomena, he noted that the low abundance of fishes and other animals in black-colored water in the Atabapo river as compared to the more productive muddy rivers of the Venezuelan llanos (von Humboldt 1833). A classification of Amazonian rivers based on water color was the first proposed by Alfred Wallace, noting the preference of local communities to build towns near the mouths of clearwater or black water rivers, such as Santarém on the Tapajós River, or Manaus on the Negro River, where there are fewer insects, but also good proximity to the muddy Amazon River with its immense productivity of fishes (Wallace 1853; 1890). These basic ecological facts were common knowledge among early Amazonian researchers (Keller 1875; Brown and Lidstone 1878), leading to the title of Gordon MacCreagh's classic narrative of travel on whitewaters and blackwaters (MacCreagh 1926).

The first scientific classification of Amazonian waters was published by Sioli (1950, 1956), focusing on water color, transparency, pH, and electrical conductivity (as a proxy for dissolved ions). Sioli's approach is the basis of the artificial classification of Amazonian rivers used in hydrological and ecological studies today, which is presented in this chapter. This classification is based on the chemistry of soils, which are derived from the chemical composition of the source bedrock and overlying regolith, and the prevailing climatic conditions over millions of years that affect the sediment weathering and mobility (Leenheer 1980; Goulding et al. 1988). From a hydrological perspective, the flowing waters of a river network drain the groundwater table of the basin-wide watershed, and the chemical properties of river water, therefore, form in soils and subsoils (Stallard and Edmond 1983; Furch 1984).

8.1.2 Classification of Waters

Amazonian waters may be classified into three broad and artificial categories, based on a combination of optical and chemical properties (Fig. 8.1; Table 8.1). The main parameters are water color based on sediment load and dissolved organic carbon (DOC) content, pH, and electrical conductivity (a proxy for dissolved cations and anions), and dissolved oxygen concentrations. These properties affect many important aspects of fish ecology and physiology, including the effects of water transparency on animal visual systems, the combined effects of electrical conductivity, pH, and water transparency on within water (i.e., autochthonous) aquatic primary productivity, the effects of low dissolved oxygen, and the effect of conductivity on the electric organ discharges of gymnotiform fishes (Crampton 1998; Val et al. 1998; Moreira-Turcq et al. 2003; Arbeláez et al. 2008; Crampton 2011; 2019; Bogotá-Gregory et al. 2020, fig. 1; Ríos-Villamizar et al. 2020b, table 1).

Clearwater rivers arise in ancient (Precambrian) and highly weathered granitic rocks of the Guiana and Brazilian Shields. Under natural conditions, clearwaters exhibit high transparency, low DOC, low sediments, and low concentrations of dissolved inorganic ions reflected by their low conductivity (typically <40 μ S cm^{-1}; Ríos-Villamizar et al. 2020a; 2020b). However, many modern clearwater rivers draining areas subjected to deforestation, gold mining, and agricultural conversion (e.g., Araguaia, Tocantins) carry a relatively higher sediment load and lower transparency than they had historically, while largely retaining other aspects of their original chemical profiles. The soils in which clearwater rivers originate contain low concentrations of carbonates or bicarbonates to buffer pH changes upon the addition of acidic or basic compounds, resulting in a near-neutral to slightly acidic pH.

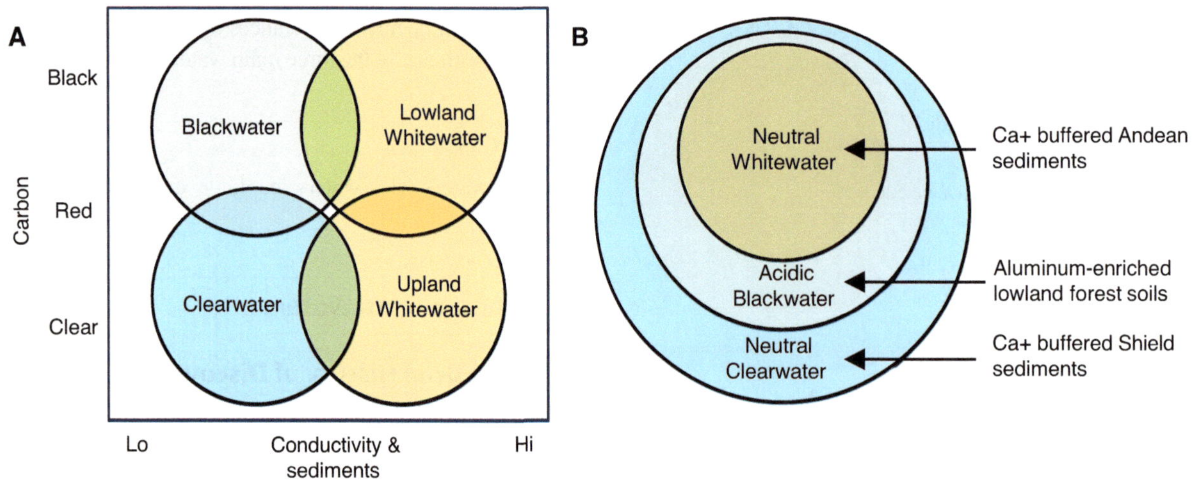

Fig. 8.1 Alternative schemes for classifying Amazonian rivers based on water conductivity and carbon load. (**A**) Two-dimensional model conceiving waters as distinct types or as intermediates. Note rivers with mixed water tributaries often have a greenish color, and oxbow lakes of lowland whitewater rivers a black color. (**B**) Nested model illustrating the cumulative roles of the two parameters in determining water chemistry profiles

Table 8.1 Summary of parameters used in the classification of Amazonian waters by optical and chemical properties

Water type	Color	Conductivity	pH	Organic load
Clearwater	Transparent	5–60	5.6–7.5	Low
Blackwater	Yellow, red, black*	5–60	3.0–5.5	High
Whitewater	Brown	80–200	5.6–7.5	High**
Estuarine	Variable	250–500	7.5–8.0	Variable

* color intensity depends on depth; ** lowland basins. Note that river water chemistry may vary seasonally, and many rivers have mixed water types depending on location in the basin and time of year.

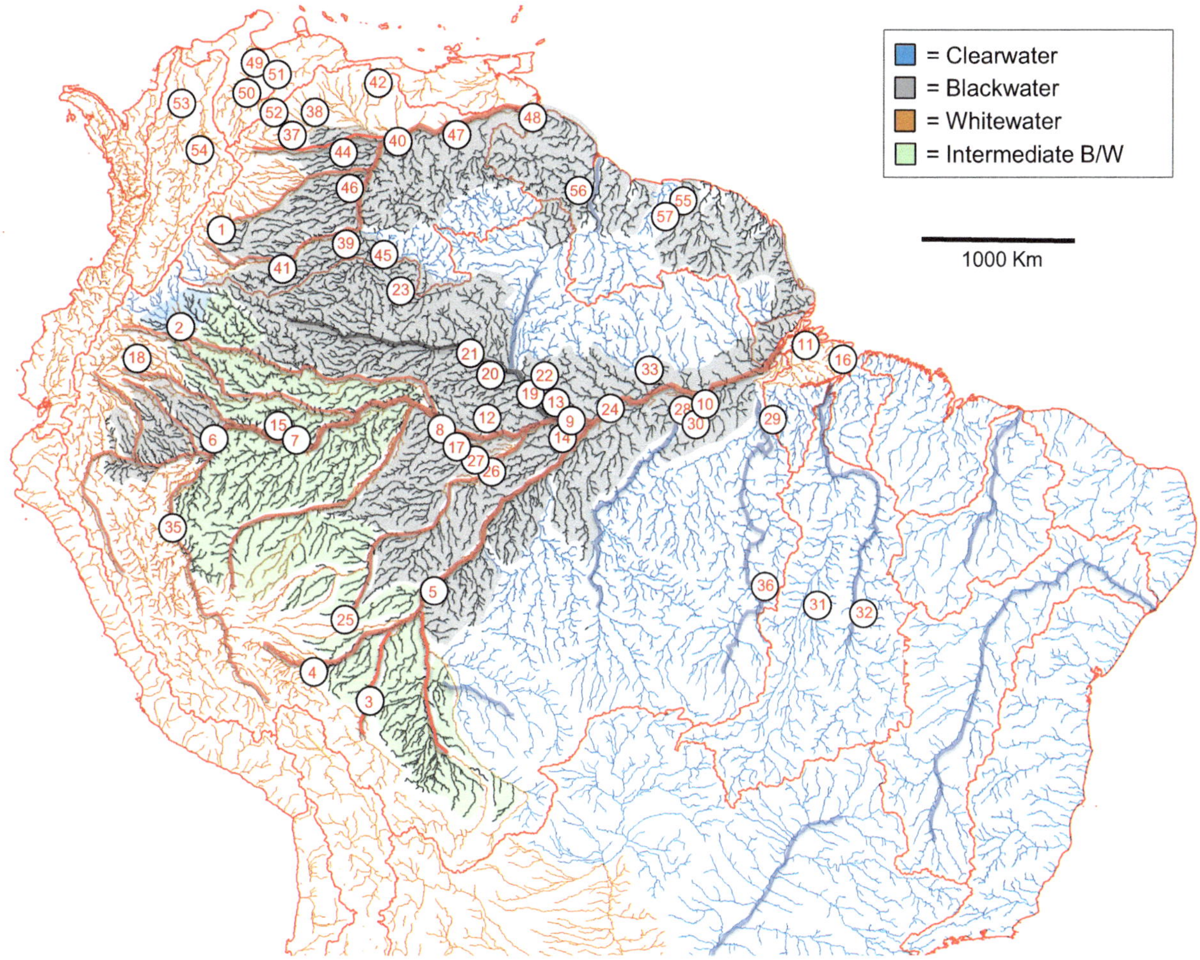

Fig. 8.2 Geographic distribution of rivers by water chemistry across northern South America. Locality and water chemistry data in Table 8.2. Note that clearwaters originate in ancient soils of the Guiana and Brazilian shields, including outcrops in the Colombian Amazon and Eastern Cordillera of Colombia, whitewaters in younger nutrient-rich soils of the Andean cordilleras and sediment fans of the foreland basin, and blackwaters from weathered acidic soils of lowland forests. Intermediate blackwater/whitewater is similar to Intermediate Type B of Ríos-Villamizar et al. (2020b)

Clearwaters are distributed in the Guiana and Brazilian Shields, including much of the São Francisco, Paraná, and Coastal Atlantic basins. Some Amazonian examples are the Araguaia (pH 6.0–7.3, 8–63.6 $\mu S\,cm^{-1}$, 22–33°C) and Tocantins (pH 7.4–8.0, 88–204 $\mu S\,cm^{-1}$, 27–28°C) basins (Fig. 8.2; Table 8.2). Large clearwater rivers with high Strahler stream orders (SO 6–7) have more length and total surface area than those of rivers with other water chemistries, and slightly higher volume (Fig. 8.3A). Large shield rivers like the Tapajós, Xingu, and Tocantins form deep mouthbay (riá) lakes where they flow into the lower Amazon. Clearwaters also drain the western portions of the Guiana Shield uplifted along the southeastern margin of the Eastern Cordillera and

Table 8.2 Data for 100 localities represented by 57 numbered circles in Fig. 8.2. Some localities are represented by a single number

Basin	Subbasin	L#	Location	Habitat	pH	Cond µ Scm–1	Temperature (°C)	References
Amazon	Andes region	1	Cauqueta	Stream	6.6–7.7	15.5–55.5	22.4–25.4	Tobes et al. (2016)
Amazon	Japurá/Caquetá	2	Apoporis	Floodplain	5.3–5.7	–	24.4–37.1	Correa (2008)
Amazon	Japurá/Caquetá	2	Paramillo	Stream	7.24–7.4	8.06–8.44	18.5	Pers. obs. JDBG
Amazon	Japurá/Caquetá	2	Aguas Calientes	Stream	6.9	8.23	25.5	Pers. obs. JDBG
Amazon	Madeira	3	Beni	Stream	4.8–8.6	21–156	16.9–34	Pouilly et al. (2006)
Amazon	Madeira	4	Madre de Dios	River	4.9–8.3	6–381	18.8–25.6	Lujan et al. (2013)
Amazon	Madeira	5	Teotônio	River	4.8–6.64	4.1–40.5	23.2–29.2	Torrente-Vilara et al. (2011)
Amazon	Main stem	6	Amazon at Iquitos	River	7.31–7.5	–	27	Stallard (1980)
Amazon	Main stem	7	Amazon at Letícia	River	4.8–7.2	4.1–32.7	24.2–26.1	Arbeláez et al. (2008)
Amazon	Main stem	8	Amazon at Tefé	River	5.7–7.4	30–113.5	27.9–31.2	Crampton (1998)
Amazon	Solimões	8	Mamiraua	Floodplain	6.2–8.79	35–378	26.5–35.1	Queiroz (2007); Affonso et al. (2011, 2015); Trevisan and Jardim-Lima (2012); Borges Pedro et al. (2013)
Amazon	Solimões	8	Manacaparu	Floodplain	5.5–7	15–60	–	Junk (1973)
Amazon	Tefé	8	Lake Tefé	Floodplain	4–6	8–14	–	Henderson and Robertson (1999)
Amazon	Tefé	8	Tefé	Stream	4.9–6.2	7.6–17.3	25.3–26.9	Crampton (1998)
Amazon	Tefé	8	Tefé	River	5.3–6.7	8.5–17.9	29.1–32.1	Crampton (1998)
Amazon	Main stem	9	Amazon at Manaus	River	6.4–6.8	20–69.6	25.9–27.6	Tundisi et al. (1984); Duncan and Fernandez (2010)
Amazon	Negro	9	Lake Cristalino	Floodplain	4.6–5.5	3.4–10.1	27–32	Hardy (1980); Tundisi et al. (1984)
Amazon	Negro	9	Lake Taruma Mirim	Floodplain	4.0–5.5	3.4–10.1	27–32	Irmler (1975); Hardy (1980)
Amazon	Solimões	9	-	Floodplain	6.6–7.6	32.7–63.2	28–30	Hardy (1980)
Amazon	Solimões	9	Ilha Curari	Floodplain	6.0–6.6	30–84	26.5–29.1	Irmler (1975)
Amazon	Solimões	9	Lake Camaleao	Floodplain	6.1–8.1	62.5–173	23.5–34.5	Ibanez (1998); Petry et al. (2003); Sánchez-Botero et al. (2003)
Amazon	Solimões	9	Lake do Rei	Floodplain	6.6–8.5	–	–	Ribeiro and Darwich (1993)
Amazon	Solimões	9	Lake Januari	Floodplain	5.7–6.4	13.7–55.9	26–28	Irmler (1975)
Amazon	Solimões	9	Lakes (in general)	Floodplain	5.7–7.5	45–320	29–40	Junk (1973); Freitas et al. (2014)
Amazon	Main stem	10	Amazon at Santarém	River	6.5–7.5	40–300	29–32	Bogotá-Gregory et al. (2020)
Amazon	Main stem	11	Curiau	Floodplain	–	27–31	–	Chellappa et al. (2005)

(continued)

Table 8.2 (continued)

Basin	Subbasin	L#	Location	Habitat	pH	Cond μ Scm–1	Temperature (°C)	References
Amazon	Main stem	12	Jacaretinga	Floodplain	6.4–7.9	25.2–54.4	27–32	Hardy (1980); Tundisi et al. (1984)
Amazon	Main stem	13	Iacoatiara (region)	Stream	1.57–6.00	6.7–10.0	23.3–25.4	Dias et al. (2008)
Amazon	Main stem	13	Manaus (region)	Stream	4.5	8.5	24.5	Fittkau (1964)
Amazon	Main stem	13	Reserva Ducke	Stream	3.7–6.2	2.99–43.7	21.8–26	Mendonça et al. (2005); Pazin et al. (2006)
Amazon	Main stem	14	Lake Catalao	Floodplain	5.8–7.4	13.4–146	28–33.5	Brito et al. (2014); Röpke et al. (2016)
Amazon	Main stem	14	Lake Grande do Jutaí	Floodplain	6.6	31–50	26–31	Santos (1980)
Amazon	Main stem	15	Leticia (region)	Stream	3.1–6.6	11.0–41.0	24.6–26.8	Mojica et al. (2009)
Amazon	Main stem	16	Marajó	Floodplain	–	50–62	25.5	Assunção and Schwassmann (1995)
Amazon	Main stem	17	Tefé to Maues (region)	Floodplain	–	25–339	25.6–29.2	Granado-Lorencio et al. (2007); Lobón-Cerviá et al. (2015)
Amazon	Napo	18	Aguarico	Floodplain	6.4–8.2	–	24–27.8	Saul (1975)
Amazon	Napo	18	Aguarico	River	7.2–7.9	–	17.3–20.7	Saul (1975)
Amazon	Napo	18	Conejo	River	6.4–6.6	–	–	Saul (1975)
Amazon	Napo	18	Jatuncocha	Floodplain	4.6–5.7	9.9–14.8	27.5–29.3	Galacatos et al. (2004)
Amazon	Napo	18	Tambococha	River	5.4–6.5	19.7–78.2	24.3–27.1	Galacatos et al. (2004)
Amazon	Napo	18	Yasuni	River	5.6–6.9	17.8–22.0	24.5–26.7	Galacatos et al. (2004)
Amazon	Negro	19	–	Floodplain	3.05–5.95	3.7–8.6	25–31	Darwich et al. (2005)
Amazon	Negro	19	Anavilhanas (region)	Stream	4.2–4.6	6.13–25.1	28.4–34.7	de Oliveira and Ferreira (2008)
Amazon	Negro	19	Lake Tupé	Floodplain	4.3–5.5	5.7–8.8	24.8–34	Reiss (1977)
Amazon	Negro	19	Negro main stem	River	3.8–6.2	6–33.6	24.8	Reiss (1977); Winemiller et al. (2008); Duncan and Fernandez (2010)
Amazon	Negro	20	Jaú	Stream	3.0–5.5	–	23–26	Kemenes and Forsberg (2014)
Amazon	Negro	21	Barcelos (region)	Stream	1.5–2.7	10.8–41.5	27.–29.1	Tavares-Dias et al. (2010)
Amazon	Negro	22	Manaus to Caracarai (region)	Stream	4.4–5.3	4.2–10.8	26–29	Schmidt (1972)
Amazon	Negro	23	Pasimoni	River	4.2–5.8	4.5–35	26–31	Winemiller et al. (1997); Jepsen and Winemiller (2007)
Amazon	Negro	24	Urubú	Stream	4.1–5.29	6.8–10.3	22.0–26.3	Barbosa et al. (2003); Buhrnheim and Fernandes (2003)
Amazon	Purus	25	Acre	River	4.5–8.0	21–76	23–33	Sendacz and Costa (1991)

(continued)

Table 8.2 (continued)

Basin	Subbasin	L#	Location	Habitat	pH	Cond μ Scm–1	Temperature (°C)	References
Amazon	Purus	26	Purus	River	6.0–8.33	60.2–77.8	20.1–33.1	Duarte et al. (2010)
Amazon	Solimões	27	Urucú	River	5.3–6.5	8.4–35.2	24–26.6	Costa and Freitas (2015)
Amazon	Tapajos	28	Arapiuns	Floodplain	4.3–6.4	7–9.8	29.3–30.8	Bogotá-Gregory et al. (2023)
Amazon	Tapajos	28	Arapiuns	River	4.8–6.2	7.5–11.2	29.0–32.5	Bogotá-Gregory et al. (2023)
Amazon	Tapajos	29	Tucurui	Floodplain	4.8–5.4	8.6–11.4	29.2–31.5	Marlier (1967)
Amazon	Tapajos	30	Tapajos	Floodplain	4.3–7.0	4.4–15.7	29.3–31.7	Bogotá-Gregory et al. (2023)
Amazon	Tapajos	30	Tapajos	River	5.6–7.4	11–38.3	26.7–31.0	Duncan and Fernandez (2010); Coelho-Souza et al. (2011); Bogotá-Gregory et al. (2023)
Amazon	Araguaia	31	Araguaia	Floodplain	6.3–7.3	25.0–63.6	27.0–30.2	Tejerina-Garro et al. (1998); Carvalho et al. (2001)
Amazon	Araguaia	31	das Mortes	River	6.07–7.00	8.0–13.0	22–33	Lima de Melo et al. (2009)
Amazon	Tocantins	32	Main	River	7.4–8.0	88–204	27.1–27.9	Vitorino Júnior et al. (2016)
Amazon	Trombetas	33	Lake Batata	Floodplain	5.1–6.6	7–30	26–37	Bozelli (1992)
Amazon	Trombetas	34	Trombetas	Floodplain	5.4–6.0	9–11	28.5–30.7	Huszar Reynolds (1997)
Amazon	Ucayali	35	Ucayali	Stream	4.5–6.2	6–14	24.2–27.0	Waddell and Crampton (2020)
Amazon	Ucayali	35	Ucayali	Floodplain	4.9–7.3	50–220	26.5–33.2	Waddell and Crampton (2022)
Amazon	Xingu	36	Xingu	River	7.7–8.3	19–21	29.1–30.5	Barbosa et al. (2015)
Orinoco	Apure	37	Apure	River	6.7–8.2	89–190	28.4–28.8	Jepsen and Winemiller (2007); Montoya et al. (2017)
Orinoco	Apure	38	Portuguesa	Floodplain			26.2–33	Winemiller (1996)
Orinoco	Main channel	38	Turbio	Stream	6.1–7.8	492–1013	24–37	Rodriguez-Olarte et al. (2017)
Orinoco	Guaviare	39	Laguna Negra	Floodplain	6.35–7.31	2–22	30.4–31.9	Kuznar et al. pers. obs.
Orinoco	Inírida	39	Inírida	River	4.33–5.0	25.7–31.3	4–8	Kuznar et al. pers. obs.
Orinoco	Main channel	40	–	Floodplain	6	13–420		Rodriguez and Lewis (1997)
Orinoco	Main channel	40	Caicara	Floodplain	3.65–7.0	–	25.5–32.5	Rodríguez and Betancourt (1999)
Orinoco	Main channel	41	Arauaca	Stream/River	6.8–7.0	70.2–82.5	28.2–30.0	Montoya et al. (2017)
Orinoco	Main channel	41	San Felipe	Stream/River	5.6	8.4	30.3	Montoya et al. (2017)
Orinoco	Main channel	41	Sarare	Stream/River	7.43	177	30	Montoya et al. (2017)
Orinoco	Main channel	43	Autana	River	3.95–5.39	5.9–12.5	24–31	Castillo et al. (2004)

(continued)

Table 8.2 (continued)

Basin	Subbasin	L#	Location	Habitat	pH	Cond µ Scm–1	Temperature (°C)	References
Orinoco	Main channel	44	Capanaparo	Stream/River	6.9	55	28.9	Montoya et al. (2017)
Orinoco	Main channel	45	Cataniapo	River	5.18–6.58	4.9–9.4	25–29	Castillo et al. (2004)
Orinoco	Main channel	45	Cunaviche	Stream/River	6	36	28.9	Montoya et al. (2017)
Orinoco	Main channel	45	La Pica	Stream/River	5.3	4.6	30.4	Montoya et al. (2017)
Orinoco	Main channel	45	Sipapo	River	4.12–5.44	4.81–9.8	25–34	Castillo et al. (2004)
Orinoco	Main channel	46	Cinaruco	River	5.5–6.3	5.9–10.5	30–32	Arrignton et al. (2006); Jepsen and Winemiller (2007); Montoya et al. (2017)
Orinoco	Main channel	46	Meta	Stream/River	7.2	113	27.8	Montoya et al. (2017)
Orinoco	Main channel	47	Cuao	River	4.92–6.53	4.8–9.5	25–31	Castillo et al. (2004)
Orinoco	Main channel	47	Lake Aguilera	Floodplain				Hamilton et al. (1990)
Orinoco	Main channel	47	Lake Tineo	Floodplain	5.56–8.13	5–479	16–36.8	Hamilton et al. (1990)
Orinoco	Main channel	48	Orinoco (delta)	River	5.0–7.7	32–37	27.2–30.4	Monente et al. (2017)
Caribbean Basin	Maracaibo	49	Guasare	Stream	6.1–9.6	260	16–30	Rincón Ramírez (2017)
Caribbean Basin	Maracaibo	50	Palmar	Stream	9.2	–	–	Rincón Ramírez (2017)
Caribbean Basin	Maracaibo	51	Santa Ana	Stream	6.11–9.6	93–176	–	Rincón Ramírez (2017)
Caribbean Basin	Maracaibo	52	Chama	Stream	6.8–8.5	20–880	–	Segnini and Chacón (2017)
Caribbean Basin	Sinú	53	Sinú	River	–	30–300	26.8–29.0	Kerguelén-Durango and Atencio-García (2015)
Caribbean Basin	Magdalena	54	Magdalena	Stream	7.8–8.6	–	–	Jaramillo-Villa et al. (2010)
Caribbean Basin	Magdalena	54	Magdalena	Floodplain	4.5–9.9	9.1–138	–	Granado-Lorencio et al. (2012)
Guiana Atlantic Basins		55	Maykabuka	Stream	5.2–6.2	29–43.4	23.1–24.9	Mol and Ouboter (2004)
Guiana Atlantic Basins		56	Mazaruni	Stream	4.4–4.8	<10	22–23.5	Maldonado-Ocampo et al. (2014)
Guiana Atlantic Basins		57	Suriname	Stream	4.7–6.4	26.4–85	24.2–25.2	Mol et al. (2000); Mol et al. (2007)
Guiana Atlantic Basins		57	Suriname	River	5.5–6.9	16–36	–	Mol et al. (2007)

Colombian Amazon, e.g., Serranía de la Macarena and Serranía de Chiribiquete (Rodríguez-Olarte et al. 2025).

Blackwaters originate in weathered arenosol, podzol, and acrisol soils of lowland tropical forests and savannas of the shield fringes and intracratonic basins. Blackwaters exhibit high DOC, low nutrients and sediment loads, low conductivity (typically $<20\ \mu S\ cm^{-1}$) and low pH; usually less than 5.5, but reaching 2.5 in some small streams from the upper Negro basin (Table 8.2). The red to black color comes from long-chain acidic polyphenols that originate from the incomplete decomposition of plants (Janzen 1974; Leenheer 1980; Bardy et al. 2011; Gonsior et al. 2016). In very acidic water, the electrical conductivity is slightly increased due to the concentration of $H+$ ions (Ríos-Villamizar et al. 2020a, 2020b).

Blackwaters drain an area of about 3.6 million km^2 of lowland rainforests that extend across the central and western

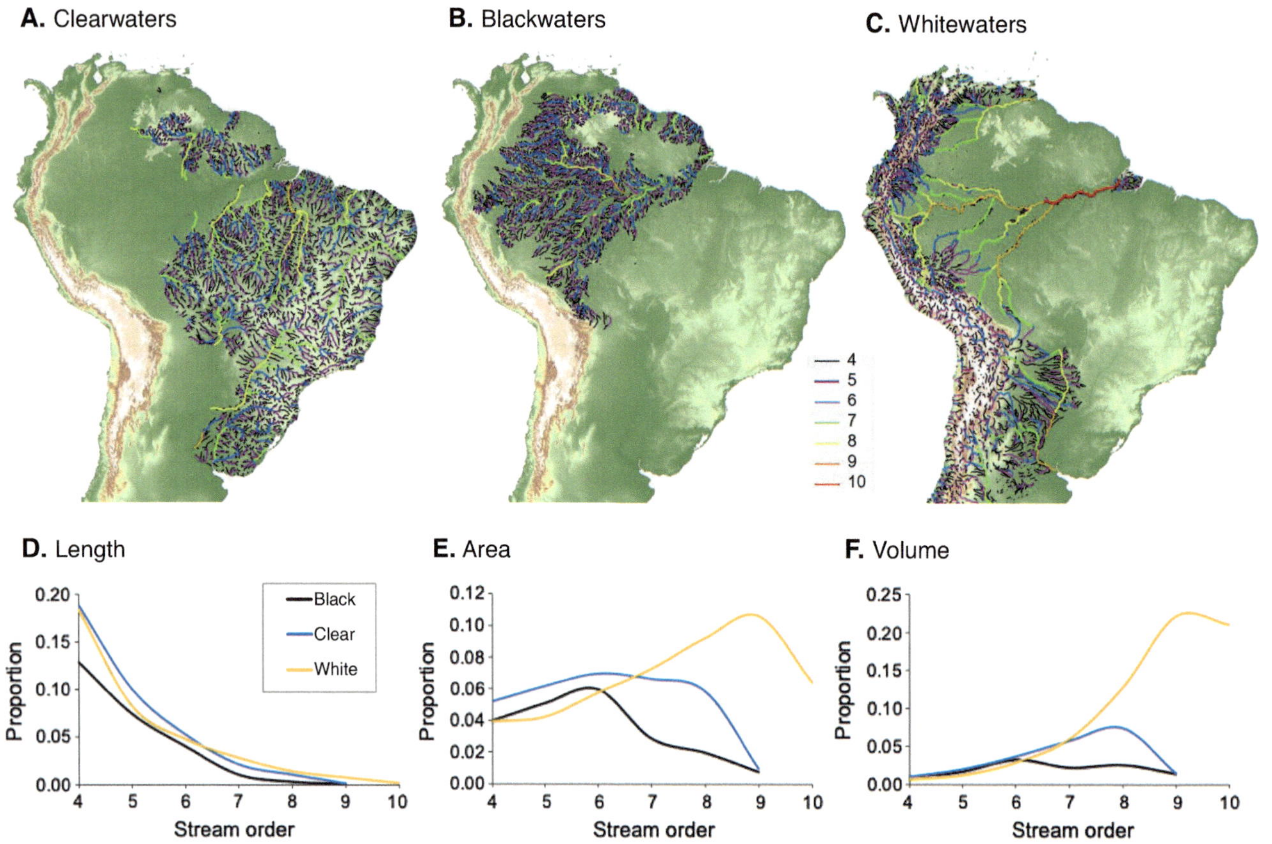

Fig. 8.3 Distributions of South American rivers by Strahler stream order (SO) among three major water types. Smaller rivers presented for SO 4–5 for clarity. Data for c. 187,000 river segments (river reaches between stream-order transitions) from HydroSHEDS. (**A**) Clearwaters of the Guiana and Brazilian Shields. (**B**) Blackwaters of Amazonian lowlands. (**C**) Whitewaters draining Andean cordilleras. Hydrological estimates of river length (**D**), river surface area (**E**), and river volume (**F**). Note that most river lengths are in low order (SO 1–5) rivers, divided approximately evenly among the three water types, while most river surface area and river volume are in large (SO 6–10) whitewater rivers. Note also the ten large whitewater tributaries flowing across the lowland blackwater source region of the Western Amazon and Llanos sedimentary basins

Amazon basins, the upper Orinoco basin, and lowland margins of the Guiana and Brazilian Shields. Besides draining several regions, blackwater rivers constitute, overall, a lower proportion in length, area, and volume when compared to large rivers with other water chemistry profiles (Fig. 8.3B). Central Amazonian blackwaters extend east of the mouth of the Purus River to the city of Óbidos and include small lowland basins on both banks. There are also blackwater basins in the lowlands around the northern and eastern margins of the Western Guiana shield. Presumably, there were blackwater rivers in the area now buried under the Fitzcarrald and Vaupes arches. The Amazonian blackwater fields are transected by six major corridors of sediment-rich rivers draining the Andes or Andean sediment fans (e.g., Madeira, Purus, Juruá, Solimões, Caquetá-Içá, Juruá), and the Orinoco blackwater field by three whitewater corridors (Guaviare, Meta, Apure).

Whitewaters originates in the Andean cordilleras, or sediment fans ultimately derived from Andean parent materials (e.g., Fitzcarrald and Vaupes Arches). Whitewaters carries a nutrient-rich sediment load (Furch and Junk 1997) and a high concentration of dissolved nutrients (particularly calcium, magnesium, nitrate, phosphate) from the erosion of biogenic sediments of originally marine provenance (e.g., limestones, cherts, and some mudstones). Whitewater rivers carry yellow to light-brown sediments and have high levels of dissolved nutrients, reflected in higher conductivity (about 40–300 µS/cm–^{1}, usually above 60). In contrast, clearwater and blackwater rivers have much lower nutrient levels and lower conductivity (around 5–40 µS/cm–^{1}) (Table 8.2).

Andean-derived whitewaters start with low DOC in the highlands, gaining high DOC after merging with lowland blackwater tributaries or receiving organic inputs from várzea floodplains along their courses. The chemistry of lowland Amazonian rivers is readily observed in whitewater floodplain lakes (including oxbow lakes and scroll-swathe lakes) where sediments have settled out, but which generally exhibit a high DOC and black color, along with the expected high conductivity and near-neutral pH. Whitewater rivers represent most of the length, area, and volume of large South American rivers, in comparison to rivers with other water chemistry profiles.

Rivers, such as lower Juruá (SO7), lower Purus (SO8), lower Madeira (SO9), and Amazon (the only SO10), are all whitewater rivers, which contribute to the predominance of whitewaters on area and volume (Fig. 8.4C). If all proportions of water length, area, and volume are summed, whitewater constitutes the preponderance of total aquatic habitat in South America.

8.1.3 Amazonian Soils

The geographic distributions of the main Amazonian soils are depicted in Fig. 8.4A. Physicochemical descriptions of these soils are provided in Quesada et al. (2011). The granitic parent rock of the Amazon Craton originally formed as magmatic arcs during the Paleoproterozoic and Mesoproterozoic Eons (*c.* 2.3–1.6 Ga) (Horbe et al. 2013; Weber et al. 2016). Much of the Amazon and other South American cratons were covered by sediments of terrestrial or marine origins during the Neoproterozoic (*c.* 1000–541 Ma), Paleozoic (541–252 Ma), and Mesozoic (252–66 Ma) Eons (Lujan and Armbruster 2011). The cratons were generally uplifted and deformed during the Mesozoic breakup of Western Gondwana, and their overlying sedimentary formations were physically eroded and chemically weathered. Throughout this lengthy time interval, the upland portions of the Guiana and Brazilian shields were denuded of their softer and more easily eroded sediments, leaving erosion-resistant rocks behind, which are now drained by clearwater rivers. Overtime, these heavily weathered sediments became

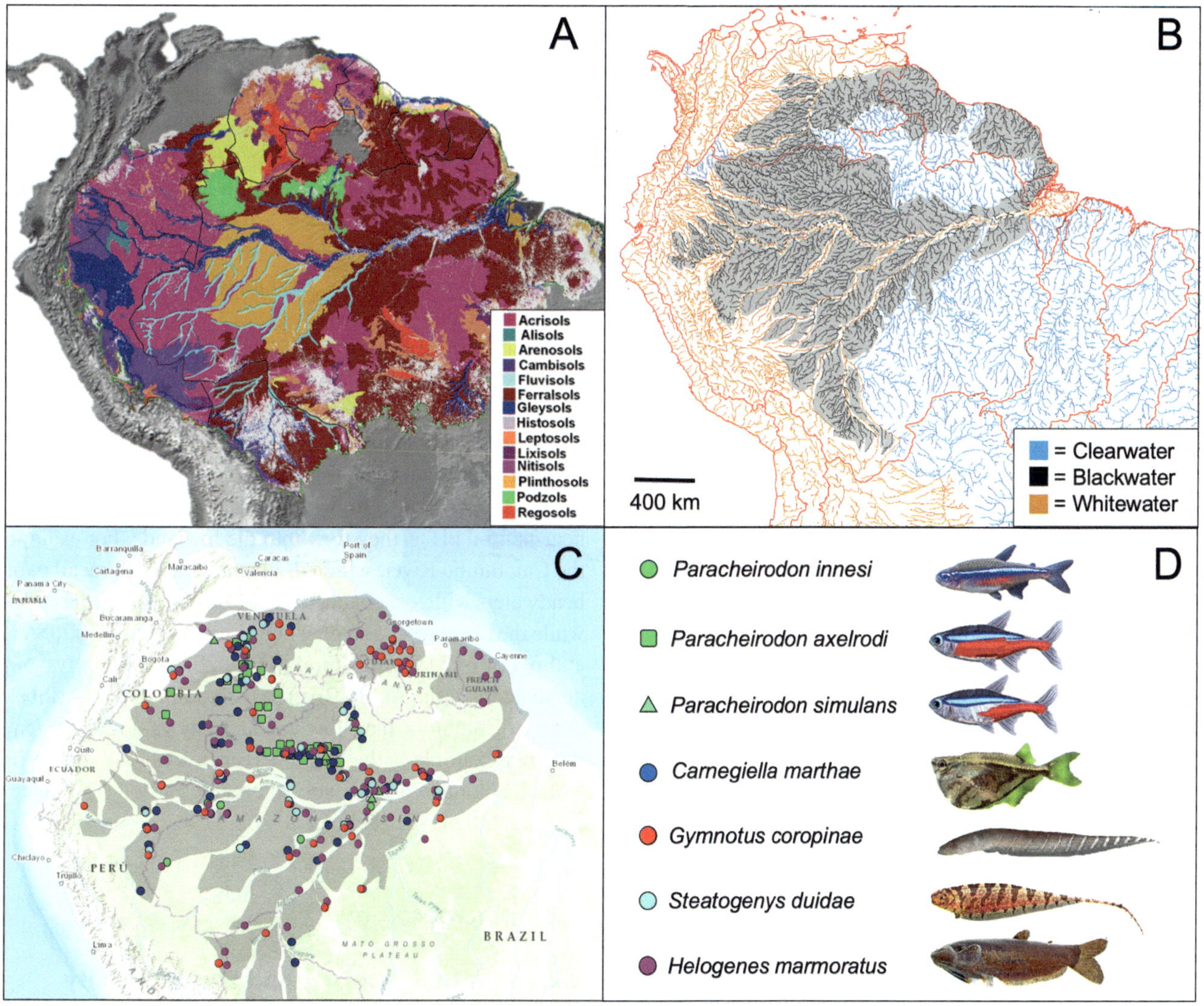

Fig. 8.4 Soil and water chemistry constrain NFF species geographic distributions. (**A**) Major soil types of northern South America. Clearwaters originate in weathered utilisols (yellow), blackwaters in acidic soils (red, pink, yellow, green), and whitewaters in geologically young inceptisols (orange). Map from Quesada et al. (2011). (**B**) Main river types of South America based on water chemistry. Rivers and watersheds (red lines) from Lehner et al. (2008). Gray shaded areas indicate zones of blackwaters and intermediate blackwaters and whitewaters. (**C**) Geographic distributions of selected NFF species ecophysiologically specialized to inhabit blackwater streams. Distribution data from Tagliacollo et al. (2021). (**D**) Images of fish species in panel **C**

enriched in aluminum and iron-bearing minerals, which are more resistant to weathering than lighter elements, such as sodium, potassium, and calcium. These soils are a main crucible for the formation of highly acidic blackwaters (pH = 3.0–5.5; Matsuo and Val 2007).

Amazonian blackwaters with high DOC and low pH originate soils with low in situ organic decomposition and high organic content (Quesada et al. 2011; Fig. 8.4A). These soils include iron-rich *ferralsols* (i.e., laterites) formed from weathering of igneous rocks under strongly oxidizing and leaching humid tropical climates, lacustrine (lake bed) clays called *acrisols* that formed under humid, tropical climates (e.g., the Solimoes/Pebas Formation), unconsolidated clay-rich floodplain sediments called *gleysols* (e.g., várzeas), and unconsolidated quartz-rich sands called *arenosols* and *podzols* (e.g., upper Negro and Atabapo basins). Ferralsols have a red or yellowish color due to high concentrations of iron and aluminum oxides and are composed of clays (e.g., kaolin) and organic matter formed by weathering and humification over tens to hundreds of millions of years. Globally, ferralsols are almost exclusively found in tropical areas on or near geologically ancient continental cratons. Acrisols are clay-rich subsoils associated with humid, tropical climates that often support tropical rainforests, with low fertility and toxic amounts of aluminum posing limitations for agricultural use. Arenosols are young soils with no diagnostic horizons and are unaltered from their parent material. Podzols derive from either quartz-rich sands and sandstone or sedimentary debris from magmatic rocks under high precipitation, are poor soils for agriculture with low moisture and nutrients, with low pH, low phosphate, and high aluminum.

The geographic distribution of water chemistry strongly predicts the geographic ranges of many NFF fish species (Figs. 8.4C, D; see Chap. 12). Clearwaters originate in soils with close to a neutral pH, including iron-rich nitisols derived from early branching parent or calcium-rich limestone parent rocks. Blackwaters originate in soils derived from erosion of the shield uplands. Extensive fields of blackwater-generating soils skirt the Guiana and Brazilian Shields, extending west into portions of the sub-Andean foreland sedimentary basins (e.g., Acre, Maranon, Llanos, etc.).

8.1.4 Limitations of the Water Classification

Although useful to help characterize fish habitats and ecotraits, this classification has several important limitations. All rivers sum the chemical attributes of their upstream tributaries. Most Amazonian rivers exhibit high variability in their physicochemical attributes, longitudinally from headwaters to mouth, regionally by soil chemistry and climate, and seasonally by local and regional precipitation patterns (Goulding et al. 2003; Seidel et al. 2016; Ríos-Villamizar et al. 2020b; Nascimento Monte et al. 2021). Many streams in the high Andes change from clearwaters to whitewaters as they descend and change again from lower to higher DOC as they merge with lowland tributaries. The lower reaches of many Amazonian rivers change dramatically between the rainy and dry seasons, depending on which contributing tributaries are in a higher flood stage (Goulding et al. 2003). Many rivers that drain the lowlands in the western Amazonian exhibit intermediate chemical profiles between blackwaters and whitewaters, and often have a greenish color at low water; e.g., Javari river (Piland et al. 2025). Some clearwater rivers change between black and yellowish to light brown in high-gradient reaches of the Yuruparí (Vaupés) and Jirijirimo (Apaporis) rivers.

Based on data from more than 400 scientific publications, Ríos-Villamizar et al. (2020a, fig. 4) classify major Amazonian rivers into five major types, including an intermediate type for clearwater waters draining heterogeneous geological formations of the Guiana and Central Brazilian shields, and another intermediate type for whitewaters draining sediments of Andean origin on paleovárzeas. The Branco River (*Rio Branco* in Portuguese) of the Negro Basin exhibits an intermediate water chemistry profile, being a little muddier than other shield rivers, but with a low to moderate pH (4.3–6.7), low electrical conductivity (7.0–36.8 μS/cm), and relatively low suspended organic material (5.0–36.0 mg/L), and is classified as a clearwater river in recent studies (Ríos-Villamizar et al. 2020a; Piedade et al. 2025).

Another limitation of the classification presented here is that it is based on the chemistry of Amazonian rivers and is less effective at classifying the rivers of other Neotropical areas. For example, many Andean rivers arise as crystal-clear nutrient-poor waters with low conductivity and a slightly acidic pH, and transition to higher conductivity whitewaters with near-neutral pH as they flow into the lowlands. For example, the Catatumbo River, which drains into Lago Maracaibo, has headwaters with low conductivity (20 μS/cm) and a pH of 5–6, while the lower reaches have a higher conductivity (100 μS/cm) and pH of 6–7 (Segnini-Flores et al. 2025). Similarly, some headwaters of the Atrato River, which drain to the Caribbean coast of Colombia, have low conductivity values (39 μS/cm). Another example is the Negro River in Uruguay, which has a black color and high dissolved organic carbon (DOC) but a near-neutral pH (de Mello et al. 2025).

8.2 Macroecology

8.2.1 Water Chemistry and Local Diversity

The chemical content of river water is derived from the vegetation and soils through rainwater and groundwater flow. The ferralsols and acrisols of the eastern Amazon have much lower concentrations of nutrient minerals, such as sodium, potassium, calcium, phosphorus, and magnesium, than the western Amazon, due to long weathering and erosive cycles (Quesada et al. 2011). Water conductivity and pH are correlated with the concentration of cations and anions present in the water. Amazonian waters may be roughly divided into those which are electrolyte-poor (e.g., blackwaters and clearwaters) or electrolyte-rich (e.g., whitewaters). The high concentrations of $H+$ ions in blackwaters strongly affect the respiratory physiology of fishes (Val and de Almeida-Val 2012). For example, fishes subjected to acidic freshwaters experience homeostasis disturbances caused by the inhibition of active $Na+$ and $Cl-$ uptake, passive diffusive ionic losses, and reversal of the transepithelial potential (TEP) at the gills (Wood 1989). However, the high DOC of blackwaters buffers some of the effects of high acidity on fish physiology, allowing specialized species to survive in these environments (Duarte et al. 2016).

Sediment content and water transparency are related to the concentration of mineral particles in the water and impact fish biology (Crampton 2011). Clearwaters and blackwaters have low sediment content, which results in high transparency, so fishes often present bright and vibrant colors in comparison to those living in whitewaters (with high sediment and low transparency). A few examples are fishes valued by the aquarium trade, such as neon tetras (*Paracheirodon*) and discus (*Symphysodon*). This indicates that fishes from high transparency rivers use vision for predation and reproduction displays (Borghezan et al. 2021). Relative eye diameter is also larger in some fish species associated with black and clearwaters.

The presence and composition of aquatic vegetation also influences water chemistry. Aquatic plants and algae provide food for fishes in many forms; i.e., aquatic leaves, fruits, seeds, benthic macroalgae, periphyton, phytoplankton, and dead organic matter (see Chapts. 16 and 17), and can substantially increase the trophic complexity of aquatic ecosystems (Thomaz and Cunha 2010; Duarte et al. 2019; Coronado-Franco et al. 2022; de Andrade et al. 2024). Whitewater rivers usually have a higher abundance and more diverse species composition of aquatic macrophytes in floodplain lakes and river margin (Junk and Piedade 1997). However, muddy waters reduce the penetration of light, inhibiting the growth of phytoplankton and periphyton in large deep whitewater rivers. In blackwater habitats, plants must be adapted to lower nutrient concentrations and lower pH.

Whitewater floodplains are characterized by extreme hypoxia during the flood season, resulting from the decomposition of organic material in flooded forests. Many fishes leave these systems during the flood season (Saint Paul and Soares 1987; Crampton 1998), while others stay—tolerating prolonged periods of hypoxic stress via a suite of physiological, anatomical, and behavioral adaptations (Junk 1983, Crampton 1998, Val et al. 1998; Crampton et al. 2008; Val and de Almeida-Val 2012). Dissolved oxygen levels in whitewater floodplain lakes are typically higher in the low-water season but decline to below 0.5 mg/L, frequently approaching anoxia for weeks to months. In contrast, large river channels with the full range of water chemistries, terra firme forest streams, and shield streams are all permanently well-oxygenated, at least by Amazonian standards, with dissolved oxygen generally in the range 2–8 mg/L (Crampton 2011, Bogotá-Gregory et al. 2020).

8.2.2 Water Chemistry and Species Distributions

Water chemistry profoundly affects the ecological conditions of Amazonian fishes, influencing local abundances, dispersal patterns, geographic distributions, and conservation status (Henderson and Crampton 1997; Junk et al. 2007; Val and de Almeida-Val 2012; Lujan et al. 2013; Goulding et al. 2019; Bogotá-Gregory et al. 2020; Borghezan et al. 2021; Bogotá-Gregory et al. 2023). Large whitewater rivers (SO 6–10) almost always have a higher *alpha diversity* (local species richness) and lower *beta diversity* (low species turnover across geographic space) than similarly sized blackwater or clearwater rivers (Henderson et al. 1998). Bogotá-Gregory et al. (2023) demonstrated that large river channels and floodplain systems (regardless of water chemistry) exhibit higher turnover (but not nestedness) than terra firme and shield streams, even when accounting for the much higher species richness of river-floodplain systems. Here, we review three nonexclusive hypotheses advanced to explain why white waters have higher fish diversity and abundance, based on differential productivity and connectivity of rivers with different water chemistry profiles. The *Productivity–Stability Hypothesis* (Colwell and Fuentes 1975; Johnson et al. 1996) proposes that the amount of available energy in local food webs constrains biomass and species diversity (Waide et al. 1999; Dodson et al. 2000). The idea is that greater trophic (i.e., food) energy supports larger population sizes and a greater metacommunity biomass, resulting in more complex food webs and redundant trophic interactions. These factors are predicted to stabilize ecosystem dynamics and population fluctuations, and to enhance local diversity by reducing the risk of stochastic extirpations. Although the connections between water chemistry, aquatic productivity, and fish biomass have long been demonstrated in Amazonian fishes (Araujo-Lima et al. 1986; Bayley 1989; Henderson and Crampton 1997; Junk 1997; Bogotá-Gregory et al. 2020), exactly how local ecological processes (e.g., productivity, trophic interactions) influence evolutionary processes (e.g., speciation and extinction) remains poorly understood (Winemiller and Jepsen 1998; Kearney et al. 2024; Schluter 2024). *The Intermediate Disturbance Hypothesis* (Huston 2014) proposes that local species richness is maximized under conditions where ecological disturbances are neither too rare nor too frequent. The idea is that occasional disturbances (e.g., hurricanes, wildfires) remove the most abundant species, allowing rarer species to persist, whereas catastrophic disturbances may extirpate even the most ecologically tolerant species. Whitewater Amazonian floodplains align well with the predictions of this hypothesis, with the annual flood pulse driving the formation and destruction of highly productive floodplain ecosystems (Furch and Junk 1997; Petsch et al. 2023). *The Species Bank Hypothesis* (Henderson et al. 1998; Albert et al. 2011) proposes that large whitewater rivers exhibit exceptionally high alpha diversity but low beta diversity due to their large surface area and high aquatic habitat volume (Fig. 1.3C), combined with strong longitudinal connectivity across the Amazon basin. This idea suggests that large whitewater river floodplains and channels function like a "bank" for species, accumulating and distributing species among similar habitats across a broad geographic range. This hypothesis also predicts that the species bank produces some species with exceptionally broad geographic distributions across river drainage networks in northern South America, some of which may give rise to daughter species (Craig et al. 2019). Such geographically widespread species that give rise to one or more daughter species without themselves becoming extinct are known as paraphyletic species, or *paraspecies*. A paraspecies is a living or fossil species that is ancestral to another species (Crisp and Chandler 1996; Albert et al. 2011). Paraspecies are predicted by evolutionary theory (Darwin 1859; De Queiroz 2007) and are empirically documented in many Amazonian taxa, both terrestrial (Russello and Amato 2004; Fernandes et al. 2014; Nascimento et al. 2021) and aquatic (Melo et al. 2018; Bernt et al. 2019; Craig et al. 2019).

Whitewater rivers often have higher alpha diversity than blackwater or clearwater rivers of similar size (see Bogotá-Gregory et al. 2020, although see Saint-Paul et al. 2000, for a contradictory conclusion), but there are more fish species in all the clearwater rivers combined (see Figs. 6.4 and 6.5). This is due to the large number of small (SO 1–5) clearwater rivers and streams that drain the expansive *c*. 4.5 million km^2 of the Brazilian Shield. The most common habitat for NFF species is in fact small, clearwater, low-gradient streams, and rivers (see Chaps. 6 and 9). Phylogenetic niche conservatism also contributes to the high diversity of NFF species in small clearwater rivers and streams. We estimated the phylogenetic origins of 79 clades (families and cichlid tribes) of the Big Nine clades (see Chap. 3), and about one-third originated in clearwaters (Fig. 8.3).

8.3 Macroevolution: Phylogenetic Transitions Among Waters

Here, we examine patterns in the evolutionary transitions of clades among rivers with different water chemistry profiles, using the Ancestral State Estimation (AEA) and the ecotrace analyses presented in Chap. 7. We conducted AEA for each of the Big Nine NFF clades and report results for most of the family- and sub-family level subclades (Fig. 8.5; Table 8.3).

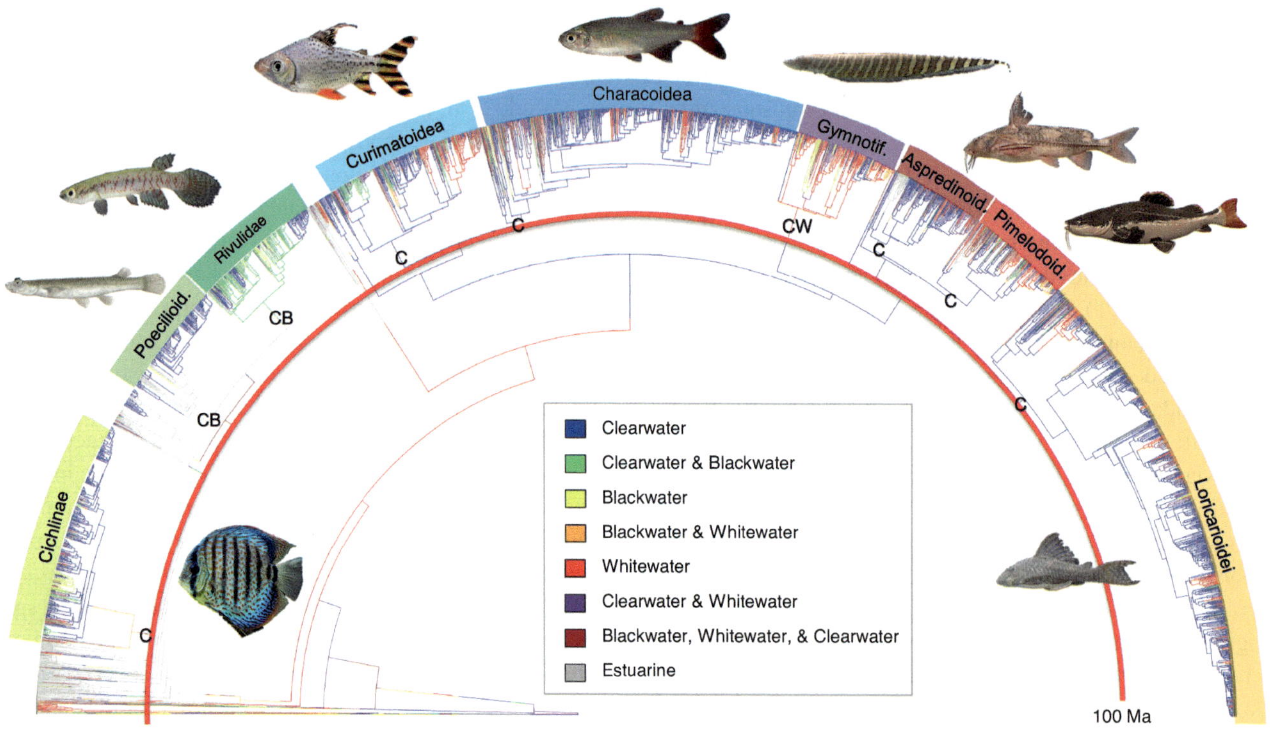

Fig. 8.5 Phylogeny of NFF species tracing ecophysiological tolerances to rivers with different water chemistry profiles. Note that all Big Nine NFF clades have estimated origins in either clearwaters or a combination of waters that includes clearwaters, on or after *c*. 100 Ma

Table 8.3 Summary of the three highest transition rates among NFF clades. Taxa arranged phylogenetically as in Fig. 8.5. Note five clades (Aspredinoidea, Cichlinae, Curimatoidea, Loricarioidei, Pimelodoidea) show high rates of exchange between combinations of multiple water types, and to a narrower range of water types. Note also that two clades (Characoidea, Cyprinodontiformes) show the highest transition rates to clearwater habitats from other water types. Note further that Gymnotiformes exhibit an evolutionary trend to a narrowing of tolerances for water types

Clade	Highest		Second highest		Third highest		Sum of absolute transition rates	General trends
	Rate	Type	Rate	Type	Rate	Type		
Big Nine	0.071	BW to W	0.044	WC to W	0.03	WC to C	0.429	Narrowing of habitat specialization from multiple water types to one
Doradioidea	14.32	W to WC	14.04	WC to W	2.8	B to BWC	32.860	High rate of exchange between whitewater–clearwater and whitewater
Cichlinae	6.518	W to BWC	4.682	BWC to W	0.65	W to WC	13.676	High rate of exchange between blackwater–whitewater–clearwater and whitewater
Curimatoidea	3.492	BWC to C	1.755	C to BWC	0.1	WC to C	5.729	High rate of exchange between blackwater–whitewater–clearwater and whitewater
Pimelodoidea	1.675	BWC to W	1.164	W to BWC	0.18	WC to BWC	3.392	High rate of exchange between blackwater–whitewater–clearwater and whitewater habitats
Loricarioidei	0.204	Other to BWC	0.187	BWC to C	0.18	B to C	1.081	High rate of exchange between blackwater–whitewater–clearwater and other water types
Characoidea	0.179	Other to C	0.065	BW to BWC	0.06	BC to BWC	0.509	Transitions to clearwater
Gymnotiformes	0.107	BW to W	0.053	BC to C	0.04	W to C	0.425	Narrowing of habitat specialization from multiple water types to one
Cyprinodontiformes	0.017	B to BC	0.016	Other to C	0.01	B to C	0.065	Transitions to clearwater

8.3.1 Big Nine NFF Clades

The ecotrace analyses of fishes inhabiting rivers with different water chemistry profiles provide several insights into the ecological origins and early diversification of the NFF fauna.

1. Each of the Big Nine NFF clades originated after 100 Ma, either in clearwaters or in a water-chemistry combination that included clearwaters (Fig. 8.5). The uncertainty about the ancestral water chemistry condition may reflect a genuine origin in both clearwaters and waters of other chemistry profiles, or it may reflect an uncertainty in our ability to estimate the ancestral condition. The fossil record does not provide information sufficient to resolve these uncertainties.
2. There has been a general trend among the Big Nine NF clades of a narrowing in the habitat utilization from waters of multiple to a single chemistry profile, especially to whitewaters, which have the highest alpha diversity, and to clearwaters, which have more total species than rivers with the other water chemistry profiles. However, this observed pattern may be a methodological artefact of the Brownian Motion (BM) model of trait evolution that averages ancestral trait values (Blomberg et al. 2020).
3. The phylogenetic origins of 75 family-level clades, or cichlid tribe-level clades, are divided about evenly between those with origins in clearwaters (38%) vs. estuarine/coastal waters (35%), with other water-chemistry combinations serving as the ancestral habitat for 25% of clades (Fig. 8.6A). However, there are pronounced differences in the subsequent fate of clades with these distinct water-chemistry origins, in which clades with clearwater origins give rise to about 4800 species, corresponding to almost 75% of all NFF species (Fig. 8.6B). The many NFF clades with estuarine/coastal origins contributed less than 10% of living NFF species (Fig. 3.3).
4. There is a high heterogeneity in evolutionary transition rates among rivers with different water chemistry profiles, both within and among the Big Nine NFF clades (Fig. 8.7; Table 8.3). Cyprinodontiformes and

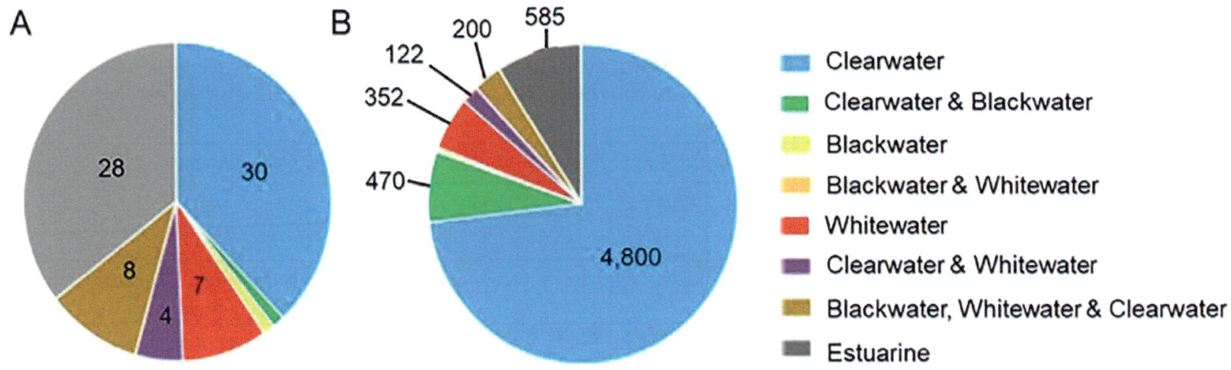

Fig. 8.6 Estimated origins of 75 South American freshwater fish clades (families and cichlid tribes) represented by two or more species among rivers with different water chemistry profiles. (**A**) Number of clades. Note that three clades are estimated with widespread geographic origins in lowlands+uplands. (**B**) Total number of species (in both uplands and lowlands) that are members of clades with upland or lowland origins. Note similar numbers of clades with origins in clearwaters, estuaries, and other water combinations combined, but species richness is dominated (72%) by clades with clearwater origins

Aspredinoidea exhibit extreme values for water chemistry evolution among the Big Nine NFF clades. The number of transitions among waters with different chemistry profiles ranges from seven combinations in the two cyprinodontiform clades combined, to 34 such combinations in Aspredinoidea, and the total rates range from 0.065 transitions Ma^{-1} in the cyprinodontiforms combined to 32.9 transitions Ma^{-1} in Aspredinoidea.

The highest transition rates occur in four clades: Aspredinoidea, Cichlinae, Curimatoidea, and Pimelodoidea. Aspredinoidea has the two highest transition rates (14.318 and 14.037 transitions Ma^{-1}) from whitewaters and clearwaters to whitewaters only, and vice versa. These high rates are due to the very high transition rates among terminal clades. The next highest rates are in Cichlinae from whitewaters to eurytopic waters, and vice versa. Curimatoidea has the fifth and seventh highest transition rates, from eurytopic waters to clearwaters and vice versa. Pimelodoidea has the eighth and ninth highest transition rates, from all inhabiting rivers with multiple chemistry profiles to whitewaters, and vice versa. The transition rates are between a more stenotopic/narrow, in most cases, whitewaters, to an eurytopic habitat (all three water chemistry profiles), which indicates increased habitat tolerance for fishes.

8.3.2 Cichlinae

Cichlinae has estimated origins in clearwaters, along with most of its tribe-level subclades; i.e., Heroini, Cichlasomatini, Guianacarina, Geophagina, Astronotini + Chaetobranchini and Retroculini. Cichlini origins are estimated in clearwaters + blackwaters, Crenicaranina has estimated eurytopic origins, and Apistogrammina has estimated origins in clearwaters with subsequent radiation in blackwaters and clearwaters + blackwaters. Cichlids show a high rate of exchange between whitewaters + clearwaters to whitewaters, with a minor transition pattern from whitewaters to clearwaters (Fig. 8.7A).

8.3.3 Poecilioidea

Poecilioidea has origins in clearwaters and blackwaters. Cyprinodontidae (Pupfishes) have clearwater origins with all species in this water chemistry profile, while Anablepidae (Four-eyed fishes) and Poeciliidae (Livebearers) are estimated to have had estuarine origins with early-branching clades maintaining this ecotrait and derived clades specializing to inhabit clearwaters.

8.3.4 Rivulidae

Rivulidae has origins in clearwaters and blackwaters with subsequent transitions to other water chemistry profiles in seven different clades (Loureiro et al. 2018). Kryptolebiasinae became specialized to estuarine waters, while Melanorivulini, Cynopoecillini, and Cynolebiasini are estimated to have originated in blackwaters, and Plesiolebiasini and Rachoviini are estimated to have clearwater origins. The cyprinodontiform Poecilioidea and Rivulidae are the only members of the Big Nine that did not undergo transitions to whitewaters (Fig. 8.7B).

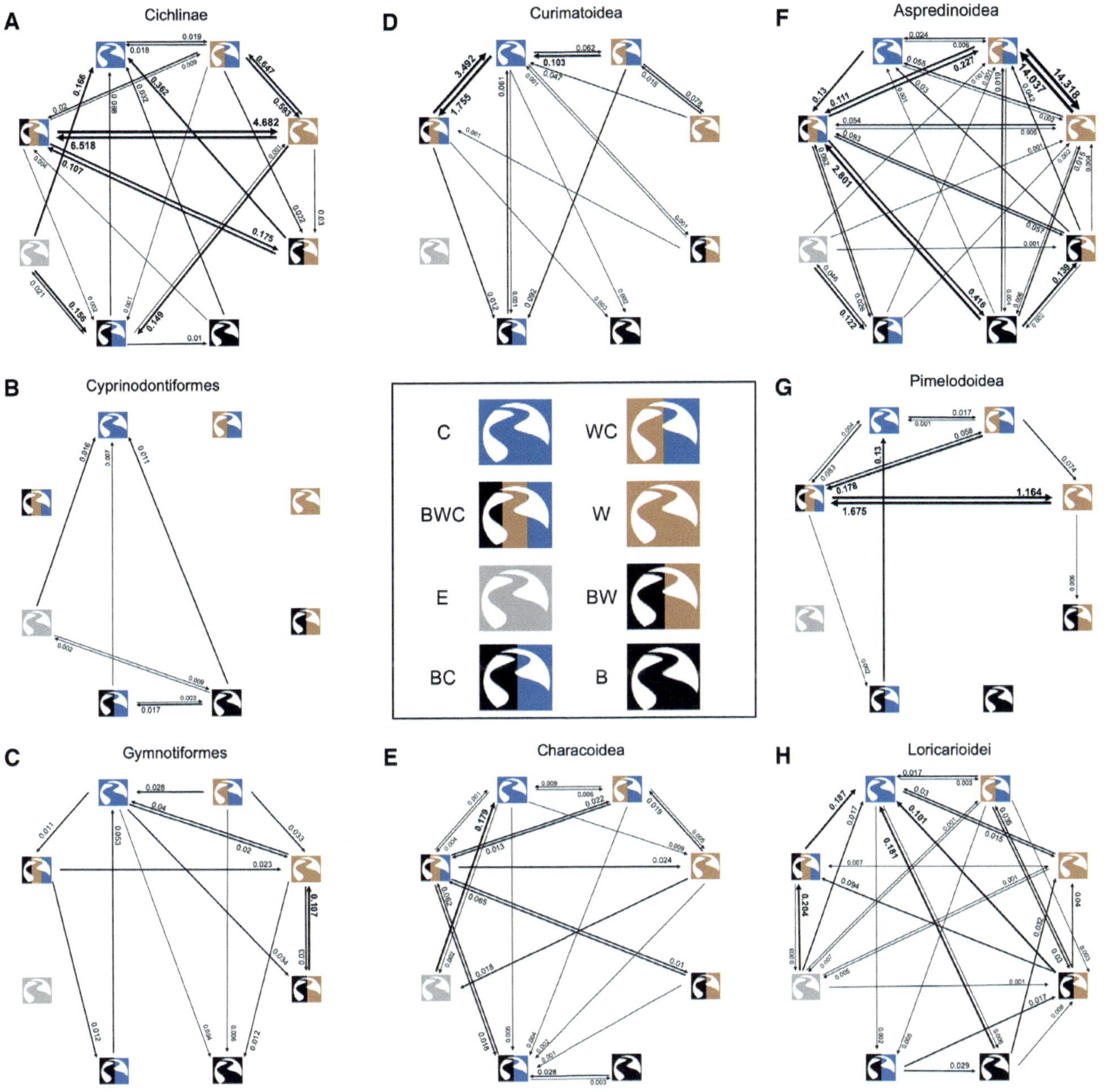

Fig. 8.7 Evolutionary transitions among NFF clades specialized to different water-chemistry profiles. Arrows indicate transitions among water chemistry, with widths proportionate to transition rates. Water chemistry legend at center; abbreviations and transition rates from Fig. 8.5. Cyprinodontiformes includes Poecilioidea and Rivulidae. Note each clade has a distinct set of evolutionary transitions; with the ranked order number of connections: Aspredinoidea (19), Loricarioidei (17), Cichlinae (16), Characoidea (14), Gymnotiformes (13), Curimatoidea (11), Pimelodoidea (8), Cyprinodontiformes (5)

8.3.5 Curimatoidea

Curimatoidea has an estimated origin in clearwaters, and this state is highly conserved in deeper nodes; only Serrasalmidae and Curimatidae underwent substantial water chemistry profile evolution. Curimatidae is estimated to have originated in whitewaters, while a few other groups underwent further transitions among water chemistry profiles. These inferences are supported by a biogeographic study that found clade origin in the lowlands (mainly whitewaters), with multiple lowland-upland river transitions (Melo et al. 2021; fig. 4A). Serrasalmidae (Piranhas and Pacus) are estimated to have originated in clearwaters with terminal clades transitioning to clear and blackwaters or whitewaters and clearwaters. Other groups, such as Chilodontidae and Anostomidae (Headstanders), and Parodontidae (Scrapetooths) derive from clearwaters, Cynodontidae (Dogtooth characins) from eurytopic, and Prochilodontidae (Flannel-mouth characiforms) estimated origins from whitewaters, with all species being whitewaters endemic. The Curimatoidea major trend is for species and clades to become more stenotopic with high exchange between eurytopic to clearwaters, the adaptation between all water chemistry profiles to clearwaters represents 60% of all relative transition rates for the group, being the highest relative transition rate for all groups (Fig. 8.7D, E).

8.3.6 Characoidei

Characoidei has estimated origins in whitewaters and/or clearwaters. There is some ambiguity in the ancestral habitat estimation due to homoplasy and a complex distribution of this trait among taxa subtending the deepest nodes of this part of the phylogeny. The early branching clades of Characoidei inhabit both of these water chemistry profiles; i.e., Ctenoluciidae, Lebiasinidae, Chalceidae, Triportheidae, Gasteropelecidae, Bryconidae; Fig 1.14C). Characidae (in the traditional sense before Melo et al. 2024) is the most diverse family-level clade of Characoidea, which arose in clearwaters, and most species (83%) retain this ancestral habitat preference. A few exceptions are blackwater endemics (three clades with relatively low diversity), clearwater and blackwaters clades (around 20 clades with less than five species each), and many whitewater endemics (20 clades, e.g., *Aphyocharax* with 11 species, *Knodus* and *Attonitus* with nine species). The largest number of transitions among water chemistry profiles with Characoidea are to clearwaters, which means most clades evolved to inhabit more eurytopic habitats, resulting in less endemism within each water chemistry profile. The second and third highest transition rates are from W to BWC, and BC to BWC (Fig. 8.7F).

8.3.7 Gymnotiformes

Gymnotiformes is estimated to have origins in whitewaters and/or clearwaters. Gymnotidae also has an estimated origin in whitewaters and/or clearwaters. The gymnotid subclade Electrophorinae (Electric eels) arose in clearwaters, while Gymnotinae (Banded electric fishes) have whitewaters origins. Subclades within Gymnotinae mainly originated in whitewaters, except for two blackwater endemic clades and two clearwater endemic clades, in the subgenera *Lamontianus* and *Tigrinus*. Rhamphichthyoidea originated in blackwaters, while its two subclades have different origins. Rhamphichthyidae originated in blackwater, while Hypopomidae (Bluntnose knifefishes) arose in clearwaters. Within Rhamphichthyidae, Steatogeninae (Painted knifefishes) had blackwaters origins, while Rhamphichthyinae (Trumpetnose knifefishes, Sand knifefishes) had eurytopic origins in blackwaters and whitewaters. Hypopomidae originated in clearwaters of the Guiana shield (e.g., *Akawaio*, *Hypopomus*, *Racenisia*), as did *Brachyhypopomus*, which also includes at least three clades (with 16 species) that subsequently evolved to inhabit high conductivity whitewaters (Crampton et al. 2016) and many species specialized to blackwater streams. Sinusoidea and both of its subclades, Sternopygidae (Glass knifefishes) and Apteronotidae (Ghost knifefishes), originated in whitewaters. Within Sternopygidae, most clades originated in whitewaters, except for two clearwater endemic groups. The same pattern can be found in Apteronotidae, with most of the clades having whitewaters origins, except nine low diversity clades in clearwaters or blackwaters + whitewaters. Gymnotiformes have a trend of narrowing habitat specialization from multiple water chemistry profiles to one, with many transitions between whitewaters and clearwaters, and few transitions among these water chemistry profiles and blackwaters (Fig. 8.7C).

8.3.8 Aspredinoidea

Aspredinoidea has estimated origins in clearwaters along with all three main subgroups: Aspredinidae, Auchenipteridae, and Doradidae. Each group underwent specific water chemistry profile evolution, for example, as Doradidae (Thorny catfishes) are extremely eurytopic, while Auchenipteridae (Driftwood catfishes) mainly inhabit clearwaters with three clades from whitewater and clearwaters, and a few small lineages endemic to whitewaters. In the case of Aspredinidae (Banjo catfishes), its biggest clade, Aspredininae, has estimated estuarine origins, and all species kept this character. Aspredinoidea groups show the same trend as cichlids with a high rate of exchange between white + clearwaters to whitewaters (Fig. 8.7G).

8.3.9 Pimelodoidea

Pimelodoidea has estimated origin in clearwaters. Two of their four families retained the same water chemistry profile origins as their parent clade (Heptapteridae and Pimelodidae). Within Heptapteridae (Three-barbeled catfishes), *Rhamdella* + *Pimelodella* expanded to clear + whitewaters, and some low-diversity clades inhabit blackwaters (e.g., *Brachyrhamdia*, *Goeldiella*, *Pariolis*). While in Pimelodidae (Long-whiskered catfishes), early branching groups inhabit whitewaters (e.g., *Aguarunichthys*, *Phractocephalus*), Pseudopimelodinae has estimated whitewater origins with minor transitions in a few species, and Pimelodinae kept the family origins, excluding one whitewater endemic clade (Pseudopimelodinae). Phreatobiidae (Cistern catfishes) have origins in hypogean ground waters, and unclassified water chemistry profile (see Chap. 15), and Pseudopimelodidae (Bumblebee and Dwarf marbled catfishes) have estimated origins in whitewaters. Pimelodoidea has a high transition rate between eurytopic and whitewaters (Fig. 8.7H).

8.3.10 Loricarioidei

Loricarioidei has estimated origins in clearwaters, along with four of the six family-level loricarioid clades analyzed. The exceptions are the early-branching clades Nematogenyidae (Chilean mountain catfish), a monotypic clade that inhabit both clearwaters and whitewaters in the southern Andes (de Pinna 2016), and Scoloplacidae (Spiny dwarf catfishes), which inhabit clearwater and blackwater habitats across much of tropical and temperate South America (Rocha et al. 2008). Astroblepidae (Stargazer catfishes), which inhabit high-elevation rivers of the northern Andes, was excluded from this analysis due to lack of species-level data on habitat, diet, and phylogenetic interrelationships (see Chap. 2). Trichomycteridae (Pencil or parasitic catfishes) has estimated clearwater origins. Their inner clades, Trichogeninae + Copiodontinae, are considered clearwater endemics, the TSVSG clade has clearwater and whitewater estimated origins, with four low-diversity clades transitioning to whitewaters, and Trichomycterinae has clearwater estimated origins with several groups specializing in different water chemistry profiles. Callichthyidae (Armored catfishes), and their two subfamilies (Callichthyinae, Corydoradinae) have estimated origins in clearwaters. Three clades of Callichthyinae have evolved into other water chemistry profiles; *Hoplosternum* + *Dianema* and *Lepthoplosternum* + *Megalechis* inhabit whitewaters and clearwaters, while species of *Callichthys* are present in all three water chemistry profiles in different combinations. Loricariidae (Suckermouth catfishes) originated in clearwaters with a few water chemistry profile transitions, such as Hypoptopomatinae with seven low-diversity clades specialized to clearwaters, and one clade to clearwaters + blackwaters, and Hypostominae with about 24 low-diversity clades specialized to several different water chemistry profile combinations. The main trend in Loricarioidei is evolving to clearwaters, with the three highest transition rates from a single water chemistry profile to clearwaters (Fig. 8.7F).

8.4 Conclusions

Amazonian rivers vary greatly in water chemistry affecting how fish live, how they behave, and how they have evolved. Amazonian rivers have different types of water, each with unique chemical and physical properties that influence fish life. Water clarity affects how fish see their surroundings for hunting and mating, while factors like acidity (pH), mineral content (conductivity), and color impact nutrient levels, temperature, and oxygen availability. Amazonian rivers may be classified in into three main categories: Clearwater rivers have clean, transparent water with little sediment or organic matter. They usually have neutral pH, low mineral content, and fewer nutrients. These rivers often flow from granite or limestone regions. Blackwater rivers are dark brown or tea-colored due to decaying plant material. They are highly acidic, low in nutrients, and have very few minerals. These rivers are common in dense rainforest areas. Whitewater rivers are muddy and full of sediment washed down from the Andes. They are rich in nutrients, have neutral pH, and contain more minerals. Some rivers mix these traits and appear greenish. Fish diversity varies across these water types, whitewater rivers support the most species because of their high nutrient levels. About 74% of Amazon fish originally came from clearwater rivers, while only 10% evolved from marine ancestors. Following chapters examine how fish evolved to survive in different Amazonian waters and how these adaptations shaped their diversity over time (Cooke et al. 2012a; Cooke et al. 2012b; Gonçalves et al. 2012; Ríos-Villamizar et al. 2013; Silva 2017).

References

Affonso AG, Queiroz HL, Novo EMLM (2011) Limnological characterization of floodplain lakes in Mamirauá Sustainable Development Reserve, central Amazon (Amazonas State, Brazil). Acta Limnol Bras 23:95–108

Affonso AG, Queiroz HL, Novo EMLM (2015) Abiotic variability among different aquatic systems of the central Amazon floodplain during drought and flood events. Braz J Biol 75:S60–S69

Albert JS, Petry P, Reis RE (2011) Major biogeographic and phylogenetic patterns. In: Albert JS, Reis RE (eds) Historical Biogeography of Neotropical Freshwater Fishes. Universality of California Press, Berkeley, pp 21–58

Andrade FS, Possamai B, Freitas CEDC, da Silva Batista J, Hoeinghaus DJ, Clements L, Siqueira-Souza FK (2024) Niche partitioning and seasonality may mediate coexistence of piranha species in Amazonian floodplain lakes. Hydrobiologia 851(18):4325–4340

Araujo-Lima CARM, Portugal LPS, Ferreira EG (1986) Fish–macrophyte relationship in the *Anavilhanas archipelago*, a blackwater system in the Central Amazon. J Fish Biol 29(1):1–11

Arbeláez F, Duivenvoorden JF, Maldonado-Ocampo JA (2008) Geological differentiation explains diversity and composition of fish communities in upland streams in the southern Amazon of Colombia. J Trop Ecol 24(5):505–515

Arbeláez JG, Orozco SG (1986) Moa. Desriptiones en lengua Cubea de los hábitos de algunos peces, 2nd edn. Editorial Townsend, Lomalina, Meta, Colombia

Arrington DA, Davidson BK, Winemiller KO, Layman CA (2006) Influence of life history and seasonal hydrology on lipid storage in three neotropical fish species. J Fish Biol 68:1347–1361

Assunção MIS, Schwassmann HO (1995) Reproduction and larval development of *Electrophorus electricus* on Marajó Island (Pará, Brazil). Ichthyol Explor Freshwaters 6:175–184

Barbosa RP, Freitas CEC, Santos SM (2003) The fish community of an upland stream in the central Amazon (Presidente Figueiredo–Amazonas–Brazil). Acta Limnol Bras 15:37–41

Barbosa TAP, Benone NL, Begot TOR, Gonçalves A, Sousa L, Giarrizzo T, Juen L, Montag LFA (2015) Effect of waterfalls and the flood pulse on the structure of fish assemblages of the middle Xingu River in the eastern Amazon basin. Braz J Biol 75:S78–S94

Bardy M, Derenne S, Allard T, Benedetti MF, Fritsch E (2011) Podzolisation and exportation of organic matter in black waters of the Rio Negro (upper Amazon basin, Brazil). Biogeochemistry 106:71–88

Bayley PB (1989) Aquatic environments in the Amazon Basin, with an analysis of carbon sources, fish production, and yield. Can Spec Publ Fish Aquat Sci 106:399–408

Bernt MJ, Tagliacollo VA, Albert JS (2019) Molecular phylogeny of the ghost knifefishes (Gymnotiformes: Apteronotidae). Mol Phylogenet Evol 135:297–307

Blomberg SP, Rathnayake SI, Moreau CM (2020) Beyond Brownian motion and the Ornstein-Uhlenbeck process: stochastic diffusion models for the evolution of quantitative characters. Am Nat 195(2):145–165

Bogotá-Gregory JD, Hurtado LFJ, Samudio JFG, Córdoba EA (2024) Ethnoecology and use of fishes by the Cubeo people from the Cuduyarí River, Colombian Amazonia. J Ethnobiol Ethnomed 20:101

Bogotá-Gregory JD, Jenkins DG, Lima FC, Magurran AE, Crampton WG (2023) Geomorphological habitat type drives variation in temporal species turnover but not temporal nestedness in Amazonian fish assemblages. Oikos 2023(11):e09967

Bogotá-Gregory JD, Lima FC, Correa SB, Silva-Oliveira C, Jenkins DG, Ribeiro FR et al (2020) Biogeochemical water type influences community composition, species richness, and biomass in megadiverse Amazonian fish assemblages. Sci Rep 10(1) :15349

Borges Pedro JP, Gomes MCRL, Jesus Trindade ME, Cavalcante DP, Oliveira JA, Hercos AP, Zucchi N, Lima CB, Pereira SA, Queiroz HL (2013) Influence of the hydrological cycle on physical and chemical variables of water bodies in the várzea areas of the middle Solimões River Region (Amazonas, Brazil). Uakari 9(2):33–47

Borghezan EDA, Pires THDS, Ikeda T, Zuanon J, Kohshima S (2021) A review on fish sensory systems and Amazon water types with implications to biodiversity. Front Ecol Evol 8:589760

Bozelli RL (1992) Composition of the zooplankton community of Batata and Mussurá lakes of the Trombetas River, state of Pará, Brazil. Amazoniana 12:239–261

Brito JG, Alves LF, Espirito Santos HMV (2014) Seasonal and spatial variations in limnological conditions of a floodplain lake (Lake Catalão) connected to both the Solimões and Negro rivers, central Amazonia. Acta Amazonica 44:121–134

Brown CB, Lidstone W (1878) Fifteen Thousand Miles on the Amazon and its Tributaries. E. Stanford, London

Bührnheim CM, Fernandes CC (2003) Structure of fish assemblages in Amazonian rain-forest streams: effects of habitats and locality. Copeia 2003:255–262

Castillo MM, Allan JD, Sinsabaugh RL, Kling GW (2004) Seasonal and interannual variation of bacterial production in lowland rivers of the Orinoco basin. Freshwater Biol 49:1400–1414

Chellappa S, Sá-Oliveira JC, Chellappa NT (2005) Fish fauna of a temporary lake in an Amazonian conservation area. Acta Limnol Bras 17:283–289

Coelho-Souza SA, Guimarães PR, Miranda MR, Poirier H, Mauro JBN, Lucotte M, Mergler D (2011) Mercury and flooding cycles in the Tapajós river basin, Brazilian Amazon: the role of periphyton of a floating macrophyte (*Paspalum repens*). Sci Total Environ 409:2746–2753

Colwell RK, Fuentes ER (1975) Experimental studies of the niche. Annu Rev Ecol Syst 6:281–310

Cooke GM, Chao NL, Beheregaray LB (2012a) Divergent natural selection with gene flow along major environmental gradients in Amazonia: insights from genome scans, population genetics and phylogeography of the characin fish *Triportheus albus*. Mol Ecol 21(10):2410–2427

Cooke GM, Chao NL, Beheregaray LB (2012b) Natural selection in the water: freshwater invasion and adaptation by water colour in the Amazonian pufferfish. J Evol Biol 25(7):1305–1320

Coronado-Franco KV, Tedesco PA, Kolmann MA, Borstein SR, Evans KO, Correa SB (2022) Feeding habits influence species habitat associations at the landscape scale in a diverse clade of Neotropical fishes. J Biogeography 49(12):2181–2192

Correa SB (2008) Fish assemblage structure is consistent through an annual hydrological cycle in habitats of a floodplain-lake in the Colombian Amazon. Neotrop Ichthyol 6:257–266

Costa ID, Freitas CEC (2015) Factors determining the structure of fish assemblages in an Amazonian river near to oil and gas exploration areas in the Amazon basin (Brazil): establishing the baseline for environmental evaluation. Zoologia 32:351–359

Craig JM, Kim LY, Tagliacollo VA, Albert JS (2019) Phylogenetic revision of Gymnotidae (Teleostei: Gymnotiformes), with descriptions of six subgenera. PLOS One 14(11):e0224599

Crampton WG (2011) An ecological perspective on diversity and distributions. Historical Biogeography of Neotropical Freshwater Fishes. University of California Press, Berkeley and Los Angeles, CA, pp 165–189

Crampton WG (2019) Electroreception, electrogenesis and electric signal evolution. J Fish Biol 95(1):92–134

Crampton WG, Chapman LJ, Bell J (2008) Interspecific variation in gill size is correlated to ambient dissolved oxygen in the Amazonian electric fish Brachyhypopomus (Gymnotiformes: Hypopomidae). Environ Biol Fishes 83:223–235

Crampton WG, De Santana CD, Waddell JC, Lovejoy NR (2016) Phylogenetic systematics, biogeography, and ecology of the electric fish genus *Brachyhypopomus* (Ostariophysi: Gymnotiformes). PLOS One 11(10):e0161680

Crampton WGR (1998) Effects of anoxia on the distribution, respiratory strategies and electric signal diversity of gymnotiform fishes. J Fish Biol 53:307–330

Crampton WGR (1998) Electric signal design and habitat preferences in a species rich assemblage of gymnotiform fishes from the upper Amazon basin. An Acad Bras Cienc 70:805–847

Crisp MD, Chandler GT (1996) Paraphyletic species. Telopea 6(4):813–844

Darwich AJ, Aprile FM, Robertson BA, Santos-Silva EN, Aprile FM, Scudeller VV, Melo S (2005) Variáveis limnológicas: contribuição ao estudo espaço-temporal de águas pretas amazônicas, BioTupé: meio físico, diversidade biológica e sociocultural do baixo rio Negro Amazônia central. Instituto Nacional de Pesquisas da Amazônia, Manaus, Brazil, pp. 20–33

Darwin C (1859) On the origin of species by means of natural selection, vol 167. John Murray, London

De Queiroz K (2007) Species concepts and species delimitation. Syst Biol 56(6):879–886

Dias MS, Magnusson WE, Zuanon J (2008) Effects of reduced-impact logging on fish assemblages in Central Amazonia. Conserv Biol 24:278–286

Dodson SI, Arnott SE, Cottingham KL (2000) The relationship in lake communities between primary productivity and species richness. Ecology 81(10):2662–2679

Duarte C, Rapp Py-Daniel LH, Pereira de Deus C (2010) Fish assemblages in two sandy beaches in lower Purus river, Amazonas, Brazil. Iheringia Série Zoologia Porto Alegre 100:319–328

Duarte C, Magurran AE, Zuanon J, Deus CP (2019) Trophic ecology of benthic fish assemblages in a lowland river in the Brazilian Amazon. Aquat Ecol 53(4):707–718

Duarte RM, Smith DS, Val AL, Wood CM (2016) Dissolved organic carbon from the upper Rio Negro protects zebrafish (*Danio rerio*) against ionoregulatory disturbances caused by low pH exposure. Sci Rep 6(1):20377

Duncan WP, Fernandes MN (2010) Physicochemical characterization of the white, black, and clearwater rivers of the Amazon Basin and its implications on the distribution of freshwater stingrays (Chondrichthyes, Potamotrygonidae). Pan-Am J Aquat Sci 5:454–464

Fernandes AM, Wink M, Sardelli CH, Aleixo A (2014) Multiple speciation across the Andes and throughout Amazonia: the case of the spot-backed antbird species complex (*Hylophylax naevius/Hylophylax naevioides*). J Biogeogr 41(6):1094–1104

Fittkau EJ (1964) Remarks on limnology of central-Amazon rain-forest streams. Verh Internat Verein Limnol 15:1092–1096

Freitas CEC, Siqueira-Souza FK, Florentino AC, Hurd LE (2014) The importance of spatial scales to analysis of fish diversity in Amazonian floodplain lakes and implications for conservation. Ecol Freshwater Fish 23:470–477

Furch K (1984) Water chemistry of the Amazon basin: the distribution of chemical elements among freshwaters. In: The Amazon: limnology and landscape ecology of a mighty tropical river and its basin. Springer Netherlands, Dordrecht, pp 167–199

Furch K, Junk WJ (1997) Physicochemical conditions in the floodplains. In: Junk WJ (ed) The Central Amazon Floodplain: Ecology of a Pulsing System. Springer-Verlag, Berlin Heidelberg, pp 69–108

Galacatos K, Barriga-Salazar R, Stewart DJ (2004) Seasonal and habitat influences on fish communities within the lower Yasuní River basin of the Ecuadorian Amazon. Environ Biol Fishes 71:33–51

Gonçalves CDS, Braga FMDS (2012) Changes in ichthyofauna composition along a gradient from clearwaters to blackwaters in coastal streams of Atlantic forest (southeastern Brazil) in relation to environmental variables. Neotrop Ichthyol 10:675–684

Gonsior M, Valle J, Schmitt-Kopplin P, Hertkorn N, Bastviken D, Luek J et al (2016) Chemodiversity of dissolved organic matter in the Amazon Basin. Biogeosciences 13(14):4279–4290

Goulding M, Barthem R, Ferreira E (2003) The Smithsonian Atlas of the Amazon. Smithsonian Books, Washington, DC

Goulding M, Carvalho ML, Ferreira EJG (1988) Rio Negro. Rich Life in Poor Water. SPB Academic Publishing, Amsterdam

Goulding M, Venticinque E, Ribeiro MLDB, Barthem RB, Leite RG, Forsberg B et al (2019) Ecosystem-based management of Amazon fisheries and wetlands. Fish Fish 20(1):138–158

Granado-Lorencio C, Cervia JL, Lima C (2007) Floodplain lake fish assemblages in the Amazon River: directions in conservation biology. Biodiversity Conserv 16:679–692

Granado-Lorencio C, Gulfo A, Alvarez F, Jiménez-Segura LF, Carvajal-Quintero JD, Hernández-Serna A (2012) Fish assemblages in floodplain lakes in a neotropical river during the wet season (Magdalena River, Colombia). J Trop Ecol 28:271–279

Hamilton SK, Sippel SJ, Lewis WM, Saunders JF (1990) Zooplankton abundance and evidence for its reduction by macrophyte nets in two Orinoco floodplain lakes. J Plankton Res 12:345–363

Hardy ER (1980) Composição do zooplâncton em cinco lagos da Amazônia Central. Acta Amazonica 10:577–609

Henderson PA, Robertson BA (1999) On structural complexity and fish diversity in an Amazonian floodplain. In: Padoch C, Ayres JM, Pinedo-Vasquez M, Henderson A (eds) Várzea: Diversity, Development and Conservation of Amazonia's Whitewater Floodplains. New York Botanical Garden Press, New York, pp 45–58

Henderson PA, Crampton WG (1997) A comparison of fish diversity and abundance between nutrient-rich and nutrient-poor lakes in the Upper Amazon. J Trop Ecol 13(2):175–198

Henderson PA, Hamilton WD, Crampton WGR (1998) Evolution and diversity in Amazonian floodplain communities. In: Dynamics of Tropical Communities: 37th Symposium of the British Ecological Society. Cambridge, UK, Cambridge University Press, p 385

Horbe AMC, Queiroz MMDA, Moura CAV, Toro MAG (2013) Geoquímica das águas do médio e baixo rio Madeira e seus principais tributários-Amazonas-Brasil. Acta Amazonica 43:489–504

von Humboldt A (1833) The travels and researches of Alexander von Humboldt. Oliver & Boyd, Edinburgh, p 424

Huston MA (2014) Disturbance, productivity, and species diversity: empiricism vs. logic in ecological theory. Ecology 95(9):2382–2396

Huszar VLM, Reynolds CS (1997) Phytoplankton periodicity and sequences of dominance in an Amazonian flood-plain lake (Lago Batata, Pará, Brazil): responses to gradual environmental change. Hydrobiologia 346:169–181

Ibanez MSR (1998) Phytoplankton composition and abundance of a central Amazon lake. Hydrobiologia 362:79–83

Irmler U (1975) Ecological studies of the aquatic soil invertebrates in three inundation forests of Central Amazonia. Amazoniana 5:337–409

Janzen DH (1974) Tropical blackwater rivers, animals, and mast fruiting by the Dipterocarpaceae. Biotropica 6(2):69–103

Jaramillo-Villa U, Maldonado-Ocampo JA, Escobar F (2010) Altitudinal variation in fish assemblage diversity in streams of the central Andes of Colombia. J Fish Biol 76:2401–2417

Jepsen DB, Winemiller KO (2007) Basin geochemistry and isotopic ratios of fishes and basal production sources in four neotropical rivers. Ecol Freshwater Fish 16:267–281

Johnson KH, Vogt KA, Clark HJ, Schmitz OJ, Vogt DJ (1996) Biodiversity and the productivity and stability of ecosystems. Trends Ecol Evol 11(9):372–377

Junk WJ (1973) Investigations on the ecology and production-biology of the "floating meadows" (Paspalo-Echinochloetum) on the middle Amazon. Part II. The aquatic fauna in the root zone of the floating vegetation. Amazoniana 4:9–102

Junk WJ (1983) Distribution on fish species in a lake of Amazon River floodplain near Manaus (Lago Camaleao), with special reference to extreme oxygen conditions. Amazoniana 7:397–431

Junk WJ, Piedade MTF (1997) Plant life in the floodplain with special reference to herbaceous plants, The Central Amazon Floodplain: Ecology of a Pulsing System. Springer Berlin Heidelberg, Berlin and Heidelberg, Germany, pp 147–185

Junk WJ, Soares MGM, Bayley PB (2007) Freshwater fishes of the Amazon River basin: their biodiversity, fisheries, and habitats. Aquat Ecosyst Health Manage 10(2):153–173

Junk WJ (1997) General aspects of floodplain ecology with special reference to Amazonian floodplains. In: Junk WJ (ed) The central Amazon floodplain: ecology of a pulsing system. Springer, Berlin, Heidelberg, pp 3–20

Kearney M, Lieberman BS, Strotz LC (2024) Tangled banks, braided rivers, and complex hierarchies: beyond microevolution and macroevolution. J Evol Biol 37(12):1402–1412

Keller F (1875) The Amazon and Madeira rivers: sketches and descriptions from the note-book of an explorer. Chapman and Hall, London, UK

Kemenes A, Forsberg BR (2014) Factors influencing the structure and spatial distribution of fishes in the headwater streams of the Jaú River in the Brazilian Amazon. Brazilian. J Biol 74:S23–S32.

Kerguelén-Durango E, Atencio-García V (2015) Environmental characterization of the reproductive season of migratory fish of the Sinú river (Córdoba, Colombia). Revista MVZ (Medicína Veterinaria e Zootecnia) Córdoba 9:143–237

Leenheer JA (1980) Origin and nature of humic substances in the waters of the Amazon River Basin. Acta Amazonica 10(3):513–526

Leenheer JA (1980) Origin and nature of humic substances in the waters of the Amazon River Basin. Acta Amazonica 10(3):513–526

Lehner B, Verdin K, Jarvis A (2008) New global hydrography derived from spaceborne elevation data. Eos Trans Am. Geophys Union 89(10):93–94

Lima de Melo T, Tejerina-Garro FL, De Melo CE (2009) Influence of environmental parameters on fish assemblage of a Neotropical river with a flood pulse regime, Central Brazil. Neotrop Ichthyol 7:421–428
Lima FCT, Ramos L, Barreto T, Cabalzar A, Tenório G, Barbosa A et al (2005) Peixes do alto Tiquié: ictiologia e conhecimentos dos tuyuka e tukano. Peixe e Gente no Alto Rio Tiquié. Instituto Socioambiental, São Paulo, Brazil, p 339
Lobón-Cerviá J, Hess LL, Melack JM, Araujo-Lima CARM (2015) The importance of forest cover for fish richness and abundance on the Amazon floodplain. Hydrobiologia 750:245–255
Loureiro M, Sá RD, Serra SW, Alonso F, Lanés LEK, Volcan MV et al (2018) Review of the family Rivulidae (Cyprinodontiformes, Aplocheiloidei) and a molecular and morphological phylogeny of the annual fish genus *Austrolebias Costa* 1998. Neotrop Ichthyol 16(03):e180007
Lujan NK, Armbruster JW (2011) Two new genera and species of Ancistrini (Siluriformes: Loricariidae) from the western Guiana Shield. Ichthyol Herpetol 2011(2):216–225
Lujan NK, Roach KA, Jacobsen D, Winemiller KO, Vargas VM, Ching VR, Maestre JA (2013) Aquatic community structure across an Andes-to-Amazon fluvial gradient. J Biogeography 40(9):1715–1728
MacCreagh G (1926) White Waters and Black. The Century Co., New York
Maldonado-Ocampo JA, López-Fernández H, Taphorn DC, Bernard CB, Crampton WGR, Lovejoy NR (2014) Akawaio penak, a new genus and species of Neotropical electric fish (Gymnotiformes, Hypopomidae) endemic to the upper Mazaruni River in the Guiana Shield. Zool Scripta 43:24–33
Marlier G (1967) Ecological studies on some lakes of the Amazon valley. Amazoniana 1:91–115
Matsuo AYO, Val AL (2007) Acclimation to humic substances prevents whole body sodium loss and stimulates branchial calcium uptake capacity in cardinal tetras Paracheirodon axelrodi (Schultz) subjected to extremely low pH. J Fish Biol 70(4):989–1000
de Mello FT, Lucas C, Inda H, Ríos M, González-Bergonzoni I (2025) Uruguay, Rivers of South America. Elsevier, Amsterdam, Netherlands, pp 539–579
Melo BF, Albert JS, Dagosta FC, Tagliacollo VA (2021) Biogeography of curimatid fishes reveals multiple lowland–upland river transitions and differential diversification in the Neotropics (Teleostei, Curimatidae). Ecol Evol 11(22):15815–15832
Melo BF, Ota RP, Benine RC, Carvalho FR, Lima FC, Mattox GM et al (2024) Phylogenomics of Characidae, a hyper-diverse Neotropical freshwater fish lineage, with a phylogenetic classification including four families (Teleostei: Characiformes). Zool J Linnean Soc 202(1):zlae101
Melo BF, Dorini BF, Foresti F, Oliveira C (2018) Little divergence among mitochondrial lineages of Prochilodus (Teleostei, Characiformes). Front Genet 9:107
Mendonça FP, Magnusson WE, Zuanon J (2005) Relationships between habitat characteristics and fish assemblages in small streams of Central Amazonia. Copeia 2005:750–763
Mojica JI, Castellanos C, Lobón-Cerviá J (2009) High temporal species turnover enhances the complexity of fish assemblages in Amazonian terra firme streams. Ecol Freshwater Fish 18:520–526
Mol JH, Ouboter PE (2004) Downstream effects of erosion from small-scale gold mining on the instream habitat and fish community of a small neotropical rainforest stream. Conserv Biol 18:201–214
Mol JH, Resida D, Ramlal JS, Becker CR (2000) Effects of El Niño-related drought on freshwater and brackish-water fishes in Suriname, South America. Environ Biol Fishes 59:429–440
Mol JH, de Mérona B, Ouboter PE, Sahdew S (2007) The fish fauna of Brokopondo Reservoir, Suriname, during 40 years of impoundment. Neotrop Ichthyol 5:351–368
Monente JA, Colonnello G, Herrera O (2017) Los ríos del Delta del Orinoco: situacíon y riesgo. In: Rodríguez-Olarte D (ed) Ríos en riesgo de Venezuela 1. Universidad Centrooccidental Lisandro Alvarado, Caracas, Venezuela, pp 127–149
Montoya JV, Osío A, Pérez MC, Pineda V (2017) Los ríos de los llanos de Apure. In: Rodríguez-Olarte D (ed) Ríos en riesgo de Venezuela 1. Universidad Centrooccidental Lisandro Alvarado, Caracas, Venezuela, pp 75–107
Moreira-Turcq P, Seyler P, Guyot JL, Etcheber H (2003) Exportation of organic carbon from the Amazon River and its main tributaries. Hydrol Processes 17(7):1329–1344
Nascimento MC, Correa SE, Costa I, Shinaigger RNT, Santos PM, Farias BL, Costa PD (2021) The physical-chemical characteristics of surface waters in the management of quality in clearwater rivers in the Brazilian Amazon. Water Policy 23(5):1303–1313
Nascimento BLS, da Silva FS, Nunes-Neto JP, de Almeida Medeiros DB, Cruz ACR, da Silva SP et al (2021) First description of the mitogenome and phylogeny of Culicinae species from the Amazon region. Genes 12(12):1983
Oliveira EC, Ferreira EJG (2008) Spawning areas, dispersion and microhabitats of fish larvae in the Anavilhanas Ecological Station, rio Negro, Amazonas State, Brazil. Neotrop Ichthyol 6:559–566
Pazin VFV, Magnusson WE, Zuanon J, Mendonca FP (2006) Fish assemblages in temporary ponds adjacent to 'terra-firme' streams in central Amazonia. Freshwater Biol 51:1025–1037
Petry P, Bayley PB, Markle DF (2003) Relationships between fish assemblages, macrophytes and environmental gradients in the Amazon River floodplain. J Fish Biol 63:547–579
Petsch DK, Cionek VDM, Thomaz SM, Dos Santos NCL (2023) Ecosystem services provided by river-floodplain ecosystems. Hydrobiologia 850(12):2563–2584
Piedade FMT, Parolin P, Lopes A, Wittmann F, Junk WJ, Mortati A, Andre T, Schoengart J, Graça MA, Latrubesse EM, Callisto M, de Mello FT, Rodríguez-Olarte D (2025) Amazonian Rivers from the Guiana Shield, Rivers of South America. Elsevier, Amsterdam, Netherlands, pp 239–277
Piland NC, Beveridge CF, Campos-Silva JV, Cañas C, Correa SB, Couto TB, Anderson EP (2025) Rivers from the Western Amazon, Rivers of South America. Elsevier, Amsterdam, Netherlands, pp 279–333
de Pinna MC (2016) The dawn of phylogenetic research on Neotropical fishes: a commentary and introduction to Baskin (1973), with an overview of past progress on trichomycterid phylogenetics. Neotrop Ichthyol 14(02):e150127
Pouilly M, Barrera S, Rosales C (2006) Changes of taxonomic and trophic structure of fish assemblages along an environmental gradient in the Upper Beni watershed (Bolivia). J Fish Biol 2006:68
Queiroz HL (2007) Classification of water bodies based on biotic and abiotic parameters at the várzeas of Mamirauá Reserve, Central Amazon. Uakari 3:19–24
Quesada CA, Lloyd J, Anderson LO, Fyllas NM, Schwarz M, Czimczik CI (2011) Soils of Amazonia with particular reference to the RAINFOR sites. Biogeosciences 8(6):1415–1440
Reiss F (1977) Qualitative and quantitative investigations on the macrobenthic fauna of central Amazon lakes. 1. Lago Tupé, a blackwater lake on the lower rio Negro. Amazoniana 6:203–235
Ribeiro JSB, Darwich AJ (1993) Produção primaria fitoplanctônica de um Iago de ilha fluvial na Amazonia Central (Lago do Rei, IIha do Careiro). Amazoniana 12:4
Rincón Ramírez JI (2017) Estado de conservacíon de los ríos en la costa occidental del Lago de Maracaibo. In: Rodríguez Olarte D (ed) Ríos en Riesgo de Venezuela 1. Universidad Centrooccidental Lisandro Alvarado, Caracas, Venezuela, pp 15–28
Ríos-Villamizar EA, Adeney JM, Junk WJ, Piedade MTF (2020a) Physicochemical features of Amazonian water typologies for water resources management. IOP Conf Ser: Earth Environ Sci 427:012003
Ríos-Villamizar EA, Adeney JM, Piedade MTF, Junk WJ (2020b) New insights on the classification of major Amazonian river water types: sustainable. Water Resour Manage 6(5):83

Ríos-Villamizar EA, Piedade MTF, Da Costa JG, Adeney JM, Junk WJ (2013) Chemistry of different Amazonian water types for river classification: a preliminary review. Water Society ii 178:1117

Rocha MS, Oliveira RRD, Py-Daniel LHR (2008) *Scoloplax baskini*: a new spiny dwarf catfish from rio Aripuanã, Amazonas, Brazil (Loricarioidei: Scoloplacidae). Neotrop Ichthyol 6:323–328

Rodriguez MA, Lewis WM (1997) Structure of fish assemblages along environmental gradients in floodplain lakes of the Orinoco River. Ecol Monograph 67:109–128

Rodríguez R JC, Betancourt L JA (1999) Caracterizacion físico-química de una laguna de inundación del tramo Orinoco medio y su relación con la biomassa de la cobertura de bora (*Eichhornia crassipes* (Mart.) Solms). Interciencia 24:243–250

Rodriguez-Olarte D, Barrios Gomez M, Marrero CJ, Marcó LM (2017) Río Turbio: un síndromo urbano en la vertienete andina del Orinoco. In: Rodríguez-Olarte D (ed) Ríos en riesgo de Venezuela 1. Universidad Centrooccidental Lisandro Alvarado, Caracas, Venezuela, pp 59–74

Rodríguez-Olarte D, Rosales J, Vásquez-Ramos JM, Ramírez-Gil H, Marrero CJ, DoNascimiento C (2025) Orinoco, Rivers of South America. Elsevier, Amsterdam, Netherlands, pp 93–152

Röpke CP, Amadio SA, Winemiller KO, Zuanon J (2016) Seasonal dynamics of the fish assemblage in a floodplain lake at the confluence of the Negro and Amazonas rivers. J Fish Biol 89:194–212

Russello MA, Amato G (2004) A molecular phylogeny of Amazona: implications for Neotropical parrot biogeography, taxonomy, and conservation. Mol Phylogenet Evol 30(2):421–437

Saint-Paul U, Soares GM (1987) Diurnal distribution and behavioral responses of fishes to extreme hypoxia in an Amazon floodplain lake. Environ Biol Fishes 20:91–104

Saint-Paul U, Zuanon J, Correa MAV, García M, Fabré NN, Berger U, Junk WJ (2000) Fish communities in central Amazonian white-and blackwater floodplains. Environ Biol Fishes 57:235–250

Sánchez-Botero JI, de Farias ML, Piedade MT, Garcez DS (2003) Ictiofauna associada às macrófitas aquáticas *Eichhornia azurea* (SW.) Kunth. e *Eichhornia crassipes* (Mart.) Solms. no lago Camaleão, Amazônia central, Brasil. Acta Sci Biol Sci 25:369–375

Santos UM (1980) Aspectos limnológicas do Lago Grande do Jutaí (Amazônia Central), face as alterações quimicas do meio hidrico da região. Acta Amazonica 10:797–822

Saul WG (1975) An ecological study of fishes at a site in upper Amazonian Ecuador. Proc Acad Nat Sci Philadelphia 127:93–134

Schluter D (2024) Variable success in linking micro- and macroevolution. Evol J Linn Soc 3(1):kzae016

Schmidt GW (1972) Chemical properties of some waters in the tropical rain-forest region of Central-Amazonia along the new road Manaus-Caracarai. Amazoniana 3:199–207

Segnini S, Chacón MM (2017) El Chama; un río andino en riesgo. In: Rodríguez Olarte D (ed) Ríos en riesgo de Venezuela 1. Universidad Centrooccidental Lisandro Alvarado, Caracas, Venezuela, pp 29–58

Segnini-Flores SE, Rodríguez-Olarte D, Rincón JE, Blanco-Libreros JF, López-Rodríguez SR (2025) Rivers of the Caribbean, Rivers of South America. Elsevier, Amsterdam, Netherlands, pp 19–56

Seidel M, Dittmar T, Ward ND, Krusche AV, Richey JE, Yager PL, Medeiros PM (2016) Seasonal and spatial variability of dissolved organic matter composition in the lower Amazon River. Biogeochemistry 131:281–302

Sendacz S, Costa SSM (1991) Caracterização do zooplâncton do rio Acre e lagos Lua Nova, Novo Andirá e Amapá (Amazônia). Rev Bras Biol 51:463–470

Silva MHC (2017) Organização da comunidade de peixes em riachos de água preta de Mata Atlântica (Doctoral dissertation, Instituto de Pesca, São Paulo, Brazil).

Silvano RA, Pereyra PE, Begossi A, Hallwass G (2022) Which fish is this? Fishers know more than 100 fish species in megadiverse tropical rivers. FACETS 7:988–1007

Sioli H (1950) Das Wasser in Amazonasgebiet. Forschung U Fortschritte, Berlin 26:274–280

Sioli H (1956) Über Natur und Mensch im brasilianischen Amazonasgebiet. Erdkunde 10:89–109

Stallard RF, Edmond JM (1983) Geochemistry of the Amazon: 2. The influence of geology and weathering environment on the dissolved load. J Geophys Res: Oceans 88(C14):9671–9688

Stallard RF (1980) Major element geochemistry of the Amazon River system. Doctoral dissertation, Massachusetts Institute of Technology

Tagliacollo VA, Dagosta FCP, Pinna MD, Reis RE, Albert JS (2021) Assessing extinction risk from geographic distribution data in Neotropical freshwater fishes. Neotrop Ichthyol 19:e210079

Tavares-Dias M, Lemos JRG, Martins ML (2010) Parasitic fauna of eight species of ornamental freshwater fish species from the middle Negro river in the Brazilian Amazon region. Revista Brasileira de Parasitologia Jaboticabal 19:103–107

Tejerina-Garro FL, Fortin R, Rodriguez MA (1998) Fish community structure in relation to environmental variation in floodplain lakes of the Araguaia River, Amazon Basin. Environ Biol Fishes 51:399–410

Thomaz SM, Cunha ERD (2010) The role of macrophytes in habitat structuring in aquatic ecosystems: methods of measurement, causes and consequences on animal assemblages' composition and biodiversity. Acta Limnol Bras 22:218–236

Tobes I, Gaspar S, Peláez-Rodríguez M, Miranda R (2016) Spatial distribution patterns of fish assemblages relative to macroinvertebrates and environmental conditions in Andean piedmont streams of the Colombian Amazon. Inland Waters 6:89–104

Torrente-Vilara G, Zuanon J, Leprieur F, Oberdorff T (2011) Effects of natural rapids and waterfalls on fish assemblage structure in the Madeira river (Amazon basin). Ecol Freshwater Fish 20:588–597

Trevisan GV, Jardim-Lima DJ (2012) Relationship between water transparency and physical-chemical variables in lakes of the Western Amazon, Brazil. Acta Sci Biol Sci 34:13–21

Tundisi JG, Forsberg BR, Devol AH, Zaret TM, Tundisi TM, Dos Santos A, Ribeiro JS, Hardy ER (1984) Mixing patterns in Amazon lakes. Hydrobiologia 108(1):3–15

Val AL, Silva MNP, Almeida-Val VMF (1998) Hypoxia adaptation in fish of the Amazon: a never-ending task. S Afr J Zool 33(2):107–114

Val AL, de Almeida-Val VM (2012) Fishes of the Amazon and their environment: physiological and biochemical aspects, vol 32. Springer Science & Business Media, Dordrecht, Netherlands

Vitorino Júnior OB, Fernandes R, Agostinho CS, Pelicice FM (2016) Riverine networks constrain β-diversity patterns among fish assemblages in a large Neotropical river. Freshwater Biol 61:1733–1745

Waddell JC, Crampton WGR (2020) Environmental correlates of circannual breeding periodicity in a multi-species assemblage of Amazonian electric fishes. Environ Biol Fishes 103:233–250

Waddell JC, Crampton WG (2022) Reproductive effort and terminal investment in a multispecies assemblage of Amazon electric fish. Ecol Monogr 92(2):e1499

Waide RB, Willig MR, Steiner CF, Mittelbach G, Gough L, Dodson SI et al (1999) The relationship between productivity and species richness. Annu Rev Ecol Syst 30(1):257–300

Wallace AR (1853) On the Rio Negro. J R Geogr Soc London 23:212–217

Wallace AR (1853) Palm trees of the Amazon and their uses. Cambridge University Press, Cambridge, UK

Wallace AR (1890) A narrative of travels on the Amazon and Rio Negro: with an account of the native tribes, and observations of the climate, geography, and natural history of the Amazon Valley. Ward, Lock and Company, London, UK

Weber B, Mota A, Helenes J, Ramírez R, Valencia Y (2016) Age and provenance of Late Miocene-Early Pliocene sedimentary rocks from the Patao high hydrocarbon reservoir offshore NE Venezuela–U-Pb detrital zircon age, Sm-Nd isotope, and biostratigraphic data. J Nat Gas Sci Eng 31:459–473

Winemiller KO, Taphorn DC, Barbarino-Duque A (1997) Ecology of *Cichla* (Cichlidae) in two blackwater rivers of southern Venezuela. Copeia 1997:690–696

Winemiller KO, Lopez-Fernandez H, Taphorn DC, Nico LG, Duque AB (2008) Fish assemblages of the Casiquiare River, a corridor and zoogeographical filter for dispersal between the Orinoco and Amazon basins. J Biogeogr 35:1551–1563

Winemiller KO (1996) Food Webs: integration of patterns & dynamics. Springer US, Boston, MA

Winemiller KO, Jepsen DB (1998) Effects of seasonality and fish movement on tropical river food webs. J Fish Biol 53:267–296

Wood CM (1989) The physiological problems of fish in acid waters. In: Morris R, Brown DJA, Taylor EW, Brown JA (eds) In Acid Toxicity and Aquatic Animals. Cambridge University Press, Cambridge, England, Society for Experimental Biology Seminar Series, pp 125–148

9 Neotropical Fishes of Upland Streams

Luciano F. A. Montag, Bruno E. Soares, Naraiana L. Benone, Lorrane G. Cantanhêde, Gabriel L. Brejão, Lilian Casatti and Roberto E. Reis

Abstract

This chapter discusses the evolutionary and ecological aspects of fishes that inhabit the upland streams of South America. Across the continent, there exist 1257 species of upland stream fish *species* that thrive in high-gradient *habitats* characterized by erosional settings with a slope exceeding 2%. This diverse faunal *assemblage* spans various clades of Neotropical fishes, showcasing pronounced convergent *traits*—both morphological and behavioral—with *adaptations* to the challenges of aquatic life in upland stream. This chapter discusses the evolutionary and ecological dimensions of survival and reproduction in upland streams. Only 16 species commonly inhabit all three of the main water chemistry profiles: clearwaters, whitewaters, and blackwaters. The majority of species exclusively inhabit the Brazilian and Guiana Shields, with 769 species in clearwaters, 377 species in whitewaters, and 33 species in blackwaters. Notably, 49% of these species predominantly occupy gravel and rock substrates, with others residing up in the water column. Predominantly, benthivorous fishes primarily consume invertebrates (40%) and followed by species that are algivores (36%). This chapter introduces four ecomorphological syndromes for upland stream fishes, emphasizing that a majority inhabit gravel bottoms in clear water environments. Despite their ecological significance, constituting almost 20% of Neotropical freshwater species, upland stream fishes often face neglect, underscoring the imperative for heightened conservation endeavors and the establishment of strategically protected areas grounded in scientific understanding.

L. F. A. Montag (✉)
Federal University of Pará, Belém, PA, Brazil
e-mail: montag@ufpa.br

B. E. Soares
University of Regina, Regina, SK, Canada
e-mail: soares.e.bruno@gmail.com

N. L. Benone
Universidade do Estado de Minas Gerais, Belo Horizonte, MG, Brazil
e-mail: nbenone@gmail.com

L. G. Cantanhêde
Instituto Chico Mendes De Conservação, São Luiz, MA, Brazil
e-mail: lorranegabrielle@hotmail.com

G. L. Brejão · L. Casatti
Universidade Estadual Paulista Rio Claro, Rio Claro, SP, Brazil
e-mail: gbrejao@gmail.com

L. Casatti
e-mail: lilian.casatti@unesp.br

R. E. Reis
Pontifícia Universidade Católica Rio Grande do Sul, Porto Alegre, RS, Brazil
e-mail: reis@pucrs.br

9.1 Introduction to Upland Streams

Upland streams of South America refer to waterways of Strahler *stream order* 1–5, a minimum slope (i.e., gradient) of 2%, and elevations between approximately 250 m and 1000 m on the Brazilian and Guiana Shields, or Andean Cordilleras (Babkin 2009; Kroonenberg et al. 2016). These streams are characterized by rapid flow and drain elevated or mountainous regions, which by the "force of water" produces the other characteristics of an upland stream as compared to lowland or floodplain streams; i.e., an incised rock-dominated riverbed with coarse sediment clasts, presence of riffles and pools, and cold water temperatures, generally below 20°C (Crampton 2011). These particularities of environmental conditions as a result of rapid flow select fish species with limited tolerance to temperature variation, high oxygen requirements, strong swimming ability, and specialized reproductive strategies to

J. S. Albert and R. E. Reis (eds.), *Ecology and Evolution of Amazonian Fishes*, Fascinating Life Sciences,
https://doi.org/10.1007/978-3-032-09328-8_9

prevent eggs or larvae from being swept away (e.g., Silveira et al. 2018; Pereira et al. 2021).

Another characteristic of upland streams under natural conditions is the presence of a dense riparian canopy, which is expected to control the light regime and limit aquatic production (Davies et al. 2008; Benone et al. 2017). These aspects of upland streams' metabolism, indicate that they are strongly dependent on terrestrial sources of carbon, supplied by vascular plants growing in the riparian zone (Davies et al. 2008), while recent large-sale syntheses suggest a large importance of autochthonous resources (Ferreira et al. 2024; Leal et al. 2023). However, the diversity of environments and conditions in tropics is so great, that the contribution of autochthonous or allochthonous sources is an inconclusive issue (e.g., Tank et al. 2010; Reis et al. 2020; Kisaka et al. 2021).

By being associated with elevated regions, upland streams are under the effect of the current speed that is responsible for carrying sediments downstream, determines the bottom deposition in different stretches of the channel (Hoover et al. 2006), and forms meanders, sandbars, rapids, backwaters, and pools (Fig. 9.1). These continuous modifications create a highly variable longitudinal river profile, forming habitat patches that have been shown to strongly affect species distributions ("Patch Dynamics," Pringle et al. 1988). Additionally, substrate composition also influences the diversity and distribution of aquatic *biodiversity*, as most aquatic animals depend on different substrates for colonization and foraging (Hoover et al. 2006). Woody debris in the bed trap litter and other larger substrates, especially in slower flowing streams, altering depth, width, and flow, increasing habitat complexity for invertebrates and fishes (Angermeier and Karr 1984; Cadol et al. 2009; Valente-Neto et al. 2015).

Despite the enormous importance of streams as maintainers of aquatic biodiversity and providers of ecosystem services (Allan and Flecker 1993; Chaves et al. 2008), they have been affected by anthropogenic impacts in several ways (Van Meter et al. 2016). Most of these ecosystems are in watersheds predominantly used for agriculture, mining, and cattle ranching, which confer a series of impacts on the abiotic and biotic environments (e.g., Brosse et al. 2011; Allard et al. 2016; Montag et al. 2019; Seabra et al. 2021).

9.2 Geomorphology

In South America, upland streams have their sources in four main geological regions, the Central Brazilian Shield, the Guiana Shield, the Atlantic Shield, and Andean cordilleras (Fig. 9.2; Schobbenhaus and Brito Neves 2003). The Guiana Shield covers the countries of Guyana, Suriname, French Guiana, and parts of northern Brazil, southern and eastern Venezuela, and eastern Colombia (Hammond 2005). The Central Brazilian Shield covers central-western Brazil and part of Bolivia. The Atlantic Shield is derived from eruptions of a post-Triassic period extending from northwestern Ceará state to the Rio de La Plata estuary in Uruguay and Argentina. Other high elevations are of modest size and do not fit into the shield designation, but only as massifs (e.g., Rio Apa, Rivera, Uruguay–Rio Grande do Sul, Maciço Central de Goiás, etc.; Schobbenhaus and Brito Neves 2003). In addition to the shields, the Andean region, which extends from the north to the south of the South American continent, is subdivided into Northern, Central, and Southern Andes. The northern part extends to 5°S and is distinguished by high ridges of folded blocks and deep depressions (Babkin 2009). In the Equatorial region, there are the Eastern and Western Cordilleras with the depression between them filled with products of volcanic activity. In Colombia, three main systems are identified: eastern, central, and western. They are divided by the basins of the Magdalena and Cauca rivers (Babkin 2009). Most of the major rivers in South America are supplied by water from the Andean Mountain range, because they receive more precipitation than the plains. In general, they also have glaciers and snow-covered areas, which provide a large water reservoir (Schoolmaster et al. 2018).

According to Dagosta and de Pinna (2017), the following drainages are part of the Guianas Shield basins: Jari, Urubu-Uatumã, Trombetas, Curuá, and Paru do Oeste. The following are part of the Central Brazil Shield: Jamaxim, Iriri, Teles Pires, Juruena, Tapajós, Xingu, and Tocantins-Araguaia. The Atlantic Shield is composed of the São Francisco Craton in its center surrounded by several orogenic fold belts which for descriptive purposes were grouped into three geotectonic provinces: Mantiqueira, Borborema, and Tocantins.

9.3 Geographic Distribution

Description of South American upland streams' water types was primarily based on coloration of Amazonian waters, being clearwaters (= blue; Sioli 1984), blackwaters, and whitewaters (Wallace 1853). Upland streams are characterized as whitewater streams (Andean nature) and clearwater streams (central Brazilian, Guiana, and Atlantic cratons). These differences in coloration result from differences in the number of suspended particles, the degree of leaching of organic material from surrounding forests, and the geological and soil characteristics of the drainage system (Leenheer 1980; Chap. 8). Soil chemistries in the Neotropical region are very diverse, due to different geological and climatic histories, with different amounts of chemical weathering between relatively younger geological formations (such as the Andes Mountain range) and older cratonic regions (Atlantic, Guiana, and Central Brazilian Shields; Quesada et al. 2011).

Fig. 9.1 Environments and fishes of upland streams from the South American region. (**A**) Salobrinha Creek located in the Parque Nacional da Serra da Bodoquena, Mato Grosso do Sul State, Central Brazil, showing *Parodon nasus* specimens foraging on the rocky substrate (Photo R. M. Romero); (**B**) general view of a rocky stream close to the headwaters of the São Francisco River basin, Minas Gerais State, Southeastern Brazil (Photo R. M. Romero); (**C**) specimens of *Characidium* sp. among the rocks in the Taquaral Creek, Parque Nacional da Serra da Bodoquena, Mato Grosso do Sul State, Brazil (Photo R. M. Romero); (**D**) general view of a rocky stream close to the headwaters of the Juruena River basin, Mato Grosso State, North Brazil (Photo N. L. Benone); (**E**) general view of a stream fall close to the headwaters of the Trombetas River basin, Pará State, North Brazil (Photo: L. B. Seabra); (**F**) General view of a rock stream of the Grande River basin, Minas Gerais State, Southeastern Brazil (Photo N. L. Benone)

Waters draining more recent geological formations often present high turbidity that limits vertical light transmission (Muntz 1978), with a typical local range of a maximum of 0.50 m depth (Sioli 1984). Soil types in these regions (e.g., volcanic andisols, clay-rich vertisols, waterlogged gleysols) favor more complete decomposition and absorption of organic matter from the surrounding forests (Leenheer 1980), resulting in a lower concentrations of dissolved organic carbon (DOC). Because of the low DOC concentration, the pH of whitewater streams is close to neutral or alkaline, ranging from 6.2 to 7.2 (Sioli 1984; Küchler et al. 2000).

Clearwater streams have their headwaters in ancient (cratonic) terrains, the Central Brazil, Guiana, and Atlantic Shields

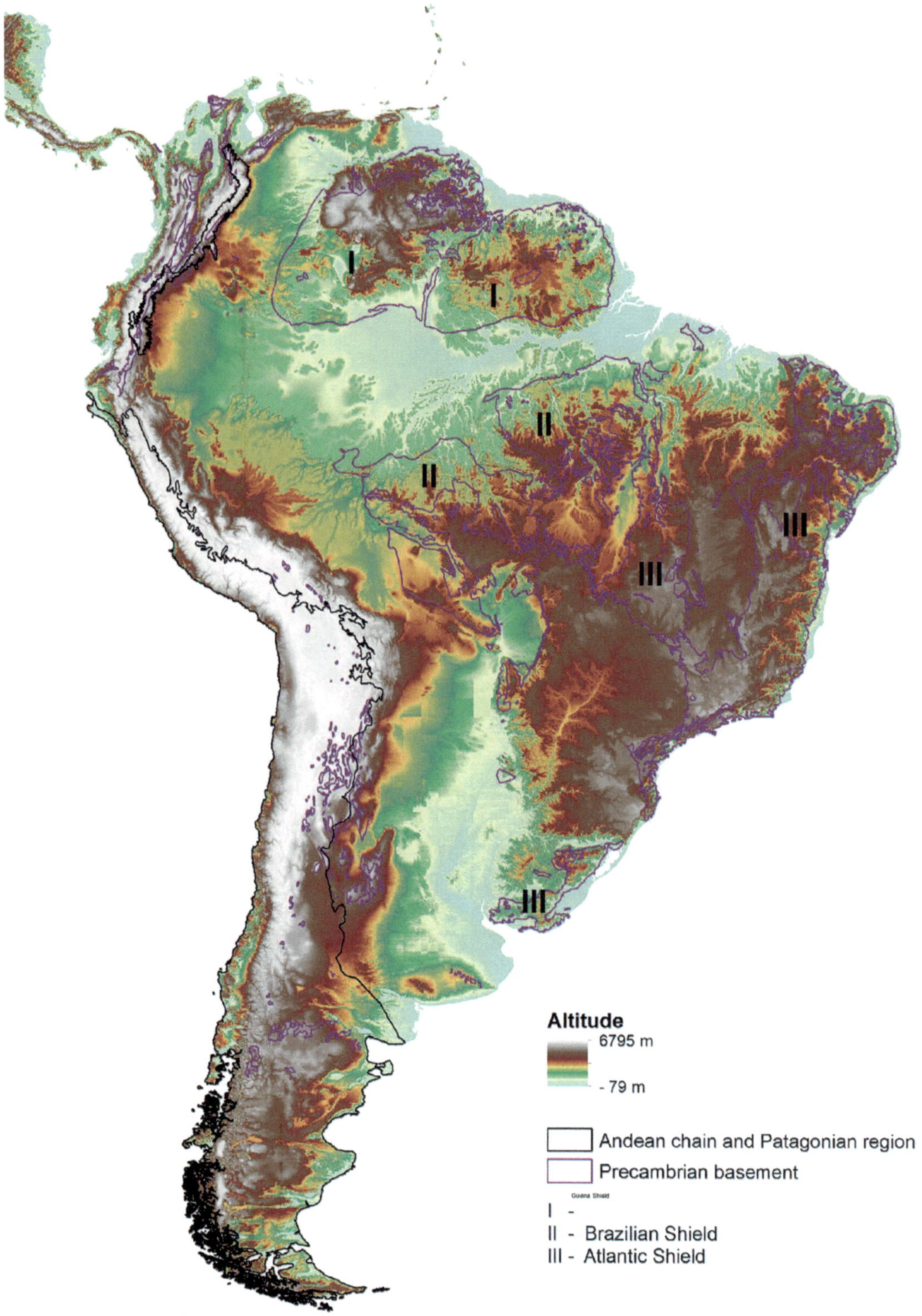

Fig. 9.2 Distribution of South American upland streams considering a slope higher than 2% in the Guiana (**I**), Brazilian (**II**), and Atlantic (**III**) Shields (Schobbenhaus and Brito Neves 2003)

(Goulding et al. 2003). Given their more ancient origins and long exposure to geochemical weathering processes, latosols (also called oxisols) lack fine sediments and various minerals that results in a higher water transparency (Furch 1984). Although more transparent than whitewaters, clearwaters are remarkably different in DOC composition. Clearwaters originate from latosolic soils that absorb soluble DOC (i.e., humic and fulvic acids), resulting in revised high transparency, ranging from 1.10 m to 4.30 m, and in a circumneutral pH between 4.50 and 7.80 (Leenheer 1980). Clearwaters also exhibit low electrical conductivity, ranging from 8.8 to 28.6 µS/cm (Küchler et al. 2000; Ríos-Villamizar et al. 2013).

Blackwaters have a lower concentration of nutrients compared to whitewaters (Furch 1984) and, as in the Rio Negro basin, have a low conductivity, ranging from 8.8 to 28.6 µS/cm (Küchler et al. 2000). Although more transparent than whitewaters, blackwaters and clearwaters are remarkably different in DOC composition. Clearwaters originate in latosolic soils that absorb soluble DOC (i.e., humic and fulvic acids) and produce high transparency ranging from 1.10 m to 4.30 m (Leenheer 1980; Sioli 1984). Blackwaters have lower transparency values, ranging from 1.3 m to 2.9 m (Sioli 1984). Blackwater are colored by tannins, fulvic, and humic acids leached from decaying vegetation (Leenheer 1980; Sioli 1984), biasing light transmission to long wavelengths (Mendonça et al. 2005). High DOC concentrations make blackwaters remarkably acidic, with a pH between 3.8 and 5.5, while the near absence of DOC in clear waters results in a circumneutral pH between 5.50 and 7.80 (Sioli 1984).

These differences in water characteristics create the stage for natural selection to act by driving local adaptations in the aquatic organisms. This set of water characteristics is of paramount influence on the biodiversity of upland stream fishes, as adaptations to the divergent pressures imposed by different habitat conditions also push populations apart that can contribute to the speciation process (Servedio and Boughman, 2017).

9.4 Fishes of Upland Streams

Across the South American continental platform, there are at least 1257 upland stream fish species encompassing seven taxonomic orders and 26 families (Fig. 9.3). The most representative orders are Siluriformes (826 species) and Characiformes (336 species), while some orders encompass only one or two species for upland stream fish species (Synbranchiformes, Gobiesociformes, and Gymnotiformes). The most representative families are Loricariidae (503 species) and Characidae (215 species), while some families encompass just one or two species, such as Erythrinidae, Pimelodidae, and Poeciliidae. Therefore, the ichthyofauna inhabiting South American upland streams encompass multiple lineages (Fig. 9.4). These lineages may encompass taxa with specialized

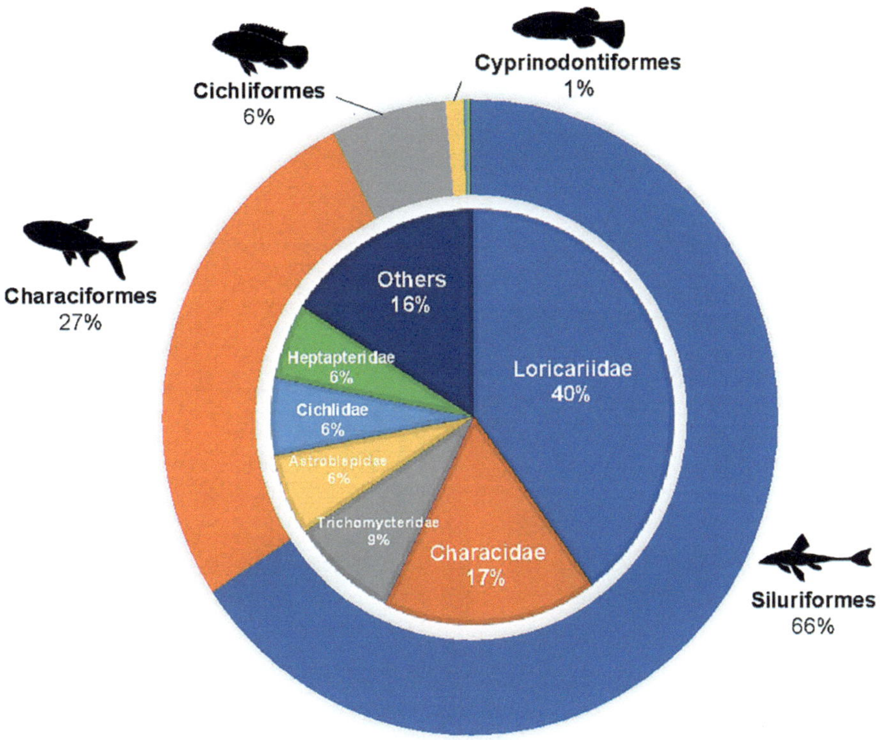

Fig. 9.3 Proportion of fish species in taxonomic orders (external circle) and families (internal circle) in upland streams in South America

traits to explore high-flow systems or generalist species. Given the high species proportion of a couple of specialized lineages (e.g., Astroblepidae and Loricariidae), upland streams might encompass a lower phylogenetic diversity than lowland systems, but these phylogenetic diversity patterns are still poorly explored in Neotropical freshwater systems (Soares and Nakamura 2021; Soares et al. 2023).

Among the upland stream fish species, only 20 are common to the three water types. Most species (934 species) inhabit clear water streams in the Atlantic, Brazilian, and Guiana Shields, while 377 species inhabit whitewater streams primarily in the Andes foothills and 113 inhabit blackwaters (Fig. 9.5). Of all the species inhabiting upland streams, 1053 species are benthic and 332 are epibenthic. Most of the benthic species utilize gravel or rocky substrate (61.4%) and submerged wood/vegetation (15.7%). In general, most upland stream fishes inhabit gravel bottoms in clear water habitats.

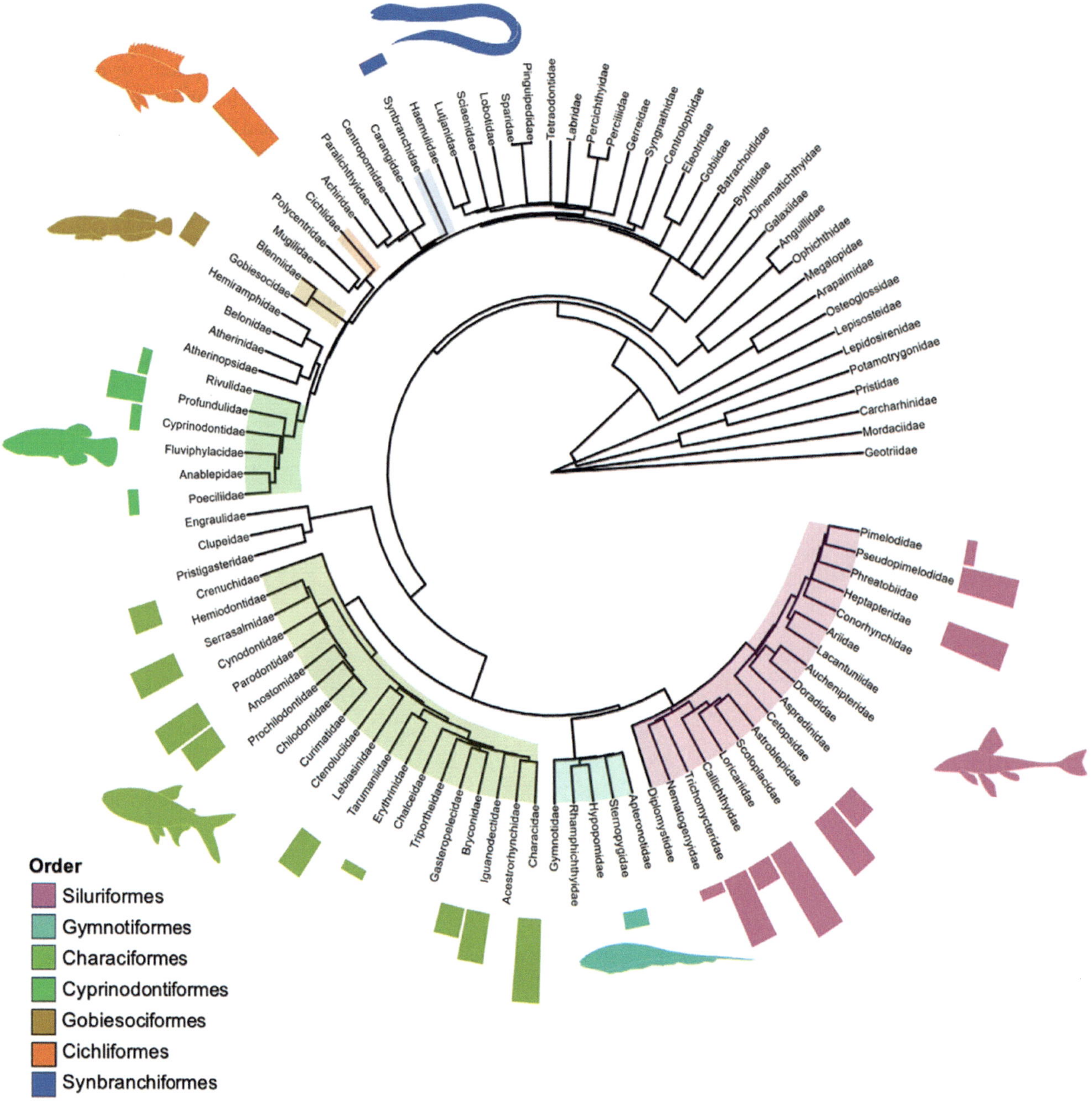

Fig. 9.4 Phylogeny for fish families inhabiting South American upland streams. The phylogeny was built using a supertree approach using the FishPhyloMaker package (Nakamura et al. 2021) merging open phylogenies and taxonomic databases

9.5 Functional Adaptations

Hydraulic energy also plays an important role in upland streams, since dispersal becomes much more cost-effective, given the strong current and the unidirectional flow. This creates a suitable scenario for isolation and vicariance, which result in elevated rates of endemism (Lujan and Conway 2015).

Upland streams have strong currents that forge a heterogeneous waterscape encompassing pools, riffles, and rapids, with bottom covered mostly by gravel, rocks, boulders, and

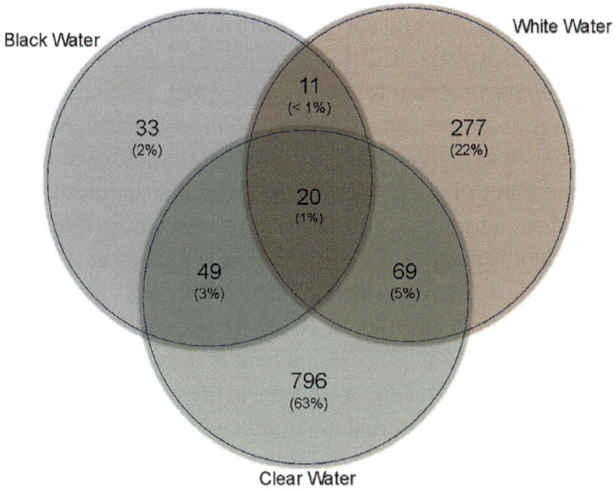

Fig. 9.5 Venn diagram showcasing the number of species of fish upland streams shared and exclusives in the three water types in the Neotropical region

cobbles. The waters are cold and well oxygenated (>5 mg/L), and turbidity varies depending on the sedimentary deposits, being high in Andean piedmont streams and low in shield streams (Crampton 2011). Due to its harsher conditions, upland streams usually have low-diversity communities encompassing a riffle-dwelling specialized fish fauna that is associated with gravel and rocks on the bottom, and riffle-avoiding species using pool habitats in the water column (Taylor 2000) (Fig. 9.6). The strong current is the most important environmental filter among the uplands distributed through the South American region. Riffle-dwelling species exhibit a few ecomorphological and behavioral syndromes to deal with the strong currents that highlight both the convergent evolution in upland streams in different biogeographic regions and the selection of different phylogenetic lineages within the Neotropical region (Dagosta and de Pinna 2019).

A first syndrome encompasses a benthic lifestyle facilitated by a dorsoventrally compressed body shape, reduced swim bladders, and long and broad paired (pectoral and/or pelvic) fins. These morphological features decrease the uplift of the water flow on the organisms' body and facilitate adherence to the bottom of the stream. In more extreme cases, the paired fins compose a suction disk that boosts adherence by the generation of negative pressure (Green and Barber 1988). Several phylogenetically unrelated genera display these traits, such as *Amphilius* (Siluriformes: Amphilidae) in Africa and *Balitora* (Cypriniformes: Balitoridae) in Asia. In the South American upland streams, this syndrome is also found in unrelated taxa, including *Parodon* (Characiformes: Parodontidae), *Sicydium* (Gobiiformes: Gobiidae), and *Gobiesox* (Gobiesociformes: Gobiesocidae).

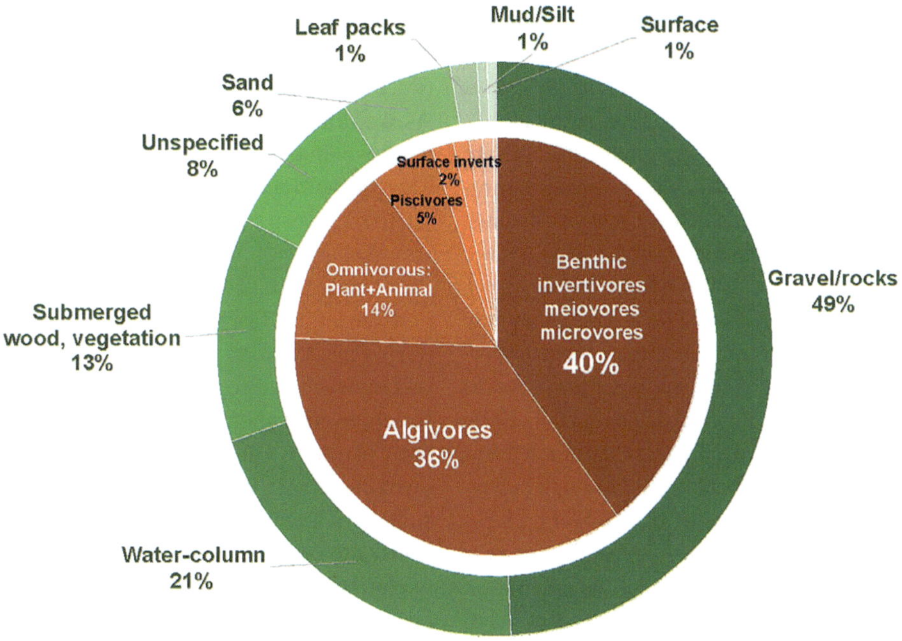

Fig. 9.6 Percent of fish species inhabiting South American upland streams in microhabitats (outer circle) and trophic guilds (inner circle)

The second syndrome comprises similar adaptations to a benthic lifestyle, but instead of the paired fins facilitating adherence to the substrate, some species exhibit fleshy mouths and oral suction to adhere to solid substrate. Neotropical representatives of the syndrome are catfishes from the families Loricariidae and Astroblepidae, and these traits are also present in the African *Chiloglanis* (Siluriformes: Mochokidae) and east Asian *Oreoglanis siamensis* (Siluriformes: Sisoridae). This syndrome has proven to be quite effective in extreme environments. An example is the genus *Astroblepus* (Siluriformes: Astroblepidae) found in high Andean streams (about 3200 masl), one of the few genera that inhabit these extreme environments (the other being *Trichomycterus*) (Jacobsen 2008).

Another ecomorphological and behavioral syndrome of riffle-dwelling species comprise the darters from North America (Percidae: Etheostomatini), the South American darters (Characiformes: Crenuchidae), as well as some cyprinids and cichlids (e.g., Levin et al. 2021). These species are able to remain motionless in flowing water due to ecomorphological and behavioral adaptations. Some morphological adaptations in this group comprise a streamlined body shape, scaleless belly, and long and large winglike paired fins that extend laterally and are steeply canted to deflect the flow of water dorsally. Streamlined bodies facilitate prolonged swimming, which is advantageous to forage in less complex environments (Pakkasmaa and Piironen 2000). The absence or partial absence of scales in the isthmus favors a closer contact of the organism belly with the substrate, improving adherence.

A fourth syndrome favors a fossorial lifestyle by body elongation, being represented by the clariid catfishes in Africa, the cobitid loaches in Asia, and the trichomycterid catfishes in South America. Trichomycterids and heptapterids have elongated bodies and dorsally located eyes, allowing them to explore the crevices between rocks (Casatti and Castro 2006). Elongated body shape along with the ability to hide in the substrate to resist flow are also common in troglobitic fish, such as *Trichomycterus conradi* and *Typhlichthys subterraneus* (Sampaio et al. 2020).

Despite these functional syndromes, upland streams are also inhabited by riffle-avoiding fishes with no specific morphological adaptations for using fast-flowing environments. Riffle-avoiding species in upland streams encompass a large range of ecomorphological traits and phylogenetic lineages, such as guppies (Cyprinodontiformes: Poeciliidae), that may or may not exhibit good swimming capability, but use marginal pools and station-keeping behaviors in upland streams to avoid being displaced by fast flowing (Chuang et al. 2008; Mohammed et al. 2012). We can also mention the small tetras (Characidae) that deal with the current mostly through behavioral strategies, by swimming downriver from larger rocks or below the surface, such as *Bryconamericus* species, or near the river margins such some *Astyanax* species, where current speed is significantly lower than in the main channel (Casatti and Castro 2006).

9.6 Conservation and Knowledge Gaps

The Guiana Shield rainforests are known to be one of the guardians of the South American climate as they have a large portion of closed forest cover remaining in the Amazon (Hammond 2005). Nonetheless, a recent research study has revealed that the deforestation of less than one-third of the regions within the Guiana Shield, presently facing the imminent threats of mining, logging, and agricultural practices, could have far-reaching consequences on biodiversity and ecosystem services throughout South America. This impact stems from disruptions in vegetation–land–atmosphere interactions, particularly affecting the initial phases of two crucial "atmospheric rivers": the Caribbean low-level Jet and the South American low-level jet (SALLJ) (Bovolo et al. 2018).

The impacts of deforestation cause structural changes not only in the forests themselves but also in associated environments, such as highland streams. Several studies have shown that the main implications for the habitat structure of rivers and streams are changes caused by agriculture (Nepstad et al. 2006; Ferreira et al. 2018), logging (Prudente et al. 2017; Jacob et al. 2021), urban development (Wang et al. 2012), and livestock (Teresa and Casatti 2012). These activities accelerate the deforestation process and tend to degrade the streams through physical–chemical changes in the water, by carrying sediment from the external environment to the interior of the streams, causing silting, in addition to contamination by toxic substances, such as pesticides and herbicides (Matthaei et al. 2010; Jinggut et al. 2012).

The factors arising from the advance of deforestation, associated with the burning of fossil fuels, intense industrial activities, and the accumulation of solid waste have intensified and accelerated climate change (Gatti et al. 2021) and contamination by emerging contaminants, such as microplastics, already recorded for streams (Ribeiro-Brasil et al. 2020). As we reflect on the main challenges to filling knowledge gaps, we first need to understand the types of gaps that exist. There are currently seven categories, according to Hortal et al. (2015): Linnean shortfall (deficits on taxonomy and systematics), Wallacean shortfall (deficits on geographic distribution), Prestonian shortfall (deficits on species abundance and population dynamics), Darwinian shortfall (deficits on evolution and phylogenetic information), Raunkiaeran shortfall (deficits on functional characteristics and ecological functions of species), Hutchinsonian shortfall (deficits in species tolerances to abiotic conditions), and Eltonian shortfall (deficits in ecological interactions). Before thinking about any efficient conservation strategy, it is extremely necessary that these shortfalls are filled in order to generate data and information about the species and to clarify the factors behind declines and extinctions (Mace and Baillie 2007). In order to do so, it is necessary to establish priorities for sampling and processing materials to prioritize the most pressing shortfalls and the areas or clades with the most prominent shortfalls (Soares et al. 2023).

The strategy adopted by environmental agencies, such as the Chico Mendes Institute for Biodiversity Conservation of Brazil, to predict the conservation status and risk of extinction of species was created by the International Union for Conservation of Nature (IUCN) and takes into account aspects arising from basic research, such as: reproductive biology, diet, population structure, geographic distribution, evolutionary history, and tolerance to external factors. Based on this information and what they represent as a risk for the perpetuity of the species, they are classified into 11 categories: not evaluated (NE), not applicable (NA), data deficient (DD), least concern (LC), near threatened (NT), vulnerable (VU), endangered (EN), critically endangered (CR), extinct in the wild (EW), and Extinct eEX) (IUCN 2022). Therefore, if all these gaps were filled, there would be essential information to design effective plans for species conservation. However, given the huge number of species already known, these assessments are still progressing timidly, so that many species are included in the DD category, especially regarding the streams' ichthyofauna. As an extremely important recommendation, it is necessary to improve knowledge about data deficient species.

Despite representing at least 20% of known Neotropical freshwater species (approximately 6200 species, Albert et al. 2020), upland stream fishes are of little economic importance and small size, reaching a maximum of 15 cm to 20 cm in length (Castro and Polaz 2020). For these reasons, these species are no longer a priority for funding ecological projects and have even greater gaps (Alkins-Koo 2000). On the other hand, the ecological importance of these species is closely related, for example, to the control of insect vectors of diseases, the maintenance of the forest through the process of seed dissemination, the nutrients cycling, and the maintenance of trophic chains (Vanni 2002; Cavalcanti et al. 2007; Horn et al. 2011; Ferreira et al. 2022). The loss of these ecosystem functions would be highly harmful to freshwaters, which justifies the call for policy makers and legislators to develop clear processes for assessing the status of smaller waters and ensuring their effective management in catchment plans (Biggs et al. 2017).

Using the largest country in South America—Brazil—as an example, to regulate and guarantee the maintenance of biological diversity and genetic resources in the national territory, Law nº 9985/2000 was created, also known as the law of the National System of Nature Conservation Units (SNUC). Based on this law, territorial spaces with significant importance for the conservation of species were specially protected and classified into two main categories: integral protection conservation unit or sustainable use conservation unit. The first category allows only the indirect use of natural resources, while the second foresees the direct use in a sustainable way (Brasil 2000).

Considering only the continental portion, 19% of the Brazilian territory (1,618,248 km²) is under some protection category: sustainable use conservation units (1,070,289 km²) represent 66.1% of the area, and fully protected conservation units (547,959 km²) account for 33.9% (MMA 2023). Despite these being significant numbers, when we take into account the territorial extension of Brazil, the distribution of these conservation units between biomes (MMA 2023), and the benefit of different groups of fauna and flora (Dias-Silva et al. 2021), it is clear that there are still some deficit. In the Brazilian Amazon, for instance, the officially protected areas do not correspond to those with high direct conservation values for stream-dwelling fishes (Frederico et al. 2018). For a reasonable protection of the upland streams and their components, not only more strategic protected areas are needed but also the selection of these areas should encompass the systematic conservation planning based on the scientific knowledge on aquatic organisms (see Frederico et al. 2018).

Acknowledgments LFAM receives a grant from CNPq (#302881/2022-0); NLB from UEMG (PQ 10/2022) and CNPq (#408299/2021-5); LGC from CNPq (#169410/2018-8); GLB received grants from FAPESP (#2018/11954-9); LC from CNPq (#304403/2021-0) and FAPESP (#2019/25445-1); and RER from CNPq (#302760/2022-9 and #400166/2016-0).

References

Albert JS, Tagliacollo VA, Dagosta F (2020) Diversification of neotropical freshwater fishes. Annu Rev Ecol Evol Syst 51:27–53

Alkins-Koo M (2000) Reproductive timing of fishes in a tropical intermittent stream. Environ Biol Fishes 57(1):49–66

Allan JD, Flecker AS (1993) Biodiversity conservation in running waters: identifying the major factors that threaten destruction of riverine species and ecosystems. BioScience 43:32–43

Allard L, Popée M, Vigouroux R, Brosse S (2016) Effect of reduced impact logging and small-scale mining disturbances on Neotropical stream fish assemblages. Aquat Sci 78(2):315–325

Angermeier PL, Karr JR (1984) Relationships between woody debris and fish habitat in a small warmwater stream. Trans Am Fish Soc 113:716–726

Babkin VI (2009) Regional distribution of rivers and streams in South America. In: Dooge JCI (Ed.)Eds. Fresh surface water. Eolsh Publishers, Oxford, 482 p

Benone NL, Esposito MC, Juen L, Pompeu PS, Montag LFA (2017) Regional controls on physical habitat structure of amazon streams. River Res Appl 33:766–776

Biggs J, von Fumetti S, Kelly-Quinn M (2017) The importance of small waterbodies for biodiversity and ecosystem services: implications for policy makers. Hydrobiologia 793:3–39

Bovolo CI, Wagner T, Parkin G, Hein-Griggs D, Pereira R, Jones R (2018) The Guiana Shield rainforests—overlooked guardians of South American climate. Environ Res Lett 13(7):074029

Brasil (2000) Lei Nº 9.985, de 18 de julho de 2000. Regulamenta o art. 225, § 1o, incisos I, II, III e VII da Constituição Federal, institui o Sistema Nacional de Unidades de Conservação da Natureza e dá outras providências. http://www.planalto.gov.br/ccivil_03/leis/l9985.htm. Accessed 9 Feb 2022

Brosse S, Grenouillet G, Gevrey M, Khazraie K, Tudesque L (2011) Small-scale gold mining erodes fish assemblage structure in small neotropical streams. Biodiversity Conserv 20(5):1013–1026

Cadol D, Wohl E, Goode JR, Jaeger KL (2009) Wood distribution in neotropical forested headwater streams of La Selva, Costa Rica. Earth Surf Process Landf 34(9):1198–1215

Casatti L, Castro R (2006) Testing the ecomorphological hypothesis in a headwater riffles fish assemblage of the Rio São Francisco, southeastern Brazil. Neotrop Ichthyol 4(2):203–214

Castro RMC, Polaz CNM (2020) Small-sized fish: the largest and most threatened portion of the megadiverse neotropical freshwater fish fauna. Biota Neotropica 20(1):e20180683

Cavalcanti LP, Pontes RJ, Regazzi AC, de Paula Júnior FJ, Frutuoso RL, Sousa EP, Dantas Filho FF, Lima JW (2007) Efficacy of fish as predators of *Aedes aegypti* larvae, under laboratory conditions. Rev Saude Publica 41(4):638–644

Chaves ML, Rieradevall M, Chainho P, Costa JL, Costa MJ, Prat N (2008) Macroinvertebrate communities of non-glacial high altitude intermittent streams. Freshwater Biol 53:55–76

Chuang LC, Shieh BS, Liu CC, Lin YS, Liang SH (2008) Effects of typhoon disturbance on the abundances of two mid-water fish species in a mountain stream of North Taiwan. Zool Stud 47(5):564–573

Crampton WGR (2011) An ecological perspective on diversity and distributions. In: Albert JS, Reis RE (Eds.)Eds. Historical biogeography of neotropical freshwater fishes. University of California, Berkeley and Los Angeles, 165–189pp

Dagosta FCP, de Pinna M (2017) Biogeography of Amazonian fishes: deconstructing river basins as biogeographic units. Neotrop Ichthyol 15(3):1–24

Dagosta FCP, de Pinna M (2019) The fishes of the Amazon: distribution and biogeographical patterns, with a comprehensive list of species. Bull Am Mus Nat Hist 431:1–163

Davies PM, Bunn SE, Hamilton SK (2008) Primary production in tropical streams and rivers. In: Dudgeon D (Ed.)Eds. Tropical stream ecology. Elsevier, London, PP23–42

Dias-Silva K, Vieira TB, Moreira FFF, Juen L, Hamada N (2021) Protected areas are not effective for the conservation of freshwater insects in Brazil. Sci Rep 11:21247

Ferreira MC, Begot TO, Prudente BS, Juen L, Montag LFA (2018) Effects of oil palm plantations on habitat structure and fish assemblages in Amazon streams. Environ Biol Fishes 101(6):547–562

Ferreira VMB, Soares BE, Leal JS, Albrecht MP (2024) Spatial factors overcome seasonality in increasing the consumption of allochthonous food resources by fishes from tropical lotic ecosystems. Freshw Biol 69(6):783–791.

Ferreira V, Albariño R, Larrañaga A, LeRoy CJ, Masese FO, Moretti MS (2022) Ecosystem services provided by small streams: an overview. Hydrobiologia 850:2501–2535

Frederico RG, Zuanon J, de Marco Jr.P (2018) Amazon protected areas and its ability to protect stream-dwelling fish fauna. Biol Cons 219(2018):12–19

Furch K (1984) Water chemistry of the Amazon basin: the distribution of chemical elements among freshwaters. In Sioli H (ed) The Amazon: limnology and landscape ecology of a mighty tropical river and its basin. Junk, Dordrecht, pp 167–169

Gatti LV, Basso LS, Miller JB, Gloor M, Domingues LG, Cassol HLG, Tejada G, Aragão LEOC, Nobre C, Peters W, Marani L, Arai E, Sanches AH, Corrêa SM, Anderson L, Randow CV, Correia CSC, Crispim SP, Neves RAL (2021) Amazonia as a carbon source linked to deforestation and climate change. Nature 595:388–393

Goulding M, Cañas C, Barthem R, Forsberg B, Ortega H (2003) Amazon headwaters: rivers, wildlife, and conservation in southeastern Peru. Asociación para la Conservación de la Cuenca Amazónica (ACCA), Lima

Green DM, Barber DL (1988) The ventral adhesive disc of the clingfish *Gobiesox maeandricus*: integumental structure and adhesive mechanisms. Can J Zool 66:1610–1619

Hammond DS (2005) Ancient land in a modern world. In: Hammond DS (ed) Tropical forests of the Guiana Shield—ancient forests in a modern world. Cambridge, Wallingford, pp 1–14

Hoover TM, Richardson JS, Yonemitsu N (2006) Flow-substrate interactions create and mediate leaf litter resource patches in streams. Freshwater Biol 51:435–447

Horn MH, Correa SB, Parolin P, Pollux BJA, Anderson JT, Lucas CM, Widmann P, Tjiu A, Galetti M, Goulding M (2011) Seed dispersal by fishes in tropical and temperate fresh waters: the growing evidence. Acta Oecol 37(6):561–577

Hortal J, Bello F, Diniz-Filho JAF, Lewinsohn TM, Lobo JM, Ladle RJ (2015) Seven shortfalls that beset large-scale knowledge of biodiversity. Annu Rev Ecol Evol System 46:523–549

IUC N (2022) The IUCN red list of threatened species. Version 2022-1. https://www.iucnredlist.org. Accessed 23 Jul 2022

Jacob LL, Prudente BS, Montag LFA, Silva RR (2021) The effect of different logging regimes on the ecomorphological structure of stream fish assemblages in the Brazilian Amazon. Hydrobiologia 848:1027–1039

Jacobsen D (2008) Tropical high-altitude streams. In: Dudgeon D (Ed.) Eds. Tropical stream ecology. Elsevier, London, 219–256pp

Jinggut T, Yule CM, Boyero L (2012) Stream ecosystem integrity is impaired by logging and shifting agriculture in a global megadiversity center (Sarawak, Borneo). Sci Total Environ 437:83–90

Kisaka TB, Almeida A, Mendonça-Galvão L, Salemi LF, Nardoto GB (2021) Understanding the factors controlling biofilm as an autochthonous resource in shaded oligotrophic neotropical streams. Aquat Sci 83:24

Kroonenberg SB, Roever EWF, Fraga LM, Reis NJ, Faraco MT, Lafon JM, Cordani UG, Wong TE (2016) Paleoproterozoic evolution of the Guiana shield in Suriname: a revised model. Netherlands J Geosci – Geologie en Mijnbouw 95:491–522

Küchler IL, Miekeley N, Forsberg R (2000) A contribution to the chemical characterization of rivers in the Rio Negro basin, Brazil. J Braz Chem Soc 11:286–292

Leal JS, González AL, Soares BE, Casa-Nova C, Marino NAC, Farjalla VF (2023) Global and local drivers of the relative importance of allochthonous and autochthonous energy sources to freshwater food webs. Ecography 2023(4):e06612

Leenheer JA (1980) Origin and nature of humic substances in the waters of the Amazon River basin. Acta Amazonica 10(3):513–526

Levin B, Simonov E, Franchini P, Mugue N, Golubtsov A, Meyer A (2021) Rapid adaptive radiation in a hillstream cyprinid fish in the East African White Nile River basin. Mol Ecol 30:5530–5550

Lujan NK, Conway KW (2015) Life in the fast lane: a review of rheophily in freshwater fishes. In: Riesch R, Tobler M, Plath M (Eds.) Eds. Extremophile fishes. Springer, Cham, Switzerland, 107–136pp

Mace GM, Baillie JEM (2007) The 2010 Biodiversity indicators: challenges for science and policy. Conserv Biol 21(6):1406–1413

Matthaei CD, Piggott JJ, Townsend CR (2010) Multiple stressors in agricultural streams: interactions among sediment addition, nutrient enrichment and water abstraction. J Appl Ecol 47(3):639–649

Mendonça FP, Magnusson WE, Zuanon J (2005) Relationships between habitat characteristics and fish assemblages in small streams of central Amazonia. Copeia 2005:751–764

Ministério do Meio Ambiente (MMA) (2023) Painel Unidades de Conservação Brasileiras. Recuperado de. https://cnuc.mma.gov.br/powerbi. Accessed 20 Nov 2023

Mohammed RS, van Oosterhout C, Schelkle B, Cable J, McMullan M (2012) Upstream guppies (*Poecilia reticulata* Peters, 1859) go against the flow. Biota Neotropica 12(3):68–72

Montag LFA, Winemiller KO, Keppeler FW, Leão H, Benone NL, Torres NR, Prudente BS, Begot TO, Bower LM, Saenz DE, Lopez'Delgado EO, Quintana Y, Hoeinghaus DJ, Juen L (2019) Land cover, riparian zones and instream habitat influence stream fish assemblages in the eastern Amazon. Ecol Freshw Fish 28:317–329

Muntz WRA (1978) A penetração de luz nas águas de rios Amazônicos. Acta Amaz 8:613–619

Nakamura G, Richter A, Soares BE (2021) FishPhyloMaker: an R package to generate phylogenies for ray-finned fishes. Ecol Inf 66:101481

Nepstad DC, Stickler CM, Almeida OT (2006) Globalization of the Amazon soy and beef industries: opportunities for conservation. Conserv Biol 20:1595–1603

Pakkasmaa S, Piironen J (2000) Water velocity shapes juvenile salmonids. Evol Ecol 14(8):721–730
Pereira LM, Dunck B, Benedito E (2021) Human impacts alter the distribution of fish functional diversity in neotropical stream system. Biotropica 53(2):536–547
Pringle CM, Naiman RJ, Bretschko G, Karr JR, Oswood MW, Webster JR, Welcomme RL, Winterbourn MJ (1988) Patch dynamics in lotic systems: the stream as a mosaic. JNABS 7:503–524
Prudente BS, Pompeu PS, Juen L, Montag LFA (2017) Effects of reduced-impact logging on physical habitat and fish assemblages in streams of Eastern Amazonia. Freshwater Biol 62:303–316
Quesada CA, Lloyd J, Anderson LO, Fyllas NM, Schwarz M, Czimczik CI (2011) Soils of Amazonia with particular reference to the RAINFOR sites. Biogeosciences 8:1415–1440
Reis AS, Albrecht MP, Bunn SE (2020) Food web pathways for fish communities in small tropical streams. Freshwater Biol 65(5):893–907
Ribeiro-Brasil DR, Torres NR, Picanço AB, Sousa DS, Ribeiro VS, Brasil LS, Montag LFA (2020) Contamination of stream fish by plastic waste in the Brazilian Amazon. Environ Pollut 266:115241
Ríos-Villamizar EA, Piedade MTF, Da Costa JG, Adeney JM, Junk WJ (2013) Chemistry of different Amazonian water types for river classification: a preliminary review. WIT Trans Ecol Environ 178:17–28
Sampaio FAC, Rufino MS, Pompeu PS, Santos HDA, Ferreira RL (2020) Hydraulic flow resistance of epigean and hypogean fish of the family Trichomycteridae (Ostariophysi, Siluriformes). Subterr Biol 35:97–110
Schobbenhaus C, Brito Neves BBde (2003) A geologia do Brasil no contexto da Plataforma Sul-Americana. In: Bizzi LA, Schobbenh aus C, Vidotti RM, Gonçalves JE (Eds.) Geology, tectonics and mineral resources of Brazil: texto, mapas & SIG. CPRM, Brasília, 56
Schoolmaster, T, Johansen, KS, Alfthan, B, Baker, E, Hesping, M, Verbist, K, 2018. The Andean Glacier and Water Atlas – the impact of glacier retreat on water resources. UNESCO and GRID-Arendal, Paris and Arendal, p. 80
Seabra LB, Benone NL, Montag LFA (2021) Assessing the effects of multiple land uses on the functional beta diversity of stream fishes in the Amazon region. Hydrobiologia 849:4515–4527
Servedio MR, Boughamn JW (2017) The role of sexual selection in local adaptation and speciation. Annu Rev Ecol Evol Syst 2017(48):85–109.
Silveira EL, Ballester ELC, Costa KA, Scheffer EWO, Vaz-dos-Santos AM (2018) Fish community response to environmental variations in an impacted neotropical basin. Ecol Freshwater Fish 27(4):1126–1139
Sioli H (1984) The Amazon and its main arteries: hydrograph, morphology of the river courses, and river types. In: Sioli H (Ed.)Eds. The Amazon. Springer Netherlands, Dordrecht, PP127–165
Soares BE, Nakamura G (2021) Ecologia filogenética de comunidades de peixes de riacho Neotropicais. Oecol Australis 25(2):433–448
Soares BE, Nakamura G, Freitas TMS, Richter A, Cadotte M (2023) Quantifying and overcoming Darwinian shortfalls to conserve the fish tree of life. Biol Cons 295:110223
Tank JL, Rosi-Marshall EJ, Griffiths NA, Entrekin SA, Stephen ML (2010) A review of allochthonous organic matter dynamics and metabolism in streams. JNABS 29(1):118–146
Taylor CM (2000) A large-scale comparative analysis of riffle and pool fish communities in an upland river system. Environ Biol Fishes 58:89–95
Teresa FB, Casatti L (2012) Influence of forest cover and mesohabitat types on functional and taxonomic diversity of fish communities in neotropical lowland streams. Ecol Freshw Fish 21(3): 433–442
Valente-Neto F, Koroiva R, Fonseca-Gessner AA, Roque FO (2015) The effect of riparian deforestation on macroinvertebrates associated with submerged woody debris. Aquat Ecol 49(1):115–125
Van Meter K, Thompson SE, Basu NB (2016) Human impacts on stream hydrology and water quality. In: Jones JB, Stanley EH (Eds.)Eds. Stream ecosystems in a changing environment. Elsevier, London, PP441–490
Vanni MJ (2002) Nutrient cycling by animals in freshwater ecosystems. Annu Rev Ecol Syst 33:341–370
Wallace, AR, 1853. A narrative of travels on the Amazon and Rio Negro: with an account of the native tribes, and observations on the climate, geology, and natural history of the Amazon valley, 1st edn,. Reeve, London
Wang B, Liu D, Liu S, Zhang Y, Lu D, Wang L (2012) Impacts of urbanization on stream habitats and macroinvertebrate communities in the tributaries of Qiangtang River, China. Hydrobiologia 680(1):39–51

10 Fishes of Blackwater Floodplains

Peter van der Sleen, Arnold Lugo-Carvajal and Jansen Zuanon

Abstract

Amazonian blackwater rivers contain some of the most extreme environmental conditions on our planet, with water that has exceptionally low electrical conductivity values and very acidic pH levels. Regardless of such environmental challenges, almost one-fifth (19%) of all Neotropical fish species inhabit blackwater habitats. One of the most pronounced adaptations that allow fish to thrive in blackwater habitats relates to their ability to maintain ion-regulation in blackwaters, but how this is achieved remains poorly understood. To a large extent, fish diversity and abundance in the blackwater rivers are subsidized by the predictable seasonal access to lateral floodplains, which provide critical feeding, spawning, and rearing grounds. Consequently, loss of floodplain habitats (i.e., igapó forests) due to increased wildfires, and diminishing floodplain access due to flood pulse change, represent major threats to the persistence of blackwater fish diversity in the twenty-first century.

P. van der Sleen (✉)
Wageningen University, Wageningen, Netherlands
e-mail: j.p.vandersleen@gmail.com

A. Lugo-Carvajal · J. Zuanon
Instituto Nacional de Pesquisas da Amazônia, Manaus, AM, Brazil

A. Lugo-Carvajal
e-mail: arnoldlugo@hotmail.com

J. Zuanon
e-mail: jzuanon3@gmail.com

10.1 Introduction to Blackwaters

"Blackwater" is a widely used term in limnology to indicate one of the three major water-chemistry profiles of South American rivers and streams; the others being "clearwater" and "whitewater" (see definitions in Chap. 8). Although the chemical properties and origins of blackwater are not uniform, it typically forms on sandy, nutrient-poor soils, where decomposition rates are slow. High levels of toxic plant compounds, an adaptation of plants in nutrient-poor conditions, may contribute to arrested decomposition rates and thus the formation of blackwaters in some regions. Slow and incomplete decomposition of leaf litter ultimately results in leaching and accumulation of humic acids and fulvic acids and other secondary plant compounds in run-off waters (Leenheer 1980; Ertel et al. 1986), staining them a yellow to reddish color, which darkens to a black color in deeper waters (Fig. 10.1).

Although blackwaters are found throughout the Neotropics, they reach their greatest volume in the Amazon Basin (Fig. 10.2a). Notably, large blackwater rivers with a Strahler *stream order* 5–10 are exclusive to the Amazon Basin, originating in the Guiana Shield—a highly eroded, ancient region spanning much of northeastern South America, between the Orinoco and Amazon rivers and the Atlantic Ocean (Ferreira et al. 2007). Among these blackwater rivers is the Negro River (hereafter Rio Negro), the second-largest tributary of the Amazon River and the largest blackwater river in the world. The Rio Negro drains an area of approximately 720,000 km^2 and encompasses the planet's most extensive blackwater floodplain habitats (Fig. 10.2b).

In terms of water chemistry, the Rio Negro presents some of the most extreme environmental conditions for fishes. Soils in the Rio Negro basin and elsewhere in the Guiana Shield have long since been stripped of their major cations, leaving the water with a very low content of dissolved minerals,

J. S. Albert and R. E. Reis (eds.), *Ecology and Evolution of Amazonian Fishes*, Fascinating Life Sciences,
https://doi.org/10.1007/978-3-032-09328-8_10

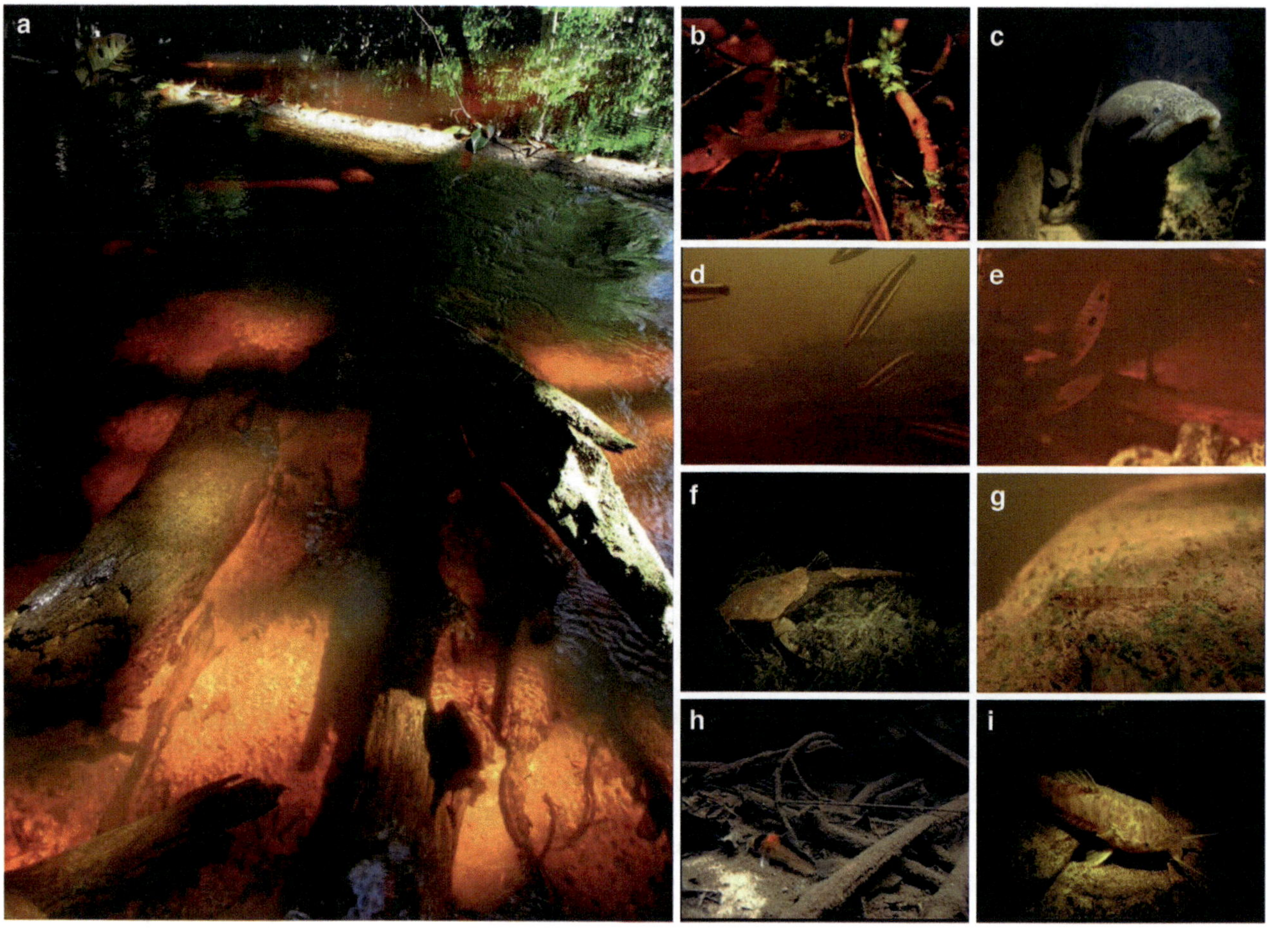

Fig. 10.1 Blackwater rainforest stream draining into the Içana river, a tributary of the upper Rio Negro, Brazil (**a**), and some of its fishes: unidentified rivulid (**b**), *Synbranchus* sp. (**c**) *Anostomus anostomus* (**d**), *Leporinus friderici* (**e**), *Bunocephalus* cf. *knerii* (**f**), *Characidium* sp. (**g**), *Corydoras adolfoi* (**h**), and *Callichthys serralabium* (**i**). Photos by PvdS

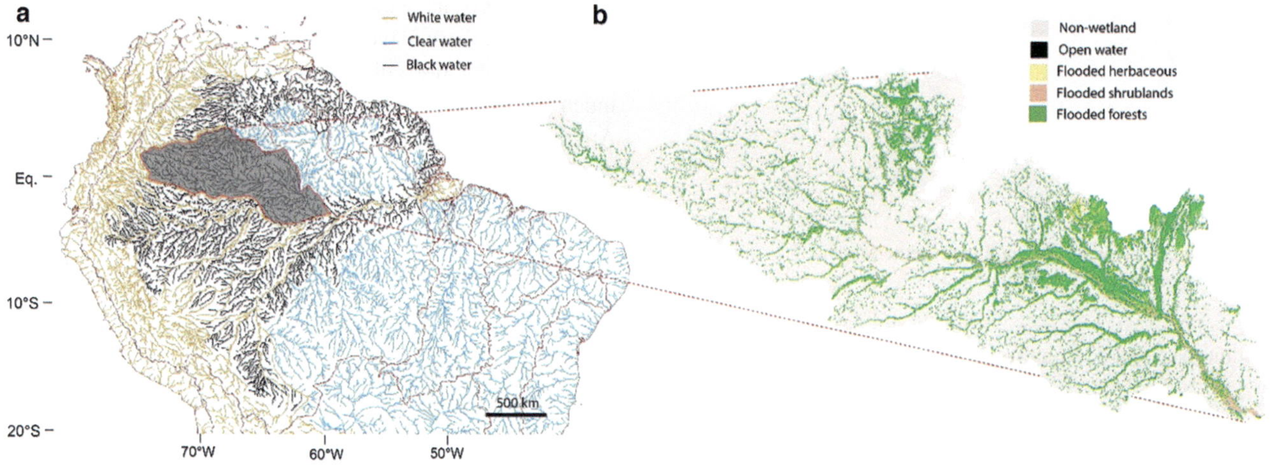

Fig. 10.2 Classification of Amazon rivers into white water, clearwater, and blackwater (**a**); base map of water types from Chap. 8. Red polygon shows area in panel (**b**); the blackwater portion of the Rio Negro basin excluding the upper Rio Branco, and vegetation types covering its floodplains; based on data from Hess et al. (2015)

resulting in exceptionally low electrical conductivity values between 5 and 20 µS cm^{-1} (Crampton 2011; Junk et al. 2015). Concentrations of Na^+ and Cl^- level in the main channel of the Rio Negro are commonly one-sixteenth and one-fifth of global river averages, respectively; water Ca^{2+} levels are even lower at about one-seventieth of global river average (Furch 1984; Gonzalez et al. 2006). The large quantities of organic acids in combination with low ion levels, produce very acidic pH levels. The pH of the main river is typically between 4.0 and 6.0, but forest streams and floodplain lakes in the Rio Negro basin can be two or more log-units lower (Walker and Henderson 1996). In addition, fish utilizing floodplains are commonly confronted with high water temperatures and very low oxygen levels.

Regardless of such challenging environmental conditions, the Rio Negro basin is home to a rich fish fauna, with >1100 species, representing almost one-fifth of all freshwater fish species in South America (Beltrão et al. 2019). To a large extent, fish diversity and abundance in the Negro basin are supported by the predictable access to lateral floodplains. Such access is provided by a large annual flood pulse, with flood heights of up to 10 m or more, causing inundation of floodplains that, depending on the topography, can last from a couple of weeks to up to more than 6 months (Goulding et al. 2003; Junk et al. 2015). In this chapter, we focus on blackwater floodplain habitats in the Rio Negro basin and the fishes that depend on them.

10.1.1 Blackwater Fishes

A preliminary survey of the occurrence of freshwater fishes among rivers with different water types indicates that almost one-fifth (19%) of all Neotropical species inhabit blackwater habitats, and of these, about one-third (396 species) are endemic to blackwaters. However, the taxonomic composition of blackwater fish faunas does not mirror that of the broader Neotropical fish fauna, with certain orders and families being either strongly overrepresented or underrepresented (Fig. 10.3). For example, the Characiformes have a much higher proportion of species in blackwater (~20%) as compared to Siluriformes (~10%), while >40% of all species of Cyprinodontiformes are found in blackwaters. Low sampling efforts in some blackwater regions and an incomplete knowledge of the distribution pattern of many species (e.g., Dagosta and de Pinna 2019; Oberdorff et al. 2019) make comparison of fish diversity across river types challenging at this point. Nonetheless, it is clear that the harsh limnological conditions in blackwater are not a constraint for fish diversity, though they have differentially affected the diversity of various fish groups (Fig. 10.3).

For the Rio Negro specifically, recent work revealed a total species richness of 1165 species distributed in 389 genera, 56 families, and 17 orders (Beltrão et al. 2019). Based on current rates of species discovery, the estimated total number may ultimately reach ~1750 species (Beltrão et al. 2019), suggesting that the Rio Negro contains a very high species diversity even for Amazonian standards, rivalling the Rio Madeira's remarkably rich ichthyofauna (Queiroz et al. 2013). Such high diversity has been related to the large variety of habitats (see section below; Figs. 10.4 and 10.5) and its connectivity to large adjacent basins, such as the upper Amazon, Orinoco, and other Guiana Shield rivers.

Almost 60% of the Rio Negro fish diversity is comprised of species in the following families: Apteronotidae, Auchenipteridae, Characidae, Cichlidae, Curimatidae, Doradidae, Loricariidae, Pimelodidae, and Serrasalmidae. Many species in the Rio Negro have a wider geographical distribution that includes parts of adjacent watersheds, such as the upper Orinoco, or other blackwater rivers in the central Amazon (Dagosta and de Pinna 2019). This group is represented mainly by medium to large-sized species, some of which make long-distance spawning migrations (Beltrão et al. 2019). Nonetheless, at least 156 species (or 13.4%) are probably endemic to the Rio Negro drainage (Beltrão et al. 2019). The region centering on the middle and upper Rio Negro appears to be an area of significant fish species endemicity. Some specific taxonomic genera, including *Coydoras*, *Apistogramma*, *Anablepsoides*, and *Laimosemion*, show a higher rate of endemism in the Rio Negro than in other adjacent drainage basins (Beltrão et al. 2019). Small size and low dispersal capacity of these clades may have favored speciation events at relatively small spatial scales (de Oliveira et al. 2009), resulting in discontinuous patterns of occurrence and high endemism in some parts of the Rio Negro basin.

10.1.2 The Importance of Floodplains

Productivity is generally much lower in blackwater habitats when compared to clearwater and whitewater habitats. In floodplains of the Rio Negro, for example, fish abundance is about five times lower than in whitewater floodplains of the main Amazon (Saint-Paul et al. 2000). Hence, a low abundance of most species and their strong dependence on allochthonous food sources is a characteristic of the fish fauna in blackwaters.

Low nutrient availability and low transparency of the water limit primary productivity in the central Rio Negro (Schmidt 1973). Carbon stable isotopes results, however, do suggest that benthic fish species living in the main river channel appear to derive their energy predominantly from algae, and that aquatic invertebrates and detritus provide the energy transfer from algae to channel fish communities (de Souza 2005). Nonetheless, for the majority of species living in the Rio Negro basin, diatoms and filamentous algae do not comprise a major food item (Goulding et al. 1988). The food chain sustaining most Rio Negro fishes is largely based on the primary productivity of floodplain forests, locally called *igapós*, and Rio Negro fishes heavily exploit floodplain forests during the high-water

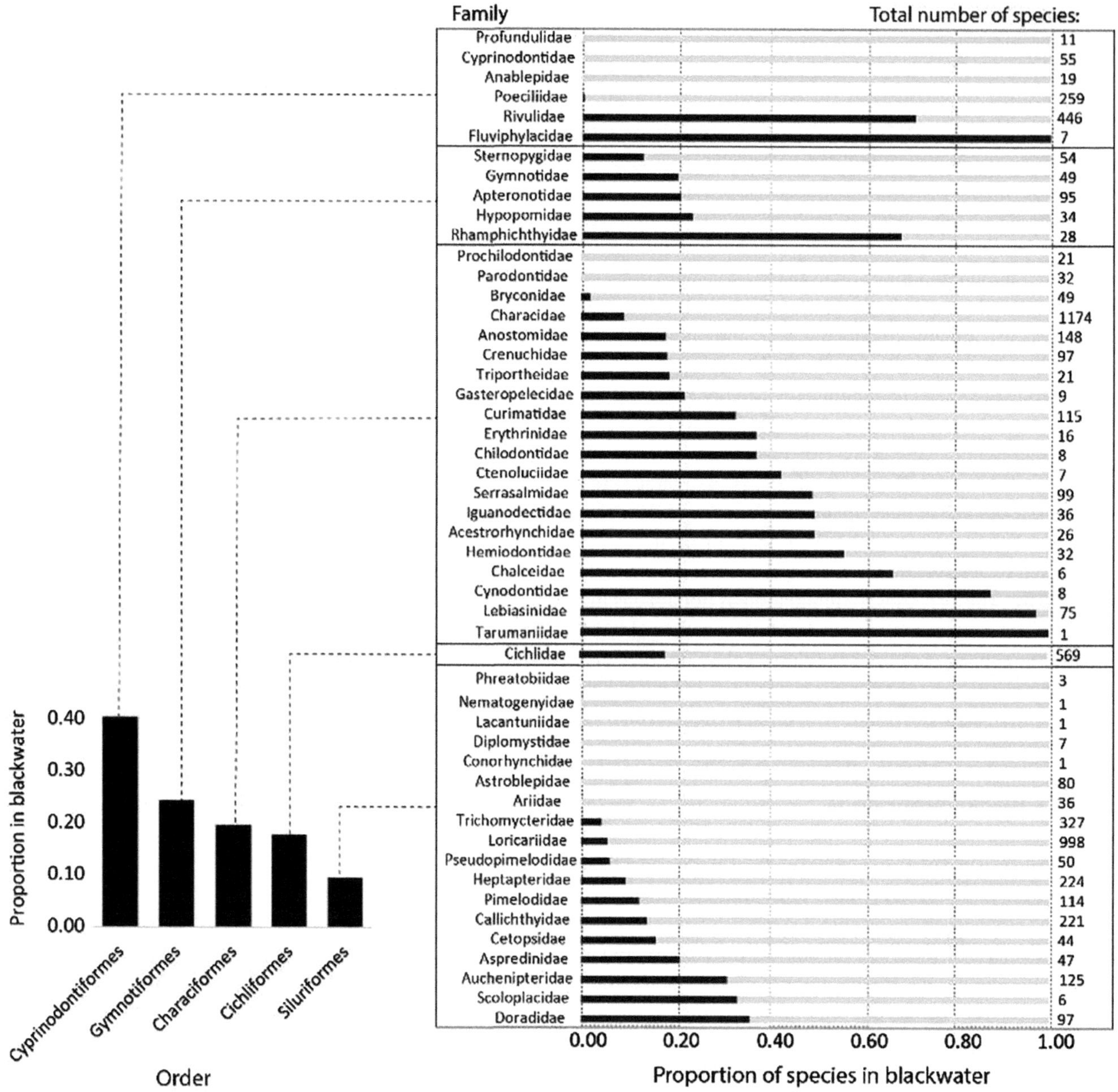

Fig. 10.3 Proportion of species found in blackwaters (dark grey bars) across fish families in five taxonomic orders that comprise 94% (or 5989 species) of the Neotropical freshwater fishes

season (e.g., Goulding et al. 1988; Noveras et al. 2012; Loebens et al. 2019; Lugo Carvajal et al. 2023). In addition, floodplain forests also function as important spawning and rearing habitats for many fish species (Araújo-Lima et al. 2001; de Oliveira and Ferreira 2008; van der Sleen et al. 2020).

While tree leaves may seem a highly abundant food source to which fish have access during the flooding season, there are no specialized folivores among the herbivorous fishes of the Rio Negro, even though many species appear to occasionally feed on leaves (Goulding et al. 1988; Chap. 15). The absence of leaf-eating specialists is likely a result of the toxicity of many tree leaves. Terrestrial invertebrates, tree seeds and fruits, and detritus are the main food items through which allochthonous energy is transferred to aquatic ecosystems. Tree fruits and seeds in particular are an important nutritional source for many Rio Negro fishes (e.g., Goulding et al., 1988 in prep).

Detritivores also depend on access to lateral floodplains because particulate organic matter settles on submerged woody vegetation during the high-water season and can hence be grazed upon. The most important commercial fish species in

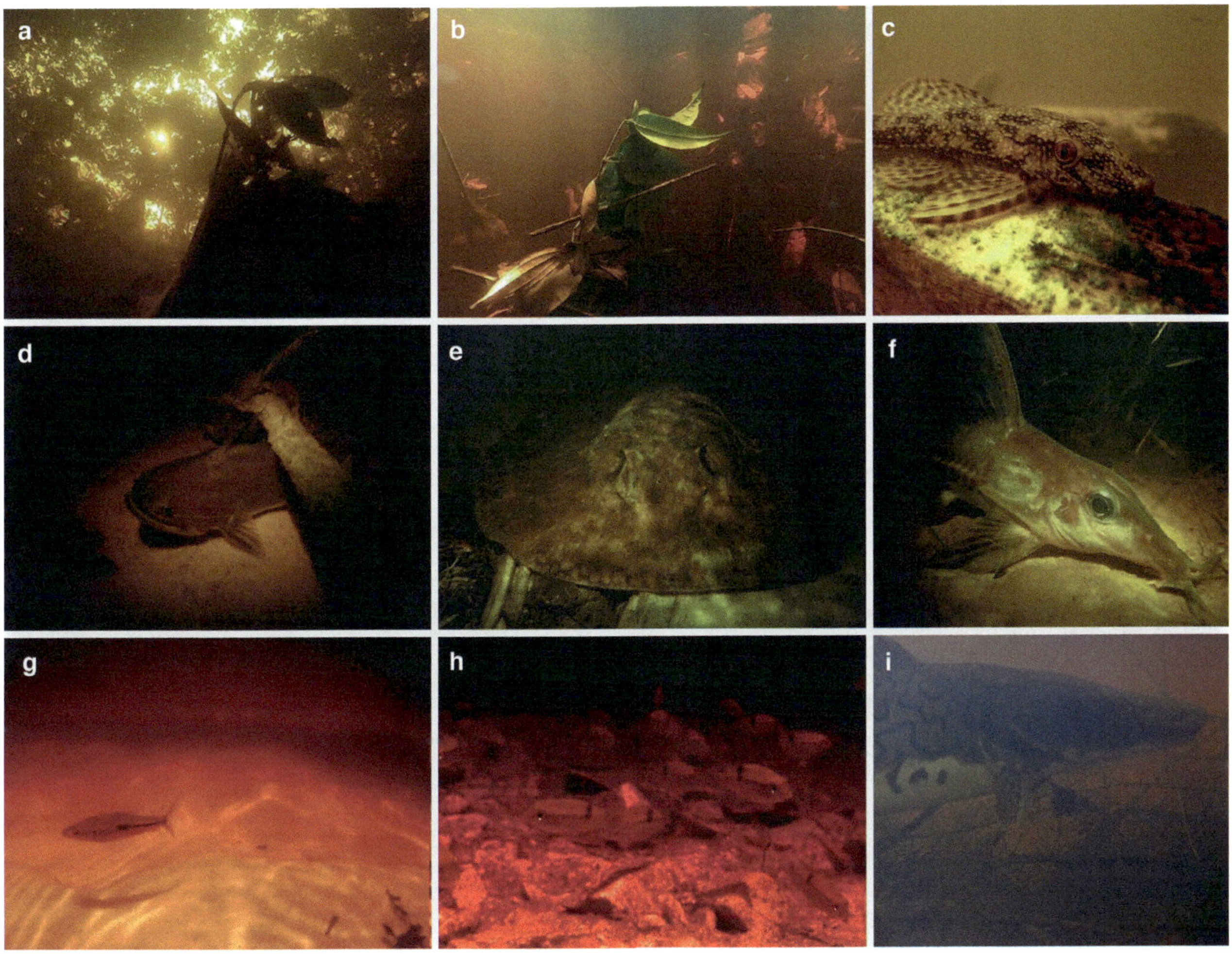

Fig. 10.4 Floodplain forest and fish from the Rio Negro basin. Flooded forest as seen just below the water surface (**a, b**); *Dekeyseria scaphirhyncha* on a submerged tree trunk (**c**); a gulper catfish (*Asterophysus batrachus*) hiding under submerged branches (**d**); the freshwater stingray *Potamotrygon wallacei* (**e**); *Anduzedoras oxyrhynchus* (**f**); *Leporinus agassizii* on a submerged beach (**g**), *Aequidens* cf. *pallidus* in a floodplain lake (**h**); and a large tiger-shovelnose catfish (*Pseudoplatystoma* cf. *reticulatum*) resting close to submerged vegetation (**i**). Photos by PvdS

the lower Rio Negro (*Semaprochilodus* spp.) are detritivores whose major food source is found on submerged vegetation in flooded forests (Barthem and Goulding 2007). The annual migrations of such detritivorous species (as well as of other migratory characiforms, such as *Anodus* spp., and *Brycon* spp.) from nutrient-rich whitewater rivers into blackwater rivers result in important energy and biomass transfers, where these species are preyed upon by large piscivores that could normally not support high population densities without these subsidies (Winemiller and Jepsen 1998; Hoeinghaus et al. 2006).

Fishes inhabiting floodplains can be divided into three broad categories in terms of their migratory behavior: (1) nonmigratory part-time residents who spend most of their lifespan in river channels and undertake lateral migrations onto floodplains during the high-water season, (2) lowland migratory species including many characiforms and some pimelodid catfishes who undertake longitudinal upstream migrations over distances of hundreds to thousands of kilometers, and which spend most of their life on floodplains, using river channels as migratory corridors (Duponchelle et al. 2021), and (3) floodplain residents who complete their entire life cycle within the floodplains and that are usually well-adapted to seasonally hypoxic conditions (Crampton 2011). Examples of the latter group in the Rio Negro include *Osteoglossum ferreirai*, *Hoplias malabaricus* and *H. curupira*, as well as many cichlids, such as *Geophagus* spp., *Uaru amphiacanthoides*, and *Symphysodon discus* (Saint-Paul et al. 2000).

10.2 Habitat Types in Blackwater Floodplains

Blackwater floodplains extend over about 119,000 km^2 in the Rio Negro basin, of which >85% are forested habitats or *igapós* (Hess et al. 2015; Fig. 10.2). Floodplains in the Rio Negro basin

Fig. 10.5 Generalized schematic cross-section of vegetation on a blackwater floodplain (**a**). The transition of deeply flooded forest (low igapó) to low flooded forests (high igapó) occurs when mean flood levels are less than 3 m and is often indicated by the presence of the palm species *Astrocaryum jauari*. Some examples of floodplain vegetation: high igapó on steep shore (**b**), island edge with *Astrocaryum jauari* palms (**c**), *Mauritia flexuosa* palm swamp, with herbaceous plants of the genus *Montrichardia* in the front (**d**), low igapó during the low-water (**e**) and high-water (**f**) season, and white-sand sclerophyllous vegetation (i.e., *campina*; **g**). Photos by PvdS

are thus dominated by forest and these has remained relatively unaffected by human activities because the low fertility of the soil does not favor agriculture and cattle ranching (Junk et al. 2011). Junk et al. (2015) recognized five main environments in blackwater floodplains: (1) permanently aquatic areas, including river channels and lakes; (2) periodically aquatic areas, including habitats not dominated by terrestrial plants, but by free-floating and rooted aquatic macrophytes; (3) periodically terrestrial areas, dominated by terrestrial plants adapted to tolerate periodic flooding, such as herbaceous plants, shrubs and trees; (4) permanently flooded or waterlogged areas, including palm swamps; and (5) permanently terrestrial areas, including different types of upland forests. Only the first four are relevant habitats for fish and will be discussed in more detail below.

10.2.1 Permanently Aquatic Areas

Main river channels in the Rio Negro basin have sandy substrate, which can form large sand bars during the low water season. Most beaches in the Rio Negro appear to have quite smooth gradients and thus provide a large and relatively homogeneous habitat for aquatic organisms. However, the highly seasonal nature of beaches and its low structural complexity have been hypothesized to result in rather unpredictable species assemblages, which feed primarily on food items drifting of falling into Rio Negro beach waters (Goulding et al. 1988; Ibarra and Stewart 1989; Jepsen 1997). Nonetheless, some studies show the presence of a few "core" species that are relatively common in beach fish faunas. These species are commonly small to medium sized and drab and light colored, including many small characins, hemiodontids, loricariinae catfishes, shoaling engraulidids and cichlids (e.g., "earth-eaters" species) (Zuanon and Ferreira 2008). Another group of fish commonly found on submerged beaches are scale eaters, such as species of the genera *Roeboides* and the wimple piranha *Catoprion mento*. At night, a different set of fish species emerges from deeper river portions to forage in shallow beaches, including large pimelodid catfishes, apteronotid knifefishes, and freshwater stingrays.

Upriver of the municipality of Santa Isabel, the bed-load of the Rio Negro is made up of gravel, and in some stretches (e.g., near São Gabriel da Cachoeira and upriver, and in several tributaries, especially on left margin of Rio Negro) rocky outcrops are a feature along river shores and in the river bed. Such rocky habitats, sometimes associated with rapids, are habitats of many endemic fish species that exhibit a variety of specialized ways to exploit local periphyton production.

Another distinct habitat in blackwaters is the accumulation of sunken trees at the lower part of streams entering the Rio Negro and its tributaries (locally known as "pauzadas"). These areas can contain dense populations of nocturnal catfishes such as *Trachycorystes trachycorystes*, *Liosomadoras oncinus*, and *Batrochoglanis* spp. Possibly, these species hide in tree logs during the day to avoid predation by giant river otters and river dolphins.

In addition to main river channels, deeper floodplain lakes constitute another type of permanent aquatic habitat and can be important refuges for aquatic organisms during the dry season. The Rio Negro basin however contains fewer floodplain lakes than typical whitewater rivers due to the much less vigorous hydrodynamics (Laraque et al. 2009; Junk et al. 2015). Many of the lakes in the Rio Negro basin are located on the islands of the Anavilhanas and Mariuá archipelagos. While some lakes are accessible to humans only during high water levels, others can be accessed year-round because they are formed in U-shaped islands that open at the downriver end (Junk et al. 2015). Lakes have lentic water conditions and as such provide a different habitat in the low-water season than main river channels. Many cichlid species, for example, favor such lentic conditions. In addition, medium-sized migratory Characiformes and Siluriformes are often the most abundant species in lakes (Saint-Paul et al. 2000; Lima et al. 2005; Beltrão et al. 2019). As lakes can trap a large quantity of organic debris from surrounding inundated forests when waters recede, and their bottoms are commonly covered with vegetation in various stages of decomposition, many species in floodplain lakes feed on detritus (Goulding et al. 1988).

10.2.2 Periodically Aquatic Areas

Habitats that are dominated by free-floating and rooted aquatic macrophytes are not abundant in the Rio Negro basin, whereas they are a dominant feature in white water floodplains (although stands of wild rice *Oryza* spp. can be found in island lakes and other lentic habitats). A main reason for this difference is the low nutrient status of the water and that the dark water prevents submerged macrophytes from establishing at depth more than 2–3 m (Goulding et al. 2003; Junk et al. 2015). Exceptions are the extended mud flats built-up by Rio Branco sediments to the left of the Anavilhanas archipelago in the lower Rio Negro, which can be covered by floating herbaceous vegetation during rising water levels. Yet, few to no fish species seem to be specifically associated with these habitats in the Rio Negro basin (Goulding et al. 1988).

10.2.3 Periodically Terrestrial Areas

This category includes habitats dominated by terrestrial plants that are adapted to tolerate periodic flooding. Most of this area in the Rio Negro basin is covered by forest (*igapó*), which in terms of structure and species composition is shaped by flood duration (Householder et al. 2021; Fig. 10.5). Consequently, floodplain forests are highly heterogeneous habitats. The most specialized plant communities are found at deeply and prolonged inundation sites (i.e., closest to the river's edge). Here, colonization of sandy substrate is dominated by shrubs that tolerate flooding over 9–10 months. A very common shrub on sand bars is "camu-camu" (*Myrciaria dubia*, Myrtaceae), which seems to be an important early colonizer of such habitats. It can form monodominant stands, and the fruits are an important food source for fish. In the middle and upper Rio Negro, the palm species, *Astrocaryum jauari* can also play an important role for island development and plant succession and is among the first colonizers of newly-formed sandy islands (Goulding and Smith 2007; Fig. 10.5). This palm species can survive inundation periods of up to 300 days (Piedade et al. 2006) and is able to grow on nearly sterile sandy soils. As soon as newly formed beaches have elevations high enough for a 2–3-month dry period, *A. jauari* palms can establish and contribute to the beach-building process by trapping sediments and fine organic matter, which improves soil fertility and allows the establishment of other plant species. Fish likely play an important role in the effective dispersal of *A. jauari* to such new areas (Piedade et al. 2006). *Astrocaryum* palms are heavily covered with spines and a yet undescribed spiny *Pseudacanthicus* species (Loricariidae) is often found associated with these palms.

On more clayish soils, monodominant forests of *Eschweilera tenuifolia* are found at the lowest topographies. These stands may be flooded for periods of up to 10 months, extremely shortening the growing season, which, in combination with very low soil fertility, causes the growth rates of these trees to be extremely slow, and trees found with a diameter of 1.75 m may be more than 1000 years old (de Resende et al. 2020).

The presence of trees with exceptional longevity at such low topographies highlights the long-lasting hydro-geomorphic stability of the Rio Negro (Junk et al. 2015).

Over time pioneer palm and shrub communities can develop into forests that, when flood duration is reduced to <8 months per year, can harbor dense stands of trees of low heights (<15 m) and generally small diameters (low igapó; Fig. 10.5). The transition of deeply flooded to low flooded forests occurs at mean flood levels of less than 3 m, where igapó forests on clayish soils can reach canopy heights of 20–25 m, with emergent trees up to 30 m tall (high igapó; Fig. 10.5). Floristic inventories have registered approximately 600 flood-tolerant tree species in igapó forests (Wittmann et al. 2010), the majority of which are also present elsewhere in the Amazon Basin, such as in *terra firme* and várzea forests (Stropp et al. 2011; Montero et al. 2014). Nonetheless, because of its periodic inundations and low fertility, the *igapó* contains many endemic tree species (Kubitzki 1989).

Prolonged flooding is a major stress factor for most tree species. Tree root systems are particularly sensitive to oxygen deprivation during prolonged flooding, and floodplain trees exhibit strategies to alleviate the potentially harmful effects of hypoxic and anoxic conditions (e.g., Parolin et al. 2004; Haase and Ratsch 2010; Piedade et al. 2010). Another remarkable adaptation of floodplain trees is that many species time their phenology to match the flooding cycle. About half of the tree species in blackwater floodplains produce fruits during the time of high water and are dispersed by water (hydrochory) or consumed by animals that exploit flooded forests (e.g., Parolin et al. 2013). Fish are an important, and likely the oldest, vertebrate group that consume tree seeds in flooded forests, and are effective (upstream) dispersal agents for trees (ichthyochory; e.g., Kubitzki and Ziburski 1994; Correa et al. 2015).

10.2.4 Campina-Igapós or "campinapós"

Studding the Rio Negro basin are clusters of white-sand vegetation (known as *Campinas* in Brazil, or *Caatinga* in Colombia and Venezuela), consisting of plant species with *sclerophylic* leaves that form forests of low stature, with slender trees, simple canopy structure, and varying structurally from almost open vegetation to shrublands or even dense forests (Goulding et al. 1988; Stropp et al. 2011; Adeney et al. 2016). The reasons for the existence of open white-sand vegetation within the closed-forest canopy remain incompletely understood and are likely multifaceted, including hydrological, anthropogenic (i.e., fires), and geological causes (Prance and Schubart 1978; Adeney et al. 2016; Rossetti et al. 2019). Fish communities in white-sand vegetation within the floodplains have not been systematically studied, but our recent inventories suggest they may be distinct (although with large overlap) from those found in floodplain forests (Lugo Carvajal et al. 2023).

10.2.5 Permanently Flooded or Waterlogged Areas

Permanently waterlogged areas, such as the borders of floodplain lakes, may be covered by herbaceous plants, mainly the arborescent *Montrichardia* spp., or a species-poor palm swamp forest (Fig. 10.5). Palm swamps are found in the headwaters of many blackwater tributaries of the Rio Negro, as well as at the confluence of many lowland tributaries in the middle and upper Rio Negro, where they can cover extensive areas (known as "*buritizais*" in Brazil and "*morichales*" in Venezuela; Goulding and Smith 2007). These palm swamps are often dominated by *Mauritiella aculeata*, which is replaced upriver by *Mauritia flexuosa* when flooding amplitude is reduced to 1–2 m (Junk et al. 2015). Palm swamps in the Rio Negro basin are also important habitats for small fishes, including many of the most important aquarium species (Goulding and Smith 2007; Montaña et al. 2008). The Cardinal Tetra (*Paracheirodon axelrodi*) and Green Neon Tetra (*P. simulans*), for example, are common in *Leopoldina pulchra* (*P. axelrodi*) and *Mauritia flexuosa* (*P. simulans*) palm swamps of the Rio Negro. Segregation of the two Neon tetra species in distinctive interfluvial habitats could be partially related to differences in thermal tolerance between the species (Marshall et al. 2011).

10.3 Adaptations of Igapó Fishes

Although thorough analyses are missing, most of the adaptations that allow the occupation of blackwater rivers and their floodplains are mirrored in clearwater and whitewater systems. In addition, many genera and species of floodplain-forest dwellers are found in different water types, and the set of endemic species in blackwaters does not seem to have a clear evolutionary link to the occupation of blackwater habitats. It has been argued that the traits that allow species to thrive in blackwaters may have emerged much earlier than the formation of the Rio Negro system (Gonzalez et al. 2018). Even though no unique *adaptations* are known for fishes occurring in blackwater floodplains, some adaptations are well pronounced in Rio Negro fishes and will be briefly discussed below.

10.3.1 Small Size, But Bright Colors

As in the rest of the Amazon basin, fishes of small size appear to dominate the Rio Negro ichthyofauna, with almost 55% of the fish species in the Rio Negro basin having an adult size of <10 cm SL (Beltrão et al. 2019; Chap. 19). A reduction of adult body size (i.e., miniaturisation, where adult fish do not surpass 25–30 mm standard length) has been hypothesized to represent an evolutionary strategy to minimize energetic demand (Weitzman and Vari 1988), allowing larger population sizes (in number of individuals) in oligotrophic waters. Given the semi-transparent conditions in shallow blackwaters, many fishes have also developed intense colors, which commonly function as species-specific visual cues used in reproduction and mate recognition (Borghezan et al. 2021), or even sophisticated anti-predatory strategies (Ikeda and Kohshima 2009).

10.3.2 Ionoregulation

Blackwaters, specifically those in the Rio Negro basin, are characterized by extremely low ion concentrations and very low pH (Furch 1984; Gaillardet et al. 1997). This combination poses challenges for ion-regulation in fishes (Chap. 12). High H+ concentrations in the water can hinder the active uptake of Na+ and Cl− (Packer and Dunson 1970; McWilliams and Potts 1978) and at the same time greatly stimulate diffusive ion loss (McWilliams 1982; McDonald 1983). Such effects can cause an ionic imbalance, leading to a drop in plasma Na+ and Cl− levels that can ultimately result in reduced blood volume and elevated blood viscosity and blood pressure (Gonzalez et al. 2006). Once the plasma ion losses surpass about 30%, cardiovascular failure can lead to fish mortality (Milligan and Wood 1982).

Consequently, Rio Negro water can be toxic for most typical freshwater fishes (Gonzalez et al. 2006). Experiments with two Amazonian species *Hoplosternum littorale* and *Brycon amazonicus* (identified as *B. erythropterum*), both not occurring in the Rio Negro, died within 24 h after exposure to water of pH 3.5, whereas *Colossoma macropomum*, naturally found in the lower Rio Negro, showed no mortality at all when exposed to the same water (Wilson et al. 1999; Gonzalez et al. 2021b). This result suggests that tolerance to exceptionally low ion concentration and pH is not shared by Amazonian fishes in general. Indeed, Goulding et al. (1988) listed several species (including *Schizodon fasciatus* and *Colomesus asellus*) that would have been expected to occur in the Rio Negro if hydrochemical barriers did not prevent them from entering it. Such barriers are also evident when comparing the ichthyofauna of the Branco River to that of the Rio Negro. The Branco River is the largest tributary of the Rio Negro, but it is richer in sediments and has a much higher cation concentration and pH. Although the two rivers are not separated by any physical barrier, each of them shares more species with adjoining non-Amazonian basins than with each other (with the Orinoco in the case of the Rio Negro and with the Essequibo River in the case of the Branco River; Dagosta and de Pinna 2019). The small number of taxa exclusively shared between the Branco and Negro rivers indicates the presence of strong physicochemical dispersal barriers for fish.

Ionoregulation has been studied in detail for only a few species, but these studies already show that there is no general mechanism to achieve tolerance to ion-poor acidic blackwaters. Measurements of ion fluxes in fish initially suggested that there may be two basic ionoregulatory strategies of fishes native to ion-poor, acidic blackwaters. Rio Negro cichlids, for example, inhibit Na+ uptake and control diffusive Na+ loss and in doing so maintain ion balance (Gonzalez et al. 2021a). In contrast, all characiforms studied to date possess very high rates of Na+ uptake, even in ion-poor water (Gonzalez and Wilson 2001, Gonzalez et al., 2017, 2018), and seem to be insensitive to water acidity down to at least pH 3.25 (Gonzalez and Preest 1999; Gonzalez and Wilson 2001). This insensitivity to low pH is different from any other freshwater fish examined, but the mechanisms through which this high tolerance is achieved remain unclear. Surprisingly, two characiforms that are not native to the Rio Negro or Rio Amazonas were found to possess the same transport characteristics, including pH insensitive uptake, as their Rio Negro relatives. This suggests that these physiological traits may be ancestral and predate colonization of the Rio Negro (Gonzalez et al. 2017). One of the species, the Congo tetra (*Phenacogrammus interruptus*), is from Africa, implying that such traits may have appeared early in the evolution of Characiformes—or, alternatively, could be an example of convergent evolution. The success of Characiforms in the Rio Negro (Fig. 10.3) may thus have resulted from pre-existing physiological traits, rather than from adaptations to Rio Negro waters (Gonzalez et al. 2018).

Yet, tolerance to ion-poor acidic blackwaters appears to be a highly complex process, potentially involving factors beyond just physiological adaptations in fish. For example, several studies suggest that *dissolved organic matter* (DOC) is involved in some way in the ionoregulatory strategies of blackwater fishes (Morris et al. 2021). Although dissociation of protons from DOC, such as humic acids, is the major cause of acidity in blackwaters, some DOCs can mitigate the inhibition of Na+ uptake and ammonia excretion, and protect against diffusive Na+ loss at low pH (Wood et al. 2003; Steinberg et al. 2006; Duarte et al. 2016; Morris et al. 2021). Such beneficial effects of DOC may however depend on adaptations related to chronic exposure to high levels of DOCs (Steinberg et al. 2006; da Costa et al. 2015). Nonetheless, the fact that there are fish able to live in waters of pH $\leq$ 3.5 may be partially possible because of the DOC that caused these acidic conditions in the first place. In addition, mucosal microbiomes are known to tightly regulate their host's epithelial barrier function (Wells et al. 2010). Recent studies examining gill samples from four native Amazonian blackwater fish species have challenged the notion that Cichlinae primarily rely on tightly regulated ion efflux and Characiformes depend on ion transport systems that support high ion uptake (François-Étienne et al. 2023). Instead, their results suggest that species-specific host–microbe interactions may be involved in ionoregulation of blackwater fish. In conclusion, our understanding of the adaptations that enable life to thrive in one of the planet's most hostile, yet species-rich, aquatic ecosystems remains highly incomplete.

10.3.3 Low Oxygen

A general adaptation of Amazonian fishes, including those inhabiting the blackwater floodplains, is associated with hypoxic conditions, which can occur in floodplains due to stagnant water and decomposition of high seasonal inputs of organic matter (Chap. 12). The extent of hypoxia is possibly smaller in blackwaters compared to other river types because of the lower abundance of macrophyte stands, which are responsible for a large amount of biomass in white water systems. Nonetheless, all igapó residents must possess a combination of morphological, physiological, or behavioral adaptations for hypoxia, and these have been intensively studied (e.g., Val and de Almeida-Val 1995). Such adaptations include air-breathing using a swim-bladder (e.g., *Arapaima*, *Gymnotus*, and several catfishes), vascularized mouth organ (e.g., *Electrophorus*), and oropharyngeal cavity (*Synbranchus*), vascularized stomach and intestines (several loricariid and callichthyid catfishes), aquatic surface-respiration (e.g., *Osteoglossum* and many characids), and physiological alterations to the way hemoglobin binds oxygen (e.g., *Hoplias*, some cichlids). Yet, fish kills occur in the Rio Negro when cold fronts blow (mostly from May to July), causing an upwelling of water with little to no oxygen and high loads of hydrogen sulfide and methane in slow-moving river stretches (Geisler 1969).

10.4 Threats to Floodplain Biotas

At the onset of the twenty-first century, the Rio Negro basin still is a "land of old rock, poor soils, much water, extensive forest and few people" (Hammond 2005). Extremely poor soils have prevented large-scale agricultural developments. At the same time, the scarcity of nutrients also makes ecosystems in the Rio Negro basin intrinsically sensitive to disturbances. One disturbance factor of concern is the increased occurrence of forest fires in blackwater floodplains (de Resende et al. 2014; Flores et al. 2014; Carvalho et al. 2021; van der Sleen et al. 2025). As climate change leads to increasingly hotter and drier conditions, forest fires are becoming more frequent and severe. Because productivity is exceptionally low in the Rio Negro basin, floodplain forests may need several centuries to recover after burning into a closed canopy forest again, and probably millennia to fully recover to their original, late-successional state (Junk et al. 2015; Flores et al. 2020). Such slow recovery makes blackwater floodplain forest particularly sensitive to reburn, initiating fire traps that could shift floodplain forest cover to a more open, savannah-like vegetation state (Flores et al. 2016; van der Sleen et al. 2025). At a large scale, such state-changes have significant consequences for forest-dependent fish (Lugo Carvajal et al. 2023) and other animal communities (Ritter et al. 2012), with loss of ecological links between the aquatic and terrestrial ecosystems and the critical services provided by freshwater fishes (Pelicice et al. 2023).

Besides forest loss, changes to the annual flood pulse—and consequently to lateral connectivity—could have significant ecological consequences, given the strong reliance of fishes in blackwater rivers on allochthonous productivity in floodplains (Correa et al. 2022). Flood pulse change has followed after the construction of large dams for the production of hydropower energy in many Amazonian tributaries. Although no large dams have yet been constructed in the Rio Negro basin, Brazil's plans for hydropower development are ambitious (Winemiller et al. 2016). Clear- and blackwater rivers have a high predisposition for use in hydropower production given their low sediment load. In fact, one of the oldest dams in the Amazon basin (the Balbina dam) is located in the Uatumã, a blackwater river running parallel to the Rio Negro. Although its impact on fish communities has not been assessed, the Balbina dam caused a radical change of hydrological regimes (Timpe and Kaplan 2017) and loss of floodplain habitat (da Rocha et al. 2019; de Resende et al. 2020; Schöngart et al. 2021).

It is also remarkable that approximately 60% of all ornamental tropical fish exported from Brazil are sourced from the Rio Negro basin (Zehev et al. 2015; Ladislau et al. 2020). The collection of ornamental fish provides an important income source for local communities and is a viable alternative compared to more destructive economic activities in blackwater floodplains. Yet, the harvest of ornamental fish has remained largely unregulated, and its impact on local fish stocks not known (Castanho et al. 2020). Improved handling practices that reduce fish mortality during transport and a fair payment of fishermen could decrease pressure on wild fish populations and increase the sustainability of the ornamental fish trade.

Floodplains in the Rio Negro have remained largely unchanged during the last centuries, making it a precious reminder of the world before the Anthropocene. The Rio Negro basin still contains an immense wilderness, with unique biodiversity of unrivalled beauty that should be preserved for future generations to enjoy and explore.

References

Adeney JM, Christensen NL, Vicentini A, Cohn-Haft M (2016) White-sand ecosystems in Amazonia. Biotropica 48(1):7–23

Araújo-Lima CARM, Da Silva VV, Petry P, Oliveira EC, Moura SML (2001) Diel variation of larval fish abundance in the Amazon and Rio Negro. Braz J Biol 61:357–362

Barthem R, Goulding M (2007) An unexpected ecosystem: the Amazon as revealed by fisheries. Amazon Conservation Association; Missouri Botanical Garden Press, Saint Louis. p 241

Beltrão H, Zuanon J, Ferreira E (2019) Checklist of the ichthyofauna of the Rio Negro basin in the Brazilian Amazon. Zookeys 881:53

Borghezan EDA, Pires THDS, Ikeda T, Zuanon J, Kohshima S (2021) A review on fish sensory systems and Amazon water types with implications to biodiversity. Front Ecol Evol 8:589760

Carvalho TC, Wittmann F, Piedade MTF, Resende AFD, Silva TSF, Schöngart J (2021) Fires in Amazonian blackwater floodplain forests: causes, human dimension, and implications for conservation. Front For Glob Change 4:755441

Castanho DG, de Deus CP, Zuanon J, Santorelli S, Leitão RP, Teresa FB (2020) Simulation of over-exploitation of ornamental fish and its consequences for the functional structure of assemblages of Amazonian streams. Ecol Freshwater Fish 29(4):533–541

Correa SB, Costa-Pereira R, Fleming T, Goulding M, Anderson JT (2015) Neotropical fish–fruit interactions: eco-evolutionary dynamics and conservation. Biol Rev 90(4):1263–1278

Correa SB, Van Der Sleen P, Siddiqui SF, Bogotá-Gregory JD, Arantes CC, Barnett AA et al (2022) Biotic indicators for ecological state change in Amazonian floodplains. BioScience 72(8):753–768

Costa STD, Gressler LT, Sutili FJ, Loebens L, Lazzari R, Baldisserotto B (2015) Effect of humic acid on survival, ionoregulation and hematology of the silver catfish, Rhamdia quelen (Siluriformes: Heptapteridae), exposed to different pHs. Zoologia (Curitiba) 32:215–224

Crampton WG (2011) An ecological perspective on diversity and distributions. In: Albert JS, Reis RE (eds) Historical biogeography of neotropical freshwater fishes. University of California Press, Berkeley and Los Angeles. pp 165–189

da Rocha M, de Assis RL, Piedade MTF, Feitosa YO, Householder JE, Lobo GDS et al (2019) Thirty years after Balbina dam: diversity and floristic composition of the downstream floodplain forest, Central Amazon, Brazil. Ecohydrology 12(8):e2144

Dagosta FC, De Pinna M (2019) The fishes of the Amazon: distribution and biogeographical patterns, with a comprehensive list of species. Bull Am Mus Nat Hist 2019(431):1–163

de Souza MJFT (2005) Fontes autotróficas de energia para peixes do canal principal e quilônios ao longo do médio rio Negro, Amazônia – Brasil [Unpublished doctoral thesis]. Instituto Nacional de Pesquisas da Amazônia, Manaus

Duarte RM, Smith DS, Val AL, Wood CM (2016) Dissolved organic carbon from the upper Rio Negro protects zebrafish (Danio rerio) against ionoregulatory disturbances caused by low pH exposure. Sci Rep 6(1):20377

Duponchelle F, Isaac VJ, Rodrigues Da Costa Doria C, Van Damme PA, Herrera-R GA, Anderson EP et al (2021) Conservation of migratory fishes in the Amazon basin. Aquat Conserv: Mar Freshwater Ecosyst 31(5):1087–1105

Ertel JR, Hedges JI, Devol AH, Richey JE, Ribeiro MDNG (1986) Dissolved humic substances of the Amazon River system. Limnol Oceanogr 31(4):739–754

Ferreira EJG, Zuanon JAS, Forsberg BR, Goulding M, Ferreira SRB (2007) Rio Branco: Peixes, ecologia e conservação de Roraima. Amazon Conservation Association (ACA); Instituto Nacional de Pesquisas da Amazônia (INPA); Sociedade Civil Mamirauá

Flores BM, Fagoaga R, Nelson BW, Holmgren M (2016) Repeated fires trap Amazonian blackwater floodplains in an open vegetation state. J Appl Ecol 53(5):1597–1603

Flores BM, Piedade MTF, Nelson BW (2014) Fire disturbance in Amazonian blackwater floodplain forests. Plant Ecol Diversity 7(1–2):319–327

Flores BM, Staal A, Jakovac CC, Hirota M, Holmgren M, Oliveira RS (2020) Soil erosion as a resilience drain in disturbed tropical forests. Plant Soil 450:11–25

François-Étienne S, Nicolas L, Eric N, Jaqueline C, Pierre-Luc M, Sidki B et al (2023) Important role of endogenous microbial symbionts of fish gills in the challenging but highly biodiverse Amazonian blackwaters. Nat Commun 14(1):3903

Furch K (1984) Water chemistry of the Amazon basin: the distribution of chemical elements among freshwaters. In: Sioli H (ed) The Amazon: limnology and landscape ecology of a mighty tropical river and its basin. Springer, Dordrecht. pp 167–199

Gaillardet J, Dupre B, Allegre CJ, Negrel P (1997) Chemical and physical denudation in the Amazon River basin. Chem Geol 142:141–173

Geisler R (1969) Investigations about free oxygen, biological oxygen demand and oxygen consumption of fishes in a tropical 'blackwater' (Rio Negro). Arch Hydrobiol 66:307–325

Gonzalez RJ, Preest MR (1999) Ionoregulatory specializations for exceptional tolerance of ion-poor, acidic waters in the neon tetra (Paracheirodon innesi). Physiol Biochem Zool 72:156–163

Gonzalez RJ, Wilson RW (2001) Patterns of ion regulation in acidophilic fish native to the ion-poor, acidic Rio Negro. J Fish Biol 58:1680–1690

Gonzalez RJ, Cradeur A, Guinnip M, Mitchell A, Reduta V (2018) South American characids share very similar ionoregulatory characteristics. Comp Biochem Physiol A Mol Integr Physiol 226:17–21

Gonzalez RJ, Hsu R, Mahaffey L, Rebagliatti D, Shami J (2021a) Examination of ionoregulatory characteristics of South American cichlids. Comp Biochem Physiol A Mol Integr Physiol Mar;253: 110854

Gonzalez RJ, Jones SL, Nguyen TV (2017) Ionoregulatory characteristics of non-Rio Negro characiforms and cichlids. Physiol Biochem Zool 90:407–414

Gonzalez RJ, Patrick ML, Duarte RM, Casciato A, Thackeray J, Day N, Val AL (2021b) Exposure to pH 3.5 water has no effect on the gills of the Amazonian tambaqui (Colossoma macropomum). J Comp Physiol B 191:493–502

Gonzalez RJ, Wilson RW, Wood CM (2006) Ionoregulation in tropical fish from ion-poor, acidic blackwaters. In: Val AL, Val VMFA, Randall DJ (eds) The physiology of tropical fishes. Academic Press, Amsterdam. pp 397–442

Goulding M, Smith N (2007) Palms: sentinels for Amazon conservation. Amazon Conservation Association; Missouri Botanical Garden Press, Saint Louis. p356

Goulding M, Barthem R, Ferreira EJG (2003) Smithsonian atlas of the Amazon. Smithsonian Books, Washington. p 253

Goulding M, Leal Carvalho M, Ferreira E (1988) Rio Negro: rich life in poor water. Amazonian diversity and foodchain ecology as seen through fish communities. SPB Academic Publishing, The Hague. p 200

Haase K, Ratsch G (2010) The morphology and anatomy of tree roots and their aeration strategies. In Junk WJ, Piedade MTF, Wittmann F, Schöngart J, Parolin P (eds) Amazonian floodplain forests. Springer, Dordrecht. pp 141–162

Hammond DS (2005) Tropical forests of the Guiana Shield: Ancient forests in a modern world. CABI Publishing, Wallingford. p 528

Hess LL, Melack JM, Affonso AG, Barbosa C, Gastil-Buhl M, Novo E (2015) Wetlands of the lowland Amazon basin: extent, vegetative cover, and dual-season inundated area as mapped with JERS-1 synthetic aperture radar. Wetlands 35:745–756

Hoeinghaus DJ, Winemiller KO, Layman CA, Arrington DA, Jepsen DB (2006) Effects of seasonality and migratory prey on body condition of Cichla species in a tropical floodplain river. Ecol Freshwater Fish 15:398–407

Householder JE, Schöngart J, Piedade MTF, Junk WJ, ter Steege H, Montero JC et al (2021) Modelling the ecological responses of tree species to the flood pulse of the Amazon Negro river floodplains. Front Ecol Evol 9:628606

Ibarra M, Stewart DJ (1989) Longitudinal zonation of sandy beach fishes in the Napo River basin, eastern Ecuador. Copeia 1989(2):364–381

Ikeda T, Kohshima S (2009) Why is the neon tetra so bright? Coloration for mirror-image projection to confuse predators? "Mirror-image decoy" hypothesis. Environ Biol Fishes 86:427–441

Jepsen DB (1997) Fish species diversity in sand bank habitats of a Neotropical river. Environ Biol Fishes 49:449–460

Junk WJ, Piedade MTF, Schöngart J, Cohn-Haft M, Adeney JM, Wittmann F (2011) A classification of major naturally-occurring Amazonian lowland wetlands. Wetlands 31:623–640

Junk WJ, Wittmann F, Schöngart J, Piedade MTF (2015) A classification of the major habitats of Amazonian black-water river floodplains and a comparison with their white-water counterparts. Wetlands Ecol Manage 23:677–693

Kubitzki K (1989) The ecogeographical differentiation of Amazonian inundation forests. Plant Syst Evol 162:285–304

Kubitzki K, Ziburski A (1994) Seed dispersal in flood plain forests of Amazonia. Biotropica 26:30–43

Ladislau DS, Ribeiro MWS, Castro PDS, Aride PHR, Paiva AJV, Polese MF et al (2020) Ornamental fishing in the region of Barcelos, Amazonas: socioeconomic description and scenario of activity in the view of "piabeiros". Braz J Biol 80:544–556

Laraque A, Guyot JL, Filizola N (2009) Mixing processes in the Amazon River at the confluences of the Negro and Solimões Rivers, Encontro das Águas, Manaus, Brazil. Hydrol Processes 23:3131–3140

Leenheer JA (1980) Origin and nature of humic substances in the waters of the Amazon River Basin. Acta Amazonica 10(3):513–526

Lima FCT, Ramos L, Barreto T, Cabalzar A, Tenório G, Barbosa A et al (2005) Peixes do alto Tiquié: Ictiologia e conhecimentos dos Tuyuka e Tukano. In: CA (ed) Peixe e Gente no Alto Rio Tiquié. Instituto Socioambiental, Manaus. pp 111–282

Loebens SD, Farias EU, Freitas CED, Yamamoto KC (2019) Influence of hydrological cycle on the composition and structure of fish assemblages in an igapó forest, Amazonas, Brazil. Bol Inst Pesca 45(4):e525

Lugo-Carvajal A, Holmgren M, Zuanon J, van der Sleen P (2023) Fish on fire: shifts in Amazonian fish communities after floodplain forest fires. J Appl Ecol 60(7):1637–1646

Marshall BG, Forsberg BR, Hess LL, Freitas CED (2011) Water temperature differences in interfluvial palm swamp habitats of Paracheirodon axelrodi and P. simulans (Osteichthyes: Characidae) in the middle Rio Negro, Brazil. Ichthyol Explor Freshwaters 22(4):377–383

McDonald DG (1983) The interaction of environmental calcium and low pH on the physiology of the rainbow trout, Salmo gairdneri. I. Branchial and renal net ion and H+ fluxes. J Exp Biol 102:123–140

McWilliams PG (1982) The effects of calcium on sodium fluxes in the brown trout, Salmo trutta, in neutral and acid media. J Exp Biol 96:439–442

McWilliams PG, Potts WTW (1978) The effects of pH and calcium concentration on gill potentials in the brown trout Salmo trutta. J Comp Physiol 126:277–286

Milligan CL, Wood CM (1982) Disturbances in hematology, fluid volume distribution, and circulatory function associated with low environmental pH in the rainbow trout, Salmo gairdneri. J. Exp. Biol. 99:397–415

Montaña CG, Layman CA, Taphorn DC (2008) Comparison of fish assemblages in two littoral habitats in a Neotropical morichal stream in Venezuela. Neotrop Ichthyol 6(4):577–582

Montero JC, Piedade MTF, Wittmann F (2014) Floristic variation across 600 km of inundation forests (Igapó) along the Negro River, central Amazonia. Hydrobiologia 729:229–246

Morris C, Val AL, Brauner CJ, Wood CM (2021) The physiology of fish in acidic waters rich in dissolved organic carbon, with specific reference to the Amazon basin: ionoregulation, acid-base regulation, ammonia excretion, and metal toxicity. J Exper Zool A Ecol Integr Physiol 335(9):843–863

Noveras J, Yamamoto KC, Freitas CEC (2012) Use of the flooded forest by fish assemblages in lakes of the National Park of Anavilhanas (Amazonas, Brazil). Acta Amazonica 42(4):567–572

Oberdorff T, Dias MS, Jezequel C, Albert JS, Arantes CC, Bigorne R et al (2019) Unexpected fish diversity gradients in the Amazon basin. Sci Adv 5(9):eaav8681

Oliveira EC, Ferreira EJG (2008) Spawning areas, dispersion and microhabitats of fish larvae in the Anavilhanas Ecological Station, Rio Negro, Amazonas State, Brazil. Neotrop Ichthyol 6(4):559–566

Oliveira RR, Feldberg E, dos Anjos MB, Zuanon J (2009) Mechanisms of chromosomal evolution and its possible relations with natural history characteristics in Ancistrus catfishes (Siluriformes: Loricariidae). J Fish Biol 75(8):2209–2225

Packer RK, Dunson WA (1970) Effects of low environmental pH on blood pH and sodium balance of brook trout. J Exp Zool 174(1):65–72

Parolin P, De Simone O, Haase K, Waldhoff D, Rottenberger S, Kuhn U et al (2004) Central Amazonian floodplain forests: Tree adaptations in a pulsing system. Bot Rev 70(3):357–380

Parolin P, Wittmann F, Ferreira LV (2013) Fruit and seed dispersal in Amazonian floodplain trees—a review. Ecotropica 19:15–32

Pelicice FM, Agostinho AA, Azevedo-Santos VM, Bessa E, Casatti L, Garrone-Neto D et al (2023) Ecosystem services generated by neotropical freshwater fishes. Hydrobiologia 850(12):2903–2926

Piedade MTF, Ferreira CS, de Oliveira Wittmann A, Buckeridge M, Parolin P (2010) Biochemistry of Amazonian floodplain trees. In: Junk WJ, Piedade MTF, Wittmann F, Schöngart J, Parolin P (eds) Amazonian floodplain forests. Springer, Dordrecht. pp 127–140

Piedade MTF, Parolin P, Junk WJ (2006) Phenology, fruit production and seed dispersal of Astrocaryum jauari (Arecaceae) in Amazonian black water floodplains. Rev Biol Trop 54(4):1171–1178

Prance GT, Schubart HOR (1978) Preliminary note on the origin of the open white sand campinas. Brittonia 30(1):60–63

Queiroz LJ, Torrente-Vilara G, Ohara WM, Pires THS, Zuanon J, Doria CRD (2013) Peixes do Rio Madeira, vols 1–3. Editora Dialeto Latin American Documentary

Resende AF, Nelson BW, Flores BM, de Almeida DR (2014) Fire damage in seasonally flooded and upland forests of the central Amazon. Biotropica 46(6):643–646

Resende AF, Piedade MTF, Feitosa YO, Andrade VHF, Trumbore SE, Durgante FM et al (2020) Flood-pulse disturbances as a threat for long-living Amazonian trees. New Phytol 227(6):1790–1803

Ritter CD, Andretti CB, Nelson BW (2012) Impact of past forest fires on bird populations in flooded forests of the Cuini River in the lowland Amazon. Biotropica 44(4):449–453

Rossetti DF, Moulatlet GM, Tuomisto H, Gribel R, Toledo PM, Valeriano MM et al (2019) White sand vegetation in an Amazonian lowland under the perspective of a young geological history. An Acad Bras Cienc 91(2):e20180583

Saint-Paul U, Zuanon J, Villacorta Correa MA, García M, Fabré NN, Berger U, Junk WJ (2000) Fish communities in central Amazonian white- and blackwater floodplains. Environ Biol Fishes 57:235–250

Schmidt GW (1973) Primary production of phytoplankton in the three types of Amazonian waters. Amazoniana 4(3):135–203

Schöngart J, Wittmann F, de Resende AF, Assahira C, Lobo GD, Neves JRDi et al (2021) The shadow of the Balbina dam: a synthesis of over 35 years of downstream impacts on floodplain forests in Central Amazonia. Aquat Conserv Mar Freshwater Ecosyst 31(5):1117–1135

Steinberg CEW, Kamara S, Prokhotskaya VY, Manusadzianas L, Karasyova TA, Timofeyev MA et al (2006) Dissolved humic substances—ecological driving forces from the individual to the ecosystem level? Freshwater Biol 51(7):1189–1210

Stropp J, van der Sleen P, Assunção PA, da Silva AL, ter Steege H (2011) Tree communities of white-sand and terra-firme forests of the upper Rio Negro. Acta Amazonica 41(4):521–544

Timpe K, Kaplan D (2017) The changing hydrology of a dammed Amazon. Sci Adv 3(11):e1700611

Val AL, de Almeida-Val VMF (1995) Fishes of the Amazon and their Environment. Springer-Verlag, Berlin. p 224

van der Sleen P, Decuyper M, Flores BM, Householder JE, Holmgren M (2025) ENSO wildfires impact Amazonian floodplains in complex ways. Ecosystems 28: 1-13

van der Sleen P, Lugo-Carvajal A, Zuanon J, Holmgren M (2020) Cats singing in the dark? Spawning aggregations of sound-producing fish in Amazonian floodplain forests. Environ Biol Fishes 103:1265–1267

Walker I, Henderson PA (1996) Ecophysiological aspects of Amazonian blackwater litterbank fish communities. In: Val AL, Almeida-Val VMF, Randall DJ (eds) Physiology and biochemistry of fishes of the Amazon. Instituto Nacional de Pesquisas da Amazônia (INPA), Manaus. pp 7–22

Weitzman SH, Vari RP (1988) Miniaturization in South American freshwater fishes; an overview and discussion. Proc Biol Soc Washington 101(2):444–465

Wells JM, Rossi O, Meijerink M, van Baarlen P (2010) Microbes and Health Sackler Colloquium: epithelial crosstalk at the microbiota-mucosal interface. Proc Natl Acad Sci U S A 108(Suppl. 1): 4607–4614

Wilson RW, Wood CM, Gonzalez RJ, Patrick ML, Bergman HL, Narahara A, Val AL (1999) Net ion fluxes during gradual acidification of extremely soft water in three species of Amazonian fish. Physiol Biochem Zool 72(3):277–285

Winemiller KO, Jepsen DB (1998) Effects of seasonality and fish movement on tropical river food webs. J Fish Biol 53 (Suppl. A):267–296

Winemiller KO, McIntyre PB, Castello L, Fluet-Chouinard E, Giarrizzo T, Nam S et al (2016) Balancing hydropower and biodiversity in the Amazon, Congo, and Mekong. Science 351(6269):128–129

Wittmann F, Schöngart J, Junk WJ (2010) Phytogeography, species diversity, community structure and dynamics of central Amazonian floodplain forests. In: Junk WJ, Piedade MTF, Wittmann F, Schöngart J, Parolin P (eds) Amazonian floodplain forests: ecophysiology, biodiversity and sustainable management. Springer, Dordrecht. pp 61–102

Wood CM, Matsuo AYO, Wilson RW, Gonzalez RJ, Patrick ML, Playle RC, Val AL (2003) Protection by natural blackwater against disturbances in ion fluxes caused by low pH exposure in freshwater stingrays endemic to the Rio Negro. Physiol Biochem Zool 76(1):12–27

Zehev BS, Vera A, Asher B, Raimundo R (2015) Ornamental fishery in Rio Negro (Amazon region), Brazil: combining social, economic and fishery analyses. Fish Aquac J 6(2):1–4

Zuanon J, Ferreira E (2008) Feeding ecology of fishes in the Brazilian Amazon—a naturalistic approach. In: Cyprino JEP, Bureau DP, Kapoor BG (eds) Feeding and digestive functions of fishes. CRC Press, Boca Raton. pp 1–35

11 Fishes of Amazonian Whitewater Floodplains

Aaron H. Fronk, Maxwell J. Bernt, Tiago P. Carvalho, William G. R. Crampton and James S. Albert

Abstract

This chapter traces the evolution of habitat utilization in South American fishes into or away from whitewater floodplains, locally called *várzeas*. Analyses using *ancestral state estimation* indicate nonfloodplain upland origins for all of the Big Nine NFF clades, and an estimated 168 transitions within seven of the Big Nine NFF clades between *várzeas* and other physically-defined lowland Amazonian aquatic macrohabitats (i.e., nonfloodplain or *terra firme* streams, deep river channels). The cyprinodontiform clades Poecilioidea and Rivulidae do not characteristically inhabit *várzea* habitats. We found approximately similar total numbers of transitions to and away from *várzeas* among NFF clades as a whole, with Curimatoidea exhibiting the greatest number of transitions and Aspredinoidea the highest transition rate. The most frequent pattern of habitat transition in these clades is among habitats of the same structural type (e.g., floodplain to floodplain) and not to the same water chemistry type (e.g., whitewater to whitewater). In most fish clades, transitions between floodplain habitats with different water chemistries (i.e., whitewaters versus blackwaters or clearwaters) were more common than between habitats with different hydrological structures (i.e., floodplains and nonfloodplain systems). Exceptions include Gymnotiformes and Pimelodoidea, both of which demonstrate a relatively higher transition rate between whitewater floodplains and nonwhitewater nonfloodplain habitats.

A. H. Fronk · J. S. Albert (✉)
University of Louisiana at Lafayette, Lafayette, LA, USA
e-mail: jsalbe@gmail.com

A. H. Fronk
e-mail: fronaar@gmail.com

M. J. Bernt
Susquehanna State University, Selinsgrove, PA, USA
e-mail: mjbernt@gmail.com

T. P. Carvalho
Universidad Nacional de Colombia, Bogotá, Colombia
e-mail: carvalho.ictio@gmail.com

W. G. R. Crampton
University of Central Florida, Orlando, FL, USA
e-mail: crampton@ucf.edu

11.1 Introduction to The Várzea

11.1.1 Geography

Seasonally-inundated floodplains flanking large lowland Amazonian whitewater rivers are an important habitat for Amazonian fishes (Junk et al. 1997; Crampton 1998a, 1999a, 2011; Henderson and Crampton 1997; Henderson et al. 1998; Petry et al. 2003; Correa et al. 2008; Saint-Paul et al. 2000; Val and de Almeida-Val 2012; Junk 2013; Bogotá-Gregory et al. 2020, 2023). More than one third (c. 1170 species or 39%) of all fish species in Greater Amazonia use these floodplains for feeding, breeding, and dispersal to complete their life cycle (van der Sleen and Albert 2017; Albert et al. 2020). The floodplains of sediment-rich whitewater rivers, called *várzeas* in Brazil, occupy about 4.2% (295,000 km^2) of the surface area of the Amazon basin (Junk 1984). The *várzea* forms contiguous ribbons of habitat that ramify across much of the Amazon basin, across a *thalweg* that extends over about 3300 km from rivers and wetlands of the sub-Andean foreland to the edge of the Marajó estuary near the Amazon river mouth (Fig. 11.1).

11.1.2 Várzea Habitats

The *várzea* includes a variety of habitats and other ecological resources that are important for fish populations. Major lowland Amazonian aquatic habitats include oxbow lakes

J. S. Albert and R. E. Reis (eds.), *Ecology and Evolution of Amazonian Fishes*, Fascinating Life Sciences,
https://doi.org/10.1007/978-3-032-09328-8_11

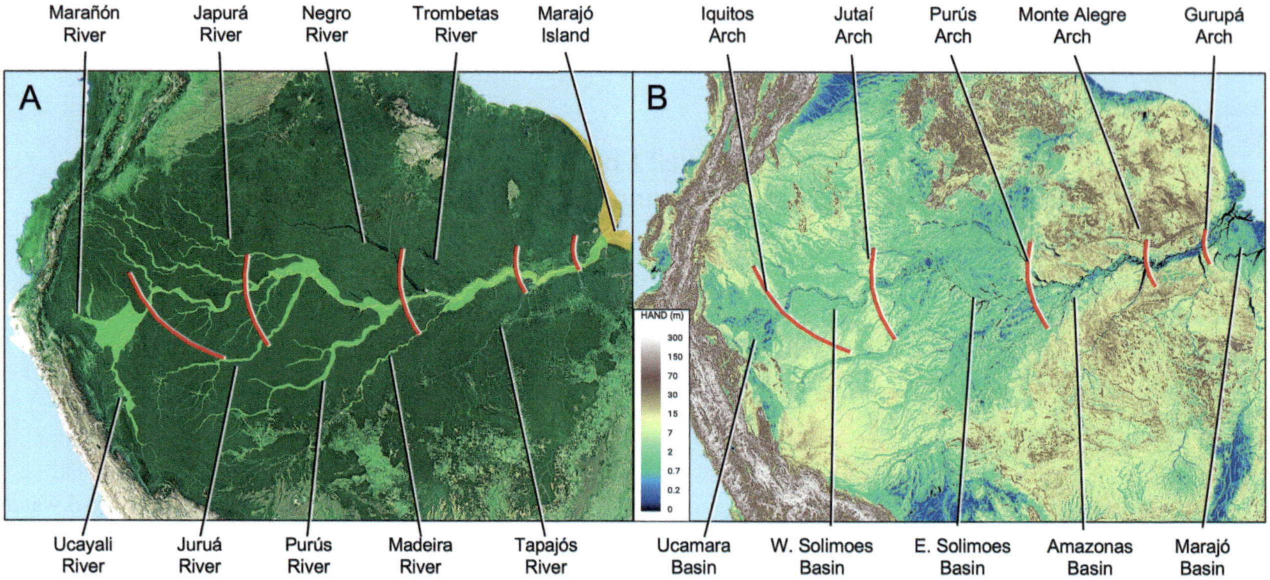

Fig. 11.1 Geographic distribution of Amazonian whitewater floodplains (*várzea*) and associated landscape features. **A** *Várzea* (light green) and tidally influenced floodplain (orange at right), with some major tributaries indicated. Structural arches are indicated by red curved lines. **B** Major geographic features governing lowland flooding, including sub-surface or surface-level structural arches (top) and sedimentary basins (bottom) along the Amazon main stem, visualized as height above nearest drainage (HAND in m, from MERIT Hydro). Areas with HAND values below 2.0 m (shades of black, blues, and greens) experience large annual flooding

(*lagoas*, *cochas*), seasonally-flooded floodplain forests like low-canopy inter-levee forests (*chavascals*) and higher-canopy forests (*restingas*) on natural river levees, and palm swamps (*morichales, aguajales*), among many others (Sioli 2012; Junk et al. 2012). Standing waters in floodplain lakes and wetlands and slack waters near the river margins are seasonally covered by mats of floating macrophytes, dominated by the C4 grasses *Echinochloa polystachya*, *Paspalum fasiculatum*, and *P. repens*, the herbaceous *Polygonum ferrugineum*, the water lettuce *Pistia stratiotes*, and the water hyacinth *Pontederia* (formerly *Eichhornia*) *crassipes*, among many others (including the Amazon lily *Victoria amazonica*) (Piedade et al. 2010). The root mats of these floating meadows constitute an important habitat for foraging and reproduction in many riverine fishes, often serving as nurseries for juvenile stages (Crampton 1996, 1998b, 1999a; Cox Fernandes 1997; Araujo-Lima and Oliveira 1998; Leite et al. 2002; Arantes et al. 2019; Goulding et al. 2019; Correa et al. 2022).

11.1.3 Annual Flood Pulse

The annual flood cycle generates a predictable expansion of habitat volume and food resources at high water, with most fish breeding during the rising-water and high-water seasons (Lowe-McConnell 1979; Crampton 1999a, 1999b; Waddell and Crampton 2020). Perhaps surprisingly, the peak of the breeding season for many species coincides with a high-water period of hypoxia (<0.5 mg/L) or anoxia (~0 mg/L), activities of which can last for weeks to several months in large floodplains. High-water hypoxia occurs due to the decomposition of organic material in the flooded forests (Crampton 1998a, 1998b; Crampton et al. 2008; Val et al. 1998; Soares et al. 2006). Despite the unfavorable oxygen levels, for which fish have evolved a range of physiological, morphological, and behavioral adaptive or phenotypically plastic responses (Crampton 1998a; Val et al. 1998; Crampton et al. 2008, 2025), high-water breeding occurs because it corresponds to the periods of lowest predator density, highest allochthonous inputs from flooded forests, and the most rapid expansion of floating meadow habitats (Waddell and Crampton 2022).

During the low-water period, the floodwaters recede into lakes, with the density of predators increasing, and prey species likely confronting severe population bottlenecks (Waddell and Crampton 2022). During the low-water period, the surface layers of *várzea* lakes and channels tend to be variable in dissolved oxygen, sometimes becoming hypoxic (especially at night, in vegetation) (Melack 1983). However, oxygen conditions are generally more benign during the low-water period when compared to high water.

11.1.4 Várzea Fishes

The *várzea* serves as an important site for reproduction in many Amazonian fishes, given their high primary productivity, seasonally-abundant food resources, and habitat shelter for juveniles (Petry et al. 2003; Crampton et al. 2004; Waddell et al. 2019, Waddell and Crampton 2022). *Várzea* fishes can occupy multiple habitats over the course of their life cycles. Some species undergo lateral migrations between floodplains and river channels (Cox Fernandes 1997; Araujo-Lima and Oliveira 1998; Crampton 1998a, 1998b; Castello 2008; Castello et al. 2015). Other species, mostly characiforms and siluriforms, undertake longitudinal migrations, some over lengthy geographic distances (>3000 km), using river channels as dispersal corridors for adults or larvae (Barthem and Goulding 1997; Herrera-R et al. 2024; Chap. 21). The complex use of floodplain habitats across geographical space, and over the course of the annual flood cycle, greatly increases the environmental complexity of Amazonian aquatic ecosystems, especially as different species exhibit distinct migration patterns (Cox Fernandes et al. 2004; Arantes et al. 2018; Carvalho et al. 2021).

11.1.5 Macroecology

The changing environmental conditions of the *várzea* in time and space promote the coexistence of numerous species, as there is little competitive exclusion and most species are limited by the extreme conditions present at low water, generating and contributing to high local species richness or *alpha diversity* (Henderson et al. 1998; Tockner et al. 2000; Hurd et al. 2016). The relatively continuous distribution of habitats across the Amazon River *várzea* over 3000 km results in relatively low turnover of species composition, referred to as low *beta diversity* (Albernaz et al. 2012; Arantes et al. 2018; Bogotá-Gregory et al. 2023). In these regards, the *várzea* may be thought of as a "species-bank," in analogy to a commercial bank that stores and lends money (Henderson et al. 1998; Albert et al. 2011b). The combination of high alpha diversity, low beta diversity, high connectivity, and high temporal species turnover allows the many fishes accumulated in local assemblages to disperse and become established across wide areas of the Amazon basin (Albert et al. 2011a; see also Siqueira-Souza et al. 2016; Carvalho et al. 2021; Duarte et al. 2022; Bogotá-Gregory et al. 2023).

11.1.6 Macroevolution

Less well understood are the consequences of the rich and variable *várzea* fish faunas on the evolution of Amazonian fishes and the accumulation of species richness at a regional scale. Has the *várzea* served an evolutionary cradle or source of new evolutionary lineages, which seeded the species composition of other habitat types? Or does the ecologically specialized *várzea* fish fauna serve more as an evolutionary museum or evolutionary graveyard, where species accumulate, but ultimately go extinct? To use a concept introduced by Henderson et al. (1999), does the *várzea* act as an evolutionary "battery" that stores a charge, where the charge is species that trickle in from surrounding *terra firme* systems? Or does the *várzea* ichthyofauna served in a combination of evolutionary roles, contributing to expansion of some clades and the demise of others?

11.1.7 Chapter Goals

This chapter documents the evolutionary history of Neotropical freshwater fish (NFF) clades among habitats based on hydrological structures (floodplains vs. others) and water chemistry (whitewater vs. others). Using the phylogeny presented in Chap. 2, and the habitat-use data in Chap. 6, we estimated the ancestral macrohabitat of 3152 species representing seven of the Big Nine NFF clades outlined in Chap. 3. We then estimate instances and frequencies of evolutionary transitions in clades that inhabit the *várzea* and other habitats, and fit alternative models of trait evolution to these phylogenetic and trait data. The goal is to estimate the habitat origins of *várzea* fishes and establish patterns of habitat transition to and from *várzea* habitats.

11.2 Methods

11.2.1 Ecotrait Categorization

Adult modal macrohabitat utilization of all NFF species was scored as present in one of the following four major categories: nonfloodplain (i.e., *terra firme*) habitats of small rivers and streams with SO 1–5; floodplain habitats of large (SO 6–10) whitewater rivers (i.e., *várzea*); floodplain habitats of large (SO 6–10) blackwater (i.e., *igapó*) or clearwater rivers; riverine habitats (i.e., nonfloodplain river channels and river margins) of large (SO 6–10) whitewater of rivers. NFF species were scored as inhabiting *várzea,* if they present adult, modal, values as present in all three of following three ecotraits of Chap. 6: whitewaters, floodplains, and large rivers with SO 6–10 (Chap. 1). Species present in whitewaters, but not the other two ecotraits, were scored as "whitewater nonfloodplain." Species present in floodplains and large rivers, but not whitewaters, were scored as "nonwhitewater floodplain." Species not found in any of the three ecotrait categories were classed as "nonwhitewater nonfloodplain". See Chap. 7 for a more detailed description of estimating ancestral states and evolutionary rates. Of the Big Nine NFF clades, the two cyprinodontiform clades were excluded from this study, as they do not occur or are rare in *várzea* and other lowland whitewater habitats.

11.2.2 Ancestral State and Evolutionary Rates Analyses

Ancestral State Estimation (ASE) was used to estimate the ancestral macrohabitat using analytical tools for discrete characters and the Mk model of trait evolution in the R package

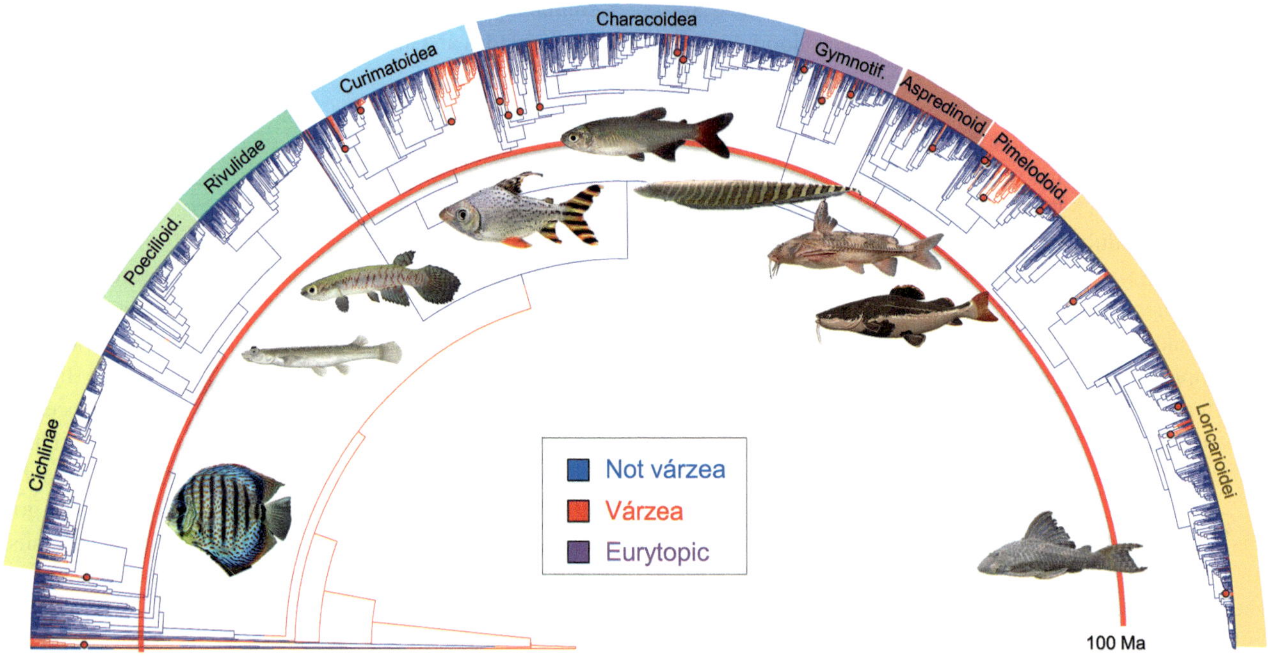

Fig. 11.2 Evolution of habitat utilization in NFF taxa among whitewater floodplains (*várzeas*) and other physical habitats. *Várzea* species defined as adult modal habitat in whitewater floodplains of large rivers (i.e., Strahler stream orders 6–10). Data for 3167 species in all NFF 880 genera, 96 families, and 40 orders. Big Nine NFF clades (colored regions) from Chap. 3. Red semicircle at c. 100 Ma marks the time of the final separation of South America and Africa and the latest origin of phylogenetically independent NFF clades. Red circles indicate *várzea* clades with five or more species. Note polyphyletic origin of *várzea* clades in seven of the Big Nine NFF clades

phytools (Revell 2014; Fig. 11.2). Evolutionary rate analyses were conducted by fitting the dataset to three different models of trait evolution: an "*equal-rates*" (ER) model in which all transition rates among character states are equal, a "*symmetric-rates*" (SYM) model in which rates of transition back and forth between two character states are equal to each other, and an "*all rates different*" (ARD) model, where all transition rates are assumed to be independent of each other. Each model-fitting analysis produced a log-likelihood value, from which was calculated an *Akaike Information Criterion* (AIC) to assess the best-fit model of trait evolution for each clade. AIC values among models were compared using the likelihood-ratio test statistic, which quantifies the difference in log likelihood scores for significance. In each case, the lowest-AIC-scoring model was found to be a significantly better fit than the other models, as assessed by >2 log likelihood score values.

11.2.3 Bayesian Stochastic Mapping

For each clade, the trait evolution results for the best-scoring model were then mapped to the phylogenetic tree using the software SIMMAP (Bollback 2006), which implements Bayesian stochastic mapping. Bayesian stochastic mapping is a method of mapping stochastic characters on trees that estimates the timing of character changes along branches. Similar to Bayesian inference, it works by sampling character histories in proportion to their posterior probabilities, due to a direct analytical approach being impractical for all but the simplest topologies. SIMMAP specifically uses a three-step algorithm for sampling character histories given a particular topology, branch lengths, and substitution model: it first calculates the conditional likelihood for each character state at each node of the tree, simulates ancestral states at each internal node by sampling from the posterior distribution of states, and finally simulates a substitution history by sampling from the posterior distribution conditioned on the aforementioned simulated ancestral states and observed tip states in the topology. One hundred stochastic maps were generated for each analysis, and from these results were derived posterior probabilities at each node of the phylogenetic tree for each character state, which appeared on the trees as color-coded pie charts at each node (Figs. 11.3–11.5). The transition rates between character states for these character histories are shown in Fig. 11.6.

11.3 Results

11.3.1 Taxonomic Composition

A total of 562 NFF species were scored as *várzea* specialists in the ecomatrix (Chap. 6), representing about 19% of the fish species in Greater Amazonia, and about 9% of all NFF species. Species of Poecilioidea and Rivulidae rarely or never inhabit

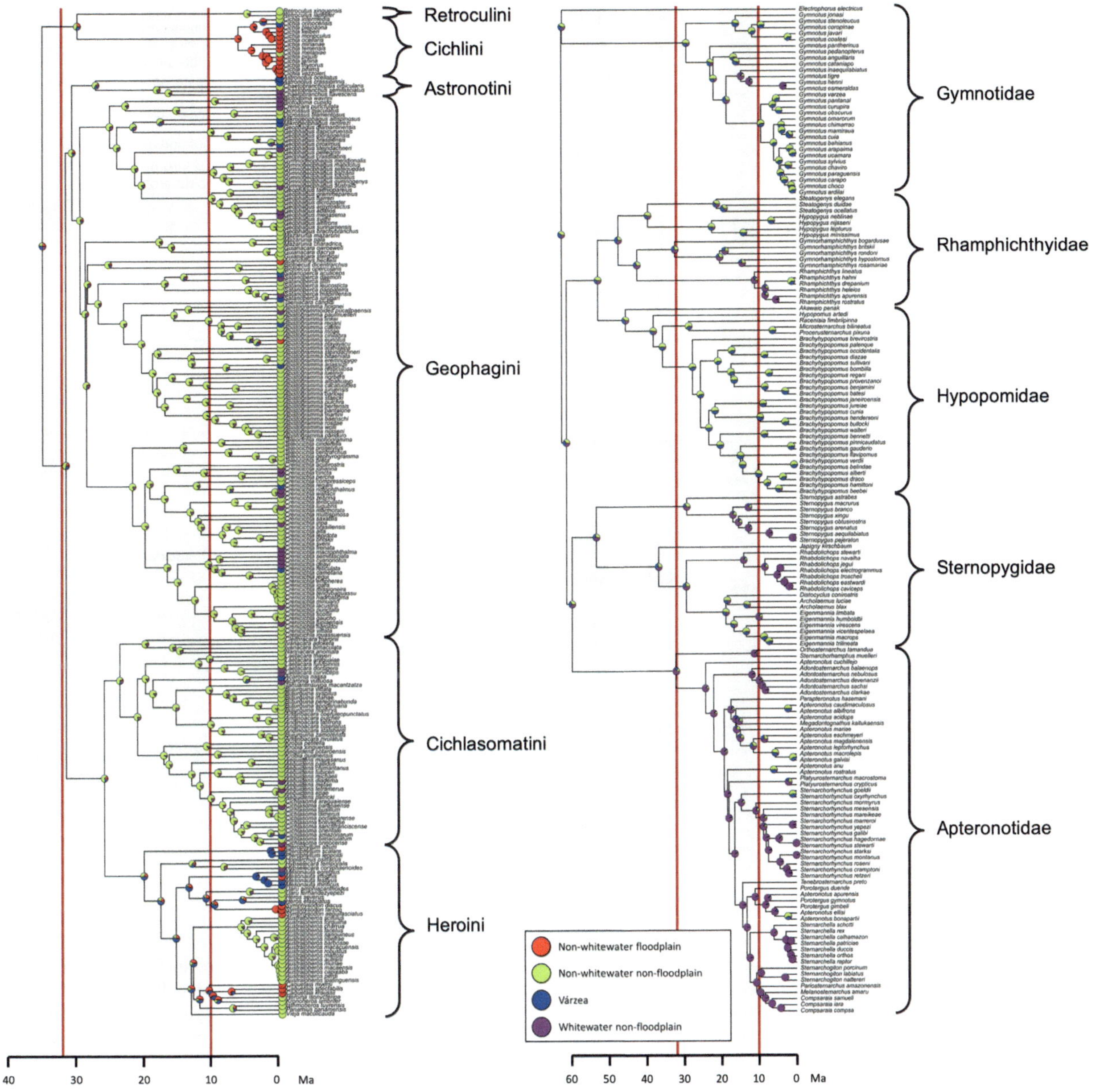

Fig. 11.3 ASE of macrohabitat utilization in whitewater floodplains for Cichlinae and Gymnotiformes. Red vertical lines indicate 34 Ma and 10 Ma. Estimated transitions to *várzea* as deepest nodes with a blue circle >50%. Note 15 separate origins of *várzea*-inhabiting clades in Cichlinae, and about seven origins in Gymnotiformes (one in Gymnotidae, two in Rhamphichthyidae, three in Hypopomidae, and one in Apteronotidae)

the *várzea*. Among the seven of the Big Nine NFF clades examined, *várzea* taxa are represented as follows:

- 22 species of Cichlinae, including many species of Apistogramma, Astronotus, Cichlasoma, Crenicichla, Mesonauta, Mikrogeophagus, Pterophyllum, and Satanoperca.
- 126 species of Curimatoidea, including many species of Amaralia, Anodus, Colossoma, Curimata, Cyphocharax, Hemiodus, Leporinus, Mylossoma, Pygocentrus, and Steindachnerina.
- 97 species of Characoidea, including many species of Acestrorhynchus, Aphyocharax, Astyanax, Charax, Ctenobrycon, Gymnocorymbus, Iguanodectes, Phenacogaster, Roeboides, Thoracocharax, and Triportheus.
- 89 species of Doradioidea, including many species of Ageneiosus, Amaralia, Amblydoras, Auchenipterus, Hemidoras, Ossancora, Tatia, Trachydoras, Tympanopleura, and Tympanopleura.
- 56 species of Gymnotiformes, including many species of Adontosternarchus, Apteronotus, Brachyhypopomus, Eigenmannia, Gymnotus, Rhamphichthys, and Sternopygus.

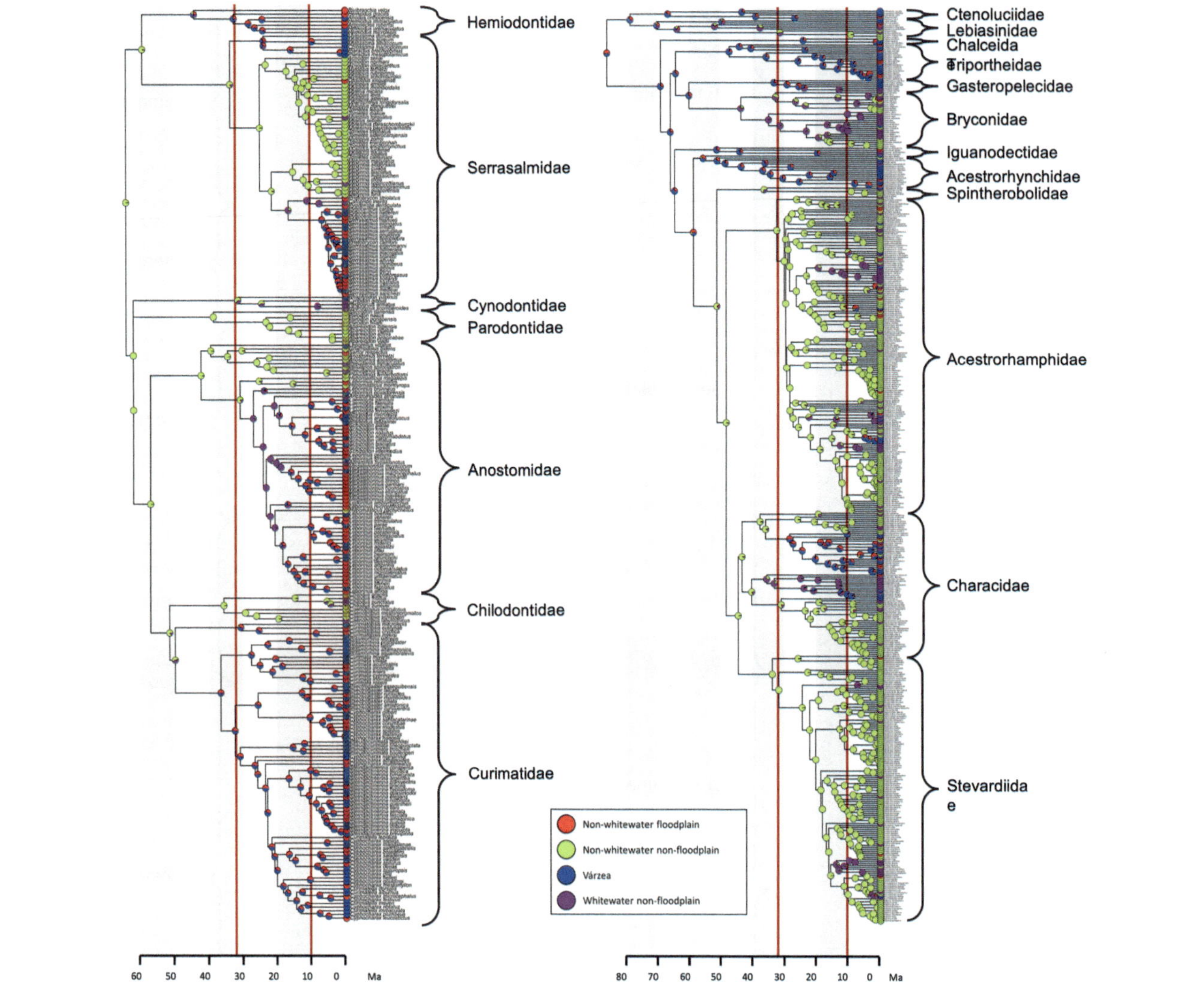

Fig. 11.4 ASE of macrohabitat utilization in whitewater floodplains for Curimatoidea and Characoidea. Red vertical lines indicate 34 Ma and 10 Ma. Note about 25 separate origins of *várzea*-inhabiting clades in Curimatoidea (two in Hemiodontidae, six in Serrasalmidae, one in Cynodontidae, 14 in Anostomidae, and two in Curimatidae with multiple reversals), and 15 clades in Characoidea (one each in Ctenolucidae, Lebiasinidae, Chalceidae, Gasteropelecidae, Bryconidae, Iguanodectidae, Acestrorhynchidae, Spintherobolidae, and Stevardiidae, and three each in Acestrorhamphidae and Characidae)

- Ten species of Pimelodoidea, including many species of Leiarius and Rhamdia.
- 101 species of Loricarioidei, including many species of Dasyloricaria, Farlowella, Hypoptopoma, Lepthoplosternum, Loricariichthys, Otocinclus, Pterygoplichthys, Rhinelepis, and Rineloricaria.

Várzea fishes are represented by at least 16 species in other clades of primary freshwater fishes, including several charismatic fishes like the bony-tongues *Arapaima* and *Osteoglossum*, Amazonian pike characins (*Boulengerella*), and some wolf fishes (*Hoplias*). *Várzea* inhabitants are also represented by 43 species of marine-derived lineages (MDL), including freshwater pufferfishes (*Colomesus*), freshwater mullets (*Mugil*), freshwater flatfishes (*Hypoclinemus*), freshwater stingrays (*Potamotrygon*), freshwater needlefishes (*Potamorrhaphis, Pseudotylosurus*), and swamp eels (*Synbranchus*).

11.3.2 Ancestral State Estimation

Ancestral state estimates and evolutionary transitions of NFF species are depicted in Fig. 11.2, in the phylogenetic sequence that they appear in Chap. 2. All the Big Nine NFF clades have estimated evolutionary origins in small rivers of the Guiana Shield and Brazilian Shield. The two cyprinodontiform clades (Poecilioidea and Rivulidae) are either rare or absent in the

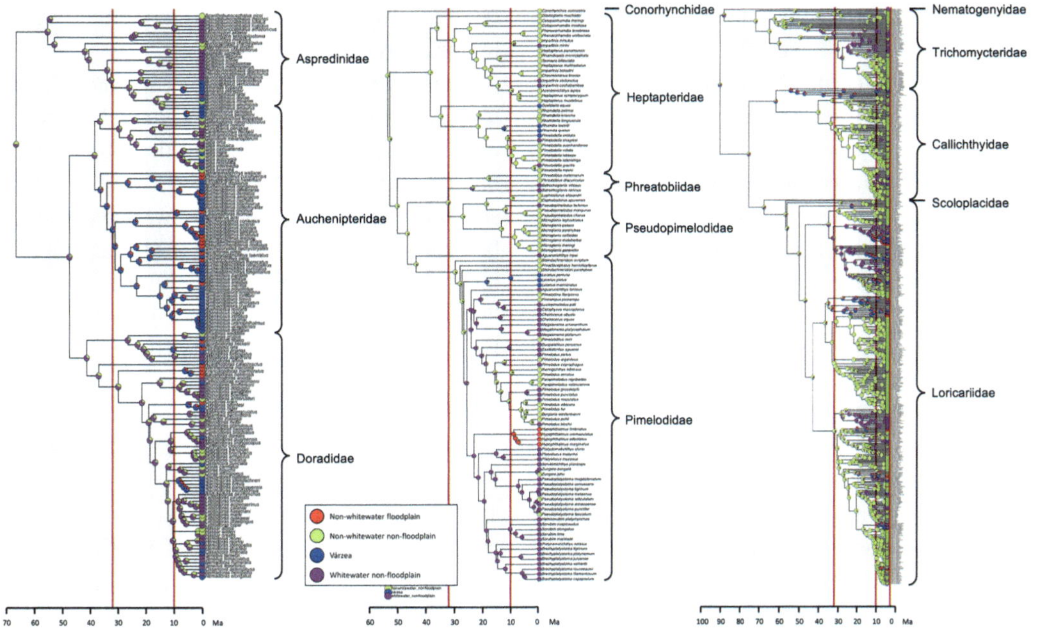

Fig. 11.5 ASE of macrohabitat utilization in whitewater floodplains for Aspredinoidea, Pimelodoidea, and Loricarioidei. Red vertical lines indicate 34 Ma and 10 Ma. Note about 23 *várzea* origins in Aspredinoidea (four clades in Aspredinidae, five in Auchenipteridae with one clade having multiple reversals, and about 14 in Doradidae), four origins in Pimelodoidea (three in Heptapteridae and one in Pimelodidae), and about 21 in Loricarioidei (five in Callichthyidae and 16 in Loricariidae with one having multiple reversals). Numbers of estimated transitions approximate due to uncertainties in ASE

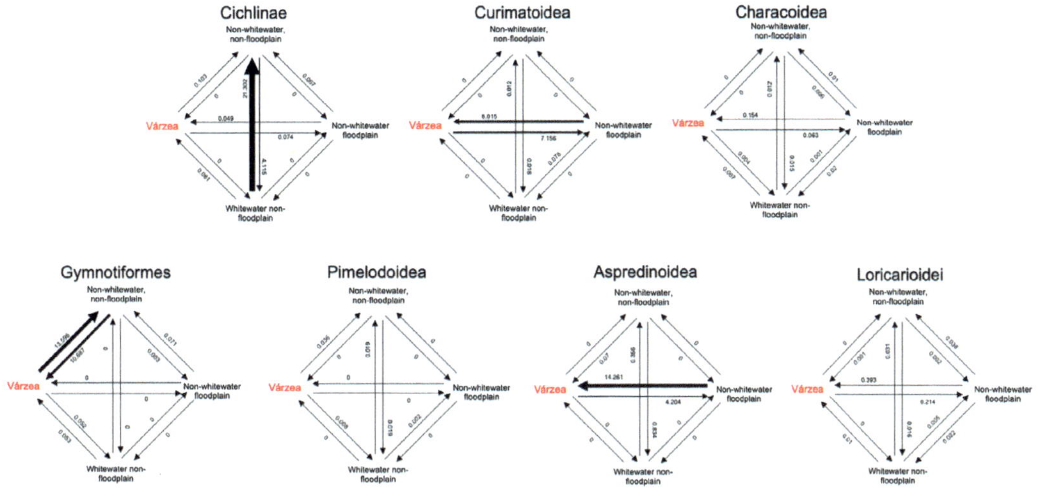

Fig. 11.6 Evolutionary transition rates among habitat types for seven NFF clades. Habitats based on physical structure (i.e., floodplains, channels, terra firme streams) and water chemistry. *Várzea* indicated in red font at left of each transition diagram. Arrow thickness indicates relative transition rate, with thicker arrows having higher rates. Note in five of the seven clades, the transitions follow an ARD pattern, and transition rates between habitats of the same hydrological structure (floodplain vs. nonfloodplain) are greater than between those of different water chemistries (whitewater vs. clearwater and/or blackwater). The exceptions are Gymnotiformes and Pimelodoidea, both of which show a SYM pattern of substitution and have the greatest transition to and from the *várzea* with nonwhitewater nonfloodplain habitats

várzea, and are not here considered further. The other seven of these clades exhibit multiple evolutionary transitions to the *várzea*.

Cichlinae was found to best follow an ARD model of trait evolution. Cichlinae is estimated to have a habitat origin in low-conductivity (clearwater or blackwater) floodplains, as observed in some early-branching species of extant Cichlini (*Cichla*), Astronotini (*Astronotus*), and Heroini (*Pterophyllum*). Cichlinae is not well represented in the *várzea* as compared with other habitats, with only 22 species being modally most abundant in this habitat, and with most of these species having origins as transitions from whitewater rivers. This analysis estimates about 15 separate transitions to the *várzea* (Fig. 11.3).

Curimatoidea was found to best follow an ARD model of trait evolution. It has an estimated habitat origin in low-conductivity (clearwater or blackwater) nonfloodplains, as observed in the majority of the Serrasalmidae and Parodontidae. The greatest number of *várzea*-dwelling species is found in Curimatidae, a family known for undertaking long-distance seasonal migrations (Lowe-McConnell 1979; Barthem and Goulding 1997; Melo et al. 2021). The majority of *várzea* species within the Curimatoidea transitioned there from blackwater or clearwater floodplain habitats, with the highest transition rates *out* of the *várzea* being back to these non-*várzea* floodplains.

Characoidea was found to best follow an ARD model of trait evolution. It has an estimated habitat origin in whitewater nonfloodplain habitats. For extant species, this is primarily found in the form of whitewater deep channel species; however, the majority of extant characoid species is found in nonwhitewater littoral habitats. Those that are found in *várzea* primarily transitioned from ancestral whitewater nonfloodplain habitats.

Unlike most of the other clades examined, Gymnotiformes was found to follow a SYM model of trait evolution, a trait shared with Pimelodoidea (see below). The habitat origin of Gymnotiformes is uncertain in this analysis, with about equal support for an estimated ancestor in whitewater river channels, *várzea*, and smaller *terra firme* (SO 1–3) low-conductivity (clear or blackwater) rivers and streams. This uncertainty arises from habitat partitioning among major gymnotiform subclades with families and subfamilies specialized to each of these major habitat types. Chapter 13 evaluates evidence that Gymnotiformes or some of its subclades originated in whitewater river channels. Two gymnotiform families (Apteronotidae and Sternopygidae) are among the most abundant and diverse animals in whitewater channels (Lundberg et al. 1987; Duarte et al. 2019). Oddly, the greatest rate of transition to and from *várzea* habitats is with nonfloodplain nonwhitewater habitats.

Aspredinoidea followed an ARD model of trait evolution, and has an estimated habitat origin within nonfloodplain habitats, which is consistent with their current association with deep channel habitats. The family Doradidae, in particular, attains among the highest animal biomasses in deep channels (Saint-Paul et al. 2000; Duarte et al. 2019). *Várzea*-dwelling doradioids transition primarily from low-conductivity, non-*várzea* (blackwater or clearwater) floodplain habitats.

Pimelodoidea followed a SYM model of trait evolution, and also has an estimated habitat origin in nonfloodplain habitats, primarily nonwhitewater nonfloodplains, which is again consistent with their extant association with primarily blackwater/clearwater terra firme streams. These same nonwhitewater nonfloodplain habitats are the primary source of *várzea* species, and vice versa, like the similarly SYM-patterned Gymnotiformes. Interestingly, no other transitions from *várzeas* to other habitat types (or vice versa) were found in this analysis for this clade.

Loricarioidei was found to best match the ARD model of trait evolution. The variegated modal habitat utilization of extant Loricarioidei made it difficult to estimate a particular habitat origin for this clade (see Chap. 13 for a more conclusive analysis, which found nonwhitewater deep channels to be their habitat of origin). The most frequent habitat of extant loricarioids is nonwhitewater nonfloodplain habitats, but those that live in *várzeas* primarily transitioned into this habitat from floodplains of low conductivity rivers.

11.3.3 Ecotrait Evolutionary Rates

The ARD model provided the most likely model of evolutionary transitions among ecotrait types for five of the seven clades examined (Fig. 11.5), while the SYM model was found to be the best model for Gymnotiformes and Pimelodoidea. For six of the clades, transition rates were higher among lineages that occupy habitats based on water chemistry (i.e., whitewater to blackwater/clearwater and vice-versa) than hydrological structure (i.e., floodplain vs. nonfloodplain). The exception is Gymnotiformes, where the highest transition was between *várzeas* and nonfloodplain nonwhitewater habitats.

ASE results were examined to find which habitats provided the greatest source of new taxa to the *várzea*, and the habitats that most frequently received taxa from the *várzea* (Table 11.1). Of the 14 most frequent transitions to and from the *várzea*,

Table 11.1 Evolutionary transitions to and from *várzea* habitats for seven NFF clades. Data for 168 transitions between *várzeas* and other physical habitats (*terra firme* streams and deep river channels) estimated on the phylogeny of Fig. 11.2. Note most transitons in Curimatoidea and the highest transition rates in Aspredinoidea, and approximately similar total numbers of transitons to and away from *várzeas* among fishes as a whole. Bolded text indicates transition to a habitat of a different structural type (i.e., floodplain to non-floodplain). Italicized text indicates transition to a habitat of a different water chemistry type (i.e., whitewater to clearwater/blackwater). Methods for estimating evolutionary rates in Chap. 7

	Number transitions		Transition rates			
Clade	To	Away	To	Rate	Away	Rate
Cichlinae	15	3	***CBNF***	0.73	**WWNF**	0.21
Curimatoidea	25	40	*CBF*	7.66	*CBF*	6.63
Characoidea	15	16	*CBF*	0.02	**WWNF**	0.01
Gymnotiformes	7	0	***CBNF***	12.40	***CBNF***	12.40
Pimelodoidea	4	0	***CBNF***	0.00	***CBNF***	0.00
Aspredinoidea	5	9	*CBF*	4.09	*CBF*	14.3
Loricarioidei	21	8	*CBF*	0.04	*CBF*	0.02
Total	92	76				

seven out of 14 (or 50%) were most frequently to or from other floodplain habitats, as opposed to three transitions to or from whitewater nonfloodplains (21.4%), and three transitions from nonwhitewater nonfloodplains (21.4%).

11.3.4 Macroevolutionary Patterns

The ASE analyses indicate nonfloodplain origins for four of the seven NFF clades examined: Characoidea, Curimatoidea, Aspredinoidea, and Pimelodoidea (Fig. 11.3). Because many deep-channel fish species use floodplains for mating and breeding (Crampton 2007), it is possible that the fish faunas of the *várzea* and other floodplain habitats were seeded from the river channels associated with those floodplains. Another common trend observed was the tendency for the estimated habitat origin to be similar to the most common habitat of the species used in the analysis. This could be an artifact of the analysis, but may also reflect phylogenetic niche conservatism.

Phylogenetic patterns in the transition rates among these habitats imply multiple transitions from and to the *várzea*, with frequent transitions to other habitats of the same physical (i.e., structural) type; i.e., blackwater or clearwater floodplains to whitewater floodplains (Fig. 11.3; Table 11.1). This pattern is most pronounced in Curimatoidea, where the only transitions to or from the *várzea* are with floodplains of other water types (Melo et al. 2021). An exception to this general pattern is observed in Gymnotiformes, which exhibit more transitions among habitats of completely differing water chemistry and structural type. This pattern is shared to some degree with Pimelodoidea, which, like Gymnotiformes, showed symmetric rates of transitions among all habitat types.

11.4 Conclusions

This chapter estimates the instances and rates of macroevolutionary transitions in freshwater fish clades adapted to whitewater floodplains (*várzeas*) and three other physically defined aquatic macrohabitats of tropical South America. The analyses use the time-calibrated phylogeny and ecomatrix of NFF species presented in earlier chapters, scoring adult, modal, habitat utilization among the following four macrohabitats: floodplain habitats of large (SO 6–10) whitewater rivers (i.e., *várzea*), floodplain habitats of large blackwater (i.e., *igapó*) or clearwater rivers, riverine habitats (i.e., nonfloodplain channels and margins) of large rivers, and nonfloodplain (i.e., *terra firme*) habitats of small (SO 1–5) rivers and streams. Among NFF clades, 562 species are *várzea* specialists, representing about 19% of all fish species in Greater Amazonia, and about 9% of all NFF species. All the Big Nine NFF clades have estimated stem-group origins in small, low-conductivity (i.e., clearwater and/or blackwater) *terra firme* rivers and streams, and this is the estimated crown-group macrohabitat for Cichlinae, Curimatoidea, and Pimelodoidea. Cichlinae is not well represented in the *várzea* as compared to other habitats, with only 22 species being modally most abundant in this habitat, with about 15 separate transitions to the *várzea*, mostly from whitewater rivers. This phylogenetic estimate is strongly influenced by the phylogenetic position of Pseudopimelodidae and Heptapteridae with mostly blackwater and clearwater *terra firme* stream species. Seven of the Big Nine NFF clades exhibit multiple evolutionary transitions into and out of the *várzea*, with an estimated 168 phylogenetic transitions with the other macrohabitats. The two cyprinodontiform clades (i.e., Poecilioidea and Rivulidae) are either rare or absent in the *várzea*. There is uncertainty in the ancestral crown-group macrohabitat estimate for four of the Big Nine NFF clades. Characoidea and its major family-level subclades have estimated origins in large river floodplains, either in low-conductivity clearwaters and/or blackwaters, or in high-conductivity whitewaters. Aspredinoidea has estimated crown-group origins in large whitewater rivers, which is consistent with the extant species-richness patterns. The ancestral crown-group macrohabitats for Gymnotiformes and Loricarioidei are uncertain, with similar likelihood values for low-conductivity (i.e., clearwater and/or blackwater) small *terra-firme* rivers and streams, and for large channels of high-conductivity whitewater rivers. These uncertainties arise from strong habitat partitioning among the extant gymnotiform and loricarioid subclades, with most family and subfamily-level taxa highly specialized to each of these two major macrohabitats. Among the Big Nine NFF clades, Curimatoidea is estimated to have undergone the most, and Pimelodoidea the fewest, evolutionary transitions between the *várzea* and other major macrohabitats.

Acknowledgments We are grateful to Milton Tan for providing the *phytools* code used in the ASE analyses, and to Michael Goulding, John Lundberg, Paulo Petry, Roberto Reis, Don Stewart, and Kirk Winemiller for many insights into the biology of *várzea* fishes.

References

Albernaz AL, Pressey RL, Costa LR, Moreira MP, Ramos JF, Assunção PA, Franciscon CH (2012) Tree species compositional change and conservation implications in the white-water flooded forests of the Brazilian Amazon. J Biogeogr 39(5):869–883

Albert JS, Carvalho TP, Petry P, Holder MA, Maxime EL, Espino J, Corahua I, Quispe R, Rengifo B, Ortega H, Reis RE (2011a) Aquatic biodiversity in the Amazon: habitat specialization and geographic isolation promote species richness. Animals 1(2):205–241

Albert JS, Petry P, Reis RE (2011b) Major biogeographic and phylogenetic patterns. In: Albert JS, Reis RE (eds) Historical biogeography of neotropical freshwater fishes. University of California Press, Berkeley, pp 3–20

Albert JS, Tagliacollo VA, Dagosta F (2020) Diversification of neotropical freshwater fishes. Annu Rev Ecol Evol Syst 51(1):27–53

Albert JS, Abrahão V, Akin DR, Allen JG, Ândrade M, Arce M et al (2025) An ecological trait matrix of Neotropical freshwater fishes. Sci Data 12(1):1127

Arantes CC, Winemiller KO, Petrere M, Freitas CE (2019) Spatial variation in aquatic food webs in the Amazon River floodplain. *Freshwater*. Science 38(1):213–228
Arantes CC, Winemiller KO, Petrere M, Castello L, Hess LL, Freitas CE (2018) Relationships between forest cover and fish diversity in the Amazon River floodplain. J Appl Ecol 55(1):386–395
Araujo-Lima CA, Oliveira EC (1998) Transport of larval fish in the Amazon. J Fish Biol 53:297–306
Barthem R, Goulding M (1997) The catfish connection: ecology, migration, and conservation of Amazon predators. Columbia University, New York
Bogotá-Gregory JD, Jenkins DG, Lima FCT, Magurran AE, Crampton WGR (2023) Geomorphological habitat type drives variation in temporal species turnover but not temporal nestedness in Amazonian fish assemblages. Oikos 11:e09967
Bogotá-Gregory JD, Lima FCT, Correa SB, Silva-Oliveira C, Jenkins D, Ribeiro FR, Lovejoy NR, Reis RE, Crampton WGR (2020) Biogeochemical water type influences community composition, species richness, and biomass in megadiverse Amazonian fish assemblages. Sci Rep 10:15349
Bollback JP (2006) SIMMAP: stochastic character mapping of discrete traits on phylogenies. BMC Bioinf 7(88). https://doi.org/10.1186/1471-2105-7-88
Carvalho RA, Teresa FB, Tejerina-Garro FL (2021) The effect of riverine networks on fish β-diversity patterns in a neotropical system. Hydrobiologia 848(2):515–529
Castello L (2008) Lateral migration of *Arapaima gigas* in floodplains of the Amazon. Ecol Freshw Fish 17(1):38–46
Castello L, Isaac VJ, Thapa R (2015) Flood pulse effects on multispecies fishery yields in the Lower Amazon. R Soc Open Sci 2(11):150299
Correa SB, Crampton WGR, Chapman LJ, Albert JS (2008) A comparison of flooded forest and floating meadow fish assemblages in an upper Amazon floodplain. J Fish Biol 72:629–644
Correa SB, Van Der Sleen P, Siddiqui SF, Bogotá-Gregory JD, Arantes CC, Barnett AA, Couto TBA, Goulding M, Anderson EP (2022) Biotic indicators for ecological state change in Amazonian floodplains. BioScience 72(8):753–768
Cox Fernandes C (1997) Lateral migration of fishes in Amazon floodplains. Ecol Freshw Fish 6(1):36–44
Cox Fernandes C, Podos J, Lundberg JG (2004) Amazonian ecology: tributaries enhance the diversity of electric fishes. Science 305(5692):1960–1962
Crampton WGR (1996) Gymnotiform fish: an important component of Amazonian floodplain fish communities. J Fish Biol 48(2):298–301
Crampton WGR (1998a) Effects of anoxia on the distribution, respiratory strategies and electric signal diversity of gymnotiform fishes. J Fish Biol 53:307–330
Crampton WGR (1998b) Electric signal design and habitat preferences in a species rich assemblage of gymnotiform fishes from the upper Amazon basin. An Acad Bras Cienc 70:805–847
Crampton WGR (1999a) Os peixes da Reserva Mamirauá: diversidade e história natural na planície alagável da Amazônia. In: Queiroz HL, Crampton WGR (eds) Estratégias para manejo de recursos pesqueiros em mamirauá. Sociedade Civil Mamirauá/CNPq, Brasília, pp 10–36
Crampton WGR (1999b) The impact of the ornamental fish trade on the discus Symphysodon aequifasciatus: a case study from the floodplain forests of Estação Ecológica Mamirauá. Adv Econ Botany 13(1999):29–44
Crampton WGR (2007) Diversity and adaptation in deep channel neotropical electric fishes. In: Sebert P, Onyango DW, Kapoor BG (eds) Fish life in special environments. Science Publishers, Enfield, NH, pp 283–339
Crampton WGR (2011) An ecological perspective on diversity and distributions. In: Albert JS, Reis RE (eds) Historical biogeography of neotropical freshwater fishes. University of California, Berkeley, pp 165–189
Crampton WGR, Chapman LJ, Bell J (2008) Interspecific variation in gill size is correlated to ambient dissolved oxygen in the Amazonian electric fish *Brachyhypopomus* (Gymnotiformes: Hypopomidae). Environ Biol Fishes 83:223–235
Crampton WGR, Castello L, Viana JP (2004) Fisheries in the Amazon várzea: historical trends, current status, and factors affecting sustainability. In: People in nature: wildlife conservation in South and Central America. Columbia University, New York, pp 76–98
Crampton WGR, Pathak LB, Janzen FH, Waddell JC, Lovejoy NR (2025) Gill evolution in neotropical electric fishes: comparative phylogenetic evidence for hypoxia-driven adaptation. Funct Ecol, in press
Duarte C, Magurran AE, Zuanon J, Deus CP (2019) Trophic ecology of benthic fish assemblages in a lowland river in the Brazilian Amazon. Aquat Ecol 53(4):707–718
Duarte C, Farago TLB, dos Anjos CS, dos Santos NR, do Nascimento LM, Cella-Ribeiro A, de Deus CP (2022) Spatial and seasonal variation of benthic fish assemblages in whitewater rivers of Central Amazon. Biota Neotropica 22(4):e20211312
Goulding M, Venticinque E, Ribeiro MLDB, Barthem RB, Leite RG, Petry P, da Silva-Junior UL, Ferraz PS, Carlos C (2019) Ecosystem-based management of Amazon fisheries and wetlands. Fish Fish 20(1):138–158
Henderson PA, Crampton WGR (1997) A comparison of fish diversity and abundance between nutrient-rich and nutrient-poor lakes in the Upper Amazon. J Trop Ecol 13:175–198
Henderson PA, Hamilton WD, Crampton WGR (1998) Evolution and diversity in Amazonian floodplain communities. In: Newbery DM, Prins HHT, Brown ND (eds) Dynamics of tropical communities. Blackwell Science, Oxford, pp 385–419
Henderson PA, Robertson BA (1999) On structural complexity and fish diversity in an Amazonian floodplain. Adv Econ Bot 13:45–58
Herrera-R GA, Heilpern SA, Couto TB, Victoria-Lacy L, Duponchelle F, Correa SB, Farah-Pérez A, López-Casas S, Cañas-Alva CM, Doria CR, Anderson EP (2024) A synthesis of the diversity of freshwater fish migrations in the Amazon basin. Fish Fish 25(1):114–133
Hurd LE, Sousa RG, Siqueira-Souza FK, Cooper GJ, Kahn JR, Freitas CE (2016) Amazon floodplain fish communities: habitat connectivity and conservation in a rapidly deteriorating environment. Biol Cons 195:118–127
Junk WJ (1984) Ecology of the *várzea*, floodplain of Amazonian whitewater rivers. In: The Amazon. Springer, Dordrecht, pp 215–243
Junk WJ, Teresa T, Piedade F, Schöngart J, Wittmann F (2012) A classification of major natural habitats of Amazonian white-water river floodplains (*várzea*s). Wetlands Ecol Manage 20:461–475
Junk WJ (ed) (2013) The Central Amazon floodplain: ecology of a pulsing system, vol. 126. Dordrecht, Netherlands, Springer Science & Business Media
Junk WJ, Soares MG, Saint-Paul U (1997) The fish. The Central Amazon floodplain: ecology of a pulsing system 385–408
Leite RG, Araujo-Lima CARM, Victoria RL, Martinellie LA (2002) Stable isotope analysis of energy sources for larvae of eight fish species from the Amazon floodplains. Ecol Freshwater Fish 11(1):56–63
Lowe-McConnell RH (1979) Ecological aspects of seasonality in fishes of tropical waters. Symp Zool Soc London 44(1):219–241
Lundberg JG, Lewis Jr WM, Saunders III JF, Mago-Leccia F (1987) A major food web component in the Orinoco River channel: evidence from planktivorous electric fishes. Science 237(4810):81–83
Melack JM (1983) Diel oxygen variations and their ecological implications in Amazon floodplain lakes. Archiv für Hydrobiologie 98:422
Melo BF, Albert JS, Dagosta FC, Tagliacollo VA (2021) Biogeography of curimatid fishes reveals multiple lowland–upland river transitions and differential diversification in the neotropics (Teleostei, Curimatidae). Ecol Evol 11(22):15815–15832
Petry P, Bayley PB, Markle DF (2003) Relationships between fish assemblages, macrophytes and environmental gradients in the Amazon River floodplain. J Fish Biol 63(3):547–579

Piedade MTF, Junk W, D'Ângelo SA, Wittmann F, Schöngart J (2010) Aquatic herbaceous plants of the Amazon floodplains: state of the art and research needed. Acta Limnol Bras 22:165–178

Revell LJ (2014) Package 'phytools.' https://cran.r-project.org/web/packages/phytools

Saint-Paul U, Zuanon J, Correa MAV, García M, Fabré NN, Berger U, Junk WJ (2000) Fish communities in central Amazonian white-and blackwater floodplains. Environ Biol Fishes 57(3):235–250

Sioli H (ed) (2012) The Amazon: limnology and landscape ecology of a mighty tropical river and its basin, vol 56. Dordrecht, Netherlands, Springer Science & Business Media

Siqueira-Souza FK, Freitas CE, Hurd LE, Petrere M (2016) Amazon floodplain fish diversity at different scales: do time and place really matter?. Hydrobiologia 776:99–110

Soares MGM, Menezes NA, Junk WJ (2006) Adaptations of fish species to oxygen depletion in a central Amazonian floodplain lake. Hydrobiologia 568:353–367

Tockner K, Malard F, Ward JV (2000) An extension of the flood pulse concept. Hydrol Processes 14(16-17):2861–2883

Val AL, Silva MNP, Almeida e Val VMF (1998) Hypoxia adaptation in fish of the Amazon: a never-ending task. South Afr J Zool 33:107–114

Val AL, de Almeida-Val VM (2012) Fishes of the Amazon and their environment: physiological and biochemical aspects, vol 32. Dordrecht, Netherlands, Springer Science & Business Media

Van der Sleen P, Albert JS (eds) (2017) Field guide to the fishes of the Amazon, Orinoco, Guianas, vol 115. Princeton, Princeton University, p 464

Waddell JC, Crampton WGR (2020) Environmental correlates of circannual breeding periodicity in a multi-species assemblage of Amazonian electric fishes. Environ Biol Fishes 103:233–250

Waddell JC, Crampton WGR (2022) Reproductive effort and terminal investment in a multi-species assemblage of Amazon electric fish. Ecol Monograph 92:e1499

Waddell JC, Njeru SM, Akhiyat YM, Schachner BI, Correa-Roldán EV, Crampton WGR (2019) Reproductive life-history strategies in a species-rich assemblage of Amazonian electric fishes. PLOS One 14:e0226095

12 Neotropical Freshwater Fishes of Seasonally Hypoxic Waters

Waldir Heinrichs-Caldas, Derek Campos, Jaqueline Custódio da Costa, Daiani Kochhann, Susana Braz-Mota, Adalberto Luis Val and Vera Maria F. Almeida-Val

Abstract

The Amazon Basin consists of diverse and complex environments, with drastic seasonal and daily variations in dissolved oxygen concentration, resulting in hypoxic and anoxic zones, with long and short duration, which have led throughout the evolutionary process to various biological adaptations. Among fishes, these adaptations occur at all levels of biological organization, including genetic, biochemical, physiological, respiratory, morphological, and behavioral traits. For example, genetic adaptations determine physiological responses, such as the strong modulation of metabolic adjustments, which can alter the O_2 uptake by the gills (MO_2) and O_2 delivery to the body tissues. Limiting oxygen in the cell can increase the lactate dehydrogenase enzyme activity, shifting the metabolism from aerobic to anaerobic, concomitantly to a metabolic depression, a strategy used by the most hypoxia-tolerant Amazon species. At the same time, the oxygen limitation presented by the Amazon environments can become limiting, and ATP production via mitochondrial oxidative phosphorylation (OXPHOS) is compromised. In addition, changes in animal physiology can lead to behavioral adaptations such as changes in dominance hierarchies, decreased activity level, use of hypoxic zones as a refuge against predators, use of aquatic surface respiration, and changes in opercular movements. Finally, responses to hypoxia can be influenced by other parameters, such as immune responses, parasitism, and temperature. Here, we will present a brief review of Amazonian fish adaptation to hypoxia, from genes to behaviors, and how immunity, parasitism, and temperature may affect these responses.

W. Heinrichs-Caldas · J. C. da Costa · S. Braz-Mota · A. L. Val (✉) · V. M. F. Almeida-Val
Instituto Nacional de Pesquisas da Amazônia, Manaus, AM, Brazil
e-mail: dalval.inpa@gmail.com

W. Heinrichs-Caldas
e-mail: dinho_c@hotmail.com

S. Braz-Mota
e-mail: susanabmota@gmail.com

V. M. F. Almeida-Val
e-mail: veraval30@gmail.com

D. Campos
College of Agricultural and Veterinary Sciences, São Paulo, SP, Brazil
e-mail: campos.derek@gmail.com

D. Kochhann
Acaraú Valley State University, Sobral, CE, Brazil
e-mail: daia.kochhann@gmail.com

12.1 Introduction

Greater Amazonia encompasses a vast mosaic of landscapes and riverscapes shaped by geomorphological processes over thousands to millions and even billions of years. Geological and climatic processes have guided the evolution of the landforms and river basins, the emergence of the forests, and the biodiversity of the aquatic and terrestrial ecosystems that we see today. The same geological and climatic processes that molded the abiotic environment also imposed adaptive pressures on the rich biodiversity of Amazonian ecosystems. When it comes to fishes, changes in the physical and chemical conditions of the waters are important drivers of evolutionary processes, affecting the processes of dispersal, adaptation, speciation, and extinction over geological time scales (Chap. 8). Extreme environmental conditions operating over millions of years have affected how some species and clades have survived while others have gone extinct. Adaptive changes in genetic, biochemical, physiological, and behavioral traits associated

J. S. Albert and R. E. Reis (eds.), *Ecology and Evolution of Amazonian Fishes*, Fascinating Life Sciences,
https://doi.org/10.1007/978-3-032-09328-8_12

with respiration, locomotion and reproduction have enabled some groups to survive and diversify under extreme abiotic environmental conditions of low pH and low oxygen.

The Amazon basin includes many extreme environments, with daily and seasonal changes in physical-chemical conditions. The aquatic environments of the Amazon can be divided into rivers, *igapós* (blackwater flooded forest), *várzeas* (whitewater floodplain), lakes, beaches, and small forest streams, with each environment having a distinct set of conditions. The water temperature, pH and concentration of dissolved oxygen differ among these environments, and vary daily and seasonally according to the annual flood pulse, which initiates several physiological responses in the animals living in these environments.

Even though there are several other stressors, this chapter will address the impacts of oxygen and temperature variation in the Amazon basin on fish adaptation and responses, starting from evolutionary processes that led to behavioral, physiological, biochemical, and genetic adaptation. Furthermore, we aim to address how oxygen availability in the Amazon can be related to the oxygen-thermal tolerance hypothesis of oxygen capacity limitation, and how immune responses can affect hypoxia-related metabolic changes.

12.2 Oxygen

When compared to fishes of the temperate zone, Amazonian fishes experience greater variations in oxygen availability. Hemoglobin is vital when it comes to oxygen and carbon dioxide transport, and several adjustments are related to its function, among which are adjustments to hemoglobin's affinity for oxygen.

12.2.1 Adaptations to Hypoxia

Tropical aquatic environments, especially freshwater ones, have complex interactions between the physical, chemical, and biological parameters. These interactions include the decomposition of organic matter, high photosynthetic rates, bacterial respiration, interactions between animals and plants, light penetration, wind incidence, and amounts of minerals in the soil (reviewed by Val and Almeida-Val 1996). This set of biotic and abiotic parameters promotes conditions of hypoxia (low oxygen concentration), or even anoxia, that are often quite prolonged, which creates a challenge for the fishes that inhabit this environment, since they involve a limiting element to aerobic metabolism; i.e., oxygen. This situation becomes even more critical with the formation of harmful gases, such as hydrogen sulfide and methane, which are formed under anaerobic environmental conditions. Associated with this, the predicted climate change events could bring about the potentiation of these hypoxic conditions via increasing temperature and carbon dioxide (CO_2; Collins et al. 2019; IPCC 2014). Together, these changes set the stage for an extremely challenging environment for freshwater tropical fishes.

Even in view of the environmental peculiarities inherent to this extreme scenario, especially with regard to hypoxic conditions, tropical fish have morphological and/or physiological/behavioral adaptations aimed at better use of the surface of the water, which is richer in oxygen, such as aquatic surface respiration (ASR) (described in more details below) whether associated or not with morphological alterations, such as lip expansion (Braum and Junk 1982), or even better capture of atmospheric air (fish with accessory or obligatory air-breathing). In addition, physiological strategies to compensate for low oxygen levels are also observed, which usually occur with exclusively aquatic-breathing fish, such as gill tissue remodeling, increased glycolytic metabolism with increased blood glucose levels, and increased anaerobic metabolism with increased plasma lactate levels.

12.2.2 Perception and Uptake

During hypoxic events, in order to activate survival strategies, fish must first perceive a lack of oxygen in the environment. For this, these animals have well-developed sensors, the chemoreceptors, responsible for monitoring the O_2 tension (PO_2) of water taken in and/or arterial blood. O_2 chemoreceptors are mainly found in the gills, but can also be located in the orobranchial cavity or arteries in other parts of the body (Milsom 2012). In fish, these chemoreceptors are mainly associated with endoderm-derived neuroepithelial cells (NECs) in the gills (Porteus et al. 2012; Hockman et al. 2017). Since they are able to detect changes in the chemical composition of the external (water) and internal (blood) environments, the activation of these chemoreceptors is the first step in a cascade of events that lead to cardiorespiratory and metabolic adjustments. The main responses resulting from the activation of these receptors are (i) increased branchial ventilation (V_G), (ii) increased heart rate (f_H), (iii) behavioral changes (escape to more oxygenated environments, ASR, facultative air-breathing), and (iv) hormonal alterations (Gilmour and Perry 2018).

Some of these receptors may also be sensitive to alterations in the levels of respiratory gases other than oxygen, such as carbon dioxide (CO_2) and ammonia (NH_3) or even H+ (Perry et al. 2009; Milsom 2012; Florindo et al. 2018). This is critical for many tropical fish that live in extreme environmental conditions. Thus, the ecological success of these animals depends, among other factors, on their ability to activate these receptors and quickly engage and sustain metabolic, cardiovascular and ventilatory responses that adjust the O_2 supply according to their demand (Fritsche et al. 1993). Gill functions play an important role in adapting the perception of the absence of

O_2 since they are controlled by NECs through the paracrine production of serotonin and synaptic relationships with the branchial nervous system (Bailly 2009; Zaccone et al. 2017). Thus, in many cases, NECs appear to respond immediately, and in a dose-dependent manner, to changes in local PO_2, and the distribution of these cells in the gill tissue may be an important adaptation mechanism for tropical species subjected to hypoxia.

For the tropical Amazonian fish *Hoplerythrinus unitaeniatus*, which has facultative air respiration, the O_2 receptors involved in inducing cardiorespiratory responses to hypoxia are present in all the gill arches (Lopes et al. 2010). In this species, there is evidence that externally and internally oriented O_2 chemoreceptors mediate branchial ventilation and cardiovascular responses, while internally oriented O_2 chemoreceptors trigger air respiration responses. Similarly, tropical fish such as pacuguaçu (*Piaractus mesopotamicus*) and tambaqui (*Colossoma macropomum*) have receptors located in all the gill arches, and also in extrabranchial regions (Milsom et al. 2002; Leite et al. 2007). Thus, it is possible that hypoxic stimulation of NECs, both branchial and extrabranchial, promotes adaptive physiological reflexes, which allow the effective uptake and delivery of O_2 to meet metabolic demands, especially in tropical fish that inhabit regions that are frequently hypoxic. As a result, these receptors are responsible for the activation of ventilatory compensation mechanisms, which will act to regulate both the flow of water through the gills and the flow of blood through the secondary lamellae and, as such, ensure an efficient ventilation/perfusion ratio through the gas exchange surface (Nilsson and Sundin 1998).

As a consequence of the activation of these hypoxia response chemoreceptors, the increase in branchial ventilation (V_G) occurs in order to allow an adaptation between the animal's O_2 demand and the adequate uptake of this gas as well as its distribution to the tissues. Tolerant fish usually show an increase in V_G (hyperventilation) associated with a greater increase in ventilatory amplitude (V_T) when compared to an increase in respiratory rate (f_R) (Rantin et al. 1992). This compensatory mechanism seems to be less costly from the energy point of view and is often used by tropical fish. This is because, in general, fish respond to environmental hypoxia with an increase in f_R and/or V_T to raise the volume of water passing through the secondary gills. Concomitantly, a reduction in heart rate occurs (f_H).

12.2.3 Transport and Delivery

The uptake of oxygen from the environment is dependent on the branchial ventilation, the perfusion capacity of the gills or the accessory air respiration organ, and the diffusion of oxygen through the branchial epithelium (Randall and Daxboeck 1984). Although the hypoxic environment is one of the main characteristics of tropical environments, the fish that live in these habitats possess a series of adaptations to maintain aerobic metabolism whenever possible (Val et al. 2016). In addition to the cardiorespiratory adjustments mentioned earlier, an increase in hematocrit is generally associated with a higher affinity to oxygen (Brauner and Val 2005; Nilsson and Östlund-Nilsson 2005). In general, lower values of P_{50} (the oxygen tension necessary to saturate 50% of hemoglobin with oxygen) are found in aquatic respirators compared to aerial respirators, and this is supported by works on amphibians and aerial-breathing fish. However, data for fish come predominantly from studies of tropical species, such as from the Amazon, where aquatic hypoxia can confound the observed relationship (Johansen et al. 1978; Morris and Bridges 1994; Powers et al. 1979; Wells et al. 2005; Yang and Albright 1992). Graham (1997) provides an excellent synthesis on the subject, and his analysis supports the above conclusion that many aquatic breathers have lower P_{50} values when compared to closely related aerial-breathing fish.

However, this relationship is less consistent if applied more broadly and with increasing phylogenetic distance. Based on limited evidence, it has been suggested that species that are likely to encounter hypoxia have lower intrinsic Hb P_{50} values compared to species that inhabit O_2-rich environments, and that athletic species have higher intrinsic Hb P_{50} values compared to slower species (Jensen 2004; Mairbaurl and Weber 2012; Mislan et al. 2017). Consequently, oxygen uptake and delivery will influence several other responses to hypoxia and changes in temperature, as follows in the next topics.

12.3 Behavioral Responses and Ecological Consequences

The physiological adaptations of Amazonian fish species to low concentrations of environmental oxygen are much better described and known than their behavioral adaptations. The activation of NECs initially triggers a series of behavioral responses. The main objective of these responses is to promote: (i) a decrease in oxygen demand and (ii) an increase in oxygen uptake.

Individual behavioral responses that aim to reduce oxygen demand are described in less detail for Amazonian species. A simple and effective way to reduce oxygen demand is to reduce swimming activity. Reduction in swimming activity during exposure to hypoxia has been described in the species *Apistogramma agassizii* (Kochhann et al. 2015), *Schizodon fasciatus, Ctenobrycon. hauxwellianus, Colossoma macropomum*, and *Parauchenipterus galeatus* (Soares et al. 2006). In *A. agassizii*, known to demonstrate strong agonistic behavior, a decrease in aggressive interactions was also observed (Kochhann et al. 2015).

12.3.1 Air and Aquatic-Surface Respiration

As reviewed by Braz-Mota and Almeida-Val (2021), in many moments, the Amazonian aquatic environment has oxygen available only in the most superficial layer, which is in contact with the aerial environment and, therefore, has oxygen transfer by diffusion. This region is very narrow, usually around two centimeters. Fish without adaptations to aerial breathing can make use of this thin layer to increase oxygen uptake when under hypoxic conditions. The use of this surface layer of water is called aquatic surface respiration (ASR) and is known in a multitude of Amazonian species. Species that perform ASR may or may not have morphological adaptations that facilitate the uptake of this surface layer. Regardless of the presence or absence of these morphological adaptations, the use of the surface layer of water improves the oxygen uptake, thus increasing the tolerance of the species to hypoxia (Kramer and McClure 1982). For example, in tambaqui, performing ASR increases blood oxygenation by 30% when the species is exposed to hypoxia (Val 1995).

The use of the ASR has as negative points the increased susceptibility to predation (mainly by aerial predators) and a decrease in foraging opportunities. Increased susceptibility to predation by aerial predators during ASR or aerial respiration was observed under laboratory experimental conditions by Kramer et al. (1983) using six species of fish with aquatic or bimodal respiration (which performed ASR or aerial respiration). The authors observed that the risk of predation by the striated heron (*Butorides striata*) increased the closer the fish was to the water's surface. They also observed antipredatory responses to the presence of the predator, such as a decrease in the rate of ASR or aerial respiration and the use of deeper regions of the aquarium. Riesch et al. (2010) also observed a higher rate of predation by birds when poecilids perform ASR. However, another study with *Poecilia mexicana* and *P. sulphuraria* reported no increase in predation risk related to increased frequency of ASR (Tobler et al. 2009). The same authors observed that the time spent on ASR is inversely proportional to the time spent foraging and, therefore, there is a trade-off between these behaviors (Tobler et al. 2009) in environments with a high concentration of hydrogen sulfide gas.

However, these hypotheses have not yet been tested in Amazonian species. It is possible that, as hypoxia exerts an important selective pressure on Amazonian species, they have developed strategies to deal with the possible problems of a longer period performing ASR or aerial breathing. An interesting example is the *Mesonauta festivus*, one of the most common and well-distributed species of Amazonian cichlids, which occupies mainly the upper layer of water bodies. Pires et al. (2015) observed that, when disturbed from its rest, the species mimics a leaf sinking into the water, keeping its body with its head facing up and its fins close to the body. This posture, along with its black-banded coloration pattern, causes the fish to look like a leaf. This behavior could occur when, for example, an aerial predator approaches during the moment in which the species is performing ASR, and could be an anti-predation strategy of the species. Another important point is that, at many times, Amazonian water bodies have low visibility, especially due to the large number of suspended solids, which is a factor that can also interfere in predator–prey relationships, and modulate prey behavior (Shingles et al. 2005) while also decreasing the visual acuity of the predator (Cezilly 1992).

One strategy that is known to reduce predation risk in social species during the use of upper part of the water column for respiration (either ASR or aerial) is synchronous ascent to the surface, which would generate a dilution effect of predation risk, similar to that which occurs in the formation of shoals (Kramer and Graham 1976; Gee 1980; Baird 1983; Chapman and Chapman 1994). For Amazonian fish, this synchronicity was observed in the species *Hoplosternum littorale* (Sloman et al. 2009), and controversial results were observed in *Corydoras aeneus* (Graham 1997; Kramer and McClure 1980; Pineda et al. 2020). It is likely that many Amazonian species use this strategy to reduce individual predation risk when using air or the air-water interface to breathe.

12.3.2 Behavior and Community Structure

Very few studies have investigated the effects of hypoxia on the behavior of Amazonian fish. Individual behavioral responses to hypoxia can lead to responses at the community level since they can alter relationships between species. One example of this is the modifications caused by hypoxia in predator-prey relationships, since hypoxia mainly interferes with the escape capacity of prey fish and decreases cohesion and coordination of fish in shoals (reviewed by Domenici et al. (2007, 2017)). However, none of these studies have been conducted with Amazonian species, and this is a vast area open to investigation.

Although physiologically challenging, hypoxic environments can offer advantages to species that manage to occupy them. One of the supposed advantages of occupying hypoxic environments would be the reduction of predation pressure since large predatory fish would be less tolerant to hypoxia (Junk et al. 1983; Reid et al. 2013). Anjos and collaborators (2008) found the same species diversity, though fewer individuals of these species, in hypoxic environments when compared to normoxic environments in a floodplain lake, thus corroborating this hypothesis. The authors argue that hypoxic environments may be especially important for the survival of juveniles of non-predatory species, which contributes to the maintenance of these species. In this way, the development of adaptations to hypoxia would end up being favorable.

Migration is a behavior that can occur in response to hypoxic environments, and is one of the main determinants of the composition of tropical fish communities of floodplain lakes (Henderson and Crampton 1997). Röpke et al. (2016) observed that the composition of the fish community changes seasonally in a floodplain lake. During the dry period, when the water becomes warmer and oxygen concentration decreases dramatically, the fish species that dominate the environment are those that present efficient strategies to survive in hypoxic environments. The behavioral mechanisms that trigger the outflow response of species less tolerant to hypoxia from floodplain lakes during the drought period have not yet been investigated.

Changes in behavior and communities' structure are sustained, and also determine modifications in the cellular metabolism, whether the animals can or cannot escape from the hypoxic environment, relying on cellular responses to survive.

12.4 Cellular Metabolism

To cope with the natural variations in dissolved oxygen, aquatic organisms have evolved diverse cellular response mechanisms. In the case of Amazonian fish, genetic adaptations, enzymatic activity, changes in metabolites, and their energy source play a fundamental role in the survival of the animal (Fig. 12.1). The expression of the *hif-1α* (hypoxia-inducible factor-1α) gene controls cellular metabolism, together with the activity of several enzymes, such as lactate dehydrogenase (LDH) and citrate synthase (CS), thus determining the energy source to be used (lipids, proteins and carbohydrates). These adjustments are species-specific in Amazonian fish and may be related to the evolution of the differentiation of the LDH enzyme.

The expression of *hif-1α* gene is linked to these deficiencies caused by a lack of oxygen since it is an immediate response to this condition of low availability of oxygen (Maxwell et al. 1993). When oxygen levels are limited, the cell signals the production of hypoxia-inducing factors (HIF) proteins. HIF-1α is the hypoxia-related subunit and, under hypoxic conditions, the HIF -1α protein is preserved and transported to the cell nucleus, where it polymerizes with the *hif-1β* subunit to produce the HIF-1 protein. This protein acts immediately in the cell and its role is to regulate several genes, including those linked to the breakdown of glucose. This gene also activates important pathways for the maintenance of oxygen homeostasis, increasing vascularization in the areas most affected by hypoxia and changing the cell's metabolism from aerobic to anaerobic to maintain energy production in the form of ATP in the absence of oxygen (Semenza 2000; Wenger 2002; Bruick 2003; Kajimura et al. 2005).

Hif-1α subunit expression is critically regulated by oxygen uptake and promotes anaerobic metabolism (Semenza et al. 2008). Studies of Amazonian fish have already demonstrated this important relationship between *hif-1α* gene expression and tolerance to hypoxia, and its control of anaerobic metabolism in different tissues. Baptista et al. (2016), when evaluating the expression of *hif-1α* in the liver of *Astronotus ocellatus*, observed that the levels of mRNA transcripts increased during hypoxia and returned to normal levels after reoxygenation,

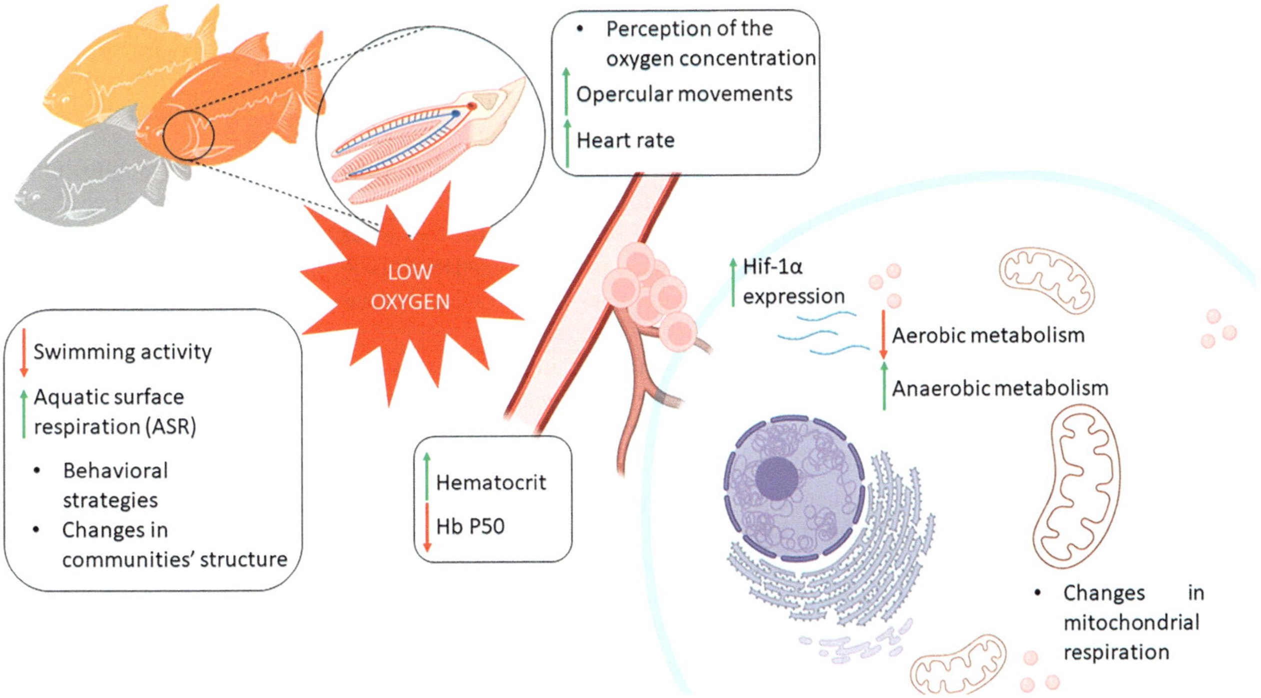

Fig. 12.1 The changes in Amazonian fishes when exposed to hypoxia go from behavioral to gene expression, changing oxygen uptake, hematocrit, gene expression and others

which suggests that the positive regulation and stabilization of this gene is the mechanism by which *A. ocellatus* regulates its homeostasis. Similar results were also observed by Heinrichs-Caldas et al. (2019) in *Astronotus crassipinis* after 5 h of exposure to hypoxia, in addition to demonstrating a direct relationship with the activity of LDH CS enzymes. In addition, Heinrichs-Caldas et al. (2019) demonstrated that there was no change in the expression of this gene in the white muscle of the animal, and the same was observed by Vasconcelos-Lima et al. (2021) in the gills of *A. ocellatus*. These results demonstrate that the expression of this gene is species- and tissue-specific, and has a strong relationship with the animal's tolerance to oxygen variations, its life history and the evolution of these species, and thus playing a fundamental role in the regulation of metabolism and has a direct relationship with the enzymes responsible for such regulation, such as the enzyme lactate dehydrogenase.

The LDH enzyme is the main controller of cellular anaerobic metabolism. It catalyzes the conversion of pyruvate to lactate and regulates the levels of these metabolites according to the availability of oxygen in the cell depending on the active isoform of this enzyme. Several studies of Amazonian fish species have already demonstrated the increased activity of this enzyme in situations of hypoxia, demonstrating a strong relationship with the tolerance of these animals to this stressor (Heinrichs-Caldas et al. 2019; Chippari-Gomes et al. 2005). Heinrichs-Caldas and Almeida-Val (2021) demonstrated that the regulation of the activity of this enzyme is related to local adaptation, life history, and tolerance of two species of Cichlidae to natural oxygen variations, which demonstrates the importance of this enzyme for the control of cellular metabolism, in addition to its direct relationship with mitochondrial respiration.

The stress naturally exerted by some environments, such as floodplain lakes, in which there is a large variation in dissolved oxygen levels, leads to higher phenotypic plasticity. Almeida-Val et al. (1995) reported that the sensitivity of 15 species to hypoxia is related to the distribution of LDH isoforms in different tissues; for example, species from hypoxic environments showed a decrease in the expression of LDH-B in the heart, while species from normoxic waters present a regular distribution of LDH. In this study, *Cichlasoma amazonarum* demonstrated both patterns, with changes after 51 days of exposure to hypoxia, which resulted in a metabolic shutdown in this species and showed that LDH gene expression is regulated according to environmental conditions (Almeida-Val et al. 1995). These works show that the phenotypic plasticity observed in Amazonian cichlids, and also in other fish species, can be considered the key that allows animals to survive chronic and periodic hypoxia, and it involves a range of mechanisms, from genetic to biochemical, that play a cascade role for the control of cellular metabolism, and, as pointed by some new research, mitochondrial respiration control plays a pivotal role in low oxygen survivability.

12.5 Mitochondrial Metabolism

Although studies on mitochondrial respiration in Amazonian fish are scarce, it is known that fish respiration and ROS production play a key role in adaptation to hypoxia. Some studies in Amazonian fish have already shown that changes in mitochondrial respiration in situations of hypoxia are species-specific, that they vary with the natural availability of dissolved oxygen, and that they are related to the animal's tolerance to hypoxia.

One recent study with Amazonian fish demonstrated that mitochondrial respiration of cardiac tissue is related to the respiratory strategy of the studied fish. High respiration is coupled to heart fibers, especially in the species *Erythrinus erythrinus*, thus indicating a high energy demand in this species of facultative air respiration (Campos et al. 2020).

With exposure to hypoxia, several metabolic mechanisms are activated, starting from the increase in aerobic or anaerobic metabolism of glycolysis and leading to the regulation of mitochondrial activity, a factor that has been shown to be determinant in the tolerance of several species to hypoxia (Solaini et al. 2010; Hickey et al. 2012; Du et al. 2016). Heinrichs-Caldas and Almeida-Val (2021), when studying two species of Cichlidae with different levels of tolerance to hypoxia (*Aequidens pallidus* and *M. festivus*), demonstrated a greater regulation of mitochondrial capacity in liver tissue in *A. pallidus*, with a lower P_{50} and higher mitochondrial catalytic efficiency, thus resulting in a greater ability of mitochondria to maintain respiration at low oxygen levels to compensate for the inability of this species to increase its anaerobic metabolism. This work also demonstrated that there is a dependence on mitochondrial respiration by *A. pallidus* in order to survive hypoxia, and that its low tolerance is due to adaptation to a normally normoxic environment.

Braz-Mota et al. (in prep.) evaluated the mitochondrial respiration of several Amazonian fish species in two phases of the flood pulse, low-water and high-water, in which dissolved natural oxygen levels range from normoxic to anoxic, depending on the period of the year. In this study, it was observed that mitochondrial respiration in cardiac tissue was lower, in general, in animals in period of low water, when compared to the high-water period, when there is less availability of dissolved oxygen in the water. Therefore, it is concluded that, for the species studied, there is an increase in oxygen delivery capacity to the tissues, which reflects in the increase in mitochondrial activity in the period in which dissolved oxygen is scarcer, that, when combined with greater control of ROS production in this period, allows the maintenance of cardiac function, increased ATP production, and survival in periods of severe hypoxia.

These mitochondrial respiration studies only scratch the surface of what can still be studied in Amazonian fish, such as the relationship with other stressors, the biochemical and physiological responses triggered by modifications in mitochondria, in addition to exploring other species and environments with different physicochemical characteristics.

12.6 Oxygen-Thermal Tolerance Hypothesis

It has been predicted that climate change will exacerbate the already extreme temperature and oxygen conditions in the Amazon region, so understanding the limits of temperature and hypoxia tolerance is extremely important in order to be able to predict the level of vulnerability of species. The critical thermal maximum (CT_{max}), typically determined as the temperature at which animals exhibit a loss of equilibrium (LOE) (Beitinger and Bennett 2000), defines the upper limit of the fundamental thermal niche of a species and is the temperature at which the collapse of one or more vital physiological functions will occur (Pörtner 2010); while the limit of tolerance to hypoxia determines the concentration or survival time when animals exhibit a LOE_{hyp}.

Both CT_{max} and LOE_{hyp} have been linked to a collapse in efficient oxygen delivery to tissues. Oxygen delivery capacity refers to the maximum capacity of the cardiorespiratory system to deliver oxygen to tissues by increasing ventilation, cardiac output, and oxygen transport in the blood. In recent decades, studies have reported a failure in the cardiorespiratory system and an accumulation of anaerobic metabolites in ectotherms when approaching their thermal and hypoxic limits (Frederich and Pörtner 2000; Mark et al. 2002; Melzner et al. 2009; Wittmann et al. 2008). This led to the hypothesis that, when they approach their CT_{max} and LOE_{hyp}, these organisms are unable to maintain sufficient oxygen supply for basic metabolism as a result of oxygen supply inability that is induced by cardiorespiratory restrictions, either by high demand at high temperatures or supply restriction in situations of hypoxia. Therefore, upper thermal limits and hypoxia may be physiologically interconnected, and a worsening of these situations as a result of climate change could be devastating for Amazonian species, since they are already close to their physiological limits (Campos et al. 2021).

Increases in temperature lead to an increase in energy demand, thus fish require a greater supply of oxygen to perform their normal physiological functions. Therefore, a reduction in tolerance to hypoxia is expected in species exposed to elevated temperatures. In fact, Jung et al. (2020), who studied tolerance to hypoxia in 37 species acclimated to 28°C, 31°C, and 33°C, found that all species reduce LOE_{hyp} with increasing temperature. On average, the species showed a 20% reduction in tolerance to hypoxia with each degree of elevated temperature. In addition, the higher the acclimation temperature, the more the tolerance to hypoxia decreases, which indicates a cumulative effect of the temperature increase. The lower tolerance to hypoxia at high temperatures may be related to increased energy demand and an inability to physiologically adjust for adequate oxygen supply.

As in hypoxia, in fish, a reduction in oxygen supply capacity at supra-optimal temperatures has been correlated with a reduction in cardiac output, which leads to tissue hypoxemia and, consequently, to the thermal limit. Therefore, in species where upper thermal limits are determined by insufficient oxygen supply for vital physiological functions, any reduction in oxygen availability should reduce CT_{max}. The upper thermal limits of these species can thus be described as oxygen-dependent. However, the link between oxygen supply capacity has recently been questioned, since some aquatic breathing species show alterations in CT_{max} only when exposed to severe concentrations of hypoxia, below the PO_{2crit} (critical oxygen tension). It has therefore been suggested that some species have cardiac systems that are more thermo-resistant and that insufficient oxygen supply is not the main determinant of thermal limits. In fact, the effect of oxygen concentration on CT_{max} in Amazonian species appears to be species-specific. Jung et al. (2020) showed that four of the seven temperatures increased thermal tolerance when exposed to 200% oxygen, which suggests that these species have a CT_{max} that is oxygen-dependent; however, three other species showed no alterations when exposed to the same situations. The mechanisms that may or may not make a species dependent on oxygen have been determined, and should be studied in order to understand which species are most susceptible.

In conclusion, the alterations as a result of climate change will affect Amazonian species in different ways. The increase in the average temperature of Amazonian waters will affect the tolerance to hypoxia of species, and a reduction in oxygen concentration will affect some, but not all species. Species with higher oxygen dependency have a higher susceptibility to common events of low oxygen concentration and high temperatures in the face of climate change. In this sense, facultative air-breathing species are less dependent on oxygen concentration and are therefore less susceptible to future alterations.

12.7 Parasitism and Immune Responses

Exposure to oxygen variation modulates oxygen uptake, animal behavior, and cellular and mitochondrial metabolism, which leads to energy demand modifications. The immune response is a factor that increases energy demand; therefore, we propose a model of possible impacts on the immunity of Amazonian fish when exposed to variations in dissolved oxygen (Fig. 12.2).

The immune system of teleost fish is well developed and has elements of both innate and adaptive immunity. As in other vertebrates, the immune system protects fish from threats by pathogens such as viruses, bacteria, and parasites that can establish themselves and cause disease (Alvarez-Pellitero 2008; Workenhe et al. 2010). However, disturbances

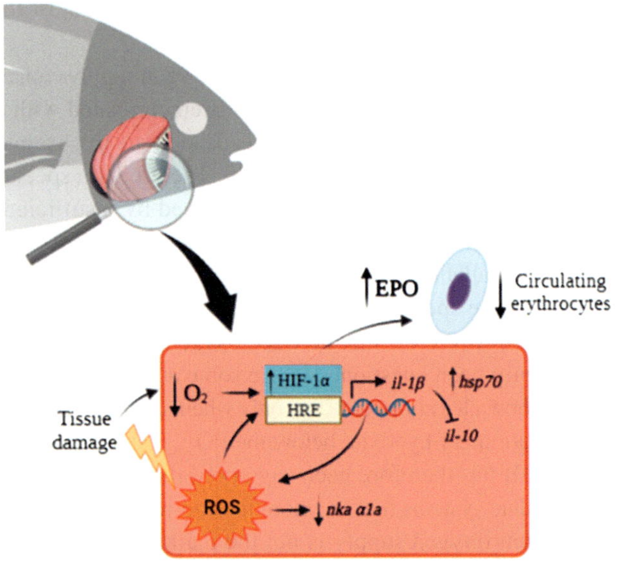

Fig. 12.2 Suggested model of the effects of hypoxia and parasitism on the immunity of Amazonian fish species. Since parasitism causes tissue damage that can lead to a hypoxic microenvironment and generation of ROS, transcription activation of the *hif-1α* gene occurs, which can both increase erythropoietin (EPO) levels and decrease the amount of erythrocytes in the circulating blood, as well as bind to hypoxia response elements (HRE) and induce transcription of *il-1β*, which, in turn, can repress the expression of its *il-10* antagonist gene, leading to inflammation. Due to the stress generated by tissue damage, *hsp70* is also overexpressed and can act as a "chaperokine" inducing transcription of *il-1β*. Inflammation can also activate ROS and restart the cycle with new activation of *hif-1α*, as well as decrease the expression of Na+, K+-ATPase genes and enzymes, leading to another cascade of biochemical events detrimental to the ionic regulation of the animals

in environmental factors, such as changes in temperature, CO_2, and oxygen in the water, can affect how the immune system of animals is able to react to pathogens (Costa et al. 2021).

As discussed earlier, variations in dissolved oxygen levels in water can induce stress, which can affect the immune responses of fish. For fish that inhabit temperate climates, such as Atlantic salmon, it is known that acute hypoxia induces the expression of genes related to antiviral defense such as *mx* and *ifnα* (Rodríguez et al. 2016). While chronic and moderate intermittent hypoxia reduces the respiratory burst of leukocytes, which is an indicator of phagocytosis in the innate immune response (Burt et al. 2013). Furthermore, in salmon, chronic hypoxia can increase susceptibility to infection, in addition to modulating the expression of genes and macrophages, which can have negative consequences on the immune response (Kvamme et al. 2013).

As a consequence of a deficient immune response, due to the energy demand being concentrated in anaerobic metabolism under hypoxic conditions, fish tend to become susceptible to infectious diseases such as parasitism. Piazzon et al. (2019) observed that *Sparus aurata* parasitized by *Sparicotyle chrysophrii* (Platyhelminthes: Monogenea) in the gills not only presented down-regulation of *hif-1α*, a gene that mediates the adaptation of cells and tissues to low oxygen concentrations, but also showed that parasitism can, by itself, create a hypoxic micro-environment in the tissue site of infection. The expression of several genes related to immunity, such as inflammasomes and others linked to apoptosis, cell proliferation, and autophagy, was observed. Similarly, parasitism by the crustacean *Argulus bengalensis* in *Labeo rohita* also led to hypoxia in the host due to the caused tissue damage, leading to changes in erythropoietin (EPO) levels and deterioration of the leukocyte profile as a consequence of damage to the immune system (Datta et al. 2022).

In general, little is known about the effects of hypoxia and climate change in Neotropical fish species and even less about these effects in Amazonian species, especially in relation to the impact of these environmental disturbances on animal health. Recent studies by Costa and Val (2020) and Costa et al. (2021) suggest that, when exposed to climate change scenarios and parasitized by monogenes in the gills, the tambaqui (*C. macropomum*), also presents tissue hypoxia, which was observed through a decrease in circulating erythrocytes in the blood and elevated levels of *hif-1α*, and which may have influenced a cascade of inflammatory events in ROS-dependent mechanisms.

In addition to all the effects on the immune system, such as leukocytosis, thrombocytosis, increased expression of pro-inflammatory genes such as interleukin—1β (*il-1β*) and heat shock protein-70 (*hsp70*), inhibition of the anti-inflammatory gene *il-10*, followed by the activation of ROS, can lead to another biochemical cascade causing damage to ion regulation of animals, as observed by decreased expression of the gene Na+, K+-ATPase subunit α1a (*nka α1a*) and the enzyme Na+, K+-ATPase (Costa and Val 2020; Costa et al. 2021).

Taking into account that species exposed to high temperatures have reduced tolerance to hypoxia and that parasitism can also lead the host to tissue hypoxia, a scenario with reduced O_2 levels in water added to parasitism tends to make animals more susceptible to other diseases and to hypoxia itself. This is due to the energy demand being directed to maintain anaerobic metabolism, although a high demand is also being requested for the immune system, thus weakening the animal's defense capacity against pathogens.

12.8 Conclusion

Many Amazonian fishes are exposed to a wide range of extreme environmental conditions, including low oxygen levels (hypoxia and anoxia), and they have evolved various adaptations and behaviors to cope with these challenges. The advent of molecular biology and new technologies for studying comparative physiology indicates that we are only touching the tip of a huge unknown field. Here, we highlight what we have learned and what we still do not know about important aspects of fish adaptations to hypoxic tropical freshwaters.

Oxygen sensing and adaptations: Many Amazonian fishes have evolved specialized oxygen-sensing cells called *peripheral oxygen chemoreceptors* (POCs). These receptors are part of the peripheral nervous system located in the gills and blood vessels where they detect changes in oxygen concentrations and can trigger responses that facilitate acclimation to hypoxic conditions (Zachar and Jonz 2012; Janes et al. 2015). Some fish species alter their blood circulation patterns and adjust the affinity of hemoglobin for oxygen. However, we are still far from understanding the processes controlling these adjustments as they are not all-or-nothing.

Behavioral responses and ecological consequences: Fish often exhibit behavioral adaptations to cope with low oxygen levels. Some species are known to migrate to shallower waters, where oxygen concentrations are higher, during periods of hypoxia. This movement can have ecological consequences, affecting predator-prey interactions and overall fish community structure. Some species are able to skim the surface of the water column which is richer in oxygen and face more risk of predation. In addition, fish may alter their feeding behaviors and reproductive behaviors in response to oxygen availability. These behavioral responses regress when the animals return to adequate oxygen availability. We do not know about the long-term consequences of the behavioral changes related to hypoxia, and very little is known about how the different behavioral traits will interact and how animals will define priorities in their behavioral performance. Also, these are therefore behaviors driven by water quality, in this case dissolved oxygen, and evidently involve the synthesis of different proteins. We still know very little about the processes regulating the expression of these proteins in Amazonian fishes.

Respiratory strategies: Fish use various respiratory strategies to extract oxygen from the air or water surface when oxygen levels are low. This includes aerial respiration and surface water respiration. Some fish species have specialized structures such as labyrinthine organs, vascularization of the stomach/intestine, among others, that allow them to extract oxygen from the air. The oxygen pressure (PO_2) in these different structures is variable and evidently the functional characteristics of the blood need to be adjusted—there is much to discover here, particularly with regard to the regulation of the functional properties of the blood for oxygen uptake as well as for the reduction of oxygen loss by these structures when the environment returns to normoxia.

Cellular metabolism and mitochondrial adaptations: In low oxygen environments, fish often undergo metabolic changes to maintain energy production. They may switch to anaerobic metabolism to generate energy without relying on oxygen. However, anaerobic metabolism can lead to the accumulation of lactate, which can be harmful if not eliminated efficiently. Fish also exhibit mitochondrial adaptations to increase their efficiency in utilizing oxygen. Recent and more sensitive technologies have shed new light on these processes. However, much remains to be described, mainly related to environmental conditions involving climate change and water pH.

Oxygen and thermal tolerances: Low oxygen levels can interact with thermal tolerance, affecting the ability of fish to tolerate high temperatures. The oxygen and capacity limitation theory of thermal tolerance (OCLTT) suggests that oxygen availability influences the upper temperature limit that a fish can tolerate. When oxygen availability is reduced, a fish's thermal tolerance may also decrease. Considering the immense diversity of fish in the Amazon basin, there is still much to uncover about the relationship between thermal sensitivity and environmental oxygen availability.

Parasitism and immune responses: Low oxygen levels can weaken the immune system of fish, making them more susceptible to disease and parasitic infections. Stress caused by hypoxia can suppress immune responses, allowing parasites to proliferate more easily. Considering that Amazonian fish species are exposed to high temperatures and low oxygen levels, parasitism can also lead the host to tissue hypoxia, making the animals susceptible to other diseases. This can have cascading effects on fish health and ecosystem dynamics.

In general, fish from the Amazon basin have evolved a number of physiological, behavioral, and biochemical adaptations to survive in extreme conditions such as hypoxia. However, rapid environmental changes due to factors such as deforestation, pollution and climate change may render these adaptations ineffective and challenge the resilience of these fish populations. Understanding these adaptations and their limitations is crucial for fish farming and for the conservation of Amazonian aquatic ecosystems and their biodiversity (Domenici et al. 2007; Fritsche et al. 1993; Jensen et al. 2017; Johansson 1978; Melzner et al. 2009; Mislan et al. 2017; Randall et al. 1981; Sloman et al. 2009; Solaini et al. 2010; Val et al. 2015; Yang and Albright 1992).

References

Almeida-Val VMF, Farias IP, Paula-Silva MN, Duncan WP, Val AL (1995) Biochemical adjustments to hypoxia by Amazon cichlids. Braz J Med Biol Res 28:1257–1263

Alvarez-Pellitero P (2008) Fish immunity and parasite infections: from innate immunity to immunoprophylactic prospects. Vet Immunol Immunopathol 126:171–198

Anjos MB, De Oliveira RR, Zuanon J (2008) Hypoxic environments as refuge against predatory fish in the Amazonian floodplains. Braz J Biol 68:45–50

Bally Y (2009) Serotonergic neuroepithelial cells in fish gills: cytology and innervation. Airw chemorecept vertebr 61–97

Baird TA (1983) Influence of social and predatory stimuli on the air-breathing behavior of the African Clawed Frog, Xenopus laevis. Copeia 1983(2):411

Baptista RB, Souza-Castro N, Almeida-Val VMF (2016) Acute hypoxia up-regulates HIF-1α and VEGF mRNA levels in Amazon hypoxia-tolerant Oscar (Astronotus ocellatus). Fish Physiol Biochem 42:1307–1318

Beitinger TL, Bennett WA (2000) Quantification of the role of acclimation temperature in temperature tolerance of fishes. Environ Biol Fishes 58(3):277–288

Braum E, Junk WJ (1982) Morphological adaptation of two Amazonian characoids (Pisces) for surviving in oxygen deficient waters. Int Rev Hydrobiol 67(6):869–886

Brauner CJ, Val AL (2005) Oxygen transfer. Fish Physiol 21:277–306

Braz-Mota S, Almeida-Val VMF (2021) Ecological adaptations of Amazonian fishes acquired during evolution under environmental variations in dissolved oxygen: a review of responses to hypoxia in fishes, featuring the hypoxia-tolerant Astronotus spp. J Exper Zool A Ecol Integr Physiol 335:771–786

Bruick RK (2003) Oxygen sensing in the hypoxic response pathway: regulation of the hypoxia-inducible transcription factor. Genes Dev 17:2614–2623

Burt K, Hamoutene D, Perez-Casanova J, Kurt Gamperl A, Volkoff H (2013) The effect of intermittent hypoxia on growth, appetite and some aspects of the immune response of Atlantic salmon (Salmo salar). Aquac Res 45(1):124–137

Campos DF, Amanajás RD, Almeida-Val VMF, Val AL (2021) Climate vulnerability of South American freshwater fish: thermal tolerance and acclimation. J Exper Zool A Ecol Integr Physiol 335:723–734

Campos DF, Braz-Mota S, Heinrichs-Caldas W, Val AL, Almeida-Val VMF (2020) Control of heart mitochondrial respiration in Amazon fish with different respiratory strategies. In: Val, A. L., & De Boeck, G. (2017). PADAPTA—adaptations of aquatic biota of the Amazon. Hydrobiologia, 789(1), 1-5., vol. 2, pp 11–15

Cezilly F (1992) Turbidity as an ecological solution to reduce the impact of fish-eating colonial waterbirds on fish farms. Colonial Waterbirds 15(2):249–252

Chapman LJ, Chapman CA (1994) Observations on synchronous air breathing in Clarias liocephalus. Copeia 1994(1):246

Chippari-Gomes AR, Gomes LC, Lopes NP, Val AL, Almeida-Val VMF (2005) Metabolic adjustments in two Amazonian cichlids exposed to hypoxia and anoxia. Comp Biochem Physiol A Mol Integr Physiol 141(1):347–355

Collins M, Tills O, Turner LM, Clark MS, Spicer JI, Truebano M (2019) Moderate reductions in dissolved oxygen may compromise performance in an ecologically-important estuarine invertebrate. Sci Total Environ 693:133444

Da Costa JC, Val AL (2020) Extreme climate scenario and parasitism affect the Amazonian fish Colossoma macropomum. Sci Total Environ 726:138628

Da Costa JC, De Souza SS, Castro JDS, Amanajás RD, Val AL (2021) Climate change affects the parasitism rate and impairs the regulation of genes related to oxidative stress and ionoregulation of Colossoma macropomum. Sci Rep 11: 22350

Datta N, Kar PK, Saha SK (2022) Circulatory physiology and erythropoiesis in freshwater fish Labeo rohita experimentally parasitized by Argulus bengalensis. J Appl Ichthyol 38:63–72

Domenici P, Lefrançois C, Shingles A (2007) Hypoxia and the antipredator behaviours of fishes. Philos Trans R Soc B: Biol Sci 362(1487):2105–2121

Domenici P, Steffensen JF, Marras S (2017) The effect of hypoxia on fish schooling. Philos Trans R Soc B 372(1727):20160236

Du SNN, Mahalingam S, Borowiec BG, Scott GR (2016) Mitochondrial physiology and reactive oxygen species production are altered by hypoxia acclimation in killifish (Fundulus heteroclitus). J Exp Biol 219:1130–1138

Florindo LH, Armelin VA, McKenzie DJ, Rantin FT (2018) Control of air-breathing in fishes: Central and peripheral receptors. Acta Histochem 120(7):642–653

Frederich M, Pörtner HO (2000) Oxygen limitation of thermal tolerance defined by cardiac and ventilatory performance in spider crab, Maja squinado. Am J Physiol Regul Integr Comp Physiol 279(5):R1531–R1538

Fritsche R, Axelsson M, Franklin CE, Grigg GG, Holmgren S, Nilsson S (1993) Respiratory and cardiovascular responses to hypoxia in the Australian lungfish. Respir Physiol 94(2):173–187

Gee JH (1980) Respiratory patterns and antipredator responses in the central mudminnow, *Umbra limi*, a continuous, facultative, air-breathing fish. Can J Zool 58:819–827

Gilmour KM, Perry SF (2018) Conflict and compromise: using reversible remodeling to manage competing physiological demands at the fish gill. Physiology 33:412–422

Graham, J, 1997. Air-breathing fishes: evolution, diversity, and adaptation. Academic Press, New York

Heinrichs-Caldas W, De Almeida-Val VMF (2021) Hypoxia tolerance in two Amazon cichlids: mitochondrial respiration and cellular metabolism adjustments are result of species environmental preferences and distribution. Fish Physiol Biochem 47:1759–1775

Heinrichs-Caldas W, Campos DF, Paula-Silva MN, Almeida-Val VMF (2019) Oxygen-dependent distinct expression of hif-1α gene in aerobic and anaerobic tissues of the Amazon Oscar, *Astronotus crassipinnis*. Comp Biochem Physiol B Biochem Mol Biol 227: 31–38

Henderson PA, Crampton WGR (1997) A comparison of fish diversity and abundance between nutrient-rich and nutrient-poor lakes in the Upper Amazon. J Trop Ecol 13:175–198

Hickey AJR, Renshaw GMC, Speers-Roesch B, Richards JG, Wang Y, Farrell AP, Brauner CJ (2012) A radical approach to beating hypoxia: depressed free radical release from heart fibres of the hypoxia-tolerant epaulette shark (*Hemiscyllum ocellatum*). J Comp Physiol B 182:91–100

Hockman D, Burns AJ, Schlosser G, Gates KP, Jevans B, Mongera A, Baker CV (2017) Evolution of the hypoxia-sensitive cells involved in amniote respiratory reflexes. eLife 6:e21231

IPCC (2014) Climate Change 2014: Impacts, Adaptation, and Vulnerability. Intergovernmental panel on climate change.

Janes TA, Xu F, Syed NI (2015) Graded hypoxia acts through a network of distributed peripheral oxygen chemoreceptors to produce changes in respiratory behaviour and plasticity. Eur J Neurosci 42(2):1858–1871

Jensen FB, Kolind RA, Jensen NS, Montesanti G, Wang T (2017) Interspecific variation and plasticity in hemoglobin nitrite reductase activity and its correlation with oxygen affinity in vertebrates. Comp Biochem Physiol A Mol Integr Physiol 206:47–53

Johansson RS (1978) Tactile sensibility in the human hand: receptive field characteristics of mechanoreceptive units in the glabrous skin area. J Physiol 281(1):101–125

Jung EH, Brix KV, Richards JG, Val AL, Brauner CJ (2020) Reduced hypoxia tolerance and survival at elevated temperatures may limit the ability of Amazonian fishes to survive in a warming world. Sci Total Environ Dec 15;748:141349

Junk WJ, Soares GM, Carvalho FM (1983) Distribution of fish species in a lake of the Amazon river floodplain near Manaus (Lago Camaleão), with special reference to extreme oxygen conditions. Amazoniana 7:397–431

Kajimura S, Aida K, Duan C (2005) Insulin-like growth factor-binding protein-1 (IGFBP-1) mediates hypoxia-induced embryonic growth and developmental retardation. Proc Natl Acad Sci U S A 102:1240–1245

Kochhann D, Campos DF, Val AL (2015) Experimentally increased temperature and hypoxia affect stability of social hierarchy and metabolism of the Amazonian cichlid *Apistogramma agassizii*. Comp Biochem Physiol A Mol Integr Physiol 190:54–60

Kramer DL, Graham JB (1976) Synchronous air breathing, a social component of respiration in fishes. Copeia 1976:689

Kramer DL, McClure M (1980) Aerial respiration in the catfish, *Corydoras aeneus* (Callichthyidae). Can J Zool 58:1984–1991

Kramer DL, McClure M (1982) Aquatic surface respiration, a widespread adaptation to hypoxia in tropical freshwater fishes. Environ Biol Fishes 7:47–55

Kramer DL, Manley D, Bourgeois R (1983) The effect of respiratory mode and oxygen concentration on the risk of aerial predation in fishes. Can J Zool 61:653–665

Kvamme BO, Gadan K, Finne-Fridell F, Niklasson L, Sundh H, Sundell K et al. (2013) Modulation of innate immune responses in Atlantic salmon by chronic hypoxia-induced stress. Fish Shellfish Immunol 34:55–65

Leite CAC, Florindo LH, Kalinin AL, Milsom WK, Rantin FT (2007) Gill chemoreceptors and cardio-respiratory reflexes in the neotropical teleost pacu, Piaractus mesopotamicus. J Comp Physiol A 193:1001–1011

Lopes JM, De Lima Boijink C, Florindo LH, Leite CAC, Kalinin AL, Milsom WK, Rantin FT (2010) Hypoxic cardiorespiratory reflexes in the facultative air-breathing fish jeju (*Hoplerythrinus unitaeniatus*): role of branchial O_2 chemoreceptors. J Comp Physiol B 180:797–811

Mairbäurl H, Weber RE (2012) Oxygen transport by hemoglobin. Compr physiol 2(2):1463–1489

Mark FC, Bock C, Pörtner HO (2002) Oxygen-limited thermal tolerance in Antarctic fish investigated by MRI and (31)P-MRS. Am J Physiol Regul Integr Comp Physiol 283(5):R1254–R1262

Maxwell PH, Pugh CW, Ratcliffe PJ (1993) Inducible operation of the erythropoietin 3'enhancer in multiple cell lines: evidence for a widespread oxygen-sensing mechanism. Proc Natl Acad Sci 90(6):2423–2427

Melzner F, Gutowska MA, Langenbuch M, Dupont S, Lucassen M, Thorndyke MC, Bleich M, Pörtner HO (2009) Physiological basis for high CO_2 tolerance in marine ectothermic animals: pre-adaptation through lifestyle and ontogeny?. Biogeosciences 6:2313–2331

Milsom WK (2012) New insights into gill chemoreception: receptor distribution and roles in water and air breathing fish. Respir Physiol Neurobiol 184(3):326–339

Milsom WK, Reid SG, Rantin FT, Sundin L (2002) Extrabranchial chemoreceptors involved in respiratory reflexes in the neotropical fish *Colossoma macropomum* (the tambaqui). J Exp Biol 205(12):1765–1774

Mislan KAS, Deutsch CA, Brill RW, Dunne JP, Sarmiento JL (2017) Projections of climate-driven changes in tuna vertical habitat based on species-specific differences in blood oxygen affinity. Global Change Biol 23(10):4019–4028

Morris S, Bridges CR (1994) Properties of respiratory pigments in bimodal breathing animals: air and water breathing by fish and crustaceans. Amer Zool 34(2):216–228

Nilsson GE, Östlund-Nilsson S (2005) Hypoxia tolerance in coral reef fishes. Fish physiol 21:583–596

Nilsson S, Sundin L (1998) Gill blood flow control. Comp Biochem Physiol A Mol Integr Physiol 119(1):137–147

Perry SF, Esbaugh A, Braun M, Gilmour KM (2009) Gas transport and gill function in water-breathing fish. In: Cardio-respiratory control in vertebrates: comparative and evolutionary aspects, pp 5–42

Piazzon MC, Mladineo I, Naya-Català F, Dirks RP, Jong-Raadsen S, Vrbatović A, Hrabar J, Pérez-Sánchez J, Sitjà-Bobadilla A (2019) Acting locally - affecting globally: RNA sequencing of gilthead sea bream with a mild *Sparicotyle chrysophrii* infection reveals effects on apoptosis, immune and hypoxia related genes. BMC Genomics 20: 200

Pineda M, Aragao I, McKenzie DJ, Killen SS (2020) Social dynamics obscure the effect of temperature on air breathing in Corydoras catfish. J Exp Biol 223(21):jeb.222133

Pires THS, Campos DF, Röpke CP, Sodré J, Amadio S, Zuanon J (2015) Ecology and life-history of *Mesonauta festivus*: biological traits of a broad ranged and abundant neotropical cichlid. Environ Biol Fish 98:789–799

Porteus CS, Brink DL, Milsom WK (2012) Neurotransmitter profiles in fish gills: putative gill oxygen chemoreceptors. Respir Physiol Neurobiol 184(3):316–325

Pörtner HO (2010) Oxygen- and capacity-limitation of thermal tolerance: a matrix for integrating climate-related stressor effects in marine ecosystems. J Exper Biol 213(6):881–893

Powers DA, Martin JP, Garlick RL, Fyhn HJ, Fyhn UE (1979) The effect of temperature on the oxygen equilibria of fish hemoglobins in relation to environmental thermal variability. Comp Biochem Physiol A: Physiol 62(1):87–94

Randall DJ, Burggren WW, Farrell AP, Haswell MS (1981) The evolution of air breathing in vertebrates. Cambridge University Press. New York, USA.

Randall D, Daxboeck C (1984) 5 Oxygen and carbon dioxide transfer across fish gills. In Fish physiology (Vol. 10, pp. 263-314). Academic Press. https://doi.org/10.1016/S1546-5098(08)60321-0

Rantin FT, Kalinin AL, Glass ML, Fernandes MN (1992) Respiratory responses to hypoxia in relation to mode of life of two erythrinid species (*Hoplias malabaricus* and *Hoplias lacerdae*). J Fish Biol 41(5):805–812

Reid AJ, Farrell MJ, Luke MN, Chapman LJ (2013) Implications of hypoxia tolerance for wetland refugia use in Lake Nabugabo, Uganda. Ecol Freshw Fish 22:421–429

Riesch RW, Oranth A, Dzienko J, Karau N, Schiebl A, Stadler S, Wigh A, Zimmer C, Arias-Rodriguez L, Schlupp I, Plath M (2010) Extreme habitats are not refuges: poeciliids suffer from increased aerial predation risk in sulphidic southern Mexican habitats. Biol J Linnean Soc 101:417–426

Rodríguez FE, Valenzuela B, Farías A, Sandino AM, Imarai M (2016) β-1,3/1,6-Glucan-supplemented diets antagonize immune inhibitory effects of hypoxia and enhance the immune response to a model vaccine. Fish Shellfish Immunol 59:36–45

Röpke CP, Amadio SA, Winemiller KO, Zuanon J (2016) Seasonal dynamics of the fish assemblage in a floodplain lake at the confluence of the Negro and Amazon Rivers: assemblage dynamics in the central amazon. J Fish Biol 89:194–212

Semenza GL (2000) HIF-1: mediator of physiological and pathophysiological responses to hypoxia. J Appl Physiol 88:1474–1480

Semenza GL, Shimoda LA, Prabhakar NR (2008) Regulation of Gene Expression by HIF-1. In: Chadwick DJ, Goode J (Eds.)Eds. Novartis foundation symposia. John Wiley & Sons, Ltd, Chichester, UK, 2–14

Shingles A, McKenzie DJ, Claireaux G, Domenici P (2005) Reflex cardioventilatory responses to hypoxia in the flathead gray mullet (*Mugil cephalus*) and their behavioral modulation by perceived threat of predation and water turbidity. Physiol Biochem Zool 78:744–755

Sloman KA, Sloman RD, De Boeck G, Scott GR, Iftikar FI, Wood CM, Almeida-Val VMF, Val AL (2009) The role of size in synchronous air breathing of *Hoplosternum littorale*. Physiol Biochem Zool 82:625–634

Soares MGM, Menezes NA, Junk WJ (2006) Adaptations of fish species to oxygen depletion in a central Amazonian floodplain lake. Hydrobiologia 568:353–367

Solaini G, Baracca A, Lenaz G, Sgarbi G (2010) Hypoxia and mitochondrial oxidative metabolism. Biochim Biophys Acta BBA - Bioenerg 1797(6-7):1171–1177

Tobler M, Riesch RW, Tobler CM, Plath M (2009) Compensatory behavior in response to sulphide-induced hypoxia affects time budgets, feeding efficiency, and predation risk. Evol Ecol Res 11:935–948

Val AL (1995) Oxygen transfer in fish: morphological and molecular adjustments. Braz J Med Biol Res 28:1119–1127

Val AL, Gomes KRM, de Almeida-Val VMF (2015) Rapid regulation of blood parameters under acute hypoxia in the Amazonian fish *Prochilodus nigricans*. Comp Biochem Physiol A Mol Integr Physiol 184:125–131

Val AL, Fearnside PM, Almeida-Val VMF (2016) Environmental disturbances and fishes in the Amazon: environment and fishes in the Amazon. J Fish Biol 89:192–193

Val AL, Almeida-Val VMF (1996) Fishes of the Amazon and their environment: physiological and biochemical aspects. Zoophysiology, vol 32. Springer-Verlag, Berlin, Heidelberg

Vasconcelos-Lima JL, Oikawa-Cardoso VL, Heinrichs-Caldas W, Almeida-Val VMF (2021) Influence of hypoxia on biochemical aspects and on expression of genes related to oxygen-homeostasis of the Amazonian cichlid *Astronotus ocellatus* (Agassiz, 1831). Genet Mol Biol 44:e20210127

Wells RM, Baldwin J, Seymour RS, Christian K, Brittain T (2005) Red blood cell function and haematology in two tropical freshwater fishes from Australia. Comp Biochem Physiol A Mol Integr Physiol 141(1):87–93

Wenger RH (2002) Cellular adaptation to hypoxia: O_2-sensing protein hydroxylases, hypoxia-inducible transcription factors, and O_2-regulated gene expression. FASEB J 16:1151–1162

Wittmann A, Schröer M, Bock C, Steeger H-U, Paul R, Pörtner HO (2008) Seasonal patterns of thermal tolerance and performance capacity in lugworm (*Arenicola marina*) populations in a latitudinal cline. Clim Res 37:227–240

Workenhe ST, Rise ML, Kibenge MJT, Kibenge FSB (2010) The fight between the teleost fish immune response and aquatic viruses. Mol Immunol 47:2525–2536

Yang CZ, Albright LJ (1992) Effects of the harmful diatom Chaetoceros concavicornis on respiration of rainbow trout Oncorhynchus mykiss . Dis Aquat Organ 14:105–114

Zaccone G, Lauriano ER, Kuciel M, Capillo G, Pergolizzi S, Alesci A, Icardo JM (2017) Identification and distribution of neuronal nitric oxide synthase and neurochemical markers in the neuroepithelial cells of the gill and the skin in the giant mudskipper, *Periophthalmodon schlosseri*. Zoology 125:41–52

Zachar PC, Jonz MG (2012) Oxygen sensitivity of gill neuroepithelial cells in the anoxia-tolerant goldfish. In: Arterial chemoreception: from molecules to systems. Springer, Netherlands, pp 167–172

Fishes of Amazonian River Channels

13

Aaron H. Fronk, Maxwell J. Bernt, Tiago P. Carvalho, William G. R. Crampton and James S. Albert

Abstract

The main channels of large lowland South American rivers support a distinctive and diverse assemblage of specialized fishes, critically important for the aquatic ecology and human economy of the region. These waterways provide essential routes for transportation and commerce, and support important fisheries upon which local communities depend for food and livelihoods. Deep Amazonian river channels are home to a diverse fish fauna of at least 844 species in 368 genera and 68 families, representing just over one-quarter (28.1%) of all fish species of Greater Amazonia, all in a total habitat footprint of only c. 3500 km^2. Many of these species exhibit phenotypes usually associated with caves and other hypogean waters, including reduced eyes and pigmentation. The tremendous volume, depth, and remoteness of Amazonian river channels make them challenging and even dangerous to sample and study, and most deep channel habitats of the great rivers of South America remain unmapped and poorly unexplored. Species were scored as inhabiting small (SO1-5) or large (SO6-10) river channels, and high-conductivity whitewater or low-conductivity black or clear waters. Ancestral state estimation and evolutionary rates for each of the Big Nine NFF clades were estimated by fitting data to three models of trait evolution (equal-rates, symmetric-rates, or all rates different models), and best-fit models were calculated using the Akaike Information Criterion. Trait evolution was mapped to the phylogenetic tree using software implementing Bayesian stochastic mapping to estimate the timing of character changes along branches. All clades have estimated evolutionary origins in small rivers of the Guiana and Brazilian shields. Cichlinae, Characoidea, and the two cyprinodontiform clades (Poecilioidea and Rivulidae) are either rare or absent in the deep channels and were not considered further. The other five of these clades exhibit multiple evolutionary transitions to and from deep channel habitats. Among the Big Nine NFF clades, Pimelodoidea and Aspredinoidea have estimated origins in nonchannel habitats, Curimatoidea and Gymnotiformes in deep river channels, and the origin of Loricarioidei is inconclusive. Transition rates between habitat character states in five species-rich NFF clades from the deep channel analysis show transitions occur most frequently among habitats of the same hydrological structure, primarily among nonchannel habitats. The most frequent habitat transitions among NFF clades are transitions among habitats of the same physical structure (e.g., among channels) rather than water chemistry (e.g., among white waters). The general pattern of macrohabitat evolution among South American fishes is transitioning among habitats of similar hydrological structure rather than similar water chemistry. These results are consistent with other studies indicating strong phylogenetic niche conservatism among NFF clades in deep river channels.

J. S. Albert (✉) · A. H. Fronk
University of Louisiana at Lafayette, Lafayette, LA, USA
e-mail: jsalbe@gmail.com

A. H. Fronk
e-mail: fronaar@gmail.com

M. J. Bernt
Susquehanna State University, Selinsgrove, PA, USA
e-mail: mjbernt@gmail.com

T. P. Carvalho
Universidad Nacional de Colombia, Bogotá, Colombia
e-mail: carvalho.ictio@gmail.com

W. G. R. Crampton
University of Central Florida, Orlando, FL, USA
e-mail: crampton@ucf.edu

J. S. Albert and R. E. Reis (eds.), *Ecology and Evolution of Amazonian Fishes*, Fascinating Life Sciences,
https://doi.org/10.1007/978-3-032-09328-8_13

13.1 Introduction to Amazonian River Channels

13.1.1 Geography

The main channels of large lowland South American rivers, such as the Amazon, Orinoco and La Plata rivers and their larger tributaries, support a diverse and unique tropical aquatic biota (Lundberg and Mago-Leccia 1986; Mago-Leccia et al. 1985; Lundberg et al. 1987; Lundberg 1991; Lundberg and Py-Daniel 1994; Lundberg et al. 1996; Barthem and Goulding 1997; Fernandes et al. 2004; Hilton et al. 2007; Ivanyisky and Albert 2014; Carvalho and Albert 2015; Carvalho et al. 2016; Bernt and Albert 2017; Bernt et al. 2018; Duarte et al. 2019; Bernt et al. 2020; Duarte et al. 2022). These channels exhibit a distinctive set of hydrological and geomorphological features that transport water and sediments, shape the riverbed and surrounding floodplains, and affect sediment deposition and water chemistry of river basins and the Atlantic Ocean (Salo et al. 1986; Mertes et al. 1996; Trigg et al. 2012; Gualtieri et al. 2020; Siddiqui et al. 2021; Jézéquel et al. 2022). The Amazon River transports an average of about 1.2 billion tons of sediment, and 25.5. billion tons of dissolved organic carbon per year to the Atlantic Ocean (Körtzinger 2003; Ternon et al. 2000). Like all rivers, deep channels interact with underlying aquifers, influencing groundwater recharge and regional hydrology (Miguez-Macho and Fan 2012; Brunner et al. 2017).

Large river channels ramify across millions of square km of tropical and subtropical South America, but the riverbed benthos itself occupies a surprisingly small total habitat footprint of only c. 3500 km^2, or about half the area of metropolitan São Paulo, Brazil (Fig. 13.1). The main channel is the central and largest waterway of any river drainage network, being deeper and wider than secondary or tributary channels, and carrying a majority of the water from headwaters to mouth (Richey et al. 1986; Flood and Damuth 1987; Fig. 1.7C). The main river channels of South America have widths of about 1000–10,000 m between the banks at low water, maximum depths commonly reaching 10–60 m (~100 m in the lower Rio Near its confluence with the Amazonas), and Strahler stream orders (SO) of 6–10 (Chap. 6). The Amazon River achieves SO10 below the confluence of the Solimões and Negro rivers, forming the river segment with the highest SO on Earth. By comparison, the lower Orinoco has SO9, and the lower Mississippi SO8.

The main channels of Amazonian rivers are also critically important for the human economy of the region, providing essential routes for transportation and commerce (Abramovay et al. 2021). All the major regional urban centers (e.g., Ciudad Bolívar, Iquitos, Leticia, Manaus) have deep-water ports to receive large oceanic cargo ships. Amazonian river channels also support important fisheries that local communities depend on for food and livelihoods (Sirén and Valbo-Jørgensen 2022). Using published data from the United Nations, World Bank, national fisheries agencies, and market studies, the total direct commercial value of Amazonian fishes, including food fish, ornamental species, and ecotourism, is estimated to range from $1.0 to 2.0 billion USD per year (Jimenez et al. 2020; Sirén and Valbo-Jørgensen 2022; Mikkola 2024). Most commercially-exploited Amazonian fishes require access to the resources of large river channels to complete their life-cycle, for food, forage, dispersal, and reproduction (Duponchelle et al. 2021; Heilpern et al. 2022; Lopes et al. 2023).

13.1.2 Habitats

Main river channels represent a distinctive set of habitats in all the major ecological regions and biogeographic areas of Greater Amazonia (Albert and Reis 2011; Albert et al. 2011, 2018; Fig. 13.2). Deep channel habitats vary greatly along their length in terms of hydrology, substrates, channel topography, and water chemistries (Crampton 2007; 2011; Junk et al. 2011; Seidel et al. 2015, 2016). Environmental properties of large river channels also vary vertically (from surface to benthos) and laterally (from channel to floodplain) in relation to water velocity, bank slope, substrate composition, and other physical variables (Barthem and Goulding 1997; Bouchez et al. 2010; Crampton 2011; Val and de Almeida-Val 2012; Gualtieri et al. 2017). Some rivers of lowland Amazonia have among highest sinuosity on Earth, with a meander index often exceeding 3.0, measured as a ratio of the channel and *thalweg* (i.e., valley way) distances (Fig. 13.3). Such high meander indices are associated with exceptional physical and ecological heterogeneity, from mesoscale features (10–100 m) like point bars and cut banks, through macroscale features (100–1,000 m) like backwater sloughs, riparian forests, oxbow lakes and abandoned meanders, to regional scale features such as floodplains and other wetlands.

The main channel of large rivers can take on a range of geomorphological configurations, based on the number, confinement, and sinuosity of the channels (Potter 1978; Ashworth and Lewin 2012; Latrubesse 2015; Jézéquel et al. 2022). Hydrologists distinguish between *meandering*, *anastomosing*, *anabranching*, and *braided* channels (Table 13.1). A meandering river has a single sinuous channel and the other channels morphologies have multiple channels, differing in their characteristics, formation processes, and environmental settings. Anabranching channels split and rejoin around stable or semistable islands, whereas braided channels form a network of unstable, intertwining channels that constantly shift around transient bars or sediment deposits. An anastomosing channel is subset of anabranching channels with relatively low-energy associated with mostly fine-grained or organic deposition. By contrast, in a *braided* river the divided

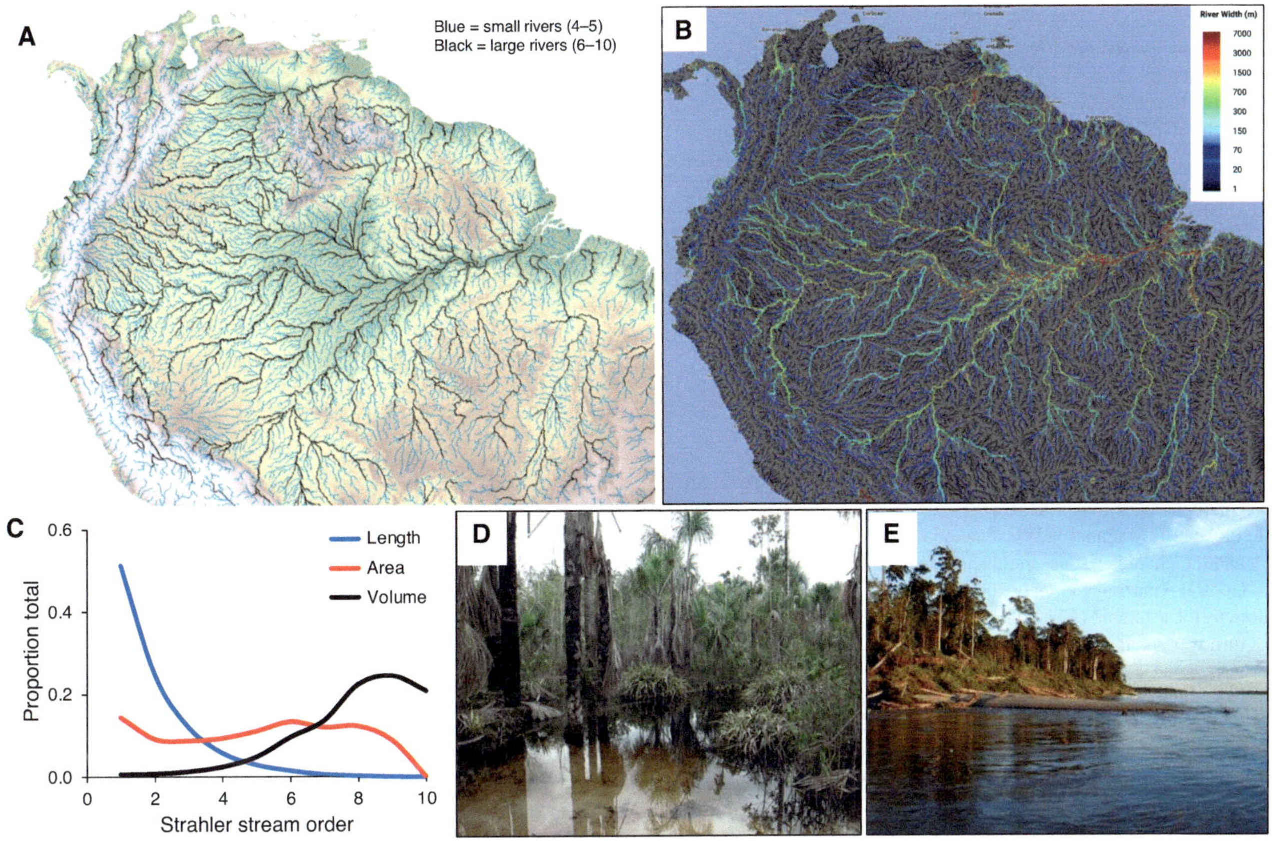

Fig. 13.1 Strahler stream order (SO) in rivers of northern South America. (**A**) Medium-sized rivers (SO = 4–5) in blue, large rivers (6–10) in black. (**B**) River width in meters. From MERIT Hydro (Yamazaki et al. 2019). Stream-order six rivers are ~200 meters wide at low water. (**C**) Fraction of South American riverine habitat by stream order, assessed as length, area, and volume. Calculations based on a total stream length of about 5.3 million km, a total stream surface area of about 764,000 km^2, and a total habitat volume of about 410 million km^3. Note larger rivers (6–10) representing 3% of total river-network length account for high proportion 53% of the total river surface area, and 90% of the total habitat volume. Data from HydroSHEDS analyzed by Jon Allen. (**D**) A first-order stream (SO = 1) in the río Arapiuns. (**E**) Rio Solimoes near Tefé, SO = 8 (photos by Will Crampton)

flow pattern is strongly stage-dependent around sand or mud bars that are unconsolidated, ephemeral, poorly vegetated, and overtopped at less than bankfull flow. The floodplains of large meandering and anabranching rivers are characterized by a "scroll-swathe" landscape of curvilinear ridges and swales that are remnants of former channels and their *levees*, or embankments that form along channel margins (Chap. 11). The floodplains are also characterized by meandering channels known as *paranás* in Portuguese and *caños* in Spanish, which can form elongate and sinuous lakes from sections of former channels. Despite this broad range of channel morphologies, there are no known examples of Amazonian fishes specialized to inhabit a subset of channel morphologies, and it appears that channel fishes do not discriminate among rivers with different channel morphologies.

Many studies have suggested a strong influence of deep river-channel habitats on the diversification and evolution of fishes (Smith 1992; Barthem and Goulding 1997; Winemiller and Adite 1997; Henderson et al. 1998). Many deep-channel Amazonian fishes exhibit a syndrome of phenotypes usually associated with caves and other hypogean waters, including reduced eyes and pigmentation (Albert 2001; Lujan and Chamon 2008; Evans et al. 2017; see Chap. 15). Large river channels act as ecological corridors, allowing long-distance migrations of many fishes over hundreds to several thousand km (see Chap 20). During high-water periods, deep channels interact with flooded forests, enabling nutrient exchanges and sustaining aquatic food webs, and channels represent a seasonal refuge for floodplain fishes during the dry season. Deep channel habitats are somewhat buffered from ecological effects of human activities, but are sensitive to changes in water chemistry following deforestation and high levels of erosion from agricultural landscapes (Figueiredo et al. 2020). Dam construction and climate change are substantially altering sediment loads and flow regimes of deep channel ecosystems in rivers of the Guiana and Brazilian Shield and Andean foothills (Winemiller et al. 2016; He et al. 2024).

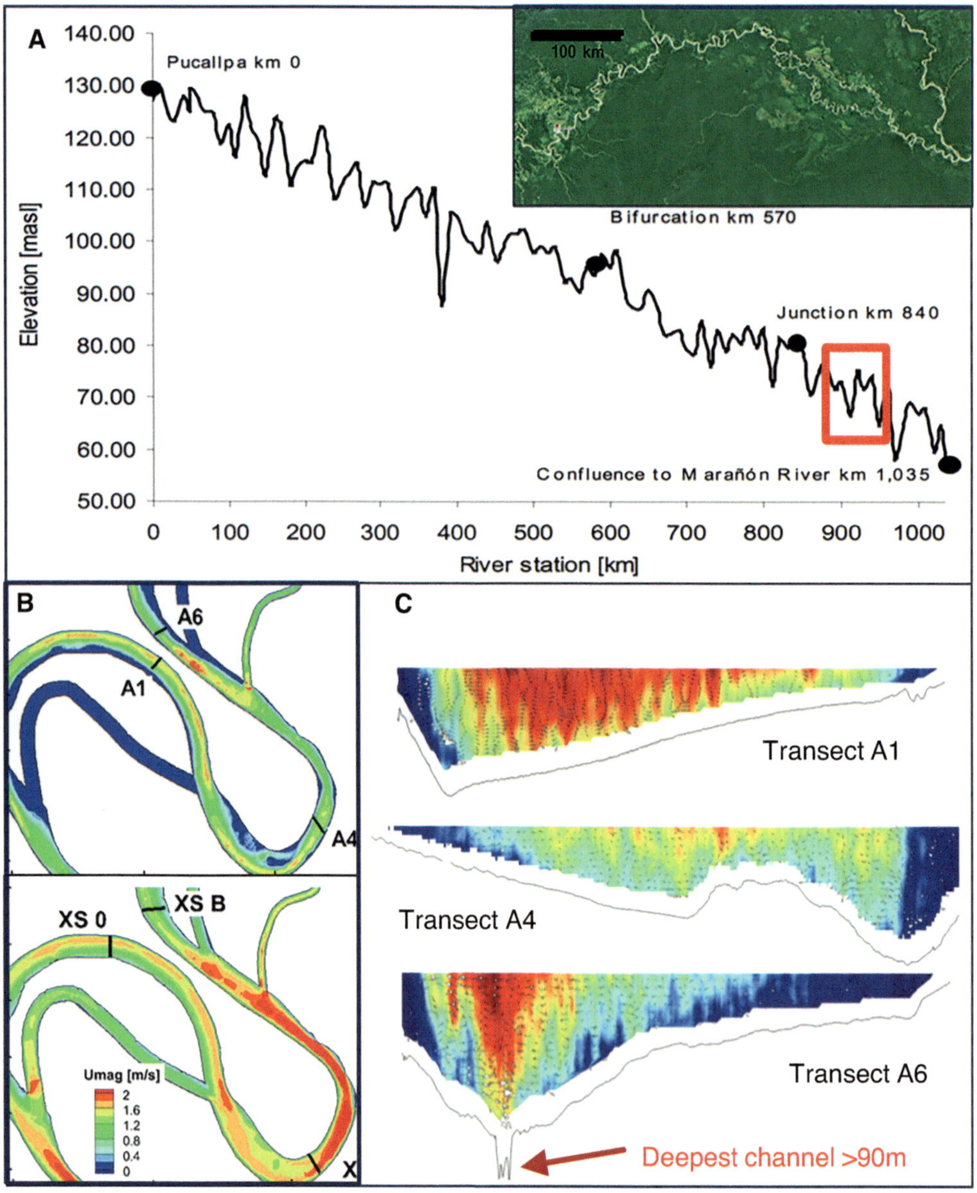

Fig. 13.2 Hydrological profile of the Rio Ucayali, Peru, a large lowland Amazonian river. (**A**) Longitudinal profile of the river bottom from Pucallpa to confluence with the Rio Marañón. River distance 1035 km; thalweg distance 550 km. Note large-scale (20–30 km) undulations of the river bottom resulting from meander cutoffs. Bifurcation and junction with Puinahua canal. Image from Ettmer and Alvarado-Ancieta (2010); inset from Google Earth. (**B**) Plan view of water velocities for river meanders near the community of Requeña (red box in A). Top: Low water (August), bottom: high water (May). From Dauer (2015). (**C**) Transects showing the fastest water velocities. Note the deepest channel at 90 m at A6 formed by crossing a former main channel. Note the fastest water flows during high water and along river runs and outer bends

The tremendous volume, depth, and remoteness of Amazonian river channels make them challenging and even dangerous to sample and study (Lopez-Rojas et al. 1984; Ceber and Montoya-Burgos 2002; Crampton 2007; Urbano-Bonilla et al. 2021). Most of the deep channel habitat of the Amazon basin remains unmapped and poorly unexplored (Galeazzi et al. 2018; Tamura et al. 2019; Marinho et al. 2022; Guerrero et al. 2024), with many ecology commercially important fishes (Zacardi et al. 2017; Arantes et al. 2018; Goulding et al. 2019; Formiga et al. 2021), and newly species of commercially important fishes still being described (e.g., Lundberg and Akama 2005; Littmann et al. 2021). Advances are being made, however, by merging data from new and traditional technologies. Sonar mapping using GPS from boats is used to generate profiles of the river bed and its substrates, and remote sensing data from satellites is used to generate digital elevation models of topography, relief, and other geomorphological attributes. These data from the physical environment are being combined

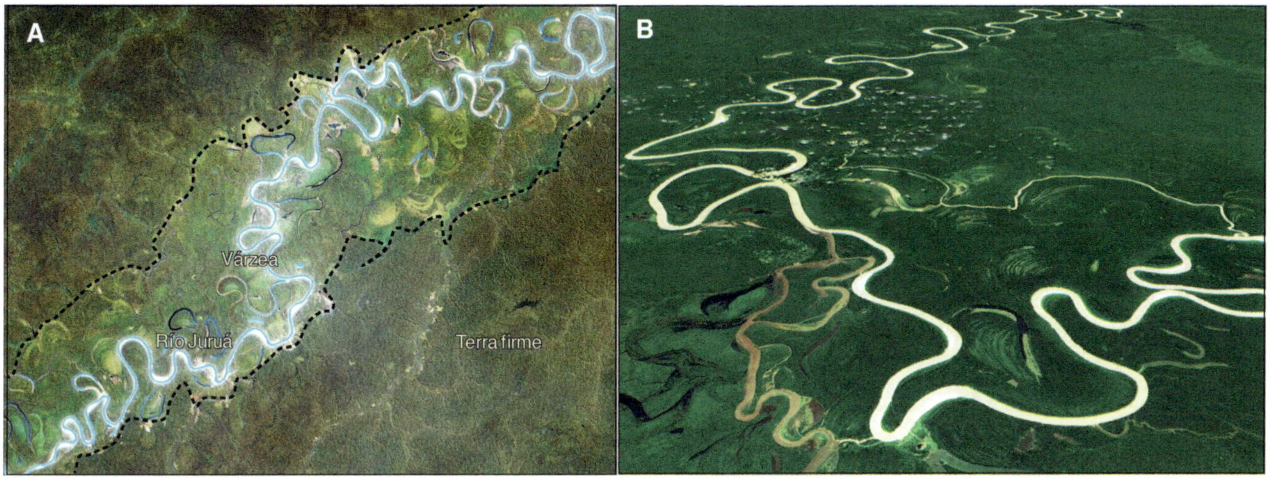

Fig. 13.3 Riverine habitats of lowland Amazonia. The Rio Juruá has one of the highest sinuosity or meander indices of any river on Earth, exceeding values of 3.0 in many places. The meander index is the ratio of channel length to valley (thalweg) length. Rivers with high sinuosity have more floodplain lakes, more seasonally flooded forest, and a larger volume of lentic aquatic habitat (Stølum 1998). (**A**) Alan view showing the boarder (dashed lines) between floodplain and nonfloodplain habitats. (**B**) Google Earth image simulating view of the Juruá valley from an airplane flying at c. 6000 m (c. 20,000 ft)

Table 13.1 Comparisons among morphologies of large Neotropical river channels

Feature	Meandering	Braided	Anabranching	Anastomosing
Definition	Single sinuous channel	Multiple channels, branches diverging and rejoining around stable islands	Subtype of anabranching, interconnected, stable channels, and vegetated islands	Multiple, intertwining, unstable channels, transient sediment bars
Energy level	Low to moderate energy, more lateral than vertical erosion	Both low- and high-energy environments	Low-energy systems	High-energy systems with steep gradients and flash flows
Island/bar formation	No islands, point bars form on inner meander bends	Stable or semistable islands, often vegetated	Permanent, vegetated islands formed from floodplain material	Temporary sediment bars that erode and reform frequently
Vegetation	Banks vegetated constraining channel migration	Vegetation may stabilize islands and banks	Dense vegetation stabilizes channels and islands	Minimal vegetation on bars due to frequent sediment shifting
Examples in South America	Paraná River (Brazil/ Argentina)	Orinoco River (Venezuela/ Colombia)	Magdalena River (Colombia)	Upper Mamore River (Bolivia)

with biological data from traditional methods (field sampling, fisheries landing data) to document the habitat affinities of deep channel Amazonian fishes (Barthem et al. 1991; Cox-Fernandes 1997; Lundberg et al. 2013; Duarte et al. 2019, 2022; Torgersen et al. 2023) (Fig. 13.4).

13.1.3 Chapter Goals

This chapter describes the taxonomic composition of large Amazonian river channels, traces the evolution of ecotraits associated with habitat utilization of riverine environments, and assesses differences in the evolutionary rates of transitions among these environments.

13.2 Methods

13.2.1 Ecotrait Categorization and Evolutionary Rates

Species were scored as inhabiting small (SO1-5) or large (SO6-10) river channels, and high-conductivity whitewater or low-conductivity black or clear waters. Of the Big Nine NFF

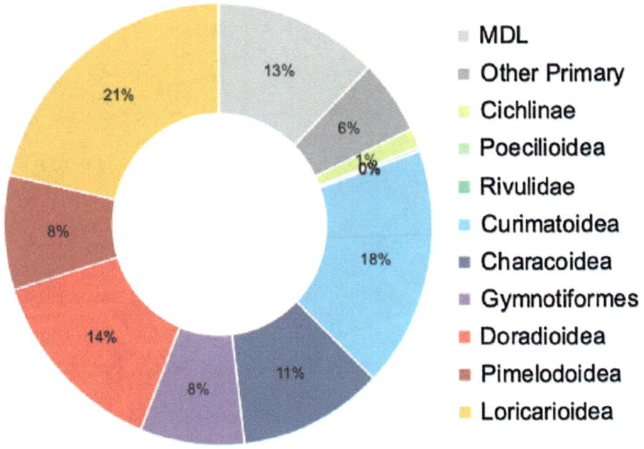

Fig. 13.4 Percentage of Big Nine NFF clades, other primary freshwater fishes, and marine-derived clade (MDC) fishes, which inhabit large Amazonian river channels

clades, Characoidea, Cichlinae, and the two cyprinodontiform clades were excluded from this study, as they do not occur or are rare in deep channels and other lowland whitewater habitats (Fig. 13.5). *Ancestral State Estimation* (ASE) was conducted using the Mk model of trait evolution in the R package *phytools* (Revell 2014; Fig. 13.5). Evolutionary rates were estimated by fitting the dataset to three models of trait evolution: an "*equal-rates*" (ER) model in which all transition rates between character states are the same, a "*symmetric-rates*" (SYM) model in which rates of transition back and forth between two character states are equal, and an "*all rates different*" (ARD) model where all transition rates are assumed to be independent of each other. Each model-fitting analysis produced a log-likelihood value, from which was calculated an *Akaike Information Criterion* (AIC) for each model in each clade. The lowest AIC values were used to select a single model for each clade, which was then compared with the other models using the likelihood-ratio test statistic, which quantifies the difference in log likelihood scores for significance. In each case, the lowest AIC-scoring model was found to be significantly better than the other models. See Chap. 7 for a more detailed description of methods for estimating ancestral states and evolutionary rates.

13.2.2 Bayesian Stochastic Mapping

For each clade, the trait evolution results for the best-scoring model were then mapped to the phylogenetic tree using the software SIMMAP (Bollback 2006), which implements Bayesian stochastic mapping. Bayesian stochastic mapping is a method of mapping stochastic characters on trees that estimates the timing of character changes along branches. Similar to Bayesian inference, it works by sampling character histories in proportion to their posterior probabilities, due to a direct analytical approach being impractical for all but the simplest topologies. SIMMAP specifically uses a three-step algorithm for sampling character histories given a particular topology, branch lengths, and substitution model: it first calculates the conditional likelihood for each character state at each node of the tree, simulates ancestral states at each internal node by sampling from the posterior distribution of states, and finally simulates a substitution history by sampling from the posterior distribution conditioned on the aforementioned simulated ancestral states and observed tip states in the topology. Hundred stochastic maps were generated for each analysis, and from these results were derived posterior probabilities at each node of the phylogenetic tree for each character state, which appeared on the trees as color-coded pie charts at each node (Figs. 13.6 and 13.7). The transition rates between character states for these character histories are shown in Fig. 13.8.

13.3 Results

13.3.1 Taxonomic Composition

Deep channel fishes are defined species in which the modal adult habitat utilization is the pelagic or benthic zones of deep river channels (Chap. 6). The deep channels of large lowland Amazonian rivers are home to a diverse fish fauna, with about 844 species representing just over one-quarter (28.1%) of all fish species of Greater Amazonia, representing 368 genera and 68 families (van der Sleen and Albert 2017; Dagosta and de Pinna 2019; Albert, Tagliacollo, and Dagosta 2020; Fig. 13.4).

Among the five NFF clades examined, deep channel specialists are represented as follows:

- 271 species of Curimatoidea, including several species of *Chilodus, Curimata, Psectrogaster, Pseudocurimata, Anodus, Hemiodus, Prochilodus, Myleus, Myloplus, Mylossoma, Pristobrycon, Semaprochilodus,* and *Serrasalmus*.
- 217 species of Doradioidea, including several species of *Bunocephalus, Xyliphius, Hoplomyzon, Auchenipterus, Hemidoras, Leptodoras, Megalodoras, Nemadoras, Ossancora, Oxydoras, Rhinodoras, Rhynchodoras, Tenellus*, and *Trachydoras*.
- 122 species of Gymnotiformes, including several species of *Adontosternarchus, Apteronotus, Sternarchorhynchus, Porotergus, Sternarchella, Sternarchogiton, Rhamphichthys, Rhabdolichops*, and *Sternopygus*.
- 128 species of Pimelodoidea, including several species of *Pimelodella, Rhamdella, Rhamdia, Brachyplatystoma, Hypophthalmus, Pimelodus, Megalonema, Sorubim*, and *Pseudoplatystoma*.
- 329 species of Loricarioidei, including several species of *Acestridium, Hypoptopoma, Otocinclus, Oxyropsis,*

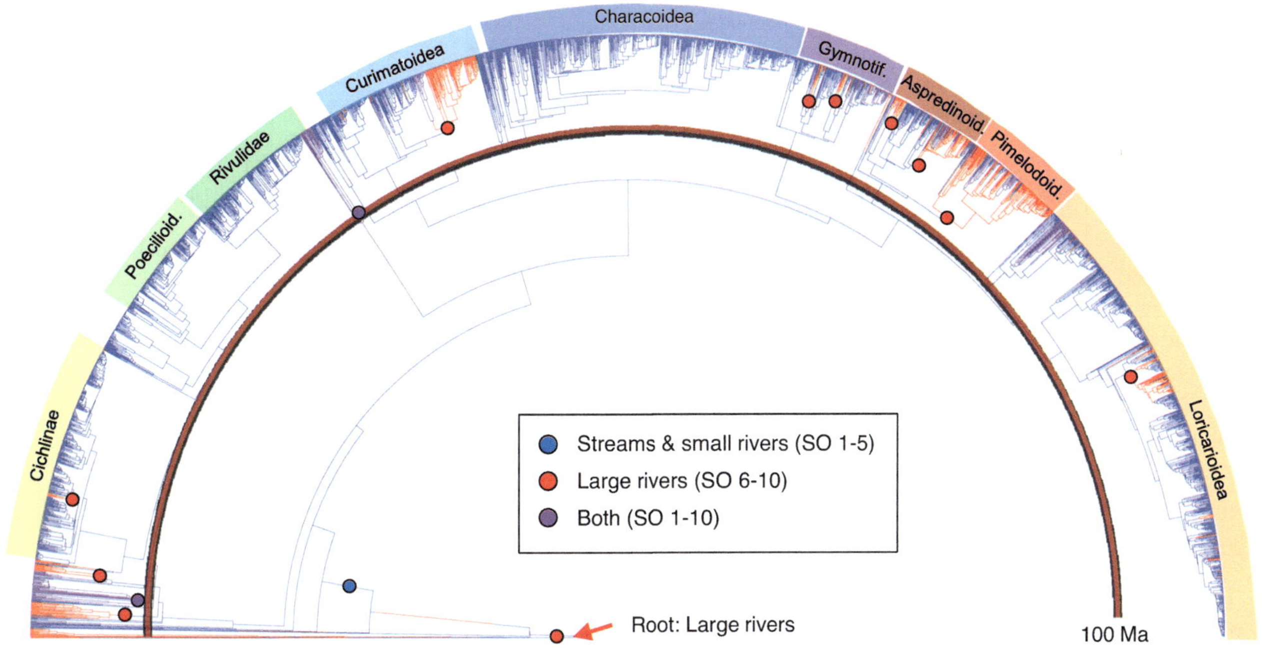

Fig. 13.5 Evolution of habitat utilization in NFF taxa along the river continuum. Species scored for adult modal habitat utilization by Strahler stream order (SO; see text). Data for 3167 species in all NFF 880 genera, 96 families, and 40 orders. Big Nine NFF clades (colored regions) from Chap. 3. Red semicircle at *c*. 100 Ma marks the time of the final separation of South America and Africa and the latest origin of phylogenetically independent NFF clades. Colored circles indicate estimated changes in clades with five or more species. Note that all Big Nine NFF clades have estimated origins in small rivers (SO 1–5). Note also that the root optimization for vertebrates is interpreted here to be a spurious result from the many marine-derived NFF clades that inhabit large rivers among early-branching taxa. Vertebrate trace origins to the Cambrian *c*. 500 Ma, long before the Upper Cretaceous (*c*. 100–66 Ma) origins of modern South American freshwater habitats or NFF taxa

Parotocinclus, Ancistomus, Dekeyseria, Isorineloricaria, Pterygoplichthys, Farlowella, Loricaria, Loricariichthys, Pseudohemiodon, Rhadinoloricaria, Rineloricaria, Henonemus, Homodiaetus, Pseudostegophilus, and *Vandellia*.

- 191 species of MDLs, including several species of *Geotria, Potamotrygon, Anchoviella, Pristigaster, Centropomus, Pachyurus, Strongylura*, and *Mugil*.

13.3.2 Ancestral State Estimation

Ancestral state estimates and evolutionary transitions of NFF species are depicted in Fig. 13.5, in the phylogenetic sequence that they appear in Chap. 2. All the Big Nine NFF clades have estimated evolutionary origins in small rivers of the Guiana and Brazilian shields. Cichlinae, Characoidea, and the two cyprinodontiform clades (Poecilioidea and Rivulidae) are either rare or absent in the deep channels, and are not here considered further. The other five of these clades exhibit multiple evolutionary transitions to and from deep channel habitats.

Curimatoidea was found to best follow an ARD model of trait evolution. It has an estimated habitat origin in deep channels, as observed in the majority of Hemiodontidae, Serrasalmidae, Chilodontidae, Cynodontidae, and Curimatidae. The greatest number of channel-dwelling species are found in Curimatidae, a family known for undertaking long-distance seasonal migrations (Lowe-McConnell 1979; Barthem and Goulding 1997; Melo et al. 2021). The majority of channel species in Curimatoidea retained their ancestral habitat preferences, with those clades living in nonchannel habitats transitioning from channel habitats between 45 and 10 million years ago. The highest rates of transition are between different channel habitats, and between nonchannel habitats, showing a pattern of high transition between habitats of similar structural type rather than water type.

Unlike most of the other clades tested, Gymnotiformes was found to follow a SYM model of trait evolution. The habitat origin of Gymnotiformes was found to be predominantly whitewater channels, the current habitat of most of the Sternopygidae and Apteronotidae, where they form two of the most abundant and diverse animal clades (Lundberg et al. 1987; Duarte et al. 2019). The greatest rate of transfer in Gymnotiformes was found to be between nonchannel habitats of differing water types, with very little exchange between any other habitat types.

Aspredinoidea followed an ARD model of trait evolution and has an estimated habitat origin within nonchannel habitats,

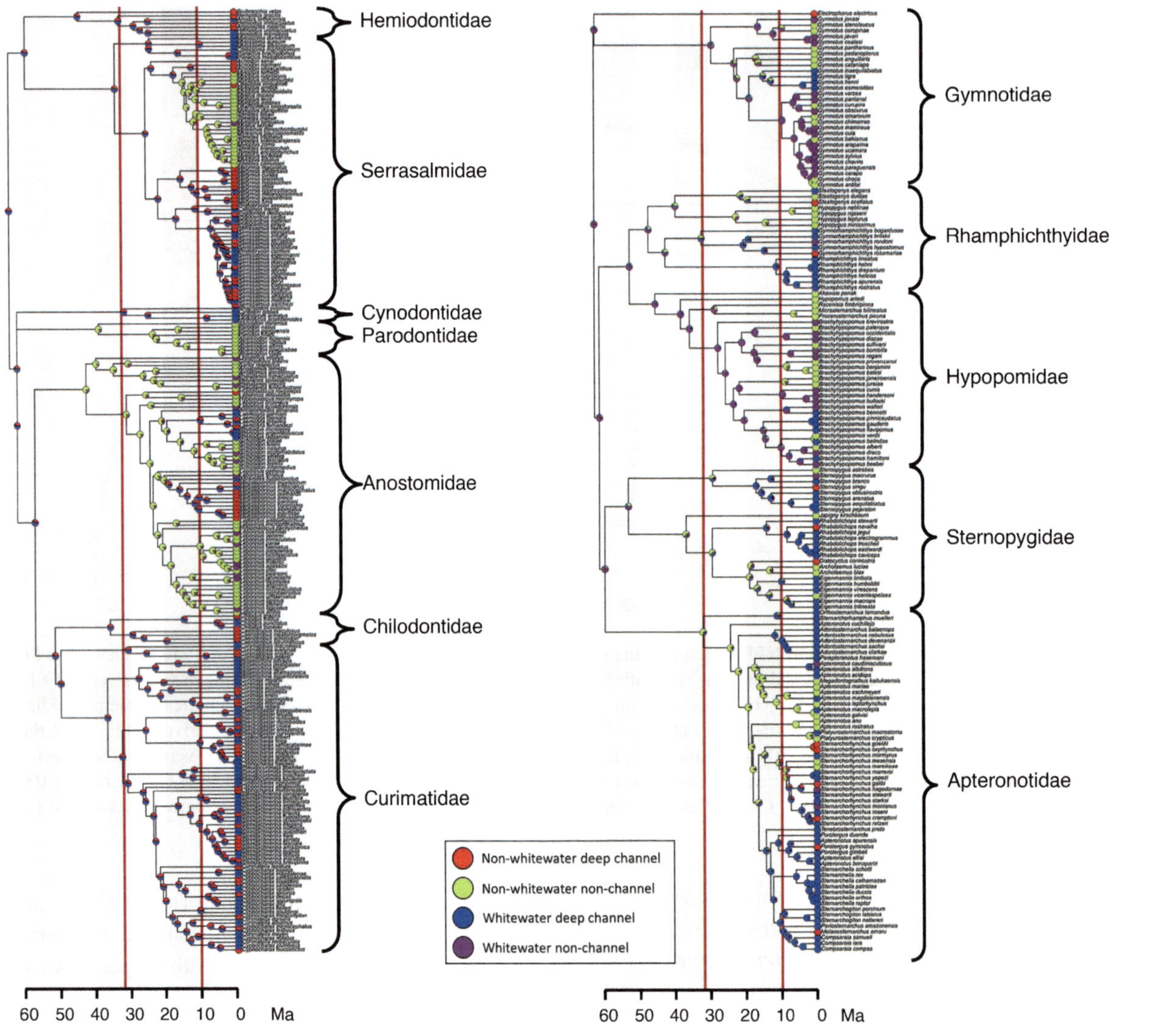

Fig. 13.6 Ancestral state estimation of macrohabitats related to water chemistry and physical habitat structure for Curimatoidea and Gymnotiformes. Time-calibrated phylogeny from Chap. 2. Vertical red lines at 34 and 10 Ma. Pie charts at internal tree nodes indicate estimated fractional probabilities from a Brownian motion model of trait evolution

which is surprising considering their current association with deep channel habitats. The family Doradidae in particular attains among the highest animal biomasses in deep channels (Saint-Paul et al. 2000; Duarte et al. 2019). This analysis suggests multiple transitions from nonchannel habitats to channel habitats amongst the various clades of Aspredinoidea, one (within the genus *Tatia*) within the last 10 Ma. Despite this, the greatest source of transition is between channel habitats of differing water types.

Pimelodoidea also has an estimated habitat origin in nonchannel habitats, primarily nonwhitewater nonchannels, which is consistent with their extant association with primarily blackwater/clearwater terra firme streams. They show a SYM model of trait evolution, similar to the Gymnotiformes, but with few transitions overall.

The variegated modal habitat utilization of extant Loricarioidei made it difficult to estimate a particular habitat origin for this clade. The most frequent habitat of extant loricarioids is nonwhitewater nonchannel habitats, but those that live in channels primarily transitioned into this habitat from whitewater nonchannel habitats. Overall, the rates of transition were very low for all of Loricarioidei.

13.3.3 Ecotrait Evolutionary Rates

Transition rates between habitat character states in five species-rich NFF clades from the deep channel analysis (Fig. 13.8) show transitions occur most frequently among habitats of the same hydrological structure, primarily among nonchannel

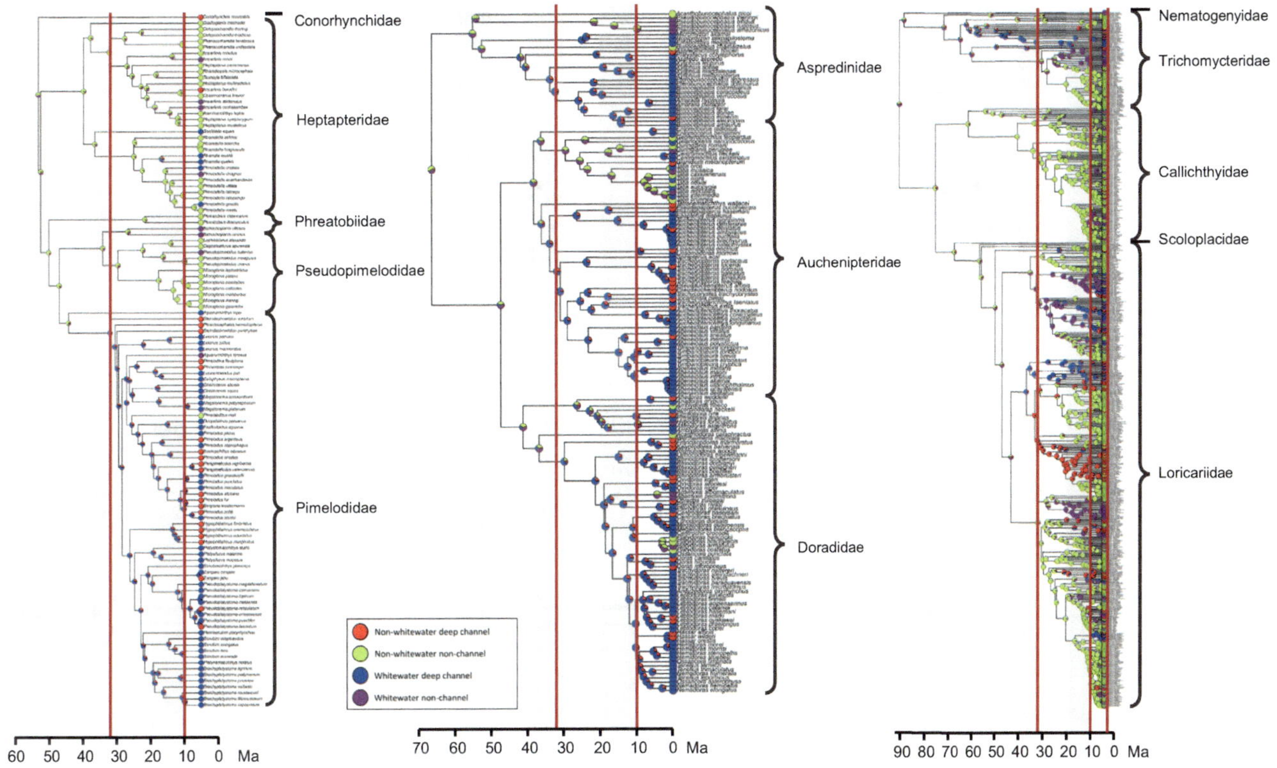

Fig. 13.7 Ancestral state estimation of macrohabitats related to water chemistry and physical habitat structure for Pimelodoidea, Aspredinoidea, and Loricarioidei. Data sources and graphical conventions as in Fig. 13.6

habitats. The only exception is Aspredinoidea, which showed the highest rate of transitions between whitewater and nonwhitewater channels. The results of the AEA analyses showed transitions occurring most frequently between habitats of the same structural type, primarily among nonchannel habitats. This pattern holds true even for Pimelodoidea and Aspredinoidea, which showed the lowest relative difference in transition rates (Table 13.2).

13.3.4 Macroevolutionary Patterns

Among the Big Nine NFF clades, Pimelodoidea and Aspredinoidea have estimated origins in nonchannel habitats, Curimatoidea and Gymnotiformes in deep river channels, and the origin of Loricarioidei is inconclusive. As many primarily deep channel species are also known to inhabit floodplains, it is possible that the original fish fauna of the *várzea* and other floodplain habitats were seeded from the deep channels associated with those floodplains. However, the transition rates for both analyses suggest a more frequent pattern is for lineages to transition to another habitat of the same hydrological structure (i.e., floodplain vs. channel) regardless of water chemistry. While most of the species examined here are primarily found in the two habitats discussed so far, it should be noted that there are also members of these lineages that are found in terra firme streams and other freshwater habitats of South America. With this in mind, it should also be noted that in the deep channel analysis, consistently higher transition rates were found between nonchannel habitats than channels, which corresponded with the high rates of transition between floodplains in the previous analysis.

13.4 Conclusions

This chapter estimates the instances and rates of macroevolutionary transitions in freshwater fish clades adapted to large Amazonian river channels and three other physically defined aquatic macrohabitats of tropical South America. The analyses use the time-calibrated phylogeny and eco-matrix of NFF species presented in earlier chapters, scoring adult, modal, and habitat utilization to each of the following four macrohabitats: deep channels of large (SO 6-10) rivers, floodplain habitats of large whitewater rivers (i.e., *várzea*), floodplain habitats of large blackwater (i.e., *igapó*) or clearwater rivers, and nonfloodplain (i.e., *terra firme*) habitats of small (SO 1-5) rivers and streams. Among NFF clades, *** species are deep-channel specialists, representing about **% of all fish species in Greater Amazonia,

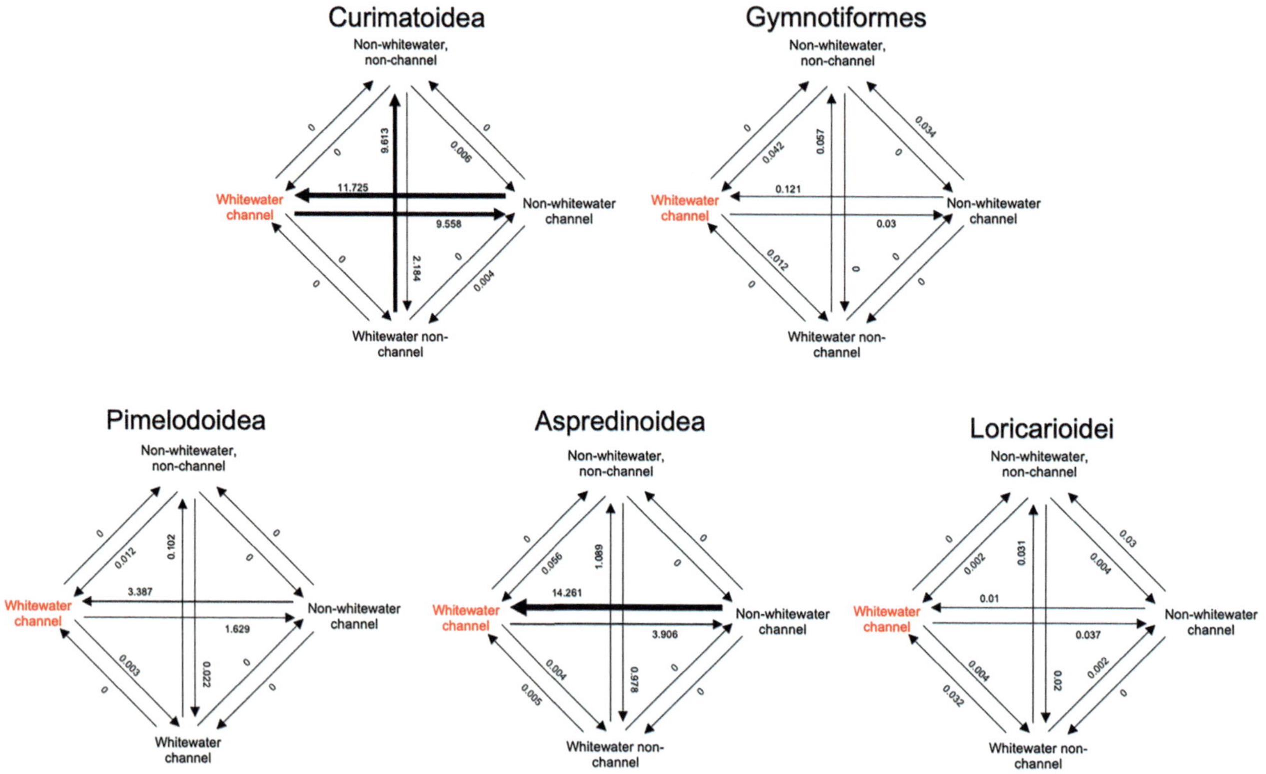

Fig. 13.8 Transition rates among four macrohabitats related to water chemistry and physical habitat structure for seven of the Big Nine NFF clades that are abundant and diverse in deep channel habitats

Table 13.2 Habitat transitions among deep channel fishes in five major NFF clades

Clade	Highest transition rate			Second highest transition rate			Overall pattern
	From	To	Rate	From	To	Rate	
Curimatoidea	*CBC*	*WWC*	11.725	*WWNC*	*CBNC*	9.613	Highest transition rates among similar structural types
Gymnotiformes	*CBNC*	*WWNC*	13.932	*WWNC*	*CBNC*	13.932	Highest transition rate between nonchannel habitats
Pimelodoidea	*CBC*	*WWC*	0.024	*WWC*	*CBC*	0.024	Highest transition rate between channel habitats
Aspredinoidea	*CBC*	*WWC*	14.261	*WWC*	*CBC*	3.906	Highest transition rates among similar structural types
Loricarioidei	*WWC*	*CBC*	0.037	**WWNC**	**WWC**	0.032	Little transition

Transitions that change habitat structure are shown in bold. Transitions that change water chemistry are italicized. Note that the most common transition is water chemistry only. CBC clear/blackwater channel, CBNC clear/blackwater nonchannel, WNC whitewater nonchannel, WWC whitewater channel

and about **% of all NFF species. Five of the Big Nine NFF clades exhibit sizeable diversity in deep Amazonian river channels. All these clades are members of the teleost superorder Ostariophysi with estimated stem-group ancestry inhabiting small (SO 1-5), low-conductivity (clearwater or blackwater) rivers, and streams of Western Gondwana during the Upper Cretaceous before *c.* 100 Ma. Crown-group members of the Curimatoidea, and five of its seven family-level subclades (Hemiodontidae, Serrasalmidae, Cynodontidae, Chilodontidae, and Curimatidae) have estimated macrohabitat origins in deep channels of large (SO 6-10) rivers, ambiguously in either low-conductivity clearwaters and/

or blackwaters, or in high-conductivity whitewaters. The other two curimatoid family-level subclades (Parodontidae and Anostomidae) have estimated macrohabitat origins in small, low-conductivity (clearwater or blackwater) rivers and streams. Crown-group Gymnotiformes and three of its family-level subclades (Rhamphichthyidae, Sternopygidae, and Apteronotidae) have estimated macrohabitat origins in the deep channels of large whitewater rivers, while the other two family-level subclades (Gymnotidae and Hypopomidae) have ambiguous macrohabitat origins, with Electrophorinae in small clearwater rivers and streams, and Gymnotinae and Hypopomidae in either small, low-conductivity rivers and streams or large whitewater river floodplains. Crown-group Pimelodoidea and three of its five family-level subclades (Heptapteridae, Phreatobiidae, and Pseudopimelodidae) have estimated macrohabitat origins in small, low-conductivity (clearwater or blackwater) rivers and streams. The other two pimelodoid family-level subclades (Conorhynchidae and Pimelodidae) have estimated macrohabitat origins in deep channels of large rivers, clearwater, and whitewater, respectively. Crown-group Aspredinoidea, and all three of its family-level subclades (Aspredinidae, Auchenipteridae, and Doradidae) have estimated macrohabitat origins in small, low-conductivity (clearwater or blackwater) rivers and streams. All the aspredinoid families exhibit a similar macroevolutionary pattern of early branches with a mix of these two macrohabitat types, and with one or more high-diversity clades in whitewater floodplains. Crown-group Loricarioidei, and its four to five family-level subclades (Nematogenyidae, Callichthyidae, Scoloplacidae, and Loricariidae) have ancestral macrohabitat origins in small, low-conductivity rivers and streams. The other loricarioid family-level subclade (Trichomycteridae) has ambiguously estimated macrohabitat origins, either in small, low-conductivity rivers and streams or high-conductivity whitewater floodplains (Wiens and Graham 2005).

Acknowledgments We are grateful to Milton Tan for providing the *phytools* code used in the ASE analysis, and to Ronaldo Barthem, Michael Goulding, Nathan Lujan, John Lundberg, Roberto Reis, Don Stewart, and Kirk Winemiller for many insights into the biology of deep channel Amazonian fishes.

References

Abramovay, R, Ferreira, J, Costa, FDA, Ehrlich, M, Euler, AMC, Young, CEF, et al (2021) The new bioeconomy in the Amazon: Opportunities and challenges for a healthy standing forest and flowing rivers, Amazon Assessment Report 2021. United Nations Sustainable Development Solutions Network, New York, p. 1300

Albert JS (2001) Species diversity and phylogenetic systematics of American knifefishes (Gymnotiformes, Teleostei). In: Miscellaneous publications museum of zoology, University of Michigan, p 190

Albert JS, Reis RE (2011) Introduction to Neotropical freshwaters. In: Albert JS, Reis RE (Eds.)Eds. Historical biogeography of neotropical freshwater fishes. University of California Press, Berkeley, pp 3–20

Albert JS, Carvalho TP, Petry P, Holder MA, Maxime EL, Espino J, Reis RE (2011) Aquatic biodiversity in the Amazon: habitat specialization and geographic isolation promote species richness. Animals 1(2):205–241

Albert JS, Tagliacollo VA, Dagosta F (2020) Diversification of Neotropical freshwater fishes. Annu Rev Ecol Evol Syst 51(1):27–53

Albert JS, Val P, Hoorn C (2018) The changing course of the Amazon River in the Neogene: center stage for Neotropical diversification. Neotrop Ichthyol 16(03):e180033

Arantes CC, Winemiller KO, Petrere M, Castello L, Hess LL, Freitas CE (2018) Relationships between forest cover and fish diversity in the Amazon River floodplain. J Appl Ecol 55(1):386–395

Ashworth PJ, Lewin J (2012) How do big rivers come to be different?. Earth-Sci Rev 114(1–2):84–107

Barthem RB, Ribeiro MCLB, Petrere Jr M (1991) Life strategies of some long-distance migratory catfish in relation to hydro-electric dams in the Amazon basin. Biol Conserv 55:339–345

Barthem R, Goulding M (1997) The catfish connection: ecology. Migration, and conservation of Amazon predators. Columbia University Press

Bernt MJ, Albert JS (2017) A new species of deep-channel electric knifefish *Compsaraia* (Apteronotidae, Gymnotiformes) from the Amazon River. Copeia 105(2):211–219

Bernt MJ, Crampton WG, Orfinger AB, Albert JS (2018) *Melanosternarchus* amaru, a new genus and species of electric ghost knifefish (Gymnotiformes: Apteronotidae) from the Amazon Basin. Zootaxa 4378(4):451–479

Bernt MJ, Fronk AH, Evans KM, Albert JS (2020) A redescription of deep-channel ghost knifefish, *Sternarchogiton preto* (Gymnotiformes: Apteronotidae), with assignment to a new genus. Neotrop Ichthyol 18(01):e190126

Bollback JP (2006) SIMMAP: Stochastic character mapping of discrete traits on phylogenies. BMC Bioinformatics 7, 88. https://doi.org/10.1186/1471-2105-7-88

Bouchez J, Lajeunesse E, Gaillardet J, France-Lanord C, Dutra-Maia P, Maurice L (2010) Turbulent mixing in the Amazon River: the isotopic memory of confluences. Earth Planet Sci Lett 290(1–2):37–43

Brunner P, Therrien R, Renard P, Simmons CT, Franssen HJH (2017) Advances in understanding river-groundwater interactions. Rev Geophys 55(3):818–854

Carvalho TP, Albert JS (2015) A new species of *Rhamphichthys* (Gymnotiformes: Rhamphichthyidae) from the Amazon Basin. Copeia 103(1):34–41

Carvalho TP, Lundberg JG, Baskin JN, Friel JP, Reis RE (2016) A new species of the blind and miniature genus Micromyzon Friel and Lundberg, 1996 (Siluriformes: Aspredinidae) from the Orinoco River: describing catfish diversity using high-resolution computed tomography. Proc Acad Nat Sci Philadelphia 165(1):37–53

Ceber C, Montoya-Burgos JI (2002) *Hypostomus fonchii* sp. n. (Siluriformes: Loricariidae) from Peru, a key species suggesting the synonymy of *Cochliodon* with *Hypostomus*. Rev Suisse Zool 109:355–368

Cox-Fernandes C (1997) Lateral migration of fishes in Amazon floodplains. Ecol Freshw Fish 6:36–44

Crampton WGR (2007) Diversity and adaptation in deep channel Neotropical electric fishes. Fish Life in Special. In: Sebert P, Onyango DW, Kapoor BG (Eds.)Eds. Environments. Science Publishers, Enfield, NH, pp 283–339

Crampton WGR (2011) An ecological perspective on diversity and distributions. In: Albert JS, Reis RE (Eds.)Eds. Historical biogeography of neotropical freshwater fishes. University of California Press, Berkeley, pp 165–189

Dagosta FC, de Pinna M (2019) The fishes of the Amazon: distribution and biogeographical patterns, with a comprehensive list of species. Bull Am Mus Nat Hist 2019(431):1–163

Duarte C, Espírito-Santo HMV, Zuanon J, Py-Daniel LR, Deus CP (2019) What happens in the darkness? Seasonal variations in tropical benthic fish assemblages. Mar Freshwater Res 71(4):419–431

Duarte C, Farago TLB, dos Anjos CS, dos Santos NR, do Nascimento LM, Cella-Ribeiro A, de Deus CP (2022) Spatial and seasonal variation of benthic fish assemblages in whitewater rivers of Central Amazon. Biota Neotropica 22(4):e20211312

Duponchelle F, Isaac VJ, Rodrigues Da Costa Doria C, Van Damme PA, Herrera-R GA, Anderson EP et al. (2021) Conservation of migratory fishes in the Amazon basin. Aquat Conserv: Mar Freshw Ecosyst 31(5):1087–1105

Evans KM, Crampton WG, Albert JS (2017) Taxonomic revision of the deep channel electric fish genus Sternarchella (Teleostei: Gymnotiformes: Apteronotidae), with descriptions of two new species. Neotrop Ichthyol 15(02), e160168.

Fernandes CC, Podos J, Lundberg JG (2004) Amazonian ecology: tributaries enhance the diversity of electric fishes. Science 305(5692):1960–1962

Figueiredo RDO, Cak A, Markewitz D (2020) Agricultural impacts on hydrobiogeochemical cycling in the Amazon: is there any solution?. Water 12(3):763

Flood RD, Damuth JE (1987) Quantitative characteristics of sinuous distributary channels on the Amazon deep-sea fan. Geol Soc Am Bull 98(6):728–738

Formiga KM, Batista JDS, Alves-Gomes JA (2021) The most important fishery resource in the Amazon, the migratory catfish *Brachyplatystoma vaillantii* (Siluriformes: Pimelodidae), is composed by an unique and genetically diverse population in the Solimões-Amazonas River System. Neotrop Ichthyol 19(01):e200082

Galeazzi CP, Almeida RP, Mazoca CE, Best JL, Freitas BT, Ianniruberto M et al. (2018) The significance of superimposed dunes in the Amazon River: implications for how large rivers are identified in the rock record. Sedimentology 65(7):2388–2403

Goulding M, Venticinque E, Ribeiro MLDB, Barthem RB, Leite RG, Forsberg B et al. (2019) Ecosystem-based management of Amazon fisheries and wetlands. Fish Fish 20(1):138–158

Gualtieri C, Ianniruberto M, Filizola N, Santos R, Endreny T (2017) Hydraulic complexity at a large river confluence in the Amazon basin. Ecohydrology 10(7):e1868

Gualtieri C, Martone I, Filizola Junior NP, Ianniruberto M (2020) Bedform morphology in the area of the confluence of the Negro and Solimões-Amazon rivers. Brazil. Water 12(6):1630

Guerrero L, Abad JD, Ortals C, Valderde H, Estrada Y, Chicchon H, Canas-Alva C (2024) Hydrogeomorphology of the origin of the Amazon River, the confluence between the Marañón and Ucayali rivers. Earth Surface Processes and Landforms 49(12):3949–3967

He F, Zarfl C, Tockner K, Olden JD, Campos Z, Muniz F, et al. (2024) Hydropower impacts on riverine biodiversity. Nat Rev Earth Environ 5(11):755–772.

Heilpern SA, Sethi SA, Barthem RB, Batista VDS, Doria CR, Duponchelle F, Flecker AS (2022) Biodiversity underpins fisheries resilience to exploitation in the Amazon river basin. Proceedings of the Royal Society B 289(1976):20220726

Henderson PA, Hamilton WD, Crampton WGR (1998) 15. Evolution and diversity in Amazonian floodplain communities. In:. Dynamics of tropical communities: 37th symposium of the British Ecological Society. Cambridge University Press, 385

Hilton EJ, Fernandes CC, Sullivan JP, Lundberg JG, Campos-da-Paz R (2007) Redescription of *Orthosternarchus tamandua* (Boulenger, 1898)(Gymnotiformes, Apteronotidae), with reviews of its ecology, electric organ discharges, external morphology, osteology, and phylogenetic affinities. Proc Acad Nat Sci Philadelphia 156(1):1–25

Ivanyisky IIISJ, Albert JS (2014) Systematics and biogeography of Sternarchellini (Gymnotiformes: Apteronotidae): diversification of electric fishes in large Amazonian rivers. Neotrop Ichthyol 12:565–584

Jézéquel C, Oberdorff T, Tedesco PA, Schmitt L (2022) Geomorphological diversity of rivers in the Amazon Basin. Geomorphology 400:108078

Jimenez EA, Amaral MT, de Souza PL, Costa MDNF, Lira AS, Fredou FL (2020) Value chain dynamics and the socioeconomic drivers of small-scale fisheries on the Amazon coast: a case study in the state of Amapá, Brazil. Marine Policy 115:103856

Junk WJ, Piedade MTF, Schöngart J, Cohn-Haft M, Adeney JM, Wittmann F (2011) A classification of major naturally-occurring Amazonian lowland wetlands. Wetlands 31(4):623–640

Körtzinger A (2003) A significant CO2 sink in the tropical Atlantic Ocean associated with the Amazon River plume. Geophys Res Lett 30:24. https://doi.org/10.1029/2003GL018841

Latrubesse EM (2015) Large rivers, megafans and other Quaternary avulsive fluvial systems: a potential "who's who" in the geological record. Earth-Sci Rev 146:1–30

Littmann MW, Lundberg JG, Rocha MS (2021) Revision of the South American catfish genus *Hypophthalmus* (Siluriformes, Pimelodidae) with descriptions of two new species from the Amazon and Orinoco Basins. Proc Acad Nat Sci Philadelphia 167(1):191–243

Lopes GCDS, Matos OF, Freitas CE (2023) Spatial dynamics of Amazonian commercial fisheries: an analysis of landscape composition and fish landings. Braz J Biol 83:e265791

Lopez-Rojas H, Lundberg JG, Marsh E (1984) Design and operation of a small trawling apparatus for use with dugout canoes. N Am J Fish Manag 4(3):331–334

Lujan NK, Chamon C (2008) Two new species of Loricariidae (Teleostei: Siluriformes) from main channels of the upper and middle Amazon Basin, with discussion of deep water specialization in loricariids. Ichthyol Explor Freshwat 19:271–282

Lundberg JG (1991) *Exallodontus aguanai*, a new genus and species of Pimelodidae (Pisces: Siluriformes) from deep river channels of South America, and delimitation of the subfamily Pimelodinae. Proc Biol Soc Washington 104:840–869

Lundberg JG, Akama A (2005) *Brachyplatystoma capapretum*: a new species of goliath catfish from the Amazon basin, with a reclassification of allied catfishes (Siluriformes: Pimelodidae). Copeia 2005(3):492–516

Lundberg JG, Mago-Leccia F (1986) A review of Rhabdolichops (Gymnotiformes, Sternopygidae), a genus of South American freshwater fishes, with descriptions of four new species. Proc Acad Nat Sci Philadelphia 138:53–85.

Lundberg JG, Py-Daniel LR (1994) Bathycetopsis oliveirai, gen. et sp. nov., a blind and depigmented catfish (Siluriformes: Cetopsidae) from the Brazilian Amazon. Copeia 1994:381–390.

Lundberg JG, Fernandes CC, Albert JS, Garcia M (1996) Magosternarchus, a new genus with two new species of electric fishes (Gymnotiformes: Apteronotidae) from the Amazon River Basin, South America. Copeia 1996:657–670.

Lundberg JG, Fernandes CC, Campos-Da-Paz R, Sullivan JP (2013) *Sternarchella calhamazon* n. sp., the Amazon's most abundant species of apteronotid electric fish, with a note on the taxonomic status of *Sternarchus capanemae* Steindachner, 1868 (Gymnotiformes, Apteronotidae). Proc Acad Nat Sci Philadelphia 162(1):157–173

Lundberg JG, Lewis Jr WM, Saunders III JF, Mago-Leccia F (1987) A major food web component in the Orinoco River channel: evidence from planktivorous electric fishes. Science 237(4810):81–83

Mago-Leccia F, Lundberg JG, Baskin JN (1985) Systematics of the South American freshwater fish genus Adontosternarchus (Gymnotiformes, Apteronotidae). Contrib Sci (Los Angel, Calif) 358:1–19

Marinho RR, Furtado AR, Dos Santos VC, Nascimento AZA, Junior NPF (2022) Riverbed morphology and hydrodynamics in the confluence of complex mega rivers-A study in the Branco and Negro rivers, Amazon basin. J South Am Earth Sci 118:103969

Melo BF, Albert JS, Dagosta FC, Tagliacollo VA (2021) Biogeography of curimatid fishes reveals multiple lowland–upland river transitions and differential diversification in the Neotropics (Teleostei, Curimatidae). Ecol Evol 11(22):15815–15832

Mertes LA, Dunne T, Martinelli LA (1996) Channel-floodplain geomorphology along the Solimões-Amazon river, Brazil. Geol Soc Am Bull 108(9):1089–1107

Miguez-Macho G, Fan Y (2012) The role of groundwater in the Amazon water cycle: 1. Influence on seasonal streamflow, flooding and wetlands. J Geophys Res: Atmos 117(D15)

Mikkola H (2024) Aquaculture and fisheries as a food source in the Amazon region-A review. Food Nutr J 9:286

Potter PE (1978) Significance and origin of big rivers. J Geol 86(1):13–33

Revell LJ (2014) Package 'phytools'. https://cran.rproject.org/web/packages /phytools

Richey JE, Meade RH, Salati E, Devol AH, Nordin Jr CF, Santos UD (1986) Water discharge and suspended sediment concentrations in the Amazon River: 1982–1984. Water Resour Res 22(5):756–764

Saint-Paul U, Zuanon J, Correa MAV, García M, Fabré NN, Berger U, Junk WJ (2000) Fish communities in central Amazonian white-and blackwater floodplains. Environ biol fishes 57(3):235–250

Salo J, Kalliola R, Häkkinen I, Mäkinen Y, Niemelä P, Puhakka M, Coley PD (1986) River dynamics and the diversity of Amazon lowland forest. Nature 322(6076):254–258

Seidel M, Dittmar T, Ward ND, Krusche AV, Richey JE, Yager PL, Medeiros PM (2016) Seasonal and spatial variability of dissolved organic matter composition in the lower Amazon River. Biogeochemistry 131(3):281–302

Seidel M, Yager PL, Ward ND, Carpenter EJ, Gomes HR, Krushe AV, Richey JE, Dittmar T, Medeiros PM (2015) Molecular-level changes of dissolved organic matter along the Amazon River-to-ocean continuum. Mar Chem 177:218–231

Siddiqui SF, Zapata-Rios X, Torres-Paguay S, Encalada AC, Anderson EP, Allaire M, Kaplan DA (2021) Classifying flow regimes of the Amazon basin. Aquat Conserv: Mar Freshw Ecosyst 31(5):1005–1028

Sirén A, Valbo-Jørgensen J (2022) Quantifying fish catches and fish consumption in the Amazon Basin. Aquat Ecosyst Health Manag 25(1):59–71

Smith GR (1992) Phylogeny and biogeography of the Catostomidae, freshwater fishes of North America and Asia. In: Mayden RL (Ed.) PP. 778-826 in Systematics, Historical Ecology, and North American Freshwater Fishes. Stanford University Press, Stanford, California, 969

Stølum HH (1998) Planform geometry and dynamics of meandering rivers. Geol Soc Am Bull 110(11):1485–1498

Tamura LN, Almeida RP, Galeazzi CP, Freitas BT, Ianniruberto M, Prado AH (2019) Upper-bar deposits in large Amazon rivers: occurrence, morphology and internal structure. Sediment Geol 387:1–17

Ternon JF, Oudot C, Dessier A, Diverres D (2000) A seasonal tropical sink for atmospheric CO2 in the Atlantic Ocean: the role of the Amazon River discharge. Mar Chem 68(3):183–201

Torgersen KT, Bogotá-Gregory JD, Acosta-Santos A, Potosí-Chuña JJ, Agudelo-Cordoba E, Albert JS (2023) Gymnotiform electric fishes of the Tres Fronteras region of the western Amazon. Check List 19(5):767–790

Trigg MA, Bates PD, Wilson MD, Schumann G, Baugh C (2012) Floodplain channel morphology and networks of the middle Amazon River. Water Resour Res 48:10

Urbano-Bonilla A, Roa-Varon A, Arce HM (2021) Dr. Javier Alejandro Maldonado-Ocampo. Ichthyology & Herpetology 109(4):1073–1074

Val, AL, de Almeida-Val, VM (2012) Fishes of the Amazon and their environment: physiological and biochemical aspects. Springer Science & Business Media, Berlin, (Vol. 32)

Van der Sleen P, Albert JS (eds) (2017) Field Guide to the Fishes of the Amazon, Orinoco, Guianas. Princeton University Press, Princeton, (Vol. 115).

Wiens JJ, Graham CH (2005) Niche conservatism: integrating evolution, ecology, and conservation biology. Annu Rev Ecol Evol Syst 36(1):519–539.

Winemiller KO, Adite A (1997) Convergent evolution of weakly electric fishes from floodplain habitats in Africa and South America. Environ Biol Fishes 49:175–186

Winemiller KO, McIntyre PB, Castello L, Fluet-Chouinard E, Giarrizzo T, Nam S et al. (2016) Balancing hydropower and biodiversity in the Amazon, Congo, and Mekong. Science 351(6269):128–129

Zacardi DM, Bittencourt SDS, Nakayama L, de Queiroz HL (2017) Distribution of economically important fish larvae (Characiformes, Prochilodontidae) in the Central Amazonia, Brazil. Fish Manag Ecol 24(4):283–291

14 Neotropical Freshwater Fishes in Hypogean Waters

Maria Elina Bichuette

Abstract

Subterranean or hypogean fishes spend their whole lives underground, either in caves or groundwaters (aquifers), and many species have adaptations that allow them to persist over evolutionary time in these specialized and often extreme environments. Hypogean fishes worldwide are moderately diverse, with about 300 species, and fish lineages representing at least 28 families have colonized subterranean habitats and phreatic environments multiple times. Of these, there are at least 57 hypogean species representing 14 families in the Neotropical realm. Hypogean environments tend to be more stable over ecological time frames in terms of physico-chemical variation (temperature and water chemistry), although they are often transient over evolutionary time scales of millions of years. Hypogean fishes in the Neotropics and worldwide exhibit a similar suite of cave-adapted phenotypes, reflecting similar and extreme phreatic environmental conditions. The most prominent influences are little to no sunlight or primary productivity, leading to lower predation and food and space restrictions. Traits most often reduced are those that are inefficient and costly with respect to fitness in conditions of permanent darkness—diurnal activity, predominant visual orientation, photoautotrophism, and some metabolic paths dependent on light (e.g., calcium absorption in most vertebrates). In addition, some foraging strategies are not as effective inside caves as they are in surface habitats. This distinctive set of characteristics found only in subterranean environments, especially loss of image-forming eyes and melanin pigments is referred to as the "blind, white, cave animal" syndrome. Other troglomorphic traits are gains or specializations related to life under nutrient-poor conditions. Examples include an expanded laterosensory system, loss of the trunk (posterior) lateral line as an adaptation to inhabit small spaces (*Stygichthys typhlops*, *Phreatobius*, and *Rhamdiopsis krugi*), body size reduction (*Ituglanis epikarsticus,* and *Rhamdiopsis krugi*), and loss of behavioral inhibition for surface swimming (*Astyanax mexicanus*).

M. E. Bichuette (✉)
Universidade Federal de São Carlos, São Carlos, SP, Brazil
e-mail: lina.cave@gmail.com

14.1 Introduction to Hypogean Fishes

Subterranean or hypogean fishes spend their whole lives underground, either in caves which have relatively large water conduit sizes (e.g., voluminous lakes reaching 60 m of depth), or in groundwater or phreatic aquifers where the conduit size is smaller. Hypogean fishes are adapted to inhabit a range of specialized and often extreme environments, such as fissures of the upper vadose zone above the water-table, or interstitial spaces of the hyporheic zone with the water table (see Juberthie 2000 for descriptions of these subterranean habitats). Hypogean fishes worldwide are moderately diverse, with about 300 species, and hypogean fishes have evolved multiple times, with species representing at least 28 families. Of these, there are at least 57 hypogean species representing 14 families in the Neotropical realm.

Hypogean species may be described as troglobitic, troglobiont, troglobiotic, or stygobiotic, all terms referring to animals that are restricted to subterranean environments. As such, these terms have both ecological and evolutionary implications (*sensu* Schiner 1854; Racovitza 1907). Hypogean environments are highly heterogeneous through time and across space (see Fig. 14.1). Obligate subterranean fishes may occupy a variety of specialized habitats at the macro (>10 m), meso (1–10 m) and micro (<1 m) scales, in different climate zones, and under varying water quality and lithological conditions (Bichuette and Gallão 2021; Trajano 2021). Caves and other subterranean waterways develop in different types

J. S. Albert and R. E. Reis (eds.), *Ecology and Evolution of Amazonian Fishes*, Fascinating Life Sciences,
https://doi.org/10.1007/978-3-032-09328-8_14

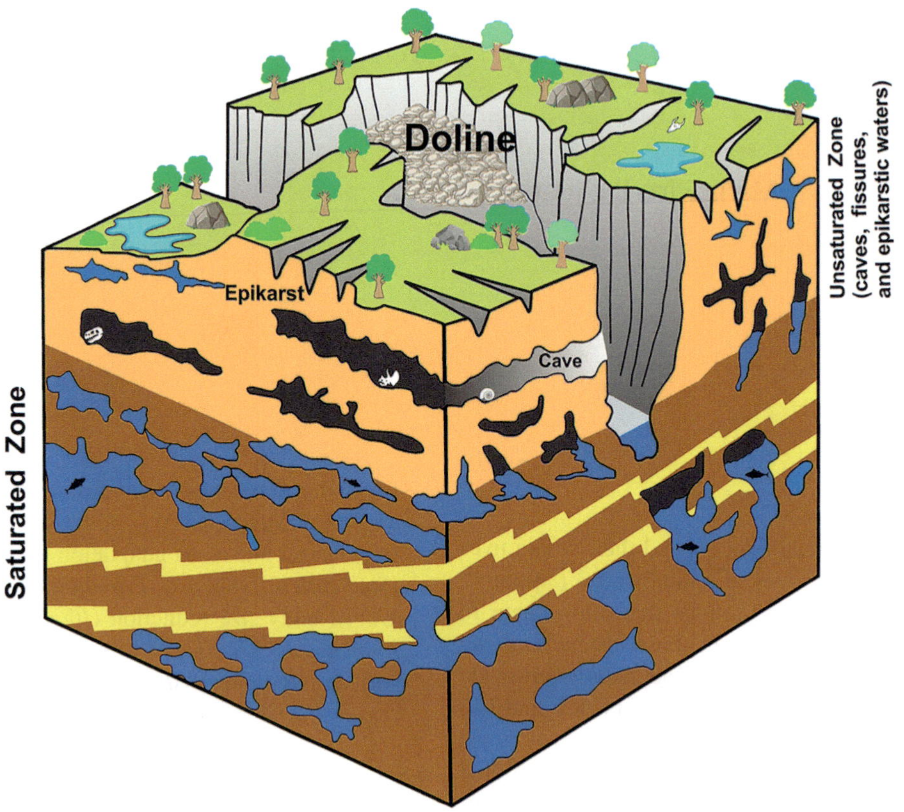

Fig. 14.1 Diagram of subterranean aquatic ecosystems showing the saturated zone (phreatic waters), unsaturated zone (dry caves, subterranean level base streams, epikarstic/upper aquifers), dolines (depression or hole in the ground caused by some form of collapse of the surface rocky layer), and other cave entrances (caused by chemical erosion of acidified water)

of rock substrates (i.e., lithology), and subterranean aquatic habitats occur in the water flowing through the sediment bed beneath rivers and streams (i.e., the under-river flow in the hyporheic zone), in phreatic outcrops (i.e., saturated zone) of flooded caves, and in the upper aquifer through sediments of the regolith (Juberthie 2000) (Figs. 14.1 and 14.2).

Groundwater from the hyporheic zone also emerges from alluvial sediments along river margins and at the transition from floodplain to upland soils. Anchialine pools (e.g., cenotes) are landlocked pools with surface freshwater and subterranean connections to the ocean, that often occur along coastlines of karstic (limestone) platforms, for example, in the Caribbean (Bahamas, Belize, Cuba, Mexico) and also on volcanic oceanic islands (e.g., Galapagos) (Bátori et al. 2019; Arroyave et al. 2021). Besides this great diversity of habitats, these fish are adapted to different ecological conditions, ranging from severe food scarcity, often seasonal, to relative climatic stability and, in some cases, abundance of food resources (tropical regions). Other specializations are often observed, as hypertrophy of some sensory structures (chemoreceptors and mechanoreceptors) and expanse of the space used (Soares and Niemiller 2013).

There is a marked bias in studies of hypogean fishes, with most research effort focused on a few well-studied species, focusing especially on evolutionary, molecular developmental, and biogeographic patterns (e.g., the tetra-characin *Astyanax mexicanus*). However, the majority of studies are more related to taxonomic than to ecological or behavioral data. Much of the subterranean Neotropical fish fauna is still unknown due to the lack of collections and the formation of researchers with expertise in subterranean fauna (e.g., Argentina, Chile, Ecuador, Peru, Venezuela, Uruguay, the Caribbean countries, and even in Brazil), which strongly impedes science-based conservation policies and actions (Mammola et al. 2021).

14.2 Taxonomic Diversity

The Neotropical subterranean ichthyofauna stands out for its taxonomic diversity, both at the family level (14 families, Fig. 14.3) and generic (22 genera, Fig. 14.3). Currently, there are records of 57 exclusively subterranean fish, occurring in caves and other subterranean habitats in 12 countries (Table 14.1). Some are monotypic, geographic relicts, and/or of marine origin (e.g., the genera *Lucifuga, Ogilbia*, and *Typhlias*). Most of these fish are threatened, mainly due to pollution and water withdrawals from rivers and aquifers (Bichuette and Trajano 2010; Bichuette and Gallão 2021; Raghavan et al. 2021).

Fig. 14.2 Examples of subterranean aquatic habitats. (**A**) Limestone flooded cave at Mato Grosso do Sul state, Brazil, a lentic habitat; (**B**) Base level stream in a sandstone cave from Bahia state, Brazil; (**C**) a phreatic lake inside a limestone cave at Bahia state, Brazil, the width reaches 40 m and the depth surpasses 60 m; (**D**) a sinkhole showing a base level stream, crossing a limestone cave at São Paulo state, Brazil

14.3 Functional Diversity

In surface habitats, climate conditions are more variable than in subterranean habitats, and this aspect can act as an environmental filter, decreasing the number of viable strategies and, ultimately, decreasing functional diversity. The physical (abiotic) conditions inside caves and aquifers are usually more stable than in surface waters, with lower annual variation in temperature, water flow, and water chemistry (e.g., DO_2, pH, etc.). However, other aspects can act as constraints for the establishment of populations, such as food and space restrictions. The traits most often filtered out are those that are inefficient and costly with respect to fitness in conditions of permanent darkness—diurnal activity, predominant visual orientation, photoautotrophism, and some metabolic paths dependent on light (e.g., calcium absorption in most vertebrates). In addition, some foraging strategies are not as effective inside caves as they are in surface habitats.

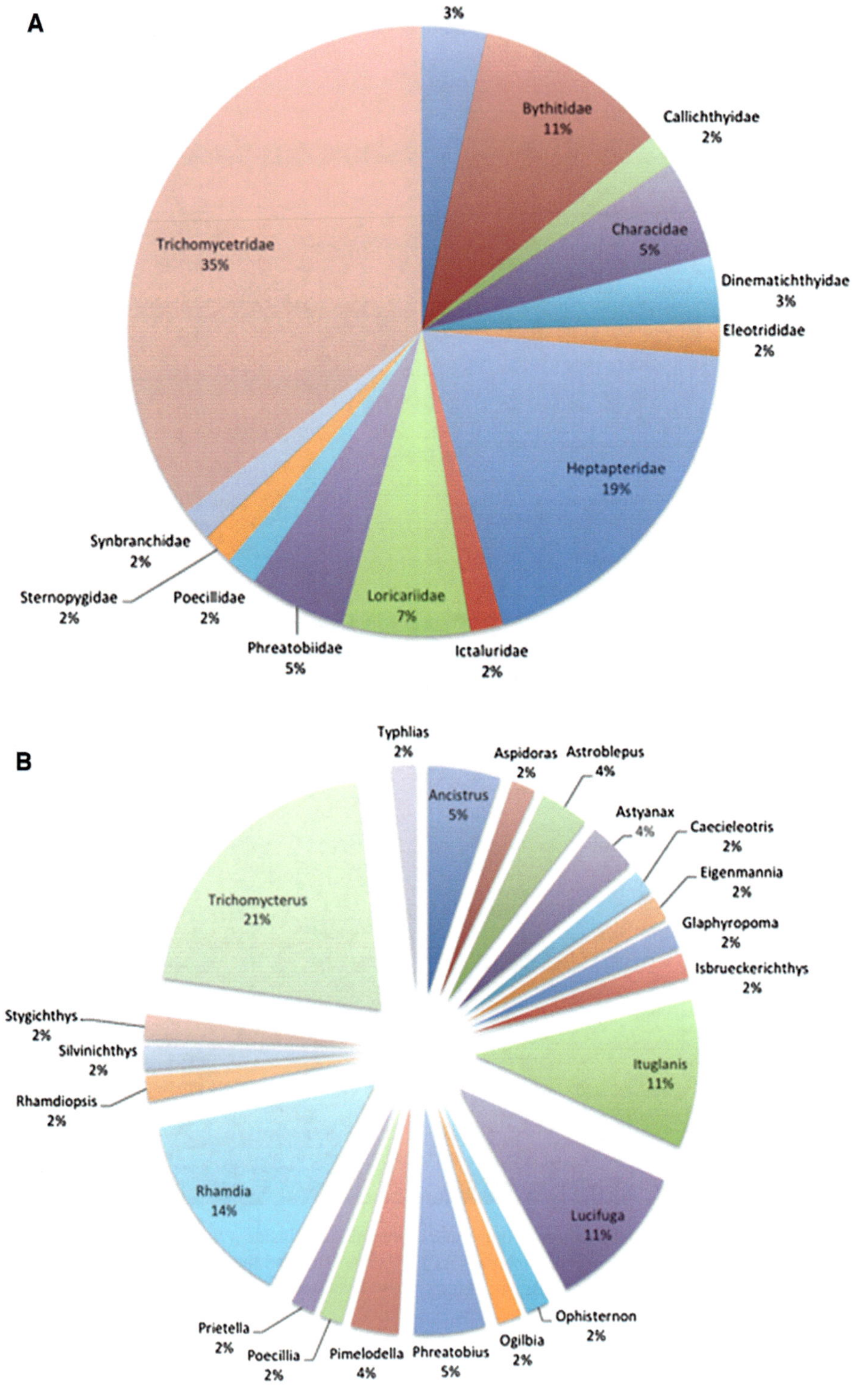

Fig. 14.3 Taxonomic distribution of Neotropical subterranean fishes by family (**A**) and genus (**B**)

The comparatively low food supply is one of the main limiting factors in caves (Hüppop 2000), and the consequent reduced metabolic rate has long been suggested as an adaptation to the general food scarcity in hypogean habitats. However, comparisons of metabolic rates between species have not often included closely related surface and cave species. Besides food scarcity, many cave animals are faced with seasonal availability of resources or even infrequent input of food and have the ability to survive under starvation periods (Culver and Pipan 2009). There are several experimental

Table 14.1 Subterranean neotropical fish species

Valid species	Country	Habitat	Observation
Ancistrus cryptophthalmus	Brazil	Cave stream	Five populations
Ancistrus formoso	Brazil	Flooded cave	
Ancistrus galani	Venezuela	Cave stream	
Aspidoras mephisto	Brazil	Cave stream	
Astroblepus pholeter	Ecuador	Cave stream	
Astroblepus riberae	Peru	Cave stream	
Astyanax jordani[a]	Mexico	Cave stream phreatic	*A. mexicanus* complex
Astyanax mexicanus[a]	Mexico	Cave stream	*A. mexicanus* complex
Caecieleotris morrisi	Mexico	Phreatic	
Eigenmannia vicentespelaea	Brazil	Cave stream	
Glaphyropoma spinosum	Brazil	Cave stream	
Isbrueckerichthys alipionis[b]	Brazil	Vadose tributary	Epigean/isolated
Ituglanis bambui	Brazil	Vadose tributary	
Ituglanis boticario	Brazil	Cave stream	
Ituglanis epikarsticus	Brazil	Epikarst	
Ituglanis mambai	Brazil	Cave stream	
Ituglanis passensis	Brazil	Cave stream	
Ituglanis ramiroi	Brazil	Vadose tributary	
Lucifuga dentata	Cuba	Anchialine habitat	Marine origin
Lucifuga gibarensis	Cuba	Anchialine habitat	Marine origin
Lucifuga lucayana	Bahamas	Anchialine habitat	Marine origin
Lucifuga simile	Cuba	Anchialine habitat	Marine origin
Lucifuga spelaeotes	Bahamas	Anchialine habitat	Marine origin
Lucifuga subterranea	Cuba	Anchialine habitat	Marine origin
Ogilbia galapagosensis	Galapagos	Anchialine habitat	Marine origin
Ophisternon infernale	Mexico	Phreatic	
Phreatobius cisternarum	Brazil	Hyporheic (alluvium)	Several populations
Phreatobius dracunculus	Brazil	Hyporheic (alluvium)	
Phreatobius sanguijuela	Bolivia, Brazil	Hyporheic (alluvium)	
Pimelodella kronei	Brazil	Cave stream	
Pimelodella spelaea	Brazil	Vadose tributary	
Poecilia mexicana	Mexico	Cave stream	Sulfurous waters
Prietella phreatophila	Mexico and USA	Phreatic	

(continued)

Table 14.1 (continued)

Valid species	Country	Habitat	Observation
Rhamdia enfurnada	Brazil	Cave stream	
Rhamdia guasarensis	Venezuela	Cave stream	
Rhamdia laluchensis	Mexico	Cave stream	
Rhamdia laticauda	Belize	Cave stream	
Rhamdia macuspanensis	Mexico	Cave stream	
Rhamdia reddelli	Mexico	Cave stream	
Rhamdia urichi	Trinidad and Tobago	Cave stream	
Rhamdia zongolicensis	Mexico	Cave stream	
Rhamdiopsis krugi	Brazil	Upper phreatic (cave)	
Silvinichthys bortayro	Argentina	Phreatic	
Stygichthys typhlops[c]	Brazil	Phreatic	Monotypic genus
Trichomycterus chaberti	Bolivia	Cave stream	
Trichomycterus dali	Brazil	Flooded cave	Several populations/lineages?
Trichomycterus donascimientoi	Colombia	Cave stream	
Trichomycterus itacarambiensis	Brazil	Cave stream	
Trichomycterus rosablanca	Colombia	Cave stream	
Trichomycterus rubbioli	Brazil	Upper phreatic (cave)	
Trichomycterus sandovali	Colombia	Cave stream	
Trichomycterus santanderensis	Colombia	Cave stream	
Trichomycterus sketi	Colombia	Cave stream	
Trichomycterus spectrum	Colombia	Cave stream	
Trichomycterus spelaeus	Venezuela	Cave stream	
Trichomycterus uisae	Colombia	Cave stream	
Typhlias pearsei	Mexico	Phreatic	Marine origin

Taxa listed alphabetically by genus and species. Details of subterranean habitat included. ?, no data about feeding habit
[a]Comprises a complex group (*Astyanax mexicanus*)
[b]Represents an epigean (surface) species with an isolated population in a single cave of southeastern Brazil
[c]Represents a monotypic genus

works on metabolic rates focusing subterranean organisms showing this pattern, since crustaceans (isopods, decapods, and amphipods), spiders, etc. For cavefishes, the first experiments were conducted for cyprinids from Africa, and amblyopsids from the USA, both considering oxygen consumption. Another aspect, growth rates, was studied for nemacheilids from India. All these experiments showed smaller growth rates and respiratory frequencies, compared to the epigean ones (see review on Hüppop 2000). Few works focused on this aspect considering the Neotropical hypogean ichthyofauna.

Some studies have shown the opposite result, such as experiments related to the production of electric fields. Electrogenesis may be energetically costly (but see Julian et al. 2003), and caves commonly have reduced food resources. Fortune et al. (2020) studied these aspects in the troglobitic gymnotiform *Eigenmannia vicentespelaea* and the epigean *E. trilineata*. The troglobitic species showed stronger electric fields than the surface fish and also larger electric organs. These differences may result from differences in movement: the cavefish appeared to maintain territories whereas the surface fish did not (at night). The fact that *E. vicentespelaea* produces such energetically costly electric fields suggests that sufficient food resources are available and accessible in the cave habitat, as observed by Bichuette and Trajano (2015).

Looking at ecological and physiological adjustments is particularly relevant to determining whether a species is adapted to living in complete darkness with attendant food restriction, for example. Some functional traits observed in cavefish are presented in Table 14.2, from habits to behavior. These traits could represent specializations or even a character-state shared with other species from epigean habitats. Because of this, it is fundamental to consider the pool of species in the community as a whole. Due to sparse information, part of the possible eco-physiological specializations is present in the text and not in Table 14.2. For example, one intrinsic eco-physiological specialization of troglobites in general is the loss of their ability to cope with environmental fluctuations, since they evolved in relatively constant abiotic conditions.

14.4 Comparisons Among Global Hypogean Faunas

The Neotropics has the greatest documented taxonomic and phylogenetic diversity of subterranean fishes of any region on Earth, with 57 species in 22 genera and 14 families. By comparison, China has at least 88 species but only 12 genera and four families. Southeast Asia has 27 species in 17 genera and eight families (Indonesia, Laos, Malaysia, Myanmar, Philippines, Thailand, Vietnam), and India has 17 species in nine genera and seven families. By comparison, the Middle East has nine species in three genera and two families (Iran, Iraq, and Oman); Continental Africa has six species in six genera and three families; Madagascar has five species in two genera and two families; United States has 11 species in eight genera and three families; Australia has three species in two genera, and for the entire Europe there is only one species of cavefish, discovered in 2016 (for an update checklist see website https://cavefishes.org.uk; Behrmann-Godel et al. 2017).

The environmental conditions are not similar among the regions, since we have different climatic conditions and biogeographical patterns, except for the occurrence of typical subterranean habitats, with total darkness in the deep zones, and food and space restrictions (with exceptions). Because of this, we note a mosaic in some aspects considered specialized attributes, such as melanin pigmentation, eye condition, morphological traits, ecological and behavioral aspects.

14.5 Synthesis of Major Diversity Patterns

This can include discussions of within-habitat radiations, parallel and convergent evolution. This can also include your own informed perceptions about the contributing roles of hypothesized intrinsic (i.e., genetic, developmental, phylogenetic) and extrinsic (local environmental, geographic, climatological) factors in producing these patterns of taxonomic and functional diversity.

During its course from deposition as rainwater to eventual arrival at the oceans, some freshwater penetrates the surface to flow through various types and various depths of rocks, sediments, and subterranean habitats. These habitats vary from the tiny interstices between sand grains, through larger caves, to vast groundwater aquifers that lie tens to hundreds of meters below the ground surface. Evolution has been at work within these subterranean places and animals that have found themselves in the darkness have adapted and evolved. They have evolved a set of very distinctive characteristics typical of, and found only in, subterranean environments. These are found in all taxa, vertebrates and invertebrates, which are now obligate inhabitants of the subterranean (or hypogean) realm. Some of these animals lose their eyes and melanin pigmentation.

Although hypogean environments often tend to be relatively more stable than epigean surface water habitats over ecological time frames, karstic and phreatic landforms are often geologically unstable and can change rapidly at evolutionary scales (Benito-Calvo et al. 2018; White 2020). As a result, extinction rates may be high for at least some hypogean species, many of which are relatively young lineages with ages of thousands (van Hengstum et al. 2019) to just a few million (Policarpo et al. 2021) years. However, some hypogean fish clades in North America have been shown to be more ancient, dating back tens

Table 14.2 Selected functional traits observed in neotropical subterranean fish species

Species	Feeding	Use of space	Social behavior	Water flow
Ancistrus cryptophthalmus	Detritivorous	Bottom	Gregarious/solitary (adults)	Lotic
Ancistrus formoso	Detritivorous	Bottom	?	Lentic
Ancistrus galani	?	Bottom	?	?
Aspidoras mephisto	Detritivorous	Bottom and midwater	Gregarious	Lentic
Astroblepus pholeter	Carnivore	?	?	Lentic
Astroblepus riberae	?	?	?	Lentic
Astyanax jordani	Generalist	Midwater	Gregarious	Lentic-lotic
Astyanax mexicanus	Generalist	Midwater	Gregarious	Lentic-lotic
Caecieleotris morrisi	?	?	?	?
Eigenmannia vicentespelaea	Carnivore	Bottom and midwater	Gregarious/solitary (adults)	Lentic
Glaphyropoma spinosum	Carnivore	Bottom and midwater	Gregarious	Lentic
Isbrueckerichthys alipionis	?	Bottom	?	Lentic
Ituglanis bambui	Carnivore	Bottom and midwater	Solitary	Lentic
Ituglanis boticario	Carnivore	Bottom	Solitary	Lentic
Ituglanis epikarsticus	Carnivore	Bottom and midwater	Solitary	Lentic
Ituglanis mambai	Carnivore	Bottom	?	Lentic
Ituglanis passensis	Carnivore	Bottom and midwater	Solitary	Lentic-lotic
Ituglanis ramiroi	Carnivore	Bottom and midwater	Solitary	Lentic
Lucifuga dentata	Carnivore	?	?	Lentic
Lucifuga gibarensis	Carnivore	?	?	Lentic
Lucifuga lucayana	Carnivore	?	?	Lentic
Lucifuga simile	Carnivore	?	?	Lentic
Lucifuga spelaeotes	Carnivore	?	?	Lentic
Lucifuga subterranea	Carnivore	?	?	Lentic
Ogilbia galapagosensis	?	?	?	?
Ophisternon infernale	Carnivore	Bottom	?	Lentic
Phreatobius cisternarum	Carnivore	Bottom	Gregarious	Lentic
Phreatobius dracunculus	Carnivore	Bottom	Gregarious	Lentic
Phreatobius sanguijuela	Carnivore	Bottom	Gregarious	Lentic
Pimelodella kronei	Carnivore	Bottom and midwater	Solitary	Lentic-lotic
Pimelodella spelaea	Carnivore	Bottom and midwater	Solitary	Lentic
Poecilia mexicana	Generalist	Midwater	Gregarious	Lentic-lotic
Prietella phreatophila	?	?	?	?

Rhamdia enfurnada	Carnivore	Bottom and midwater	Gregarious/solitary (adults)	Lentic
Rhamdia guasarensis	?	?	?	?
Rhamdia laluchensis	?	?	?	?
Rhamdia laticauda	Carnivore	?	?	?
Rhamdia macuspanensis	?	?	?	?
Rhamdia reddelli	?	?	?	?
Rhamdia urichi	?	?	?	?
Rhamdia zongolicensis	?	?	?	?
Rhamdiopsis krugi	Carnivore	Midwater and surface	Solitary	Lentic
Silvinichthys bortayro	?	?	?	?
Stygichthys typhlops	Carnivore	Midwater and surface	Solitary	Lentic
Trichomycterus chaberti	Carnivore	Bottom	?	?
Trichomycterus dali	Carnivore	Bottom and midwater	Solitary	Lentic
Trichomycterus donascimientoi	?	?	?	?
Trichomycterus itacarambiensis	Carnivore	Bottom	Solitary	Lentic
Trichomycterus rosablanca	?	?	?	?
Trichomycterus rubbioli	Carnivore	Bottom	?	Lentic
Trichomycterus sandovali	?	?	?	?
Trichomycterus santanderensis	?	?	?	?
Trichomycterus sketi	?	?	?	?
Trichomycterus spectrum	?	?	?	?
Trichomycterus spelaeus	?	?	?	?
Trichomycterus uisae	?	?	?	?
Typhlias pearsei	Carnivore	?	?	?

?, no available data

of millions of years; e.g., in Ictaluridae (Arce-H et al. 2017) and Amblyopsidae (Ghezelayagh et al. 2021).

These losses have led to the oft-used phrase "blind white cave animals." In addition to these losses, there is a set of gains, which are also specializations related to life without light and most often very food-poor conditions. For example, the lateral line can be expanded, besides behavioral, physiological, and ecological specializations, referred to in the literature as troglomorphic traits (Christiansen 2012). However, on the contrary, life in small spaces can also be the cause for loss of the lateral line in the body observed in *Stygichthys typhlops*, the species of *Phreatobius*, and *Rhamdiopsis krugi* (Bichuette and Trajano 2010). Evolutionary reduction in body size is discussed for some Brazilian subterranean fish (e.g., *Ituglanis epikarsticus* and *Rhamdiopsis krugi*) (Bichuette and Trajano 2004; Bichuette et al. 2015). Also, diversity in the degree of reduction of eyes and melanin pigmentation is commonly observed, and this variation has no clear taxonomic correlation (Fig. 14.4).

Morphological and physiological studies indicate that the diversity in modes of pigmentation variation results from a combination of different mechanisms of pigmentation reduction and the ability to synthesize melanin. Apparently, a polygenic system is involved in the differentiation of melanophores, determining their density in the skin as observed for *Astyanax mexicanus* cavefish and their surface-dwelling counterparts, model system for studying the role of development in the evolution of cave-related phenotypes. The loss of melanophore pigmentation is caused by a specific block in the ability of undifferentiated melanoblasts to convert L-tyrosine to L-DOPA, a precursor of melanin (Borowsky 2010).

Some authors state that the relative degree of troglomorphy is related directly to the time of evolution—the phylogenetic age—of the particular taxon in question (e.g., Porter et al. 2007). More than 80% of Neotropical subterranean fish species are otophysans. The explanation for this is that otophysans make up the majority of the primary freshwater fishes, and it is mainly freshwater fishes that can get into the subterranean domain. Only the families Bythitidae and Eleotridae have invaded caves from the sea.

The high diversity of Neotropical subterranean fishes when compared to other regions worldwide (e.g., Australia, USA, Africa, India) may be due to one or all the following factors: (1) general high ichthyofaunal diversity in the basins of Neotropical region and also the special habitats in the Caribbean region (anchialine caves; (2) high research efforts than in most other regions (Brazil and Mexico represent the best studied regions); and/or (3) the essentially comparative approach by researchers, using phylogenetic criteria for the recognition of the troglobitic status, which is primarily a distributional attribute (the restriction to the subterranean environment).

Isolation during past drier climates seems to be the case of the blind catfish, *Pimelodella kronei*, currently in secondary contact with its epigean relative, *P. transitoria*. The Alto Ribeira karst area is presently situated in a wet zone, but there is good evidence of dry, isolation phases in subtropical Brazil during the last 116,200 years (Cruz-Jr. et al. 2005). On the other hand, part of the karst areas in Bahia State, Northeastern Brazil,

Fig. 14.4 Examples of cavefish showing homogenous patterns of depigmentation and anophthalmy (**A**, **B**, and **C**) and heterogeneity in these characters (**D**, **E**, and **F**). (**A**) *Rhamdiopsis* new species; (**B**) *Stygichthys typhlops*; (**C**) *Rhamdiopsis krugi*; (**D**) *Trichomycterus* new species; (**E**) *Ancistrus cryptophthalmus*; (**F**) *Aspidoras mephisto*

are located in currently semiarid regions, a possible phase favoring differentiation of the species of *Rhamdiopsis* and *Trichomycterus*, besides the cave catfish *Rhamdia enfurnada*. All these species live in water bodies without direct epigean connections, and in some localities, there is evidence pointing to accentuate paleoclimatic fluctuations in the past 210,000 years (Wang et al. 2004). The phreatic habitats of *Stygichthys typhlops* have no permanent connections with epigean drainages. In this case, lowering of the water table due to extensive water pumping for irrigation projects in the last seven decades is speeding up the disconnection process and threatening the species by habitat loss (Moreira et al. 2010; pers. obs.).

Few works have focused on the evolution of traits considering the tempo of genetic decay for Neotropical subterranean fishes. We can cite the work of Porter and collaborators (2007) for *Astyanax mexicanus* and, recently, that of Policapo et al. (2020), who studied some general principles on the extent and tempo of the decay of genes involved in vision, circadian clock, and pigmentation in cavefishes. The authors considered two species of Cuban cavefish of the genus *Lucifuga*, and compared to the considered recently evolved cavefish, *Astyanax mexicanus*, and three species belonging to the Chinese genus *Sinocyclocheilus* (Cyprinidae). Their analyses suggested that blind cavefishes for which genomes are available are not very ancient and that they all lost their eyes during the Pleistocene, and the limited extent of gene decay in these cavefishes and the very small number of loss-of-function mutations per pseudogene suggested that their eye degeneration may not be very ancient, ranging from early to late Pleistocene.

Acknowledgments I thank the Laboratório de Estudos Subterrâneos/UFSCar team for support of field work, to the financial agencies who supported the projects about cavefishes in Brazil (Conselho Nacional de Desenvolvimento Científico e Tecnológico/CNPq; FAPESP, Fundação de Amparo à Pesquisa do Estado de São Paulo/FAPESP; Fundação o Boticário de Proteção à Natureza); to J.S. Albert and R.E. Reis for the invitation to collaborate with this subject, and to J.S. Albert, R.E. Reis, and J.E. Gallão for all suport and the critical suggestions in the work.

References

Arce-H M, Lundberg JG, O'Leary MA (2017) Phylogeny of the North American catfish family Ictaluridae (Teleostei: Siluriformes) combining morphology, genes and fossils. Cladistics 33(4):406–428

Arroyave J, Martinez CM, Martínez-Oriol FH, Sosa E, Alter SE (2021) Regional-scale aquifer hydrogeology as a driver of phylogeographic structure in the neotropical catfish *Rhamdia guatemalensis* (Siluriformes: Heptapteridae) from cenotes of the Yucatán Peninsula, Mexico. Freshw Biol 66(2):332–348

Bátori Z, Vojtkó A, Maák IE, Lőrinczi G, Farkas T, Kántor N et al (2019) Karst dolines provide diverse microhabitats for different functional groups in multiple phyla. Sci Rep 9(1):1–13

Behrmann-Godel J, Nolte AW, Kreiselmaier J, Berka R, Freyhof J (2017) The first European cave fish. Curr Biol 27(7):257–258. https://doi.org/10.1016/j.cub.2017.02.048 PMID: 28376329

Benito-Calvo A, Ortega AI, Navazo M, Moreno D, Pérez-Gonzalez A, Parés JM et al (2018) Pleistocene geodynamic evolution of the Arlanzón valley: implications for the formation of the endokarst system and the open-air archaeological sites of the Sierra de Atapuerca (Burgos, España). Bol Geol Min 129(1/2):59–82

Bichuette ME, Trajano E (2010) Conservation of subterranean fishes. In: Trajano E, Bichuette ME, Kapoor BG (eds) Biology of subterranean fishes. Science Publishers, New Hampshire, pp 65–80

Bichuette ME, Trajano E (2015) Population density and habitat of an endangered cave fish *Eigenmannia vicentespelaea* Triques, 1996 (Ostariophysi: Gymnotiformes) from a karst area in central Brazil. Neotrop Ichthyol 13:113–122. https://doi.org/10.1590/1982-0224-20140095

Bichuette ME, Gallão JE (2021) Under the surface: what we know about the threats to subterranean fishes in Brazil. Neotrop Ichthyol 19(3):e210089. https://doi.org/10.1590/1982-0224-2021-0089

Bichuette ME, Trajano E (2004) Three new subterranean species of Ituglanis from Central Brazil (Siluriformes:Trichomycteridae). Ichthyol Expl Fresh 15:243–256

Bichuette ME, Rantin B, Hingst-Zaher E, Trajano E (2015) Evolution of a subterranean catfish. Biol J Linn Soc Lond 114:136–151. https://doi.org/10.1111/bij.12405

Borowsky R (2010) The evolutionary genetics of cave fishes:convergence, adaptation and pleiotropy. In: Trajano E, Bichuette ME, Kapoor BG (eds) Biology of subterranean fishes. Science Publishers, New Hampshire, pp 115–140

Christiansen K (2012) Morphological adaptations. In: White WB, Culver DC (eds) Encyclopedia of Caves. Academic Press, Waltham, pp 517–528. https://doi.org/10.1016/b978-0-12-383832 2.00075-x

Cruz JrFW, Burns SJ, Karmann I, Sharp WD, Vuille M, Cardoso AO, Ferrari JA, Dias PLS, Viana JrO (2005) Insolation-driven changes in atmospheric circulation over the past 116,000 years in subtropical Brazil. Nature 434:63–66

Culver DC, Pipan T (2009) The biology of caves and other subterranean habitats. Oxford University Press, Oxford, p 256

Fortune ES, Andanar N, Madhav M, Jayakumar RP, Cowan NJ, Bichuette ME, Soares D (2020) Spooky interaction at a distance in cave and surface dwelling electric fishes. Front Integr Neurosci 14:561524. https://doi.org/10.3389/fnint.2020.561524

Ghezelayagh A, Harrington RC, Burress ED, Campbell MA, Buckner JC, Chakrabarty P et al (2021) Prolonged morphological expansion of spiny-rayed fishes following the end-Cretaceous. bioRxiv

van Hengstum PJ, Cresswell JN, Milne GA, Iliffe TM (2019) Development of anchialine cave habitats and karst subterranean estuaries since the last ice age. Sci Rep 9(1):1–10. https://doi.org/10.1038/s41598-019-48058-8

Hüppop K (2000) How do cave animals cope with the food scarcity in caves?. In: Wilkens H, Culver DC, Humphreys WF (eds) Ecosystems of the world, vol. 30: subterranean ecosystems. Elsevier Press, Amsterdam, pp 159–188

Juberthie C (2000) The diversity of the karstic and pseudokarstic hypogean habitats in the world. In: Wilkens H, Culver DC, Humphreys WF (eds) Ecosystems of the world, vol. 30: subterranean ecosystems. Elsevier Press, Amsterdam, pp 17–39

Julian D, Crampton WG, Wohlgemuth SE, Albert JS (2003) Oxygen consumption in weakly electric neotropical fishes. Oecologia 137:502–511

Mammola S, Lunghi E, Bilandžija H, Cardoso P, Grimm V, Schmidt SI et al (2021) Collecting eco-evolutionary data in the dark: impediments to subterranean research and how to overcome them. Ecol Evol 11(11):5911–5926

Moreira CR, Bichuette ME, Oyakawa OT, de Pinna MCC, Trajano E (2010) Rediscovery and redescription of the unusual subterranean characiform *Stygichthys typhlops*, with notes on its life history. J Fish Biol 76(7):1815–1824. https://doi.org/10.1111/j.1095-8649.2010.02625.x

Policarpo M, Fumey J, Lafargeas P, Naquin D, Thermes C, Naville M et al (2021) Contrasting gene decay in subterranean vertebrates:

insights from cavefishes and fossorial mammals. Mol Biol Evol 38(2):589–605. https://doi.org/10.1093/molbev/msaa249

Porter ML, Dittmar K, Perez-Losada M (2007) How long does evolution of the troglomorphic form take? Estimating divergence times in *Astyanax mexicanus*. In: Kranjc A, Gabrovsek F, Culver DC, Sasowsky ID (eds) Time in Karst, Karst Water Institute special publication 12. Karst Waters Institute, Leesburg, pp 173–182

Racovitza EG (1907) Essai sur les problèmes biospéologiques. Arch Zool Exp Gén 6:371–488

Raghavan R, Britz R, Dahanukar N (2021) Poor groundwater governance threatens ancient subterranean fishes. Trends Ecol Evolut 36(10):875–878

Schiner JR (1854) Fauna der Adelsberger-, Lueger- und Magdalen-Grotte. Verh Zool-Bot Ges 3:1–40

Soares D, Niemiller ML (2013) Sensory adaptations of fishes to subterranean environments. BioScience 63(4):274–283

Trajano E (2021) Diversity of Brazilian troglobitic fishes: models of colonization and differentiation in subterranean habitats. Diversity 13(3):106. https://doi.org/10.3390/d13030106

Wang X, Auler AS, Edwards RL, Cheng H, Cristalli PS, Smart PL et al (2004) Wet periods in northeastern Brazil over the past 210 kyr linked to distant climate anomalies. Nature 432:740–743

White WB (2020) Rate processes: chemical kinetics and karst landform development. Routledge, London, pp 227–248

15 Folivorous and Frugivorous Neotropical Freshwater Fishes

Joshua P. Egan and Matthew A. Kolmann

Abstract

The Neotropics are home to a rich diversity of folivorous and frugivorous fishes, such as the enigmatic folivorous Discus *Symphysodon* spp., a popular aquarium fish, and the frugivores Tambaqui *Colossoma macropomum* and Red Belly Pacu *Piaractus brachypomus*, which are important food fishes for human consumption. Folivorous fishes consume the roots, stems, leaves, and flowers of aquatic and terrestrial plants, and frugivores consume fruits and seeds. Early in ontogeny, folivorous and frugivorous fishes consume plankton, small nekton, and in the case of Discus, mucus exuded from the skin of their parents, before transitioning to diets containing large quantities of plant material. The ecology of frugivorous and folivorous fishes has not been extensively studied, but research has revealed that these fishes generally exhibit seasonal differences in the amount and types of plants and fruits and seeds they consume. Folivorous and frugivorous fishes have also been shown to facilitate key food web linkages and frugivores play major roles in seed dispersal (ichthyochory) within Neotropical floodplain ecosystems, a function that is relatively rare amongst fishes. Ancestral state reconstructions inferred that folivory independently evolved at least six times and frugivory at least five times in Neotropical fishes. The timing of these transitions during the Eocene and Oligocene corresponds to the diversification of many of the plant clades that are important in the diets of these fishes. There is preliminary evidence of morphological adaptations associated with folivory and frugivory involving a variety of structures, including digestive tracts, abdominal cavities, teeth, skulls, lips, and cranial musculature, but additional work is needed to better understand adaptations that have enabled folivorous and frugivorous diets in Neotropical fishes. Folivorous and frugivorous fishes provide food for people living along the Amazon and other rivers, and the dispersal of seeds from plants used in ceremonial practices means that these fishes provide both provisioning and cultural services. Dramatic declines in the abundances of folivorous and frugivorous fishes have been documented and are expected to correspond to reductions in ecosystem health and functioning.

15.1 What Are Folivores and Frugivores?

Fishes that eat plant material may be divided among folivore, frugivore, and algivore trophic guilds. In this chapter, we consider folivores and frugivores (see Chap. 16 for information about algivores). Folivorous fishes consume the roots, stems, leaves, and flowers of aquatic and terrestrial plants (Mérona and Rankin-de-Mérona 2004; Da Silva et al. 2005; Rejas 2018) and frugivores consume fruits and seeds (Correa et al. 2007; Correa and Winemiller 2014). In this chapter, we refer to plant matter comprised of fruits and seeds as "fruits and seeds" and we call any non-fruit and non-seed plant material "plants" or "plant material" for brevity. Frugivores and folivores generally exhibit seasonal differences in the amount and types of plants and fruits and seeds they consume (Mérona and Rankin-de-Mérona 2004; Correa et al. 2007; Anderson et al. 2009; Correa and Winemiller 2014). Each trophic guild is associated with unique morphologies, behaviors, and ecosystem functions, while also harboring substantial intra-guild phenotypic and ecological diversity (Gottsberger 1978; Correa et al. 2007; Huie et al. 2019; Carvalho et al. 2022).

J. P. Egan (✉)
University of Idaho, Moscow, USA
e-mail: eganx149@umn.edu

M. A. Kolmann
University of Louisville, Louisville, KY, USA
e-mail: mkolmann@gmail.com

J. S. Albert and R. E. Reis (eds.), *Ecology and Evolution of Amazonian Fishes*, Fascinating Life Sciences,
https://doi.org/10.1007/978-3-032-09328-8_15

The Neotropics is home to a rich diversity of folivorous and frugivorous fishes. Enigmatic examples include the folivorous Discus *Symphysodon* spp., a popular aquarium fish, and the frugivorous Tambaqui *Colossoma macropomum*, an important food fish (Goulding and Carvalho 1982; Crampton 2008). Folivorous and frugivorous fishes facilitate key food web linkages and frugivores play major roles in seed dispersal (ichthyochory) within Neotropical floodplain ecosystems, a function that is relatively rare amongst fishes (Mannheimer et al. 2003; Horn et al. 2011; Bueno et al. 2013; Correa et al. 2015b; Weiss et al. 2024). In this chapter, we explore the taxonomic diversity of Neotropical folivores and frugivores, compare the diets of folivores and frugivores, and describe ontogenetic and seasonal dietary patterns exhibited by these species. We also discuss the evolution of folivory and frugivory in the Neotropics, highlight morphological traits (e.g., teeth and jaw shapes and gut morphology) associated with frugivory and folivory, and discuss the functional roles of folivorous and frugivorous fishes in Neotropical freshwater ecosystems.

15.2 Taxonomic Diversity

Trophic guilds were assigned to Neotropical fishes based on the dominant prey types in adult diets, and species may have been assigned to multiple trophic guilds (see ecotraits in Chap. 6). There are additional Neotropical fishes that consume plant and fruit materials, but only in minor quantities, so they were not assigned to the folivore or frugivore trophic guilds. Nine species were assigned to both frugivory and folivore trophic guilds. An additional 50 species were designated as folivorous and 25 species as frugivorous (Table 15.1). Together, these 82 species represent three orders: Characiformes (four families, 74 species), Cichliformes (one family, four species), and Siluriformes (two families, four species). Frugivory occurs in Characiformes (two families, six genera, 29 species) and Siluriformes (two families, three genera, four species). Folivory is exhibited by Characiformes (three families, ten genera, 57 species) and Cichliformes (one family, one genus, four species).

15.3 Feeding Modes

15.3.1 Substrate Scraping

Substrate scrapers rasp plants attached to the surfaces of hard substrates using oral teeth (da Silva Gonçalves and Cestari 2018). *Ossubtus xinguense* consumes allochthonous plant material (not algae) by scraping it from rocky substrates (Huie et al. 2019). The folivore *Deuterodon iguape*, and possibly other *Deuterodon* species, also exhibits scraping behavior, but preliminary research suggests this species appears to primarily use substrate scraping behavior to feed on algae and employs other feeding modes when consuming plants (da Silva Gonçalves and Cestari 2018).

15.3.2 Kleptophagy

Kleptophagy is a form of parasitism in which one individual steals food that has already been procured by another individual (Iyengar 2008). Carvalho et al. (2022) observed the frugivore *Brycon falcatus* exhibiting kleptophagous behavior in the Teles Pires River basin, Mato Grosso state, Brazil. *Brycon falcatus* individuals stole fruit from the mouths of *Leporinus friderici* after the latter found small fruits hidden in the leaf litter.

Table 15.1 Summary of neotropical folivorous and frugivorous fishes

Order	Family	Species	Fol	Fru	Citations
Characiformes	Anostomidae	*Schizodon altoparanae*	1		
Characiformes	Anostomidae	*Schizodon australis*	1		
Characiformes	Anostomidae	*Schizodon borellii*	1		Manetta et al. (2003)
Characiformes	Anostomidae	*Schizodon corti*	1		Galvis et al. (1997)
Characiformes	Anostomidae	*Schizodon dissimilis*	1		
Characiformes	Anostomidae	*Schizodon fasciatus*	1		Dos Santos (1981)[a], Forsberg et al. (1993), Mérona and Rankin-de-Mérona (2004)[a,b], Da Silva et al. (2005), Oliveira et al. (2006), Mortillaro et al. (2015), and Soares et al. (1986)
Characiformes	Anostomidae	*Schizodon intermedius*	1		Yabe and Bennemann (1994)[a]
Characiformes	Anostomidae	*Schizodon isognathus*	1		Winemiller and Jepsen (1998)
Characiformes	Anostomidae	*Schizodon jacuiensis*	1		
Characiformes	Anostomidae	*Schizodon knerii*	1		Alvim and Peret (2004)

Table 15.1 (continued)

Order	Family	Species	Fol	Fru	Citations
Characiformes	Anostomidae	*Schizodon nasutus*	1		Andrade and Braga (2005), Villares Junior et al. (2011)[a], de Avila-Simas et al. (2019)[a], and López-Rodríguez et al. (2019)[a]
Characiformes	Anostomidae	*Schizodon platae*	1		López-Rodríguez et al. (2019)[a]
Characiformes	Anostomidae	*Schizodon rostratus*	1		
Characiformes	Anostomidae	*Schizodon scotorhabdotus*	1		Sidlauskas et al. (2007)
Characiformes	Anostomidae	*Schizodon succinctus*	1		
Characiformes	Anostomidae	*Schizodon vittatus*	1		de Mérona et al. (2001)[a] and Zuanon and Ferreira (2008)
Characiformes	Bryconidae	*Brycon falcatus*		1	Menezes (1969), Albrecht et al. (2009)[a], Correa and Winemiller (2014)[a,b], de Matos et al. (2016), and Carvalho et al. (2022)
Characiformes	Bryconidae	*Brycon melanopterus*		1	Mérona and Rankin-de-Mérona (2004)[a], Correa and Winemiller (2014)[a,b], Correa and Winemiller (2018), and Soares et al. (1986)
Characiformes	Characidae	*Deuterodon iguape*	1		Sabino and Corrêa e Castro (1990) and da Silva Gonçalves and Cestari (2018)
Characiformes	Characidae	*Deuterodon langei*	1		Vitule et al. (2008)[a] and Wolff et al. (2013)[a]
Characiformes	Characidae	*Deuterodon longirostris*	1		
Characiformes	Characidae	*Deuterodon pedri*	1		
Characiformes	Characidae	*Deuterodon rosae*	1		
Characiformes	Characidae	*Deuterodon singularis*	1		
Characiformes	Characidae	*Deuterodon stigmaturus*	1		Dala-Corte et al. (2016)[a]
Characiformes	Characidae	*Deuterodon supparis*	1		
Characiformes	Serrasalmidae	*Acnodon normani*	1	1	Gosline (1951), Leite and Jégu (1990)[a], de Mérona et al. (2001)[a], and Pereira et al. (2007)[a]
Characiformes	Serrasalmidae	*Acnodon oligacanthus*	1	1	Planquette et al. (1996)
Characiformes	Serrasalmidae	*Acnodon senai*	1	1	Jégu and dos Santos (1990) and Leite and Jégu (1990)[a]
Characiformes	Serrasalmidae	*Colossoma macropomum*		1	Roubach and Saint-Paul (1994), Mérona and Rankin-de-Mérona (2004)[a,b], Lucas (2008), Anderson et al. (2009)[a,b], and Soares et al. (1986)
Characiformes	Serrasalmidae	*Mylesinus paraschomburgkii*	1		Jégu et al. (1989)[a]
Characiformes	Serrasalmidae	*Mylesinus paucisquamatus*	1		Vitorino Júnior et al. (2016)[a]
Characiformes	Serrasalmidae	*Mylesinus schomburgkii*	1		Nico (1991)
Characiformes	Serrasalmidae	*Myleus altipinnis*	1		
Characiformes	Serrasalmidae	*Myleus knerii*	1		de Carvalho et al. (2020)
Characiformes	Serrasalmidae	*Myleus micans*	1		de Mérona et al. (2001)[a] and Alonso et al. (2020)
Characiformes	Serrasalmidae	*Myleus pachyodus*	1		
Characiformes	Serrasalmidae	*Myleus pacu*	1		de Mérona et al. (2001)[a]
Characiformes	Serrasalmidae	*Myleus setiger*	1		Andrade et al. (2019a)[a] and Andrade et al. (2023)
Characiformes	Serrasalmidae	*Myloplus arnoldi*		1	Andrade et al. (2023)
Characiformes	Serrasalmidae	*Myloplus asterias*		1	Andrade et al. (2019b)[a] and Correa and Winemiller (2014)[a,b]
Characiformes	Serrasalmidae	*Myloplus levis*		1	Silveira et al. (2020)
Characiformes	Serrasalmidae	*Myloplus lobatus*		1	Anjos et al. (2019)
Characiformes	Serrasalmidae	*Myloplus lucienae*		1	
Characiformes	Serrasalmidae	*Myloplus planquettei*		1	
Characiformes	Serrasalmidae	*Myloplus rubripinnis*		1	Alexander (1964), Correa and Winemiller (2014)[a,b] and Andrade et al. (2019b)[a]

(continued)

Table 15.1 (continued)

Order	Family	Species	Fol	Fru	Citations
Characiformes	Serrasalmidae	*Myloplus schomburgkii*		1	Andrade et al. (2019b)[a]
Characiformes	Serrasalmidae	*Myloplus taphorni*		1	
Characiformes	Serrasalmidae	*Myloplus ternetzi*		1	Boujard et al. (1990)[a,b] and Planquette et al. (1996)
Characiformes	Serrasalmidae	*Myloplus tiete*		1	Correa et al. (2015b)
Characiformes	Serrasalmidae	*Myloplus torquatus*		1	Correa and Winemiller (2014)[a,b]
Characiformes	Serrasalmidae	*Myloplus tumukumak*		1	
Characiformes	Serrasalmidae	*Myloplus zorroi*		1	Andrade et al. (2016b)
Characiformes	Serrasalmidae	*Mylossoma acanthogaster*	1	1	
Characiformes	Serrasalmidae	*Mylossoma albiscopum*	1	1	
Characiformes	Serrasalmidae	*Mylossoma aureum*	1	1	Mérona and Rankin-de-Mérona (2004)[a] and Soares et al. (1986)
Characiformes	Serrasalmidae	*Mylossoma duriventre*	1	1	de Mérona et al. (2001)[a], Mérona and Rankin-de-Mérona (2004)[a,b], Rejas (2018), Melo et al. (2019)[a], Barbarino Duque and Winemiller (2003), Araujo et al. (2020), and Soares et al. (1986)
Characiformes	Serrasalmidae	*Mylossoma unimaculatum*	1	1	
Characiformes	Serrasalmidae	*Ossubtus xinguense*	1		Andrade et al. (2019a, 2019b)[a]
Characiformes	Serrasalmidae	*Piaractus brachypomus*		1	de Mérona et al. (2001)[a], Mérona and Rankin-de-Mérona (2004)[a], Anderson et al. (2009)[a], and Correa et al. (2014)[a]
Characiformes	Serrasalmidae	*Piaractus mesopotamicus*		1	Galetti et al. (2008) and Araujo et al. (2020)
Characiformes	Serrasalmidae	*Piaractus orinoquensis*		1	
Characiformes	Serrasalmidae	*Prosomyleas rhomboidalis*		1	Boujard et al. (1990)[a,b] and Andrade et al. (2019b)
Characiformes	Serrasalmidae	*Serrasalmus nigricauda*	1		
Characiformes	Serrasalmidae	*Tometes ancylorhynchus*	1		Andrade et al. (2016a)
Characiformes	Serrasalmidae	*Tometes camunani*	1		Andrade et al. (2013)
Characiformes	Serrasalmidae	*Tometes kranponhah*	1		Andrade et al. (2016a) and Andrade et al. (2019b)[a]
Characiformes	Serrasalmidae	*Tometes lebaili*	1		
Characiformes	Serrasalmidae	*Tometes makue*	1		
Characiformes	Serrasalmidae	*Tometes siderocarajensis*	1		Andrade et al. (2017)
Characiformes	Serrasalmidae	*Tometes trilobatus*	1		Sa-Oliveira et al. (2014)[a]
Characiformes	Serrasalmidae	*Utiaritichthys esguiceroi*	1		Pereira and Castro (2014)
Characiformes	Serrasalmidae	*Utiaritichthys longidorsalis*	1		
Characiformes	Serrasalmidae	*Utiaritichthys sennaebragai*	1		de Mérona et al. (2001)[a]
Cichliformes	Cichlidae	*Symphysodon aequifasciatus*	1		Crampton (2008)
Cichliformes	Cichlidae	*Symphysodon discus*	1		Crampton (2008)
Cichliformes	Cichlidae	*Symphysodon haraldi*	1		Bleher (2006)[a] and Crampton (2008)[a]
Cichliformes	Cichlidae	*Symphysodon tarzoo*	1		
Siluriformes	Auchenipteridae	*Auchenipterichthys longimanus*		1	de Mérona et al. (2001)[a], Mannheimer et al. (2003), Freitas et al. (2011)[a,b], Freitas et al. (2017)[a], and Weiss et al. (2024)
Siluriformes	Doradidae	*Lithodoras dorsalis*		1	Barbosa et al. (2015)[a]
Siluriformes	Doradidae	*Pterodoras granulosus*		1	de Souza-Stevaux et al. (1994), de Mérona et al. (2001)[a], Mérona and Rankin-de-Mérona (2004)[a], Agostinho et al. (2009)[a,b], and González-Bergonzoni et al. (2010)
Siluriformes	Doradidae	*Pterodoras rivasi*		1	Echevarría et al. (2018)[a]

A "1" in the folivore (Fol) and frugivore (Fru) columns shows which species belong to each trophic guild [a]Citations containing quantitative diet data used to conduct hierarchical agglomerative cluster analysis and generate bar plots [b]References used for Fig. 15.2

15.3.3 Cropping and Browsing

Cropping is a feeding mode exhibited by folivores and involves biting off pieces of growing plants, such as grasses, close to the substrate. For example, *Schizodon nasutus* consumes large quantities of grasses (Poaceae) via cropping (Sazima 1980; Barbarino Duque and Winemiller 2003; Villares Junior et al. 2011), as do pacus like *Mylesinus* (Vitorino et al. 2016).

Many folivorous and frugivorous fishes *browse* on attached or unattached plant and fruit materials that may be resting on the substrate, suspended in the water column, or floating on the water surface (Souza et al. 2019). Browsing is accomplished by biting pieces off of plants and fruits or ingesting whole pieces of plants (e.g., unattached flowers and leaves) and whole fruits and seeds via biting and/or suction feeding. For example, *Brycon* spp. frequently consume whole fruits floating on the surface of the water (Costa-Pereira et al. 2011; Carvalho et al. 2022), and the catfish *Auchenipterichthys longimanus* eats whole fruits resting on the substrate (Mannheimer et al. 2003). Some species crush fruits and seeds using their oral jaws before they pass into the gastrointestinal tract (Kubitzki and Ziburski 1994; Galetti et al. 2008; Anderson et al. 2009; Horn et al. 2011; Correa and Winemiller 2014; Correa et al. 2015b; see Sects. 4.1, 6.2, and 7.4 for additional discussion about seed crushing).

15.4 Trophic Diversity

15.4.1 Interspecific Variation in Prey Types

To quantitatively compare the prey consumption of folivorous and frugivorous fishes, We conducted hierarchical agglomerative cluster analysis based on the proportions of different prey categories (Table 15.2) in diets. We collected diet data from scientific articles that reported the proportions of prey types in diets (e.g., % volume, % number, % weight, Alimentary Index; Hyslop 1980; Table 15.3). We did not include diet data reported as indices that do not sum to 100%, such as frequency of occurrence (percentage of fish that consume each prey type). To standardize interspecific comparisons, We only collected diet data from large juvenile and adult-sized individuals. In some cases, articles reported diet composition information in figures, but not in text or tables. In such cases, We extracted data from figures using ImageJ software (http://www.imagej.nih.gov/ij). To incorporate data from studies reporting diet data at varying levels of resolution, We assigned prey types to one of eleven coarse prey categories that were not based on taxonomy, but instead contained functionally and morphologically similar prey (Table 15.2). We calculated mean percentages of diet composition for each prey category in diets when diet data were available from multiple locations, time points, or studies for a single species. When studies included "unidentified materials" categories in % diet composition calculations, We removed this category and recalculated the percentages using only the identifiable materials. We calculated Bray-Curtis dissimilarity between each species, then grouped species using hierarchical agglomerative clustering with complete linkage in the vegan R package (Bray and Curtis 1957; Legendre and Legendre 2012; Oksanen et al. 2016).

Among the 34 species included in cluster analysis, the proportions of fruits and seeds in diets ranged from lows of 0% in several folivorous species to a high of 81% in *Mylossoma aureum* (Fig. 15.1). Nine species had diets composed of more than 30% fruits and seeds. The percentages of plants in

Table 15.2 Prey types comprising each prey category used for hierarchical agglomerative cluster analysis

Prey category	Category composition
Algae/periphyton (ALGA)	Algae, filamentous algae, periphyton
Detritus (DETR)	Detritus, microphytobenthos, sediment
Fish (FISH)	Fish, scales
Fruits and seeds (FRUI)	Whole fruits, fruit material, seeds
Invertebrates (INVE)	Arthropoda, Crustacea, Insecta, invertebrates, allochthonous insects/invertebrates, autochthonous insects/invertebrates
Mollusca (MOLL)	Bivalva, Gastropoda, Mollusca
Phytoplankton (PHYT)	Diatoms, phytoplankton
Plant material (PLAN)	Allochthonous plants, autochthonous plants, flowers, leaves, plant material, roots
Protozoa (PROT)	Protozoa
Terrestrial vertebrates (VERT)	Amphibia, Mammalia, terrestrial vertebrates
Zooplankton (ZOOP)	Copepoda, zooplankton

Prey category codes are shown in parentheses. Prey categories are based on the functional and morphological similarity of prey and are not taxonomic

Table 15.3 Percent consumption of prey types by folivorous and frugivorous neotropical fishes

Species	PLAN	FRUI	FISH	ZOOP	PROT	MOLL	VERT	DETR	ALGA	PHYT	INVE
Deuterodon langei	0.438	0.030	0.000	0.000	0.000	0.000	0.000	0.086	0.013	0.000	0.433
Deuterodon stigmaturus	0.092	0.000	0.031	0.000	0.000	0.000	0.000	0.276	0.357	0.000	0.245
Schizodon fasciatus	0.700	0.095	0.010	0.000	0.000	0.000	0.000	0.043	0.089	0.000	0.063
Schizodon intermedius	0.694	0.000	0.000	0.000	0.069	0.000	0.000	0.000	0.000	0.000	0.237
Schizodon nasutus	0.900	0.010	0.000	0.000	0.000	0.010	0.000	0.010	0.050	0.000	0.030
Schizodon vittatus	0.752	0.000	0.035	0.014	0.000	0.000	0.000	0.199	0.000	0.000	0.000
Acnodon normani	0.389	0.000	0.142	0.000	0.000	0.001	0.000	0.002	0.000	0.000	0.465
Schizodon platae	1.000	0.000	0.000	0.000	0.000	0.000	0.000	0.000	0.000	0.000	0.000
Acnodon senai	0.980	0.000	0.000	0.000	0.000	0.000	0.000	0.000	0.000	0.000	0.020
Colossoma macropomum	0.095	0.558	0.015	0.025	0.000	0.000	0.000	0.232	0.000	0.000	0.077
Mylesinus paraschomburgkii	1.000	0.000	0.000	0.000	0.000	0.000	0.000	0.000	0.000	0.000	0.000
Mylesinus paucisquamatus	0.703	0.050	0.000	0.000	0.000	0.000	0.000	0.152	0.086	0.000	0.009
Myloplus asterias	0.242	0.250	0.000	0.000	0.000	0.000	0.000	0.001	0.010	0.484	0.014
Myloplus rhomboidalis	0.070	0.430	0.100	0.000	0.000	0.100	0.000	0.000	0.000	0.210	0.100
Myloplus rubripinnis	0.406	0.214	0.121	0.000	0.000	0.000	0.000	0.007	0.049	0.160	0.044
Myloplus schomburgkii	0.448	0.009	0.014	0.000	0.000	0.000	0.000	0.000	0.028	0.255	0.247
Myloplus ternetzi	0.830	0.170	0.000	0.000	0.000	0.000	0.000	0.000	0.000	0.000	0.000
Myloplus torquatus	0.325	0.393	0.016	0.000	0.000	0.000	0.000	0.020	0.000	0.236	0.010
Mylossoma aureum	0.130	0.810	0.000	0.000	0.000	0.000	0.000	0.000	0.000	0.000	0.060
Mylossoma duriventre	0.305	0.419	0.000	0.001	0.000	0.000	0.000	0.000	0.007	0.000	0.268
Ossubtus xinguense	0.532	0.042	0.124	0.000	0.000	0.000	0.000	0.000	0.156	0.055	0.092
Piaractus brachypomus	0.289	0.392	0.185	0.000	0.000	0.000	0.000	0.014	0.000	0.000	0.120
Tometes kranponhah	0.288	0.171	0.082	0.000	0.000	0.000	0.000	0.000	0.173	0.178	0.108
Tometes trilobatus	1.000	0.000	0.000	0.000	0.000	0.000	0.000	0.000	0.000	0.000	0.000
Utiaritichthys sennaebragai	0.941	0.000	0.000	0.000	0.000	0.000	0.000	0.006	0.000	0.000	0.053
Auchenipterichthys longimanus	0.023	0.188	0.042	0.000	0.000	0.000	0.000	0.000	0.000	0.000	0.748
Lithodoras dorsalis	0.143	0.638	0.000	0.000	0.000	0.033	0.000	0.083	0.010	0.000	0.092
Pterodoras granulosus	0.606	0.134	0.172	0.000	0.000	0.018	0.000	0.035	0.007	0.000	0.029
Pterodoras rivasi	0.612	0.000	0.112	0.017	0.000	0.000	0.000	0.103	0.000	0.000	0.156
Symphysodon haraldi	0.130	0.000	0.000	0.010	0.000	0.000	0.000	0.140	0.460	0.000	0.260
Brycon falcatus	0.094	0.370	0.132	0.000	0.000	0.000	0.031	0.010	0.000	0.000	0.370
Brycon melanopterus	0.077	0.466	0.000	0.000	0.000	0.000	0.000	0.036	0.000	0.000	0.421
Myleus micans	0.908	0.000	0.001	0.000	0.000	0.000	0.000	0.052	0.000	0.000	0.039
Myleus setiger	0.421	0.000	0.137	0.000	0.000	0.000	0.000	0.209	0.103	0.000	0.129

Data sources in Table 15.1. Diet classification in Table 15.2

diets ranged from 0.2% in the frugivore *Auchenipterichthys longimanus* to 100% in *Mylesinus paraschomburgkii* and *Tometes trilobatus*. Twenty-two species had diets composed of more than 30% plant material. Invertebrates were important in the diets of several folivore and frugivore species, particularly *Auchenipterichthys longimanus*, *Acnodon normani*, *Deuterodon langei*, *Brycon melanopterus*, and *Brycon falcatus*. Allochthonous and autochthonous insects were the dominant types of invertebrates consumed (e.g., de Merona et al. 2001; Vitule et al. 2008). A few species consumed small amounts of fish, most notably *Brycon falcatus* and *Piaractus brachypomus*, which had diets composed of 13% and 19% fish, respectively. Organic detritus was consumed in small quantities by several species.

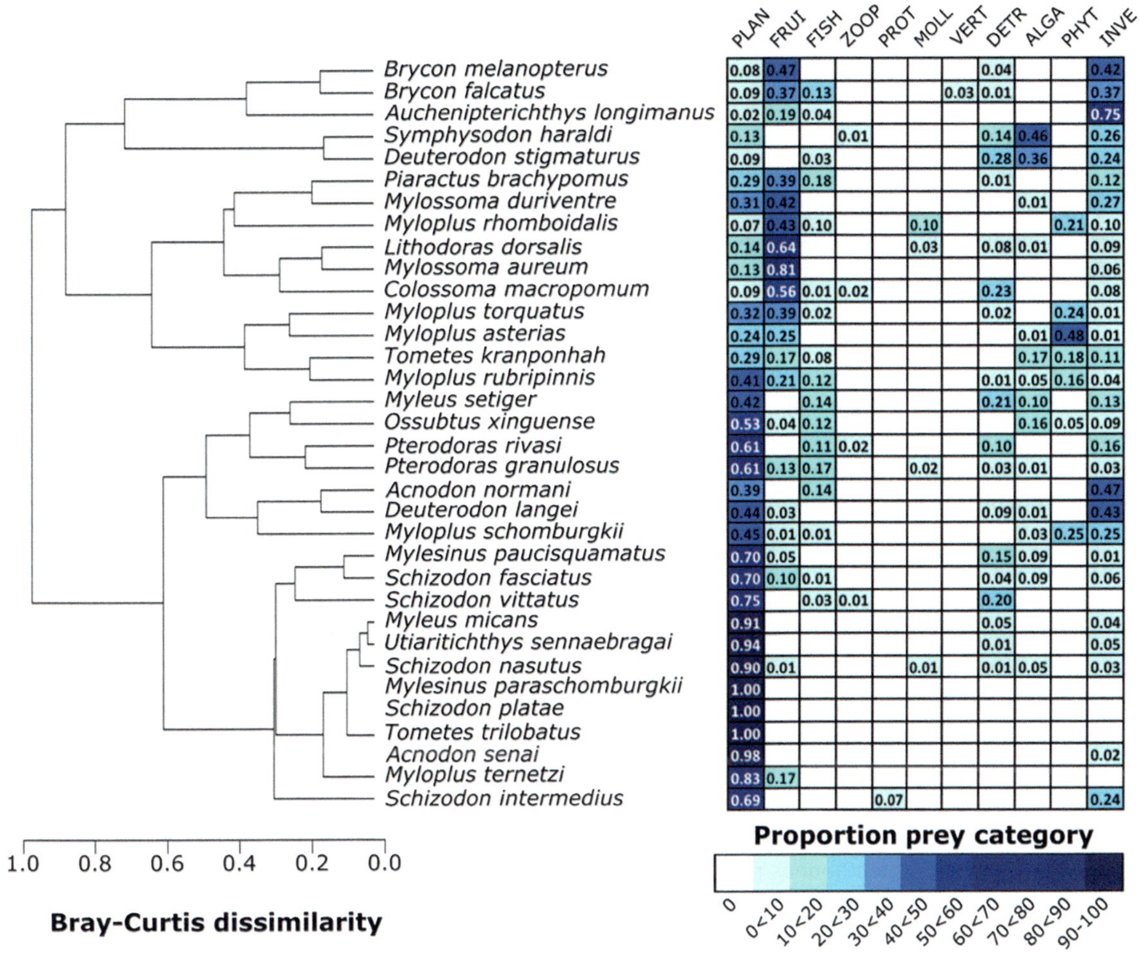

Fig. 15.1 Left: Dendrogram showing dietary similarty among folivorous and frugivorous Neotropical fishes. This dendrogram was generated using complete linkage hierarchical agglomerative cluster analysis based on Bray-Curtis dissimilarity indices of non-taxonomic prey categories consumed. Right: Heat map showing the percentages of prey categories in the diets of each species. Percentage diet composition data are rounded to two decimal places, which in some cases resulted in total prey consumption slightly different than 100%. Sources of diet data are reported in Table 15.1. Definitions of prey category codes are in Table 15.2. Unrounded diet data (three decimal places) used as input for cluster analysis are in Table 15.3.

15.4.2 Diversity of Plant Parts Consumed

Within frugivore and folivore trophic guilds, there is interspecific diversity in the diets of large juvenile- and adult-sized fishes (See Sect. 15.4.3 for discussion about interspecific variation in the diets of earlier life stages; Fig. 15.1). Assemblages of frugivores often exhibit interspecific differences in the types and components of fruits and seeds they consume (Lucas 2008; Anderson et al. 2009; Correa and Winemiller 2014; Araujo et al. 2020). Fruit and seed size (Kubitzki and Ziburski 1994; Agostinho et al. 2009) and hardness (Lucas 2008; Anderson et al. 2009; Horn et al. 2011; Correa et al. 2014) represent key axes of frugivore trophic niche variation. For example, the catfish *Auchenipterichthys longimanus* consumes relatively large fruits whole (e.g., fruits from *Licania* spp. and Bombacaceae spp.; Mannheimer et al. 2003), while some frugivorous serrasalmid species (e.g. *Myloplus* spp.) with smaller mouth gapes generally eat greater numbers of small seeds and fruits, and must process large fruits and seeds using their oral jaws before they are ingested (see Sect. 15.6.2 for discussion about seed crushing; Kubitzki and Ziburski 1994; Galetti et al. 2008; Anderson et al. 2009; Horn et al. 2011; Correa and Winemiller 2014; Correa et al. 2015b).

Most Neotropical frugivores appear to have some capacity for breaking down hard seeds before they enter the esophagus, and there is evidence of interspecific diversity in seed processing performance (see Sects. 15.6.2 and 15.7.4 for further discussion about seed crushing and bite performance; Alexander 1964; Lucas 2008; Correa and Winemiller 2014; Huby et al. 2019). For example, *Brycon* spp. and *Piaractus* spp. appear to be particularly effective at crushing seeds (Correa and Winemiller 2014; Correa et al. 2014). However, seed processing performance is not well characterized in Neotropical frugivores, which hinders interspecific comparisons (Alexander 1964; Lucas 2008; Correa and Winemiller 2014; Huby et al. 2019).

There are differences in the types of plant material consumed by Neotropical folivores, although many folivores appear to be quite opportunistic and unselective, often consuming plant materials in proportion to their availability. One major axis of interspecific trophic variation among folivores is the relative contributions of autochthonous (aquatic) versus allochthonous plant materials to diets. Autochthonous macrophytes tend to more frequently dominate the diets of some species, such as *Schizodon* spp., *Myleus* spp., *Tometes* spp., and *Ossubtus xinguense*. *Schizodon nasutus* consumes large quantities of plant roots and aquatic macrophytes, including grasses (Poaceae) via cropping (Sazima 1980; Barbarino Duque and Winemiller 2003; Villares Junior et al. 2011; López-Rodríguez et al. 2019). Andrade et al. (2019a) found that the diets of all members of an assemblage of rheophilic folivores (*Myleus setiger, Ossubtus xinguense,* and *Tometes kranponhah*) inhabiting a fast-flowing river primarily consumed epilithic macrophytes (Podostemaceae). However, Andrade et al. (2019a) did identify interspecific differences in the relative proportions of autochthonous and allochthonous plant materials in the diets of these fishes, with *Myleus setiger* eating more allochthonous material (e.g., leaves and flowers) than *Ossubtus xinguense* and *Tometes kranponhah*. Allochthonous plant material is frequently more important in the diets of folivores, such as *Mylossoma* spp. and *Deuterodon* spp., that browse on leaves and flowers floating on the surface or suspended in the water column (Barbarino Duque and Winemiller 2003; Mérona and Rankin-de-Mérona 2004; Vitule et al. 2008; Dala-Corte et al. 2016).

15.4.3 Ontogenetic Dietary Changes

There are two general ontogenetic diet trajectories exhibited by folivores and frugivores. Within each trajectory, there is interspecific diversity in the timing of diet transitions and in a variety of functionally relevant attributes of prey consumed (e.g., prey sizes). In the first trajectory, which is exhibited by nearly all frugivores and folivores, small fishes feed on microscopic planktonic invertebrates (e.g., Copepoda) when they initiate exogenous feeding. As fish grow, they transition to consuming larger-sized plankton (e.g., aquatic insect larvae) and sometimes small nekton (e.g., Amphipoda) before finally transitioning to diets containing large quantities of plant material and fruits and seeds (Vitule et al. 2008; Agostinho et al. 2009; Anderson et al. 2009; Dala-Corte et al. 2016; Andrade et al. 2019a; Santos et al. 2023). Many species continue feeding on small invertebrates and nekton after they transition to folivory and/or frugivory, either incidentally or purposefully. For example, in Rio Ribeirão, Paraná State, Brazil, *Deuterodon langei* individuals less than 3 cm total length consumed 73% insects, 3% plants, and 1% seeds and fruits, individuals ranging from 3 to 8 cm total length consumed 52% insects, 16% plants, and 2% seeds and fruits, and individuals more than 8 cm total length consumed 31% insects, 37% plants, and 6% seeds and fruits (Vitule et al. 2008). Agostinho et al. (2009) characterized the diets of *Pterodoras granulosus* in the Lajeado Reservoir, Tocantins State, Brazil belonging to two length classes: 6.0–22.5 and 22.5–60.2 cm standard length. Individuals in the shorter length class ate greater quantities of insects (8.0%) and zooplankton (1.1%), and fewer plants (50.9%) than individuals in the longer length class, which ate fewer insects (3.6%) and zooplankton (0.4%) and more plants (79.6%).

Following ontogenetic transitions to frugivory, there are often additional ontogenetic changes in the sizes of fruits, and by proxy seeds, consumed, because as fish grow, they generally have the capacity to consume larger prey (Goulding 1980; Galetti et al. 2008; Correa et al. 2015a; Egan et al. 2017; Egan et al. 2018; Egan et al. 2026). This is likely driven by increases in mouth gape (Costa-Pereira and Galetti 2015). Seed dispersal capacity also tends to be positively correlated with fish size, in part because larger mouth gapes permit more intact seeds to proceed further down the gastrointestinal tract without being damaged, and because larger fish tend to consume greater numbers of seeds (Kubitzki and Ziburski 1994; Pollux 2011; Correa et al. 2015a). Larger frugivores tend to disperse more seeds from more species of plants than smaller frugivores (Correa et al. 2015b), as well as disperse these seeds over longer distances (Anderson et al. 2011). The tendency of humans to target larg fishes for food makes it so that the most proficient seed-dispersing species and individuals are at high risk of overexploitation (Costa-Pereira et al. 2018; Araujo et al. 2020). For example, Costa-Pereira and Galetti (2015) summarize how in the city of Manaus in the 1970s, 26,000 tons of *Colossoma macropomum* were sold annually.

The second ontogenetic diet trajectory is only exhibited by folivorous *Symphysodon* spp. Rather than microscopic invertebrates representing the most important prey item for during early ontogeny when they initiate exogenous feeding, *Symphysodon* spp. fry feed on a proteinaceous mucus that is exuded from the skin of both parents (Hildemann 1959; Buckley et al. 2010), a strategy that is very uncommon among fishes (Satoh and Sowersby 2021). After feeding exclusively on this mucus for a period of three to four weeks, *Symphysodon* spp. transition to feeding on microscopic plankton such as rotifers and copepods (Buckley et al. 2010; Satoh et al. 2017; Satoh and Sowersby 2021). The diets of juveniles after transitioning from mucus feeding to consuming small plankton until the transition to the folivorous diets of large juveniles and adults are not well documented in *Symphysodon* spp. (Fig. 15.1, Table 15.3; Bleher 2006; Crampton 2008).

15.4.4 Seasonal Dietary Variation

There is temporal variation in the diets of folivores and frugivores associated with hydrological seasonality (Fig. 15.2; Winemiller 1996; Arantes et al. 2019; Alonso et al. 2020). Many species of Neotropical frugivorous and folivorous fishes inhabit floodplain ecosystems, which exhibit major seasonal hydrologic changes (Chaps. 10 and 11). During the high-water season, river water levels rise and inundate floodplains, allowing fish to seasonally access novel spaces and resources (Goulding 1980; Junk 1997). Fruiting in many Neotropical plant species inhabiting floodplains is synchronized with the high-water season, contributing to a pulse of fruit biomass available to fishes (Correa and Winemiller 2014; Parolin and Wittmann 2011). In this section, we discuss intraspecific variation in the diets of fishes inhabiting floodplain ecosystems in terms of hydrologic seasons: the low-water and high-water seasons and the transitional periods from the low to the high-water season (rising-water season) and the high to low water season (receding-water season). We also obtained diet data from frugivorous *Lithodoras dorsalis* individuals inhabiting a tidal delta ecosystem that did not exhibit seasonal flood pulse dynamics (Barbosa et al. 2015). We therefore discuss variation in the diet of this species in terms of pluviometric seasons (dry-wet season, wet season, wet-dry season, and dry season).

Floodplain frugivores: Frugivorous fishes tend to consume greater quantities of fruits and seeds during the high-water season and other prey types, such as plant material, detritus, and aquatic invertebrates, comprise greater proportions of diets during other seasons (Fig. 15.2; Boujard et al. 1990; Mérona and Rankin-de-Mérona 2004; Lucas 2008; Anderson et al. 2009; Agostinho et al. 2009; Freitas et al. 2011; Correa and Winemiller 2014). For example, *Colossoma macropomum* primarily consumes fruit during the high-water season and plant material, insects, aquatic invertebrates, and detritus during other seasons (Mérona and Rankin-de-Mérona 2004; Oliveira et al. 2006; Lucas 2008; Anderson et al. 2009). In the Amazon River, Amazonas State, Brazil, *Colossoma macropomum* consumed 88% fruit during high-water season, but only 15% during the low-water season (Mérona and Rankin-de-Mérona 2004). Oliveira et al. (2006) described dramatic fluctuations in the amount of fruit consumed by *Mylossoma duriventre* in Lake Camaleão, a floodplain lake on the Solimões River, Amazonas State, Brazil, with ~50%, ~80%, ~90%, ~0% of examined individuals (i.e., diet data expressed as frequency of occurrence) having consumed fruit during the rising- high-, receding- and low-water seasons, respectively. Plant material and insects made up large portions of the diets of *M. duriventre* when less fruit was consumed.

Tidal delta frugivores: Barbosa et al. (2015) characterized the diet of the frugivorous *Lithodoras dorsalis* in the Marajó delta, Pará State, Brazil. In the study area, water level changes were primarily driven by tides, and there were no distinct hydrological seasons. Consequently, Barbosa et al. (2015) documented temporal diet variation associated with pluviometric seasons. Daily flooding driven by the tides allowed

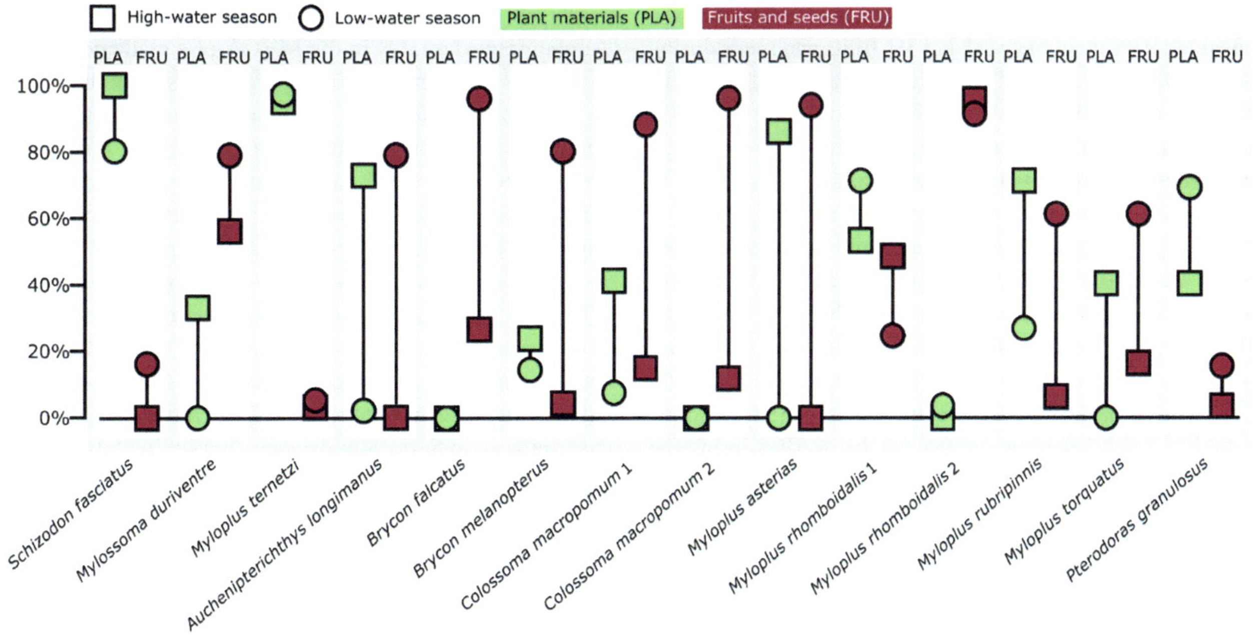

Fig. 15.2 Plots showing the magnitude of change in the consumption of fruits and seeds (FRU) and other plant materials (PLA) between high-water and low water seasons by folivores (*Myloplus ternetzi* and *Schizodon fasciatus*), the folivorous and frugivorous *Mylossoma duriventre*, and frugivores (remaining species) in river-floodplain ecosystems. Sources of diet data are denoted in Table 15.1 with a superscript "C"

temporary access to tidal floodplain areas and plant material, fruits, and seeds. During the wet season, *Lithodoras dorsalis* primarily consumed fruit, especially fruits from the açaí palm (*Euterpe oleracea*) and aninga (*Montrichardia linifera*), during all seasons, but consumed the greatest amount of fruit in the dry season (82%). *Lithodoras dorsalis* consumed less fruit during the wet-dry (58%), dry-wet (58%), and wet (57%) seasons. The greatest amount of plant material was consumed during the wet season (25%), with smaller amounts consumed during the wet-dry (9%), dry (10%), and dry-wet (14%) seasons.

Floodplain folivores: The magnitudes of seasonal changes in the diets of folivores appear to generally be less than frugivores, although seasonal diet changes have not been characterized in many species (Fig. 15.2). This may be because access to the plant materials consumed by folivores is less dependent on flooding. However, there is evidence that some tree species exhibit increased leaf shedding during high water periods (Parolin and Wittmann 2011). This, along with access to floodplains during the high-water season, may result in some seasonality in the availability of plant materials that could correspond to seasonal variation in the diets of folivores, with more plant consumption during the high-water season. We found mixed support for this hypothesis. In the Amazon River, Amazonas State, Brazil, the folivore *Schizodon fasciatus* consumed 80% plant material and 16% fruit during the high-water season and 100% plant material during the low-water season (Mérona and Rankin-de-Mérona 2004). Oliveira et al. (2006) found that in Lake Camaleão, Amazonas State, Brazil, *Schizodon fasciatus consumed* ~100%, 90%, 90%, and 100% plant material and ~0%, 10%, 10%, and 0% fruit, during the rising-, high-, receding- and low-water seasons, respectively. *Myloplus ternetzi* from the Approuague River, French Guiana consumed 97% plant material and 3% fruits and seeds during the high-water season and 95% plant material and 5% fruits and seeds during the low-water season (Boujard et al. 1990). In some species of Serrasalmidae, folivore distributions are tied to seasonal changes in water levels and habitat, as well as resource availability (Coronado-Franco et al. 2022).

Trophic niche overlap: In some assemblages there is seasonal variation in trophic niche overlap among frugivores and folivores (Correa and Winemiller 2014). Correa and Winemiller (2014) found that in an assemblage of five frugivores (*Brycon falcatus, Brycon melanopterus, Myloplus asterias, M. rubripinnis*, and *M. torquatus*) inhabiting the Apaporis River, Colombia, there was greater niche overlap during the high-water season when fruits and seeds were readily available, and presumably competition for these resources was low. During the low water-season, when there were fewer fruits and seeds available, trophic niche overlap was reduced among these co-occurring frugivores as a result of interspecific differences in the most important non-fruit and seed components of diets.

Seasonal variation in trophic niche breadth: Most studies report wider realized trophic niche breadths in frugivores inhabiting floodplain ecosystems during low-water than high-water seasons (Mérona and Rankin-de-Mérona 2004; Freitas et al. 2011; Correa and Winemiller 2014). For example, during the low-water versus high-water season, niche breadths estimated using Levin's Index (Levin's 1968) were 48% wider in *Mylossoma duriventre* (Mérona and Rankin-de-Mérona 2004), 167% wider in *Colossoma macropomum*, and 1273% wider in *Auchenipterichthys longimanus* (Freitas et al. 2011). Barbosa et al. (2015) found that the niche breadth of *Lithodoras dorsalis* from a tidal delta ecosystem did not exhibit clear associations with seasons. In contrast to floodplain frugivores, Mérona and Rankin-de-Mérona (2004) found that the niche breadth of the folivorous *Schizodon fasciatus* was 50% wider during the high-water season than the low-water season.

15.5 Phylogenetic Diversification

Ancestral state reconstructions (see Chap. 7 for methodological details) inferred that folivory evolved independently at least six times in Neotropical fishes: (1) in the most recent common ancestor (MRCA) of *Symphysodon* (Cichlidae), (2) the MRCA of Serrasalmidae, (3) the MRCA of *Schizodon* (Anostomidae), (4) in *Deuterodon iguape* (Acestrorhamphidae), (5) in *Deuterodon pedri* (Acestrorhamphidae), and (6) in the MRCA of a clade containing *Deuterodon singularis, D. stigmaturus, D. supparis*, and *D. langei* (Acestrorhamphidae). Note that *Deuterodon* was not recovered as monophyletic (Chap. 2). Frugivory was estimated to have evolved independently at least five times: (1) in the MRCA of Serrasalmidae, (2) in the MRCA of a clade containing *Brycon falcatus* and *Brycon melanopterus* (Bryconidae), (3) in *Auchenipterichthys longimanus* (Auchenipteridae), (4) in the MRCA of a clade containing *Pterodoras rivasi* and *Pterodoras granulosus* (Doradidae), and (5) in *Lithodoras dorslis* (Doradidae). These reconstructions are our best attempts and first approximations at quantifying the frequency with which folivory and frugivory have evolved in Neotropical fishes. However, it is important to remember that other Neotropical lineages consume fruit and plant material, but not sufficient quantities to be assigned to frugivore or folivore trophic guilds. Consequently, our ancestral state reconstructions are conservative and more nuanced analyses can reveal additional information about the evolution of fruit- and plant-eating fishes. Interestingly, the transitions to folivory and frugivory in Serrasalmidae were approximately contemporaneous. In other words, the MRCA of Serrasalmidae was inferred to consume substantial quantities of both seeds and fruits and other plant materials (see also, Kolmann et al. 2021).

The transitions to folivory and frugivory in Serrasalmidae were the earliest transitions to both of these trophic guilds in Neotropical fishes. Kolmann et al. (2021) estimated an age for the MRCA of Serrasalmidae of 38.0–64.6 Ma, which suggests this is the approximate timing of the transitions to folivory and frugivory in Serrasalmidae (Fig. 15.3). Likewise, the ancestors of Bryconidae, another family containing frugivorous and folivorous species), evolved around 37–47 million years ago (Abe et al. 2014). This timing corresponds to the diversification of many of the plant clades that are important in the diets of these fishes (Horn et al. 2011; Correa et al. 2015a; Kolmann et al. 2021). For example, the fruits and seeds of rubber trees (*Hevea spruceana*), tucumã palms (*Astrocaryum* spp.), Capparaceae spp. (e.g., the tapia tree *Crateva tapia* and fava tree *Vicia faba*) are important in the diets of *Piaractus* spp. and *Colossoma macropomum* (Lucas 2008; Anderson et al. 2009). In the Neotropics, these plant clades generally diversified in the Eocene and Oligocene (Mercado-Gómez et al. 2022), timing that corresponds to the rise of folivorous and frugivorous serrasalmids and proceeding the origins of folivory and frugivory in other Neotropical fish clades (Wojciechowski 2003; Correa et al. 2007; Schaefer et al. 2009; Bartish et al. 2011; Correa et al. 2018; Kolmann et al. 2021).

Resource availability appears to structure habitat affiliation, at least in serrasalmids, where frugivorous species are strongly associated with flooded lowlands with high floodplain extent and increased landscape heterogeneity (Coronado-Franco et al. 2022). Unlike frugivores, folivorous serrasalmids are associated with lesser floodplain extent and reduced landscape heterogeneity, possibly resulting from many species being associated with upland, rheophilic habitats (Andrade et al. 2019a; Coronado-Franco et al. 2022).

15.6 Functional Roles in Neotropical Ecosystems

15.6.1 Food Web Linkages

Neotropical aquatic food webs are spatially and temporally variable, but there are some food web linkages that are consistently formed by frugivores and folivores (Winemiller 1996; Wolff et al. 2013; Arantes et al. 2019; Alonso et al. 2020). In some floodplain and coastal delta ecosystems, frugivores and folivores are conduits of energy and material derived from C3 aquatic macrophytes, particulate organic matter, terrestrial plant material, fruits and seeds, and phytomicrobenthos (Soares et al. 1986; Alvim and Peret 2004; Barbosa et al. 2015; Rejas 2018; Arantes et al. 2019; Alonso et al. 2020; Andrade et al. 2023). Folivorous and frugivorous fishes channel the flow of material and energy derived from these production sources both by consuming primary consumers that feed on these materials (e.g., aquatic insects and zooplankton) and, most notably, by directly consuming these producers (see Sects. 15.4.1–15.4.4 for detailed descriptions of diet composition as well as temporal

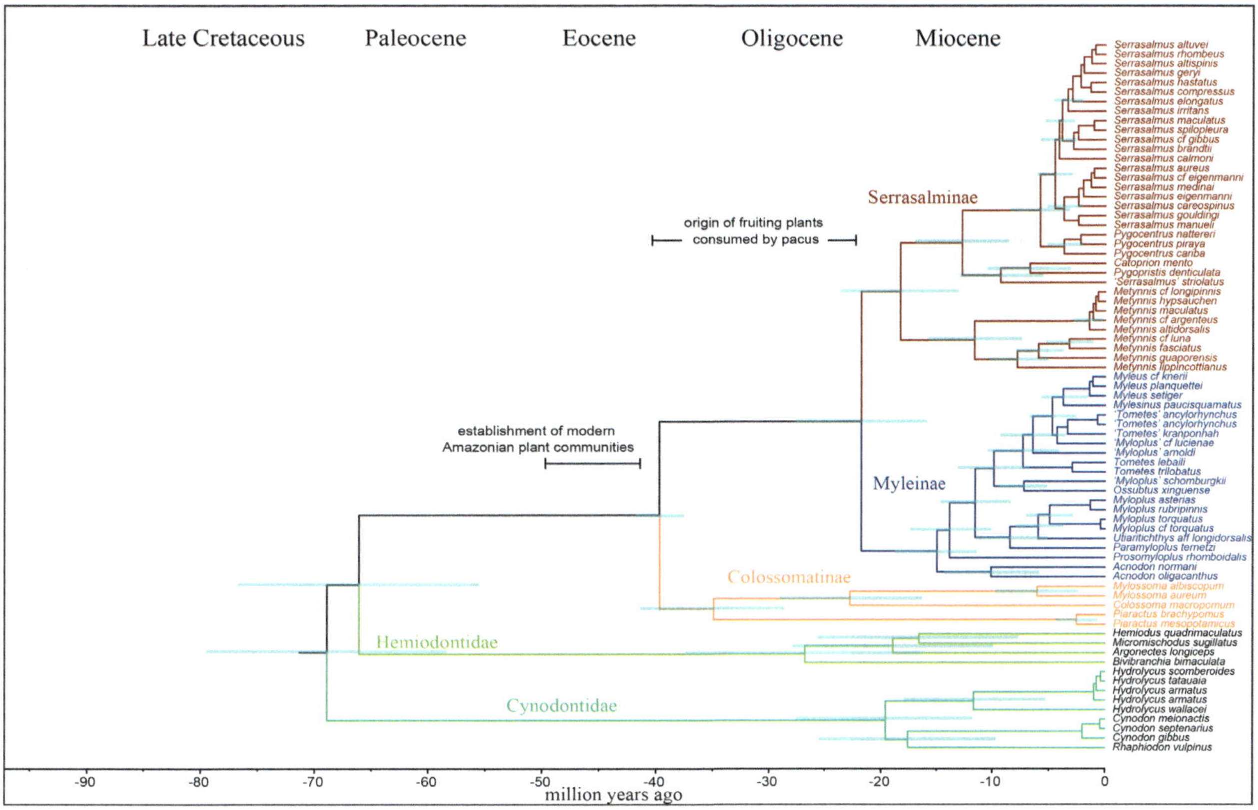

Fig. 15.3 Figure reproduced from Kolmann et al. (2021) illustrating the timing of serrasalmid diversification relative to the origin of many of the fruiting plant lineages consumed by these fishes. Ultrametric phylogeny time-calibrated using a set of 15 priors based on fossils

and ontogenetic diet variation), which is relatively uncommon in Neotropical fishes (see Chap. 6). Upland rivers and streams are largely depauperate of frugivores, although some taxa (e.g., *Brycon* spp.) occur in small numbers and their ecology in highlands is poorly understood and warrants further study (Chap. 9; M.A. Kolmann personal observation). Folivores are common in upland rivers (e.g., *Myleus* spp., *Mylesinus* spp., *Ossubtus xinguense*, and *Tometes* spp. in the Tocantins and Madeira rivers) where they facilitate the transfer of energy and materials from plants, particularly aquatic macrophytes, to higher trophic levels (Vitorino et al. 2016; Andrade et al. 2016a, 2016b).

A large proportion of Neotropical folivores and frugivores directly consume and readily assimilate C3 and CAM plant material and fruits and seeds (Tables 15.1 and 15.3; Forsberg et al. 1993; Manetta et al. 2003; Alvim and Peret 2004; Barbosa et al. 2015; Arantes et al. 2019; Alonso et al. 2020; Andrade et al. 2023). In contrast, only a handful of species consume large quantities of C4 plants (primarily grasses): *Schizodon borellii* (Mortillaro et al. 2015), *Schizodon fasciatus* (Andrade and Braga 2005; Oliveira et al. 2006; Mortillaro et al. 2015; Rejas 2018; Arantes et al. 2019), *Schizodon nasutus* (Andrade and Braga 2005), and *Mylossoma duriventre* (Oliveira et al. 2006; Rejas 2018). Presumably, other *Schizodon* and *Mylossoma* species consume C4 plants, but this is unconfirmed. Overall, C4 plants are responsible for much of the primary production in floodplain ecosystems, but are only minor basal production sources in aquatic Neotropical food webs due to their low digestibility by most consumers (Hamilton et al. 1992; Forsberg et al. 1993; Melack and Forsberg 2001; Mortillaro et al. 2015; Rejas 2018; Arantes et al. 2019). Consequently, folivorous fishes that assimilate energy and material from C4 grasses form highly unique trophic linkages that warrant further investigation.

Neotropical folivores and frugivores are thought to be consumed by a variety of aquatic and terrestrial predators, including piscivorous fishes, birds, mammals, and reptiles, and thus serve as important links between lower and higher trophic levels (Barbosa et al. 2018; Terborgh and Davenport 2021; Costa and Nunes 2022; Lubich et al. 2022). Although diet studies based on examination of gut contents and stable isotopes are generally unable to identify prey material to the species level, this limits knowledge about the transfer of material and energy from foliviores and frugivores to other Neotropical consumers. Smaller folivore and frugivore individuals and species (e.g., *Schizodon fasciatus*) are likely consumed by many species of aquatic and terrestrial piscivores (Barbosa et al. 2018; Terborgh and Davenport 2021; Costa and Nunes 2022; Lubich et al. 2022). The adults of large species (e.g., *Colossoma macropomum*) are likely primarily consumed by large predators such as river dolphins (*Inia geoffrensis*), caimans (subfamily Caimaninae), and giant otters (*Pteronura brasiliensis*; Goulding and Carvalho 1982; Silva et al. 2014; Moldowan et al. 2016; Terborgh and Davenport 2021; Echeverria et al. 2022). Diet studies using approaches such as DNA metabarcoding are needed to provide a higher-resolution understanding of Neotropical food webs (de Sousa et al. 2019; Bonato et al. 2022).

15.6.2 Seed Dispersal

In Neotropical floodplain ecosystems, frugivores fishes act as important agents of seed dispersal for many plant species (de Souza-Stevaux et al. 1994; Anderson et al. 2009; Correa et al. 2015b). Much of the fruit consumption occurs during the high-water season, when the fruits of many species ripen and frugivorous fishes move into flooded forests and savannahs (Fig. 15.2; Gottsberger 1978; Goulding 1980; Goulding and Carvalho 1982; Kubitzki and Zibursk 1994; Roubach and Saint-Paul 1994; Saint-Paul et al. 2000; Anderson et al. 2009, 2011; Horn et al. 2011; Weiss et al. 2016, 2024). See Sect. 15.4.4 for a detailed discussion of seasonal variation in the consumption of fruits and seeds. Frugivorous fishes, particularly large-bodied species, are highly effective seed dispersers because whole seeds can pass through their digestive tracts intact, and seeds that do pass through digestive tracts tend to have high germination rates (Anderson et al. 2011), and because frugivorous fishes can travel considerable distances (Anderson et al. 2011; Horn et al. 2011; Correa et al. 2015b). For example, Anderson et al. (2011) modeled seed dispersal distances by adult *Colossoma macropomum* for five plant species. Mean dispersal distances ranged from 337 to 552 m, and the maximum predicted dispersal distance was 5495 m. Seed dispersal effectiveness increases with fish size in characiforms because smaller individuals are more likely to crush and destroy seeds with their molariform oral teeth, whereas larger individuals generally swallow fruits and seeds whole and intact and can consume a wider range of seed sizes (Kubitzki and Ziburski 1994; Galetti et al. 2008; Anderson et al. 2009; Horn et al. 2011; Correa et al. 2015b). There is not a strong association between size and dispersal effectiveness in siluriform frugivores because juveniles and adults from these species tend to swallow fruits and seeds whole, although like Characiformes, large individuals do have the capacity to ingest larger fruits and seeds (de Souza-Stevaux et al. 1994; Mannheimer et al. 2003; Correa et al. 2007).

Frugivorous fishes vary with respect to the diversity of fruits and seeds they consume (and potentially disperse), with some specialist frugivores (e.g., *Myloplus, Mylossoma*) feeding on only a few plant species, while other frugivores consume seeds from many different species of plants (Correa et al. 2015a, 2015b; Costa-Pereira and Galetti 2015; Araujo et al. 2020). For example, large-bodied pacus like *Colossoma* and *Piaractus* each consume (and disperse) seeds from over 70 species of plants in their respective regions (Correa et al. 2015b; Araujo et al. 2020). Other studies found that eleven different species of fishes had ingested seeds from 49 different species of plants (Costa et al. 2024). Another study found that eight species of fishes dispersed seeds from 16 different species of plants along the Uatumã River (Weiss et al. 2024). The difference in the diversity of plants consumed by frugivores highlights how the diversity of the plant community and frugivore community can vary greatly by region.

What plants do frugivorous fishes target and what fruit characteristics are attractive to fishes? Commonly targeted fruits by fishes belong to the families Arecaceae, Convolvulaceae, Euphorbiaceae, Melastomataceae, Phyllanthaceae, Rubiaceae, and Violaceae, for example (Goulding 1980; Araujo et al. 2020; Weiss et al. 2024). Fruits in these families vary in shape, size, and texture, but fruits spread by fishes share several traits in common. Namely, these fruits tend to be fleshy, float easily, and have low seed to pulp ratios (i.e., fish-eaten fruits tend to be less dense; Correa et al. 2018). Evolutionarily, reliance on ichthyochory by plants is thought to have followed hydrochory, whereby fleshy fruits facilitated buoyancy first, and then were exapted for ichthyochory as fruit-frugivore networks evolved (Correa et al. 2018).

15.7 Morphological Adaptations

In this section, we discuss morphological traits that have been associated with folivory and frugivory and contrast these traits with non-folivorous and frugivorous taxa. Interspecific comparisons are often used to gain insight into the adaptive significance of phenotypic traits (Harvey and Pagel 1991). However, there are numerous factors that can lead to spurious inferences about adaptation when using this approach, particularly when shared evolutionary history (i.e., phylogeny) is not accounted for (Garland et al. 1992; Garland and Adolph 1994). Therefore, we caution readers to view the associations discussed in this section as preliminary hypotheses of adaptive significance only, particularly when associations have not been formally established using phylogenetic comparative statistical approaches.

15.7.1 Digestive Tracts and Abdominal Cavities

Several studies reported long digestive tracts in folivores and frugivores and hypothesized that they allow for the digestion of tough seeds and plant materials, such as thick cellulose cell walls (Alexander 1964; Crampton 2008; Vitule et al. 2008; Andrade et al. 2013; Santos et al. 2015). The *intestinal quotient* (IQ = digestive tract length/fish standard length) is used to compare digestive tract lengths. In Neotropical folivores and frugivores, We found IQs ranging from 0.97 in *Deuterodon stigmaturus* to 4.90 in *Tometes camunani* (Table 15.4). The IQs of folivores and frugivores were generally greater than those of planktivores, omnivores, benthic invertivores, surface invertivores, piscivores, and ectoparasites, similar to algivores, and shorter than detritivores, which had IQs as great as 21.60. Burns (2021) investigated the evolution of intestine length and abdominal cavity size in Characiformes using phylogenetic comparative methods. This study included the frugivore *Colossoma macropomum* and the folivores *Deuterodon iguape, Schizodon fasciatus, Schizodon intermedius, Schizodon knerii*, and *Schizodon nasutus*, but included these species in a combined detritivore/folivore trophic guild for statistical analyses. Burns (2021) identified positive correlations between gut length and detritivory/folivory and gut length and abdominal cavity size. However, a subsequent study with greater representation of folivorous and frugivorous species (strictly Serrasalmidae) did not find that herbivores, including frugivores and folivores, had larger abdominal cavities than non-herbivorous species (Kolmann et al. 2024). More research into the evolution of digestive phenotypes in herbivorous Neotropical fishes, accounting for both morphological and physiological adaptations, is a ripe area of inquiry.

15.7.2 Teeth in Extant Folivores and Frugivores

The oral jaws of frugivorous characiforms tend to have heterodont dentitions, with at least some molariform teeth, which are thought to aid in husking, shelling, or crushing seeds (Lucas 2008; Andrade et al. 2019b; Huie et al. 2019). Kolmann et al. (2019) found that in Serrasalmidae, the evolution of interlocking teeth and simultaneously-replaced dentitions (where all teeth on one side of the head are lost and replaced at the same time) were associated with the evolution of herbivory. Interlocking teeth are hypothesized to disperse stress evenly along the tooth row, promoting even tooth wear. Together, interlocking teeth and simultaneous tooth replacement may allow folivorous serrasalmids to maintain a consistently sharp cutting edge for processing tough plant material containing cellulose and silica (Kolmann et al. 2019). Frugivorous serrasalmids also have iron-laden teeth for maintaining a hard cutting edge (Velasco-Hogan et al. 2021).

Folivorous characiforms tend to have flat, sharp, and sometimes multicuspid oral teeth, which may also aid in folivory (Lucena and Lucena 2002; Andrade et al. 2013, 2016a, 2016b, 2019b; Pereira and Castro 2014; Santos et al. 2015; Huie et al. 2019; Fig. 15.4). Multiple authors have noted that tooth height may be associated and the percentage of plant material in the diets of folivorous serrasalmids (Andrade et al. 2013; Huie et al. 2019), as has been documented in herbivorous mammals (Webb 1977; Clauss et al. 2008; Codron et al. 2019). In fact, morphological convergence among folivorous and frugivorous species of serrasalmids, with respect to feeding traits, appears widespread (Kolmann et al. 2024). Recent research has shed some light on some potential biomechanical hallmarks of teeth in frugivorous fishes. Rosen et al. (2025) proposed

Table 15.4 Intestinal quotient (IQ) values for Neotropical fishes representing a range of trophic guilds

Order		Species	Guild(s)	IQ	Citation
		Frugivores and folivores			
Characiformes	Anostomidae	*Schizodon altoparanae*	FOL	1.41	Ferretti et al. (1996)
	Anostomidae	*Schizodon borellii*	FOL	1.35	Ferretti et al. (1996)
	Anostomidae	*Schizodon fasciatus*	FOL	2.08	Pouilly et al. (2003)
	Anostomidae	*Schizodon knerii*	FOL	1.30	Santos et al. (2015)
	Characidae	*Deuterodon stigmaturus*	FOL	0.97	Dala-Corte et al. (2016)
	Serrasalmidae	*Colossoma macropomum*	FRU	2.00	Goulding and Carvalho (1982)
	Serrasalmidae	*Myleus rubripinnis*	FOL	2.60	Alexander (1964)
	Serrasalmidae	*Mylossoma duriventre*	FOL, FRU	1.96	Pouilly et al. (2003)
	Serrasalmidae	*Tometes camunani*	FOL	4.90	Andrade et al. (2013)
		Nonfrugivores and folivores			
Characiformes	Acestrorhynchidae	*Acestrorhynchus falcirostris*	PLA	0.06	Ramírez et al. (2015)
	Anostomidae	*Leporinus friderici*	OMN	1.25	Albrecht et al. (2001)
	Characidae	*Charax condei*	BEN, SUR, PIS	0.19	Ramírez et al. (2015)
	Characidae	*Ctenobrycon hauxwellianus*	BEN, OMN	0.94	Ramírez et al. (2015)
	Characidae	*Gymnocorymbus thayeri*	BEN	0.69	Ramírez et al. (2015)
	Characidae	*Hemigrammus analis*	BEN, OMN	0.60	Ramírez et al. (2015)
	Characidae	*Hyphessobrycon copelandi*	BEN, OMN	0.30	Ramírez et al. (2015)
	Characidae	*Phenacogaster pectinata*	BEN	0.42	Ramírez et al. (2015)
	Characidae	*Tetragonopterus argenteus*	OMN, ECT	0.89	Ramírez et al. (2015)
	Crenuchidae	*Characidium pellucidum*	BEN	0.26	Ramírez et al. (2015)
	Curimatidae	*Potamorhina altamazonica*	DET	11.26	Pouilly et al. (2003)
	Curimatidae	*Potamorhina latior*	DET	10.21	Pouilly et al. (2003)
	Curimatidae	*Steindachnerina elegans*	DET	16.22	Ibañez et al. (2007)
	Curimatidae	*Steindachnerina guentheri*	DET	15.84	Ibañez et al. (2007)
	Iguanodectidae	*Bryconops melanurus*	OMN	0.53	Ramírez et al. (2015)
	Parodontidae	*Apareiodon affinis*	ALG	1.20	Ibañez et al. (2007)
	Parodontidae	*Parodon hilarii*	ALG	1.20	Ibañez et al. (2007)
	Serrasalmidae	*Serrasalmus rhombeus*	BEN, PIS, OMN, ECT	1.22	Pouilly et al. (2003)
Gymnotiformes	Apteronotidae	*Adontosternarchus sachsi*	PLA	0.50	Pouilly et al. (2003)
	Hypopomidae	*Hypopygus lepturus*	BEN	0.34	Ramírez et al. (2015)
	Rhamphichthyidae	*Gymnorhamphichthys rondoni*	BEN	0.27	Ramírez et al. (2015)
	Sternopygidae	*Eigenmannia virescens*	BEN, PLA	0.36	Pouilly et al. (2003)
	Sternopygidae	*Eigenmannia virescens*	BEN, PLA	0.27	Ramírez et al. (2015)
Siluriformes	Auchenipteridae	*Auchenipterus nuchalis*	BEN	1.13	Pouilly et al. (2003)
	Auchenipteridae	*Entomocorus benjamini*	SUR	0.99	Pouilly et al. (2003)
	Loricariidae	*Hypostomus oculeus*	DET, ALG	21.60	Ramírez et al. (2015)
	Loricariidae	*Limatulichthys griseus*	ALG	1.10	Ramírez et al. (2015)
	Pimelodidae	*Hypophthalmus marginatus*	PLA	1.46	Pouilly et al. (2003)

Neotropical frugivores (FRU), folivores (FOL), and nonfrugivore and folivore trophic guilds for comparison: PLA = planktivore, OMN = omnivore, BEN = benthic invertivore, SUR = surface invertivore, PIS = piscivore, ECT = ectoparasite, ALG = algivore, DET = detritivore. We report mean IQ values if available, but in a small number of cases, values are based on a single measurement

Fig. 15.4 Different phenotypes of folivorous and frugivorous serrasalmids (pacus). (**A**) *Ossubtus xinguense*, the endangered folivorous parrot pacu from the Rio Xingu. (**B**) *Prosomyleus rhomboidalis*, a frugivorous pacu. Note the narrow, cusped, and blade-like teeth in the folivore (**A**) versus the blunt, robust molariform teeth in the frugivore (**B**). CT scans on the right show the unique inferior mouth position in *Ossubtus*, a species that grazes on Podostemaceae in rapids. Figure modified from Huie et al. (2019) and pictures of live fishes courtesy of Mark Sabaj

that frugivorous fish dentitions should be sharp enough to indent fruit rinds, to gain purchase on floating food, and once engaged deeper into fruit tissues, these teeth would pulverize sugar-containing cells. As expected, frugivorous pacu teeth distribute stress more evenly, while piranha teeth concentrate stress at tooth tips for better penetration into prey. This "Low Force (for puncture), High Work (for cell wall pulverizing)" approach describes the behavior of dentitions in frugivores like *Colossoma*, as well as other vertebrate frugivores like primates and bats. However, Rosen et al. (2025) also suggest, based on the performance of the teeth in allegedly "frugivorous" piranhas (e.g., former genus *Pristobrycon*), that these fishes may be targeting harder foods like seeds instead of fruit pulp. Recent work (Huskey et al. 2026) that may support these observations demonstrated that piranhas which consume seeds (e.g., *Serrasalmus eigenmanni, Pygopristis denticulata*) also tend to have shorter, high leverage jaws, and exert high tooth pressures when biting, perhaps reflecting the demands of their granivorous diets.

15.7.3 Functional Morphology of Frugivorous Fishes

Understanding the form and function of frugivore dentitions can aid in reconstructing the evolution of animal-mediated seed dispersal in the Neotropics, as frugivorous fishes may be the earliest vertebrates to act as seed dispersers (Correa et al. 2015a). Comparative anatomists studying terrestrial vertebrates, especially mammals, have taken their understanding of dental morphology in extant herbivores and used this knowledge to hypothesize about the ecological roles of extinct species (Erickson et al. 1996; Lautenschlager 2020). Here, we do something similar with frugivorous fishes, specifically with the extinct large-bodied piranha, *Megapiranha paranensis*, known from upper jaw bones and attached teeth from the upper Miocene of Argentina (Cione et al. 2009). Previous studies have assumed that this extinct piranha was a carnivore, based mostly on its phylogenetic placement as sister to piranhas, rather than pacus (Cione et al. 2009; Grubich et al. 2012). However, the earliest branching lineages within piranhas are omnivorous or herbivorous, not carnivoreous (Kolmann et al. 2021). Taxa like "*Serrasalmus*" *striolatus* (formerly *Pristobrycon*) and *Pygopristis* are omnivores and herbivores, respectively (Nico and Taphorn 1988; Kolmann et al. 2024), while *Catoprion* species are ectoparasites (Vieira and Gery 1979; Chap. 18). Taking a strict parsimony perspective, *Megapiranha* could be expected to be an herbivore, given that it is phylogenetically bracketed by herbivorous and omnivorous lineages like *Metynnis*, *Pygopristis*, and *S. striolatus* (Fig. 15.5).

An additional line of evidence used to explore whether *Megapiranha* was likely an herbivore or carnivore could be

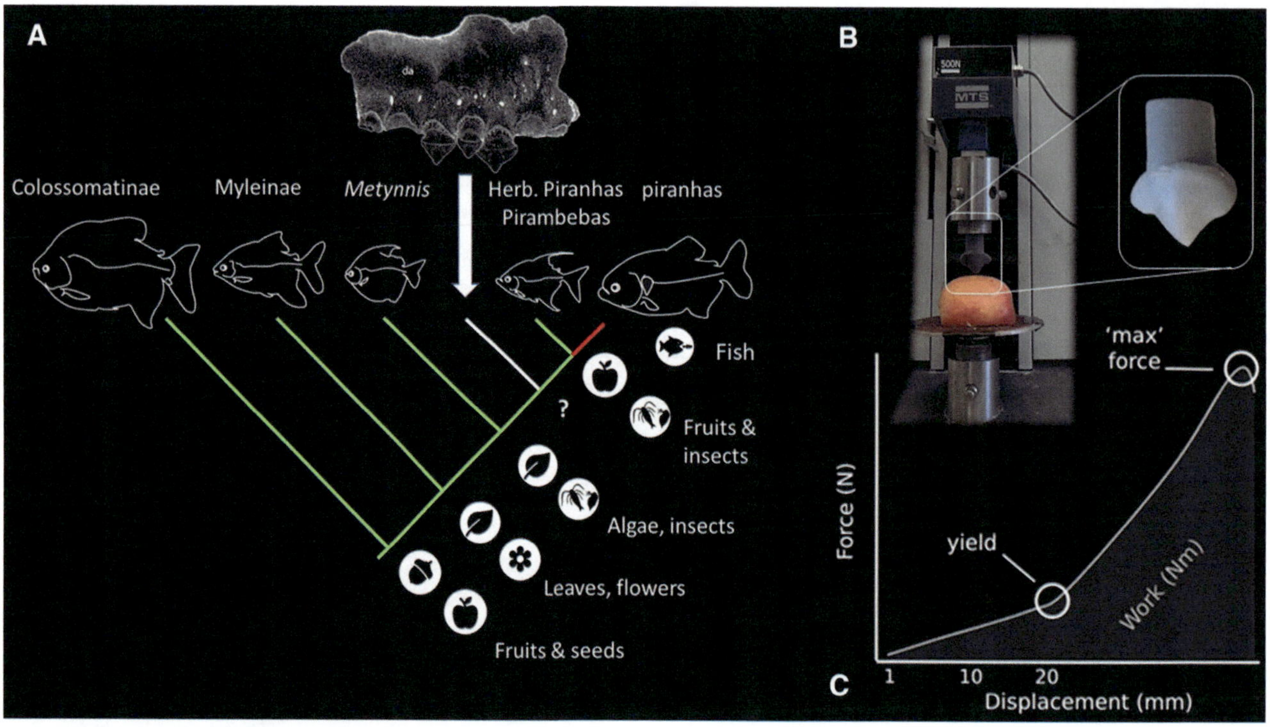

Fig. 15.5 Phylogeny of serrasalmids and an experiment for measuring tooth indentation performance in †*Megapiranha*. (**A**) Simplified phylogeny of piranhas and pacus (from Kolmann et al. 2021) with diets represented at nodes by icons. Placement of †*Megapiranha* indicated by white arrow (image from Cione et al. 2009). (**B**) Universal testing system and 3D printed piranha tooth model used for indentation tests. (**C**) Example force-displacement curve measured by a universal testing system

biomechanical: one could evaluate the performance of the dentition in *Megapiranha* to determine if it is more like herbivorous rather than carnivorous cousins. We did just that, by 3D printing piranha and pacu teeth, including tooth models from *Megapiranha*, and indenting these models into fruit (apples and oranges). We used surface renderings from microCT-scanned teeth to make 3D-printed models (FormLabs2 resin printer) and then measured the force (N) and work (Nm) required to indent these tooth models into fruit using a universal testing station (MTS Systems Corporation, Eden Prairie, MN). We then used analysis of variance to test for differences in the performance (force to puncture and work to puncture) of tooth models among three species of serrasalmids, *Pristobrycon maculipinnis* (a piranha), *Piaractus brachypomus* (a pacu), and *Megapiranha*. We found that teeth in *Megapiranha* performed more like pacu teeth (*Piaractus*) than modern piranhas (*Pristobrycon*). With respect to force and work, the behavior of *Megapiranha* teeth was indistinguishable from *Piaractus* (force, $p = 0.855$; work, $p = 0.242$). Tooth models from *Piaractus* and *Megapiranha* both exerted higher forces and higher work than piranha teeth. In contrast, *Megapiranha* tooth indentation performance almost always differed statistically from *Pristobrycon* (force, $p = 0.001$; work, $p < 0.001$). This research suggests that inferring the ecological role of *Megapiranha* based solely on its presumed inclusion within piranhas is dubious, especially when considering dietary diversity within piranhas (Nico and Taphorn 1988; Kolmann et al. 2021). Finally, we suggest that *Megapiranha* was likely a frugivore or generalist herbivore, based on phylogenetic bracketing by herbivorous taxa and the behavior of its teeth.

15.7.4 Skulls, Lips, and Cranial Musculature

Robust skulls, distinct lower-jaw morphology, terminal mouths, and large *pars rictalis* subdivisions of the adductor mandibulae muscles are proposed adaptations for folivory and frugivory in Neotropical fishes (Alexander 1964; Lucas 2008; Huby et al. 2019; Huie et al. 2019). Differences in adductor mandibulae morphology and attachment between folivores and frugivores have also been identified (Huby et al. 2019). The folivore *Ossubtus xinguense* is the only serrasalmid with a subterminal mouth, which has been suggested to be an adaptation associated with the distinct substrate scraping foraging mode exhibited by this species (see Sect. 15.3.1; Jégu and Dos Santos 1990; Huie et al. 2019).

Huby et al. (2019) used phylogenetic comparative approaches to characterize differences in cranial morphology and bite performance among 27 serrasalmid species grouped into herbivore (frugivores + folivores) and carnivore trophic guilds. This study included the folivore *Myleus micans*, the frugivores *Colossoma macropomum*, *Prosomyleus rhomboidalis*, *Myloplus rubripinnis*, *Myloplus schomburgkii*, *Myloplus ternetzi*, *Piaractus brachypomus*, and *Piaractus mesopotamicus*, and *Acnodon oligacanthus*, which is assigned to both the folivore and frugivore guilds. Huby et al. (2019) determined that adductor mandibulae muscle mass (relative to body mass) was significantly heavier in carnivorous than in herbivorous species and that among herbivorous species, frugivores tended to have larger adductor mandibulae muscles with more attachment points than folivores. The *pars rictalis* was significantly larger in herbivorous than in carnivorous species, while the *pars malaris* subdivision of the adductor mandibulae was significantly larger in carnivorous than in herbivorous species. These morphological differences corresponded to three major predicted functional differences between herbivores and carnivores. First, Huby et al. (2019) predicted that the adductor mandibulae muscles transmit relatively more force to the front of the lower jaw and less force to the back of the lower jar in herbivorous species relative to carnivorous species. Second, herbivorous species were found to have a "vice-like" pattern of dental occlusion, in contrast to carnivorous species, which have a more "scissor-like" bite. Although this pattern is more obvious in differentiating frugivores from carnivores: folivores also tend to have a more scissor-like bite (Huie et al. 2019). Finally, predicted bite force relative to body size was greater in carnivores than herbivores, although this difference was not statistically significant. Interestingly, frugivores tended to have higher predicted bite forces than folivores, suggesting that folivory and frugivory may impose different selective pressures on bite performance and that future research may benefit from designating separate frugivore and folivore categories in statistical analyses (Huby et al. 2019). In piranhas, however, frugivorous/granivorous species like, *Serrasalmus eigenmanni*, tended to exhibit higher bite forces and bite stresses than carnivores (like the black piranha, *Serrasalmus rhombeus*), pound-for-pound (Huskey et al. 2026).

Serrasalmids also have soft tissue modifications of the feeding apparatus that may aid in processing fruits and other plant materials. Cohen et al. (2023) found a tight correlation between lip morphotypes and feeding ecology. Namely, that folivorous species have thinner lips than frugivores, and lips protected from abrasion by layers of connective tissue and mucus-producing goblet cells. Frugivores, on the other hand, have larger lips filled with collagen and fat, as well as being richly covered in sensory structures lip taste buds and nerves. Frugivores were also proposed to have more mobile lips than other taxa, so much so that Cohen et al. (2023) proposed that the lips in these taxa functions like an oral appendage, i.e., like a trunk or tongue. Similarly, observations on feeding in *Brycon cephalus* (Carvalho et al. 2021) demonstrate that considerable food manipulation might be expected in frugivores who engage in seed predation. In this study, individual *Brycon cephalus* intraorally manipulated seeds, even regurgitating seeds for later manipulation, during which the seed coat was deftly removed (Carvalho et al. 2021). More investigations of feeding behavior are necessary in frugivorous fishes to determine how widespread these behaviors are and what separates seed predators from more generic frugivores.

Roberts-Hugghis et al. (2023) investigated the evolution of oral jaw morphology in Neotropical cichlids using phylogenetic comparative methods. This study found that *Symphysodon* spp. have small mouths and thin, elongate jaws relative to other Neotropical cichlids (Roberts-Hugghis et al. 2023). This project did not test for correlations between diet and jaw morphology. Thus, additional research is needed to determine if the evolution of small mouths and thin, elongate jaws is correlated with the evolution of folivory and to test for associations between other anatomical attributes and folivory in Neotropical cichlids. In some herbivorous cichlids, like in serrasalmids, hypertrophied lips are thought to aid in herbivory. Specifically, in cichlids these lips are thought to aid in scraping algae and plant materials from the substrate (Ford et al. 2016).

15.8 Future Directions

Folivorous and frugivorous fishes play major roles in structuring lowland plant communities in the Neotropics by grazing on plant material and dispersing seeds (Galetti et al. 2008; Anderson et al. 2009, 2011) and channeling energy from primary producers to higher tropic levels. Despite frugivorous fishes being perhaps the earliest vertebrate seed dispersers (Correa et al. 2007; Kolmann et al. 2021), the study of ichthyochory and frugivory in fishes in general lags behind that of terrestrial vertebrates like birds and bats. In fact, most studies on frugivory and seed dispersal in fishes have been published only in the last two decades (Correa et al. 2007), although Goulding (1980) brought attention to the idea over 40 years ago. Likewise, the ecological roles of folivores, including potential top-down impacts on aquatic plant communities and the roles of folivores in the flow of energy through both aquatic and terrestrial food webs, have not been well studied. Furthermore, we still have little understanding of what traits and behaviors are shared among folivorous and frugivorous fishes. For example, how do frugivorous fishes find, select for, and handle fruits, and how do some folivorous fishes digest C4 plant material?

Seed-dispersing fishes like pacus provide food for people living along the Amazon and other rivers, as well as aiding in the spread of the trees and plants that humans use for building materials and textiles (Goulding 1980). Fishes like *Piaractus* spp. and *Colossoma macropomum* also spread seeds from plants used in ceremonial practices (Goulding 1980), expanding the instrumental services these animals provide beyond provisioning to also include cultural services. Declines in the abundances of large fishes, particularly in pimelodid catfishes, and bryconid and serrasalmid characins, are expected to reduce both plant biodiversity and ecosystem health and functions, by interrupting seed-dispersal networks (Costa-Pereira and Galetti 2015). Given the long-distance migrations fishes like *Colossoma macropomum* are known to make, even localized reductions in fish diversity by overexploitation could have wide-ranging consequences (López-Fernández and Taphorn 2004; López-Fernández et al. 2012).

Acknowledgments M.A.K. thanks Coco Dawkins for her role in performing the *Megapiranha* tooth experiments and Dr. Stephanie Crofts for helping brainstorm these experiments and for the use of *Megapiranha* tooth models.

References

Abe KT, Mariguela TC, Avelino GS, Foresti F, Oliveira C (2014) Systematic and historical biogeography of the Bryconidae (Ostariophysi: Characiformes) suggesting a new rearrangement of its genera and an old origin of *Mesoamerican ichthyofauna*. BMC Evol Biol 14(1):152. https://doi.org/10.1186/1471-2148-14-152

Agostinho CS, Marques EE, Oliveira RJD, Braz PS (2009) Feeding ecology of *Pterodoras granulosus* (Siluriformes, Doradidae) in the Lajeado Reservoir, Tocantins, Brazil. Iheringia Sér Zool 99:301–306

Albrecht MP, Ferreira MFN, Caramaschi EP (2001) Anatomical features and histology of the digestive tract of two related neotropical omnivorous fishes (Characiformes; Anostomidae). J Fish Biol 58(2):419–430

Albrecht MP, Caramaschi ÉP, Horn MH (2009) Population responses of two omnivorous fish species to impoundment of a Brazilian tropical river. Hydrobiologia 627:181–193

Alexander RM (1964) Adaptation in the skulls and cranial muscles of South American characinoid fish. Zool J Linnean Soc 45(305):169–190

Alonso MB, de Carvalho DR, Alves CBM, Pompeu PS (2020) Trophic structure of a fish assemblage in a reference condition river located in a polluted watershed. Environ Biol Fishes 103(11):1437–1452

Alvim MCC, Peret AC (2004) Food resources sustaining the fish fauna in a section of the upper São Francisco River in Três Marias, MG, Brazil. Braz J Biol 64:195–202

Anderson JT, Saldaña Rojas J, Flecker AS (2009) High-quality seed dispersal by fruit-eating fishes in Amazonian floodplain habitats. Oecologia 161:279–290

Anderson JT, Nuttle T, Saldaña Rojas JS, Pendergast TH, Flecker AS (2011) Extremely long-distance seed dispersal by an overfished Amazonian frugivore. Proc R Soc B 278(1723):3329–3335

Andrade MC, Giarrizzo T, Jégu M (2013) *Tometes camunani* (Characiformes: Serrasalmidae), a new species of phytophagous fish from the Guiana Shield, rio Trombetas basin, Brazil. Neotrop Ichthyol 11:297–306

Andrade MC, Jégu M, Giarrizzo T (2016a) *Tometes kranponhah* and *Tometes ancylorhynchus* (Characiformes: Serrasalmidae), two new phytophagous serrasalmids, and the first *Tometes* species described from the Brazilian Shield. J Fish Biol 88:1–28

Andrade MC, Jégu M, Giarrizzo T (2016b) A new large species of *Myloplus* (Characiformes, Serrasalmidae) from the Rio Madeira basin, Brazil. Zookeys 571:153

Andrade MC, Machado VN, Jégu M, Farias IP, Giarrizzo T (2017) A new species of *Tometes* Valenciennes 1850 (Characiformes: Serrasalmidae) from Tocantins-Araguaia River Basin based on integrative analysis of molecular and morphological data. PLOS One 12(4):e0170053

Andrade MC, Fitzgerald DB, Winemiller KO, Barbosa PS, Giarrizzo T (2019a) Trophic niche segregation among herbivorous serrasalmids from rapids of the lower Xingu River, Brazilian Amazon. Hydrobiologia 829:265–280

Andrade MC, Winemiller KO, Barbosa PS, Fortunati A, Chelazzi D, Cincinelli A, Giarrizzo T (2019b) First account of plastic pollution impacting freshwater fishes in the Amazon: ingestion of plastic debris by piranhas and other serrasalmids with diverse feeding habits. Environ Pollut 244:766–773

Andrade MC, Keppeler FW, Zuluaga-Gómez MA, Conceição JW, Lisboa OP, Andrades R, Winemiller K, Giarrizzo T (2023) Food web structure in the Xingu River rapids prior to operation of the Amazon's largest hydropower plant. Aquat Sci 85(3):1–11

Andrade PM, Braga FMDS (2005) Diet and feeding of fish from Grande River, located below the Volta Grande reservoir, MG-SP. Braz J Biol 65(3):377–385

Anjos MRD, Machado NG, Pedersoli MA, Pedersoli NRB, Barros BS, Lourenço IH, Barreiros JP (2019) Survey of fish species from the Lower Roosevelt River, Southwestern Amazon basin. Biota Neotropica 19:e20180717

Arantes CC, Winemiller KO, Petrere M, Freitas CE (2019) Spatial variation in aquatic food webs in the Amazon River floodplain. Freshw Sci 38(1):213–228

Araujo JM, Correa SB, Anderson J, Penha J (2020) Fruit preferences by fishes in a Neotropical floodplain. Biotropica 52(6):1131–1141

de Ávila-Simas S, Reynalte-Tataje DA, Zaniboni-Filho E (2019) Fish predators of the golden mussel *Limnoperna fortunei* in different environments in a South American subtropical river. Bol Inst Pesca 45(2):e484

Barbarino DA, Winemiller KO (2003) Dietary segregation among large catfishes of the Apure and Arauca Rivers, Venezuela. J Fish Biol 63(2):410–427

Barbosa TA, Rosa DC, Soares BE, Costa CH, Esposito MC, Montag LF (2018) Effect of flood pulses on the trophic ecology of four piscivorous fishes from the eastern Amazon. J Fish Biol 93(1):30–39

Barbosa TAP, Barthem RB, Montag LFDA (2015) Feeding ecology of immature *Lithodoras dorsalis* (Valenciennes, 1840) (Siluriformes: Doradidae) in a tidal environment, estuary of the Rio Amazonas. Neotrop Ichthyol 13:341–348

Bartish IV, Antonelli A, Richardson JE, Swenson U (2011) Vicariance or long-distance dispersal: historical biogeography of the pantropical subfamily Chrysophylloideae (Sapotaceae). J Biogeography 38(1):177–190

Bleher H (2006) Bleher's discus, vol I. Aquapress, Pavia, Italy

Bonato KO, Silva PC, Carvalho FR, Malabarba LR (2022) Trophic interactions of vampire catfishes (Siluriformes: Vandelliinae) revealed by metabarcoding analysis of stomach contents. Freshw Biol 67(3):542–548

Boujard T, Sabatier D, Rojas-Beltran R, Prevost MF, Renno JF (1990) The food habits of three allochthonous feeding characoids in French Guiana. Rev Ecol Terre Vie 45(3):247–258

Bray JR, Curtis JT (1957) An ordination of the upland forest communities of southern Wisconsin. Ecol Monograph 27(4):326–349

Buckley J, Maunder RJ, Foey A, Pearce J, Val AL, Sloman KA (2010) Biparental mucus feeding: a unique example of parental care in an Amazonian cichlid. J Exp Biol 213(22):3787–3795

Bueno RS, Guevara R, Ribeiro MC, Culot L, Bufalo FS, Galetti M (2013) Functional redundancy and complementarities of seed dispersal by the last neotropical megafrugivores. PLOS One 8(2):e56252

Burns MD (2021) Adaptation to herbivory and detritivory drives the convergent evolution of large abdominal cavities in a diverse freshwater fish radiation (Otophysi: Characiformes). Evolution 75(3):688–705. https://doi.org/10.1111/evo.14178

de Carvalho DR, Alves CBM, Moreira MZ, Pompeu PS (2020) Trophic diversity and carbon sources supporting fish communities along a pollution gradient in a tropical river. Sci Total Environ 738:139878

Carvalho LN, dos Santos Júnior JB, Correa SB (2022) Fish thieves: an alternative tactic of food capture in a neotropical frugivorous species (*Brycon falcatus*). Austral Ecol 47(5):1140–1143

Carvalho LN, dos Santos Júnior JB, Correa SB (2021) Uncovering mechanisms of seed predation by fish. Biotropica 202:1460–1464

Cione AL, Dahdul WM, Lundberg JG, Machado-Allison A (2009) *Megapiranha paranensis*, a new genus and species of Serrasalmidae (Characiformes, Teleostei) from the upper Miocene of Argentina. J Vertebr Paleontol 29(2):350–358

Clauss M, Kaiser T, Hummel J (2008) The morphophysiological adaptations of browsing and grazing mammals. In: The ecology of browsing and grazing. Springer Berlin Heidelberg, Berlin, Heidelberg, pp 47–88

Codron D, Hofmann RR, Clauss M (2019) Morphological and physiological adaptations for browsing and grazing. In: The ecology of browsing and grazing II. Springer, Cham, pp 81–125

Cohen KE, Lucanus O, Summers AP, Kolmann MA (2023) Lip service: histological phenotypes correlate with diet and feeding ecology in herbivorous pacus. Anat Rec 306(2):326–342

Coronado-Franco KV, Tedesco PA, Kolmann MA, Borstein SR, Evans KO, Correa SB (2022) Feeding habits influence species habitat associations at the landscape scale in a diverse clade of neotropical fishes. J Biogeogr 49(12):2181–2192

Correa SB, Winemiller KO (2014) Niche partitioning among frugivorous fishes in response to fluctuating resources in the Amazonian floodplain forest. Ecology 95(1):210–224

Correa SB, Winemiller K (2018) Terrestrial–aquatic trophic linkages support fish production in a tropical oligotrophic river. Oecologia 186(4):1069–1078

Correa SB, Mérona BD, Armbruster JW (2014) Diet shift of Red Belly Pacu *Piaractus brachypomus* (Cuvier, 1818) (Characiformes: Serrasalmidae), a neotropical fish, in the Sepik-Ramu River Basin, Papua New Guinea. Neotrop Ichthyol 12:827–833

Correa SB, Winemiller KO, López-Fernández H, Galetti M (2007) Evolutionary perspectives on seed consumption and dispersal by fishes. BioScience 57:748e756

Correa SB, Costa-Pereira R, Fleming T, Goulding M, Anderson JT (2015a) Neotropical fish–fruit interactions: eco-evolutionary dynamics and conservation. Biol Rev 90(4):1263–1278

Correa SB, de Oliveira PC, Nunes da Cunha C, Penha J, Anderson JT (2018) Water and fish select for fleshy fruits in tropical wetland forests. Biotropica 50(2):312–318

Correa SB, Araujo JK, Penha JM, da Cunha CN, Stevenson PR, Anderson JT (2015b) Overfishing disrupts an ancient mutualism between frugivorous fishes and plants in neotropical wetlands. Biol Cons 191:159–167

Costa IDD, Nunes NNDS (2022) Niche partitioning of two piscivorous fish species in a river in the western Brazilian Amazon. Iheringia Sér Zool 112:e2022002

Costa G, Weiss B, Piedade MTF, Ferreira EJ, Maltchik L (2024) Structure and vulnerability of the ichthyochory network in wetland forests of Central Amazonia. Ecol Freshwater Fish 33:e12780

Costa-Pereira R, Correa SB, Galetti M (2018) Fishing-down within populations harms seed dispersal mutualism. Biotropica 50(2):319–325

Costa-Pereira R, Severo-Neto F, Yule TS, Pereira APT (2011) Fruit-eating fishes of *Banara arguta* (Salicaceae) in the Miranda River floodplain, Pantanal wetland. Biota Neotropica 11:373–376

Costa-Pereira R, Galetti M (2015) Frugivore downsizing and the collapse of seed dispersal by fish. Biol Cons 191:809–811

Crampton WG (2008) Ecology and life history of an Amazon floodplain cichlid: the discus fish *Symphysodon* (Perciformes: Cichlidae). Neotrop Ichthyol 6:599–612

Dala-Corte RB, Silva ERD, Fialho CB (2016) Diet-morphology relationship in the stream-dwelling characid *Deuterodon stigmaturus* (Gomes, 1947) (Characiformes: Characidae) is partially conditioned by ontogenetic development. Neotrop Ichthyol 14(2):e150178

Echevarría G, Rodríguez JP, Machado-Allison A (2018) Seasonal fluctuations in taxonomic and functional diversity in assemblages of catfishes in the Venezuelan Arauca River Floodplain. Stud Neotrop Fauna Environ 53(1):38–53

Echeverria A, Botta S, Marmontel M, Melo-Santos G, Fruet P, Oliveira-da-Costa M, Pouilly M, Di Tullio J, Van Damme PA (2022) Trophic ecology of Amazonian River dolphins from three rivers in Brazil and Bolivia. Mamm Biol 102(5):1687–1696

Egan JP, Gibbs S, Simons AM (2018) Trophic niches through ontogeny in 12 species of Indo-Pacific marine Clupeoidei (herrings, sardines, and anchovies). Mar Biol 165(10):153

Egan JP, Chew U-S, Kuo C-H, Villarroel-Diaz V, Hundt PJ, Iwinski NG, Hammer MP, Simons AM (2017) Diets and trophic guilds of small fishes from coastal marine habitats in western Taiwan. J Fish Biol 91(1):331–345. https://doi.org/10.1111/jfb.13355

Egan JP, Hundt PJ, Rude NP, DeBoer JA, Lubinski BJ, Maxson KA, Solomon LE, Whitten Harris AL, Simons AM (2026) Diet of juvenile skipjack herring *Alosa chrysochloris*: ontogenetic shifts predator–prey size ratios and mouth gape allometry. J Fish Biol 1–16. https://doi.org/10.1111/jfb.70390

Erickson GM, Kirk SDV, Su J, Levenston ME, Caler WE, Carter DR (1996) Bite-force estimation for *Tyrannosaurus rex* from tooth-marked bones. Nature 382(6593):706–708

Ferretti CML, Andrian IDF, Torrente G (1996) Dieta de duas espécies de *Schizodon* (Characiformes, Anostomidae) na planície de inundação do alto Rio Paraná, e sua relação com aspectos morfológicos. Bol Inst Pesca 23:171–186

Ford AG, Rüber L, Newton J, Dasmahapatra KK, Balarin JD, Bruun K, Day JJ (2016) Niche divergence facilitated by fine-scale ecological partitioning in a recent cichlid fish adaptive radiation. Evolution 70(12):2718–2735

Forsberg BR, Araújo-Lima CARM, Martinelli LA, Victória RL, Bonassi JA (1993) Autotrophic carbon sources for fish of the central Amazon. Ecology 74(3):643–652

Freitas T, Montag LF, Barthem RB (2017) Distribution, feeding and ecomorphology of four species of *Auchenipteridae* (Teleostei: Siluriformes) in Eastern Amazonia, Brazil. Iheringia Sér Zool 107:e2017008

Freitas TMDS, Almeida VHDC, Valente RDM, Montag LFDA (2011) Feeding ecology of *Auchenipterichthys longimanus* (Siluriformes: Auchenipteridae) in a riparian flooded forest of Eastern Amazonia, Brazil. Neotrop Ichthyol 9:629–636

Galetti M, Donatti CI, Pizo MA, Giacomini HC (2008) Big fish are the best: seed dispersal of *Bactris glaucescens* by the pacu fish (*Piaractus mesopotamicus*) in the Pantanal, Brazil. Biotropica 40(3):386–389

Galvis G, Mojica JI, Camargo M (1997) Peces del Catatumbo. Asociación Cravo Norte

Garland T Jr, Adolph SC (1994) Why not to do two-species comparative studies: limitations on inferring adaptation. Physiol Zool 67(4):797–828

Garland T Jr, Harvey PH, Ives AR (1992) Procedures for the analysis of comparative data using phylogenetically independent contrasts. Syst Biol 41(1):18–32

González-Bergonzoni I, Teixeira-de Mello F, Vidal N, D'Anatro A, Masdeu M (2010) Reappearance and diet of juvenile armado catfish (*Pterodoras granulosus*) in Lower Uruguay River, (Rio Negro, Uruguay). Bol Soc Zool Urug 19:42–46

Gosline WA (1951) Notes on the characid fishes of the subfamily Serrasalminae. Proc Calif Acad Sci 4(27):17–64

Gottsberger G (1978) Seed dispersal by fish in the inundated regions of Humaitá, Amazonia. Biotropica 10(3):170–183

Goulding M (1980) The fishes and the forest: explorations in Amazonian natural history. University of California Press, Berkeley

Goulding M, Carvalho ML (1982) Life history and management of the tambaqui (*Colossoma macropomum*, Characidae): an important Amazonian food fish. Rev Bras Zool 1(2):107–133

Grubich JR, Huskey S, Crofts S, Orti G, Porto J (2012) Mega-bites: extreme jaw forces of living and extinct piranhas (Serrasalmidae). Sci Rep 2(1):1009

Hamilton SK, Lewis WM, Sippel SJ (1992) Energy sources for aquatic animals in the Orinoco River floodplain: evidence from stable isotopes. Oecologia 89:324–330

Harvey PH, Pagel MD (1991) The comparative method in evolutionary biology. Oxford, England Oxford University Press

Hildemann WH (1959) A cichlid fish *Symphysodon discus* with unique nurture habits. Am Nat 93:27–34

Horn MH, Correa SB, Parolin P, Pollux BJA, Anderson JT, Lucas C, Widmann P, Tjiu A, Galetti M, Goulding M (2011) Seed dispersal by fishes in tropical and temperate fresh waters: the growing evidence. Acta Oecol 37(6):561–577

Huby A, Lowie A, Herrel A, Vigouroux R, Frédérich B, Raick X, Kurchevski G, Godinho AL, Parmentier E (2019) Functional diversity in biters: the evolutionary morphology of the oral jaw system in pacus, piranhas and relatives (Teleostei: Serrasalmidae). Biol J Linnean Soc 127(4):722–741

Huie JM, Summers AP, Kolmann MA (2019) Body shape separates guilds of rheophilic herbivores (Myleinae: Serrasalmidae) better than feeding morphology. Proc Acad Nat Sci Philadelphia 166(1):1–15

Huskey S, Fletcher K, Kolmann M, Seiler S, Kitchen G, Hemraj-Naraine D, Dinan B, Westneat MW (2026) Biomechanics muscle modeling and the elevated bite force and tooth stress of piranhas. Sci Nat 113(2):22. https://doi.org/10.1007/s00114-026-02071-w

Hyslop EJ (1980) Stomach contents analysis-a review of methods and their application. J Fish Biol 17(4):411–429

Ibañez C, Tedesco PA, Bigorne R, Hugueny B, Pouilly M, Zepita C, Zubieta J, Oberdorff T (2007) Dietary-morphological relationships in fish assemblages of small forested streams in the Bolivian Amazon. Aquat Living Resour 20(2):131–142

Iyengar EV (2008) Kleptoparasitic interactions throughout the animal kingdom and a re-evaluation, based on participant mobility, of the conditions promoting the evolution of kleptoparasitism. Biol J Linnean Soc 93(4):745–762

Jégu M, dos Santos GM (1990) Description d'Acnodon senai n. sp. du Rio Jari (Brésil, Amapà) et redescription d'A. normani (Teleostei, Serrasalmidae). Cybium 14(3):187–206

Jégu M, Santos GD, Ferreira E (1989) Une nouvelle espèce du genre Mylesinus (Pisces, Serrasalmidae), *M. paraschomburgkii*, décrite des bassins du Trombetas et du Uatumã (Brésil, Amazonie). Rev Hydrobiol Trop 22:49–62

Junk WJ (1997) General aspects of floodplain ecology with special reference to Amazonian floodplains. In: The central Amazon floodplain: ecology of a pulsing system. Springer Berlin Heidelberg, Berlin, Heidelberg, pp 3–20

Kolmann MA, Poulin E, Rosen J, Hemraj-Naraine D, Burns MD (2024) Phenotypic convergence is more frequent and stronger in herbivorous freshwater fishes. Integr Comp Biol 64(5):1467–1483

Kolmann MA, Cohen KE, Bemis KE, Summers AP, Irish FJ, Hernandez LP (2019) Tooth and consequences: Heterodonty and dental replacement in piranhas and pacus (Serrasalmidae). Evol Dev 21(5):247–262

Kolmann MA, Hughes LC, Hernandez LP, Arcila D, Betancur-R R, Sabaj MH, López-Fernández H, Ortí G (2021) Phylogenomics of piranhas and pacus (Serrasalmidae) uncovers how dietary convergence and parallelism obfuscate traditional morphological taxonomy. Syst Biol 70(3):576–592

Kubitzki K, Ziburski A (1994) Seed dispersal in flood plain forests of Amazonia. Biotropica 26:30–43

Lautenschlager S (2020) Multibody dynamics analysis (MDA) as a numerical modelling tool to reconstruct the function and palaeobiology of extinct organisms. Palaeontology 63(5):703–715

Legendre P, Legendre L (2012) Numerical ecology. Amsterdam, The Netherlands, Elsevier

Leite RG, Jégu M (1990) Régime alimentaire de deux espèces d'Acnodon (Characiformes, Serrasalmidae) et habitudes lépidophages de *A. normani*. Cybium 14(4):353–360

Levins R (1968) Evolution in changing environments: some theoretical explorations, no. 2. Princeton, New Jersey, USA Princeton University Press

López-Rodríguez A, Silva I, de Ávila-Simas S, Stebniki S, Bastian R, Massaro MV, Pais J, Tesitore G, Teixeira de Mello F, D'Anatro A, Vidal N (2019) Diets and trophic structure of fish assemblages in a large and unexplored subtropical river: the Uruguay River. Water 11(7):1374

Lubich C, Aguiar-Santos J, Corrêa F, Freitas CE, Siqueira-Souza FK (2022) Trophic ecology of *Acestrorhynchus falcirostris* Cuvier, 1819 in island lakes on the lower stretch of the Solimões River, Amazon Basin. Braz J Biol 84:e253852

Lucas CM (2008) Within flood season variation in fruit consumption and seed dispersal by two characin fishes of the Amazon. Biotropica 40(5):581–589

Lucena CAS, Lucena ZMS (2002) Redefinição do gênero Deuterodon Eigenmann (Ostariophysi: Characiformes: characidae). Comun Mus Ciênc Tecnol PUCRS, Sér Zool 15(1):113–135

Manetta GI, Benedito-Cecilio E, Martinelli M (2003) Carbon sources and trophic position of the main species of fishes of Baía River, Paraná River floodplain, Brazil. Braz J Biol 63:283–290

Mannheimer S, Bevilacqua G, Caramaschi EP, Scarano FR (2003) Evidence for seed dispersal by the catfish *Auchenipterichthys longimanus* in an Amazonian lake. J Trop Ecol 19(2):215–218

de Matos LS, Silva JOS, de Andrade PSM, Carvalho LN (2016) Diet of characin, *Brycon falcatus* (Muller and Troschel, 1844) in the Amazon Basin: a case study on an attractant for fish in the Teles Pires River. J Appl Ichthyol 32(6):1080–1085

Melack JM, Forsberg BR (2001) Biogeochemistry of Amazon floodplain. In: McClain ME, Victoria R, Richey JE (eds) The biogeochemistry of the Amazon basin, 1st edn. Oxford University Press, Oxford, pp 235–274

Melo T, Torrente-Vilara G, Röpke CP (2019) Flipped reducetarianism: a vegan fish subordinated to carnivory by suppression of the flooded forest in the Amazon. Forest Ecol Manag 435:138–143

Menezes NA (1969) The food of *Brycon* and three closely related genera of the tribe Acestrorhynchini. Pap Avulsos Zool 22(20): 217–223

Mercado-Gómez JD, Morales-Puentes ME, Gonzalez MA, Velasco JA (2022) Seasonal droughts during the Miocene drove the evolution of Capparaceae towards Neotropical seasonally dry forests. Rev Biol Trop 70(1):132–148

de Mérona B, Mendes dos Santos G, Gonççalves de Almeida R (2001) Short term effects of Tucuruí Dam (Amazonia, Brazil) on the trophic organization of fish communities. Environ Biol Fishes 60:375–392

Mérona BD, Rankin-de-Mérona J (2004) Food resource partitioning in a fish community of the central Amazon floodplain. Neotrop Ichthyol 2:75–84

Moldowan PD, Laverty TM, Emmans CJ, Stanley RC (2016) Diet, gastric parasitism, and injuries of caimans (*Caiman*, *Melanosuchus*, and *Paleosuchus*) in the Peruvian Amazon. South Am J Herpetol 11(3):176–182

Mortillaro JM, Pouilly M, Wach M, Freitas CEC, Abril G, Meziane T (2015) Trophic opportunism of central Amazon floodplain fish. Freshw Biol 60(8):1659–1670

Nico LG (1991) Trophic ecology of piranhas (Characidae: Serrasalminae) from savanna and forest regions in the Orinoco River basin of Venezuela. Doctoral dissertation, University of Florida

Nico LG, Taphorn DC (1988) Food habits of piranhas in the low llanos of Venezuela. Biotropica 20(4):311. https://doi.org/10.2307/2388321

Oksanen J, Blanchet FG, Friendly M, Kindt R, Legendre P, McGlinn D, Minchin PR, O'Hara RB, Simpson GL, Solymos P, Stevens MHH (2016) Package 'vegan' 2016. Community ecology package, R package version, 2.4-0

Oliveira ACB, Soares MGM, Martinelli LA, Moreira MZ (2006) Carbon sources of fish in an Amazonian floodplain lake. Aquat Sci 68:229–238

Parolin P, Wittmann F (2011) Tree phenology in Amazonian floodplain forests. In: Junk WJ, Piedade MTF, Wittmann F, Schöngart J, Parolin (eds) Amazonian floodplain forests: ecophysiology, biodiversity and sustainable management. pp 105–126

Pereira PR, Agostinho CS, Oliveira RJD, Marques EE (2007) Trophic guilds of fishes in sandbank habitats of a neotropical river. Neotrop Ichthyol 5(3):399–404

Pereira TNA, Castro RMC (2014) A new species of *Utiaritichthys* Miranda Ribeiro (Characiformes: Serrasalmidae) from the Serra dos Parecis, Tapajós drainage. Neotrop Ichthyol 12(2):397–402

Planquette P, Keith P, Le Bail PY (1996) Atlas des poissons d'eau douce de Guyane. Tome 1. Collection du Patrimoine Naturel, vol 22, MNHN, Paris and INRA, Paris, p 429

Pollux BJA (2011) The experimental study of seed dispersal by fish (ichthyochory). Freshw Biol 56(2):197–212

Pouilly M, Lino F, Bretenoux JG, Rosales C (2003) Dietary–morphological relationships in a fish assemblage of the Bolivian Amazonian floodplain. J Fish Biol 62(5):1137–1158

Ramírez F, Davenport TL, Mojica JI (2015) Dietary–morphological relationships of nineteen fish species from an Amazonian terra firme blackwater stream in Colombia. Limnologica 52:89–102

Rejas D (2018) Trophic structure of a floodplain fish assemblage in the upper Amazon basin, Bolivia. Rev Biol Trop 66(3): 1258–1271

Roberts-Hugghis AS, Burress ED, Lam B, Wainwright PC (2023) The cichlid pharyngeal jaw novelty enhances evolutionary integration in the feeding apparatus. Evolution 77(9):1917–1929

Rosen J, Cohen K, Donatelli CM, Summers A, Crofts S, Kolmann M (2025) Something to sink your teeth into: the mechanics of tooth indentation in frugivorous fishes. J R Soc Interface 22:20240725

Roubach R, Saint-Paul U (1994) Use of fruits and seeds from Amazonian inundated forests in feeding trials with *Colossoma macropomum* (Cuvier, 1818) (Pisces, Characidae). J Appl Ichthyol 10(2–3):134–140

Sabino J, Corrêa e Castro RM (1990) Feeding biology, activity period and spatial distribution of fishes in an Atlantic Forest stream in southeastern Brazil. Rev Bras Biol 50(1):23–36

Saint-Paul U, Zuanon J, Correa MAV, García M, Fabré NN, Berger U, Junk WJ (2000) Fish communities in central Amazonian white-and blackwater floodplains. Environ Biol Fishes 57:235–250

dos Santos GM (1981) Estudos de alimentação e hábitos alimentares de *Schizodon fasciatus* Agassiz, 1829, *Rhytiodus microlepis* Kner, 1859 e *Rhytiordus argenteofuscus* Kner, 1859, do lago Janauacá-AM. (Osteichthyes, Characoidei, Anostomidae). Acta Amazonica 11:267–285

Santos MLD, Arantes FP, Santiago KB, Santos JED (2015) Morphological characteristics of the digestive tract of *Schizodon knerii* (Steindachner, 1875), (Characiformes: Anostomidae): an anatomical, histological and histochemical study. An Acad Bras Cienc 87:867–878

Santos Z, Oliveira LS, Cajado RA, Silva FK, Silva LV, Zacardi DM (2023) Early ontogeny of the commercially valuable fish red-bellied pacu *Piaractus brachypomus* (Characiformes, Serrasalmidae) from the Amazon, Brazil. J Fish Biol 102(1):214–224

Sá-Oliveira JC, Angelini R, Isaac-Nahum VJ (2014) Diet and niche breadth and overlap in fish communities within the area affected by an Amazonian reservoir (Amapá, Brazil). An Acad Bras Cienc 86:383–406

Satoh S, Sowersby W (2021) Mucus provisioning behavior in teleost fishes: a novel model system for the evolution of secretory provisioning in vertebrates. Ichthyol Res 68:1–10

Satoh S, Tanoue H, Ruitton S, Mohri M, Komatsu T (2017) Morphological and behavioral ontogeny in larval and early juvenile discus fish *Symphysodon aequifasciatus*. Ichthyol Res 64:37–44

Sazima I (1980) Behavior of two Brazilian species of parodontid fishes, *Apareiodon piracicabae* and *A. ibitiensis*. Copeia 1980(1):166–169

Schaefer H, Heibl C, Renner SS (2009) Gourds afloat: a dated phylogeny reveals an Asian origin of the gourd family (Cucurbitaceae) and numerous oversea dispersal events. Proc R Soc B 276(1658):843–851

Sidlauskas BL, Garavello JC, Jellen J (2007) A new *Schizodon* (Characiformes: Anostomidae) from the Río Orinoco system, with a redescription of *S. isognathus* from the Rio Paraguay system. Copeia 2007(3):711–725

Da Silva DS, Lucotte M, Roulet M, Poirier H, Mergler D, Santos EO, Crossa M (2005) Trophic structure and bioaccumulation of mercury in fish of three natural lakes of the Brazilian Amazon. Water Air Soil Pollut 165:77–94

Silva RE, Rosas FCW, Zuanon J (2014) Feeding ecology of the giant otter (*Pteronura brasiliensis*) and the neotropical otter (*Lontra longicaudis*) in Jaú National Park, Amazon, Brazil. J Nat Hist 48(7–8):465–479

da Silva Gonçalves C, Cestari C (2018) Grazing of the lambari fish *Deuterodon iguape* is associated with mouth morphology. Zool Anz 274:127–130

Silveira R, Leão-Neto WM, Barbosa da Silva FH (2020) Small-sized fish as possible seed dispersers: disclosing novel fish and plant species interactions in the Pantanal wetland. Stud Neotrop Fauna Environ 55(1):36–43

Soares MGM, Almeida RG, Junk WJ (1986) The trophic status of the fish fauna in Lago Camaleão, a macrophyte dominated floodplain lake in the middle Amazon. Amazonia 9(4):511–526

de Sousa LL, Silva SM, Xavier R (2019) DNA metabarcoding in diet studies: unveiling ecological aspects in aquatic and terrestrial ecosystems. Environ DNA 1(3):199–214

de Souza-Stevaux MC, Negrelle RR, Citadini-Zanette V (1994) Seed dispersal by the fish *Pterodoras granulosus* in the Paraná River Basin, Brazil. J Trop Ecol 10(4):621–626

Terborgh J, Davenport L (2021) Mobile piscivores and the nature of top-down forcing in Upper Amazonian floodplain lakes. Hydrobiologia 848(2):431–443

Velasco-Hogan A, Huang W, Serrano C, Kisailus D, Meyers MA (2021) Tooth structure, mechanical properties, and diet specialization of Piranha and Pacu (Serrasalmidae): a comparative study. Acta Biomater 134:531–545

Vieira I, Géry J (1979) Crescimento diferencial e nutrição em *Catoprion mento* (Characoidei). Peixe lepidófago da Amazônia. Acta Amaz 9(1):143–146. https://doi.org/10.1590/1809-43921979091143

Villares Junior GA, Gomiero LM, Goitein R (2011) Biological aspects of *Schizodon nasutus* Kner, 1858 (Characiformes, Anostomidae) in the low Sorocaba River basin, São Paulo state, Brazil. Braz J Biol 71(3):763–770

Vitorino Júnior OB, Agostinho CS, Pelicice FM (2016) Ecology of *Mylesinus paucisquamatus* Jégu and Santos, 1988, an endangered fish species from the Rio Tocantins basin. Neotrop Ichthyol 14(2):e150124

Vitule JR, Braga MR, Aranha JM (2008) Ontogenetic, spatial and temporal variations in the feeding ecology of *Deuterodon langei* Travassos, 1957 (Teleostei: Characidae) in a Neotropical stream from the Atlantic rainforest, southern Brazil. Neotrop Ichthyol 6:211–222

Webb SD (1977) A history of savanna vertebrates in the new world. Part I: North America. Annu Rev Ecol Syst 8(1):355–380

Weiss B, Zuanon JA, Piedade MT (2016) Viability of seeds consumed by fishes in a lowland forest in the Brazilian Central Amazon. Trop Conserv Sci 9(4):1940082916676129

Weiss B, Santana FD, Petene Calvi G, Costa G, Zuanon J, Piedade MTF (2024) Effectiveness of fish assemblage as seed dispersers in Amazon oligotrophic flooded forests. Austral Ecol 49(1):e13330
Winemiller KO (1996) Factors driving temporal and spatial variation in aquatic floodplain food webs. In: Polis GA, Winemiller KO (eds) Food webs: integration of patterns and dynamics. Chapman and Hall, New York, pp 298–312
Winemiller KO, Jepsen DB (1998) Effects of seasonality and fish movement on tropical river food webs. J Fish Biol 53:267–296
Wojciechowski MF (2003) Reconstructing the phylogeny of legumes (Leguminosae): an early 21st century perspective. Adv Legume Syst 10:5–35
Wolff LL, Carniatto N, Hahn NS (2013) Longitudinal use of feeding resources and distribution of fish trophic guilds in a coastal Atlantic stream, southern Brazil. Neotrop Ichthyol 11:375–386
Yabe RDS, Bennemann ST (1994) Regime alimentar de Schizodon intermedius Garavello and Britski do rio Tibagi, Paraná, e sua relação com algumas características morfológicas do trato digestivo (Osteichthyes, Anostomidae). Rev Bras Zool 11:777–788
Zuanon J, Ferreira E (2008) Feeding ecology of fishes in the Brazilian Amazon—a naturalistic approach. In: Feeding and digestive functions of fishes. Science Publishers Inc, USA, pp 1–34

Benthic Algivorous and Detritivorous Neotropical Freshwater Fishes

16

Daniel R. Akin, Jonathan W. Armbruster, Bruno F. Melo, Nathan K. Lujan, Kirk O. Winemiller and James S. Albert

Abstract

Algae and detritus are somewhat loosely defined. Algae vary greatly in taxonomy, nutrition, and ecological distribution. while detritus varies greatly in size, origin, level of decomposition, and microbial community. Periphyton is a mixture of algae, detritus, and microbes that is likewise variable in composition. Algae and detritus are foundational to tropical aquatic food webs, and although detritus is more abundant at any given time, algae are generally more bioavailable for consumption and assimilation by fishes. The relative biomass of algivorous and detritivorous fishes is not a trivial amount, with some detritivores making up large fisheries in South America. However, there is still much work to be done for large-scale studies quantifying the compositions and biomass of algivores and detritivores in Neotropical fish assemblages. There are numerous morphological specializations that are associated with algivory and detritivory, but there is a large amount of disparity between phenotypes, suggesting that there are many different ways to efficiently feed on algae and detritus. Although wood-eating catfishes have been shown not to digest and assimilate the fibrous components of wood, this highly specialized form of grazing helps to break down large woody debris for other detritivores and decomposers. The modal adult diet classification employed for this book is a first step in understanding the evolution of detritivores and algivores, but it is far from resolute. Algae and detritus are obviously broad categories that can be subdivided into finer categories and are often intertwined (e.g., periphyton), and chemical analyses are beginning to shed more light on the assimilation and transfer of specific source nutrients and energy within a food web.

D. R. Akin (✉) J. S. Albert
University of Louisiana at Lafayette, Lafayette, LA, USA
e-mail: jsalbe@gmail.com

D. R. Akin
e-mail: daniel.akin1@louisiana.edu

J. W. Armbruster
Auburn University, Auburn, AL, USA
e-mail: armbrjw@auburn.edu

B. F. Melo
American Museum of Natural History, New York, NY, USA
e-mail: brunfmelo@gmail.com

UNICAMP Universidade Estadual de Campinas, Campinas, Brazil

N. K. Lujan
Royal Ontario Museum, Toronto, ON, Canada
e-mail: nlujan@rom.on.ca

K. O. Winemiller
Texas A&M University, College Station, TX, USA
e-mail: k-winemiller@tamu.edu

16.1 Introduction

16.1.1 Benthic Foragers: Defining Algae and Detritus

This chapter describes the diversity and evolution of algivorous and detritivorous fishes that inhabit Neotropical freshwaters (Fig. 16.1). An important part of this discussion concerns what ichthyologists and ecologists precisely mean by the terms algae and detritus. Some ichthyologists broadly conceive of algae as phytoplankton and periphyton (or even worse—as "living-slimy-fish-food"), while detritus is defined as nonliving organic material from undetermined sources (i.e., "nonliving-slimy-fish-food"). However, most ecological research involving fish feeding requires discrimination based on finer categories, such as phytoplankton, periphyton, benthic microalgae, filamentous algae, fine or coarse particulate organic matter, and dissolved organic matter. Feeding on algae and detritus is further complicated by nutritional differences between taxa in terms of ingestion,

J. S. Albert and R. E. Reis (eds.), *Ecology and Evolution of Amazonian Fishes*, Fascinating Life Sciences,
https://doi.org/10.1007/978-3-032-09328-8_16

Fig. 16.1 Morphological disparity amongst grazer oral jaws. From left to right; top row—Sicydium hildebrandi (ROM 10638), Parodon buckleyi (AUM 51159), Chaetostoma cf. loborhynchus (ROM 109181); middle row—Ancistrus macrophthalmus (ROM 93577), Hypostomus pyrineusi (ROM uncatalogued, field code ECU21-11), Panaqolus nocturnus (ROM uncatalogued, field code ECU21-11); bottom row—Aphanotorulus unicolor (ROM uncatalogued, field code ECU21-02), Loricaria sp. (ROM uncatalogued, field code ECU21-02), Pseudohemiodon lamina (ROM uncatalogued, field code ECU21-17)

digestion, and assimilation. Many fishes are found with significant quantities of algae or detritus in their gastrointestinal tract, but actual digestion and assimilation of this material may be minimal or nil. For example, carnivorous fishes may incidentally ingest small amounts of mostly indigestible algae or detritus during prey capture. It is helpful to define a few key terms:

Detritus comprises a wide range of nonliving organic matter, varying from wood and leaf litter to shed insect casings and fecal matter from fish or other animals. Some researchers include living microbial communities that grow on nonliving organic matter as part of the detritus pool, while others consider decomposer communities as separate entities that can only be classified as detritus after they die (Moore et al. 2004). Detritus in aquatic systems can be divided into two broad categories: *particulate organic matter* (POM) and *dissolved organic matter* (DOM) or *dissolved organic carbon* (DOC), with POM being further subdivided by size into large woody debris (LWD), coarse particulate organic matter (CPOM), and fine particulate organic matter (FPOM). It should now be clear that not all detritus is equal, as it varies according to origin, particle size, level of decomposition, type of microbial community, and longitudinal distribution in lotic ecosystems (Moore et al. 2004). In rivers generally, most detrital biomass is of terrestrial origin, including woody debris and fallen leaves, and this detritus changes along the longitudinal river continuum as it is processed by various specialized consumers, with particle size decreasing and nutrient-rich microbial communities increasing with increasing surface area as detrital particles are transported downstream. Ultimately, only indigestible recalcitrant detritus remains. Tropical rivers are generally distinguished from temperate rivers by having a majority of their fish biomass supported by detrital carbon (Bayley 1989; Cebrian 1999; Wallace et al. 1997).

Algae are a polyphyletic assemblage of plants, protists, and bacteria that use chlorophyll-α as their primary photosynthetic pigment and lack traditional terrestrial plant characteristics such as roots, leaves, and stems (Lee 2018). Algae in freshwater and marine ecosystems include distantly related lineages such as cyanobacteria (blue-green algae), green and red algae that are known for harmful algal blooms, protists like euglenoids and dinoflagellates, diatoms (Bacillariophyceae), macroscopic kelps and other brown algae, and many other taxa. *Plankton* is a term for a different polyphyletic assemblage of organisms suspended, drifting, or swimming within the water column. *Phytoplankton* are algae suspended within the water column. Whereas phytoplankton are sometimes referred to as microalgae, ecologists additionally distinguish benthic microalgae, or *microphytobenthos* as the microscopic algae that inhabit the thin layer of substratum between sediments and the water column. The unassuming and unvegetated appearance of the microphytobenthos in juxtaposition with its relatively important role in shallow water food webs is discussed in depth in the "Secret Garden" papers (MacIntyre et al. 1996; Miller et al. 1996). We further differentiate *periphyton* as a heterogenous community of algae (phytobenthos), detritus, and microbial decomposers (biofilm) overlying submerged surfaces, following the nomenclature provided by Azim et al. (2005). Furthermore, periphyton attached to different substrata are termed as follows: plants—epiphyton, sand—epipsammon, sediment/mud/silt—epipelon, wood—epixylon and rock—epilithon. These substrata differences may have impacts on nutritional composition and energy cycling, as demonstrated by differences in relationships of chlorophyll-α concentrations and total phosphorus between hard substrates (epixylon and epilithon) and soft substrates (epipelon) (Vadeboncoeur et al. 2006). Even at small spatial scales, periphyton varies widely in taxonomic composition (Chételat et al. 1999). Despite our efforts to separate algivores and detritivores in the modal adult database (Chap. 6), the difficulty in determining gut contents in numerous species paired with a lack of studies to determine sources of assimilated nutrients, means that many so-called algivores may obtain a greater portion of their nutrition from detritus than algae.

16.1.2 Basal Production Sources of Tropical Riverine Food Webs

Algae and detritus are major basal energy and nutritional sources in Neotropical freshwater ecosystems (Hoeinghaus et al. 2007; Vadeboncoeur and Power 2017; Wallace et al. 1997). *Allochthonous* inputs refer to energy and nutrients that enter a habitat or ecosystem from an adjacent habitat or ecosystem (e.g., leaf litter and runoff), while *autochthonous* inputs originate within the ecosystem (e.g., algae). Because detritus and algae provide most of the matter and energy at the base of food webs of Neotropical rivers, detritivores and algivores are integral to ecological energetics and ecosystem dynamics. Shrimps, aquatic insects, and other aquatic invertebrates consume algae and detritus, and the ubiquity of these organisms in tropical freshwater ecosystems probably explains the widespread occurrence of invertivory among Neotropical Freshwater Fish (NFF) species (see Chap. 8). Nonetheless, no analysis of food web ecology in Neotropical rivers would be complete without considering the crucial roles of algivorous and detritivorous fishes (Flecker 1996; Flecker et al. 2010; Winemiller et al. 2014).

Algae are the base of "green food webs" and support an inverted trophic pyramid, meaning that algae usually represent a very small proportion of standing biomass and carbon at any given time when compared to the organisms that consume them (Vadeboncoeur and Power 2017). This is partially due to the rapid turnover rate and productivity of algae and the high efficiency of algae consumption and assimilation by specialized consumers. Harvesting and assimilation of algae by algivores is roughly twice as efficient when compared to rates for animals that feed on terrestrial grasses and forty times more efficient compared to animals foraging on terrestrial dicot leaves (Cebrian 1999). Grazing removes superficial layers of algae, which allows penetration of sunlight into deeper strata of algal mats, and the nutrients excreted by grazing insects and fishes enrich microbial communities, potentially increasing the productivity of grazed algal patches (André et al. 2003; Munubi et al. 2018). Because of the rapid turnover in algae compared to the consumers they sustain, the biomass at higher trophic levels begins to accumulate, resulting in an inverted trophic pyramid of standing biomass.

Detritus is the base of "brown food webs" and generally is abundant in streams and rivers around the world (Brett et al. 2017; Cebrian 1999; Moore et al. 2004). Detritus flows downstream, settling and pooling as it goes, with smaller particles generally traveling further and aggregating further downstream than larger particles. Detritus of all sizes is found everywhere, but vast coarse woody debris dams (*palisadas* in Spanish) are more common immediately downstream of erosional montane uplands such as the Andes, whereas fine particles are more apt to accumulate in channel pools and margins downstream as well as floodplains where flow velocities decrease. During high flow events, detritus is resuspended, transported, and redistributed (Bayley 1995). Despite this abundance, the energy locked up in POM is not efficiently assimilated into riverine food webs because few animals possess the enzymes needed to break down cellulose, making it a less bioavailable energy source when compared to algae (Brett et al. 2017). Nevertheless, detritus and decomposers are a foundation of allochthonous inputs to riverine food webs and support the majority of fish biomass in tropical rivers. Exclusion studies have demonstrated that removal of leaf litter from streams results in negative effects for dissolved organic carbon, as well as declines in abundance and biomass of taxa at multiple trophic levels (Meyer et al. 1998; Wallace et al. 1997). The dynamics of detritus and algae inputs to river food webs is influenced by a host of environmental factors, and the two sources are probably complementary for many primary consumers. Brett and colleagues (2009) demonstrated a higher level of productivity in a zooplankter when fed a mixture of algae and detritus than would have been expected by ingesting algae and detritus independently of one another. Decomposing detritus may also stimulate algal growth. Conversely, higher DOM concentrations generally decrease light penetration, which hinders algal growth.

16.1.3 Biomass

Algae are approximately twice as calorically dense with roughly twice as much digestible protein as detritus, and algae have greater nitrogen and phosphorus content than terrestrial plants (Cebrian 1999; Elser et al. 2000). There is variation in assimilation among different algae corresponding to the relative abundance of cellulose in their cell walls (see Vadeboncoeur and Power 2017 for a comprehensive review). Despite consuming a stoichiometrically superior food source, algivores must graze almost constantly and respond to changes in algal patch quality to increase in biomass (Power 1983). There is currently a shortage of published literature that directly examines algivorous fish biomass within Neotropical fish communities. A monthly survey of the Guaritico River in Venezuela from June 1990 to January 1991 reported algivore biomass ranging from roughly 1% to 26% of the total fish assemblage biomass, with the highest algivore relative biomasses (~24%–26%) during one month with lowest overall biomass (June, rising water) and during another month with the second highest overall biomass (December, falling water) (Lasso and Castroviejo 1992).

Many Neotropical fishes are indirectly linked to detritus via direct consumption of fishes or aquatic invertebrates that feed on detritus. Detritivorous fishes in the families Chilodontidae, Curimatidae, and Prochilodontidae that consume detritus directly are often among the most abundant species in local fish assemblages. *Prochilodus lineatus* represented nearly 60% of the total fish biomass in the Paraná River of Brazil, and *Prochilodus nigricans* comprised 39.8% (±12.1%) of the total annual catch in the Loreto region of Peru over a 22-year period (Bowen 1983; Garcia et al. 2009). Dominance of detritivores in tropical regions may be altered by ongoing and future anthropogenic impacts. In a subtropical region of the United States, detritivorous shad (*Dorosoma* spp.) have been recorded as dominant species in eutrophic reservoirs (Bachmann et al. 1996), which may suggest that detrital food sources increase in relative importance within impounded systems. However, it should be noted that *Dorosoma* is also an effective phytoplanktivore, suggesting these effects may not be attributable to detritivory. Regardless, detritus supports valuable fisheries throughout the Neotropics, and research is needed to estimate the immediate and long-term effects of river impoundment on detritivore populations both within and downstream of the impoundment.

16.2 Taxonomic Composition

This section describes the taxonomic composition of benthic algivorous Neotropical freshwater fishes. These results are summarized in Fig. 16.2.

Ostariophysi:

Primary Freshwater Fishes Chilodontidae: Head-stander characins contain eight species that are omnivorous with diets that include variable fractions of detritus (Melo and Sidlauskas 2017). Chilodontids and prochilodontids have relatively small teeth that are flexibly affixed to fleshy lips rather than the jaw bones. Chilodontids, curimatids, and prochilodontids all possess an enlarged muscular epibranchial organ that aids the concentration and ingestion of FPOM from the water column and its separation from inorganic substrates (Bertmar et al. 1969; Vari et al. 1983)
Curimatidae: Toothless characins have 120 detritivorous species, which are also likely to ingest and assimilate algae (Frable, 2017). Curimatids and prochilodontids have various soft anatomical features of the mouth and buccopharyngeal cavity that allow for sifting of fine particles prior to passage to the gastrointestinal tract. Adult curimatids lack dentition but most genera possess fleshy lips that facilitate suction of dislodged FPOM and modifications on the roof of the buccopharyngeal region allowing for the processing of ingested material (Vari 1991).
Prochilodontidae: Flannel-mouth characiforms consist of 21 species, which are algivores/detritivores (Melo and Sidlauskas 2017). Curimatids and prochilodontids have various soft anatomical features of the mouth and buccopharyngeal cavity that allow for sifting and concentration of fine particles prior to passage to the gastrointestinal tract. Like chilodontids, prochilodontids have relatively small teeth that are affixed to the fleshy lips as opposed to the jaw bones. Prochilodontids have modified upper and lower jaws supporting the expanded lips that are used to scrape and suck periphyton and FPOM (Bowen 1983; Castro and Vari 2004) and have a muscular, gizzard-like structure used for mechanical reduction of detritus. Prochilodontids are migratory and often among the most abundant large fishes in river communities, where they have been shown to have a large influence on benthic ecology (Flecker 1996; Flecker et al. 2010; Winemiller et al. 2014).
Parodontidae: Scrapetooths include 32 species distributed primarily in mid-elevation waters from Panama to Argentina, with higher diversity in the northern Andes, upper Paraná

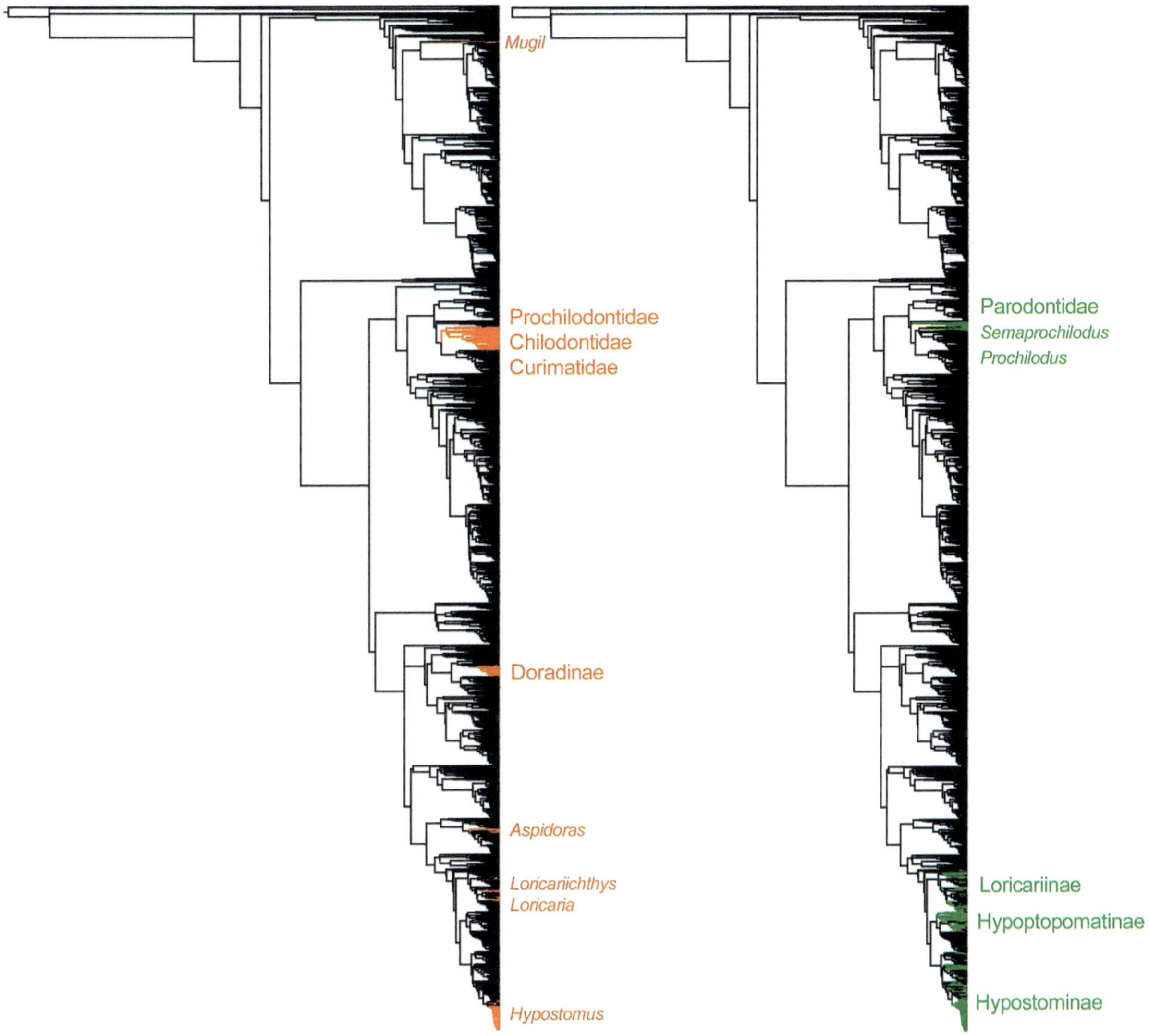

Fig. 16.2 Ecotrait phylogeny of detritivores (brown) and algivores (green). Phylogeny from Chap. 2; data from Albert et al. (2025)

River, and São Francisco River basins. Straight cutting teeth are present in the upper jaw, while the toothless lower jaw is used to scrape algae from submerged surfaces (Roberts 1974; Restrepo-Gómez et al. 2014). The species are rheophilic and generally found only in swiftly flowing waters.

Callichthyidae: The 226 species of Chevron-Armored Catfishes are mostly benthic feeding invertivores or omnivores; however, *Corydoras habrosus* is partially algivorous, and the species of the genus *Aspidoras* (24/226 species, 10.6%) are at least partially detritivorous (Cassatti et al. 2009). Callichthyids have a small mouth, jaws, and oral teeth, as well as moderately robust pharyngeal jaws and teeth. They consume small particles located on or within loose sediments.

Doradidae: Thorny Catfishes consist of 96 species and are largely omnivorous/carnivorous, but 53/96 species (55%) are thought to have detritus as a considerable portion of their diets. Sabaj (2005) suggested that the oral hood of *Leptodoras* facilitates feeding on invertebrates and detritus from sandy streams. With the variation present in the mouth ornamentation of doradids consisting of barbels and simple to branched papillae, there is likely also a broad number of food items that doradids ingest, and more research would be needed to determine the degree of algivory/detritivory in the group. The genus *Pterodoras* is notable as one of the few Neotropical fish taxa regularly encountered with enlarged guts full of dead macrophytes and other coarse detritus.

Loricariidae: Suckermouth Armored Catfishes consists of over 1050 species and are predominant periphyton grazers in Neotropical freshwaters from Costa Rica to Argentina. With most species (>80%) scored as benthic algivores or detritivores, it is easier to discuss the groups that are not grazers. Within the large subfamily Hypostominae, the tribes Lithoxini (*Avalithoxus*, *Exastilithoxus*, *Lithoxus*, and *Paralithoxus*) and Acanthicini (excluding *Acanthicus*; i.e., *Leporacanthicus*, *Megalancistrus*, *Pseudacanthicus*) as well as some Peckoltini (*Hypancistrus*, *Pseudoqolus*, *Panaqolus albomaculatus*, *Scobinancistrus*, *Spectracanthicus*) are predominantly invertivorous (Armbruster 2004; Lujan et al. 2018). Within subfamily Loricariinae, the Hartiini are mostly algivorous, whereas the Loricariini are granivorous and insectivorous (Armbruster 2004; Thomas and Rapp Py-Daniel 2008; Provenzano and Sabaj 2024). Loricariids vary tremendously in jaw shape and angle (from the jaw rami meeting at angles of ~30° to ~180° and the jaws being slight and scarcely ossified to robust; Lujan and Armbruster 2012) and number of teeth (from two to over 200 per jaw ramus, JWA pers. obs.). Loricariid jaws and teeth do not occlude against themselves but rather against substrates as part of an oral disk. The lower jaws of most species rotate around their own long axis (i.e., about the anguloarticular-quadrate joint such that they move dorsoventrally), and the premaxillae are adducted around a condyle of the mesethmoid via a novel insertion of the adductor mandibulae muscle. The two lower jaw rami operate independently of one another, and the upper jaw operates independently of the lower jaws (Schaefer and Lauder 1986, 1996). Feeding scars are often seen on rocks in streams where loricariids are located. Interspecific differences in jaw bone, tooth, and soft-tissue anatomy correspond to differences in the microhabitats where species forage (riffles, runs, pools), the types of substrates they feed on (rocks, sand, mud, woody debris, etc.), and how deeply they may feed (surface vs. gouging into the substrate) (Lujan et al. 2011, 2012; Krings et al. 2023). Species of the genus *Loricariichthys* have thin fleshy lips that protrude as a tube that functions in a manner like those of sturgeons to suck particles from the substrate and then sift it by moving water in and out of the orobranchial chamber (winnowing). Analysis of gut contents of *Loricariichthys brunneus* (as *L. typus*) from the Venezuelan Llanos (Orinoco River Basin) revealed a diet comprised of 57% detritus (mostly CPOM), with the remainder consisting of chironomid (Diptera) larvae and other aquatic invertebrates (Winemiller 1990). The precise diet of loricariids is often difficult to interpret due to the irregularity of their stomach contents and the presence of gray sludge in their intestines. As an example of diet in loricariids, Abilhoa et al. (2016) found a variety of detritus, blue-green algae, green algae, filamentous algae, diatoms, and animal remains within the digestive tracts of species of *Hypostomus*.

Acanthomorpha: Spiny-Ray Fishes

Cichlidae: Many Palaeotropical cichlids (Etroplinae and Tilapiinae) feed extensively on detritus and algae (Winemiller and Kelso Winemiller 2003), which, along with their efficient rates of food conversion, have made these fishes globally important for aquaculture. Among New World cichlids (Cichlinae), several species routinely consume large fractions of detritus (with associated microbiota) and benthic algae (diatoms, filamentous algae) with variable fractions of aquatic invertebrates, macrophyte tissues, and other items ingested depending on fish size and environmental conditions (Winemiller et al. 1995; Winemiller and Kelso-Winemiller 2003). For example, examination of gut contents from large samples of *Vieja maculicauda* from Costa Rica revealed that the dry season diet consisted of 31% detritus and 29% algae by volume, and the wet season diet was comprised of 44% detritus and 13% algae (Winemiller 1990). Other Neotropical cichlids with diets documented to contain large portions of detritus include *Chuco intermedia* (91% detritus by volume) and *Vieja melanura* (64%) (Pease et al. 2018; Soria-Barreto et al. 2019).

Eleotridae: Sleepers encompass a wide range of trophic niches ranging from detritivores, such as *Dormitator* species, to piscivores, such as *Gobiomorus* species. In Costa Rica, analysis of a large sample of *Dormitator maculatus* (n = 222) revealed 86% detritus and 10% algae in the diet (Winemiller and Ponwith 1998).

Gobiidae: American Rock-Climbing Gobies (genus *Sicydium*) is a clade of 31 anadromous fishes found throughout the Caribbean. The species largely feed on diatoms and have a broad, ventrally oriented mouth that serves multiple purposes, including scraping biofilms and assisting the fused pelvic fins in holding position on substrates in swift streams (Erdman 1961; Monti et al. 2018). Different tooth types among *Sicydium* species have been correlated with variations in the types of diatoms consumed (Monti et al. 2018). Six species—*S. condotense*, *S. hildebrandi*, *S. montanum*, *S. punctatum*, *S. rosenbergii*, and *S. salvini*—are located in the study area and were scored as algivores within the ecomatrix. Stomach contents analysis of *Evorthodus lyricus* in Costa Rica revealed 93% diatoms and 5% filamentous algae in the diet (Winemiller 1990).

Mugilidae: The 16 mullets of the genus *Mugil* separate invertebrates, algae, and detritus from the mud using their gill rakers (Thompson et al. 2015; Cardona 2016). Three species—*M. brevirostris*, *M. curema*, and *M. liza*—are located in the study area and were scored as detritivores.

Poeciliidae: The majority of poeciliids are invertivorous or omnivorous, but analyses of gut contents have shown detritus and algae being dominant components of the diets of several species. In Venezuela, the Guppy *Poecilia reticulata* consumed 32% detritus (mostly FPOM) and 34% algae (mostly diatoms) by volume (Winemiller 1990). In Costa Rica, *Poecilia gilli* was found to consume 49% detritus and 49% algae by volume, and *Phallichthys amates* consumed 55% detritus and 44% algae (Winemiller 1990).

16.2.1 Benthic Algivores

Algivores transfer energy to higher trophic positions within green food webs. Because autochthonous production is dynamic in response to light, water depth, turbidity, nutrients, etc. (Roach et al. 2014; Vadeboncoeur and Power 2017), fishes may have an advantage over larval insects when it comes to colonizing newly forming algal mats (see Power 1983). As discussed earlier, more than 900 species of Loricariidae are at least partially algivorous; however, it is more accurate to refer to them as grazers that feed on both algae and detritus (i.e., microphytobenthos). Nevertheless, algivory has played at least a complementary role in the diversification of one of the greatest radiations of freshwater fishes. Many algivorous fishes have modified jaws and teeth suitable for grazing (Fig. 16.1), as well as long intestines to accommodate a diet of mostly algal matter. See the Loricariidae account above for more examples of grazing specializations.

Although it would seem more straightforward to compare benthic algivores directly to phytoplanktivores, we contrast them here with all planktivores due to a lack of evidence that any Neotropical freshwater fishes are phytoplankton specialists (Frau 2022). Benthic algae as a food resource are typically concentrated, providing grazing benthic algivores with a resource that can be assessed and, in some cases, defended (Power 1983). In contrast, plankton are suspended in the water column, and many planktivores use a more indiscriminate, mechanical filter-feeding strategy, e.g., straining particles with comb-like gill rakers as employed by anchovies (*Anchoviella* spp.). However, food selection via this mechanism is primarily based on particle size rather than food type, e.g., algal phytoplankton versus animal zooplankton. Other planktivores are suspension feeders that consume particles that are discriminately located, often by vision but also by other senses such as electroreception, e.g., in *Rhabdolichops* (Lundberg 1987). However, these suspension feeders typically target more conspicuous zooplankton and rarely target phytoplankton (but see *Cetengraulis* as a marine example of a phytoplanktivore; Gay et al. 2002). Thus, although phytoplankton may be an important basal production source in some freshwater ecosystems, they are mostly indirectly linked to fish via zooplankton (Lazzaro 1987), and although there are many planktivorous Neotropical fishes, there are unlikely to be any Neotropical planktivorous fishes that specialize on phytoplankton alone. Therefore, most algivorous Neotropical fishes consume algae that are attached to substrates, and represent a distinct foraging strategy from fishes that feed on suspended zooplankton and phytoplankton.

16.2.2 Benthic Detritivores

Detritivorous fishes help to cycle nutrients vertically to higher trophic levels as part of brown food webs and play an important role in moving nutrients horizontally across a hypothetical landscape from one area to another. Most prochilodontids, including all species of *Prochilodus* and *Semaprochilodus*, migrate hundreds of kilometers upstream to reproduce in river channels near seasonally inundated floodplains (*várzeas* and *igapós*) into which their eggs and larvae drift. Larvae and juveniles feed and grow rapidly within productive nursery habitats within floodplains before dispersing into river networks (Fernandes 1997). Migrating and dispersing prochilodontids transporting nutrients in the forms of living biomass as well as excreted waste can also affect local food webs via consumption of basal resources in the form of microphytobenthos and detritus (Winemiller and Jepsen 1998). In addition, migrating adults and juveniles provide a spatial food web subsidy when they enter tributaries where they are preyed upon by resident piscivores (Winemiller and Jepsen 2004).

Many detritivorous fishes have highly modified jaws and fleshy lips (Fig. 16.1). Species of the detritivorous Prochilodontidae are often called *bocachicos* in Spanish, referencing sucking movements that resemble a nursing human baby. Detritivores often possess elongated intermeshing gill rakers that function as a filtration mechanism suitable for separating inorganic matter like sand. *Prochilodus* has two stomachs, with the anterior stomach acting as a reservoir (crop-like structure) and the posterior stomach (gizzard-like structure) being rigid and muscular for grinding food (Bowen 1983). Similar stomach combinations are found in many detritivores, but exceptions exist, such as the Nile tilapia (*Oreochromis niloticus*), an African cichlid that secretes acid to break down food. Detritivore intestines are also well developed to manage the slow assimilation rates of detritus, increasing surface area by adding length and coiling, or by increasing width and internal folding (Bowen 1983; Farago et al. 2020).

16.2.3 An X-Treme Case: Xylivory

Three lineages of loricariids have independently evolved specializations for xylivory (i.e., wood-eating), such as using spoon-shaped teeth and strength optimized jaws: the *Hypostomus cochliodon* group, *Panaqolus*, and *Panaque* (Schaefer and Stewart 1993; Armbruster 2003). However, because vertebrates lack the biochemical pathways needed to digest cellulose, vertebrate digestion of large woody debris requires digestive tract specializations that slow the passage of ingested wood particles, nurture microbes capable of digesting a portion of the cellulose, and absorb the released nutrients. The alimentary canals of wood-eating catfishes were examined and found to lack these specializations, and fishes fed wood in a lab setting rapidly passed the ingested particles within four hours of consumption, even losing weight when fed an all-wood diet (German 2009). German and Bittong (2009) found that wood-eating *Panaqolus nocturnus*, *P. schaeferi* (as *P.* cf. *nigrolineatus* "Marañon"), and *Hypostomus pyrineusi*, as well as the algivorous/detritivorous *Pterygoplichthys pardalis* (as *P. disjunctivus*), had low cellulase and xylanase activity, suggesting that they were digesting soluble polysaccharides as opposed to the plant matter itself. All of these species had low levels of microbial-produced enzymes that would aid in digestion, suggesting digestion was endogenous to the species and not dependent on symbiotic microbes (McCauley et al. 2020).

Although wood-eating catfishes are often observed to have large proportions of wood in their stomachs and show a carbon isotope signature matching that of the wood on which they graze, they receive little nutrition directly from the wood (Lujan et al. 2011). Rather, xylivorous catfishes mostly assimilate the microbes (bacteria and fungi), soluble detritus, and algae present within the ingested wood (German and Miles 2010). Nonetheless, this wood-eating behavior ecologically distinguishes xylivorous species from other fishes by allowing them to utilize a readily available resource substrate that cannot be accessed by other fishes in the community, thereby making them highly specialized detritivores. In this regard, xylivorous catfishes act as ecosystem engineers by facilitating the decomposition of woody debris and making these materials more readily consumed by other organisms.

16.3 Ecotrait Matrix

Of the 6342 species scored for adult modal diet in the ecotrait matrix (Chap. 6), 171 of these species or 2.7% were classified as algivores and detritivores, 832 species or 13.1% as algivores but not detritivores, and 247 species or 3.9% as detritivores but not algivores (Fig. 16.2). This scoring of species is more likely to contain errors than the other feeding categories of the ecomatrix, as the boundary between algivory and detritivory is not always clear, and many species contain algae or detritus in their guts from incidental consumption.

In very few cases, some clades are unambiguously viewed as either modally detritivorous or algivorous. For example, parodontids are rheophilic and likely only consume algae off rocks. They display low morphological disparity within their family and are convergent in morphology with cypriniform grazers of other continents, such as the North American Stonerollers (*Campostoma;* Catostomidae) and the African and Asian labeonines (e.g., *Labeo* and *Garra*; Cyprinidae). Additionally, all species of Curimatidae were coded as detritivorous (115 of 115 species), yet the finer feeding aspects of this family and function of the known morphological diversity of buccal cavity soft tissues remain largely unexplored. Farago et al. (2020) found that four sympatric species of curimatids feed on detritus in different habitats at different times. Winemiller (1990) examined gut contents of 97 *Steindachnerina argentea* (as *Curimata argentea*) and found 57.5% detritus (mostly FPOM) and 17% algae. Detritus is likely not a limiting resource in most aquatic environments, suggesting detritivorous feeding is facultative and opportunistic for many fishes, as has been observed in some catfishes of the family Loricariidae (Black and Armbruster 2022). Species scored as both algivores and detritivores also come with a level of uncertainty, as 19 of 21 species of the Prochilodontidae were scored as both detritivores and algivores, yet many aspects of feeding in this family remain poorly studied. Interspecific and ontogenetic differences in gut length and other qualities may further affect how these fishes assimilate materials (Kramer and Bryant 1995) allowing for multiple species to utilize the same resources.

To emphasize the precariousness of assigning modal adult diets between detritivory and algivory, we will highlight the Loricariidae, a hyper-diverse family with over 1050 species. Of all the NFF species considered in the ecomatrix (6345 spp) that were classified as strictly algivorous (832 spp), strictly detritivorous (247 spp), and both detritivorous and algivorous (171 spp), the loricariid catfishes accounted for 95% (792/832), 15% (37/247), and 89% (152/171) of each dietary group, respectively, and totaled for 78% (981/1250) of benthic grazers when combining these categories. Within the Loricariidae, this comes out to a relative distribution of 81% algivory, 4% detritivory, and 15% algivory + detritivory. However, although dietary data are scarce for most species, the information available suggests that the majority of loricariids most likely consume some combination of algae and detritus. As observed among curimatids, the assimilation of ingested materials is influenced by gut length and other characteristics, and there are significant differences in relative gut length within the more robust-bodied hypostomines versus the more dorsoventrally flattened loricariines (Kramer and Bryant 1995). Likewise, the global and local diversity (i.e., between and within sub-basins) of the Loricariidae is great, with many species having disparate jaw and tooth morphologies (Fig. 16.1).

The wide range of jaw morphologies observed among loricariid species has been hypothesized to correspond with specializations to their varied diets (Lujan et al. 2012). For example, consider the aforementioned specialized teeth and jaws, which are associated with eating wood. Furthermore, one hypothesis for high local diversities of loricariids is that algae and detritus are not limiting; therefore, species vary in where, when, and how they feed, based on different numbers, shapes, and strengths of teeth and differently shaped jaws. Contrary to this line of reasoning, jaw morphology and diet have not always been demonstrated to be strongly correlated (e.g., Black and Armbruster 2022); however, jaw morphological specialization could correspond to poorly recognized axes of resource variation, such as elemental nutrient concentrations or ratios (Vanni et al. 2002; Moody et al. 2019). For example, Hood et al. (2005) found that the foregut contents of two loricariid genera (*Ancistrus*, *Chaetostoma*) in an Andean piedmont river were up to four times more concentrated in phosphorus (a potentially growth-limiting resource in loricariids) than average biofilm, suggesting that they were selectively feeding on nutrient-rich subsets of the available resource base. Similarly, loricariids have been observed to respond to variation in periphyton cover at relatively small spatial scales (i.e., between pools) (Power 1983), indicating a behavioral component to their selective foraging. The tremendous variability observed in both form and habitat suggests that loricariids employ numerous specialized feeding behaviors and mechanisms. Clearly, greater refinement is needed to determine not only what loricariids ingest, but what they assimilate.

This section demonstrates the need for more nuanced distinctions among algal and detrital food resources in addition to the behavioral and physiological traits used by fishes to consume these resources. Studies of loricariids in the wild have shown that consumption of food items does not necessarily mean assimilation, and this is especially true of algae and detritus. The finer scales of resource partitioning amongst benthic algae and detritus may lead to a deeper understanding of ecological and evolutionary theory necessary to explain the taxonomic richness of NFFs, but this will require finer-scale data on food consumed and assimilated.

16.4 Problems and Future Directions

There are several long-standing challenges in studying fish diets in the wild. Traditionally, fish diets have been inferred from direct examination of stomach contents. Stomach contents analysis (SCA) is relatively inexpensive and is widely employed by fish ecologists today. However, SCA is time-consuming and requires sufficient taxonomic expertise to identify food items removed from guts, often as fragments and in various states of decomposition. Other limitations of this approach are that stomach contents represent only recently ingested materials and may exhibit a bias towards food items that undergo slower digestion (e.g., fish bones and arthropod carapaces). For example, SCA would poorly discriminate modal dietary differences among the fishes of a local assemblage in the days immediately following a transient mayfly bloom, when many species will appear to feed on Ephemeroptera. SCA also suffers from difficulty in separating partially digested food items, which can strongly affect how we interpret the absolute or relative abundance of food items (Baker et al. 2014; Hyslop 1980). Another important limitation of SCA is the presence of rare food items that may represent opportunistic omnivory or unintentional prey capture. As discussed above, periphyton is a mixture of algae, detritus, and microbes, and benthic feeding fishes may also ingest other small animals (molluscs, arthropods, meiofauna, zooplankton) and their eggs that are in the wrong periphyton at the wrong time. Grazing algivores with large mouths (on the order of centimeters in diameter) likely cannot browse selectively on periphyton or algal mats that house so many small prey items (tens of microns to millimeters in diameter) (Power 1983).

Nevertheless, the future of studying algivores and detritivores is far from bleak. Studies using stable isotopes (SI) alongside SCA are providing a more complete picture of energy and nutrient origins and pathways throughout food webs (González-Bergonzoni et al. 2019; Layman et al. 2007; Reis et al. 2020). For NFF communities, SI has recently been used in an effort to elucidate poorly recognized axes of resource variation through the assemblage centroid-standardized isotope vector analysis (ACSIVA) method proposed by Lujan and others (2012), which interprets the carbon and nitrogen depletion or enrichment of consumers from an assemblage centroid as the partitioning of resources within a community in ways that cannot otherwise be noticeably observed. Another NFF study attempted to directly compare individuals from communities that are disjunct in time and space by first standardizing all data points for an assemblage by a translation factor deterλmined by the translation of the centroid of shared basal resources in each community to the Euclidean origin of the biplot (i.e., 0 δC13‰, 0 δN15‰). This is the baseline-standardized isotopic vector analysis (BaSIVA) method by Black and Armbruster (2021), which does allow for a more direct comparison of individuals within separate communities, but at the cost of reducing variation assumed to be vital to the puzzle of resource partitioning in ACSIVA. Another tool called compound-specific SI, in which researchers analyze the stable isotope signatures of specific organic compounds (e.g., different fatty acids or amino acids), has recently been promoted as a means to achieve greater resolution for identifying dietary sources (Twining et al. 2020). In addition to SI, a promising method for estimating the transfer of material through food webs is the analysis of essential fatty acid (EFA) profiles as bio-markers. Because animals cannot make these EFAs from precursor molecules, they must obtain them in their diet (Iverson 2009; Torres-Ruiz and Wehr 2020). Researchers are also attempting to understand the flow of nutrients throughout food webs by use of nutritional stoichiometry, or ratios of carbon and potentially limiting nutrients such as nitrogen and phosphorous, at different trophic levels (Elser et al. 2000; Evans-White and Halvorson 2017). Employing studies using some combinations of SCA, SI, EFAs, and nutritional stoichiometry will improve our understanding of Neotropical fish diets—especially diets consisting of living and non-living slimy-fish-foods.

References

Abilhoa V, Valduga MO, Frehse FdeA, Vitule JRS (2016) Use of food resources and resource partitioning among five syntopic s pecies of Hypostomus (Teleostei: Loricariidae) in an Atlantic Forest river in southern Brazil. Zoologia 33:e20160062

Albert JS et al (2025) An ecological trait matrix for neotropical freshwater fishes, the most diverse continental vertebrate fauna. Sci Data 12(1):1127

André ER, Hecky RE, Duthie HC (2003) Nitrogen and phosphorus regeneration by cichlids in the littoral zone of Lake Malawi, Africa. J Great Lakes Res 29:190–201

Armbruster JW (2003) The species of the *Hypostomus cochliodon* group (Siluriformes: Loricariidae). Zootaxa 249:1–60

Armbruster JW (2004) Phylogenetic relationships of the suckermouth armored catfishes (Loricariidae) with emphasis on the Hypostominae and the Ancistrinae. Zool J Linn Soc 141:1–80

Azim ME, Beveridge MCM, van Dam AA, Verdegem MC (2005) Periphyton and aquatic production: an introduction. In: Azim ME, Verdegem MC, van Dam AA, Beveridge MC (eds) Periphyton: ecology, exploitation and management. Wallingford (UK): CABI, pp 1–13

Bachmann RW, Jones BL, Fox DD, Hoyer M, Bull LA, Canfield JrDE (1996) Relations between trophic state indicators and fish in Florida (U.S.A.) lakes. Can J Fish Aquat Sci 53(4):842–855

Baker R, Buckland A, Sheaves M (2014) Fish gut content analysis: robust measures of diet composition. Fish Fish 15(1):170–177

Bayley PB (1989) Aquatic environments in the Amazon Basin, with an analysis of carbon sources, fish production, and yield. Can Spec Publ Fish Aquat Sci 106:399–408

Bayley PB (1995) Understanding large river: floodplain ecosystems. BioScience 45(3):153–158

Bertmar G, Kapoor BG, Miller RV (1969) Epibranchial organs in lower teleostean fishes—an example of structural adaptation. Int Rev Gen Exp Zool 4:1–48

Black CR, Armbruster JW (2021) New method of isotopic analysis: baseline-standardized isotope vector analysis shows trophic partitioning in loricariids. Ecosphere 12(5):e03503

Black CR, Armbruster JW (2022) Chew on this: oral jaw shape is not correlated with diet type in loricariid catfishes. PLoS One 17(11):e0277102

Bowen SH (1983) Detritivory in neotropical fish communities. Environ Biol Fishes 9(2):137–144

Brett MT, Bunn SE, Chandra S, Galloway AWE, Guo F, Kainz MJ, Kankaala P, Lau DCP, Moulton TP, Power ME, Rasmussen JB, Taipale SJ, Thorp JH, Wehr JD (2017) How important are terrestrial organic carbon inputs for secondary production in freshwater ecosystems?. Freshw Biol 62(5):833–853

Brett MT, Kainz MJ, Taipale SJ, Seshan H (2009) Phytoplankton, not allochthonous carbon, sustains herbivorous zooplankton production. Proc Natl Acad Sci U S A 106(50):21197–21201

Cardona L (2016) Food and feeding of Mugilidae. CRC Press, Boca Raton, pp 165–195

Casatti L, Veronezi Júnior JL, Ferreira CDP (2009) Diet of the armored catfish Aspidoras fuscoguttatus (Ostariophysi, Callichthyidae) in streams with different limnological and structural features. Biota Neotropica 9:113–121

Castro RMC, Vari RP (2004) Detritivores of the South American fish family Prochilodontidae (Teleostei: Ostariophysi: Characiformes): A phylogenetic and revisionary study. SCtZ 622:1–189

Cebrian J (1999) Patterns in the fate of production in plant communities. Am Nat 154(4):449–468

Chételat J, Pick FR, Morin A, Hamilton PB (1999) Periphyton biomass and community composition in rivers of different nutrient status. Can J Fish Aquat Sci 56(4):560–569

Elser JJ, Fagan WF, Denno RF, Dobberfuhl DR, Folarin A, Huberty A, Interlandi S, Kilham SS, McCauley E, Schulz KL, Siemann EH, Sterner RW (2000) Nutritional constraints in terrestrial and freshwater food webs. Nature 408(6812):578–580

Erdman DS (1961) Notes on the biology of the gobiid fish *Sicydium plumieri* in Puerto Rico. Bull Mar Sci 11(1):448–456

Evans-White MA, Halvorson HM (2017) Comparing the ecological stoichiometry in green and brown food webs – a review and meta-analysis of freshwater food webs. Front Microbiol 8:1184

Farago T, Borba G, Amadio S, Oliveira J, Santos G, Val A, Ferreira E (2020) Feeding strategies differentiate four detritivorous curimatids in the Amazon. Web Ecol 20(2):133–141

Fernandes CC (1997) Lateral migration of fishes in Amazon floodplains. Ecol Freshw Fish 6(1):36–44

Flecker AS, McIntyre PB, Moore JW, Anderson JT, Taylor BW, Hall Jr RO (2010) Migratory fishes as material and process subsidies in riverine ecosystems. Am Fish Soc Symp 73:559–592

Flecker AS (1996) Ecosystem engineering by a dominant detritivore in a diverse tropical stream. Ecology 77:1845–1854

Frable B (2017) Family curimatidae-toothless characins. In: van der Sleen P, Albert JS (eds) Field guide to the fishes of the Amazon, Orinoco, and Guianas. Princeton University Press, Princeton (New Jersey), pp 148–153

Frau D (2022) Grazing impacts on phytoplankton in South American water ecosystems: a synthesis. Hydrobiologia 849(4):833–860

Garcia A, Tello S, Vargas G, Duponchelle F (2009) Patterns of commercial fish landings in the Loreto region (Peruvian Amazon) between 1984 and 2006. Fish Physiol Biochem 35(1):53–67

Gay D, Bassani C, Sergipense S (2002) Diel variation and selectivity in the diet of Cetengraulis edentulus (Cuvier 1828)(Engraulidae-Clupeiformes) in the Itaipu Lagoon, Niterói, Rio de Janeiro. Atlântica Rio Grande 24(2):59–68

German DP (2009) Inside the guts of wood-eating catfishes: can they digest wood?. J Comp Physiol B 179(8):1011–1023

German DP, Bittong RA (2009) Digestive enzyme activities and gastrointestinal fermentation in wood-eating catfishes. J Comp Physiol B 179(8):1025–1042

German DP, Miles RD (2010) Stable carbon and nitrogen incorporation in blood and fin tissue of the catfish Pterygoplichthys disjunctivus (Siluriformes, Loricariidae). Environ Biol Fishes 89(2):117–133

González-Bergonzoni I, D'Anatro A, Vidal N, Stebniki S, Tesitore G, Silva I, Teixeira de Mello F (2019) Phylogenetic tree diagram showing evolutionary relationships among various fish families and subfamilies. The left side highlights groups such as Prochilodontidae, Chilodontidae, Curimatidae, and Doradinae in orange. The right side features Parodontidae, Semaprochilodus, Prochilodus, Loricariinae, Hypoptopomatinae, and Hypostominae in green. The tree structure illustrates branching patterns and connections between these groups. Ecosystems 22(8):1736–1753

Hoeinghaus DJ, Winemiller KO, Agostinho AA (2007) Landscape-scale hydrologic characteristics differentiate patterns of carbon flow in large-river food webs. Ecosystems 10(6):1019–1033

Hood JM, Vanni MJ, Flecker AS (2005) Nutrient recycling by two phosphorus-rich grazing catfish: the potential for phosphorus-limitation of fish growth. Oecologia 146:247–257

Hyslop EJ (1980) Stomach contents analysis—a review of methods and their application. J Fish Biol 17(4):411–429

Iverson SJ (2009) Tracing aquatic food webs using fatty acids: from qualitative indicators to quantitative determination. In: Kainz M, Brett MT, Arts MT (eds) Lipids in aquatic ecosystems. Springer, pp 281–308

Kramer DL, Bryant MJ (1995) Intestine length in the fishes of a tropical stream: 2. Relationships to diet—the long and short of a convoluted issue. Environ Biol Fishes 42:129–141

Krings W, Konn-Vetterlein D, Hausdorf B, Gorb SN (2023) Holding in the stream: convergent evolution of suckermouth structures in Loricariidae (Siluriformes). Front Zool 20: 37

Lasso C, Castroviejo J (1992) Composition, abundance and biomass of the benthic fish fauna from the Guaritico river of a Venezuelan floodplain. Ann Limnol-Int J Lim 28(1):71–84

Layman CA, Arrington DA, Montaña CG, Post DM (2007) Can Stable isotope ratios provide for community-wide measures of trophic structure?. Ecology 88(1):42–48

Lazzaro X (1987) A review of planktivorous fishes: their evolution, feeding behaviours, selectivities, and impacts. Hydrobiologia 146(2):97–167

Lee RE (2018) Phycology. Cambridge University Press, Cambridge (UK)

Lujan NK, German DP, Winemiller KO (2011) Do wood-grazing fishes partition their niche?: morphological and isotopic evidence for trophic segregation in Neotropical Loricariidae. Funct Ecol 25(6):1327–1338

Lujan NK, Winemiller KO, Armbruster JW (2012) Trophic diversity in the evolution and community assembly of loricariid catfishes. BMC Evol Biol 12:1–13

Lujan NK, Armbruster JW, Lovejoy NR (2018) Multilocus phylogeny and generic revision of the Guiana Shield endemic Lithoxus clade of suckermouth armored catfishes (Loricariidae: Hypostominae). Zool J Linn Soc 184(4):1169–1186

Lundberg JG, Lewis JrWM, Saunders III JF, Mago-Leccia F (1987) A major food web component in the Orinoco River channel: evidence from planktivorous electric fishes. Science 237(4810):81–83

MacIntyre HL, Geider RJ, Miller DC (1996) Microphytobenthos: the ecological role of the "secret garden" of unvegetated, shallow-water

marine habitats. I. Distribution, abundance and primary production. Estuaries 19(2):186–201
McCauley M, German DP, Lujan NK, Jackson CR (2020) Gut microbiomes of sympatric Amazonian wood-eating catfishes (Loricariidae) reflect host identity and little role in wood digestion. Ecol Evol 10(14):7117–7128
Melo BF, Sidlauskas BL (2017a) Family Chilodontidae-Headstanders. In: van der Sleen P, Albert JS (eds) Field guide to the fishes of the Amazon, Orinoco, and Guianas. Princeton University Press, Princeton (New Jersey), pp 141–142
Melo BF, Sidlauskas BL (2017b) . Family prochilodontidae-flannel mouth characiformes. In: van der Sleen P, Albert JS (eds) Field Guide to the fishes of the Amazon, Orinoco, and Guianas. Princeton University Press, Princeton (New Jersey), pp 170–172
Meyer JL, Wallace JB, Eggert SL (1998) Leaf litter as a source of dissolved organic carbon in streams. Ecosystems 1(3):240–249
Miller DC, Geider RJ, MacIntyre HL (1996) Microphytobenthos: the ecological role of the "secret garden" of unvegetated, shallow-water marine habitats II. Role in sediment stability and shallow-water food webs. Estuaries 19(2):202–212
Monti D, Lefrancois E, Lord C, Mortillaro JM, Lopez P, Keith P (2018) Selectivity on epilithic diatom consumption for two tropical sympatric gobies: Sicydium punctatum Perugia, 1986 and Sicydium plumieri (Bloch, 1786). Cybium: Revue Internationale d'Ichtyologie 42:365–373
Moody EK, Lujan NK, Roach KA, Winemiller KO (2019) Threshold elemental ratios and the temperature dependence of herbivory in fishes. Funct Ecol 33(5):913–923
Moore JC, Berlow EL, Coleman DC, de Ruiter PC, Dong Q, Hastings A, Johnson NC, McCann KS, Melville K, Morin PJ, Nadelhoffer K, Rosemond AD, Post DM, Sabo JL, Scow KM, Vanni MJ, Wall DH (2004) Detritus, trophic dynamics and biodiversity. Ecol Lett 7(7):584–600
Munubi RN, McIntyre PB, Vadeboncoeur Y (2018) Do grazers respond to or control food quality? Cross-scale analysis of algivorous fish in littoral Lake Tanganyika. Oecologia 188(3):889–900
Lujan NK, Armbruster JW (2012) Morphological and functional diversity of the mandible in suckermouth armored catfishes (Siluriformes: Loricariidae). Abstract J Morphol 273(1):24–39. https://doi.org/10.1002/jmor.11003
Pease AAM, Mendoza-Carranza KO, Winemiller (2018) Feeding ecology and ecomorphology of cichlid assemblages in a large Mesoamerican river delta. Environ Biol Fishes 101:867–879
Power ME (1983) Grazing responses of tropical freshwater fishes to different scales of variation in their food. Environ Biol Fishes 9(2):103–115
Provenzano RFP, Sabaj MH (2024) Description of a new species of Loricaria Linnaeus 1758 (Siluriformes: Loricariidae) from the Orinoco Basin. Proc Acad Nat Sci Philadelphia 168:211–220
Reis A, da S, Albrecht MP, Bunn SE (2020) Food web pathways for fish communities in small tropical streams. Freshw Biol 65(5):893–907
Restrepo-Gómez A, Mancera-Rodríguez N (2014) Trophic ecology of Saccodon dariensis (Pisces: Parodontidae) in Guatapé River tributaries, Magdalena River Basin, Colombia. Rev MVZ Córdoba 19(1):3930–3943
Roach KA, Winemiller KO, Davis IIISE (2014) Autochthonous production in shallow littoral zones of five floodplain rivers: effects of flow, turbidity and nutrients. Freshw Biol 59(6):1278–1293
Roberts TR (1974) Osteology and classification of the neotropical characoid fishes of the families Hemiodontidae (including Anodontinae) and Parodontidae. Bull Mus Comp Zool 146:411–472
Sabaj MH (2005) Taxonomic assessment of Leptodoras (Siluriformes: Doradidae) with descriptions of three new species. Neotrop Ichthyol 3:637–678
Schaefer SA, Stewart DJ (1993) Systematics of the *Panaque dentex* species group (Siluriformes: Loricariidae), wood-eating armored catfishes from tropical South America. Ichthyol Explor Freshw 4:309–342
Schaefer SA, Lauder GV (1986) Historical transformation of functional design: evolutionary morphology of feeding mechanisms in loricarioid catfishes. Syst Zool 35:489–508
Schaefer SA, Lauder GV (1996) Testing historical hypotheses of morphological change: biomechanical decoupling in loricarioid catfishes. Evolution 50:1661–1675
Soria-Barret MR, Rodiles-Hernández, Winemiller KO (2019) Trophic ecomorphology of cichlid fishes of Selva Lacandona, Usumacinta, Mexico. Environ Biol Fishes 102(7):985–996
Thomas MR, Py-Daniel LHR (2008) Three new species of the armored catfish genus Loricaria (Siluriformes: Loricariidae) from river channels of the Amazon basin. Neotrop Ichthyol 6:379–394
Thompson G, Fortunato RC, Chiesa I, Volpedo A (2015) Trophic ecology of Mugil liza at the southern limit of its distribution (Buenos Aires, Argentina). Braz J Oceanogr 63:271–277
Torres-Ruiz M, Wehr JD (2020) Complementary information from fatty acid and nutrient stoichiometry data improve stream food web analyses. Hydrobiologia 847(2):629–645
Twining CW, Taipale SJ, Ruess L, Bec A, Martin-Creuzburg D, Kainz MJ (2020) Stable isotopes of fatty acids: current and future perspectives for advancing trophic ecology. Philos Trans R Soc B 375(1804):20190641
Vadeboncoeur Y, Power ME (2017) Attached algae: the cryptic base of inverted trophic pyramids in freshwaters. Annu Rev Ecol Evol Syst 48(1):255–279
Vadeboncoeur Y, Kalff J, Christoffersen K, Jeppesen E (2006) Substratum as a driver of variation in periphyton chlorophyll and productivity in lakes. JNABS 25(2):379–392
Vanni MJ, Flecker AS, Hood JM, Headworth JL (2002) Stoichiometry of nutrient recycling by vertebrates in a tropical stream: linking species identity and ecosystem processes. Ecol Lett 5(2):285–293
Vari RP (1991) Systematics of the neotropical characiform genus Steindachnerina Fowler (Pisces: Ostariophysi). Smithson Contrib Zool 507:1–118
Vari RP (1983) Phylogenetic relationships of the families Curimatidae, Prochilodontidae, Anostomidae, and Chilodontidae (Pisces, Characiformes). Smithson Contrib Zool 378:1–60
Wallace JB, Eggert SL, Meyer JL, Webster JR (1997) Multiple trophic levels of a forest stream linked to terrestrial litter inputs. Science 277(5322):102–104
Winemiller KO, Jepsen DB (1998) Effects of seasonality and fish movement on tropical river food webs. J Fish Biol 53(sA):267–296
Winemiller KO, Kelso-Winemiller LC, Brenkert AL (1995) Ecomorphological diversification and convergence in fluvial cichlid fishes. Environ Biol Fish 44:235–261
Winemiller KO (1990) Spatial and temporal variation in tropical fish trophic networks. Ecol Monogr 60:331–367
Winemiller KO, Kelso-Winemiller LC (2003) Food habits of tilapiine cichlids of the Upper Zambezi River and floodplain during the descending phase of the hydrologic cycle. J Fish Biol 63:120–128
Winemiller KO, Ponwith JB (1998) Comparative ecology of eleotrid fishes in Central American coastal streams. Environ Biol Fishes 53:373–384
Winemiller KO, Jepsen DB (2004) Migratory neotropical fish subsidize food webs of oligotrophic blackwater rivers. In: Polis GA, Power ME, Huxel GR (eds) Food webs at the landscape level. University of Chicago Press, Chicago, pp 115–132
Winemiller KO, Roelke DL, Cotner JB, Montoya JV, Sanchez L, Castillo MM, Montaña CG, Layman CA (2014) Pulsing hydrology determines top-down control of basal resources in a tropical river-floodplain ecosystem. Ecol Monogr 84(4):621–635

17 Piscivorous Neotropical Freshwater Fishes

Kirk O. Winemiller and James S. Albert

Abstract

Piscivorous fishes are diverse components of freshwater ecosystems throughout the humid Neotropics, representing all the major Neotropical freshwater fish (NFF) clades, and characterizing a range of distinct predatory strategies. We identify at least 550 species of piscivorous NFF fishes, in 113 genera and 29 families, accounting for just under 10% of the whole NFF fauna. These species occupy numerous specialized niches, with many forms frequently coexisting and possibly partitioning the general piscivorous niche space. Piscivorous NFF clades are heterogeneously distributed among higher NFF taxa, having evolved at least 37 times independently. Less than half (16% or 43%) of these clades are obligate piscivores, with a majority being facultative piscivores who largely eat nonvertebrate animals, mainly arthropods (e.g., insects and crustaceans). A piscivorous diet has evolved in at least nine cichlid genera, but just a single cyprinodontiform species (*Belonesox belizanus* in Central America, away from the Amazonian center of NFF diversity). At the genus level, the most diverse piscivorous clades are riverine-adapted pimelodid (24 genera) and trichomycterid (20 genera) catfishes. Species in these two families exhibit the extreme ends of the NFF body size spectrum, from the giant piraíba (*Brachyplatystoma filamentosum*) that grows to 2.5 m and swallows large food items whole, to the tiny candiru (*Vandellia sanguinea*) that grows to 5.3 cm, and feeds by parasitizing blood from the gills of larger fishes. Both trophic niche conservatism and evolution are widely observed in the NFF fauna, with many cases of phylogenetic stasis, adaptive divergence, and niche convergence observed at multiple levels of the phylogenetic hierarchy. Most piscivorous NFF fishes have a generalist foraging strategy and broad diets, and fewer clades are dietary specialists with extreme morphologies adapted to feed in certain microhabitats or on certain types of fish. Some piscivorous NFF fishes have enhanced sensory modalities (e.g., acute olfaction in many catfishes, high-frequency hearing in many ostariophysans, electroreception in catfishes, and gymnotiform electric fishes) used to detect prey, and others specialized methods for capturing prey. Most piscivores feed using their oral jaws for suction or envelopment of prey, and some others use a biting motion to grasp or impale prey (e.g., *Cynodon*), while yet others like the piranhas (*Pygocentrus* and *Serrasalmus*) or the piracatinga catfish (*Calophysus macropterus*) have sharp teeth used to clip off pieces of flesh. Highly specialized forms of piscivory include feeding on fish scales, fins, mucus, and blood. The various specialized modes of piscivory reviewed in this study are hypothesized to have evolved adaptively in response to ecological opportunities of omnipresent prey species, and under the perennial influences of interspecific competition during seasonal (i.e., annual flood cycle) and interannual (i.e., longer term climate oscillations) intervals of depressed prey availability. Like other predatory species with high importance in trophic webs, piscivorous fishes are generally less abundant than other species, and some have small and geographically fragmented populations. From these attributes, piscivorous species may be expected to be at greater conservation risk under human-caused habitat changes and overfishing. Research along multiple avenues (e.g., functional morphology, behavior, phylogenetic systematics, field studies, and ecological modeling) is needed to improve our understanding of the ecology, evolution, and conservation of Neotropical freshwater piscivorous fishes.

K. O. Winemiller (✉)
Texas A&M University, College Station, TX, USA
e-mail: k-winemiller@tamu.edu

J. S. Albert (✉)
University of Louisiana at Lafayette, Lafayette, LA, USA
e-mail: jsalbe@gmail.com

J. S. Albert and R. E. Reis (eds.), *Ecology and Evolution of Amazonian Fishes*, Fascinating Life Sciences,
https://doi.org/10.1007/978-3-032-09328-8_17

17.1 Introduction

This chapter examines various modes of piscivory (i.e., feeding on fishes) and its evolution among clades of Neotropical freshwater fishes (NFF). Piscivory first evolved with the origin of jawed fishes (Gnathostomata) in the Silurian Period *c.* 460 Ma, and is observed in all major living clades of cartilaginous fishes (Chondrichthyes) and boney fishes (Osteichthyes (Friedman and Sallan 2012; Brazeau and Friedman 2015). Among freshwater ecosystems globally, piscivores are among the first species that appear during local community assembly. For example, fish communities at high latitude and high elevation typically contain a few species (e.g., salmonid trouts and charrs in the Northern Hemisphere, percichthyids in the Southern Hemisphere) that feed on benthic macroinvertebrates but also fish, including conspecifics (Amundsen et al. 1995). As more species are added to fish communities at lower latitudes and elevations, more trophic niches are observed, including more specialized forms of piscivory (Winemiller 1991). Cannibalism appears to be common among freshwater fishes (Pereira et al. 2017a), and obviously fish are an ever-present food resource within any freshwater habitat occupied by one or more fish species. Cannibalism appears to be more common among freshwater fishes at higher latitudes where species diversity is low, especially in the Northern Hemisphere (Pereira et al. 2017b).

Fish flesh is highly nutritious and easily digestible food resource with high energy density, which allows piscivores, especially those capable of ingesting large prey items, to feed less often than fishes specializing on plant materials or invertebrates (Arrington et al. 2002). Among piscivorous fishes, certain foraging strategies are effective for exploiting specific types of prey, e.g., prey of different body sizes, body dimensions, movement capabilities, or defense tactics. Nonetheless, there is scant evidence that Neotropical piscivores specialize on restricted fish taxa, and most species feed on a diversity of other fish species within the hyperdiverse Neotropical ichthyofauna. The NFF fauna encompass highly diverse modes of feeding with regards to functional morphology, behavior, and physiology.

Here, we classify NFF taxa based on four aspects of feeding: degree of piscivory, mode of ingestion, attack mode, and activity period. We map piscivores onto a phylogeny of all NFF genera (Fig. 17.1) abstracted from the species-level tree presented in Chap. 2. We interpret the evolution of piscivory to be a derived state, and count the number of unique derivations from an alternative trophic niche, as well as interpretations of niche conservatism, adaptive divergence, and niche convergence. Using these ecological and phylogenetic data, we propose that many piscivorous species are at high conservation risk under human-caused habitat changes and overfishing, and advocate for renewed research into the ecology, evolution, and conservation of Neotropical freshwater piscivorous fishes.

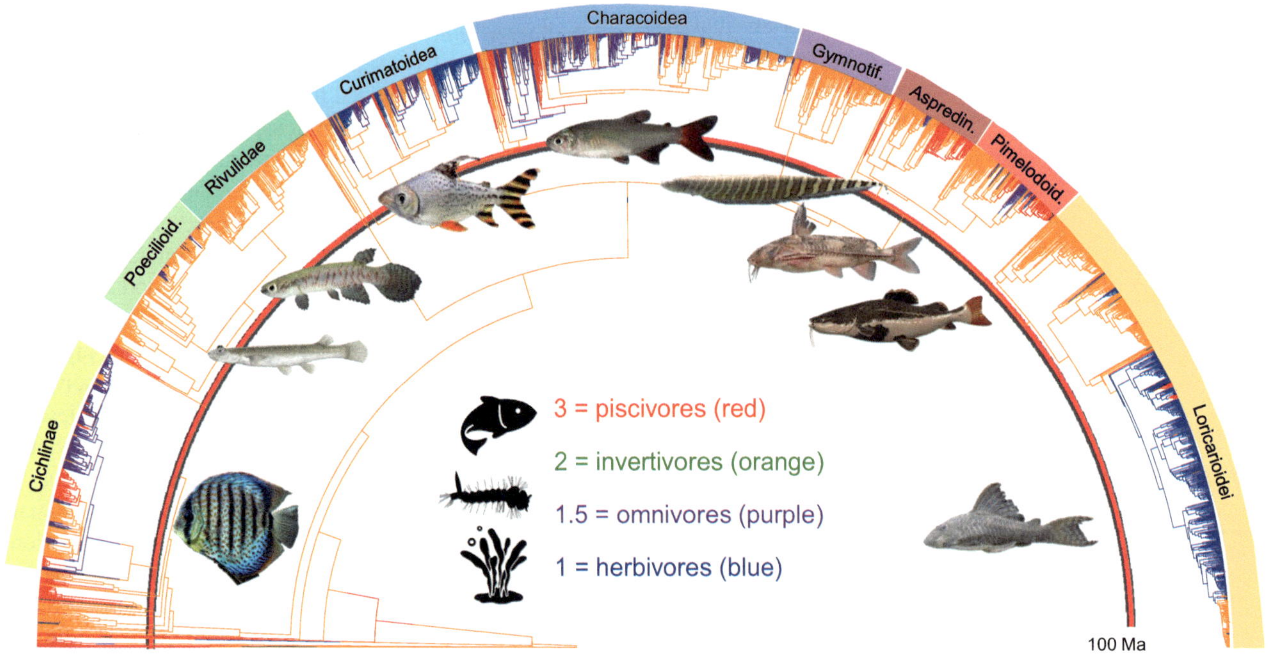

Fig. 17.1 Evolution of PC2 values from the ecospace analysis of Chap. 6 representing 8.5% of the total variance. Ecotraits loading most heavily on PC2 are related to diet. Graphical conventions as in Fig. 7.2). Phylogeny of 3167 species (49.6% of NFF total) in 865 genera, 96 families, and 40 orders from Chap. 2; Big Nine NFF clades (colored regions) from Chap. 3. Red semicircle at *c.* 100 Ma marks time of the final separation of South America and Africa and the latest origin of phylogenetically independent NFF clades (Albert et al. 2011)

17.2 Materials and Methods

Piscivorous fishes are defined as those that normally consume fish flesh, by eating whole fish or by biting off pieces of flesh, during at least some portion of the year (Winemiller 1990a; Jepsen and Winemiller 2002). It is important to note that many if not most fish will at least occasionally consume fish eggs or larvae when available, but are not here considered as piscivorous. We further distinguish between *obligate piscivores* with more than half the diet comprised of fish, and *facultative piscivores* that feed on fish occasionally but with diets comprised mainly of invertebrate prey or other food items (Table 17.1; Burress 2016).

We assessed adult model diet of all NFF species, including some peripheral freshwater fishes (*sensu* Myers 1938) belonging to otherwise marine clades that are essentially restricted to freshwater throughout the duration of their life cycle (e.g., *Paratrygon*, *Pellona*, *Potamorrhaphis*, and *Colomesus*). We excluded marine taxa that occasionally or regularly invade freshwater ecosystems, even if they occasionally are found in freshwater habitats over 100 km inland (e.g., *Megalops atlanticus, Centropomus undecimalis, Lutjanus colorado,* and *Strongylura timucu*), or if they are diadromous species (e.g., *Gobiomorus dormitor*).

Classifying the diet of NFF fish species is complicated because many species vary substantially in the proportion and size-distribution of other fishes in their diet, and in their relative consumption of fish and other animal proteins during ontogeny or throughout the seasonal water cycle. Further, quantitative or even qualitative data not available for many NFF species (Layman et al. 2005). We relied on the limited dietary information reported in wild and captive fishes (e.g., natural history and behavioral observations, stomach contents), behavioral and dietary data from the aquaculture and aquarium trades, and functional interpretations from the morphology of the oral and pharyngeal jaws and the intestine. Data were compiled from numerous published reports (Albert et al. 2025, table S1), unpublished dietary data from stomach contents analysis performed by the first author on tens of thousands of specimens (Winemiller 1990a), and collaborators (Williams et al. 1998; Barbarino-Duque 2003; Montaña and Winemiller 2009; Cochran-Biederman and Winemiller 2010; Montaña et al. 2011; Winemiller et al. 2011; Peterson et al. 2017; Pease et al. 2018; Soria-Barreto et al. 2019), and qualitative information reported in FishBase (https://www.fishbase.se/search.php) and other internet sources. We limited our analysis to genera, classifying a genus as piscivorous if at least one species was reported to consume fish either exclusively or facultatively. This genus-level analysis allows us to estimate the minimum number of evolutionary transformations among dietary modes, but does not have sufficient taxonomic resolution to estimate niche differences among taxa.

We also classified genera according to mode of ingestion: bite/impale, suction/engulf, and dislodge/scrape, the latter for species that feed on fins, scales, mucus, or blood. It is important to note that these classifications are not entirely discrete, and some fish are capable of performing more than one ingestion mode or expressing modes to varying degrees. For example, some lepidophagous fishes dislodge rows of scales by raking their lower jaw teeth across the body of their prey (*Catoprion* spp.), whereas others dislodge a few scales at a time by ramming their prey with two external teeth located on the tip of the snout (*Roeboides* spp.).

Piscivorous fish species exhibit two distinct modes of "biting." The jaw movements involved in suction versus engulfing of prey involve differences in the timing of a sequence of movements by jaw and head elements (Westneat 2006; Wainwright and Day 2007). For example, the cichlid *Petenia splendida* feeds using highly protrusible jaws and generates strong suction

Table 17.1 Definitions for functional categories associated with piscivory among Neotropical freshwater fishes

Foraging mode	Category	Definition
Piscivory	Exclusive	Fish normally comprise more than half the diet of adult size classes
	Facultative	Fish frequently comprise less than half the diet of adult size classes
Ingestion	Bite/Impale	Prey item is initially grasped within jaws, flesh is removed by biting, or prey is impaled on jaw teeth
	Suction/Engulf	Prey item is drawn into the oral chamber via suction and/or enveloped within the oral chamber as the predator passes over the prey item with jaws agape (ram feeding)
	Dislodge/Scrape	Scales are dislodged then ingested, or scales or mucus is scraped from the prey
Attack strategy	Pursue	Prey are actively pursued and captured
	Probe	Relatively sedentary prey are located by probing the environment
	Sit-and-wait	Active prey are ambushed by a relatively sedentary predator
Activity period	Diurnal	Activity during daylight
	Crepuscular	Activity in dim light during dawn and dusk
	Nocturnal	Activity at night

that draws small prey into their buccal cavity (Wainwright et al. 2001), whereas the gulper catfish *Asterophysus batrachus* has nonprotrusible jaws but a huge mouth gape that it uses to envelop very large prey, even some larger than itself (Zuanon and Sazima 2005). We also classified genera based on the principal attack strategy (pursue, probe, sit-and-wait/stalk) and the activity period (diurnal, crepuscular, nocturnal). These classifications likewise are not discrete in nature, with many fishes employing more than one attack strategy or adjusting their activity period depending on life stage, environmental conditions, and characteristics of potential prey.

The qualitative trait "adult modal diet = piscivory" was traced on the phylogeny of Chap. 2 using methods described in Chap. 6. In cases involving an ancestral node leading to a bifurcation, if one of the derived lineages was piscivorous based on the above criterion, then the node was classified as piscivorous. The mapped phylogenies were then used to estimate cases of independent evolution of piscivory from a nonpiscivorous ancestor.

17.3 Results

Among the 6345 species and 741 genera included within the NFF meta-phylogeny, 601 species (nearly 10%) were classified as piscivores, in 138 genera, 42 families, and 24 orders. When marine vagrants and diadromous forms were excluded, the dataset contained 550 piscivorous species (113 genera, 29 families, 13 orders). The most species-rich piscivorous family is Cichlidae (121 spp., nine genera), followed by Pimelodidae (85 spp., 24 genera), Trichomycteridae (47 spp., 20 genera), and Serrasalmidae (37 spp., five genera). The piscivorous genus with the most species is *Crenicichla* (Cichlidae), however, detailed dietary data are lacking for most of these species, and many of the smaller species probably consume mostly aquatic macroinvertebrates, with fish being a minor dietary component. For example, examination of gut contents of 111 *Crenicichla geayi* specimens collected over an entire year at Caño Volcán, Venezuela, yielded only 1.8% fish (guppies, *Poecilia reticulata*) by volume, whereas most of the diet consisted of macroinvertebrates, being 32% mayfly larvae (Ephemeroptera), 15% shrimp (Decapoda), 11.3% dragonfly nymphs (Odonata); Winemiller (1990a). Interestingly, some Neotropical fishes considered by aquarium hobbyists to be strongly piscivorous actually reveal limited fish consumption based on dietary analysis of wild stocks. The oscar (*Astronotus ocellatus*), a popular aquarium fish that often is fed goldfish, consumed only about 20% fish (volumetric percentage based on gut contents analysis of 99 specimens surveyed over one year at Caño Maraca, Venezuela; Winemiller 1990b), with aquatic and terrestrial macroinvertebrates comprising the bulk of the diet. Nineteen of the piscivorous trichomycterid catfish genera were blood feeders (known locally as *candiru*; Spotte 2002), and *Ochmacanthus* (5 spp.) are mucus feeders (Winemiller and Yan 1989).

Independent evolution of piscivory was identified for at least 37 NFF clades (Table 17.2). Among these, 34 genera were determined to be independently derived from ancestors that in nearly all cases were inferred to be invertivores or omnivores that feed heavily on aquatic invertebrates, especially insects and crustaceans. Three clades with long branch lengths (genera of Potamotrygonidae, Lepisosteidae, and Osteoglossidae) were considered inconclusive with regards to the principal diet of ancestors. A recent analysis of the evolution of trophic diversity in Neotropical cichlids inferred that piscivory evolved independently at least nine times (Arbour et al. 2020), and in nearly all cases, the ancestral condition was inferred to be invertivory. Here, we also identify nine independently evolved lineages of piscivory among the Neotropical Cichlidae. However, our trophic classification of genera differs slightly from that of Arbour et al. (2020), with *Nandopsis* and *Kronoheros* absent in their dataset, and we classified *Amphilophus* and *Mayaheros* as macroinvertebrate feeders rather than piscivores. In a study of the evolution of trophic diversity in the Serrasalmidae, Correa et al. (2007) inferred that piscivorous genera (piranhas) evolved from an ancestral form that fed on both invertebrate prey and plant material.

Only 16 of the 37 piscivorous NFF clades were classified as exclusive piscivores (Table 17.2). Most Neotropical freshwater piscivores are trophic generalists, in terms of the species on which they feed, and few piscivores consume one or a few fish prey species in Amazonian waters, presumably reflecting the species richness and evenness of these systems. Indeed most Neotropical piscivores, for which reliable dietary data are available, mainly consume macroinvertebrates. For example, *Roeboides* species have specialized morphology for scale feeding—external snout teeth used to dislodge scales. However, they probably feed on scales to a great extent only when availability of invertebrate prey is low. Analysis of gut contents of *Roeboides dayi* from two streams in Venezuela revealed that scales were prevalent in the diet during the early stages of the dry season, and mayfly larvae and other macroinvertebrates dominated their diet throughout the year (Peterson and Winemiller 1997).

Examples of fish that capture prey by biting and/or impaling are provided in Fig. 17.2. Equal numbers of clades were classified as capturing prey primarily using a bite/impale (Fig. 17.3) versus suction/engulf (Fig. 17.4) mode of prey capture, and only two clades were classified as dislodging or scraping tissue (Fig. 17.5; Table 17.2). Two clades had genera that captured prey using one or another mode (*Arapaima*—suction/engulf, *Osteoglossum*—bite; blood feeding trichomycterids—bite, *Ochmacanthus*—scrape). Pursuit is the predominant mode of attack observed among piscivorous clades, and probing is used by many of the catfish genera that hunt at night aided by a keen sense of taste (taste buds on barbels and lips) and ampullary electroreceptors on the snout (Fig. 17.6). Other clades classified as probers were the nocturnal freshwater stingrays

(*Heliotrygon*, *Paratrygon*) and the cichlid genus *Astronotus*. The latter has large bulging eyes used to searches for aquatic macroinvertebrates among the roots of dense mats of floating vegetation (e.g., *Pontederia*, *Paspalum*, *Hydrocotyle*, *Pistia*) and for terrestrial arthropods (spiders, grasshoppers, beetles) that venture into the interstitial spaces between floating aquatic plants (Winemiller 1990b). Only three clades were classified as sit-and-wait/stalk predators—gars (*Atractosteus*), wolffish or trahiras (three erythrinid genera), and leaffishes (two polycentrid genera). Many fishes classified as pursuers or probers also may adopt a sit-and-wait or stalking mode of foraging under certain conditions. Twenty-five clades were classified as diurnal foragers, and the main clades of nocturnal foragers are catfishes and stingrays. Three clades were classified as crepuscular foragers—gars, payaras (Cynodontidae), and croakers (Sciaenidae); but it is important to note that many

Table 17.2 Neotropical freshwater fish genera categorized according to piscivory modes

Clade	Piscivory independently derived?	Piscivory exclusive or facultative	Predominant ingestion mode	Predominant attack	Predominant activity period
Heliotrygon, Paratrygon, Plesiotrygon	?	Facultative	S/E	Probe	Nocturnal
Atractosteus	?	Facultative	B/I	Sit-and-wait	Crepuscular
Arapaima, Osteoglossum	?	Facultative	B/I, S/E	Pursue	Diurnal
Lycengraulis	Y	Facultative	B/I	Pursue	Diurnal
Pellona	Y	Exclusive	S/E	Pursue	Diurnal
Catoprion, Pristobrycon, Pygocentrus, Pygopristis, Serrasalmus	Y	Exclusive	B/I	Pursue	Diurnal
Cynodon, Hydrolycus, Rhaphiodon	Y	Facultative	B/I	Pursue	Crepuscular
Probolodus, Serrabrycon	Y	Exclusive	D/S	Pursue	Diurnal
Boulengerella, Ctenolucius	Y	Exclusive	B/I	Pursue	Diurnal
Erythrinus, Hoplias, Hoplerythrinus	Y	Facultative	B/I	Sit-and-wait	Diurnal
Agoniates	Y	Exclusive	B/I	Pursue	Diurnal
Acestrorhynchus, Gilbertolus, Gnathocharax, Heterocharax, etc	Y	Exclusive	B/I	Pursue	Diurnal
Salminus	Y	Exclusive	B/I	Pursue	Diurnal
Exodon, Roeboexodon	Y	Exclusive	D/S	Pursue	Diurnal
Oligosarcus	Y	Exclusive	B/I	Pursue	Diurnal
Acestrocephalus, Charax, Cynopotamus, Galeocharax, Roeboides, etc	Y	Facultative	B/I	Pursue	Diurnal
Electrophorus, Gymnotus	Y	Facultative	S/E	Probe	Nocturnal
Cetopsis, Cetopsidium	Y	Exclusive	B/I	Probe	Nocturnal
Lacantunia	Y	Facultative	S/E	Probe	Nocturnal
Brachyplatystoma, Leiarius, Phractocephalus, Pseudoplatystoma, etc	Y	Exclusive	S/E	Probe	Nocturnal
Ageneiosus	Y	Exclusive	S/E	Pursue	Nocturnal
Paracanthopoma, Paravandellia, Vandellia	Y	Exclusive	B/I	Probe	Nocturnal
Potamoglanis	Y	Exclusive	B/I	Probe	Nocturnal
Miuroglanis, Pareidon, Stegophilus, Ochmacanthus, Acanthopoma, etc	Y	Exclusive	B/I, D/S	Probe	Nocturnal
Percilia, Percichthys	Y	Facultative	S/E	Pursue	Diurnal
Plagioscion, Pachyurus, Pachypops	Y	Facultative	S/E	Pursue	Crepuscular
Monocirrhus, Polycentrus	Y	Facultative	S/E	Sit -and-wait	Diurnal
Cichla	Y	Exclusive	S/E	Pursue	Diurnal
Astronotus	Y	Facultative	S/E	Probe	Diurnal
Crenicichla	Y	Facultative	S/E	Pursue	Diurnal
Acaronia	Y	Facultative	S/E	Pursue	Diurnal
Nandopsis	Y	Facultative	S/E	Pursue	Diurnal
Caquetaia	Y	Facultative	S/E	Pursue	Diurnal
Petenia	Y	Facultative	S/E	Pursue	Diurnal
Kronoheros	Y	Facultative	S/E	Pursue	Diurnal
Parachromis	Y	Facultative	S/E	Pursue	Diurnal
Potamorhaphis, Pseudotylosaurus	Y	Facultative	B/I	Pursue	Diurnal
Belonesox	Y	Facultative	B/I	Pursue	Diurnal

Y, yes; ?, undetermined; *S/E*, suction/englulf; *B/I*, bite/impale; *D/S*, dislodge/scrape

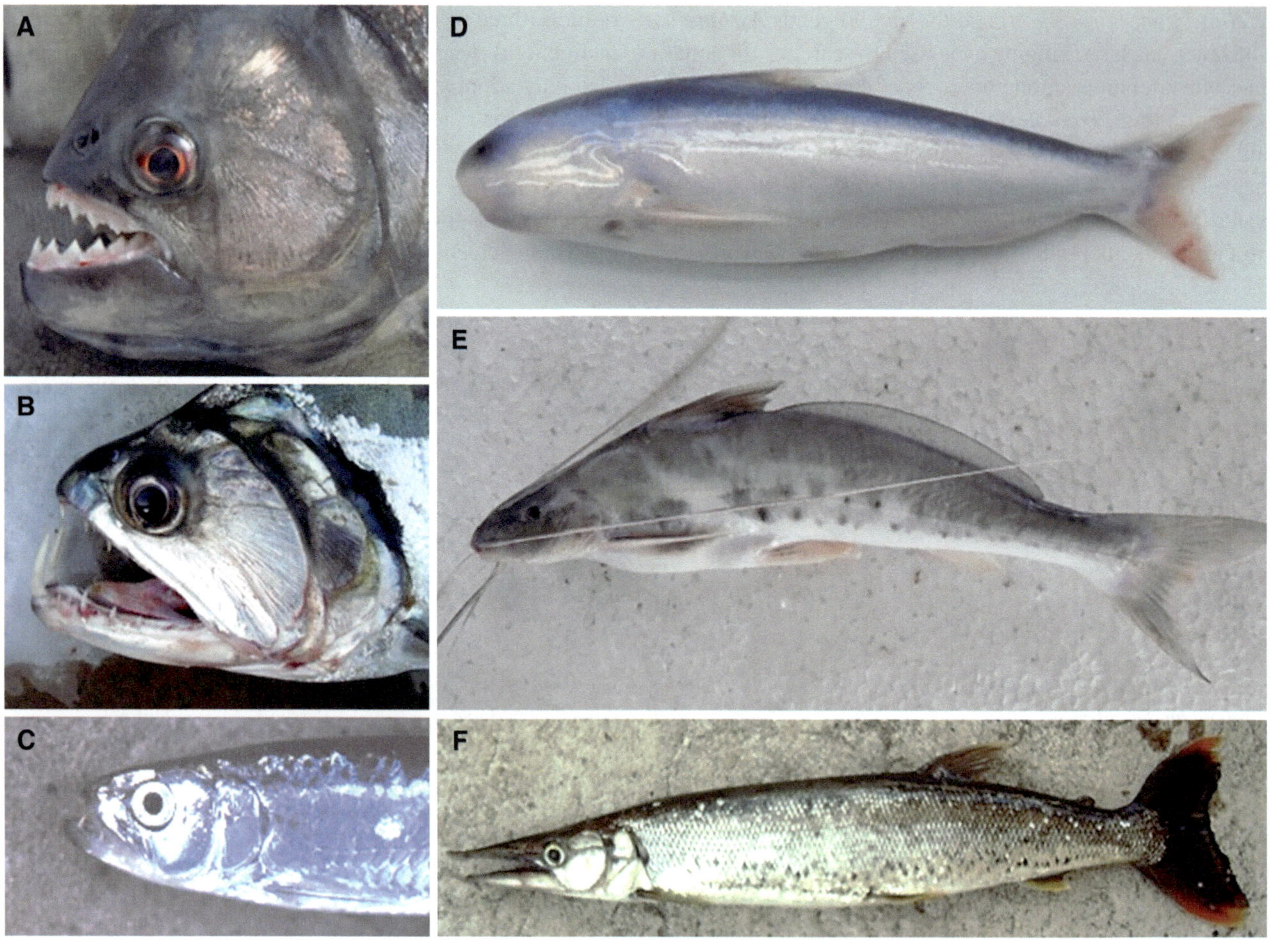

Fig. 17.2 Examples of fish that capture prey by biting and/or impaling: (**A**) *Serrasalmus rhombeus*, Serrasalmidae (lips of specimen removed to reveal sharp, triangular teeth used to bite out pieces of flesh); (**B**) *Hydrolycus armatus*, Cynodontidae (long canine teeth on lower jaw impale prey); (**C**) *Agoniates halecinus*, Triportheidae (grasps prey with conical teeth); (**D**) *Cetopsis coecutiens*, Cetopsidae (bites and spins/rotates the body to remove chunks of flesh); (**E**) *Calophysus macropterus*, Pimelodidae (bites repeatedly to remove chunks of flesh); (**F**) *Boulengerella cuvieri*, Ctenolucidae (grasps prey with numerous, tightly-spaced sharp teeth)

diurnal and nocturnal species increase foraging effort during dawn and dusk, presumably because prey can be more easily ambushed but light intensity remains sufficient for prey detection. Moreover, virtually all fishes classified here as principally crepuscular feeders often feed during day or night.

To illustrate contrasting foraging strategies, 26 piscivorous genera were positioned within a two-dimensional space defined by a gradient of ingestion mode (bite/impale vs. suction/engulf) and a gradient of prey size. Prey size varies from microphagous species feeding mainly on small prey items less than 1 mm largest dimension (e.g., plankton, periphyton, meiofauna) to macrophagous species primarily targeting larger food items with a body size more than 1 mm) (Fig. 17.6). Species were positioned within this space by the first author subjectively and based upon observation of fish feeding behavior and examination of morphology and gut contents involving hundreds of species and thousands of specimens. Several genera represent extremes within this two-dimensional space that contrasts prey size and modes of prey ingestion. *Atractosteus* (gars) and *Hydrolycus* (payaras) are specialists for grasping or impaling large fish, whereas *Arapaima* (pirarucu) and *Brachyplatystoma* (piraíba) use powerful suction to capture large fish. *Belonesox* (pike livebearer) bites to grasp small fish, and *Vandellia* (candiru) bites the gill filaments of larger fishes to ingest blood and gill tissue (microphagy) (Kelley and Atz 1964). *Monocirrhus* (leaffish), *Acaronia* (basketmouth cichlid), and a great many piscivorous fishes use suction to capture small fishes, as well as aquatic macroinvertebrates.

17.4 Discussion

17.4.1 Diversity and Abundance

How does the diversity and abundance of piscivorous fishes in the Neotropics compare with other regions? The Neotropics is home to the highest diversity of freshwater fishes on Earth (Levêque et al. 2008; Albert et al. 2020), but the percentage of

Fig. 17.3 Examples of fish that ingest prey by suction and/or engulfing: (**A**) *Electrophorus electricus*, Gymnotidae; (**B**) *Pellona castelnaeana*, Pristigasteridae; (**C**) *Plagioscion squamosissimus*, Sciaenidae; (**D**) *Ageneiosus dentatus*, Auchenipteridae. (**E**) *Caquetaia kraussii*, Cichlidae; (**F**) *Crenicichla lenticulata*, Cichlidae; (**G**) *Paratrygon aiereba*, Potamotrygonidae

fishes that can be classified as piscivores (about 10%) appears to be comparable to other species-rich regions (Kopf et al. 2021). High latitude regions with low species richness appear to have larger percentages of piscivorous fishes, with cannibalism being common. These high latitude fish faunas have much less functional morphological diversity than tropical freshwater fish faunas, an indication of less ecological diversification (Winemiller 1991, Toussaint et al. 2016).

Some Neotropical clades are entirely piscivorous, e.g., the family Cynodontidae within the Characiformes, genus *Pellona* within the Pristigasteridae, and genus *Cichla* within the Cichlidae. Among orders, the Siluriformes (catfishes) has the highest number of piscivorous genera (49) and species (167). Among families, the Pimelodidae has the most genera (24), followed by the Trichomycteridae (20), and the Cichlidae has the most species (121, which includes many *Crenicichla* spp. that, as mentioned previously, probably consume mostly aquatic macroinvertebrates). Again, we note that it is difficult to classify most NFF according to a binary piscivore versus non-piscivore classification, because many fishes that are largely invertivorous also feed on fish occasionally. For example, studies of stingrays in the genus *Potamotrygon* indicated diets comprised mostly of decapod crustaceans and other aquatic macroinvertebrates, with fish being rarely consumed (Shibuya et al. 2009, Shibuya 2022). We, therefore, excluded *Potamotrygon* species, despite the fact that they are frequently caught on hooks baited with fish. The genera *Heliotrygon*, *Paratrygon*, and *Plesiotrygon* were classified as piscivores based on reports (Shibuya et al. 2009, Shibuya and Zuanon 2013).

17.4.2 Prevalence of Piscivory

Why is piscivory more prevalent in some families than others? In the case of the Pimelodidae and other siluriform families, there appears to be a high degree of niche conservatism with

Fig. 17.4 Examples of fish that dislodge or scrape tissue from other fish: (**A**) *Exodon paradoxus*, Characidae (scale feeder); (**B**) *Roeboides dayi*, Characidae (scale feeder); (**C**) *Ochmacanthus alternus*, Trichomycteridae (mucus feeder)

regards to trophic ecology. Most pimelodids, for example, are piscivores that forage nocturnally using a probing or stalking mode and capturing prey using suction. However, there are some pimelodid genera and species that are invertivorous (e.g., *Hypophthalmus*, *Pimelodus*) and others that, although piscivores, evolved a different mode of prey capture (e.g., *Calophysus macropterus* removes chunks of flesh by biting in a manner similar to that used by piranhas). Pimelodids generally have a large and wide mouth efficient for ingestion of large prey items, and nocturnal habits allowing the ingestion of resting (nearly motionless) prey. By contrast, diurnal piscivores like cichlids have to contend with the evasive actions of active and alert prey fishes.

17.4.3 Niche Conservatism and Evolution

Examples of trophic niche conservatism and trophic niche evolution are widespread throughout the NFF fauna, with examples of phylogenetic stasis, adaptive divergence, and niche convergence observed at multiple levels of the phylogenetic hierarchy (López Fernández et al. 2010, Arbour et al. 2020; Winemiller et al. 2021). Some studies have suggested trophic niche partitioning among coexisting piscivorous fishes in the Neotropics (e.g., Winemiller 1989, Jepsen et al. 1997, Barbarino-Duque and Winemiller 2003, Andrade et al. 2019). Trophic niche convergence is also apparent among piscivorous cichlids that independently evolved among and within tribes, particularly among the Heroini genera in Central America and the Caribbean (Arbour et al. 2020). Convergence in morphology and behavior associated with piscivory also is observed among Neotropical fishes from deeply divergent lineages, as observed in the species ordination in Fig. 17.6 (see also Winemiller 1990a, Winemiller et al. 1995, Montaña and Winemiller 2013; Winemiller et al. 2015).

17.5 Conclusions: Future Studies

Many aspects of piscivory by freshwater fishes remain poorly documented, and among the most basic information that is lacking is dietary data for freshwater fishes in the Neotropics. Only a small fraction of species has been studied, and much of the reported information for others is qualitative and vague. More research is needed that quantifies and compares diet composition (e.g., prey taxa, size, frequency of occurrence,

Fig. 17.5 Examples of fish that pursue, probe, or sit-and-wait to capture prey: (**A**) *Cichla ocellaris monoculus*, Cichlidae (pursue); (**B**) *Acestrorhynchus falcirostris*, Acestrorhynchidae (pursue); (**C**) *Astronotus ocellatus*, Cichlidae (probe); (**D**) *Phractocephalus hemioliopterus*, Pimelodidae (probe); (**E**) *Monocirrhus polyacanthus*, Polycentridae (sit-and-wait); (**F**) *Hoplias aimara*, Erythrinidae (sit-and-wait)

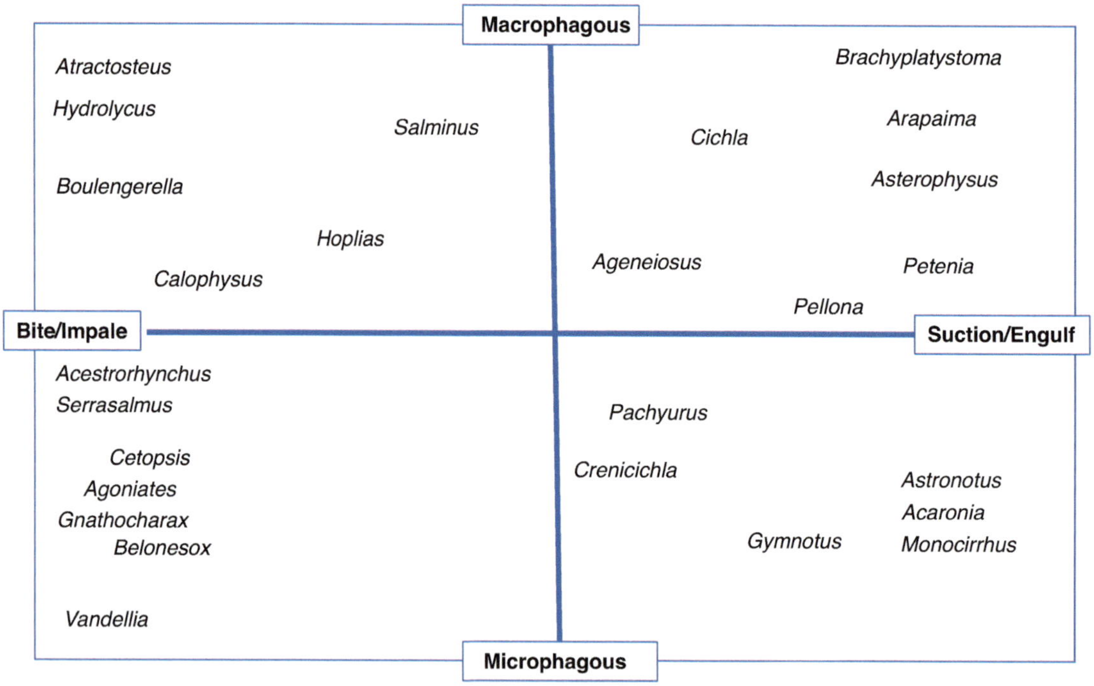

Fig. 17.6 Subjective ordination of piscivorous Neotropical freshwater fish genera in relation to ingestion mode and size of food items ingested. Note the polyphyletic taxonomic composition of each quadrant, indicating assemblages with convergent origins. Note also all these taxa (except *Belonesox*) are present in the Amazon Basin, and that most of these taxa exhibit broad geographic distributions across tropical South America

volumetric proportion, dry weight, etc.) among coexisting species, spatial gradients, time periods, and ontogenetic stages. Estimating diet selectivity based on the relative availability of alternative prey (i.e., functional response of predation) is key to understanding the trophic ecology and dynamics of piscivore populations and local species assemblages. Also needed is more laboratory research on behavior, functional morphology, and mechanics associated with prey capture and ingestion (Gidmark et al. 2019), and developmental constraints during ontogeny and evolution (Hulsey et al. 2005).

Piscivores represent some of the most ecologically and economically important fishes inhabiting Neotropical freshwaters. Large-bodied migratory pimelodid catfishes (*Brachyplatystoma*), the pirarucu or paiché (*Arapaima*), and the tucunaré or peacock bass (*Cichla*) all have high market value in commercial and subsistence fisheries in the Amazon basin. Piscivorous cichlids (*Cichla*, *Petenia*, and *Parachromis*) support important recreational fisheries and international tourism in South and Central America. As apex predators that integrate trophic dynamics throughout the entire food web, piscivorous fishes are useful indicators for the status of communities and ecosystems. Given that apex predator populations tend to be smaller than those of fishes at lower trophic levels, and also tend to be exploited for food or recreation, they are particularly prone to overharvest (Winemiller et al. 2016). Much more research is needed to document stock health and for effective management and conservation of Neotropical freshwater piscivorous fishes.

References

Albert J, Abrahão V, Akin D, Allen J, Ândrade M, Arce M, Armbruster J, Benine R, Bernt M, Bichuette M, Birindelli J, Bragança P, Britto M, Buckup P, Burns M, Calegari B, Carvalho T, Casatti L, Costa W, Crampton W, Dagosta F, Egan J, Fernández L, Figueiredo J, Fitzgerald D, Fontenelle J, Haddad E, Kolmann M, Langeani F, Lehmann P, Leitão RP, Loeb M, Lucena CS, Lucinda P, Lujan N, Malabarba L, Matamoros W, Melo B, Netto-Ferreira A, Pavanelli C, Pelicice F, Pereira E, Plasencia III X, Ramos R, Rocha M, Roxo F, Sant'Anna V, Shibatta O, Slobodian V, Stange M, Tagliacollo V, Tan M, Torgersen K, Winemiller K, Reis R, da Silva R (2025) An ecological trait matrix of Neotropical freshwater fishes. Scientific Data 12:1127. https://doi.org/10.1038/s41597-025-04674-w

Albert JS, Petry P, Reis RE (2011) Major biogeographic and phylogenetic patterns. In: Albert JS, Reis RE (eds) Historical biogeography of neotropical freshwater fishes. University of California Press, Berkeley, pp 21–58

Albert JS, Tagliacollo VA, Dagosta F (2020) Diversification of Neotropical freshwater fishes. Annu Rev Ecol Evol Syst 51:27–53

Amundsen P-A, Damsgaard B, Arnesen AM, Jobling M, Jørgensen E (1995) Cannibalism and prey specialization in Arctic charr, *Salvelinus alpinus* (L.). Environ Biol Fishes 43:285–293

Andrade M, Fitzgerald DB, Winemiller KO, Barbosa PS, Giarrizzo T (2019) Trophic niche segregation among three serrasalmids from rapids of the lower Xingu River, Brazilian Amazon. Hydrobiologia 829:265–280

Arbour JH, Montaña CG, Winemiller KO, Pease AA, Soria-Barreto M, Cochran-Biederman JL, López-Fernández H (2020) Macroevolutionary analyses indicate that repeated adaptive shifts towards predatory diets affect functional diversity in Neotropical cichlids. Biol J Linnean Soc 129(4):844–861

Arrington DA, Winemiller KO, Loftus W, Akin S (2002) How often do fishes "run on empty"? Ecology 83:2145–2151

Barbarino Duque A, Winemiller KO (2003) Dietary segregation among large catfishes of the Apure and Arauca rivers, Venezuela. J Fish Biol 63:410–427

Brazeau MD, Friedman M (2015) The origin and early phylogenetic history of jawed vertebrates. Nature 520(7548):490–497

Burress ED (2016) Ecological diversification associated with the pharyngeal jaw diversity of Neotropical cichlid fishes. J Anim Ecol 85(1):302–313

Cochran-Biederman JL, Winemiller KO (2010) Relationships among habitat, ecomorphology and diets of cichlids in the Bladen River, Belize. Environ Biol Fishes 88:143–152

Correa SB, Winemiller KO, López-Fernández H, Galetti M (2007) Seed consumption and dispersal by fishes–evolutionary perspectives. BioScience 57:748–756

Friedman M, Sallan LC (2012) Five hundred million years of extinction and recovery: a phanerozoic survey of large-scale diversity patterns in fishes. Palaeontology 55(4):707–742

Gidmark NJ, Pos K, Matheson B, Ponce E, Westneat MW (2019) Functional morphology and biomechanics of feeding in fishes. In: Bels V, Whishaw I (eds) Feeding in vertebrates. Springer, Cham, Switzerland, pp 297–332

Hulsey CD, Fraser GJ, Streelman JT (2005) Evolution and development of complex biomechanical systems: 300 million years of fish jaws. Zebrafish 2(4):243–257

Jepsen DB, Winemiller KO (2002) Structure of tropical river food webs revealed by stable isotope ratios. Oikos 96(1):46–55

Jepsen DB, Winemiller KO, Taphorn DC (1997) Temporal patterns of resource partitioning among *Cichla* species in a Venezuelan black-water river. J Fish Biol 51:1085–1108

Kelley WE, Atz JW (1964) A pygidiid catfish that can suck blood from goldfish. Copeia 1964(4):702–704

Kopf RK, Yen JD, Nimmo DG, Brosse S, Villéger S (2021) Global patterns and predictors of trophic position, body size and jaw size in fishes. Global Ecol Biogeogr 30(2):414–428

Layman CA, Winemiller KO, Arrington DA, Jepsen DB (2005) Body size and trophic position in a diverse tropical food web. Ecology 86(9):2530–2535

Levêque C, Oberdorff T, PAUGy D, Stiassny MLJ, Tedesco PA (2008) Global diversity of fish (Pisces) in freshwater. Freshwater Animal Diversity Assessment 595(1):545–567

López Fernández H, Winemiller KO, Honeycutt RL (2010) Multilocus phylogeny and rapid radiations in Neotropical cichlid fishes (Perciformes: Cichlidae: Cichlinae). Mol Phylogenet Evol 55(3):1070–1086

Montaña CG, Winemiller KO (2009) Comparative feeding ecology and habitats use of *Crenicichla* species (Perciformes: Cichlidae) in a Venezuelan floodplain river. Neotrop Ichthyol 7(2):267–274

Montaña CG, Winemiller KO (2013) Evolutionary convergence in Neotropical cichlids and Nearctic centrarchids: evidence from morphology, diet and stable isotope analysis. Biol J Linnean Soc 109:146–164

Montaña CG, Layman CA, Winemiller KO (2011) Gape size influences seasonal patterns of piscivore diets in three Neotropical rivers. Neotrop Ichthyol 9:647–655

Myers GS (1938) Fresh water fishes and West Indian zoogeography. Annual Report of the Board of Regents of the Smithsonian Institution 92:339–364

Pease AA, Mendoza-Carranza M, Winemiller KO (2018) Feeding ecology and ecomorphology of cichlid assemblages in a large Mesoamerican river delta. Environ Biol Fishes 101:867–879

Peterson CC, Winemiller KO (1997) Ontogenetic diet shifts and scale-eating in *Roeboides dayi*, a Neotropical characid. Environ Biol Fishes 49:111–118

Peterson CC, Keppeler FW, Saenz DE, Bower LM, Winemiller KO (2017) Seasonal variation in fish trophic networks in two clear-water streams in the Central Llanos region, Venezuela. Neotrop Ichthyol 15(2):e160125

Shibuya A (2022) A review of the ecological role of the Neotropical freshwater stingrays (Chondrichthyes: Potamotrygonidae). Food Webs 32:e00244

Shibuya A, Zuanon J (2013) Catfishes as prey items of potamotrygonid stingrays in the Solimões and Negro rivers, Brazilian Amazon. Biota Neotropica 13(1):376–379. http://www.biotaneotropica.org.br/v13n1/en/abstract?short-communication+bn03213012013

Shibuya A, Araújo MLG, Zuanon JAS (2009) Analysis of stomach contents of freshwater stingrays (Elasmobranchii, Potamotrygonidae) from the middle Negro River, Amazonas, Brazil. Pan-Am J Aquat Sci 4(4):466–475

Soria-Barreto M, Rodiles-Hernández R, Winemiller KO (2019) Trophic ecomorphology of cichlid fishes of Selva Lacandona, Usumacinta, Mexico. Environ Biol Fishes 102(7):985–996

Spotte S (2002) Candiru: life and legend of the bloodsucking catfishes. Creative Arts Book Co., Berkeley, CA

Pereira LS, Agostinho AA, Winemiller KO (2017a) Revisiting cannibalism in fishes. Rev Fish Biol Fish 27:499–513. https://doi.org/10.1007/s11160-017-9469-y

Pereira LS, Keppeler FW, Agostinho AA, Winemiller KO (2017b) Is there a relationship between fish cannibalism and latitude or species richness? PLOS One 12(1):e0169813. https://doi.org/10.1371/journal.pone.0169813

Toussaint A, Charpin N, Brosse S, Villéger S (2016) Global functional diversity of freshwater fish is concentrated in the Neotropics while functional vulnerability is widespread. Sci Rep 6:22125. https://doi.org/10.1038/srep22125

Wainwright PC, Day SW (2007) The forces exerted by aquatic suction feeders on their prey. J R Soc Interface 4(14):553–560

Wainwright PC, Ferry-Graham LA, Waltzek TB, Carroll AM, Hulsey CD, Grubich JR (2001) Evaluating the use of ram and suction during prey capture by cichlid fishes. J Exp Biol 204:3039–3051

Westneat MW (2006) Skull biomechanics and suction feeding in fishes. In: Shadwick RE, Lauder G (eds) Fish biomechanics. Fish physiology. Academic Press, San Diego, pp 29–75

Williams JD, Winemiller KO, Taphorn DC, Balbas L (1998) Ecology and status of piscivores in Guri, an oligotrophic tropical reservoir. North Am J Fish Manage 18:274–285

Winemiller KO (1989) Ontogenetic diet shifts and resource partitioning among piscivorous fishes in the Venezuelan llanos. Environ Biol Fishes 26:177–199

Winemiller KO (1990a) Spatial and temporal variation in tropical fish trophic networks. Ecol Monograph 60:331–367

Winemiller KO (1990b) Caudal eyespots as deterrents against fin predation in the neotropical cichlid *Astronotus ocellatus*. Copeia 1990:665–673

Winemiller KO (1991) Ecomorphological diversification of freshwater fish assemblages from five biotic regions. Ecol Monograph 61:343–365

Winemiller KO, Yan HY (1989) Obligate mucus feeding in a South American trichomycterid catfish. Copeia 1989:511–514

Winemiller KO, Fitzgerald DB, Bower LM, Pianka ER (2015) Functional traits, convergent evolution, and periodic tables of niches. Ecol Lett 18:737–751

Winemiller KO, Kelso Winemiller LC, Montaña CG (2021) Peacock bass: diversity, ecology and conservation. Academic Press/Elsevier, London, p 305

Winemiller KO, Kelso-Winemiller LC, Brenkert AL (1995) Ecomorphological diversification and convergence in fluvial cichlid fishes. Environ Biol Fishes 44:235–261

Winemiller KO, Humphries P, Pusey BJ (2016) Protecting large apex predators. In: Closs G, Krkosek M, Olden JD (eds) Conservation of freshwater fishes. Cambridge University Press, Cambridge, UK, 361–398

Winemiller KO, Zeug SC, Robertson CR, Winemiller BK, Honeycutt RL (2011) Food web structure of coastal streams in Costa Rica revealed by dietary and stable isotope analyses. J Trop Ecol 27:463–476

Zuanon J, Sazima I (2005) The ogre catfish: prey scooping by the auchenipterid *Asterophysus batrachus*. J Ichthyol Aquat Biol 10(1):15–22

Ectoparasitic Neotropical Freshwater Fishes 18

Matthew A. Kolmann, Karine O. Bonato, Kory M. Evans, Jonathan M. Huie and Luiz R. Malabarba

Abstract

The study of parasitic behaviors in Neotropical freshwater fishes, particularly hematophagy, began with early nineteenth-century descriptions of trichomycterid catfishes. These fishes, notably *Vandellia* species, became notorious for their reported habit of entering human excretory openings, although scientific evidence was scarce. Documented cases of blood-feeding by these catfishes have since been confirmed, albeit are rare occurrences. Other Neotropical fishes, including characiforms and gymnotiforms, also display various ectoparasitic behaviors, such as eating scales, fins, and mucus from other fishes. These feeding strategies have evolved multiple times across and within different fish lineages and offer insights into the ecological complexity and functional diversity of ectoparasitic fishes. Sazima (1983) conducted one of the first comparative studies on the behaviors employed by scale-feeding fishes and proposed several evolutionary pathways that could lead to scale-eating (lepidophagy) in fishes. Hypotheses regarding trophic-mediated origins for ectoparasitism, particularly where ectoparasites evolve from piscivores, do not seem well supported. Multiple different evolutionary pathways are probable given the diversity of lineages engaged in ectoparasitism; however, intraspecific aggression may be a putative origin for scale- and fin-feeding in characiforms. Study into the nuances of ectoparasite ecology, behavior, and phenotypes is still in its infancy and remains an area ripe for further study. Advances in genomic techniques, specifically metabarcoding methods for analyzing ectoparasite diet preferences, could offer new insights into the evolutionary dynamics of these specialized feeding strategies.

M. A. Kolmann (✉)
University of Louisville, Louisville, KY, USA
e-mail: mkolmann@gmail.com

K. O. Bonato · L. R. Malabarba
Universidade Federal do Rio Grande do Sul, Porto Alegre, RS, Brazil
e-mail: kakabio2005@yahoo.com.br

L. R. Malabarba
e-mail: malabarb@ufrgs.br

K. M. Evans
Rice University, Houston, TX, USA
e-mail: Kory.Evans@rice.edu

J. M. Huie
George Washington University, Washington, USA
e-mail: jonathanmhuie@gmail.com

18.1 What is an Ectoparasite?

Fishes that engage in feeding behaviors like pterygophagy (fin-feeding), lepidophagy (scale-feeding), mucophagy (mucus-feeding), and hematophagy (blood-feeding) (Fig. 18.1) are frequently referred to as either ectoparasites or micropredators. The Neotropical fishes that engage in any of the above feeding modes are all generally smaller than their prey/host and while they are certainly detrimental to the overall health of their prey, rarely (if ever) do these fishes kill their prey outright. Advocates of the "micropredator" label typically seize on the former phenomenon (predator smaller than prey), while those using "ectoparasite" usually point to the latter (parasites do not kill prey) to support their respective monikers. This latter caveat is particularly important because it separates lepidophages, pterygophages, etc. from more "normal" predators and parasitoids, which kill their prey or hosts, thereby leaving parasite as a remaining option for those with discerning tastes in ecological nomenclature.

Specifically, **ectoparasites** are defined popularly as *any parasites which attach or feed from the outside of the host (or live on the outside of the host)* while **parasites** are more broadly defined as: *an organism that lives in or on an organism of another species (its host) and benefits by deriving nutrients at the other's expense.* Arguments against the ectoparasite label for fishes like scale-feeders would point out several caveats: (1) parasites are commonly thought to use only a single host

J. S. Albert and R. E. Reis (eds.), *Ecology and Evolution of Amazonian Fishes*, Fascinating Life Sciences,
https://doi.org/10.1007/978-3-032-09328-8_18

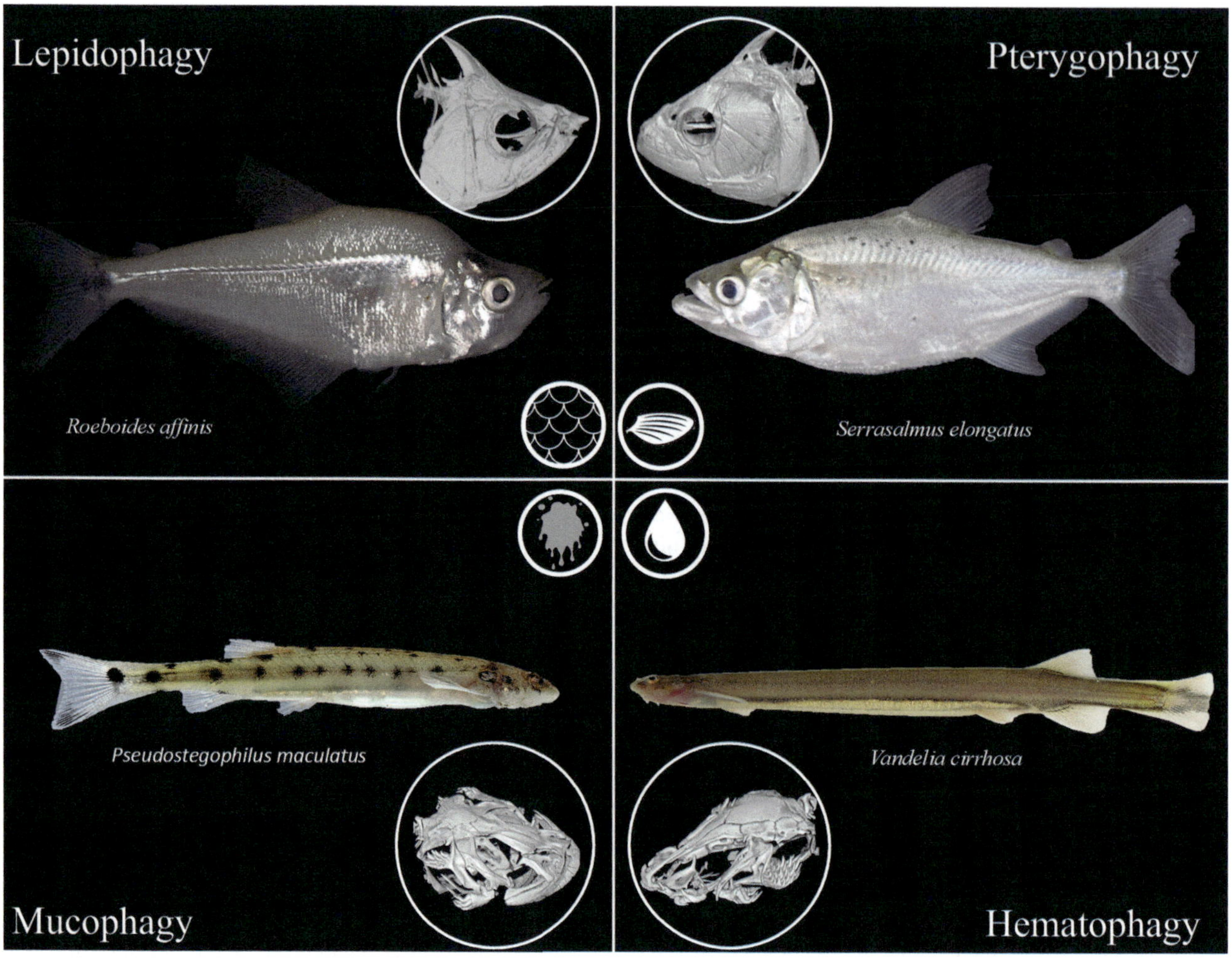

Fig. 18.1 Diverse cranial and postcranial phenotypes in Neotropical freshwater ectoparasitic fishes. Top left: hunchback or scale-eating characin *Roeboides* (Characidae: Characiformes). Top right: pike piranha *Serrasalmus elongatus* (Serrasalmidae: Characiformes). Bottom left: pencil catfish *Acanthopoma annectens* (Trichomycteridae: Siluriformes). Bottom right: the vampire catfish or candiru *Vandellia cirrhosa* (Trichomycteridae: Siluriformes). Insets are volume renders of the skull from each taxon, using microcomputed tomographic imaging (microCT). Little overt convergence is obvious in ectoparasitic fishes, except perhaps modifications to the dentition

individual to complete some distinct phase of its life cycle, and relatedly (2) that parasites consume nutrients from a single host individual (or different hosts at distinct life cycle stages), while a predator feeds on multiple prey. Finally, (3) ectoparasites are assumed to typically spend their entire lives living on their (single) host. Fin-, mucus-, blood-, and scale-feeding fishes obviously do not fit any of these three prerequisites, but neither do many widely acknowledged parasites, like ticks (Acaridae), for example. Curiously, authors like Leung (2014) derive hematophagous catfishes as parasites, but not scale-feeders (although they are not consistent; Leung 2014, p. 7). Leung's categorization also ignores that trichomycterid catfishes feed on multiple prey throughout their lives and do not required a host to aid in the completion of some life cycle phase.

So, are lepidophagous or hematophagous fishes either ectoparasites or micropredators? To avoid controversy, we acknowledge that the differences among parasites, parasitoids, and predators are a matter of gradation. The idea that micropredators are smaller than their prey is vague to the point of being useless—stoats (Mustelidae), for example, hunt and kill prey some four to five-times greater in mass than themselves, yet are not considered micropredators. What is a suitable body size threshold for distinguishing between micropredators and predators? Conversely, is it then reasonable to call whales or basking sharks "macropredators" because these animals feed on prey much smaller than themselves? Alternatively, we and others proposed the idea that fishes like pterygophages and lepidophages are essentially carnivorous "grazers" (Northcote et al. 1987; Nico and Taphorn 1988; Kolmann et al. 2018). After all, these fishes forage by removing superficial fleshy layers, leaving the underlying substrate untouched, making the target prey resource replenishable and abundant, like herbivorous grazers. For convenience, we will refer to those fishes that engage in mucus-, scale-, fin-, or blood-feeding as ectoparasites throughout the rest of the chapter.

18.2 Hematophagy or Sanguinivory

First, it is worth noting just how rare blood-feeding vertebrates are—to our knowledge, this ecology has only evolved about four to six times across some 70,000+ species of vertebrates; namely, in birds (vampire finches and oxpeckers), mammals (vampire bats), jawless fishes (lampreys), and some catfishes. Scientific study of parasitic Neotropical freshwater fishes begins with narratives of trichomycterid catfishes by Martius (in Spix and Agassiz 1829), which describe little fishes that enter the excretory openings of humans or other animals. Documentation of scale-feeding by Neotropical freshwater fishes came after descriptions of parasitism in catfishes, with Kner (1860, p. 34) observing scales ingested by the Wimple piranha *Catoprion* (Roberts 1970). In this section, we first discuss the history of blood-feeding in catfishes and then the history of work on scale- and fin-feeding, given the historical precedence of the former.

After Martius first discussed parasitism in catfishes, the few published accounts in the nineteenth and beginning of the twentieth centuries gave attention to anecdotal but undocumented reports of fishes entering and ascending the urethra of persons while bathing, causing inflammation, amputation, and sometimes death (Castelnau 1855; Günther 1868; Boulenger 1897; Pellegrin 1909; Haseman 1911; Eigenmann 1918). However, (1) the lack of captured specimens showing this "urinophilous habit" (Eigenmann 1918) and the (2) assignment of the local common name "Candiru" for both large species of the genus *Cetopsis* (Cetopsidae) as well as to the small species of *Vandellia* (Pellegrin 1909; Eigenmann 1918), led Martius to consider these fishes as possible juvenile cetopsids or an unnamed smaller species. The description of these catfishes' habits of entering the excretory opening of humans or other animals is a clear reference to vandelliine trichomycterids, which were undescribed in Martius' time. The anecdotal report of penetration into urethral openings of unprotected bathers has made these fishes notorious in medical literature, despite the lack of documented scientific reports (Bauer 2013; Breault 1991; Gudger 1930; Toro et al. 2001). Documented reports of urethra penetration and of blood feeding on humans by vandelliines have been provided recently by Haddad (2016) and Haddad Junior et al. (2021), respectively. However, there is only one report of urethral penetration (with some controversy over how the fish entered the urethra) and then only infrequent (but well-documented) reports of blood feeding on humans by catfishes.

After Martius, some reports were made of *Vandellia* collected inside the branchial chamber of large catfishes. Based on these reports and on the morphology of the teeth and opercular and interopercular spines of *Vandellia wieneri* [= *Plectrochilus wieneri* (Pellegrin 1909)], Pellegrin (1909) hypothesized that these various spines serve several purposes for vandelliines: (1) for attachment to the gills of their host, (2) as well as facilitating the locomotion of the animal in a narrow duct (for example, between the gill lamellae of *Pseudoplatystoma*), and finally, (3) to make such wounds that an abundant flow of blood for feeding is possible. Pellegrin (1909) then hypothesized that *Plectrochilus wieneri* are not, as Günther thought, mere commensals of the great catfishes on which they usually live, stating that *The vandellines therefore represent, among the catfishes, the last term of specialization with a view to a most characteristic parasitism* [Les Vandellies representent done, chez les Silurides, le dernier terme de la specialisation en vue d'un parasitisme des plus caracterises.]

The first documented reports of parasitic blood-feeding behavior were made by Kelley and Atz (1964) who demonstrated, in aquaria, that *Vandellia* invading a goldfish's gill chamber and blood-feeding. Similarly, Machado and Sazima (1983) reported *Branchioica bertonii* (= *Paravandellia oxyptera* Miranda Ribeiro 1912) entering and feeding on blood inside the gill chambers of the large catfish *Pseudoplatystoma* spp. Later, *Vandellia cirrhosa* and *V. sanguinea* have been demonstrated in laboratory experiments to bite mostly at the ventral or dorsal arteries inside the branchial chamber of *Colossoma macropomum*, hypothesizing that the catfishes rapidly engorged themselves by using the hosts' own blood pressure—whereby blood is pumped directly into the gut of the parasite (Zuanon and Sazima 2004). Most of the information on ecological interactions of vandelliines has been based on incidental observations of vandelliines in the gill chambers or partly buried into the skin of large freshwater fish host species, such as pimelodid and doradid catfishes, serrasalmids, and freshwater stingrays (Lasso et al. 2015; Schmidt 1993; Zuanon and Sazima 2005). Additional reports document catfishes found in the wound of a caiman, an electric eel, and two small characid fishes (Ribeiro 1954). Based on DNA barcoding analysis of stomach contents in *Paravandellia oxyptera* obtained from natural environments, Bonato et al. (2022) have demonstrated that this vandelliine shows a much larger spectra of ecological interactions than previously reported. Bonato et al. (2022) confirmed that catfishes fed on eight different host species, putatively selected by host body size, from five families in the orders Characiformes, Gymnotiformes, and Siluriformes.

18.3 Scale-, Fin-, and Mucus-Feeders

After Kner's initial reports on scale-feeding, later records by Breder (1927) and Géry (1963) described scale-feeding by characiforms like *Roeboides* and *Exodon*, but Géry at least believed scales to be ingested incidentally or infrequently, rather than as a primary source of nutrients (Roberts 1970). Scales are a relatively common item found in stomach contents in several groups of fishes but as an occasional and less relevant item in most fishes' diets. Among 250 species from the Rio Negro, for example, Goulding et al. (1988) found scales in the stomachs of 100 or 40% of these species. Scales or fins may appear in fish stomachs as part of a fish that was swallowed entirely or in part, removed from carrion, or scavenged from the detritus layer (Goulding et al. 1988). Scales are commonly found in the stomachs of stethaprionine and stevardiine characids, but these incidences have been considered the results of aggressive encounters among schooling individuals (Sazima 1983; Arcifa et al. 1991; Vilella et al. 2002). It is necessary to separate these records from specialized lepidophagous fishes adapted to remove and feed on scales of other living fishes. In *Bryconexodon juruenae*, for example, scales may represent more than 80% of the stomach volume, along with smaller contributions of insect, algae, and leaves (Godoi 2004). While vandelliines have been known for their hematophagous habit, the representatives of the trichomycterid subfamily Stegophilinae have been reported as mucus and scale-feeders, as well as outright flesh-feeders (Eigenmann and Allen 1942; Roberts 1972; Baskin et al. 1980; Machado and Sazima 1983; Winemiller and Yan 1989; Datovo and Bockmann 2010; de Pinna 1998, 2016). The analysis of gut contents of *Homodiaetus anisitsi* and *Pseudotegophilus maculatus* has shown these stegophilines are obligatory mucus feeders, which has also been observed in *Ochmacanthus* by Winemiller and Yan (1989).

Here, we attempt to discriminate between species for which there is anecdotal evidence for ectoparasitism (i.e., fins, scales, etc., present in stomachs), from those species that are demonstrably ectoparasitic, based on considering both ecological (e.g., gut content data) and behavioral (in vivo observations of ectoparasitism) evidence. For instance, scale-feeding has been reported in Characidae (Characinae, Probolodini, Exodontinae, Stethaprioninae, and putatively Stevardiinae), Serrasalmidae, Anostomidae, and Erythrinidae, while fin-feeding has been reported in Serrasalmidae and Characidae (as well as African Alestidae and Distichodontidae). More recently, lepidophagy and pterygophagy have been confirmed in some gymnotiforms in the family Apteronotidae (Roberts 1970; Evans et al. 2017, 2019). However, isolated reports of scale-feeding behaviors have also been documented in some species of Cichlidae (Kramer and Bryant 1995), Ariidae (Hoese 1966; Szelistowski, 1989), Heptapteridae (Kramer and Bryant 1995), Callichthyidae (Zardo and Behr 2013), Rivulidae (Keppeler et al. 2013), and mucus-feeding in Doradidae (Zaret 2013). In this chapter, we do our best to discriminate between anecdotal or incidental "ectoparasitism" (for instance, where scales have been reported in stomach contents) versus those species that have been observed actively parasitizing other fishes (in addition to having ectoparasitic habits confirmed by diet studies). This distinction should highlight the necessity for both ecological and behavioral studies on the natural history of Neotropical freshwater fishes.

18.4 Taxonomic Diversity

The number of ectoparasitic fishes listed in the diet database (Chap. 6) is 117 species, including the estuarine species *Oligoplites saurus* (Carangidae; Sazima and Uieda 1980). We chose to treat reports of ectoparasitism (exclusively reports of scale-feeding) in Ariidae, Callichthyidae, Cichlidae, Doradidae, Rivulidae, and Tetragonopterinae as incidental, without behavioral evidence or further replication. Considering only those lineages for which we have multiple lines of evidence for ectoparasitism (behavioral, morphological, and/or ecological), these trophic niches have evolved some 13–20 times in Neotropical fishes (and ten to15 times in characiforms alone). This includes scale-, fin-, mucus-, or blood-feeding in orders, such as Gymnotiformes (e.g., *Sternarchella*), Characiformes (e.g., *Serrabrycon*, *Landonia*), Siluriformes (e.g., Vandelliinae), and Cyprinodontiformes (e.g., *Belonesox*). Of these 138 total, the vast majority (97 species) are lepidophages, while 21 species (all piranhas) are pterygophages, and 20 species, all pencil catfishes, are hematophagous. Diagnosing how many of these lineages also feed on mucus is difficult, given that most lepidophages and pterygophages are also consuming mucus (and probably quite intentionally).

Outside the Neotropics, there are documented freshwater (and marine) ectoparasites arising not only in many of the same orders as described above, like Characiformes (*Distichodus*, *Phago*, *Belonophago*, etc.), Siluriformes, Cyprinodontiformes (*Cyprinodon*, *Belonesox*; Martin and Wainwright 2011, 2013; Marchio and Piller 2013), but also in Carangiformes (*Echeneis*, *Oligoplites*; Strasburg 1959; Major 1973), Cichliformes (*Perissodus*, *Plecodus*; Fryer and Iles 1955), Centrarchiformes (*Terapon*), Gobiiformes (*Elacatinus*, *Tigrigobius*; Huie et al. 2020), Labriformes (*Labroides*), Blenniiformes (*Plagiotremus*; Bshary and Bshary 2012; *Aspidontus*, fin-feeder; Fujisawa et al. 2020), Tricanthodidae like *Tydemania* (Mok 1978) and *Macrorhamphosodes* (Nakae and Sasaki 2002), as well as some clades ambiguous in their higher-level taxonomic affiliations like Ambassidae (*Chanda*; Grubh and Winemiller 2004). For the most part, these lineages are tropical and freshwater, begging the question as to whether this bias is biological or methodological. Similarly, it would appear that the majority of freshwater ectoparasites follow a Gondwanan distribution, with the greatest diversity of these taxa found in South America and Africa, although this is heavily biased toward otophysans. In any case, ectoparasitic fishes are known from faunas worldwide and contribute to the ecological complexity of these regions and communities. These generalities pose a few questions: Why are ectoparasitic fishes so uncommon? How worthwhile are ectoparasitic niches for fishes in general?

18.5 Energetic Benefits

Fins and scales appear to provide very similar nutrient profiles, which perhaps is no surprise given they share similar skeletal constituents (acellular bone) and are coated with mucus (Nico and Morales 1994; Fig. 18.2). Many characiforms occasionally feed on scales as juveniles, to supplement their diets, perhaps due to the mineral-limited nature of blackwater and clearwater rivers (Goulding 1980; Goulding et al. 1988). Considering that mucosal dry weight is high in both carbohydrates and proteins, while scales (and by proxy, fins) are predominantly calcium phosphate (hydroxyapatite), these prey items are presumably nutritious while being readily digestible and perhaps more easily digested than chunks of fish flesh or whole bones (Nico and Morales 1994). Finally, fins, scales, and mucus are abundant and renewable resources, given the diversity of fishes in South American river systems (Northcote et al. 1987; Nico and Taphorn 1988).

Comparable research on the digestive or ecophysiological adaptations for blood-feeding in catfishes is lacking. However, if these catfishes are expected to be at least partially convergent with other hematophagous vertebrates, like vampire bats, blood-drinking finches, or oxpecker birds, we might expect to see similar solutions to the shared challenges posed by hematophagy, like: (1) blood viscosity and coagulation, (2) low nutritional value of blood, and relatedly (3) the difficulty in refining sugars and fats chemically bound in blood, (4) osmoregulatory and renal stresses, due to high salt and protein concentrations in blood, (5) iron assimilation and excretion, and (6) disease and immunological stress (Zepeda Mendoza et al. 2018). Based on these challenges, we might expect similar adaptations for blood-feeding in catfishes regarding increased gill and renal function, anticoagulant production, and metabolic or behavioral modifications to combat nutrient starvation and increase lipid and glucose assimilation (Zepeda Mendoza et al. 2018).

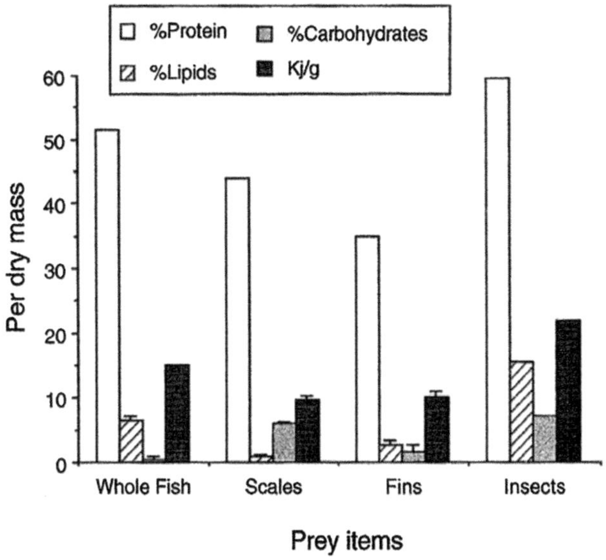

Fig. 18.2 Macromolecular constituents of an ectoparasitic diet vs. other common freshwater fish prey items. From Nico and Morales (1994). Note how feeding on scales and/or fins still provides considerable nutritional benefits to ectoparasitic fishes, similar to feeding on whole fishes or insects

18.6 Trophic Position

Parasites are rarely included in food-web studies, but community ecology research has demonstrated the importance of these organisms in understanding ecological interactions (Hernandez and Sukhdeo 2008; Lafferty et al. 2006; Marcogliese 2003; Sukhdeo 2010, 2012). Lafferty et al. (2008) argue that parasites can alter food-web topology, changing chain length, connectivity, and robustness. For example, species known as top predators are not the highest trophic level, if they are parasitized. Parasites can alter the process and function of food webs, affecting host survival, or by inducing stress that affects biological processes, such as reproduction and growth, and by increasing host vulnerability to predators (Lafferty et al. 2006). So, the exclusion of parasites from food-web studies leaves the evolutionary and ecological implications of parasite mediation in trophic interactions unknown, which also makes it difficult to understand parasite–host interactions in a community context (Sukhdeo and Hernandez 2005; Sukhdeo 2012). In order to mitigate these deficits in our understanding and investigate these complex relationships, it is first necessary to answer the question "Who is the host?" for any given parasite or community of parasites.

The lack of information about host identities limits our understanding of the role of parasites in food webs; however, recent studies have begun to use modern molecular tools to identify hosts. The origin of mucus, scales, and blood typically cannot be traced by a simple visual inspection; therefore, precise molecular techniques are needed, as well as the use of DNA extraction and amplification protocols for stomach contents that minimize DNA contamination between food sample and the ectoparasite itself. An efficient identification of hosts was obtained in some studies with Neotropical fishes (Bonato et al. 2018; Bonato et al. 2022). These studies indicated that mucophagous, scale-feeding, and blood-feeding fishes interact with a wide range of species in their community, although mucus- and scale-feeders showed a predilection for characiform hosts.

Ectoparasitism may also be a seasonal feeding strategy as fishes face the wax and wane of different stressors in their environments. Neotropical freshwaters experience seasonal fluctuations in water level that can increase available foraging habitat (wet season) or severely restrict foraging habitat and compress fishes together (dry season). In the genus *Roeboides*, individuals may facultatively feed on scales as they lose access to marginal floodplain habitat during dryer months (Peterson and McIntyre 1998; Peterson and Winemiller 1997). These seasonal fluctuations in lepidophagy may temporarily inflate the trophic positions of any species that engage in ectoparasitic behavior. Species that facultatively engage in parasitic behaviors also possess derived traits that allow them to feed on scales while also retaining traits that allow them to switch back to more general prey resources. This pattern was observed in the gymnotiform genus *Sternarchella* (Apteronotidae), in which some species feed facultatively on scales while also incorporating invertebrate prey (Evans et al. 2019). These observations on seasonality and facultative feeding in ectoparasites recalls Liem's Paradox—that even species with specialized traits can exploit a wide range of ecological niches or resources (Robinson & Wilson, 1998)—making ectoparasites a potent model for testing the "Paradox" in situ.

While dietary modes like lepidophagy and hematophagy have received a fair amount of attention, comparatively little has been written regarding fin- or mucus-feeding, especially among Neotropical freshwater fishes. For the rest of this chapter, we chose not to discuss scale-feeding separate from mucus-feeding, as mucus presumably contributes much, if not most, of the energetic gains from scale-feeding (Whitfield and Blaber 1978; Vieira and Gery 1979; Nico and Morales 1994; Bonato et al. 2018). Similarly, where we discuss lepidophagous clades that also contain known fin-feeders (e.g., in piranhas), we will discuss both these topics in parallel. Whether discussing fin-feeders, blood-feeders, or mucophages—we acknowledge that we know very little regarding how these and other parasites participate in energy transfer across and within communities, even whole landscapes.

18.7 Functional Diversity

From blood- to scale-feeding, ectoparasitic fishes are regarded as some prime examples of ecomorphological "specialization," arising from their seemingly nuanced niches and behavioral ecologies. To this end, the morphological taxonomy of many ectoparasitic fishes has been obfuscated by autapomorphies: unique, or novel phenotypes which leave little information for traditional systematics to work with (Vari 1986; Mattox and Toledo-Piza 2012). Moreover, many taxa are phenotypically and ethologically convergent, which misled early morphological attempts to categorize taxa like *Serrabrycon* and *Bryconexodon*, for example. However, these trends do not mean that ectoparasites are less diverse in terms of morphology or behavior. Rather, in comparison with their nearest, non-parasitic relatives, ectoparasites exhibit wonderful examples of phenotypic novelty (Vari 1986; Adriaens et al. 2010; Kolmann et al. 2018).

Are scale feeders diverse with regard to whom or what they eat? Some anecdotal evidence has suggested selectivity in prey resources used by scale-feeders. For example, fin-feeding piranhas (e.g., *Serrasalmus irritans, S. altuvei*) ate fins from a wide variety of prey, (cichlids, characids, and curimatids, even other serrasalmids like *Pygocentrus*), but predominantly ate fins that were similar in size to their own (Nico and Taphorn 1988; Nico and Morales 1994). Wimple piranhas (*Catoprion*) appear to feed on large scales, or more appropriately, fishes with proportionately larger scales, while other serrasalmids and characiform lepidophages feed on smaller scales (Kolmann et al. 2018; Kolmann unpublished data). Some lepidophagous characiforms, e.g., *Roeboides* or *Probolodus*, sometimes stalk, mimic, and prey upon relatives which they closely resemble (aggressive mimicry) (Sazima 1977; Sazima 1983). Similarly, African lepidophagous cichlids predominantly consume the scales of other cichlids, sometimes aggressively mimicking species that appear similar only in deeper waters (Nshombo et al. 1985; Kovac et al. 2019).

Mucus-, scale-, and blood-feeding catfishes of the family Trichomycteridae reportedly feed on other, larger catfishes (Baskin et al. 1980; Adriaens et al. 2010; Zuanon and Sazima 2005)—although blood-feeding vandellines are also known to feed on piranhas and pacus (Zuanon and Sazima 2003). However, these findings are biased from only a few field observations, while recent studies using molecular data have demonstrated that these catfish predators feed widely within their community (Bonato et al. 2018). There appears to be a selectivity behavior to scale-feeders, which is related to the size of the prey. Bonato et al. (2018) found that the mucus in the stomach of *Homodiaetus anisitsi* and *Pseudostegophilus maculatus* belonged to both large and small-size species, but the presence of scales was restricted to the small species. In other words, small scales of small hosts are ingested along with mucus, but scales of large hosts are never ingested—only the mucus is found in the stomachs. Therefore, these species were found to be obligatory mucus-feeders and occasionally (or incidentally) lepidophagous (Bonato et al. 2018). We find it interesting that the feeding targets of scale-/mucus- feeding characiforms appear to differ from scale-/mucus- feeding catfishes. While the former appear to target scales and incidentally ingest mucus and skin (akin to scales as "buttered toast" covered with mucus), the latter have smaller, scraping teeth to target mucus and incidentally ingest small scales (more similar to eating butter that is littered with breadcrumbs). Future studies using DNA barcoding methods (see Kovac et al. 2019) could elucidate prey selectivity in Neotropical freshwater fishes and help to explore if prey preference trends vary with season, community assemblage, or other ecological factors.

Lepidophagous otophysans share perhaps one ubiquitous trait among them: specialized teeth, which are often robust, raptorial, or projecting at odd angles from the jaws and skull (Sazima 1983; Fig. 18.3). For example, characid lepidophages like the Exodontinae (*Exodon*, *Bryconexodon*, and *Roeboexodon*), *Roeboides* (a characin), and *Serrabrycon,* share in common large, procumbent teeth used to scrape scales from prey (Hahn et al. 2000; Novakowski et al. 2004; Kolmann et al. 2018; Fig. 18.1). Prey are typically collided with at an angle perpendicular to their flank, with robust, anteriorly (or aborally) oriented teeth (Hata et al. 2011). These teeth presumably resist the compressive impact of the strike as well as the shearing forces that come afterward as the ectoparasite scrapes the dentition along the prey's flanks (Sazima 1983; Sazima and Machado 1990; Janovetz 2005). Similarly, the wimple piranha *Catoprion mento*, relative to other piranhas, has squat, anteriorly angled, multicuspid teeth that are used to impact and then leverage large scales from prey (Janovetz 2005). These teeth look drastically different from the triangular, blade-like teeth in other piranhas; furthermore, *Catoprion* is the only piranha that lacks interlocking teeth or a unilateral tooth replacement mechanism (Kolmann et al. 2019). Most of these scale-feeding characiforms have elongate lower jaws, perhaps for increasing gape height (Whitfield and Blaber 1978; Sazima 1983; Kolmann et al. 2018), as well as what appear to be particularly robust skulls (Huie and Kolmann, personal observation). Similarly, scale-feeding gymnotiforms have elongate lower jaws and large, raptorial teeth, relative to their non-ectoparasitic confamilials (Evans et al. 2017).

To what extent are the phenotypes of lepidophagous fishes fine-tuned to match increased dependencies on scale-feeding? That is a tricky question to answer. Many lepidophagous lineages are only loosely convergent, and rarely diversify beyond one or two species (Fig. 18.4). However, notable genera with multiple species of variable scale feeders (e.g., *Perissodus* and *Roeboides*) lend themselves to comparative investigations. Currently, there are 21 recognized species of *Roeboides* that vary greatly in their morphology and dependency on lepidophagy (Lucena et al. 2003, 2007; Souza et al. 2022). Adult *Roeboides affinis* that consume more scales (based on volumetric proportions) have larger and more external teeth, and a more elongated snout, compared to their sympatric and more facultative scale-feeding congenerics (Novakowski et al. 2004). Similarly, as the proportion of scale-feeding increases over the ontogeny of many *Roeboides* species, their feeding morphology also changes to reflect their more specialized diet (Peterson and Winemiller 1997; Peterson and McIntyre 1998; Hahn et al. 2000; Novakowski et al. 2004; Albrecht et al. 2013). Higher degrees of specialized scale-feeding may also be correlated with more pronounced behavioral and morphological "handedness"—asymmetry in the predator's mouth attributed to consistently using either the right or left side of its jaws to remove scales from a consistent side of its prey. Morphological and behavioral handedness have been observed in *Perissodus* cichlids, *Chanda nama* (Ambassidae), and *Exodon paradoxus* (Characidae) (Grubh and Winemiller 2004; Hata et al. 2011; Lee et al. 2012). Behavioral and morphological handedness were observed in *Roeboides* (Sazima 1983) and *Macrorhamphosodes uradoi* (Triacanthodidae; Nakae and Sasaki 2002), respectively, but additional work is needed to assess the potential correlation between trophic specialization and the degree of handedness.

However, piranhas, which feed on scales only as juveniles, do not share the traits listed above with more obligate lepidophages (Kolmann et al. 2018). Many of these facultative scale-feeding piranhas are also frequent fin-feeders, for example, species like *Serrasalmus elongatus* and *S. altuvei*, which feed on fins as juveniles through to adulthood (Nico 1991; Nico and Taphorn 1988; Röpke et al. 2014). The shearing, saw-like

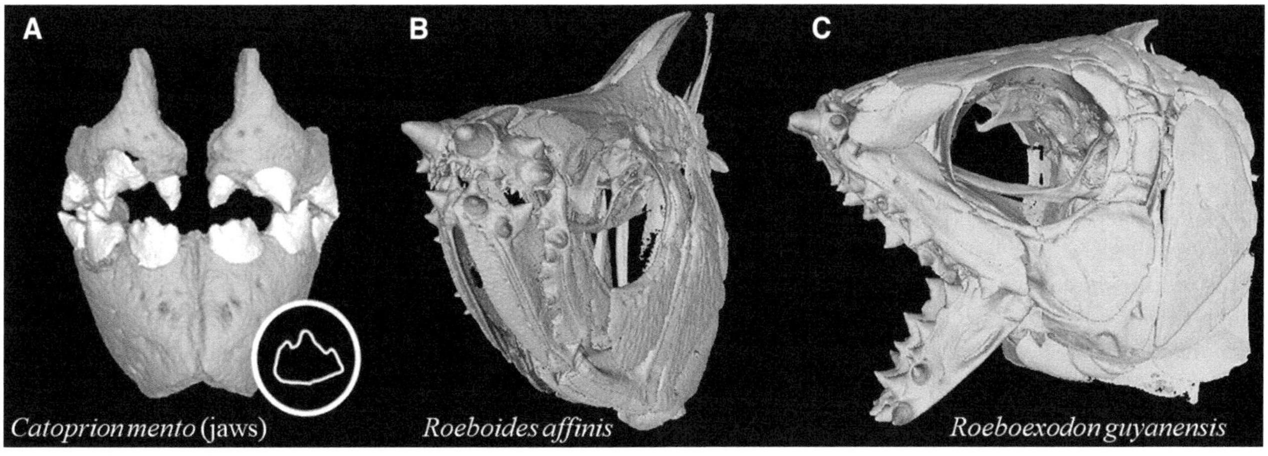

Fig. 18.3 Examples of the robust dentition characteristic of many lepidophagous characiforms. (**A**) A piranha (Serrasalmidae), the wimple piranha *Catoprion mento* (from Kolmann et al. 2018). (**B**) The scale-feeding characin, *Roeboides affinis* (Characinae). (**C**) Another scale-feeding characin, but from another characid subfamily, *Roeboexodon guyanensis* (Exodontinae)

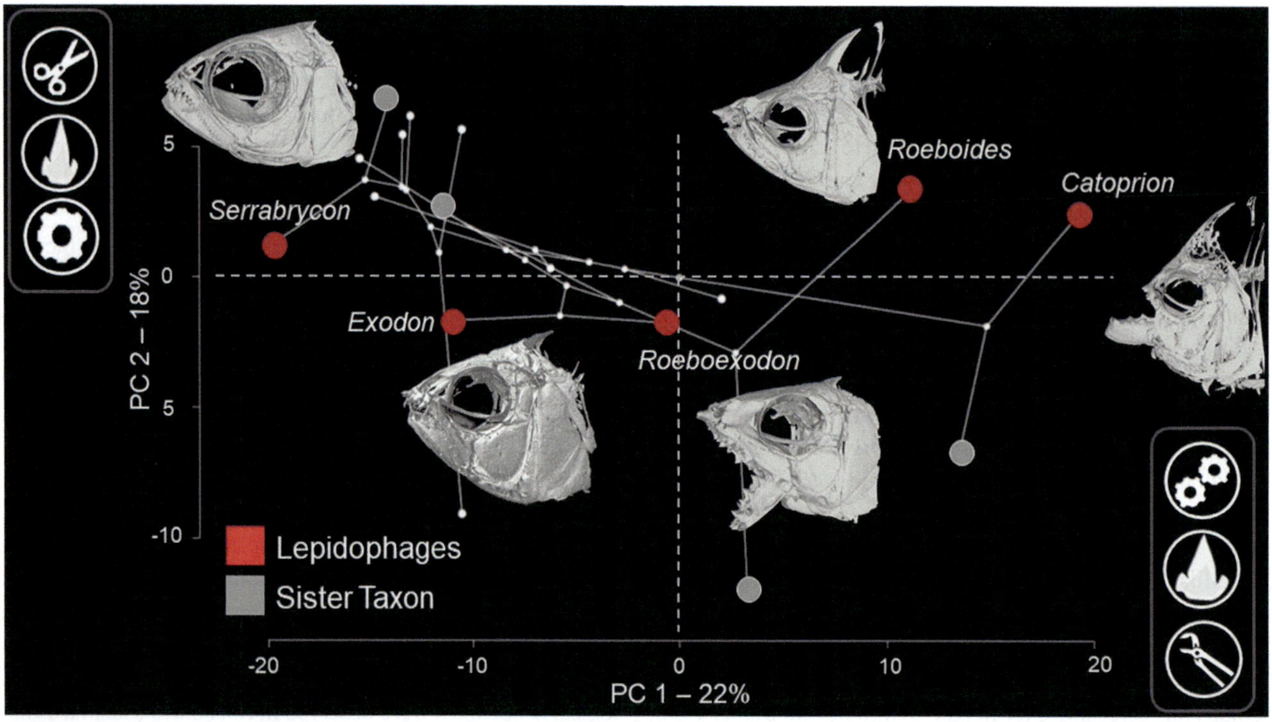

Fig. 18.4 The diversity of feeding mechanics in lepidophagous characiforms and their non-ectoparasitic relatives. Scale-feeders denoted by red circles, non-ectoparasitic confamilials denoted by gray circles. Note the considerable divergence among sister taxa, yet little overall similarity among lepidophage feeding mechanisms. Skulls are microCT renders. Tool icons denote more scissor-like tooth occlusion, weaker jaw leverage, and taller teeth in the top right quadrant of the phylomorphospace. Lower left quadrant: greater jaw leverage, more robust teeth, and more precisely occluding dentition

dentitions of piranhas would appear to be ideally suited for slicing through fins, but also work well enough for removing scales from prey (Northcote et al. 1987). Similarly, the skulls and teeth of more facultative scale-feeders, like *Odontostilbe* (now *Holoshesthes*) *pequira* and *Probolodus,* seem more similar to their close relatives (Cheirodontinae and Stethaprioninae, respectively; Fig. 18.4), rather than other characiform scale-feeders (Roberts 1970; Lima et al. 2012; Santos and Castro 2014). So, although there is certainly morphological and functional evidence for feeding specialization in characiform scale-feeders, scale-feeding does not always beget specialized phenotypes, but rather some anatomical aspects are simply modified from ancestral *bauplans* (Fig. 19.4).

Hematophagous catfishes also exhibit an interesting mix of autapomorphies and synapomorphies, although there appear to be fewer independent transitions to blood-feeding in siluriforms than there are for lepidophagous characiforms. This makes it difficult to disentangle which traits are shared in common among ectoparasitic catfishes that result primarily from inherited characteristics (phylogenetic inertia or even exaptation) *versus* traits that are truly adaptive, either for blood-feeding specifically, or some other aspect of natural history (i.e., shared habitat or phoresis). For example, in most trichomycterids, a modified opercular apparatus with elongated odontodes (integumentary teeth) appears ancestral for the family but has been subsequently modified for either (1) external attachment to a host or (2) as a bracing mechanism for clinging to substrates in high-flow regimes (Eigenmann 1918; Arratia 1983; Adriaens et al. 2010; Fig. 18.5). However, whereas similar odontodes on the interopercle see a reduction in size among some of the ectoparasitic lineages like stegophilines (Baskin 1973; de Pinna 1992), the opercular odontodes are noticeably enlarged in other stegophilines and vandelliines, which is useful for implantation within the host's flesh. As in lepidophages, stegophilines and vandelliines also have specialized, needle-like teeth for cutting through flesh or scraping mucus and scales from prey (Ihering 1930; Baskin 1973; Spotte et al. 2001; Adriaens et al. 2010).

Among all of the stunning ecological and morphological diversity in ectoparasitic fishes, there is perhaps an even more stunning lack of convergence (Figs. 18.4 and 18.5; Kolmann et al. 2024). Our analysis in this chapter suggests that ectoparasites tend to more closely resemble their non-ectoparasitic sister taxa than they do other ectoparasites (Fig. 18.4). This may be due to phylogenetic constraints that restrict the range of variation produced by a species' developmental program. These constraints would produce a pattern of morphological diversification in which ectoparasitic species elaborate

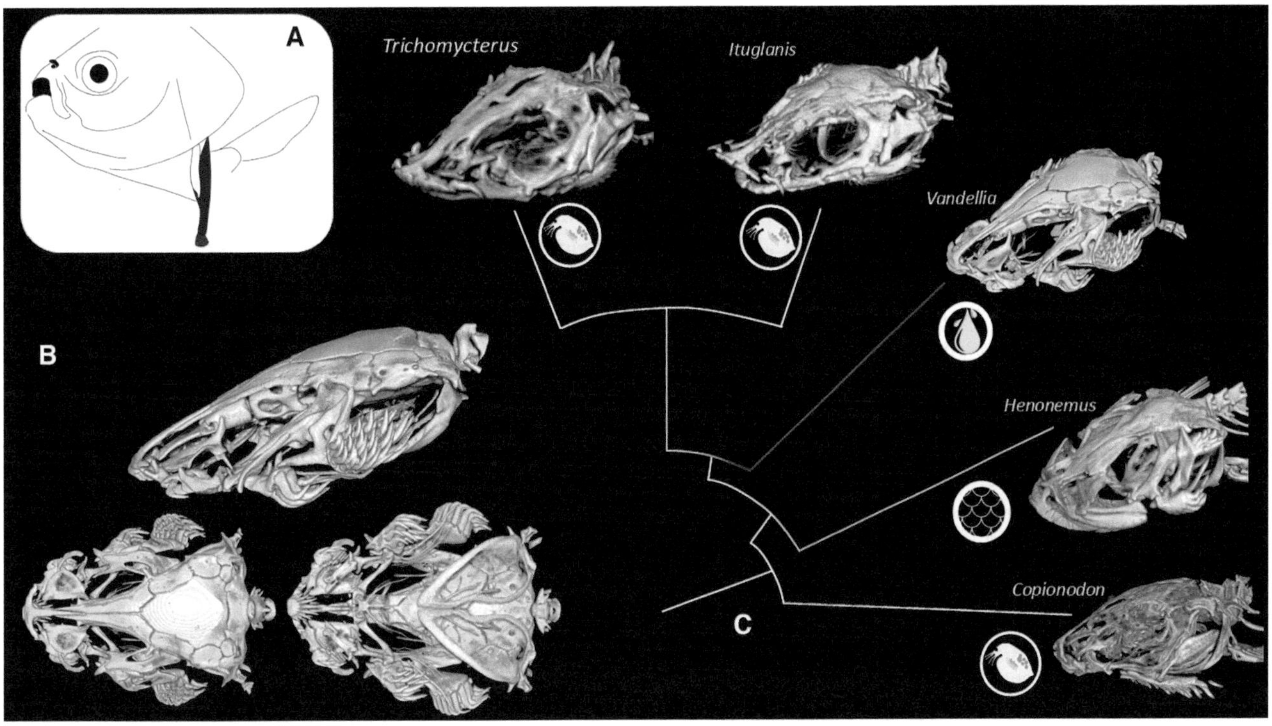

Fig. 18.5 MicroCT renders of skulls from ectoparasitic and non-ectoparasitic trichomycterids. Insets show images of a vampire catfish or candiru lodged in the gill chambers of a (**A**) large pacu or tambaqui, *Colossoma macropomum*, in a natural setting (Zuanon and Sazima 2004). (**B**) Close-up of specialized feeding morphology, notably large opercular spines, in *Vandellia cirrhosa* (FLMNH 77839); lateral image top, dorsal and ventral bottom right and left, respectively. (**C**) Phylogeny of trichomycterids (Chap. 1) and demonstrating the relatively conserved nature of overall cranial phenotype (except *Vandellia*)

and modify an ancestral body-plan, rather than innovating novel morphologies. In other words, some phenotypes that are critical to ectoparasitism, like odontodes in vandellines, are mosaics modified from ancient evolutionary forms to reflect more recent ecological functions.

18.8 Origins of Ectoparasitic Fishes

Sazima (1983) provided one of the first truly comparative approaches to understanding the evolutionary pathways and prerequisites that fishes might navigate to arrive at ectoparasitism by examining the literature on scale-feeders from multiple ecosystems. There are two hypothetical domains that have been proposed as the ecological precursors or routes to ectoparasitism; however, both are specific to scale-feeding. Sazima (1983) summarized earlier hypotheses based on typically single-taxon observations, which proposed either a trophic or behavioral origin for scale-feeding (Sazima 1983). This first hypothesis (trophic) was built on observations of scale-feeders in other parts of the tropics, relative to what were considered their nearest relatives at the time. Conversely, authors working in the Neotropics (and studying piranhas), proposed that scale-feeders arose from "degenerate" or perhaps more eloquently, "opportunistic" variations on either piscivory or scavenging (i.e., necrophagy) (Gery 1972; Sazima and Uieda 1980; Sazima and Machado 1990).

Alternatively, observations on scale-feeding cichlids from East Africa noted that the sister taxon to scale-feeding taxa (e.g., *Perissodus*), were algae-scraping species (Fryer and Iles 1955; Greenwood 1965; Fryer and Iles 1972). Ecologically, the shift from outright piscivory or scavenging to parasitism seems like a logical jump (Gery 1972), while from a functional standpoint, the scraping behavior of cichlids resembles how lepidophages remove prey items (scales) from a substrate (fish). However, the former hypotheses appear unfounded given our more recent comprehensive understanding of the serrasalmid phylogeny (Thompson et al. 2014), which shows the most obligate scale-feeding piranha, *Catoprion*, as sister to omnivorous or herbivorous piranhas (*Pygopristis* and *"Pristobrycon" striolatus*, respectively), which in turn form a clade sister to all other piranhas (Sazima 1983; Correa et al. 2007). If we consider this character state in the sister group to all piscivorous piranhas, and that many piranhas scale-feed as juveniles, one might propose that scale-feeding may have given rise to outright piscivory and not *vice versa* (Goulding 1980). Conversely, more recent African cichlid phylogenies suggest that scale-feeding cichlids are indeed sister to more carnivorous or omnivorous taxa; however, the authors of these studies also found significant amounts of plant material and plankton in the stomachs of some of the nearest relatives to obligate scale-feeders, as well as the lepidophagous species themselves (Nshombo et al. 1985; Takahashi et al. 2007; Koblmüller et al. 2007).

Mucus-feeding has been proposed as the ecological precursor to scale- and blood- feeding in trichomycterids (Machado and Sazima 1983), based on the assumption that one of the earliest-branching lineages, *Ochmacanthus*, is known to be an obligate mucophage (Roberts 1970; Winemiller and Yan 1989; Cochran et al. 2015). Bonato et al. (2018) found mucus in the stomachs of *Homodiaetus* and *Pseudostegophilus*, both stegophiline catfishes. Relatedly, Zuanon and Sazima (2005) demonstrated some degree of mucus-feeding among more aggressive scale- and flesh- feeding catfishes, such as *Paracanthopoma*, one of the earliest branching lineages within Vandelliinae. However, given more recent and comprehensive phylogenetic hypotheses for Trichomycteridae (Ochoa et al. 2020), the evolution of diets in the "TSVSG" clade (Tridentinae, Stegophilinae, Vandelliinae, Sarcoglanidinae, and Glanapteryginae) merits more investigation, particularly with quantitative methods. The immediate sister clades to more ectoparasitic catfishes, lineages like *Pygidianops*, *Listrura*, and *Microcambeva*, are either psammophilic or leaf-litter taxa which feed on invertebrates, suggesting a similar diet for the more distant ancestors of tridentine, vandelline, and stegophiline catfishes (Carvalho et al. 2014; Costa et al. 2020). However, without basic natural history and dietary ecology information for Tridentinae (Adriaens et al. 2010), our ability to predict the ancestral diet of the ectoparasitic trichomycterids remains contentious.

Finally, Sazima (1983) and others have also hinted that mutualisms like cleaning behaviors could lead to scale-feeding. This mode has been recognized in marine fishes like grunters (e.g., *Terapon*; Whitfield and Blaber 1978) and wrasses (*Labroides*; Losey 1972), which are facultative cleaners that supplement their diet with scales and mucus, similar to echeneids (Cressey and Lachner 1970). Unfortunately, it may be that too little is known about cleaning behaviors in tropical freshwaters, where it is harder to observe such behaviors for any length of time, although Sazima and Machado (1990) noted cleaning behaviors in juvenile piranhas, which were also fin-feeders. Investigation of mutualisms among freshwater vertebrates is surely an area for which greater research is needed.

In opposition to a trophic impetus for scale-feeding, Sazima (1983) also proposed a social origin for scale-feeding behaviors, stemming from interspecific aggression and perhaps in conjunction with trophic benefit. Sazima (1983) pointed to characiforms like *Probolodus*, which form moderate–large schools of small, surly fishes. The social hierarchy of these schools involves frequent (but usually not deadly) acts of chasing, agonistic display, and most importantly, biting of the fins, flanks, and tails of conspecifics (Breder 1927; Sazima 1977, 1983). Relatedly, the armored keel of serrasalmids, formed of bony, individual scutes called "serrae," has been proposed as defense against just this sort of aggressive "negotiation" among shoal members (Kolmann et al. 2020). More recently, work by St. John, Martin, and colleagues found that the advent of scale-feeding in *Cyprinodon* pupfishes is coincident with behavioral and genomic underpinnings of increased aggression (St. John et al. 2019, 2020). In gymnotiform electric fishes, tail biting is a well-documented form of aggression across many of the major families (Albert 2001). It is possible thatv this behavior became evolutionarily fixed within some members of the genus *Sternarchella* who feed almost exclusively on the scales and tails of other electric fishes (Evans et al. 2017, 2019).

Finally, another behavioral hypothesis for the origin of ectoparasitism, in trichomycterids at least, was proposed by Zuanon and Sazima (2005) and was based on observations of *Paracanthopoma*, which burrows into the flesh of large catfishes. Zuanon and Sazima (2005) point out that *Paracanthopoma* do not embed themselves in anatomical regions of their hosts, which would be preferential for blood-sucking (i.e., near blood supply, like the gill arches), as the dorsal surface of pimelodid catfishes is sparsely vascularized. Instead, these vampire catfishes might be using their larger host brethren as long-distance transport, i.e., hitch-hiking (phoresis) in order to increase their dispersal capability. Phoresis may have acted synergistically with the tendency of vandelliine catfishes to wedge or embed themselves in substrates with their modified opercular apparatus (Adriaens et al. 2010), which is reminiscent of sharksuckers (Echeneidae) and their chondrichthyan hosts (Kenaley et al. 2019). Juvenile sharksuckers (*Echeneis*) practice facultative phoresis with elasmobranchs and other large aquatic fauna, while their sister group, *Remora*, forms more obligate associations with specific host lineages (Strasburg 1959; Cressey and Lachner 1970). Sharksuckers like *Echeneis* will also remove bits of flesh from their hosts, whereas remoras scavenge and perform mutualistic cleaner functions (but see Sazima et al. 1999). Regardless of its origins, and unlike scale-feeding in characiforms, blood-feeding and ectoparasitism in general are hypothesized to have arisen only once in catfishes, with blood-feeders nested within a clade of mucus- and scale-feeding trichomycterids (Fernández and Schaefer 2009). However, if catfishes in the family Tridentinae are truly hematophagous (Azpelicueta 1990; Adriaens et al. 2010), this would suggest that blood-feeding has evolved twice in ectoparasitic catfishes.

18.9 Testing the Trophic Origin Hypothesis

To investigate the trophic origins of scale-feeding in characiforms and ectoparasitism (more broadly) in trichomycterid catfishes, we used a tried and true method in phylogenetics: ancestral state reconstruction, through Bayesian stochastic character mapping (Bollback 2006; Huelsenbeck et al. 2003). This method can be easily assessed with discrete traits, obviating the need for particularly detailed diet data, and can be extended specifically to pterygophagy, mucophagy, or hematophagy, wherever a comprehensive phylogeny exists for a given clade. We used the recent, species-dense cladograms introduced in Chap. 2 which build on well-sampled phylogenies like Betancur et al. (2019) and Ochoa et al. (2020b) for characiforms and trichomycterids, respectively. We first tested which of three possible models of discrete trait evolution (ER; equal transition rates among state, SYM; symmetrical rates, or ARD; all-rates-different among transitions among states) best fit the data, using the model with the lowest AICc as our criterion moving forward. For characiforms, this was the SYM model (log-likelihood = −235.306, AICc = 518.017) and for trichomycterids this was the ER model (log-likelihood = −30.488, AICc = 63.018). We then simulated the evolution of these diet categories 500 times. Likewise, we repeated these methods for trichomycterids using a finer-grained diet guild breakdown, separating ectoparasitic guilds into three categories (general ectoparasites, e.g., tridentines; sanguinivores, e.g., vandelliines; as well as mucus- and scale- feeders, e.g., stegophilines—Adriaens et al. 2010). In this finer-grained analysis, the ER model was also selected (log-likelihood = −36.166, AICc = 74.374) once again.

Our analyses suggest that lepidophagy has arisen some 10–15 times independently across the characiform phylogeny (*sans* Crenuchoidea and Citharinoidea; Fig. 18.6). The uncertainty in the number of reconstructions stems predominantly from whether these behaviors have arisen multiple times within single clades like Serrasalmidae or Characinae, with lepidophagy potentially arising in piranhas and pacus some two to four times, although this is unlikely. These findings also suggest that scale-feeding has arisen some six to nine times independently in Characidae and is congruent with phylogenetic niche conservatism, with each isolated instance associated with its own subfamily (except the Exodontinae). We also coded lepidophages, to the best of our ability, as either obligate or facultative, as well as distinguishing these diet modes from taxa that transition from scale-feeding over ontogeny. We find that this latter category is the most prevalent across the Neotropical characiform phylogeny and denotes taxa like *Exodon*, *Roeboexodon*, *Bryconexodon*, *Catoprion*, *Serrabrycon*, and some species of *Roeboides* as obligate lepidophages. Finally, we find that invertivory (feeding on insects, crustaceans, and other arthropods) is the most common trophic precursor to scale-feeding, with only a single instance of piscivory leading to scale-feeding in the Characinae. The results suggest that piscivory or algae-feeding

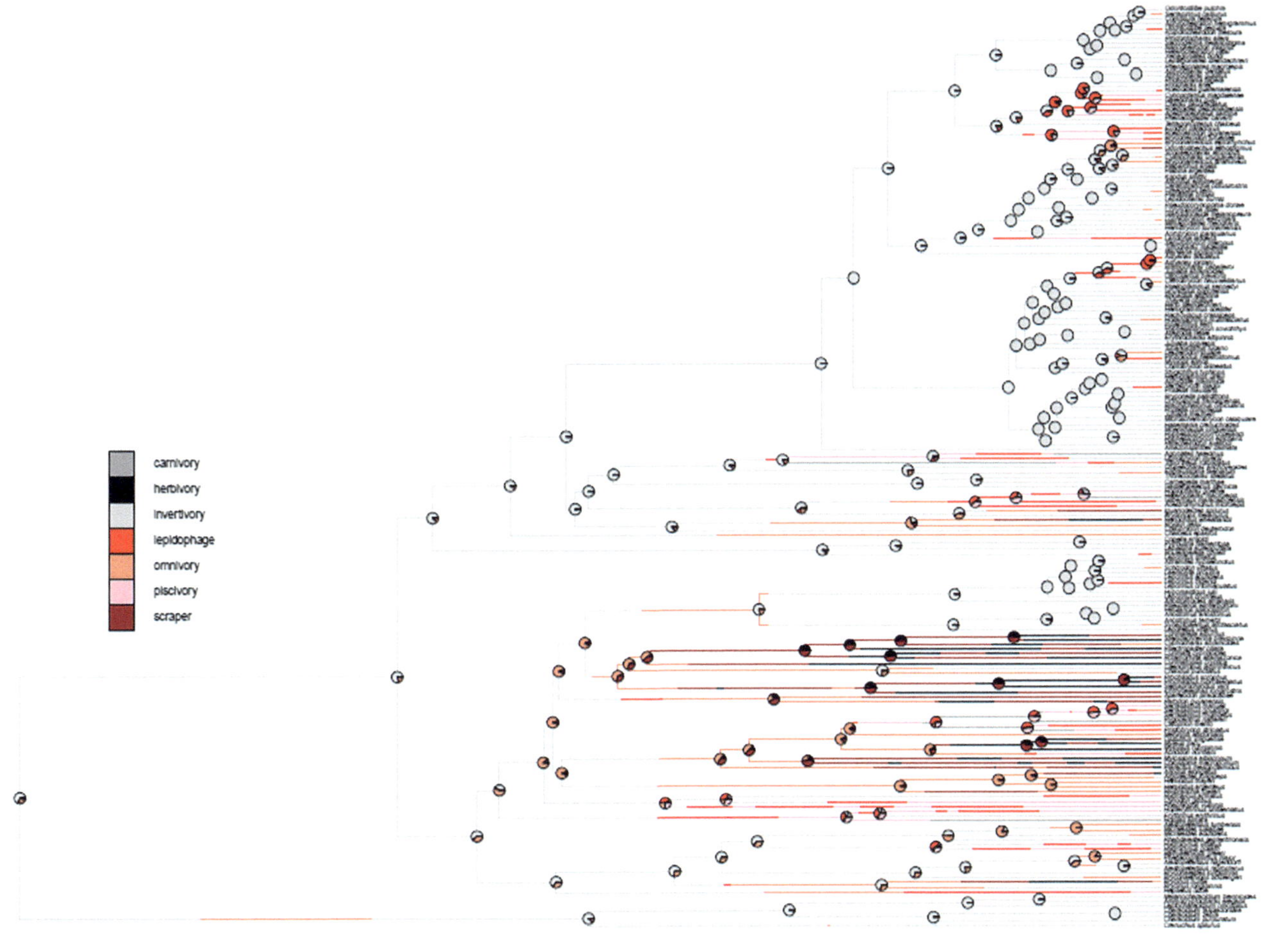

Fig. 18.6 Ancestral state reconstruction of discrete diet categories in the Characiformes. Red denotes lepidophagous taxa. Gray and black shaded lines denote some manner of invertivory (arthropod specialists). Results from 500 simulations using the all NFF phylogeny (Chap. 2)

do not appear to lend themselves to shifts to scale-feeding, at least in Neotropical characoids and their allies. This method is easily adapted to other Neotropical freshwater fishes and their respective phylogenies; moreover, one could foreseeably code discrete characters as proxies for interspecific aggression, such as school size, number of conspecific attacks per minute, etc., in order to test behavioral hypotheses regarding the origin of ectoparasitism in characiforms or other Neotropical fishes.

These analyses recovered a similar pattern for trichomycterids—with some manner of ancestral invertebrate feeder (an arthropod specialist) giving rise to ectoparasitic lineages (Fig. 18.7). According to the NFF phylogeny (see Chap. 2), this occurred twice, once in the ancestor of Stegophilinae + Tridentinae and independently in the common ancestor of Vandellinae. This analysis also suggests that ectoparasitism is perhaps lost, once in the ancestor of *Pygidianops* + [*Sarcoglanis* + *Typhlobelus*] and again in the lineage leading to *Stauroglanis*. However, the loss of adaptations associated with parasitism seems unlikely and correspondingly, the support for this hypothesis is lower than in a situation where the ancestor of *Potamoglanis* and other tridentine catfishes was an invertebrate feeder of some variety. Identification of the primary prey of various species within Tridentinae is of paramount importance for understanding the evolution of parasitism in trichomycterid pencil catfishes.

According to the finer reconstruction of trichomycterid diet evolution—some variety of ectoparasitism might have evolved four separate times: once in vandelliines (sanguinivory), another time in stegophilines (mucophagy and lepidophagy), and two instances of more general ectoparasitism in Tridentinae (Fig. 18.8). This scenario suggests that ectoparasitism arose in the common ancestors of [*Miuroglanis* + *Tridensimilis*] and again in the ancestor of *Potamoglanis*. In all four of these lineages, the ancestors prior to ectoparasitic lineages were suggested to be benthic invertebrate (i.e., arthropod) feeders. These results should be interpreted with considerable caution, however, as the topology of the NFF phylogeny differs in some regards from that of other published phylogenies for trichomycterids (Ochoa et al. 2020). Although our analyses were not extended to the Ochoa et al. (2020) phylogeny, the topological arrangements of this phylogeny and the NFF phylogeny would appear to agree that non-ectoparasitic can be nested within ectoparasitic clades (e.g., *Stauroglanis*). Whether ectoparasitism has arisen multiple times (NFF phylogeny) or

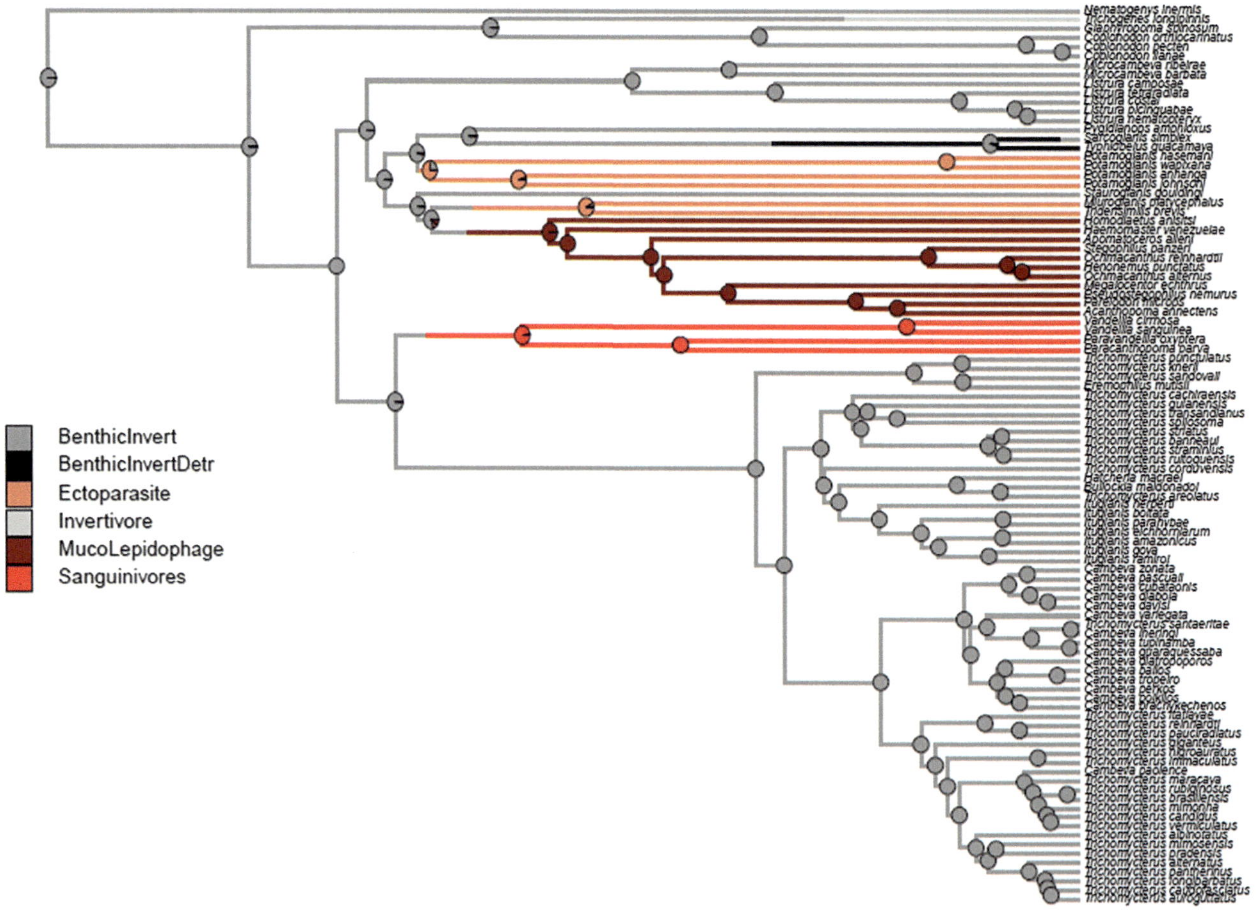

Fig. 18.7 Ancestral state reconstruction of discrete diet categories in the Trichomycteridae. Red denotes some matter of ectoparasite, be they hematophages, mucus, or scale-feeders. Gray and black shaded lines denote some manner of invertivory (arthropod specialists). Results from 500 simulations using the all NFF phylogeny (Chap. 2)

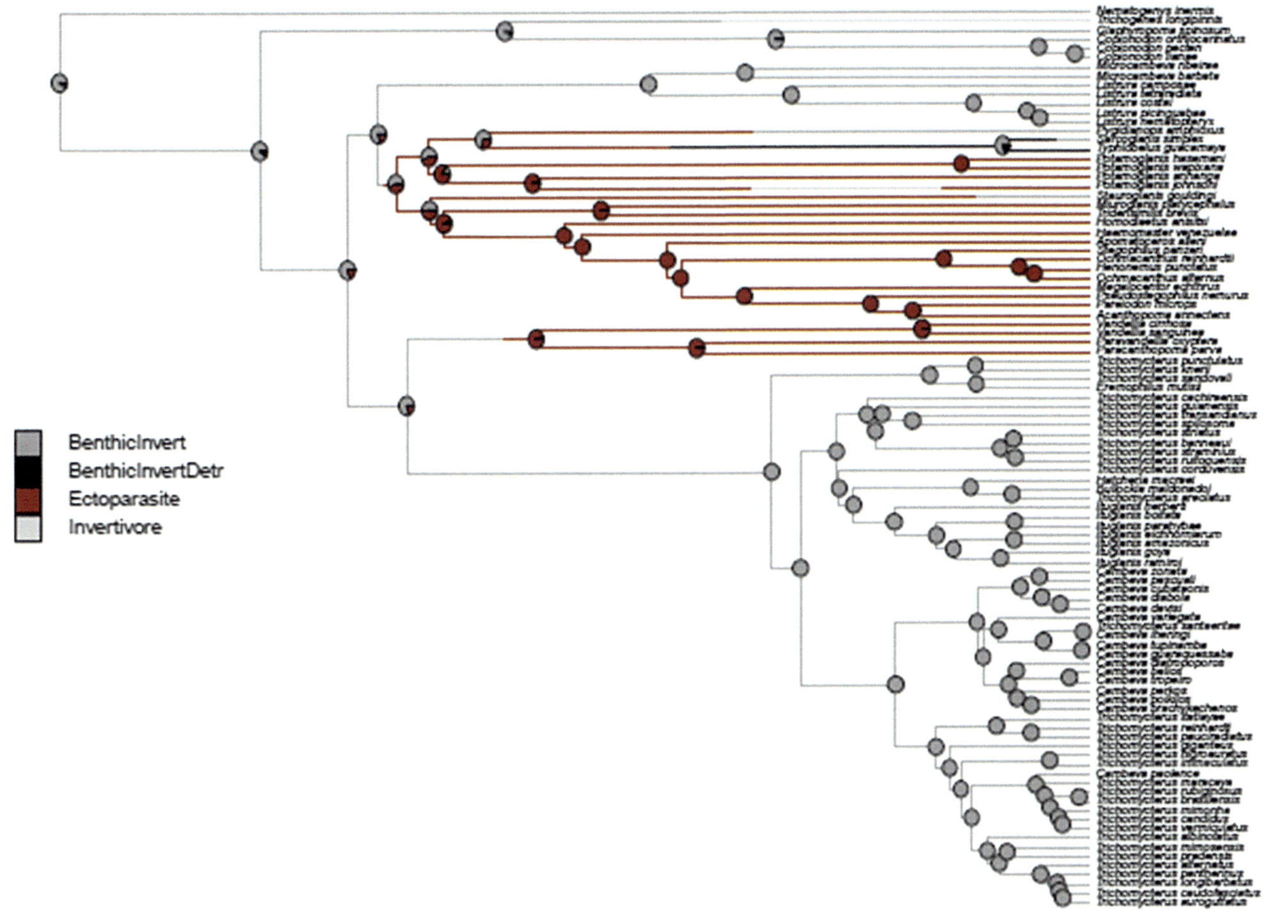

Fig. 18.8 Ancestral state reconstruction of discrete diet categories in the Trichomycteridae, with more granular classifications for different ectoparasites. Red and orange shading denotes some matter of ectoparasite, be they hematophages (bright red), mucus, or scale-feeders (maroon), or a generalist ectoparasite (salmon). Gray and black shaded lines denote some manner of invertivory (arthropod specialists). Results from 500 using the all NFF phylogeny (Chap. 2)

arose once with later diversification into hematophagous and lepidophages niches merits further investigation.

18.10 Conclusions: Future Studies

Although ectoparasitic fishes are relatively rare in the context of over 37,000 marine and freshwater fishes in general, Neotropical freshwaters are undoubtedly a hotspot for these fishes. Given the myriad examples of behavioral, ecological, and morphological diversity in ectoparasitic fishes, they offer compelling study systems for the evolution of complex phenotypes, coevolution (of predator and prey), and various other topics in ethology. In the genomics era, some notable examples of studies using ectoparasitic fishes have explored the underlying genetic machinery under selection for producing specific morphological (Stewart and Albertson 2010) and behavioral (St. John et al. 2019, 2020) phenotypes unique to scale-feeding fishes, with similar studies underway for gymnotiform fishes. Likewise, molecular tools offer considerable promise for unveiling the role of parasites in mediating trophic interactions (Bonato et al. 2018, 2022). Fruitful areas of research remain in identifying the preferred host of some ectoparasitic fishes, or whether these resources change with ontogeny, seasonality, or host availability. The eco-evolutionary origins of ectoparasitism, be they trophic or social, also remain unexplored (Sazima 1983). However, here we found little (if any support) for a trophic-mediated origin for scale-feeding. Future research could and should examine intraspecific aggression as a potential evolutionary origin for scale-feeding (see section above, Sazima 1983; St. John et al. 2019, 2020).

We suggest that adaptations toward many (but not all) ectoparasitic trophic strategies are shifts of convenience: most scale- and fin-feeders seem little differentiated, morphologically speaking, from their close relatives. Instead, slight changes to the dentition seem to be enough, for characiforms and gymnotiforms at least, to engage in raptorial feeding on scales, fins, and mucus. Therefore, shifting to an ectoparasitic bauplan might require few changes overall, especially if ectoparasitism is present in juveniles of the same species (and then paedomorphosis might be suspected as mechanism for producing morphological "novelty"; Kolmann et al. 2018). Likewise, most ectoparasites seem to be evolutionary singletons—scale- or fin-feeding taxa like *Probolodus*, *Landonia*, *Odontostilbe*, or *Serrasalmus elongatus* are isolated examples of obligate ectoparasites in their clades. Others, like *Roeboides* or the exodontines (e.g., *Exodon*, *Bryconexodon*), are small radiations of ectoparasitic fishes (Roberts 1970; Novakowski et al. 2004; Meza-Vargas et al. 2019). Perhaps, an ectoparasitic feeding strategy is a way to avoid competing with ecomorphologically similar confamilials, surely a reasonable hypothesis given the great diversity of small characoid fishes. However, the great diversity within ectoparasitic trichomycterid lineages might suggest that hematophagy or mucophagy holds some promise for diversification, perhaps even adaptive radiation. A combined morphological, ecological, and phylogenetic comparative approach is needed to test these questions in the future.

The digestive ecophysiology of ectoparasitic fishes, particularly hematophagous catfishes, and the adaptations that allow for and protect against the confounds of a poor nutrient diet, are also ripe for exploration. Advances in genomic techniques and metabarcoding, hold particular promise in this area (e.g., Bonato et al. 2018, 2022). Finally, and perhaps most importantly, the basic natural history, taxonomy, and phylogenetic positon for most Neotropical ectoparasitic fishes is still lacking. Information pertaining to the diets of most of the stegophiline, vandelline, and tridentine catfishes is still unknown, and while the exact phylogenetic positions of many characiform scale-feeders are ambiguous. The specific hosts of scale- and fin-feeding fishes are largely unknown as well—do these ectoparasites target specific prey species or are they opportunists? Future research should also spend considerable time observing the behaviors of ectoparasites and evaluate whether such symbioses, and related mutualisms like cleaning behaviors, are really that rare—or a function of the difficulty of observing such behaviors in Neotropical freshwater ecosystems. Finally, ectoparasitic fishes likely exercise some degree of top-down control on their hosts, in turn perhaps even influencing direct and indirect ecological interactions among their community—but these ideas remain suppositions until the basic biology, and then manipulative experiments, can explore them.

References

Adriaens D, Baskin JN, Coppens H (2010) Evolutionary morphology of trichomycterid catfishes: about hanging on and digging in. In: Nelson JS, Schultze HP, Wilson MV (eds) Origin and phylogenetic interrelationships of teleosts. New York, Friedrich Pfeil, pp 337–362

Albert JS (2001) Species diversity and phylogenetic systematics of American knifefishes (Gymnotiformes, Teleostei) (vol 190). Ann Arbor, MI, Division of Ichthyology, Museum of Zoology, University of Michigan, pp 1–127

Albrecht MP, Reis VC, Caramaschi ÉP (2013) Resource use by the facultative lepidophage *Roeboides affinis* (Günther, 1868): a comparison of size classes, seasons and environment types related to impoundment. Neotrop Ichthyol 11(2):387–394

Arcifa MS, Northcote TG, Froehlich O (1991) Interactive ecology of two cohabiting characin fishes (*Astyanax fasciatus* and *Astyanax bimaculatus*) in an eutrophic Brazilian reservoir. J Trop Ecol 7(2):257–268

Arratia G (1983) Preferencias de habitat de peces siluriformes de aguas continentales de Chile (Fam. Diplomystidae y Trichomycteridae). Stud Neotrop Fauna Environm 18(4):217–237

Azpelicueta M (1990) *Tridentopsis cahuali* n. sp.(Siluriformes, Trichomycteridae), a new miniature tridentine from Paraguay System, in Argentina. Rev Suisse Zool 97(4):981–988

Baskin JN (1973) Structure and relationships of the Trichomycteridae. Unpublished PhD thesis, City University, New York, p 398

Baskin JN, Zaret TM, Mago-Leccia F (1980) Feeding of reportedly parasitic catfishes (Trichomycteridae and Cetopsidae) in the Rio Portuguesa Basin, Venezuela. Biotropica 182–186

Bauer IL (2013) Candiru—A little fish with bad habits: need travel health professionals worry? A review. J Travel Med 20(2) 119–124. 10.1111/jtm.12005

Betancur RR, Arcila D, Vari RP, Hughes LC, Oliveira C, Sabaj MH, Ortí G (2019) Phylogenomic incongruence, hypothesis testing, and taxonomic sampling: the monophyly of characiform fishes. Evolution 73(2):329–345

Bollback JP (2006) SIMMAP: stochastic character mapping of discrete traits on phylogenies. BMC Bioinform 7(1):88

Bonato KO, Silva PC, Malabarba LR (2018) Unrevealing parasitic trophic interactions—A molecular approach for fluid-feeding fishes. Front Ecol Evol 6. 10.3389/fevo.2018.00022

Bonato KO, Silva PC, Malabarba LR (2018) Unrevealing parasitic trophic interactions—a molecular approach for fluid-feeding fishes. Front Ecol Evol 6:22

Bonato KO, Silva PC, Carvalho FR, Malabarba LR (2022) Trophic interactions of vampire catfishes (Siluriformes: Vandelliinae) revealed by metabarcoding analysis of stomach contents. Freshw Biol 67(3):542–548

Boulenger GA (1897) Exhibition of specimens, and remarks upon the habits of the Siluroid *Vandellia cirrhosa*. Proc Zool Soc Lond 901

Breault JL (1991) Candirú: Amazonian parasitic catfish. J Wilderness Med 2(4):304–312. 10.1580/0953-9859-2.4.304

Breder CM Jr (1927) The fishes of the Rio Chucunaque drainage, eastern Panama. Bull Amer Mus Nat Hist 57:91–176

Bshary A, Bshary R (2012) Individual differences in foraging strategies of parasitic sabre-tooth blennies. PLOS One 7(9):e45998

Carvalho MS, Zuanon J, Ferreira EJ (2014) Diving in the sand: the natural history of *Pygidianops amphioxus* (Siluriformes: Trichomycteridae), a miniature catfish of Central Amazonian streams in Brazil. Environ Biol Fishes 97(1):59–68

Castelnau FL (1855) Poissons. In: Animaux nouveaux or rares recueillis pendant l'expédition dans les parties centrales de l'Amérique du Sud, de Rio de Janeiro a Lima, et de Lima au Para; exécutée par ordre du gouvernement Français pendant les années 1843 a 1847. Part 7. Zoologie, v. 2: i-xii + 1–112. P. Bertrand, Paris, pp 1–50

Cochran PA, Wagner RJ, Golden EJ (2015) The feeding behavior of a South American parasitic catfish (*Ochmacanthus alternus*). J Fresh Ecol 30(3):417–424

Correa SB, Winemiller KO, LóPez-Fernández H, Galetti M (2007) Evolutionary perspectives on seed consumption and dispersal by fishes. BioScience 57(9):748–756. 10.1641/B570907

Costa WJ, Vilardo P, Katz AM (2020) Sympatric sister species with divergent morphological features of psammophilic catfishes of the south-eastern Brazilian genus *Microcambeva* (Siluriformes: Trichomycteridae). Zool Anz 285:12–17

Cressey RF, Lachner EA (1970) The parasitic copepod diet and life history of diskfishes (Echeneidae). Copeia (2):310–318

Datovo A, Bockmann FA (2010) Dorsolateral head muscles of the catfish families Nematogenyidae and Trichomycteridae (Siluriformes: Loricarioidei): comparative anatomy and phylogenetic analysis. Neotrop Ichthyol 8(2):193–246. 10.1590/S1679-62252010000200001

Eigenmann CH (1918) The Pygidiidae, a family of South American catfishes. Mem Carnegie Mus 7(5):259–398

Eigenmann CH, Allen WR (1942) Fishes of western south America. I. The intercordilleran and Amazonian lowlands of Peru. II.-The high pampas of Peru, Bolivia, and northern Chile. With a revision of the Peruvian Gymnotidae, and of the genus Orestias. University of Kentucky

Evans KM, Kim LY, Schuber BA, Albert JS (2019) Ecomorphology of Neotropical electric fishes: an integrative approach to testing the relationships between form, function, and trophic ecology. Integr Org Biol 1(1):obz015

Evans KM, Crampton WGR, Albert JS (2017) Taxonomic revision of the deep channel electric fish genus *Sternarchella* (Teleostei: Gymnotiformes: Apteronotidae), with descriptions of two new species. Neotrop Ichthyol 15(2):e160168

Fernández L, Schaefer SA (2009) Relationships among the Neotropical Candirus (Trichomycteridae, Siluriformes) and the evolution of parasitism based on analysis of mitochondrial and nuclear gene sequences. Mol Phylogenet Evol 52(2):416–423

Fryer G, Greenwood PH, Trewavas F (1955) Scale-eating habits of African cichlid fishes. Nature 175:1089–1090

Fryer G, Iles TD (1955) Predation pressure and evolution in Lake Nyasa. Nature 176(4479):470

Fryer G, Iles TD (1972) The cichlid fishes of the Great Lakes of Africa: Their biology and evolution. Oliver & Boyd, Edinburgh, UK

Fujisawa M, Sakai Y, Kuwamura T (2020) The false cleanerfish relies on aggressive mimicry to bite fish fins when benthic foods are scarce in their local habitat. Sci Rep 10(1):8652. 10.1038/s41598-020-65304-6

Géry J (1963) A L'Étude des foisons characoides-27: systematique et évolution de some piranhas (Serrasalmus). Vie et Milieu, pp 597–618

Géry J (1969) The fresh-water fishes of South America. Biogeogr Ecol S Am 2:828–848

Géry J (1972) Poissons characoïdes des Guyanes. I. Généralités. II. Famille des Serrasalmidae. Zoologische Verhandelingen 122(1):1–250

Godoi DSD (2004) Diversidade e hábitos alimentares de peixes de um córrego afluente do rio Teles Pires, Carlinda, MT, drenagem do rio Tapajós. Unpublished thesis. Universidade Estadual Paulista

Gorlick DL, Atkins PD, Losey GS (1987) Effect of cleaning by *Labroides dimidiatus* (Labridae) on an ectoparasite population infecting *Pomacentrus vaiuli* (Pomacentridae) at Enewetak Atoll. Copeia 41–45

Goulding M (1980) The fishes and the forest: explorations in Amazonian natural history. Univ of California Press

Goulding M, Carvalho ML, Ferreira EG (1988) Río Negro, rich life in poor water: Amazonian diversity and foodchain ecology as seen through fish communities. SPB Academic Publishing, The Hague, p 200

Greenwood PH (1965) The cichlid fishes of Lake Nabugabo, Uganda. Bull Br Mus (Nat Hist) Zool 12(9):315–357

Grubh AR, Winemiller KO (2004) Ontogeny of scale feeding in the Asian glassfish, *Chanda nama* (Ambassidae). Copeia 2004(4):903–907

Grubh AR, Winemiller KO (2004) Ontogeny of scale feeding in the Asian glassfish, Chanda nama (Ambassidae). Ichthyol Herpetol 2004(4):903–907

Gudger EW (1930) On the alleged penetration of the human urethra by an Amazonian catfish called candiru with a review of the allied habits of other members of the family Pygidiidae part I. Am J Surg 8(1):170–188

Günther ACLG (1868) Catalogue of the fishes in the British Museum, vol 7. London, Wheldon & Wesley

Haddad V (2016) Aquatic animals inducing mechanical injuries. In: Bonamonte D, Angelini G (eds) Aquatic dermatology: biotic, chemical and physical agents. Springer, 217–226

Haddad V, Zuanon, J, Sazima I (2021) Medical importance of candiru catfishes in Brazil: a brief essay. Rev Soc Bras Med Trop 54:1–3

Hahn NS, Pavanelli CS, Okada EK (2000) Dental development and ontogenetic diet shifts of *Roeboides paranensis* Pignalberi (Osteichthyes, Characinae) in pools of the upper Rio Paraná floodplain (state of Paraná, Brazil). Rev Bras Biol 60(1):93–99

Haseman JD (1911) Descriptions of some new species of fishes and miscellaneous notes on others obtained during the expedition of the Carnegie Museum to central South America. Ann Carnegie Mus 7(17):315–328

Hata H, Yasugi M, Hori M (2011) Jaw laterality and related handedness in the hunting behavior of a scale-eating characin, *Exodon paradoxus*. PLOS One 6(12):e29349

Hernandez AD, Sukhdeo MVK (2008) Parasites alter the topology of a stream food web across seasons. Oecologia 156(3):613–624. 10.1007/s0044 2-008-0999-9

Hoese HD (1966) Ectoparasitism by juvenile sea catfish, *Galeichthys felis*. Copei (4):880

Huelsenbeck JP, Nielsen R, Bollback JP (2003) Stochastic mapping of morphological characters. Syst Biol 52(2):131–158

Huie JM, Thacker CE, Tornabene L (2020) Co-evolution of cleaning and feeding morphology in western Atlantic and eastern Pacific gobies. Evolution 74(2):419–433. 10.1111/evo.v74.2 10.1111/evo.13904

Ihering RV (1930) Notas ecológicas referentes a peixes d'água doce do Estado de S. Paulo e descripção de 4 espécies novas. *Archivos do Instituto Biológico*. São Paulo 3:93–103

Janovetz J (2005) Functional morphology of feeding in the scale-eating specialist *Catoprion mento*. J Exp Biol 208(24):4757–4768

St. John ME, McGirr JA, Martin CH (2019) The behavioral origins of novelty: did increased aggression lead to scale-eating in pupfishes?. Behav Ecol 30(2):557–569

Kelley WE, Atz JW (1964) A pygidiid catfish that can suck blood from goldfish. Copeia 1964(4):702–704

Kenaley CP, Stote A, Ludt WB, Chakrabarty P (2019) Comparative functional and phylogenomic analyses of host association in the remoras (Echeneidae), a family of hitchhiking fishes. Integr Organ Biol 1(1):obz007

Keppeler FW, Lanés LEK, Rolon AS, Stenert C, Maltchik L (2013) The diet of *Cynopoecilus fulgens* Costa, (Cyprinodontiformes: Rivulidae) in Southern Brazil wetlands. Ital J Zool 80(2):291–302

Kner R (1860) Zur Familie der Characinen, II. Denkschriften der mathematisch-naturwissenschaftlichen Classe der kaiserlichen Akademie der Wissenschaften, Wien, 18:9–62

Koblmüller S, Egger B, Sturmbauer C, Sefc KM (2007) Evolutionary history of Lake Tanganyika's scale-eating cichlid fishes. Mol Phylogenet Evol 44(3):1295–1305

Kolmann MA, Urban P, Summers AP (2020) Structure and function of the armored keel in piranhas, pacus, and their allies. Anat Rec 303(1):30–43

Kolmann MA, Huie JM, Evans K, Summers AP (2018) Specialized specialists and the narrow niche fallacy: a tale of scale-feeding fishes. R Soc Open Sci 5(1):171581

Kolmann MA, Cohen KE, Bemis KE, Summers AP, Irish FJ, Hernandez LP (2019) Tooth and consequences: heterodonty and dental replacement in piranhas and pacus (Serrasalmidae). Evol Dev 21(5):247–262

Kolmann MA, Poulin E, Rosen J, Hemraj-Naraine D, Burns MD (2024) Phenotypic convergence is more frequent and stronger in herbivorous freshwater fishes. Integr Comp Biol 64(5):1467–1483

Kovac R, Boileau N, Muschick M, Salzburger W (2019) The diverse prey spectrum of the Tanganyikan scale-eater *Perissodus microlepis* (Boulenger, 1898). Hydrobiologia 832(1):85–92

Kramer DL, Bryant MJ (1995) Intestine length in the fishes of a tropical stream: 2. Relationships to diet—the long and short of a convoluted issue. Environ Biol Fishes 42(2):129–141

Lafferty KD, Allesina S, Arim M, Briggs CJ, De Leo G, Dobson AP, Dunne JA, Johnson PT, Kuris AM, Marcogliese DJ, Martinez ND, Memmott J, Marquet PA, McLaughlin JP, Mordecai EA, Pascual M, Poulin R, Thieltges DW (2008) Parasites in food webs: the ultimate missing links. Ecol Lett 11(6):533–546. 10.1111/ele.2008.11.issue-6. 10.1111/j.1461-0248.2008.01174.x

Lafferty KD, Dobson AP, Kuris AM (2006) Parasites dominate food web links. Proc Natl Acad Sci USA 103(30):11211–11216

Lasso CA, DoNascimiento C, Morales-Betancourt MA, Lasso-Alcalá OM (2015) Parasitism of freshwater stingrays (Potamotrygonidae) by hematophagous catfishes (Vandelliinae). Ichthyol Explor Freshw 26(1):83–86

Lee HJ, Kusche H, Meyer A (2012) Handed foraging behavior in scale-eating cichlid fish: its potential role in shaping morphological asymmetry. PLoS One 7(9):e44670

Leung TLF (2014) Fish as parasites: an insight into evolutionary convergence in adaptations for parasitism. J Zool 294(1):1–12. 10.1111/jzo.2014.294.issue-1. 10.1111/jzo.12148

Lima MRL, Bessa E, Krinski D, Carvalho LN (2012) Mutilating predation in the Cheirodontinae *Odontostilbe pequira* (Characiformes: Characidae). Neotrop Ichthyol 10(2):361–368

Losey Jr GS (1974) Cleaning symbiosis in Puerto Rico with comparison to the tropical Pacific. Copeia 960–970

Lucena CAS (2003) Revisão taxonômica e relações filogenéticas das espécies de Roeboides grupo-microlepis (Ostariophysi, Characiformes, Characidae). Iheringia, Sér Zool 93(3):283–308

Lucena CASD (2007) Revisão taxonômica das espécies do gênero Roeboides grupo-affinis (Ostariophysi, Characiformes, Characidae). Iheringia, Sér Zool 97:117–136

Major PF (1973) Scale feeding behavior of the leatherjacket, *Scomberoides lysan* and two species of the genus *Oligoplites* (Pisces: Carangidae). Copeia 1973(1):151–154

Marchio EA, Piller KR (2013) Scale-stacking in the live-bearer *Belonesox belizanus* (Poeciliidae). J Fish Biol 83(5):1387–1390

Marcogliese DJ (2003) Food webs and biodiversity: are parasites the missing link? J Parasitol 89(6):S106–S113

Martin CH, Wainwright PC (2011) Trophic novelty is linked to exceptional rates of morphological diversification in two adaptive radiations of *Cyprinodon* pupfish. Evolution 65:2197–2212

Martin CH, Wainwright PC (2013) On the measurement of ecological novelty: scale-eating pupfish are separated by 168 my from other scale-eating fishes. PLoS One 8:e71164

Mattox GM, Toledo-Piza M (2012) Phylogenetic study of the Characinae (Teleostei: Characiformes: Characidae). Zool J Linn Soc 165(4):809–915

Meza-Vargas V, Faustino-Fuster D, Marchena J, Ortega H (2019) Geographic distribution extension of *Landonia latidens* Eigenmann & Henn, 1914 (Characidae, Stevardiinae) in coastal drainages of Peru. Check List 15:851

Miranda Ribeiro Ade (1912) Loricariidae, Callichthyidae, Doradidae e Trichomycteridae. In: Commissão de Linhas Telegraphicas Estrategicas de Matto-Grosso ao Amazonas. Annexo 5:1–31

Mok HK (1978) Scale-feeding in *Tydemania navigatoris* (Pices: Triacanthodidae). Copeia 78:338

Nakae M, Sasaki K (2002) A scale-eating triacanthodid, *Macrorhamphosodes uradoi*: prey fishes and mouth "handedness" (Tetraodontiformes, Triacanthoidei). Ichthyol Res 49:2–5

Nico LG (1991) Trophic ecology of piranhas (Characidae: Serrasalminae) from savanna and forest regions in the Orinoco River basin of Venezuela. Unpublished Ph.D. dissertation. University of Florida, Gainesville

Nico LG, de Morales M (1994) Nutrient content of piranha (Characidae, Serrasalminae) prey items. Copeia (2):524–528

Nico LG, Taphorn DC (1988) Food habits of piranhas in the low llanos of Venezuela. Biotropica 20(4):311–321

Northcote TG, Arcifa MS, Froehlich O (1987) Fin-feeding by the piranha (*Serrasalmus spilopleura* Kner): the cropping of a novel renewable resource. Proc 5th Congr Europ Ichthyol Stockholm 1985:133–143

Novakowski GC, Fugi R, Hahn NS (2004) Diet and dental development of three species of *Roeboides* (Characiformes: Characidae). Neotrop Ichthyol 2(3):157–162

Nshombo M, Yanagisawa Y, Nagoshi M (1985) Scale-eating in *Perissodus microlepis* (Cichlidae) and change of its food habits with growth. Jpn J Ichthyol 32(1):66–73

Ochoa LE, Datovo A, DoNascimiento C, Roxo FF, Sabaj MH, Chang J, Melo BF, Silva GS, Foresti F, Alfaro M, Oliveira C (2020a) Phylogenomic analysis of trichomycterid catfishes (Teleostei: Siluriformes) inferred from ultraconserved elements. Sci Rep 10(1):1–15

Ochoa LE, Melo BF, García-Melo JE, Maldonado-Ocampo JA, Souza CS, Albornoz-Garzón JG, Oliveira C (2020b) Species delimitation reveals an underestimated diversity of Andean catfishes of the family Astroblepidae (Teleostei: Siluriformes). Neotrop Ichthyol 18(4):e200048

Pellegrin J (1909) Les Poissons du genre "Vandellia" CV. au siège de la Société philomatique 1(4-6):197–204

Peterson CC, Winemiller KO (1997) Ontogenetic diet shifts and scale-eating in *Roeboides dayi*, a Neotropical characid. Environ Biol Fishes 49(1):111–118

Peterson CC, McIntyre P (1998) Ontogenetic diet shifts in *Roeboides affinis* with morphological comparisons. Environ Biol Fishes 53(1):105–110

de Pinna MC (1992) A new subfamily of Trichomycteridae (Teleostei, Siluriformes), lower loricarioid relationships and a discussion on the impact of additional taxa for phylogenetic analysis. Zool J Linn Soc 106(3):175–229

de Pinna MCC (1998) Phylogenetic relationships of Neotropical Siluriformes (Teleostei: Ostariophysi): historical overview and synthesis of hypotheses. In: Malabarba LR, Reis RE, Vari RP, Lucena ZM, Lucena CAS (eds) *Phylogeny and classification of Neotropical fishes*. EDIPUCRS, pp 279–330

de Pinna MC (2016) The dawn of phylogenetic research on Neotropical fishes: a commentary and introduction to Baskin (1973), with an overview of past progress on trichomycterid phylogenetics. Neotrop Ichthyol 14(02):e150127. 10.1590/1982-0224-20150127

Ribeiro PM (1954) Catálogo dos Peixes do Museu Nacional. I. Pygidiidae Eigenmann & Eigenmann, 1888. Publicações Avulsas do Museu Nacional 15:1–17

Roberts TR (1970) Scale-eating American characoid fishes, with special reference to *Probolodus heterostomus*. Proc Calif Acad Sci 38:383–390

Robinson, B.W. and Wilson, D.S., 1998. Optimal foraging, specialization, and a solution to Liem's paradox. The American Naturalist, 151(3), pp.223-235.

Roberts TR (1972) Ecology of fishes of the Amazon and Congo Basins. Bull Mus Comp Zool 143:117–149

Röpke CP, Ferreira E, Zuanon J (2014) Seasonal changes in the use of feeding resources by fish in stands of aquatic macrophytes in an Amazonian floodplain, Brazil. Environ Biol Fishes 97(4):401–414

Santos O, Castro R (2014) Taxonomy of Probolodus Eigenmann, 1911 (Characiformes: Characidae) with description of two new species, and comments about the phylogenetic relationships and biogeography of the genus. Neotrop Ichthyol 12:403–418

Sazima I (1977) Possible case of aggressive mimicry in a neotropical scale-eating fish. Nature 270(5637):510–512

Sazima I (1983) Scale-eating in characoids and other fishes. Environ Biol Fishes 9(2):87–101

Sazima I, Machado FA (1990) Underwater observations of piranhas in western Brazil. Environ Biol Fishes 28(1–4):17–31

Sazima I, Moura RL, Rodrigues MCM (1999) A juvenile sharksucker, *Echeneis naucrates* (Echeneidae), acting as a station-based cleaner fish. Cybium 23(4):377–380

Sazima I, Machado FA (1982) Hábitos e comportamento de Roeboides prognathus, um peixe lepidófago (Osteichthyes, Characoidei). Bol Zool 7(7):37–56

Sazima I, Uieda VS (1980) Scale-feeding behavior of *Oligoplites saurus* and recording of scale feeding in *Oligoplites palometa* and *Oligoplites saliens* (Pisces, Carangidae). Rev Bras Biol 40:701–710

Schmidt RE (1993) Relationships and notes on the biology of *Paracanthopoma parva* (Pisces: Trichomycteridae). Ichthyol Explor Freshw 4(2):185–191

Souza CS, Melo BF, Mattox GM, Oliveira C (2022) Phylogenomic analysis of the Neotropical fish subfamily Characinae using ultraconserved elements (Teleostei: Characidae). Mol Phylogenet Evol 171:107462

Spix JB von, Agassiz L (1829) Selecta genera et species piscium quos in itinere per *Brasiliam annis*. Selecta genera et species piscium quos in itinere per *Brasiliam annis* MDCCCXVII-MDCCCXX : jussu et auspiciis Maximiliani Josephi I. *Bavariae regis augustissimi* peracto. (1):1–82

Spotte S, Petry P, Zuanon JA (2001) Experiments on the feeding behavior of the hematophagous candiru, *Vandellia cf. plazaii*. Environ Biol Fishes 60(4):459–464

St. John MES, Holzman R, Martin CH (2020) Rapid adaptive evolution of scale-eating kinematics to a novel ecological niche. J Exp Biol 223(6). 10.1242/jeb.217570

Stewart TA, Albertson RC (2010) Evolution of a unique predatory feeding apparatus: functional anatomy, development and a genetic locus for jaw laterality in Lake Tanganyika scale-eating cichlids. BMC Biol 8(1):8

Strasburg DW (1959) Notes on the diet and correlating structures of some central Pacific echeneid fishes. Copeia 1959(3):244–248

Sukhdeo MV (2010) Food webs for parasitologists: a review. J Parasitol 96(2):273–284

Sukhdeo MVK (2012) Where are the parasites in food webs? Paras Vect 5:239

Sukhdeo MVK, Hernandez AD (2005) Food web patterns and the parasite's Perspective. In: Thomas F, Guegan JF, Renaud F (eds) *Parasitism and ecosystems*. Oxford University Press, pp 54–67

Szelistowski WA (1989) Scale-feeding in juvenile marine catfishes (Pisces: Ariidae). Copeia 1989(2):517–519

Takahashi R, Moriwaki T, Hori M (2007) Foraging behaviour and functional morphology of two scale-eating cichlids from Lake Tanganyika. J Fish Biol 70(5):1458–1469

Takahashi R, Watanabe K, Nishida M, Hori M (2007) Evolution of feeding specialization in Tanganyikan scale-eating cichlids: a molecular phylogenetic approach. BMC Evol Biol 7(1):195

Thompson AW, Betancur-RR, López-Fernández H, Ortí G (2014) A time-calibrated, multi-locus phylogeny of piranhas and pacus (Characiformes: Serrasalmidae) and a comparison of species tree methods. Mol Phylogenet Evol 81:242–257

Toro OA, Arango SA, Miranda EF (2001) La vandellia cirrhosa un parásito urológico poco conocido. Actas Urol Esp 25(4):325–326. 10.1016/S0210-4806(01)72626-5

Vari RP (1986) Serrabrycon magoi, a new genus and species of scale-eating characid (Pisces: Characiformes) from the upper Rio Negro. Proc Biol Soc Wash 99(2):328–334

Vieira I, Géry J (1979) Crescimento diferencial e nutrição em *Catoprion mento* (Characoidei). Peixe lepidófago da Amazônia. Acta Amaz 9(1):143–146

Vitule JRS, Aranha JMR (2002) Ecologia alimentar do lambari, Deuterodon langei Travassos, 1957 (Characidae, Tetragonopterinae), de diferentes tamanhos em um riacho da Floresta Atlântica, Paraná (Brasil). Acta Biológica Paranaense 31(4), 137-150.

Vilella FS, Becker FG, Hartz SM (2002) Diet of *Astyanax* species (Teleostei, Characidae) in an Atlantic forest river in Southern Brazil. Braz Arch Biol Technol 45:223–232

Whitfield AK, Blaber SJM (1978) Scale-eating habits of the marine teleost *Terapon jarbua*. J Fish Biol 12:61–70

Winemiller KO, Yan HY (1989) Obligate mucus-feeding in a South American trichomycterid catfish (Pisces: Ostariophysi). Copeia 1989(2):511–514

Zardo ÉL, Behr ER (2013) Dieta e Ritmo Circadiano da Atividade Alimentar do Tamoatá *Hoplosternum littorale* (Hancock, 1828) (Siluriformes, Callichthyidae) Capturados no Rio Vacacaí, Rio Grande do Sul, Brasil. Rev Bras Multidiscip 16(2):55–65

Zaret TM (ed) (2013) Evolutionary ecology of neotropical freshwater fishes. Springer Science & Business Media

Zepeda Mendoza ML, Xiong Z, Escalera-Zamudio M, Runge AK, Thézé J, Streicker D, Frank HK, Loza-Rubio E, Liu S, Ryder OA, Samaniego Castruita JA, Katzourakis A, Pacheco G, Taboada B, Löber U, Pybus OG, Li Y, Rojas-Anaya E, Bohmann K, Carmona Baez A, Arias CF, Liu S, Greenwood AD, Bertelsen MF, White NE, Bunce M, Zhang G, Sicheritz-Pontén T, Gilbert MPT (2018) Hologenomic adaptations underlying the evolution of sanguivory in the common vampire bat. Nat Ecol Evol 2(4):659–668

Zuanon J, Sazima I (2004) Vampire catfishes seek the aorta not the jugular: Candirus of the genus Vandellia (Trichomycteridae) feed on major gill arteries of host fishes. J Ichthyol Aquat Biol 8(1):31–36

Zuanon J, Sazima I (2005) Free meals on long-distance cruisers: the vampire fish rides giant catfishes in the Amazon. Biota Neotrop 5(1):109–114

19 Miniaturization in Neotropical Freshwater Fishes

Pedro H. N. Bragança and Michael D. Burns

Abstract

This chapter reviews the evidence for the evolution of miniaturization in Amazonian fishes. First, we present an alternative criterion to identify miniature fish *species* that incorporates *phylogenetic* relationships instead of an arbitrary size threshold. We then update the list of miniature Neotropical Freshwater Fish (NFF) species. We also show that most miniature NFF species feed primarily on benthic invertebrates and occupy acidic blackwater *habitats*. We then show that NFF species occupying blackwater habitats have significantly smaller adult body sizes than species inhabiting rivers with other water-chemistry profiles, and show that phylogenetic transitions to blackwater habitats are associated with a significant reduction in body size across all clades. Lastly, we show that the conservation status of most miniature Amazonian species is poorly known, but that most miniature species are likely at a high risk of *extinction* due to their small geographic range sizes and narrow habitat tolerances.

19.1 Introduction to Miniature Neotropical Freshwater Fishes

Body size is a key biological and ecological trait that varies substantially across fishes, and which influences almost all aspects of their physiological and ecological performances, including metabolic rate (Kleiber 1932), dispersal capacity (Roff 1988, 1991; Bloom et al. 2018; Burns and Bloom 2020), foraging capacity and population sizes (Romanuk et al. 2011; Burress et al. 2016; Keppeler et al. 2020), and fecundity (Roff 1992; Barneche et al. 2018). Because of these intimate relationships between adult body size (hereafter body size) and ecological performances, fishes have adapted sizes under the influences of multiple and often conflicting ecological and evolutionary pressures (Knope and Scales 2013; Fronk et al. 2019; Kolmann et al. 2020; Egan et al. 2021; de Brito et al. 2022).

The different pressures of *natural selection* across ecological gradients have led to an astounding range of body sizes of fishes in the Amazon basin and other regions of South America. Neotropical Freshwater Fish (NFF) species range in size from just under 1.0 cm (0.4 in) standard length (SL) in species of *Leptophilypnion* (Eleotridae) and *Trichomycterus* (Trichomycteridae), to the Largetooth sawfish *Pristis pristis* (Pristidae) with a maximum recorded size of 610 cm (about 20 ft), and the Pirarucu *Arapaima gigas* (Arapaimidae) at 450 cm in SL (about 15 ft) and weighing over 200 kg (440 lbs). While much is known about growth and body size diversity of fishes generally (Knouft and Page 2003; Albert and Johnson 2012; Steele and López-Fernández 2014), little is known about the ecological and evolutionary processes underlying body-size evolution in the NFF fauna (but see Cerezer et al. 2023).

The major *macroevolutionary* patterns of size-evolution among fishes generally are: (1) to maintain a similar body sizes across speciation events (i.e., evolutionary stasis), and (2) to stochastically increase the range of body sizes, eventually generating a right-skewed frequency distribution of log-body sizes, with a longer tail towards the larger end of the size spectrum (Albert et al. 2009; Albert and Johnson 2012). The macroevolutionary causes of these patterns are hypothesized to be (1) inheritance, phylogenetic niche conservatism, and elevated extinction risk of species with extreme body sizes serving to keep species closer to ancestral trait values (Crisp and Cook 2012; Pyron et al. 2015; Ripple et al. 2017; Chichorro et al. 2019), and (2) entropy and adaptation driving species away from ancestral trait values (Brooks and Wiley 1988; Albert et al. 2020; Burns 2021; Melo et al. 2022).

P. H. N. Bragança
American Museum of Natural History, New York, NY, USA
e-mail: pedrobra88@gmail.com

M. D. Burns (✉)
Oregon State University, Corvallis, OR, USA
e-mail: mdburns01@gmail.com

J. S. Albert and R. E. Reis (eds.), *Ecology and Evolution of Amazonian Fishes*, Fascinating Life Sciences,
https://doi.org/10.1007/978-3-032-09328-8_19

These empirically observed macroevolutionary patterns contradict the expectations of the widely cited Cope's rule, which posits a directional tendency to evolve larger body sizes within individual species lineages, under microevolutionary pressures of natural selection (Hone and Benton 2005). However, larger species are also predicted to have: (1) higher viability costs due to extended development, as well as higher predation, parasitism, and energy requirements, and (2) decreased reproductive success to delayed achievement of the reproductive stage (Blanckenhorn 2000). In general, there is a lack of conclusive results from experimental and empirical studies on the selective pressures affecting adult body size in fishes (Gagliano et al. 2007; Ahti et al. 2020).

This chapter focuses on the different evolutionary and ecological aspects favoring the selection of extremely reduced body sizes in Neotropical fishes, a process known as *miniaturization*. From an evolutionary perspective, a miniature may be defined as a species or clade that has undergone extreme phylogenetic reduction in adult body size, impacting phenotypes across the full integrated hierarchy of phenotypic organization (see Fig. 7.1), and affecting functionally important aspects of metabolism, physiology, morphology, and behavior related to locomotion, feeding, dispersal and reproduction (Hanken and Wake 1993; Britz et al. 2009, Britz and Conway 2009, Britz et al. 2014; Britz and Conway 2016). Evolutionary miniaturization may arise from developmental truncation or developmental acceleration resulting in the reduction, simplification, or loss of these phenotypes (Weitzman and Vari 1988; Hanken and Wake 1993).

Among vertebrates, evolutionary miniaturization is generally associated with the reduction and loss of bones. Structural simplification of the skeleton can arise from reduced mineralization or non-ossification of cartilaginous or membranous bone-precursor tissues (Hanken and Wake 1993, Johnson and Brothers 1993; Britz and Kottelat 2003). Among fishes, miniaturization frequently results in simplification of the cephalic laterosensory canal system, reduction in the number of vertebrae, scales, and fin rays, reduced or lack of separation of the median fins, and reduced sculpturing and ornamentation of dermal bones (Weitzman and Vari 1988). For example, miniatures species of the characiform genus *Priocharax* share several pedomorphic features of the laterosensory canal system, numbers of fin rays, and presence of a single series of teeth in the premaxilla, maxilla, and dentary bones (Mattox et al. 2024). Similarly, species of the cyprinodontiform genus *Fluviphylax* (Fluviphylacidae) are characterized by the absence of the vomer and interhyal bones, a reduced number of caudal-fin rays, and a simplified laterosensory canal system (Bragança 2018).

Historically, little attention was directed to miniature NFF species during the first expeditions of the nineteenth century. Many specimens were regarded as juveniles of larger-bodies species, or just ignored and forgotten on museum collection shelves (Roberts 1984; Weitzman and Vari 1987, 1988; Costa and Le Bail 1999). However, the past four decades have seen a dramatic increase in the discovery rate and description of small-sized NFF species (Weitzman and Vari 1988; Costa and Le Bail 1999; Toledo-Piza et al. 2014). Although little is known about the habitat and ecological conditions associated with miniatures, Weitzman and Vari (1988) suggested that most miniature NFF species are habitat specialists, appearing in specific and restricted microhabitats, mainly in slow-flowing or lentic clearwater or blackwater rivers and streams. The overall lack of data on the ecology of miniature NFF species hampers a clearer understanding of the possible evolutionary pathways favoring a highly reduced body size.

The goals of this chapter are to: (1) present alternative criteria to identify miniature NFF species, (2) update the list of miniatures described for the Neotropical region; (3) provide a historical overview of the description rate of Neotropical miniatures; (4) present data on their ecological requirements and diet; and (5) comment on the conservation status of miniature NFF species.

19.2 Definitions of Miniaturization

Almost four decades have passed since Weitzman and Vari (1988) published the first compilation of miniature fish species of the Neotropical region, proposing two arbitrary size thresholds to define miniature species: (1) adults not reaching a maximum SL over 26 mm; or (2) individuals reaching sexual maturity below 20 mm SL. The justification for these size thresholds was that most fishes that mature at such small sizes exhibit a characteristic syndrome of reductive morphological characters, especially reduction and loss of bones used in phylogenetic analyses. Knowledge of phylogenetic relationships among Neotropical fishes was at that time far from complete, thus not allowing the use of relative body size, defined as size differences from close relatives, into their definition of evolutionary miniaturization. Weitzman and Vari (1988) recognized a weakness of their proposed arbitrary thresholds in the presence of species that slightly surpassed the size thresholds but nonetheless exhibit apparent pedomorphic and reductive traits typical of miniatures. For instance, the *Tomeurus gracilis* (Poeciliidae) and many trichomycterid catfishes (e.g., *Listrura* species) are pedomorphic with reductive characteristics typical of miniatures, but were not classified as such due to their size slightly surpassing the arbitrary thresholds. Unfortunately, in many cases, such species have been systematically removed from miniature lists without further information on the presence or absence of reductive or pedomorphic traits (e.g., Toledo-Piza et al. 2014).

Subsequent advances in the study of evolutionary relationships among Neotropical fishes have made it clear that the field needs to include knowledge of the phylogenetic relationships to identify miniature species. Therefore, we propose a broader definition of miniatures to include species belonging to lineages that have undergone a reduction of body size relative to their closely-related lineages, coupled with morphological reductions and losses of structures. In doing that, the central criteria defining a miniature is no longer restricted to body size *per se*, but with the reduction of body size in a given lineage or species within a particular phylogenetic context. Thus, according to the proposed method, identifying a miniature species does not rely solely on the species maximum size, but on whether the lineage the species belongs to has undergone a reduction in body size relative to its closely related and sister lineages. Recognizing miniatures based on the presence of morphological simplifications, reductions, and losses. This definition will expand the study of miniaturization in groups currently underrepresented or missing from the lists of NFF miniatures (e.g., cichlid *Apistogramma*; gymnotiform *Hypopygus*).

19.3 Evolution of Body Size

For this chapter, we compiled a comprehensive dataset on the phylogenetic distribution and evolution of reductive traits and losses for NFF species showing size reductions. Body size data were extracted from the ecomatrix presented in Chap. 6, and methods of ancestral state estimation were adapted from those presented in Chap. 7. Following the size-based criteria presented in Weitzman and Vari (1988), there are a total of 279 NFF miniature species, belonging to the following orders: Characiformes ($n = 104$), Cichliformes ($n = 4$), Clupeiformes ($n = 2$), Cyprinodontiformes ($n = 86$), Gobiiformes ($n = 8$), and Siluriformes ($n = 75$) (Fig. 19.1; Table 19.1).

Analyses of dates of species descriptions confirm an increase in the discovery and description rates from 1981 to 2020 (Tagliacollo et al. 2021). During this time, about 60% of all valid miniature species were described (Fig. 19.2). Two orders contain the highest number of miniature species descriptions during this time, including Cyprinodontiformes with about 84%, and Siluriformes with 64% (Fig. 19.2). Curiously, Characiformes, which includes the highest number of miniatures, did not show an increase in the rate of miniature species descriptions, with an average of about ten species described per decade (Fig. 19.2). The increased number of miniature NFF species described is a direct result of more specialized and directed sampling of the unusual biotopes where most miniature species occur (Costa and Le Bail 1999).

19.4 Habitat Utilization

Previous studies have shown an association between specific water types, such as clearwater and blackwaters, and a greater diversity of miniature fishes (Roberts, 1984; Weitzman and Vari, 1988; Perkin et al. 2022). However, until now, this relationship has not been clearly understood, with few studies looking at the ecology of miniature species (Henderson and Walker, 1990; Carvalho et al. 2006; Perkin et al. 2022). The different chemical and physical properties known for each water type likely play a significant role in selecting smaller body sizes in clearwater and blackwater habitats, and may explain the relatively low diversity observed in whitewaters. For example, whitewater rivers, which are muddy with a brownish coloration, contrast with clearwater and blackwater rivers by the presence of a high sediment load, which can damage and block the ability of fish gills to filter particles and extract oxygen (Chap. 8). The increased sediment load is especially problematic for small-sized species, significantly limiting their distribution and movement and, in some cases, causing death (Kjelland et al. 2015). Additionally, a high concentration of suspended sediments in whitewater affects fishes' visual, chemical, and mechanical sensorial responses (Borghezan et al. 2021). These sediments probably reduce the feeding and communication capabilities of miniature fish inhabiting the water column, the most common microhabitat among miniatures (49%, Fig. 19.3).

In addition to the suspended sediment load and the differences in nutrient availability, the acidity (pH) of the different water types is likely a primary determinant of the characteristic fish fauna of a given locality. In contrast to clearwater and whitewater rivers with pH values in the range 6.2–7.2, blackwater rivers are more acidic, with values ranging from 3.8 to 4.9 (Sioli 1984; Borghezan et al. 2021). More acidic water implies a significant challenge for freshwater fishes in maintaining their ionic balance, influencing the gill's functional ability (Kwong et al. 2014). Studies indicate that some small characids, including *Paracheirodon axelrodi* (Schultz, 1956), *P. innesi* (Myers, 1936) (a miniature species), *Gymnocorymbus ternetzi* (Boulenger, 1895), and *Hemigramus* sp., from the Negro River have developed a unique way of counterbalancing the ionic Na–K osmotic losses in poorly mineralized and ion-poor blackwater (Gonzalez et al. 2005). Apparently, their $Na+$ and $Cl-$ mechanism uptake is not sensible to changes in water pHs as low as 3.25, instead they show an ion uptake mechanism that is highly specialized for operation in ion-poor and low pH waters. However, no other studies have approached the mechanisms that might play an essential role in the ionic balance in other miniature blackwater species and if a reduction in the body size could make ionic balance less energy costly. Similarly, there are no studies investigating how the low calcium concentration in blackwaters might affect the osteogenesis of different fish species or if it could be related to a decrease in fish body size.

According to the present data, most miniature species are known to occur in clearwater (46.9%), clearwater and blackwater (22.6%), blackwater (16.8%), and whitewater (9%) rivers (Fig. 19.3). The Cyprinodontiformes species are known from

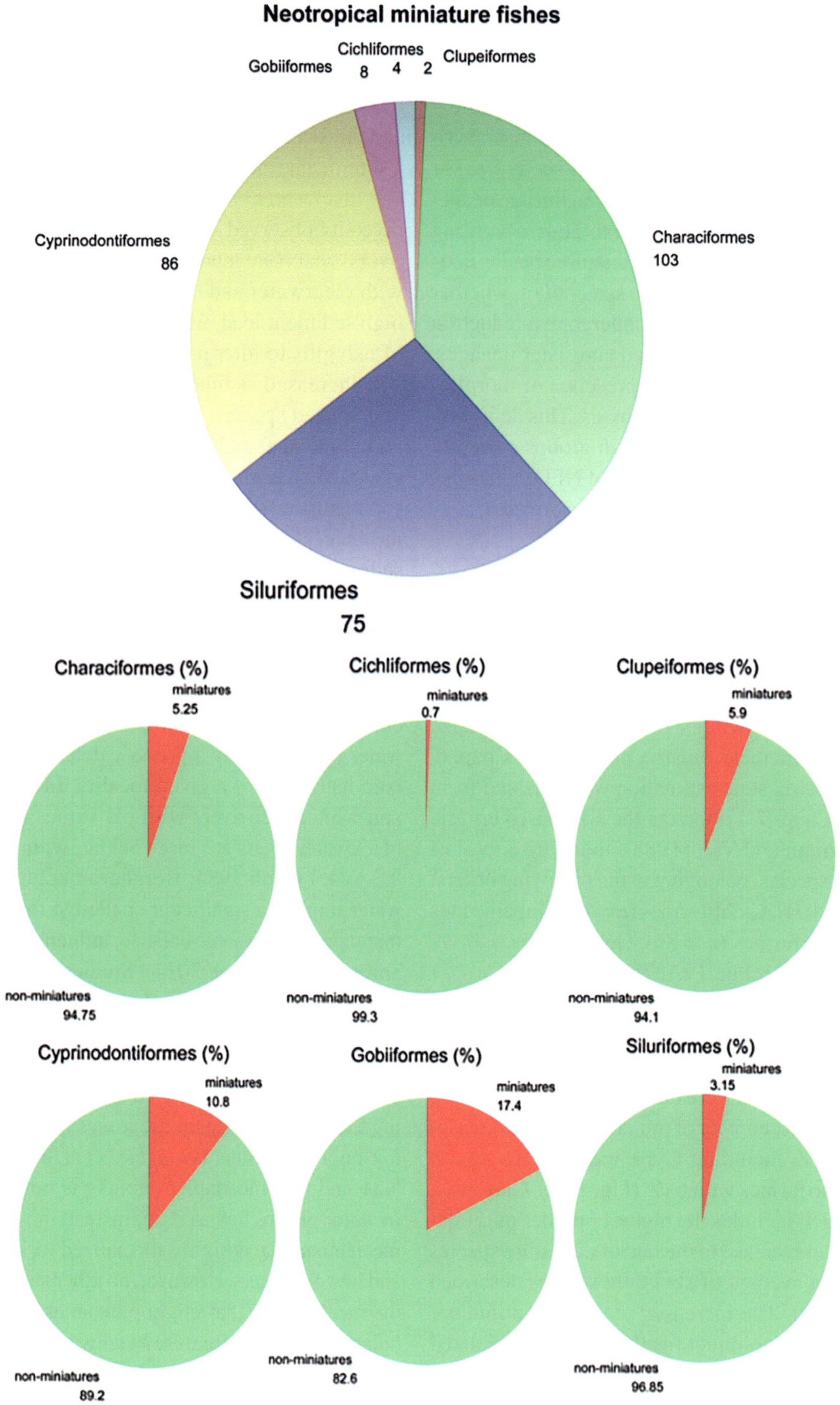

Fig. 19.1 Taxonomic composition of miniature species of Neotropical freshwater fishes. Top: Species per order. Bottom: Proportions of miniature species in each taxonomic order

clearwater and blackwater (38.4 %), clearwater (31.4%) and blackwater (38.4%) rivers. In contrast, despite Siluriformes and Characiformes miniature species also occurring in the same habitats as cyprinodontiform species, they are also present in whitewater rivers. Most miniature catfishes occur in clearwater (55.4%), whitewater (21.5%), blackwater (7.7%), and clearwater or blackwater (7.7%) rivers, and the Characiformes can be found in clearwater (58.1%), clearwater and blackwater

Table 19.1A List of miniature NFF species, with information on their ecology and conservation. NFF miniature species in four studies. Reference abbreviations: W&V88 = Weitzman and Vari (1988); C&L99 = Costa and Le Bail (1999); T-P14 = Toledo-Piza et al. (2014); and B&B26 = Bragança and Burns (2026, this chapter). Species arranged in phylogenetic sequence by Order, Family, Subfamily, and Genus; only family names shown

Family	Genus species	SL	W&V88	C&L99	T-P14	B&B26
Engraulidae	*Amazonsprattus scintilla*	19.5	X			
	Anchoviella manamensis	25.0	X			
Characidae	*Axelrodia lindeae*	20.6	X			
	Axelrodia riesei	16.7	X			
	Axelrodia stigmatias	20.5	X			
	Microschemobrycon cryptogrammus	23.6				X
	Microschemobrycon elongatus	25.0	X			
	Microschemobrycon meyburgi	22.0			X	
	Oxybrycon parvulus	15.7	X			
	Tyttobrycon dorsimaculatus	20.5	X			
	Tyttobrycon hamatus	16.9	X			
	Tyttobrycon marajoara	22.1			X	
	Tyttobrycon shibattai	19.0				X
	Tyttobrycon spinosus	20.5	X			
	Tyttobrycon xeruini	22.6	X			
	Amazonspinther dalmata	19.6			X	
	Cheirodon luelingi	17.6	X			
	Nanocheirodon insignis	24.4			X	
	Odontostilbe gracilis	23.5			X	
	Odontostilbe littoris	18.0			X	
	Serrapinnus malabarbai	24.5				X
	Serrapinnus kriegi	23.6			X	
	Spintherobolus broccae	25.6	X			
	Priocharax ariel	17.1	X			
	Priocharax britzi	14.1				X
	Priocharax nanus	15.4			X	
	Priocharax pygmaeus	16.4	X			X
	Priocharax varii	14.0				
	Creagrutus maracaiboensis	22.0			X	
	Creagrutus nigrostigmatus	23.4			X	
	Cyanogaster noctivaga	17.4			X	
	Iotabrycon praecox	21.8	X			
	Pterobrycon landoni	25.1			X	
	Scopaeocharax atopodus	22.0	X			
	Scopaeocharax rhinodus	25.0	X			
	Trochilocharax ornatus	17.0			X	
	Tyttocharax cochui	22.0	X			
	Tyttocharax madeirae	17.5	X			
	Tyttocharax metae	15.8			X	
	Tyttocharax tambopatensis	15.5		X		
	Xenurobrycon coracoralinae	15.8			X	
	Xenurobrycon heterodon	20.1	X			
	Xenurobrycon macropus	19.8	X			

(continued)

Table 19.1A (continued)

Family	Genus species	SL	W&V88	C&L99	T-P14	B&B26
	Xenurobrycon polyancistrus	13.8	X			
	Xenurobrycon varii	16.8				X
	Xenurobrycon pteropus	13.8	X			
	Hemigrammus aereus	24.0			X	
	Hemigrammus iota	21.0	X			
	Hemigrammus tridens	20.0			X	
	Hyphessobrycon albolineatus	25.8			X	
	Hyphessobrycon amandae	19.5	X			
	Hyphessobrycon arianae	24.0			X	
	Hyphessobrycon axelrodi	22.0			X	
	Hyphessobrycon catableptus	18.0	X			
	Hyphessobrycon caru	25.4				X
	Hyphessobrycon ecuadorensis	20.5			X	
	Hyphessobrycon eilyos	24.7				X
	Hyphessobrycon elachys	17.9	X			
	Hyphessobrycon frickei	19.6				X
	Hyphessobrycon gracilior	21.5	X			
	Hyphessobrycon griemi	25.7	X			
	Hyphessobrycon haraldschultzi	21.0				X
	Hyphessobrycon heteresthes	17.0				X
	Hyphessobrycon hildae	18.8				X
	Hyphessobrycon loretoensis	24.0	X			
	Hyphessobrycon minimus	21.0	X			
	Hyphessobrycon myrmex	25.6				X
	Hyphessobrycon parvellus	21.6			X	
	Hyphessobrycon piranga	23.1				X
	Hyphessobrycon roseus	19.3			X	
	Hyphessobrycon saizi	23.0			X	
	Hyphessobrycon taphorni	24.1				X
	Hyphessobrycon tenuis	26.0			X	
	Hyphessobrycon tropis	21.3			X	
	Hyphessobrycon tukunai	20.6	X			
	Hyphessobrycon weitzmanorum	25.5				X
	Paracheirodon innesi	22.2	X			
	Paracheirodon simulans	20.2	X			
Characidae *insertae sedis*	*Bryconella pallidifrons*	23.2			X	
	Tucanoichthys tucano	16.6			X	
Crenuchidae	*Ammocryptocharax minutus*	19.7		X		
	Characidium bahiense	23.3	X			
	Characidium chicoi	18.7				X
	Characidium geryi	24.0			X	
	Characidium heinianum	25.0			X	
	Characidium mirim	20.2			X	
	Characidium nana	21.5				X
	Characidium pteroides	21.1			X	

Table 19.1A (continued)

Family	Genus species	SL	W&V88	C&L99	T-P14	B&B26
	Elachocharax geryi	19.0	X			
	Elachocharax junki	23.1	X			
	Elachocharax mitopterus	13.9	X			
	Elachocharax pulcher	22.1	X			
	Klausewitzia ritae	25.0	X			
	Microcharacidium bombioides	23.2				X
	Microcharacidium eleotrioides	21.0	X			
	Microcharacidium gnomus	22.3		X		
	Microcharacidium weitzmani	12.0		X		
	Odontocharacidium aphanes	16.5	X			
	Odontocharacidium varii	16.6				X
Gasteropelecidae	*Carnegiella myersi*	21.5			X	
	Carnegiella schereri	26.0			X	
Lebiasinidae	*Nannostomus anduzei*	16.2	X			
	Nannostomus britskii	24.0			X	
	Nannostomus minimus	23.0	X			
	Nannostomus rubrocaudatus	25.7				X
Aspredinidae	*Acanthobunocephalus nicoi*	19.7		X		
	Acanthobunocephalus scruggsi	22.1				X
	Hoplomyzon atrizona	25.3				X
	Hoplomyzon cardosoi	18.5				X
	Hoplomyzon papillatus	16.9	X			
	Dupouyichthys sapito	23.0				X
	Micromyzon akamai	15.8		X		
	Micromyzon orinoco	15.6				X
	Ernstichthys taquari	22.8				X
Astroblepidae	*Astroblepus chimborazoi*	25.0			X	
Auchenipteridae	*Gelanoglanis nanonocticolus*	22.2			X	
	Gelanoglanis pan	24.7				X
	Gelanoglanis vari	17.7				X
	Tatia marthae	23.1			X	
Callichthyidae	*Aspidoras brunneus*	21.3	X			
	Aspidoras carvalhoi	25.4	X			
	Corydoras boehlkei	25.7			X	
	Corydoras cochui	25.0			X	
	Corydoras gracilis	23.2	X			
	Corydoras habrosus	20.1	X			
	Corydoras hastatus	24.0	X			
	Corydoras pygmaeus	23.7	X			
Cetopsidae	*Denticetopsis royeroi*	18.0			X	
	Denticetopsis sauli	21.0			X	
Doradidae	*Physopyxis ananas*	22.0			X	
	Physopyxis cristata	22.7			X	
Heptapteridae	*Horiomyzon retropinnatus*	24.0	X			
Loricariidae	*Corumbataia liliai*	24.7				X

(continued)

Table 19.1A (continued)

Family	Genus species	SL	W&V88	C&L99	T-P14	B&B26
	Hisonotus bocaiuva	25.6				X
	Rhinotocinclus bockmanni	23.5				X
	Rhinotocinclus dinizae	23.3				X
	Nannoplecostomus eleonorae	22.2			X	
	Otocinclus tapirape	24.1			X	
	Parotocinclus amazonensis	23.0			X	
	Rhinotocinclus collinsae	25.5			X	
	Rhinotocinclus halbothi	19.9			X	
	Rhinotocinclus pentakelis	23.5				X
	Rhinotocinclus hardmani	25.1				X
Pseudopimelodidae	*Microglanis robustus*	23.3				X
	Microglanis zonatus	20.0			X	
Scoloplacidae	*Scoloplax baileyi*	15.4			X	
	Scoloplax baskini	16.1			X	
	Scoloplax dicra	13.8	X			
	Scoloplax distolothrix	17.9		X		
	Scoloplax dolicholophia	12.0		X		
	Scoloplax empousa	19.9		X		
Trichomycteridae	*Pygidianops cuao*	18.7			X	
	Pygidianops eigenmanni	23.0	X			
	Pygidianops magoi	12.6			X	
	Typhlobelus macromycterus	21.9			X	
	Typhlobelus guacamaya	22.3			X	
	Ammoglanis amapaensis	17.9			X	
	Ammoglanis diaphanus	18.7		X		
	Ammoglanis multidentatus	19.7				X
	Ammoglanis natgeorum	12.9				X
	Ammoglanis obliquus	15.5				X
	Ammoglanis pulex	14.9			X	
	Malacoglanis gelatinosus	19.9	X			
	Microcambeva draco	24.6			X	
	Microcambeva mucuriensis	25.2				X
	Stenolicmus ix	19.0				X
	Sarcoglanis simplex	21.0	X			
	Schultzichthys gracilis	26.0			X	
	Ituglanis compactus	25.8				X
	Miuroglanis platycephalus	12.3	X			
	Potamoglanis anhanga	13.1			X	
	Potamoglanis hasemani	18.0	X			
	Potamoglanis johnstoni	16.0	X			
	Potamoglanis wapixana	15.7				X
	Tridensimilis venezuelae	25.0	X			
	Tridentopsis cahuali	22.2			X	
	Tridentopsis pearsoni	23.0	X			
	Tridentopsis tocantinsi	23.0	X			

Table 19.1A (continued)

Family	Genus species	SL	W&V88	C&L99	T-P14	B&B26
	Paracanthopoma saci	21.8				X
	Paracanthopoma cangussu	15.7				X
Fluviphylacidae	*Fluviphylax gouldingi*	14.6				X
	Fluviphylax obscurus	17.3		X		
	Fluviphylax palikur	13.9		X		
	Fluviphylax pygmaeus	22.0	X			
	Fluviphylax simplex	15.5		X		
	Fluviphylax wallacei	14.4				X
	Fluviphylax zonatus	15.9		X		
Poeciliidae	*Cnesterodon raddai*	23.2			X	
	Neoheterandria elegans	18.0	X			
	Phalloptychus eigenmanni	22.8			X	
	Phallotorynus dispilos	25.3			X	
	Phallotorynus psittakos	25.3			X	
	Phallotorynus victoriae	23.0			X	
	Poecilia (Pamphorichthys)	23.8				X
	Poecilia (Pamphorichthys)	24.5			X	
	Poecilia (Pamphorichthys)	22.4			X	
	Poecilia (Pamphorichthys)	15.0	X			
	Poecilia (Pamphorichthys)	20.0			X	
	Poecilia (Pamphorichthys)	23.0	X			
	Poecilia (Pamphorichthys)	24.8			X	
	Poecilia (Micropoecilia)	19.8				X
	Poecilia (Micropoecilia)	24.4				X
	Poecilia (Micropoecilia)	20.0				X
	Poecilia (Micropoecilia)	23.6		X		
	Poecilia (Micropoecilia)	22.4				X
	Poecilia (Micropoecilia)	14.9				X
	Poecilia (Micropoecilia)	20.0			X	
	Pseudopoecilia austrocolumbiana	20.0			X	
	Scolichthys iota	21.8			X	
Rivulidae	*Anablepsoides gamae*	22.2				X
	Anablepsoides henschelae	22.5				X
	Anablepsoides ottonii	22.4				X
	Atlantirivulus jurubatibensis	25.7				X
	Austrolebias paranaensis	25.4				X
	Campellolebias intermedius	23.9				X
	Hypsolebias fasciatus	20.0				X
	Hypsolebias marginatus	23.1				X
	Hypsolebias notatus	25.2				X
	Laimosemion jauaperi	18.9			X	
	Laimosemion kirovskyi	22.7			X	
	Laimosemion leticia	23.1				X
	Laimosemion nicoi	19.1				X
	Laimosemion romeri	21.7			X	

(continued)

Table 19.1A (continued)

Family	Genus species	SL	W&V88	C&L99	T-P14	B&B26
	Laimosemion uatuman	22.2			X	
	Laimosemion ubim	18.0			X	
	Leptolebias marmoratus	23.3			X	
	Leptopanchax itanhaensis	21.8			X	
	Leptopanchax sanguineus	20.9				X
	Leptopanchax splendens	22.7				X
	Maratecoara gesmonei	24.5				X
	Melanorivulus atlanticus	17.8				X
	Melanorivulus bororo	24.6				X
	Melanorivulus decoratus	23.2				X
	Melanorivulus egens	26.0			X	
	Melanorivulus jalapensis	20.1				X
	Melanorivulus javaeh	24.0				X
	Melanorivulus modestus	23.8				X
	Melanorivulus nigropunctatus	24.1				X
	Melanorivulus paracatuensis	23.9			X	
	Melanorivulus paresi	25.8				X
	Melanorivulus pictus	25.8				X
	Melanorivulus rossoi	21.5			X	
	Melanorivulus rubroreticulatus	25.4				X
	Melanorivulus rutilicaudus	22.9				
	Melanorivulus schuncki	24.4				X
	Mucurilebias leitaoi	22.0				X
	Notholebias cruzi	22.9		X		
	Notholebias fractifasciatus	21.0				X
	Notholebias vermiculatus	23.7			X	
	Papiliolebias francescae	21.8				X
	Plesiolebias altamira	18.1				X
	Plesiolebias aruana	19.9		X		
	Plesiolebias canabravensis	19.5				X
	Plesiolebias filamentosus	20.7				X
	Plesiolebias fragilis	20.7				X
	Plesiolebias glaucopterus	24.1		X		
	Plesiolebias lacerdai	19.7		X		
	Simpsonichthys cholopteryx	23.5			X	
	Simpsonichthys nigromaculatus	25.6			X	
	Simpsonichthys parallelus	23.5			X	
	Spectrolebias costae	19.8		X		
	Spectrolebias gracilis	20.8				X
	Spectrolebias reticulatus	20.1			X	
	Spectrolebias semiocellatus	22.2		X		
	Stenolebias bellus	16.5		X		
	Stenolebias damascenoi	24.4		X		
Eleotridae	*Leptophilypnion fittkaui*	9.7			X	
	Leptophilypnion pusillus	9.1			X	

Table 19.1A (continued)

Family	Genus species	SL	W&V88	C&L99	T-P14	B&B26
	Microphilypnus acangaquara	18.5			X	
	Microphilypnus hypolyrasimeion	14.4				X
	Microphilypnus macrostoma	20.0	X			
	Microphilypnus tapajosensis	22.7			X	
	Microphilypnus ternetzi	23.2	X			
Gobiidae	*Gobiosoma yucatanum*	26.0			X	
Cichidae	*Apistogramma agayuara*	24.7				X
	Apistogramma inornata	20.7				X
	Apistogramma minima	19.8				X
	Apistogramma staecki	20.8			X	
23	278		69	23	95	90

Table 19.1B List of miniature NFF species with ecological attributes. B = blackwater, C = clearwater, E = estuarine, Gen = Generaist, and W = whitewater.

Genus species	Water	Gradient	Mesohabitat	Microhabitat
Amazonsprattus scintilla	Gen	Lo	Pelagic channel	Water-column
Anchoviella manamensis	Gen	Lo	Pelagic channel	Water-column
Axelrodia lindeae	C, B	Lo	Lotic	Water-column
Axelrodia riesei	W, B	Lo	Lotic	Water-column
Axelrodia stigmatias	W, B	Lo	Lotic	Water-column
Microschemobrycon cryptogrammus	C	Lo	Pelagic, littoral, floodplains	Water-column
Microschemobrycon elongatus	C, B	Lo	Pelagic, littoral, floodplains	Water-column
Microschemobrycon meyburgi	C, B	Lo	Pelagic, littoral, floodplains	Water-column
Oxybrycon parvulus	W	Lo	Lotic	Water-column
Tyttobrycon dorsimaculatus	W	Lo	Lotic	Water-column
Tyttobrycon hamatus	W	Lo	Lotic	Water-column
Tyttobrycon marajoara	C	Lo	Lotic	Water-column
Tyttobrycon shibattai	W	Lo	Lentic	Water-column
Tyttobrycon spinosus	W	Lo	Lotic	Water-column
Tyttobrycon xeruini	B	Lo	Lotic	Water-column
Amazonspinther dalmata	W	Lo	Lotic	Wood, vegetation
Cheirodon luelingi	C, W	Lo	Lotic	Water-column
Nanocheirodon insignis	W	Lo	Lotic	Water-column
Odontostilbe gracilis	C	Lo	Lotic	Water-column
Odontostilbe littoris	C, W	Lo	Lotic	Water-column
Serrapinnus malabarbai	C	Lo	Lotic	Water-column
Serrapinnus kriegi	C	Lo	Lotic	Water-column
Spintherobolus broccae	C	Lo	Lotic	Water-column
Priocharax ariel	C	Lo	Lentic	Water-column
Priocharax britzi	W	Lo	Lentic	Water-column
Priocharax nanus	B	Lo	Lentic	Water-column
Priocharax pygmaeus	C	Lo	Lentic	Water-column
Priocharax varii	C	Lo	Lentic	Water-column
Creagrutus maracaiboensis	C	Hi	Lotic	Gravel/rocks
Creagrutus nigrostigmatus	C	Hi	Lotic	Gravel/rocks

(continued)

Table 19.1B (continued)

Genus species	Water	Gradient	Mesohabitat	Microhabitat
Cyanogaster noctivaga	C, B	Lo	Lotic	Gravel/rocks
Iotabrycon praecox	C	Lo	Lotic	Water-column
Pterobrycon landoni	C	Lo	Lotic	Water-column
Scopaeocharax atopodus	C	Lo	Lotic	Water-column
Scopaeocharax rhinodus	C	Lo	Lotic	Water-column
Trochilocharax ornatus	C	Lo	Lotic	Water-column
Tyttocharax cochui	C, B	Lo	Lotic	Water-column
Tyttocharax madeirae	C, B	Lo	Lotic	Water-column
Tyttocharax metae	C, B	Lo	Lotic	Water-column
Tyttocharax tambopatensis	C, B	Lo	Lotic	Water-column
Xenurobrycon coracoralinae	C	Lo	Lotic	Water-column
Xenurobrycon heterodon	C	Lo	Lotic	Water-column
Xenurobrycon macropus	C	Lo	Lotic, floodplains	Sand, submerged wood/vegetation
Xenurobrycon polyancistrus	C	Lo	Lotic	Water-column
Xenurobrycon varii	C	Lo	Lotic	Water-column
Xenurobrycon pteropus	C	Lo	Lotic	Water-column
Hemigrammus aereus	C	Lo	Lotic	Water-column
Hemigrammus iota	C	Lo	Lotic	Water-column
Hemigrammus tridens	C	Lo	Lotic, lentic	Water-column
Hyphessobrycon albolineatus	C	Lo	Lotic	Water-column
Hyphessobrycon amandae	C	Lo	Lotic, lentic	Water-column
Hyphessobrycon arianae	C	Lo	Lotic, lentic	Water-column
Hyphessobrycon axelrodi	W	Lo	Lentic	Water-column
Hyphessobrycon catableptus	C	Lo	Lotic	Water-column
Hyphessobrycon caru	C	Lo	Lotic	Water-column
Hyphessobrycon ecuadorensis	C	Lo	Lotic	Water-column
Hyphessobrycon eilyos	C	Lo	Lotic	Water-column
Hyphessobrycon elachys	C	Lo	Lotic	Water-column
Hyphessobrycon frickei	C	Lo	Lotic	Water-column
Hyphessobrycon gracilior	C	Lo	Lotic	Water-column
Hyphessobrycon griemi	C	Lo	Lotic	Water-column
Hyphessobrycon haraldschultzi	C	Lo	Lotic	Water-column
Hyphessobrycon heteresthes	C	Lo	Lotic	Water-column
Hyphessobrycon hildae	C	Lo	Lotic	Water-column
Hyphessobrycon loretoensis	C	Lo	Lotic	Water-column
Hyphessobrycon minimus	C	Lo	Lotic	Water-column
Hyphessobrycon myrmex	C	Lo	Lotic	Water-column
Hyphessobrycon parvellus	C	Lo	Lotic	Water-column
Hyphessobrycon piranga	C	Lo	Lotic	Water-column
Hyphessobrycon roseus	C	Lo	Lotic	Water-column
Hyphessobrycon saizi	C	Lo	Lotic	Water-column
Hyphessobrycon taphorni	C	Lo	Lotic	Water-column
Hyphessobrycon tenuis	C	Lo	Lotic	Water-column
Hyphessobrycon tropis	B	Lo	Lotic	Water-column
Hyphessobrycon tukunai	C	Lo	Lotic	Water-column
Hyphessobrycon weitzmanorum	C	Lo	Lotic	Submerged wood/vegetation, leaf packs

Table 19.1B (continued)

Genus species	Water	Gradient	Mesohabitat	Microhabitat
Paracheirodon innesi	C, B	Lo	Lotic	Water-column
Paracheirodon simulans	C, B	Lo	Lotic	Water-column
Bryconella pallidifrons	W, B	Lo	Lotic/lentic	Water-column
Tucanoichthys tucano	B	Lo	Lotic	Water-column
Ammocryptocharax minutus	C, B	Lo	Lotic	Sand/submerged wood, vegetation
Characidium bahiense	C	Lo	Lotic	Sand
Characidium chicoi	C	Lo	Lotic	Submerged wood, vegetation
Characidium geryi	C	Lo	Lotic	Sand
Characidium heinianum	C	Lo	Lotic/lentic	Sand
Characidium mirim	C	Lo	Lotic	Sand
Characidium nana	C	Lo	Lotic	Sand
Characidium pteroides	C, B	Lo	Lotic/lentic	Sand
Elachocharax geryi	C	Lo	Lotic	Sand
Elachocharax junki	C, B	Lo	Lotic	Sand/leaf packs
Elachocharax mitopterus	C, B	Lo	Lotic	Sand/leaf packs
Elachocharax pulcher	C, B	Lo	Lotic	Sand/leaf packs
Klausewitzia ritae	C	Lo	Lotic	Sand
Microcharacidium bombioides	C, B	Lo	Lotic	Sand
Microcharacidium eleotrioides	C, B	Lo	Lotic	Sand/leaf packs
Microcharacidium gnomus	C, B	Lo	Lotic	Sand/leaf packs
Microcharacidium weitzmani	C, B	Lo	Lotic	Sand/leaf packs
Odontocharacidium aphanes	C, B	Lo	Lotic	Sand
Odontocharacidium varii	B	Lo	Lotic	Sand/leaf packs
Carnegiella myersi	W	Lo	Lotic	Surface
Carnegiella schereri	W	Lo	Lotic	Surface
Nannostomus anduzei	B	Lo	Lotic	Water-column
Nannostomus britskii	B	Lo	Lotic/pelagic channel	Water-column
Nannostomus minimus	B	Lo	Lotic/pelagic channel	Water-column
Nannostomus rubrocaudatus	B	Lo	Lotic/pelagic channel	Water-column
Acanthobunocephalus nicoi	C, B	Lo	Lentic	Sand/leaf packs
Acanthobunocephalus scruggsi	C, B	Lo	Lentic	Sand/leaf packs
Hoplomyzon atrizona	W	Lo	Lotic	Sand
Hoplomyzon cardosoi	W	Lo	Lotic	Sand/gravel-rocks
Hoplomyzon papillatus	W	Lo	Benthic channel	/Mud-silt/sand
Dupouyichthys sapito	C, W	Lo	Lotic	Mud-silt/sand/gravel-rocks
Micromyzon akamai	W	Lo	Benthic channel	Mud-silt/sand
Micromyzon orinoco	W	Lo	Benthic channel	Mud-silt/sand
Ernstichthys taquari	W	Lo	Benthic channel	Sand/gravel-rocks
Astroblepus chimborazoi	C	Hi	Lotic	Gravel-rocks
Gelanoglanis nanonocticolus	W, B	Lo	Benthic channel/pelagic channel	Sand/gravel-rocks
Gelanoglanis pan	C	Lo	Benthic channel/pelagic channel	Sand/gravel-rocks
Gelanoglanis vari	C	Lo	Benthic channel/pelagic channel	Sand/gravel-rocks
Tatia marthae	C	Lo	Lotic	Surface
Aspidoras brunneus	C	Lo	Lotic	Submerged wood, vegetation

(continued)

Table 19.1B (continued)

Genus species	Water	Gradient	Mesohabitat	Microhabitat
Aspidoras carvalhoi	C	Lo	Lotic	Submerged wood, vegetation
Corydoras boehlkei	B	Lo	Lotic	Gravel-rocks
Corydoras cochui	W	Hi	Floodplains	Gravel-rocks
Corydoras gracilis	C	Lo	Lotic	Gravel-rocks
Corydoras habrosus	W	Lo	Lotic	Gravel-rocks
Corydoras hastatus	C	Lo	Lotic	Gravel-rocks
Corydoras pygmaeus	C	Lo	Lotic	Gravel-rocks
Denticetopsis royeroi	B	Lo	Pelagic-channel	Water-column
Denticetopsis sauli	B	Lo	Pelagic-channel	Water-column
Physopyxis ananas	G	Lo	Lentic (littoral channels, floodplains)	Sand/Submerged wood, vegetation
Physopyxis cristata	B	Lo	Lotic	Sand
Horiomyzon retropinnatus	W	Lo	Lotic (benthic channel)	Sand/gravel-rocks
Corumbataia liliai	C	Hi	Lotic	Submerged wood, vegetation
Hisonotus bocaiuva	C	Lo	Lotic	Gravel-rocks/submerged wood, vegetation
Rhinotocinclus bockmanni	C	Lo	Lotic	Gravel-rocks/submerged wood, vegetation
Rhinotocinclus dinizae	C	Hi	Lotic	Submerged wood, vegetation
Nannoplecostomus eleonorae	C	Hi	Lotic	Gravel-rocks
Otocinclus tapirape	C	Lo	Lotic (floodplains)	Submerged wood, vegetation
Parotocinclus amazonensis	G	Hi/Lo	Lotic (littoral channel)	Submerged wood, vegetation
Rhinotocinclus collinsae	C	Hi/Lo	Lotic (littoral channel)	Submerged wood, vegetation
Rhinotocinclus halbothi	C	Hi/Lo	Lotic (littoral channel)	Gravel-rocks/submerged wood, vegetation
Rhinotocinclus pentakelis	C	Hi/Lo	Lotic (littoral channel)	Submerged wood, vegetation
Rhinotocinclus hardmani	B	Hi	Lotic (littoral channel)	Gravel-rocks
Microglanis robustus	C	Hi	Lotic	Submerged wood, vegetation
Microglanis zonatus	C	Hi	Lotic	Submerged wood, vegetation
Scoloplax baileyi	C, B	Lo	Lotic	Leaf packs
Scoloplax baskini	C	Lo	Lotic	Leaf packs
Scoloplax dicra	C	Lo	Lotic	Leaf packs
Scoloplax distolothrix	C	Lo	Lotic	Leaf packs
Scoloplax dolicholophia	C, B	Lo	Lotic	Leaf packs
Scoloplax empousa	C	Lo	Lotic	Leaf packs
Pygidianops cuao	C	Lo	Lotic	Sand
Pygidianops eigenmanni	B	Lo	Lotic	Sand
Pygidianops magoi	C	Lo	Lotic	Sand
Typhlobelus macromycterus	C	Lo	Lotic	Sand
Typhlobelus guacamaya	C	Lo	Lotic	Sand
Ammoglanis amapaensis	C	Lo	Lotic	Sand
Ammoglanis diaphanus	C	Lo	Lotic	Sand
Ammoglanis multidentatus	B	Lo	Lotic	Sand
Ammoglanis natgeorum	B	Lo	Lentic	Sand/leaf packs
Ammoglanis obliquus	C	Lo	Lotic	Sand
Ammoglanis pulex	C	Lo	Lotic	Sand

Table 19.1B (continued)

Genus species	Water	Gradient	Mesohabitat	Microhabitat
Malacoglanis gelatinosus	W	Lo	Lotic	Sand
Microcambeva draco	C	Lo	Lotic	Sand
Microcambeva mucuriensis	C	Lo	Lotic	Sand
Stenolicmus ix	C	Lo	Lotic	Sand
Sarcoglanis simplex	C, B	Lo	Lotic	Sand
Schultzichthys gracilis	C, W	Lo	Pelagic channel	Sand
Ituglanis compactus	C	Lo	Lotic	Sand
Miuroglanis platycephalus	C	Lo	Pelagic channel	Sand
Potamoglanis anhanga	W	Lo	Pelagic channel	Sand
Potamoglanis hasemani	C, B	Lo	Pelagic channel	Sand/submerged wood vegetation
Potamoglanis johnstoni	C	Lo	Pelagic channel	Sand
Potamoglanis wapixana	C	Lo	Pelagic channel	Sand
Tridensimilis venezuelae	W	Lo	Pelagic channel	Sand
Tridentopsis cahuali	W	Lo	Lentic/pelagic channel	Sand
Tridentopsis pearsoni	W	Lo	Pelagic channel	Sand
Tridentopsis tocantinsi	W	Lo	Pelagic channel	Sand
Paracanthopoma saci	W, C	Lo	Pelagic channel	Sand
Paracanthopoma cangussu	C	Lo	Lentic/pelagic channel	Sand
Fluviphylax gouldingi	B	Lo	Littoral,floodplains	Sand/ wood, vegetation/surface
Fluviphylax obscurus	B	Lo	Littoral,floodplains	Sand/ wood, vegetation/surface
Fluviphylax palikur	C	Lo	Littoral,floodplains	Sand/ wood, vegetation/surface
Fluviphylax pygmaeus	B	Lo	Littoral,floodplains	Sand/ wood, vegetation/surface
Fluviphylax simplex	C, B	Lo	Littoral,floodplains	Sand/ wood, vegetation/surface
Fluviphylax wallacei	B	Lo	Littoral,floodplains	Sand/ wood, vegetation/surface
Fluviphylax zonatus	B	Lo	Littoral,floodplains	Sand/ wood, vegetation/surface
Cnesterodon raddai	C	Lo	Lotic	Water-column
Neoheterandria elegans	C	Lo	Lotic	Water-column
Phalloptychus eigenmanni	C	Lo	Lotic	Water-column
Phallotorynus dispilos	C	Lo	Lotic	Water-column
Phallotorynus psittakos	C	Lo	Lotic	Water-column
Phallotorynus victoriae	C	Lo	Lotic	Water-column
Poecilia (Pamphorichthys)	C	Lo	Lotic	Water-column
Poecilia (Pamphorichthys)	C	Lo	Lotic	Water-column
Poecilia (Pamphorichthys)	C	Lo	Lotic	Water-column
Poecilia (Pamphorichthys)	C	Lo	Lotic	Water-column
Poecilia (Pamphorichthys)	C	Lo	Lotic	Water-column
Poecilia (Pamphorichthys)	C	Lo	Lotic	Water-column
Poecilia (Pamphorichthys)	C	Lo	Lotic	Water-column
Poecilia (Micropoecilia)	C	Lo	Lotic	Water-column
Poecilia (Micropoecilia)	C	Lo	Lotic	Water-column
Poecilia (Micropoecilia)	C	Lo	Lotic	Water-column
Poecilia (Micropoecilia)	C	Lo	Lotic	Water-column
Poecilia (Micropoecilia)	C	Lo	Lotic	Water-column
Poecilia (Micropoecilia)	C	Lo	Lotic	Water-column
Poecilia (Micropoecilia)	C	Lo	Lotic	Water-column

(continued)

Table 19.1B (continued)

Genus species	Water	Gradient	Mesohabitat	Microhabitat
Pseudopoecilia austrocolumbiana	C	Lo	Lotic	Water-column
Scolichthys iota	C	Lo	Lotic	Water-column
Anablepsoides gamae	C, B	Lo	Lentic	Surface
Anablepsoides henschelae	B	Lo	Lentic	Surface
Anablepsoides ottonii	C, B	Lo	Lentic	Surface
Atlantirivulus jurubatibensis	B	Lo	Lentic	Surface
Austrolebias paranaensis	C	Lo	Ephemeral	Water-column (mud/silt)
Campellolebias intermedius	B	Lo	Ephemeral	Water-column
Hypsolebias fasciatus	C, B	Lo	Ephemeral	Water-column (mud/silt)
Hypsolebias marginatus	C, B	Lo	Ephemeral	Water-column (mud/silt)
Hypsolebias notatus	C, B	Lo	Ephemeral	Water-column (mud/silt)
Laimosemion jauaperi	B	Lo	Lentic	Surface, submerged wood, vegetation
Laimosemion kirovskyi	B	Lo	Lentic	Surface, submerged wood, vegetation
Laimosemion leticia	C	Lo	Lentic	Surface, submerged wood, vegetation
Laimosemion nicoi	C	Lo	Lentic	Surface, submerged wood, vegetation
Laimosemion romeri	B	Lo	Lentic	Surface, submerged wood, vegetation
Laimosemion uatuman	B	Lo	Lentic	Surface, submerged wood, vegetation
Laimosemion ubim	B	Lo	Lentic	Surface, submerged wood, vegetation
Leptolebias marmoratus	B	Lo	Ephemeral floodplains	Water-column, mud/silt
Leptopanchax itanhaensis	B	Lo	Ephemeral	Water-column, mud/silt
Leptopanchax sanguineus	B	Lo	Ephemeral	Water-column, mud/silt
Leptopanchax splendens	B	Lo	Ephemeral	Water-column, mud/silt
Maratecoara gesmonei	C, B	Lo	Ephemeral	Submerged wood, vegetation
Melanorivulus atlanticus	C, B	Lo	Lentic	Surface
Melanorivulus bororo	C, B	Lo	Lentic	Surface
Melanorivulus decoratus	C, B	Lo	Lentic	Surface
Melanorivulus egens	C, B	Lo	Lentic	Surface
Melanorivulus jalapensis	C, B	Lo	Lentic	Surface
Melanorivulus javaeh	C, B	Lo	Lentic	Surface
Melanorivulus modestus	C, B	Lo	Lentic	Surface
Melanorivulus nigropunctatus	C, B	Lo	Lentic	Surface
Melanorivulus paracatuensis	C, B	Lo	Lentic	Surface
Melanorivulus paresi	C, B	Lo	Lentic	Surface
Melanorivulus pictus	C, B	Lo	Lentic	Surface
Melanorivulus rossoi	C, B	Lo	Lentic	Surface
Melanorivulus rubroreticulatus	C, B	Lo	Lentic	Surface
Melanorivulus rutilicaudus	C, B	Lo	Lentic	Surface
Melanorivulus schuncki	C, B	Lo	Lentic	Surface
Mucurilebias leitaoi	B	Lo	Ephemeral	Water-column (mud/silt)
Notholebias cruzi	B	Lo	Ephemeral	Water-column (mud/silt)

Table 19.1B (continued)

Genus species	Water	Gradient	Mesohabitat	Microhabitat
Notholebias fractifasciatus	B	Lo	Ephemeral	Water-column (mud/silt)
Notholebias vermiculatus	C, B	Lo	Ephemeral	Water-column (mud/silt)
Papiliolebias francescae	C	Lo	Ephemeral	Water-column (submerged wood, vegetation)
Plesiolebias altamira	C, B	Lo	Ephemeral	Water-column (submerged wood, vegetation)
Plesiolebias aruana	C, B	Lo	Ephemeral	Water-column (submerged wood, vegetation)
Plesiolebias canabravensis	C, B	Lo	Ephemeral	Water-column (submerged wood, vegetation)
Plesiolebias filamentosus	C, B	Lo	Ephemeral	Water-column (submerged wood, vegetation)
Plesiolebias fragilis	C, B	Lo	Ephemeral	Water-column (submerged wood, vegetation)
Plesiolebias glaucopterus	C, B	Lo	Ephemeral	Water-column (submerged wood, vegetation)
Plesiolebias lacerdai	C, B	Lo	Ephemeral	Water-column (submerged wood, vegetation)
Simpsonichthys cholopteryx	B	Lo	Ephemeral	Water-column (mud/silt)
Simpsonichthys nigromaculatus	B	Lo	Ephemeral	Water-column (mud/silt)
Simpsonichthys parallelus	B	Lo	Ephemeral	Water-column (mud/silt)
Spectrolebias costae	C, B	Lo	Ephemeral	Water-column (mud/silt)
Spectrolebias gracilis	C, B	Lo	Ephemeral	Water-column (mud/silt)
Spectrolebias reticulatus	C	Lo	Ephemeral	Water-column (mud/silt)
Spectrolebias semiocellatus	B	Lo	Ephemeral	Water-column (mud/silt)
Stenolebias bellus	C, B	Lo	Ephemeral floodplains	Submerged wood, vegetation
Stenolebias damascenoi	C, B	Lo	Ephemeral floodplains	Submerged wood, vegetation
Leptophilypnion fittkaui	B	Lo	Benthic channel	Leaf packs/sand
Leptophilypnion pusillus	B	Lo	Benthic channel	Leaf packs/sand
Microphilypnus acangaquara	C, B	Lo	Benthic channel	Leaf packs/sand
Microphilypnus hypolyrasimeion	B	Lo	Benthic channel	Leaf packs/sand
Microphilypnus macrostoma	C, B	Lo	Benthic channel	Leaf packs/sand
Microphilypnus tapajosensis	C, B	Lo	Benthic channel	Leaf packs/sand
Microphilypnus ternetzi	C, B	Lo	Benthic channel	Leaf packs/sand
Gobiosoma yucatanum	E	Lo	Benthic channel	Sand
Apistogramma agayuara	C	Lo	Lotic	Leaf packs
Apistogramma inornata	C	Lo	Lotic	Leaf packs
Apistogramma minima	B	Lo	Lotic	Leaf packs
Apistogramma staecki	C	Lo	Lotic	Leaf packs

We also found that most miniature NFF species (95.6%) occur in low-gradient waterways, and only a few appear in high-gradient waterways (3.0%) or both habitats (1.4%). These results support the hypothesis (Fig. 19.4) that physical challenges related to locomotion in a high water velocity system likely restrict miniature species from these habitats (Weitzman and Vari 1988). Among all NFF species, only 3% occur in high-gradient waterways. These include two Characiformes, *Creagrutus maracaiboensis* (Schutz 1944) and *C. nigrostigmatus* (Dahl 1960) and nine Siluriformes, *Astroblepus chimborazoi* (Fowler 1915), *Corydoras cochui* (Myers and Weitzman 1954), *Corumbataia liliai* (Silva, Roxo, Souza and Oliv 2018), *Nannoplecostomus eleonorae* (Ribeiro, Lima and Pereira 2012), *Rhinotocinclus collinsae* (Schmidt and Ferraris 1985),

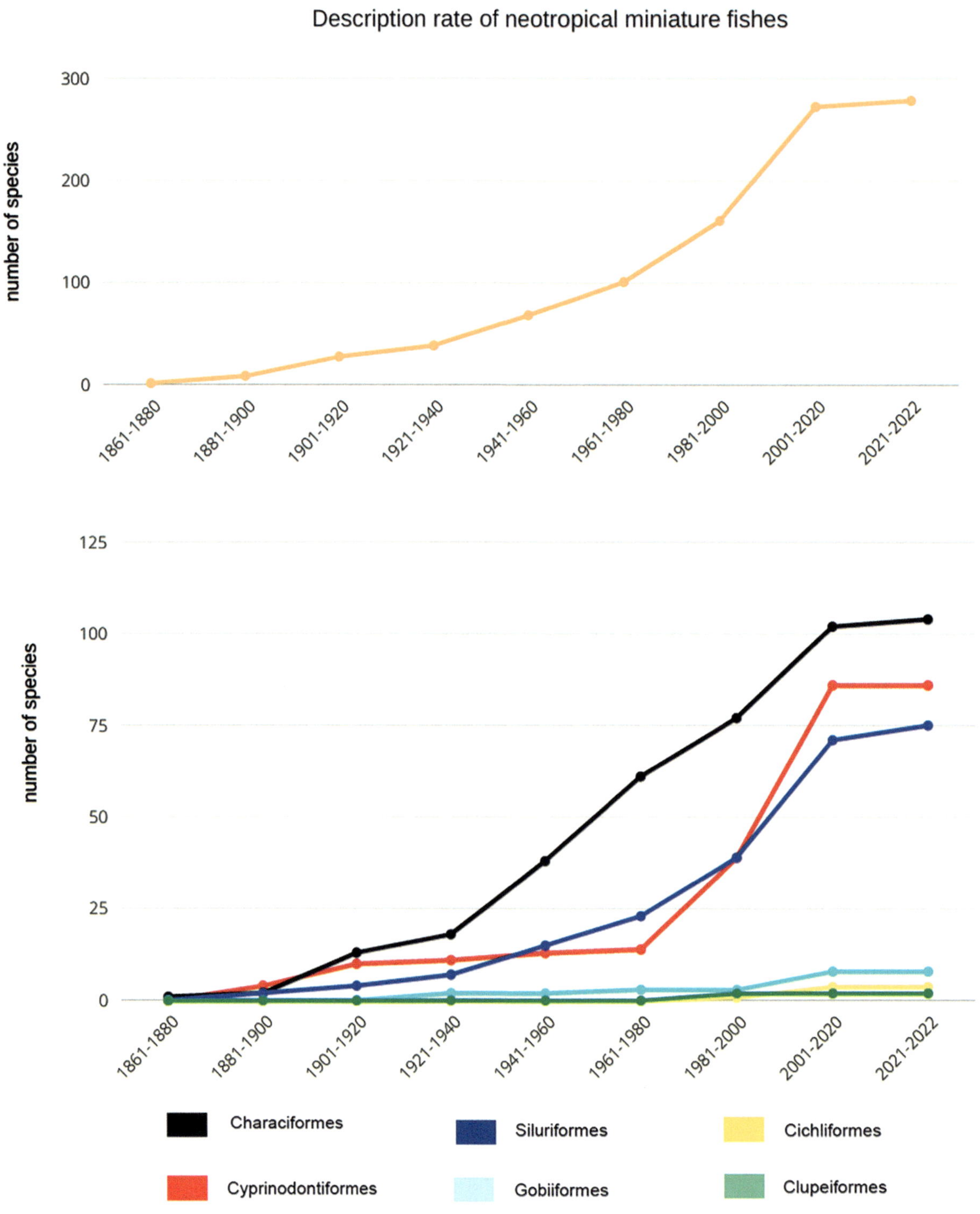

Fig. 19.2 Cumulative numbers of miniature NFF species. Data from 1861 to 2022. Top: All Neotropical freshwater fish species. Bottom: Species by taxonomic order

R. halbothi (Lenhmann Lazzarotto and Reis 2014), *R. pentakelis* (Roxo, Messias and Silva 2019), *Microglanis robustus* (Ruiz and Shibatta 2010), and *M. zonatus* (Eigenmann and Allen 1942). The occurrence of the aforementioned catfishes in high-gradient habitats might be due to their evolutionary histories, since they belong to lineages that typically occupy fast-flowing environments (Chap. 9).

Some miniature NFF species occupy distinct microhabitats. Most miniatures in the Neotropical region occur in the water column (49%). In comparison, fewer inhabit sandy

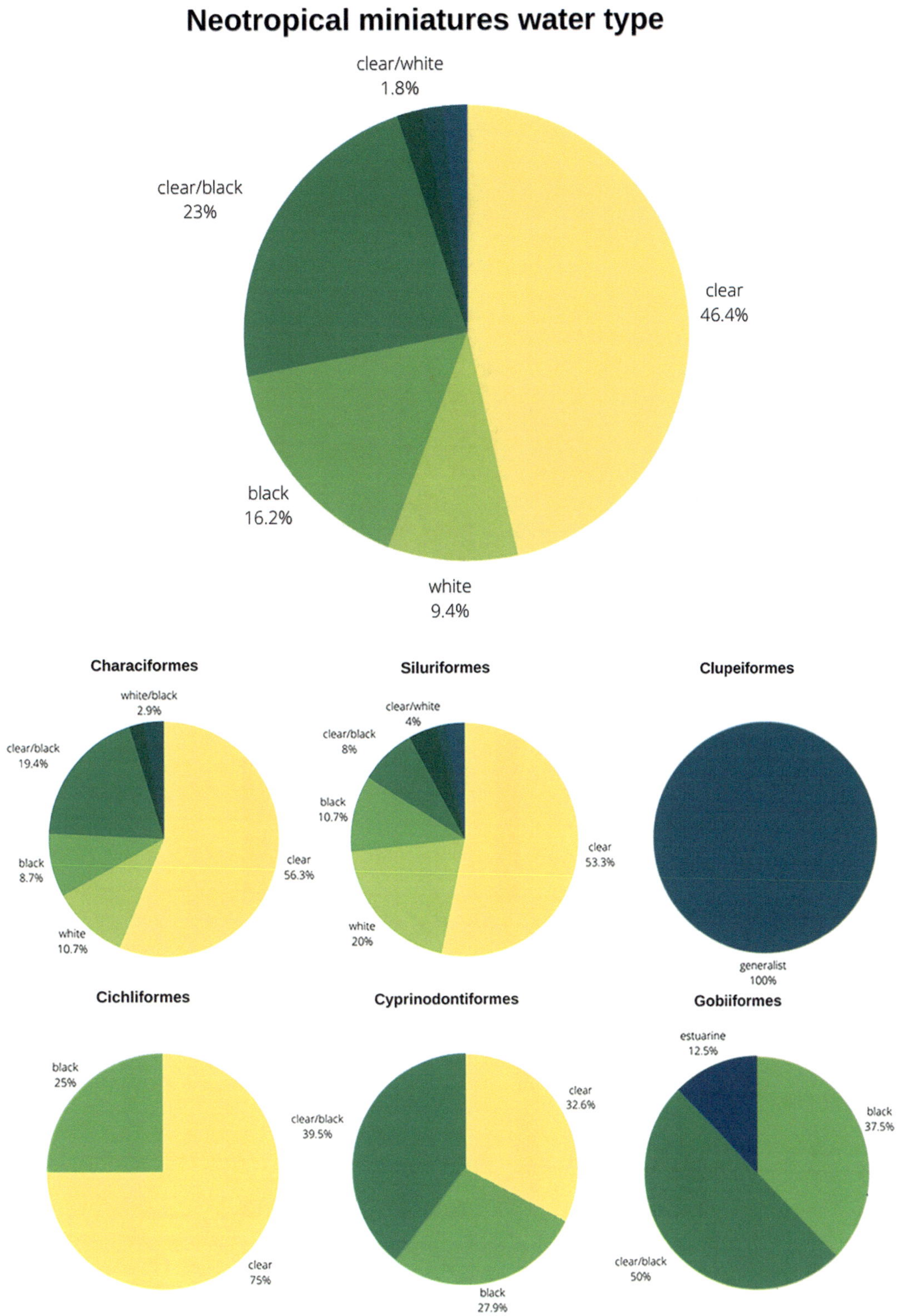

Fig. 19.3 Distribution of miniatures NFF species by water chemistry. Clearwaters, blackwaters, and whitewaters are defined in Chap. 10. Top: All NFF species. Bottom: Species by taxonomic order

(20.4%), whitewater (8.6%), and blackwater (8.6%) rivers. Very few species occur in both clearwater and whitewater ($n = 3$) or in blackwater and whitewater ($n = 4$) rivers.

bottoms (20.5%), submerged wood (13.5%), or the water surface (11.2%) (Fig. 19.5). The Characiformes, for example, are generally found up in the water-column (Weitzman and

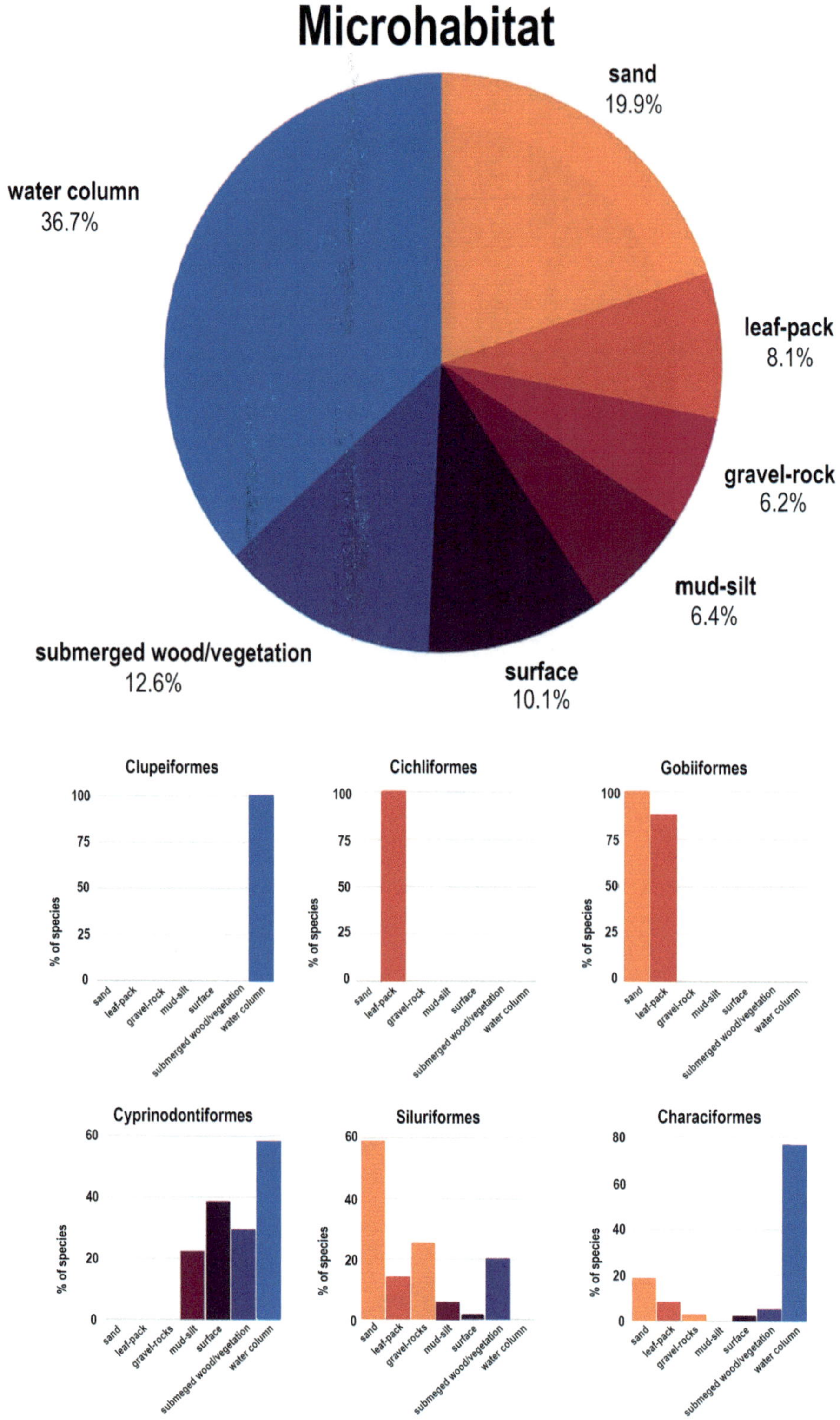

Fig. 19.4 Distribution of miniatures NFF species by microhabitats. Habitat definitions in Chap. 6. Top: All NFF species. Bottom: Species by taxonomic order

Vari 1988), whereas trichomycterid catfishes inhabit fine sandy bottoms (Costa et al. 2020). Henderson and Walker (1990) were the first to report a high diversity of miniatures in the leaf litter of the Tarumã-Mirim River, a blackwater tributary of the Negro River in Brazil. Herein, leaf pack substrates are occupied by 7.4% of all miniature species, especially by the characiform

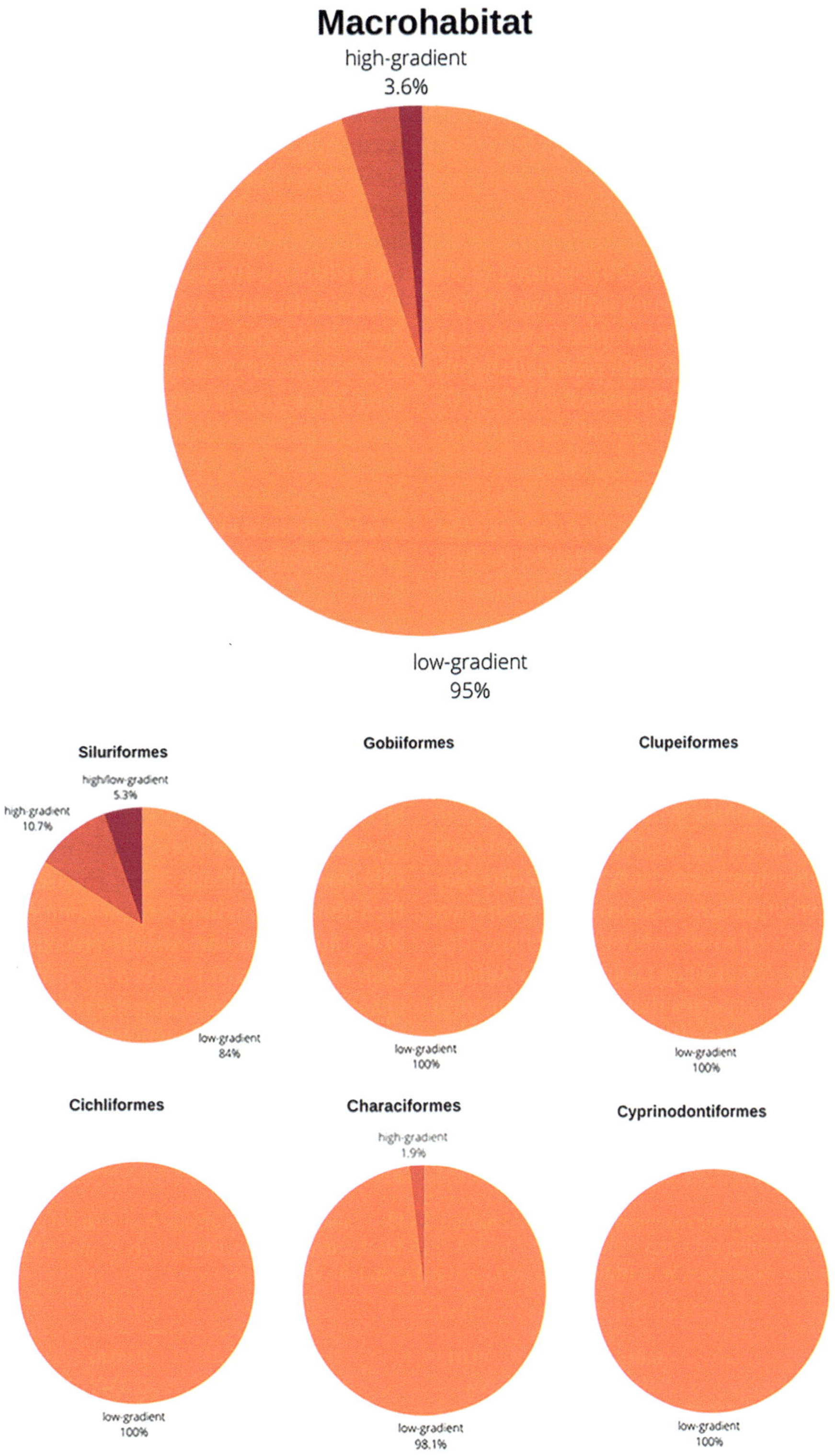

Fig. 19.5 Distribution of miniatures NFF species by macrohabitats. Habitat definitions in Chap. 6. Top: All NFF species. Bottom: Species by taxonomic order

family Crenuchidae, by the siluriform species belonging to the genus *Scoloplax* (Bailey and Baskin 1976) and the cichlid genus *Apistogramma* (Regan 1913).

Recently, Perkin et al. (2022) investigated the distribution and diversity of miniature fishes in an Amazonian stream, supporting previous suggestions that miniatures specialize in specific micro-habitats to avoid competition (Carvalho et al. 2006; Costa et al. 2020; Henderson and Walker 1990; Henschel et al. 2020; Weitzman and Vari 1988). Within the psamophilic (sand dwelling) catfishes of the Trichomycteridae family, recent studies revealed sympatric miniature species associated with entirely different microhabitats (e.g., different sand granulometry or the presence of patches of organic material mixed with the sand). For example, *Ammoglanis obliquus* (Henschel, Bragança, Rangel-Pereira and Costa 2020) occurred in the same sandbank as a *Potamoglanis* species in the Preto da Eva River. Still, the former was generally sampled in finer sand with no organic material, whereas the latter preferred slightly thicker sand with organic material (Henschel et al. 2020). Another example of a specialization in miniature catfishes occurs in *Microcambeva filamentosa* (Costa, Katz and Vilardo 2020), from a coastal river in the Mata Atlântica of southeastern Brazil. This species is sympatric to its putative sister species *M. ribeirae* (Costa et al. 2004) but has a strikingly divergent morphology, suggesting that the former is nocturnal and the latter diurnal, a potential strategy for resource partitioning.

19.5 Habitat and Body Size Evolution

The above section discussed how miniature fishes often inhabit blackwater and clearwater habitats. We also discussed the ecological factors of acidic waters that might drive reduced body sizes in fishes. This section investigated how water acidity may influence body size evolution across Neotropical freshwater fishes. Specifically, we tested whether Neotropical fishes that occupied the most acidic (blackwater) habitats have smaller body sizes on average than species that inhabit pH neutral habitats. Furthermore, we tested whether the repeated transition to blackwater habitat resulted in a reduced body size, indicating a predictable adaptive response to the invasion of blackwater environments.

We employed comparative phylogenetic methods to test these predictions using a new supertree of Neotropical fishes (see Chap. 2 for details) and a complete ecological dataset (see Chap. 5 for more information). First, we classified species as occupying black, white, clear, both clear and black, and both clear and whitewater types. Next, we classified generalists as species occupying all three water types. We then assembled maximum body size data for each species in the phylogeny from FishBase (Froese and Pauly 2000) using the R package rfishbase (Boettiger et al. 2012). We then log10-transformed the body size data to correct for normality, as many statistical tests assume an underlying normal distribution. Finally, we used a phylogenetic ANOVA to test for differences in body size between the different habitat classifications. We performed the phylogenetic ANOVA in the R package geomorph using the function procD.pgls, which uses permutation procedures to assess statistical hypotheses under a Brownian motion model of trait evolution (Adams and Collyer 2018; Baken et al. 2021). If the overall phylogenetic ANOVA was considered significant, we performed pairwise comparisons between habitat groups using the function pairwise from the R package RRPP (Adams and Collyer 2018).

Lastly, we used phylogenetically independent contrasts (PICs) to determine if there was an increase or decrease in body size associated with habitat transitions; see Friedman et al. (2020) for details. First, we inferred the ancestral habitat types across the phylogeny using stochastic character mapping (SIMMAP) (Bollback 2006; Huelsenbeck et al. 2003). We then obtained the independent contrasts and ancestral states of the nodes that immediately preceded habitat transitions in each SIMMAP and recorded the daughter branch where the change occurred, including the type of habitat transition. We then calculated the average PIC values of each clade by type of transition, where negative values indicated a decrease and positive values showed an increase in body size. We then used a *t*-test on the average PICs per habitat transition type. If an

Table 19.2 Results of the phylogenetic independent contrasts analysis. Bolded values were statistically significant at $p < 0.05$. Subclades that did not have transitions between ecological states were labeled with NA. B = blackwater, C = clearwater, C/B = mix of blackwater and clearwater, C/W = mix of clearwater and whitewater, W = whitewater, and G = water chemistry generalists

	C to B	C/B to B	C/W to B	G to B	W to B
All lineages	**−0.0142**	**−0.0138**	0.0020	**−0.0044**	**−0.0057**
Aspredinoidea	0.0459	NA	−0.0009	−0.0047	−0.0025
Characoidea	**−0.0088**	0.0019	**-0.0090**	0.0007	**−0.0057**
Cichlidae	−0.0012	**0.0120**	0.0024	−0.0061	0.0062
Curimatoidea	**−0.0088**	**0.0053**	0.0004	−0.0026	−0.0036
Cyprinodontiformes	**−0.0248**	**−0.0046**	NA	NA	NA
Gymnotiformes	**−0.0172**	**−0.0075**	−0.0035	**−0.0168**	**−0.0082**
Loricariidae	**−0.0101**	0.0027	−0.0156	−0.0083	0.0017
Pimelodoidea	NA	NA	NA	NA	NA

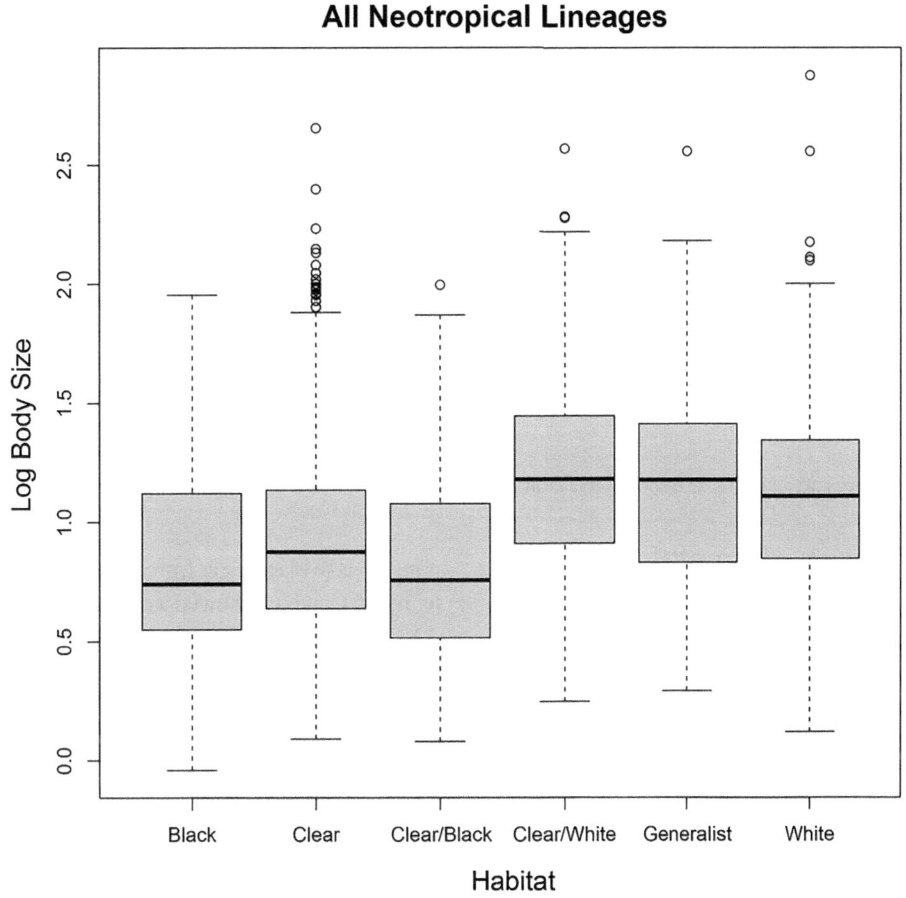

Fig. 19.6 Boxplot of log body size of species occupying each habitat type. Points represent outliers

Table 19.3 Mean pairwise differences for each water type across all Neotropical freshwater fishes

	d	*Z*	*P*-value
Black:clear	0.10	2.25	0.01
Black:clear/black	0.10	2.21	0.02
Black:clear/white	0.22	3.51	0.00
Black:generalist	0.30	3.86	0.00
Black:white	0.19	3.33	0.00
Clear:clear/black	0.00	-0.11	0.57
Clear:clear/white	0.31	6.63	0.00
Clear:generalist	0.20	4.32	0.00
Clear:white	0.29	6.47	0.00
Clear/black:clear/white	0.32	5.45	0.00
Clear/black:generalist	0.20	3.94	0.00
Clear/black:white	0.29	5.33	0.00
Clear/white:generalist	0.52	6.08	0.00
Clear/white:white	0.02	1.87	0.03
Generalist:white	0.49	5.90	0.00

independent contrast differed significantly from zero, that indicated that the respective habitat transition was associated with directional morphological change (Friedman et al. 2020).

We ran all the analyses across eight major Neotropical subclades, namely the Aspredinoidea, Characoidea, Cichlinae, Curimatoidea, Cyprinodontiformes, Gymnotiformes,

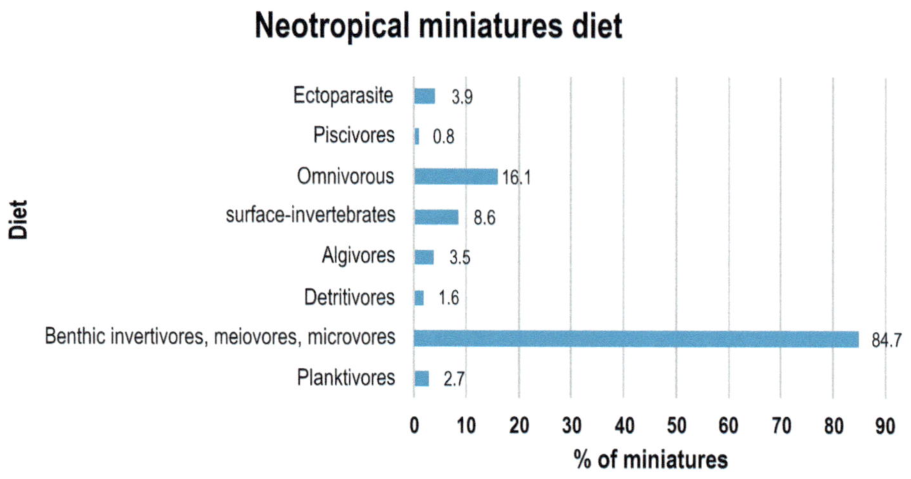

Fig. 19.7 Dietary preferences of miniature NFF species by taxonomic order

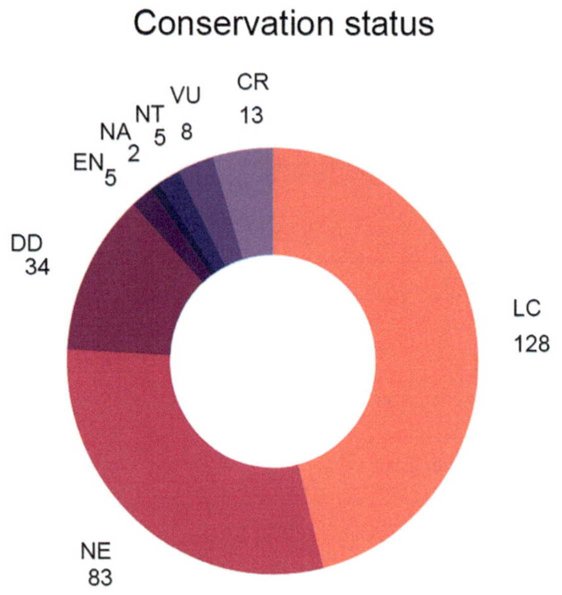

Fig. 19.8 Conservation status of miniature NFF species according to IUCN criteria

Loricarioidei, and Pimelodoidea. We did this to determine whether individual clades of Neotropical fishes exhibited different patterns from the entire dataset.

Our results showed that the invasion of blackwater habitats was significantly associated with a decrease in body sizes across all Neotropical fishes, except for the invasion of blackwaters from lineages that occupy a mix of blackwater and clearwater habitats, which tended to cause an increase in body size across all Neotropical lineages (Table 19.2; Fig. 19.6). The reduction in body size during transitions to blackwater is remarkably consistent across all subclades, except for transitions to blackwater from a mix of blackwater and clearwater habitats, which showed a mix of patterns. The phylogenetic ANOVA found that body sizes were significantly smaller in lineages that occupied more acidic waters, with blackwater fishes having the smallest body sizes and whitewater fishes the largest (Table 19.3; Fig. 19.8). However, the phylogenetic ANOVA analyses on each subclade showed no significant results, and subclade patterns were not examined further. Our results show that blackwater fishes have significantly smaller body sizes than other water types, and most transitions to blackwater drive a reduction in body size across Neotropical freshwater fishes.

Blackwaters are characterized as nutrient-poor, acidic habitats, which can strongly influence aspects of body size in fishes. There are strong correlations between water chemistry, productivity, and fish body size, with fish in nutrient-poor and acidic water habitats tending to have smaller body sizes (Fernandes et al. 2021; Santos et al. 2019). Fishes in these environments are thought to have smaller body sizes because of the increased metabolic stresses and lower food availability in the more acidic habitat. Acidic water disrupts the balance of electrolytes and ions across gill filament and other epithelial surfaces (Camargo et al. 2009), and smaller fish have a higher surface-area-to-volume ratio, which allows for more efficient and faster exchange of gases and waste products (Fernandes et al. 2012). Furthermore, smaller fishes have a lower absolute volume of body fluids, which may allow for faster osmoregulatory adjustments. Lastly, smaller fishes have a lower total metabolic demand than larger fishes, which may enable them to survive with lower food consumption in low-nutrient acidic waters.

Our results show that invasion of blackwater habitat significantly reduces body size across Neotropical freshwater fishes, a result shared across different subclades. Additionally, the acidity of water significantly affects the body size of Neotropical fishes. In highly acidic blackwater habitats, fishes have the smallest body sizes, followed by species in slightly less acidic clearwater habitats, and the largest body sizes are found in species inhabiting whitewater environments. While we cannot definitively state that the invasion of acidic habitats is the primary driver of miniaturization, our results show that the consistent reduction in body size during the invasion of highly acidic blackwater environments might explain the prevalence of miniature species in more acidic habitats and their general absence from non-acidic whitewater habitat.

19.6 Position in Food Webs

The most common diets among the Neotropical miniatures are benthic invertivores, meiovores, and microvores (84.7%); omnivorous, a combination of plant and animal material (16.1%); and water-surface invertebrates (8.6%) (Fig. 19.7). In a nutrient-poor system such as clearwaters and blackwaters, fishes are often more dependent on biomass originated from the terrestrial environment, with non-aquatic invertebrates representing a primary source of allochthonous biomass. This biomass is only available to the aquatic system after predation by small/miniature fishes (Goulding et al. 1988).

To date, there is little direct evidence from population studies demonstrating selective pressures for smaller body sizes in nutrient-poor Amazonian river systems. More detailed investigations of the diet and a more specific identification/distinction of each food item into smaller categories are needed to document the mechanistic connections between dietary specialization and evolutionary miniaturization. However, there are data emerging from studies at the ecological scale supporting these general interpretations. A clear example is the Peacock bass (*Cichla ocellaris*; Cichlidae), which inhabits all three major Amazonian water types, but which grows significantly larger and faster in nutrient-rich whitewater habitats than blackwater systems (Winemiller et al. 1997). These size differences reflect how environmental constraints shape growth and size within a single species across river with different water chemistry profiles.

There is as yet little direct evidence that Amazonian fishes evolved smaller body sizes as an anti-predatory defense against piscivorous fishes and birds. Yet acidic blackwaters with lower in situ rates of organic decomposition often have more structurally complex substrates with submerged leaf litter, tangled roots, and woody debris. Many fishes have evolved smaller body sizes and cryptic phenotypes to navigate and exploit narrow spaces, avoid predators, or feed on macroinvertebrates (Olentino et al. 2024). Small body size in fishes can also reduce the likelihood of being detected or pursued by larger fish and bird predators (Peters, 1986; Macchiusi and Baker, 1991), as is possibly the case for the surface-dwelling species of *Fluviphylax* (Olentino et al. 2024). Future studies on predatory behavior and ecology networks are needed to better understand the main predators of miniature NFF species.

19.7 Conservation

Many miniature NFF species have not yet been evaluated by the IUCN for their conservation status, and 34 species (12.2%) are data deficient (Fig. 19.8). Among the threatened categories, 1.8% of miniatures are near threatened, 2.5% vulnerable, 2.1% endangered, and 4.7% critically endangered. The family with the most critically endangered miniature species is Rivulidae, the fish family with the most threatened species in the Neotropics. Despite the high number of species with the least concern conservation status (44.4%), most new miniatures occur in only a few locations. Thus, future research on the NFF fauna will presumably keep revealing more species with restricted geographic distributions, which is a significant criterion in defining a species' conservation status. The lack of information is discouraging because of the current biodiversity crisis and the unique vulnerability of freshwater taxa.

Small-sized species (maturing at less than 15 cm SL) are the most threatened of all Neotropical fishes (Castro and Polaz 2020; Tagliacollo et al. 2021). Among the major threats, the destruction of natural habitats, such as small streams and swamps, is the most common. Furthermore, recent studies investigating the ecological importance of rare species suggest that they might play a significant role in ecosystem functionality (Basile, 2022; Dee et al. 2019; Leitão et al. 2016). Considering both points, we believe that the general findings from rare species studies might apply to most miniature Neotropical freshwater lineages. According to these studies, a rare species is generally more likely to be lost in the system, but at the same time, it is the one performing unique functional roles due to a high specialization. Lastly, the lack of information on the conservation status of an increased number of miniature species and their potentially critical role in the ecosystem raises concern about their conservation status. This concern is especially applicable when considering small-sized fish species' higher vulnerability to extinction (Olden et al. 2007; Ripple et al. 2017).

19.8 Conclusions

Body size is a key organismal trait that varies over about 2.5 orders of magnitude among Amazonian fishes (~1–450 cm standard length), with profound consequences for almost all aspects of physiological and ecological performances and tolerances, evolutionary adaptations, and biodiversity patterns. This chapter reviews the evolution of miniaturization in Amazonian fishes, defined as a species or clade that has undergone extreme phylogenetic reduction in adult body size, impacting phenotypes across the full integrated hierarchy of phenotypic organization. This chapter presents phylogenetic criteria to identify miniature NFF species, updates the list of miniaturized NFF species, and reviews the ecological requirements and conservation status of miniature NFF species. Fishes of blackwater habitats generally grow to smaller adult body sizes than their close relatives in whitewater and clearwater habitats, due to low sediments, nutrients, and pH of the waters that limit food availability and favoring species with smaller body sizes and lower total metabolic demands.

References

Adams DC, Collyer ML (2018) Multivariate phylogenetic comparative methods: evaluations, comparisons, and recommendations. Syst Biol 67(1):14–31

Ahti PA, Kuparinen A, Uusi-Heikkilä S (2020) Size does matter—the eco-evolutionary effects of changing body size in fish. Environ Rev 28(3):311–324

Albert JS, Johnson DM (2012) Diversity and evolution of body size in fishes. Evol Biol 39(3):324–340

Albert JS, Johnson DM, Knouft JH (2009) Fossils provide better estimates of ancestral body size than do extant taxa in fishes. Acta Zool 90:357–384

Albert JS, Tagliacollo VA, Dagosta F (2020) Diversification of Neotropical freshwater fishes. Annu Rev Ecol Evol Syst 51(1):27–53

Baken EK, Collyer ML, Kaliontzopoulou A, Adams DC (2021) *geomorph* v4.0 and *gmShiny*: enhanced analytics and a new graphical interface for a comprehensive morphometric experience. Methods Ecol Evol 12(12):2355–2363

Barneche DR, Robertson DR, White CR, Marshall DJ (2018) Fish reproductive-energy output increases disproportionately with body size. Science 360(6389):642–645

Basile M (2022) Rare species disproportionally contribute to functional diversity in managed forests. Sci Rep 12(1):1

Blanckenhorn WU (2000) The evolution of body size: what keeps organisms small?. Q Rev Biol 75(4):385–407

Bloom DD, Burns MD, Schriever TA (2018) Evolution of body size and trophic position in migratory fishes: a phylogenetic comparative analysis of Clupeiformes (anchovies, herring, shad and allies). Biol J Linn Soc 125(2):302–314

Boettiger C, Lang DT, Wainwright PC (2012) rfishbase: exploring, manipulating and visualizing FishBase data from R. J Fish Biol 81(6):2030–2039

Bollback JP (2006) SIMMAP: stochastic character mapping of discrete traits on phylogenies. BMC Bioinform 7(1):88. https://doi.org/10.1186/1471-2105-7-88

Borghezan E, Pires TH, Ikeda T, Zuanon J, Kohshima S (2021) A review on fish sensory systems and Amazon water types with implications to biodiversity. Front Ecol Evol 8:589760

Bragança PHN (2018) *Fluviphylax gouldingi* and *F. wallacei*, two new miniature killifishes from the middle and upper Rio Negro drainage, Brazilian Amazon. Spixiana 41(1):133–146

de Brito V, Betancur-R R, Burns MD, Buser TJ, Conway KW, Fontenelle JP, Kolmann MA, McCraney WT, Thacker CE, Bloom DD (2022) Patterns of phenotypic evolution associated with marine/freshwater transitions in fishes. Integr Comp Biol 62(2):406–423

Britz R, Conway KW (2009) Osteology of *Paedocypris*, a miniature and highly developmentally truncated fish (Teleostei: Ostariophysi: Cyprinidae). J Morphol 270(4):389–412

Britz R, Conway KW (2016) *Danionella dracula*, an escape from the cypriniform Bauplan via developmental truncation?. J Morphol 277(2):147–166

Britz R, Conway KW, Rüber L (2009) Spectacular morphological novelty in a miniature cyprinid fish, *Danionella dracula* n. sp. Proc R Soc B 276(1665):2179–2186

Britz R, Conway KW, Rüber L (2014) Miniatures, morphology and molecules: Paedocypris and its phylogenetic position (Teleostei, Cypriniformes). Zool J Linn Soc 172(3):556–615

Britz R, Kottelat M (2003) Descriptive osteology of the family Chaudhuriidae (Teleostei, Synbranchiformes, Mastacembeloidei), with a discussion of its relationships. Am Mus Novit 3418:1–62

Brooks DR, Wiley EO (1988) Evolution as entropy: toward a unified theory of biology. Chicago, University of Chicago Press

Burns MD (2021) Adaptation to herbivory and detritivory drives the convergent evolution of large abdominal cavities in a diverse freshwater fish radiation (Otophysi: Characiformes). Evolution 75(3):688–705

Burns MD, Bloom DD (2020) Migratory lineages rapidly evolve larger body sizes than non-migratory relatives in ray-finned fishes. Proc R Soc B 287(1918):20192615

Burress ED, Holcomb JM, Bonato KO, Armbruster JW (2016) Body size is negatively correlated with trophic position among cyprinids. R Soc Open Sci 3(5):150652

Camargo MM, Fernandes MN, Martinez CB (2009) How aluminium exposure promotes osmoregulatory disturbances in the neotropical freshwater fish *Prochilodus lineatus*. Aquat Toxicol 94(1):40–46

Carvalho LN, Zuanon J, Sazima I (2006) The almost invisible league: crypsis and association between minute fishes and shrimps as a possible defense against visually hunting predators. Neotrop Ichthyol 4(2):219–224

Castro RMC, Polaz CNM (2020) Small-sized fish: the largest and most threatened portion of the megadiverse Neotropical freshwater fish fauna. Biota Neotrop 20(1):e20180683

Cerezer FO, Dambros CS, Coelho MT, Cassemiro FA, Barreto E, Albert JS et al (2023) Accelerated body size evolution in upland environments is correlated with recent speciation in South American freshwater fishes. Nat Commun 14(1):6070

Chichorro F, Juslén A, Cardoso P (2019) A review of the relation between species traits and extinction risk. Biol Cons 237:220–229

Costa WJEM, Le Bail P-Y (1999) *Fluviphylax palikur*: a new poeciliid from the Rio Oiapoque Basin, Northern Brazil (Cyprinodontiformes: Cyprinodontoidei), with comments on miniaturization in *Fluviphylax* and other Neotropical freshwater fishes. Copeia 1999(4):1027–1034

Costa WJEM, Vilardo P, Katz AM (2020) Sympatric sister species with divergent morphological features of psammophilic catfishes of the south-eastern Brazilian genus *Microcambeva* (Siluriformes: Trichomycteridae). Zool Anz 285:12–17

Costa WJEM, Lima SMQ, Bizerril CRSF (2004) *Microcambeva ribeirae* sp. n. (Teleostei: Siluriformes: Trichomycteridae): a new sarcoglanidine catfish from the Rio Ribeira do Iguape basin, southeastern Brazil. Zootaxa 563:1–10

Crisp MD, Cook LG (2012) Phylogenetic niche conservatism: what are the underlying evolutionary and ecological causes?. New Phytol 196(3):681–694

Dee LE, Cowles J, Isbell F, Pau S, Gaines SD, Reich PB (2019) When do ecosystem services depend on rare species?. Trends Ecol Evol 34(8):746–758

Egan JP, Buser TJ, Burns MD, Simons AM, Hundt PJ (2021) Patterns of body shape diversity and evolution in intertidal and subtidal lineages of combtooth blennies (Blenniidae). Integr Org Biol 3(1):obab004

Fernandes IM, Sacoman KDJ, Farias-Neto JPD, Silva HPD, Vendruscolo J, Lourenco LDS (2021) Effect of environmental and spatial factors on small-sized fish assemblages in a tropical river. Acta Amazon 51:129–138

Fernandes MN, da Cruz AL, da Costa OTF, Perry SF (2012) Morphometric partitioning of the respiratory surface area and diffusion capacity of the gills and swim bladder in juvenile Amazonian air-breathing fish, *Arapaima gigas*. Micron 43(9):961–970

Friedman ST, Price SA, Corn KA, Larouche O, Martinez CM, Wainwright PC (2020) Body shape diversification along the benthic–pelagic axis in marine fishes. Proc R Soc B 287(1931):20201053

Froese R, Pauly D (2000) FishBase2000: concepts designs and data sources. WorldFish, Penang

Fronk AH, Kim LY, Craig JM, Crampton WG, Albert JS (2019) Sexual size dimorphism in the Macana Tigrina, *Gymnotus javari* (Gymnotidae, Gymnotiformes). Copeia 107(2):305–313

Gagliano M, McCormick MI, Meekan MG (2007) Survival against the odds: ontogenetic changes in selective pressure mediate growth-mortality trade-offs in a marine fish. Proc R Soc B 274(1618):1575–1582

Gonzalez RJ, Wilson RW, Wood CM (2005) Ionoregulation in tropical fishes from ion-poor, acidic blackwaters. In: Val A, Almeida-Val VMF, Randall D (eds) Fish physiology, vol 21. San Diego, Academic Press, pp 397–442

Goulding M, Carvalho M, Ferreira EG (1988) Rio Negro, rich life in poor water. The Hague, Netherlands, SPB Academic Publishing

Hanken J, Wake DB (1993) Miniaturization of body size: organismal consequences and evolutionary significance. Annu Rev Ecol Syst 24(1):501–519

Henderson PA, Walker I (1990) Spatial organization and population density of the fish community of the litter banks within a central Amazonian blackwater stream. J Fish Biol 37(3):401–411

Henschel E, Bragança PHN, Rangel-Pereira F, Costa WJEM (2020) A new psammophilic species of the catfish genus *Ammoglanis* (Siluriformes, Trichomycteridae) from the Amazon River basin, northern Brazil. Zoosyst Evol 96(1):1

Hone DWE, Benton MJ (2005) The evolution of large size: how does Cope's rule work?. Trends Ecol Evol 20(1):4–6

Huelsenbeck JP, Nielsen R, Bollback JP, Schultz T (2003) Stochastic mapping of morphological characters. Syst Biol 52(2):131–158

Johnson DG, Brothers EB (1993) *Schindleria*: a paedomorphic goby (Teleostei: Aobioidei). Bull Mar Sci 52(1):441–471

Keppeler FW, Montaña CG, Winemiller KO (2020) The relationship between trophic level and body size in fishes depends on functional traits. Ecol Monogr 90(4):e01415

Kjelland ME, Woodley CM, Swannack TM, Smith DL (2015) A review of the potential effects of suspended sediment on fishes: potential dredging-related physiological, behavioral, and transgenerational implications. Environ Syst Decis 35(3):334–350

Kleiber M (1932) *Body Size and Metabolism*. Hilgardia 6(11):315–353

Knope ML, Scales JA (2013) Adaptive morphological shifts to novel habitats in marine sculpin fishes. J Evol Biol 26(3):472–482

Knouft JH, Page LM (2003) The evolution of body size in extant groups of North American freshwater fishes: speciation, size distributions, and Cope's rule. Am Nat 161(3):413–421

Kolmann MA, Burns MD, Ng JYK, Lovejoy NR, Bloom DD (2020) Habitat transitions alter the adaptive landscape and shape phenotypic evolution in needlefishes (Belonidae). Ecol Evol 10(8):3769–3783

Kwong RWM, Kumai Y, Perry SF (2014) The physiology of fish at low pH: the zebrafish as a model system. J Exp Biol 217(5):651–662

Leitão RP, Zuanon J, Villéger S, Williams SE, Baraloto C, Fortunel C, Mendonça FP, Mouillot D (2016) Rare species contribute disproportionately to the functional structure of species assemblages. Proc R Soc B 283(1828):20160084

Macchiusi F, Baker RL (1991) Prey behaviour and size-selective predation by fish. Freshw Biol 25(3):533–538

Mattox GM, Lima FC, Britz R, Souza CS, Oliveira C (2024) Two new miniature species of the fish genus *Priocharax* from the Rio Tapajós and Amazonas drainages, Pará, Brazil (Teleostei: Characiformes: Characidae). Vertebr Zool 74:533–550

Melo BF, Sidlauskas BL, Near TJ, Roxo FF, Ghezelayagh A, Ochoa LE et al (2022) Accelerated diversification explains the exceptional species richness of tropical characoid fishes. Syst Biol 71(1):78–92

Olden JD, Hogan ZS, Zanden MJV (2007) Small fish, big fish, red fish, blue fish: size-biased extinction risk of the world's freshwater and marine fishes. Glob Ecol Biogeogr 16(6):694–701

Olentino D, Lubich C, Yamamoto K (2024) Fish associated with microhabitats of submerged fallen logs in a lake in the Brazilian Amazon. Acta Sci Biol Sci 46:e70913

Perkin JS, Montaña CG, Nogueira EJ, Brandão BB, Mattox GMT, Conway KW (2022) Estimated richness and environmental correlates of miniature fish assemblages in the Rio Jacundá, Brazil. Neotrop Ichthyol 20(2):e210051

Peters RH (1986) The ecological implications of body size, vol 2. Cambridge, Cambridge University Press

Pyron RA, Costa GC, Patten MA, Burbrink FT (2015) Phylogenetic niche conservatism and the evolutionary basis of ecological speciation. Biol Rev 90(4):1248–1262

Ripple WJ, Wolf C, Newsome TM, Hoffmann M, Wirsing AJ, McCauley DJ (2017) Extinction risk is most acute for the world's largest and smallest vertebrates. Proc Natl Acad Sci 114(40):10678–10683

Roberts TR (1984) *Amazonsprattus scintilla*, new genus and species from the Rio Negro, Brazil, smallest known clupeomorph fish. Proc Calif Acad Sci 43(20):317–321

Roff DA (1988) The evolution of migration and some life history parameters in marine fishes. Environ Biol Fishes 22(2):133–146

Roff DA (1991) Life history consequences of bioenergetic and biomechanical constraints on migration. Amer Zool 31(1):205–216

Roff D (1992) Evolution of life histories: theory and analysis. London, Chapman and Hall

Romanuk TN, Hayward A, Hutchings JA (2011) Trophic level scales positively with body size in fishes. Glob Ecol Biogeogr 20(2):231–240

Santos LL, Benone NL, Soares BE, Barthem RB, Montag LF (2019) Trait–environment relationships in Amazon stream fish assemblages. Ecol Freshw Fish 28(3):424–433

Sioli H (1984) Introduction: history of the discovery of the Amazon and of research of Amazonian waters and landscapes. In: Sioli H (ed) The Amazon: limnology and landscape ecology of a mighty tropical river and its basin. Springer, Netherlands, pp 1–13

Steele SE, López-Fernández H (2014) Body size diversity and frequency distributions of Neotropical cichlid fishes (Cichliformes: Cichlidae: Cichlinae). PLOS One 9(9):e106336

Tagliacollo VA, Dagosta FCP, Pinna MD, Reis RE, Albert JS (2021) Assessing extinction risk from geographic distribution data in Neotropical freshwater fishes. Neotrop Ichthyol 19:e210079

Toledo-Piza M, Mattox GMT, Britz R (2014) *Priocharax nanus*, a new miniature characid from the Rio Negro, Amazon basin (Ostariophysi: Characiformes), with an updated list of miniature Neotropical freshwater fishes. Neotrop Ichthyol 12:229–246

Weitzman SH, Vari RP (1987) Two new species and a new genus of miniature characid fishes (Teleostei: Characiformes) from northern South America. Proc Biol Soc Washington 101:640–652

Weitzman SH, Vari RP (1988) Miniaturization in South American freshwater fishes; an overview and discussion. Proc Biol Soc Washington 101:444–465

Winemiller KO, Taphorn DC, Barbarino-Duque A (1997) Ecology of *Cichla* (Cichlidae) in two blackwater rivers of southern Venezuela. Copeia 1997(4):690–696

20 Migratory Fishes of Neotropical Freshwaters

Bruno F. Melo and James S. Albert

Abstract

The upstream migration of fishes in large Neotropical rivers during the spawning season is triggered by rising water levels near the equator and temperature changes in subtropical regions. This event is crucial for the reproduction of fish *species* and holds ecological and economic significance, particularly for local fishing communities. Notable NFF clades exhibiting migratory behaviors include members of four characiform families (Anostomidae, Bryconidae, Curimatidae, and Prochilodontidae), and the siluriform family Pimelodidae. Migratory fishes include nine clades of Characiformes, six of Gymnotiformes, and 12 of Siluriformes. River basins exhibit distinct physical attributes that impact directly on the migratory behavior, and the establishment of numerous dams across the Neotropical region has affected the lifecycle of migratory fishes and fish-dependent human communities. Evidence from paleontology indicates that migratory fishes occupied western portions of the proto-Amazonas basin as far back as the Paleocene *c.* 60 Ma. Time-calibrated *phylogenies* of multiple freshwater fish groups indicate that initial diversifications occurred during the Eocene, and that anostomids, bryconids, curimatids, and prochilodontids colonized upland habitats during the Miocene (23–5.3 Ma) and Pliocene (5.3–2.6 Ma). Several migratory fish species have been included in governmental Red List assessments, including nine characiform species in genera *Brycon* and *Prochilodus*, six siluriform species in the genera *Conorhynchos, Kalyptodoras, Rhyncodoras*, and *Steindachneridion*, and two gymnotiform species in the genera *Sternarchogiton* and *Sternarchorhynchus*. Management plans for migratory freshwater fishes are challenging because their geographic ranges span international borders.

20.1 Introduction to Neotropical Fish Migrations

Migratory species of Neotropical freshwater fish (NFF) play an important role in the ecology and *biodiversity* of the world's largest and most diverse aquatic ecosystems, including the Amazon, Orinoco, and Paraná Rivers (Carolsfeld et al. 2003; Herrera et al. 2024). For many fishes, migrations are essential for reproduction, feeding, and maintaining *genetic diversity*, supporting both natural ecosystems and the livelihoods of local human communities. Some of the more conspicuous long-distance migrants are prochilodontid fishes of the genera *Prochilodus* ("coporo" in northern South America, "sábalo" in the south, and "curimbatá" in Brazil) and *Semaprochilodus* ("bocachico" or "sapuara" in Spanish and "jaraquí" in Portuguese). Migratory behaviors are known in many large-bodied riverine fishes, but little information is available regarding the timing or extents; e.g., river stingrays (Potamotrygonidae; Shibuya 2022); *Pellona* (Pristigasteridae) is known to undergo lateral migrations from the main river channel (Moreira-Hara et al. 2009; Ximenes et al. 2017).

20.1.1 Ecological Drivers

Migratory species may undergo longitudinal movements upstream and downstream, or lateral movements between the river channel and the surrounding floodplains (Fernandes 1997; Silva et al. 2021). Some species undertake extraordinary

B. F. Melo (✉)
American Museum of Natural History, New York, NY, USA
e-mail: brunfmelo@gmail.com

UNICAMP Universidade Estadual de Campinas, Campinas, Brazil

J. S. Albert
University of Louisiana at Lafayette, Lafayette, LA, USA
e-mail: jsalbe@gmail.com

J. S. Albert and R. E. Reis (eds.), *Ecology and Evolution of Amazonian Fishes*, Fascinating Life Sciences,
https://doi.org/10.1007/978-3-032-09328-8_20

journeys of hundreds to thousands of kilometers driven by seasonal cues, such as rainfall, temperature changes, and water-level fluctuations. Goliath catfishes (*Brachyplatystoma* spp., Pimelodidae) undertake the longest recorded freshwater migration in the world, with juveniles swimming upstream from the Amazon River estuary in a multistaged journey toward their adult spawning grounds in the Andean piedmont (Barthem and Goulding 1997; Barthem et al. 2017). Many herbivorous and detritivorous species (e.g., Prochilodontidae, Serrasalmidae, Bryconidae) in nutrient-poor blackwater and clearwater tributaries move downstream to whitewater river channels to spawn (e.g., Araujo-Lima and Goulding 1997; Silva and Stewart 2017). Many species couple longitudinal with lateral movements at smaller spatial scales to complete their life cycles. During rising waters, larvae may move into adjacent floodplain habitats, such as flooded forests and oxbow lakes, which provide shelter and feeding grounds (De Lima and Araujo-Lima 2004).

Massive, long-distance, fish migrations serve as a major source of upstream nutrient transport in Neotropical aquatic ecosystem (Ribeiro and Petrere 1990; Barthem et al. 1991; Winemiller and Jepsen 1998). The biomass of migratory fishes reaches up to thousands of kilograms per river kilometer during peak migration, depending on drainage basin and region (Winemiller and Jepsen 2004; Barradas et al. 2012; González-Bergonzoni et al. 2019). Many other fishes undertake lateral or short longitudinal migrations (Cajado et al. 2023; Chap. 13). However, the intricate patterns of these migrations face increasing threats from habitat fragmentation, overfishing, and climate change, underscoring the need for conservation efforts and sustainable management practices to ensure their survival and the health of their habitats (Barthem and Goulding 1997; Duponchelle et al. 2016).

20.1.2 Terminology

Migration refers to the purposeful and active movement among two or more distinct aquatic habitats, which occurs on an ontogenetic (i.e., adult or juvenile stages) or seasonal (e.g., dry–wet periods) basis (Dingle and Drake 2007). In migratory species, movements are critical to individual fitness and population persistence because it enables specialized use of different habitats for growth and reproduction (McIntyre et al. 2015). Where their migration routes are blocked or key habitats are lost, migratory fishes often suffer rapid and catastrophic losses. However, human appropriation and degradation of the Earth's freshwater ecosystems have transformed this reliance on multiple habitats into a detriment for many migratory fishes (Albert et al. 2021a).

Numerous terms are associated with fish migration, such as *anadromy*, *catadromy*, amphidromy, oceanodromy, and potamodromy. *Anadromous* fishes migrate to oceans, where they spend the majority of their time for maturation before returning to freshwater to spawn. In contrast, *catadromous* fishes migrate from the ocean to freshwaters for feeding and growing, and then return to the ocean for reproduction. In a particular case, *amphidromy*, the movements are not indicative of reproductive processes; this can be either freshwater amphidromy, whereby the breeding occurring in freshwaters, or marine amphidromy, wherever the breeding takes place in the ocean (Myers 1949; Gross 1987; McDowall 1992).

An alternative categorization of migratory fishes exists based on the specificity of their habitats: *oceanodromy* refers to the migrations of exclusively marine fishes between spawning grounds and ocean currents, while *potamodromy* refers to the migrations of exclusively freshwater fishes. Potamodromy can be subdivided into two categories: the *longitudinal migration* when movements of fishes occur upstream for spawning and downstream for feeding and growing, and the *lateral migration*, when movements of fishes occur between marginal lagoons and rivers. Partial migration occurs when a given population comprises both migratory and resident individuals (Fernandes 1997; Chapman et al. 2012). The focus of this chapter centers on potamodromous fishes of South America, which engage in both lateral and longitudinal migration for feeding or spawning.

20.1.3 Goals of This Chapter

This chapter summarizes the *species richness*, *taxonomic composition*, phylogenetic history, and conservation status of migratory Neotropical freshwater fishes.

20.2 The Piracema

The predominant migration pattern observed in Neotropical migratory fishes is upstream migration. This event is popularly referred to as *piracema* in Brazil, from the Indigenous Tupi word *pirá* meaning fish, and *sema* meaning exit, and also denotes the time of the year during which fish reproduce. The term can be more broadly applied to annual breeding migrations in large lowland river basins of tropical South America, such as the Amazon, Orinoco, Tocantins, Magdalena, Paraná-Paraguay, and São Francisco River basins. However, it is not uncommon for migrations to also be observed in other coastal rivers, including the Doce, Paraíba do Sul, and Uruguai rivers (Carolsfeld et al. 2003; Godinho and Godinho 2003). The annual arrival of fishes is referred to by different terms in Spanish-speaking regions, e.g., *arribazón, ribazón, subida de los peces*, or *subienda*.

In general terms, piracema fishes perform two distinct forms of migration: (1) during the rainy season, adults migrate upstream from foraging channels to spawning grounds, where they lay eggs and develop into larvae and juveniles, primarily in floodplains or marginal lagoons; (2) during the dry season, adults migrate downstream from upstream or tributaries to feeding in large river channels. Longitudinal migrations are observed consistently across many basins and taxa, while the supplementary lateral movement primarily takes place in the várzeas of the floodplains in the central Amazonia (Araújo-Lima and Ruffino 2003; Fernandes 1997; Godinho and Kynard 2009).

The duration of the piracema is critical for the fishes and the populations that depend on them. In the upper Paraná basin near the Itaipu Reservoir, migratory fishes constituted 39%–57% of the artisanal fish landings (Agostinho et al. 2004). The piracema of southeastern Brazil from October to March coincides with the abundance of precipitation and the reproductive season of numerous migratory fishes, including *pintados* (*Pseudoplatystoma*), *dourados* (*Salminus*), and *curimbatás* (*Prochilodus*; de Magalhães Lopes et al. 2018). During this time, fishing activities cease in preparation for the reproduction season, populations form large aggregations or schools, and migrate upstream for spawning, while exotic and allochthonous species emerge as the predominant targets of fishing efforts.

River basins exhibit distinct attributes that have a direct impact on the migratory behavior. The region upstream of the Itaipu Dam in the upper Paraná River contains an enormous and contiguous lowland area and connections to tributaries that serve as nursery areas for several migratory fishes. The fish migration in the middle São Francisco River is entirely influenced by marginal lagoons, during which juveniles mature throughout the dry season and become prepared to swim in the main river during the rainy season (Pompeu and Godinho 2006; Sato et al. 2005). This natural process has been impacted by the establishment of numerous powerplants across the Neotropical region (e.g., Jirau dam on the Madeira River, Tucuruí dam on the Tocantins River, Itaipu dam on the Paraná River, Três Marias dam on the São Francisco River) that has had a discernible influence on the water pulse, resulting in severe consequences for migratory fish populations and the natural aquatic lifecycle (Barletta et al. 2010; Fitzgerald et al. 2018; Godinho and Kynard 2009; Winemiller et al. 2016).

20.2.1 Taxonomic Composition

The *taxonomic composition* of migratory Neotropical freshwater fishes is exceptionally diverse, including at least 805 species in 153 genera, 30 families, and 15 taxonomic orders (Table 20.1). These include 258 characiform, 105 gymnotiform, 263 siluriform, and 164 species of *marine derived lineages* (MDLs).

20.2.2 Marine-Derived Clades

Migratory Neotropical freshwater fishes of relatively recent (i.e., Neogene or Quaternary) marine origins constitute a diverse *assemblage*, both in terms of *species richness* and *phenotypic disparity*. This fauna includes at least 164 species in 42 genera, 14 families, and 11 taxonomic orders, and represents three of the five fish clades traditionally ranked as Classes; Hyperoartia (lampreys), Chondrichthyes (cartilaginous fishes), and Actinopterygii (ray-finned fishes). There are no freshwater hagfishes (Hyperotreti), and the extant Neotropical lungfish species *Lepidosiren* (Sarcopterygii) is a primary freshwater fish (*sensu* Myers 1938) that only undertakes small lateral migrations onto floodplains, and is not known to undertake long-distance longitudinal migrations.

20.2.3 Characiformes

Characiformes species have been documented to exhibit both lateral and longitudinal migrations throughout the rainy season. Migration behavior is present in Curimatoidea, such as Serrasalmidae (particularly Colossomatinae), Cynodontidae, Anostomidae (particularly *Megaleporinus*), Prochilodontidae, Curimatidae (particularly *Curimata*, *Potamorhina*, and *Psectrogaster*), and in families of Characoidea such as Bryconidae. Considerable scientific effort has been dedicated to deciphering the many aspects of migratory behavior in Prochilodontidae and Bryconidae due to their relative relevance in industrial and artisanal fisheries; in contrast, the knowledge base about the remaining characiform taxa remains less understood.

The Neotropical fish family Prochilodontidae has been the subject of much research in the fields of evolution, genetics, biogeography, reproduction, and ecology. The majority of the 21 species migrate in schools, with some performing movements of more than 1000 kilometers upstream for spawning. Biogeographic reconstructions hypothesize that the earliest diversification of prochilodontids occurred in the Proto-Amazonas basin during the Eocene (Frable et al. 2022). Population genetic studies have shown lack of structure, indicating high *gene flow* among populations and that the molecular pools can be shared between individuals from Argentina to southeastern Brazil (Table 20.1). The genetic homogeneity can be observed in most species, with direct evidence for curimbatás *Prochilodus argenteus* and *P. costatus* from the São Francisco, *P. lineatus* from the La Plata and Paraíba do Sul, *P. nigricans* from the Amazon River Basin (hereafter, Amazon Basin), and *P. magdalenae* from the Cauca-Magdalena river basins of Colombia. The lack of population genetic structuring in which a single population generally is widespread and have little molecular variation, likely due to the migratory behavior, has been repeatedly found in almost 30 years of genetic research (Revaldaves et al. 1997; Sivasundar et al. 2001; Turner et al. 2004; Hatanaka et al. 2006; Aguirre-Pabón et al. 2013; Melo et al. 2013; Rueda et al. 2013; Berdugo and Narváez Barandica 2014; Chagas et al. 2015; Machado et al. 2017b; Melo et al. 2018a; Santos et al. 2021).

Table 20.1 Population genetic structure of principal migratory NFF species

Order	Species	Basin	F_{ST} or equivalent	Population genetic structure	Reference
Characiformes	*Brycon amazonicus*	Amazon	0.009	Low	de Oliveira et al. 2018
	Brycon henni	Cauca and Magdalena	0.179	High	Landínez-García and Márquez 2020
	Brycon insignis	Paraíba do Sul	0.07	Moderate	Matsumoto and Hilsdorf 2009
	Brycon opalinus	Paraíba do Sul	0.043	Low	Barroso et al. 2005
	Brycon orbignyanus	La Plata	0.234–0.258	High	Ashikaga et al. 2015
	Brycon orthotaenia	São Francisco	0.002	Low	Coimbra et al. 2020
	Colossoma macropomum	Amazon	0.062	Moderate	Santos et al. 2018
	Curimata mivartii	Cauca and Magdalena	0.001	Low	Landínez-García and Márquez 2018
	Ichthyoelephas longirostris	Cauca and Magdalena	0.01	Low	Landínez-García and Márquez 2016
	Megaleporinus obtusidens	São Francisco	0.073	Moderate	Oliveira-Farias et al. 2022
	Piaractus brachypomus	Amazon	0.379	High	Jorge et al. 2018
	Piaractus mesopotamicus	La Plata	0.001–0.014	Low	Calcagnotto and DeSalle 2009; Iervolino et al. 2010
	Prochilodus argenteus	São Francisco	0.008	Low	Hatanaka et al. 2006; Melo et al. 2013
	Prochilodus costatus	São Francisco	0.008–0.031	Low	Melo et al. 2013; Braga-Silva and Galetti 2016
	Prochilodus magdalenae	Cauca	0.045	Low	García-Castro and Márquez 2023
	Prochilodus nigricans	Amazon	0.153	High	Machado et al. 2017b
	Salminus brasiliensis	Uruguay	0.053	Moderate	Ribolli et al. 2018
	Salminus franciscanus	São Francisco	0.007	Low	Coimbra et al. 2020
	Salminus hillari	Paraná	0.019	Low	Silva et al. 2015
	Semaprochilodus insignis	Amazon	0.003–0.008	Low	Nunes et al. 2023
Siluriformes	*Brachyplatystoma flavicans*	Amazon	0.078	Moderate	Coronel et al. 2004
	Brachyplatystoma platynemum	Amazon	0.157	High	Ochoa et al. 2015
	Brachyplatystoma rousseauxii	Amazon	0.028–0.043	Low	Carvajal-Vallejos et al. 2014
	Pimelodus maculatus	Uruguay	0.022	Low	Ribolli et al. 2012
	Pimelodus yuma	Cauca	<0.00–0.019	Low	Joya et al. 2021
	Pseudoplatystoma corruscans	La Plata	0.034–0.163	Moderate	Pereira et al. 2009; Rueda et al. 2023
	Pseudoplatystoma fasciatum	Amazon	0.01	Low	S. Coronel et al. 2004
	Pseudoplatystoma magdaleniatum	Cauca and Magdalena	0.003	Low	García-Castro et al. 2021
	Pseudoplatystoma punctifer	Amazon	0.003	Low	Cunha-Machado et al. 2022
	Steindachneridion scriptum	Uruguay	0.068	Moderate	Paixão et al. 2018

F_{ST} genetic differentiation threshold: 0–0.05 = low; 0.05–0.25 = moderate; >0.25 = high (Wright 1965)

The flooded *várzea* forest in the Amazon Basin serves as habitats for juvenile and adult members of numerous migratory fish families, including Serrasalmidae. During periods of low water level, adults migrate to the main river channel and upstream to spawn in woody shores of large sediment-rich rivers, with larvae drifting downriver to várzea nursery habitats (Araújo-Lima and Ruffino 2003). Genetic studies have identified a single population of the tambaqui *Colossoma* (Serrasalmidae) in the Amazonas-Solimões, as well as genetic structure in the upper sections of major tributaries, including Juruá, Madeira, Purus, and Tapajós Rivers (Santos et al. 2007; 2018). The absence of population structure in pacus-caranha *Piaractus* between and within the Paraguay and Paraná rivers indicates that migratory behavior contributed to genetic homogeneity (Calcagnotto and DeSalle 2009; Iervolino et al. 2010). However, population genetic investigations in serrasalmids have encountered difficulties as a result of species delineation uncertainty within genera; several studies have identified cryptic diversity and species complexes, as well as underestimation of species richness (Machado et al. 2018; Nico et al. 2018; Kolmann et al. 2020; Mateussi et al. 2020). In addition, the expansion of aquaculture programs in southern and southeastern Brazil has catalyzed the viability of hybrids among native and nonnative serrasalmid genera *Colossoma* × *Piaractus* and the different hybrid combinations of *C. macropomum*, *P. brachypomus*, and *P. mesopotamicus* (patinga, tambacu, tambatinga), which represent competitors against native populations (Hashimoto et al. 2012).

Certain genera of Curimatidae have been reported to conduct short lateral migrations in the Amazon Basin, such as *Curimata*, *Potamorhina*, and *Psectrogaster* (Araújo-Lima and Ruffino 2003; Fernandes 1997; Landínez-García and Marquez 2018). Although much research has centered on the systematics, biogeography, and species delimitation, very few studies have addressed population genetics and the correlation with migratory behavior (Melo et al. 2016a, 2018b, 2021). Specimens of the short-migratory sardina *Curimata mivartii* present high levels of genetic diversity and gene flow in the lower Cauca-Magdalena basin of Colombia (Landínez-García and Marquez 2018). Similarly, little mitochondrial variation has been observed for branquinha *Potamorhina*, such as *P. altamazonica* and *P. latior*, across vast expanses of central-western Amazonia suggesting that migration may account for genetic homogeneity (Dorini et al. 2020). In molecular phylogenies, *Curimata* and *Potamorhina* form a clade, indicating a potential single origin for the migratory behavior within the family. However, short-distance migration is observed in species of *Psectrogaster* and rarely in *Curimatella*, which complicates the evolutionary understanding of curimatid migration. Recently, a nuclear-based population genetic study detected two genetic groups of *Cyphocharax magdalenae* structured by rainy and dry seasons in the middle and lower Cauca River (Ochoa-Aristizábal and Márquez 2024). The migratory behavior reported for *C. magdalenae* probably accounts for the presence of three distinct mitochondrial lineages distributed along the Magdalena-Cauca-Ranchería, León-Atrato, and Chucunaque-Tuira, Santa María, and Chiriquí basins of Panama and Costa Rica (Melo et al. 2024).

As observed in fishes generally, Neotropical freshwater fishes have larger modal adult body sizes as compared with nonmigratory fishes (Fig. 20.1; Burns and Bloom 2020). Anostomidae is another curimatoid family with some genera achieving large body sizes (e.g., *Hypomasticus*, *Megaleporinus*, *Schizodon*). Although genetic studies have shown certain population isolation, little is known about the impact of the migratory behavior on this genetic structure. The piapara (*Megaleporinus obtusidens*) is an example of Anostomidae that migrates in the piracema of the upper Paraná and São Francisco basins. Analyses of microsatellite DNA markers indicate high levels of genetic diversity and low levels of population structure in the upper Paraná river basin (Ramos et al. 2012). Another study indicates three main genetic clusters of *M. obtusidens*

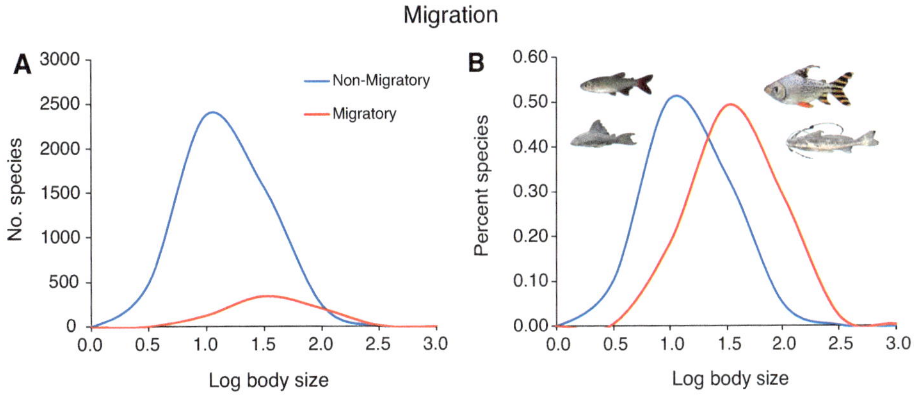

Fig. 20.1 Body size diversity among migratory and nonmigratory NFF species. Data for 5406 NFF species scored by adult modal phenotype. Absolute (**A**) and relative (**B**) size-frequency distributions. Note that migratory NFF species tend to be larger than nonmigratory NFF species

along the upper and lower sections of the São Francisco, and a significant reduction in genetic diversity in stocked populations when compared to wild specimens (Oliveira-Farias et al. 2022). The piava-vermelha *Hypomasticus copelandii* is another important component of the local fisheries in the Paraíba do Sul and adjacent coastal rivers of eastern Brazil. Molecular data indicate two genetic lineages along Brazilian coastal rivers, suggesting that Plio-Pleistocene river disconnection resulted in population structure (Mendes et al. 2022). Although these findings deliver an indication of population structure, more research is required to understand the function of migration in anostomid evolution.

Bryconidae contains several threatened species mostly within the genus *Brycon* and population genetic studies are extremely crucial to determine genetic variability and conserve natural populations. Several studies in the Bryconidae have shown instances of genetic differentiation. Nuclear and mitochondrial data revealed a lack of population structure and high levels of genetic variability in pirapitinga (*Brycon nattereri*) populations from the heavily impacted upper Paraná basin (Galindo et al. 2019). On the other hand, molecular analyses of the Endangered piracanjuba (*Brycon orbignyanus*) reveal significant levels of population structure, implying that the Ivinhema, Ilha Grande, Sucuriú, and Verde rivers of the upper Paraná should be priority for their conservation (Ashikaga et al. 2015). Microsatellite data from dourados (*Salminus*) revealed three distinct populations isolated by time along the Uruguay river basin (Ribolli et al. 2017). These results are extremely relevant for future management plans aiming the conservation of *Brycon and Salminus*.

20.2.4 Siluriformes

Migratory catfishes are mostly represented by members of the family Pimelodidae. They are commonly known as long-whiskered catfishes and comprise a diverse group of Siluriformes characterized by distinctive barbels, elongated bodies, and robust pectoral and dorsal fins. Many pimelodids perform long-distance migrations upriver to reach specific spawning or feeding sites. Giant catfishes of the piraíba (*Brachyplatystoma*) perform migrations of over 8000 km in the Amazon Basin (Barthem and Goulding 1997; Duponchelle et al. 2016). A recent radio-tagging study in the Xingu River basin has shown that pirarara (*Phractocephalus hemioliopterus*) migrates up to 347 km and sorubim (*Pseudoplatystoma punctifer*) up to 164 km upriver (Hahn et al. 2019). A genetic study with pintado (*Pseudoplatystoma corruscans*) in the La Plata system evidenced multiple genetically-structured populations in the Paraná-Paraguay rivers and the occurrence of homing behavior, i.e., tendency of adults in returning to their nursery tributaries (Pereira et al. 2009). A genetic study conducted with *Brachyplatystoma platynemum* indicate at least two structured populations in the Amazon and Madeira Rivers (Ochoa et al. 2015). These data indicate that migration plays an important role in the genetic structure of pimelodids.

Migration is also present in other catfish families. The medium-sized catfishes mapará (*Hypophthalmus*, Auchenipteridae) from the floodplains of the central Amazon basin and Tocantins-Araguaia rivers are reported to perform migrations in the Tocantins River (Carvalho and de Mérona 1986). Jundiá (*Rhamdia*) has been reported migrating up to tributaries during the rainy peak in the Uruguay basin (Zaniboni-Filho and Shulz 2003). Armados (*Pterodoras*) occur above the Itaipu Reservoir where they migrate upstream in tributaries of the Paraná River to reproduce; tagging studies indicate upstream movements of up to 150 km (Agostinho et al. 2004; Makrakis et al. 2007). The *pirá Conorhynchos*, an early-diverged genus related to Heptapteridae (Silva et al. 2021), has historical records of migration upstream in the São Francisco River basin, and is currently threatened to extinction (ICMBIO 2018). Genetic data show that *Conorhynchos* has suffered reductions in population size and loss of heterozygosity (Carvalho and Beheregaray 2018).

The cascudo-preto (*Rhinelepis aspera;* Loricariidae) migrates upriver for spawning from October to January in the upper Paraná basin (Agostinho et al. 2004; Suzuki et al. 2000). Short migrations for spawning in *Loricariichthys* were detected during the rainy season of Rio Ibicuí of the Uruguay River basin (Zardo and Behr 2015). An exceptional and unusual observation of massive migration of schools of *Trichomycterus barbouri* (Trichomycteridae) of over 370 km was observed in February and March in the Río Beni of Bolivia (Miranda-Chumacero et al. 2015). Species of other groups might perform short trophic migrations or occasional movements (gymnotiforms, cichlids, sciaenids) but are not here considered actual piracema fishes.

20.3 Phylogenetic History and Evolution of Piracema Fishes

Paleontological evidence, along with molecular phylogenetics and fossil-dated trees, has provided some evidence of when and where piracema fishes evolved in the Neotropics. Edentolous dentaries intermediate to curimatids, chilodontids, and prochilodontids have been found in Late Cretaceous formations of West Texas (Wick 2021). Fossils of serrasalmids are reported from the Maastrichtian El Molino Formation through the Paleocene in Contamana Formation and Cayara Formation in Peru and Bolivia (Gayet et al. 2001; Gayet et al. 2003; Antoine et al. 2016). Paleocene fossils of anostomids, serrasalmids, and pimelodids are also found in the Contamana Formation of the Peruvian Amazon (Antoine et al. 2016). Despite the incomplete fossil record, these combined data indicate that early forms of migratory fish groups occupied western portions of proto-Orinoco-Amazonas during the Paleocene from 66 to 56 million years ago.

Many phylogenies involving Neotropical migratory fish groups have been published recently. Molecular phylogenies exist at ordinal levels for migratory Characiformes (Melo et al. 2022) and Siluriformes (Sullivan et al. 2006), and at family-level for Anostomidae (Ramirez et al. 2017; Sidlauskas et al. 2022), Auchenipteridae (Calegari et al. 2019), Bryconidae (Abe et al. 2014), Curimatidae (Melo et al. 2018b), Doradidae (Arce-H. et al. 2013), Heptapteridae (Silva et al. 2021), Pimelodidae (Lundberg et al. 2011), Prochilodontidae (Melo et al. 2016b), and Serrasalmidae (Kolmann et al. 2020; Mateussi et al. 2020). These datasets are extremely useful to estimate the timing of diversification and biogeographic range evolution across multiple protowatersheds (Albert et al. 2021b; Frable et al. 2022).

Substantial molecular dating hypotheses have converged for the diversification of Neotropical migratory fish lineages during the Eocene (56.0–33.9 Ma). A dated phylogeny of Bryconidae indicates that the most recent common ancestor (MRCA) of the family that includes *Chilobrycon*, *Salminus*, *Henochilus*, and cis- and trans-Andean *Brycon* originated during the Eocene at approximately 46.7 Ma, suggesting the occurrence in the Andean foreland basin of western Amazonia (Abe et al. 2014). The timetree of Curimatidae indicates that the lineage leading to the migratory branquinhas *Potamorhina* split from the sister genus *Curimata* during the Eocene at 45 Ma (Melo et al. 2021). The origins of the two migratory prochilodontid genera *Prochilodus* and *Semaprochilodus* dates to approximately 34 Ma (Frable et al. 2022). The MRCA of Serrasalmidae that is consequently the split of Colossomatinae and Serrasalminae dates from the Eocene at around 42–38 Ma (Kolmann et al. 2020), and the origins of migratory anostomids seem to be around the Eocene-Oligocene boundary at approximately 37 Ma (Sidlauskas et al. 2022). Time-calibrated trees for siluriforms seem to follow the same pattern with the early diversification of Pimelodidae during the Paleocene-Eocene transition at around 56.4 Ma (Tagliacollo et al. 2015). Therefore, the current evidence from dated molecular phylogenetics indicates that the lineage diversification in the Eocene led to the origin of modern clades of migratory fishes in the Neotropics.

Historical biogeography studies in the Neotropical biota indicate the presence of the Sub-Andean foreland basin disconnected from the eastern side of the Amazonia by the Eocene. The paleogeographic conditions at the time suggest connectivity between the western Proto-Amazon with the Orinoco River basin and adjacent Pacific versant rivers (Hoorn et al. 2010). Although some studies have consistently demonstrated that many fish clades had origins in the Proto-Amazonas basin (Machado et al. 2018; Fontenelle et al. 2021; Melo et al. 2021), little research has investigated Cenozoic patterns of lineage diversification (Tagliacollo et al. 2015; Frable et al. 2022). Some questions remain active, such as when and where did the behaviors evolve: western Amazon in the Miocene? In which paleogeographic conditions: Pebas mega-wetland system? What are the roles of floodplains and estuaries as food sources and nurseries in migratory channel fishes such as *Brachyplatystoma* and *Semaprochilodus*?

Subsequent diversification within those families promoted later colonization of upland habitats mostly during the Miocene and Pliocene. For example, the curimbatás and *jaraquis Prochilodus rubrotaeniatus* and *Semaprochilodus varii* dispersed to Guiana Shield, *P. britskii*, *P. nigricans*, and *S. brama* dispersed to Amazonian Brazilian Shield, whereas *P. lineatus*, *P. argenteus*, and *P. costatus* colonized the uplands of the upper Paraná and São Francisco basins (Melo et al. 2016b; Santos et al. 2021). The migratory pimelodid catfishes also experienced events of upland geodispersal, for example, the pintado *Pseudoplatystoma corruscans* in La Plata and São Francisco, *Hemisorubim*, *Sorubim*, *Iheringichthys*, and *Pimelodus* in the La Plata system (Tagliacollo et al. 2015). Several lineages of migratory anostomids transitioned from Amazonian lineages to uplands of the Brazilian Shield, such as piapara *Megaleporinus obtusidens* and piavuçu *Megaleporinus macrocephalus* (Ramirez et al. 2017; Sidlauskas et al. 2022). Some clades achieved eastern Atlantic drainages where they perform small-scale migrations in medium-sized Jequitinhonha River, Doce River, and Paraíba do Sul River, as for example *Prochilodus harttii* and the threatened *P. vimboides* (Frable et al. 2022).

20.4 Phylogenetic Distribution of Piracema Fishes

We plotted the migratory behavior in the phylogenetic tree, including all Neotropical freshwater fishes. Migratory clades can be found across the teleost tree of life (Fig. 20.2; red branches). Although migrating behavior is apparent in early Neotropical lampreys, such as *Geotria* and *Mordacia* and chondrichthyans *Carcharhinus* and *Pristis*, the marine elopomorphs *Anguilla* and *Megalops* are the earliest actinopterygian lineages to exhibit such behavior. Other migratory lineages include Sciaenidae, *Centropomus*, Achiridae, Mugilidae, *Cichla*, Atherinopsidae, Clupeiformes, and Ariidae. However, in this chapter, the "piracema fishes" refer to otophysan potamodromous taxa, such as Bryconidae, Apteronotidae, and Pimelodidae.

There have been about 83 evolutionary transitions to long-distance (>100 km) migratory behaviors in members of the NFF fauna, in at least 31 family-level clades representing all of the Big Nine NFF clades (Fig. 20.2). The predominance of migratory otophysan fishes concentrates in the characiform superfamily Curimatoidea, including Anostomidae, Curimatidae, Cynodontidae,

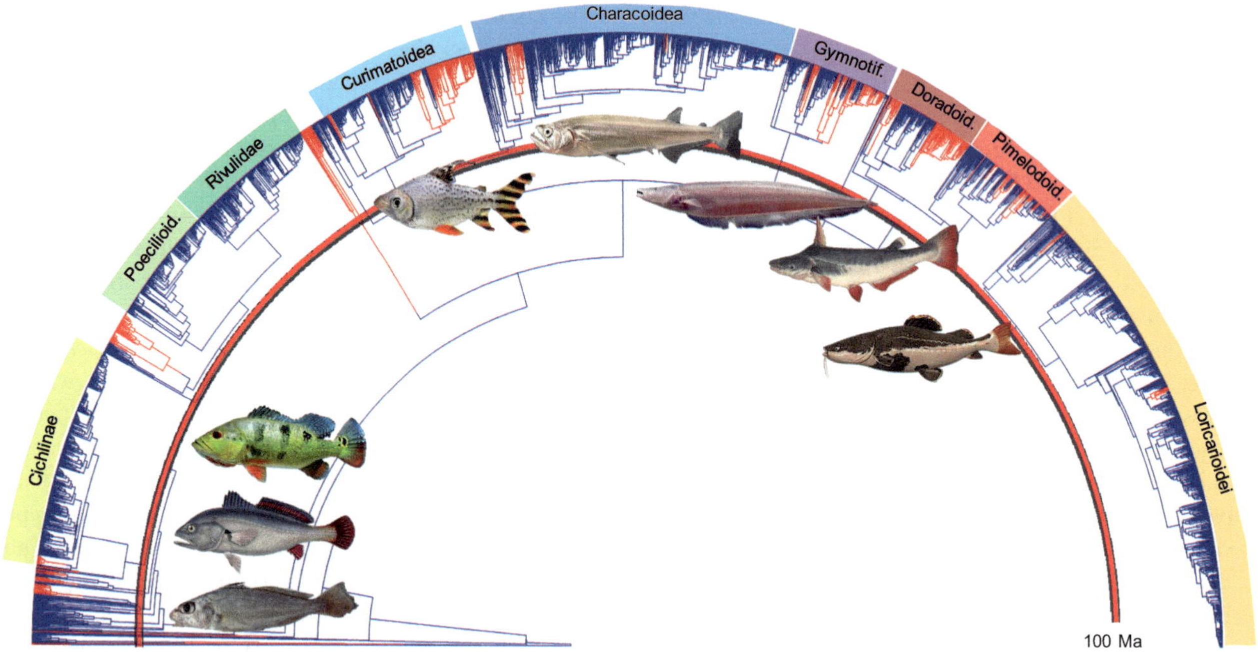

Fig. 20.2 Phylogenetic distribution of long-distance (>100 km) migratory behavior NFF species. Migratory lineages in red. Phylogeny from Chap. 2. Stochastic mapping of trait evolution with 1000 replicates resulted in about 83 evolutionary transitions to, and 48 transitions away, from migratory behavior on this tree

Prochilodontidae, and Serrasalmidae, in the gymnotiform family Apteronotidae, and in the siluriform suborder Siluroidei, mainly in Pimelodidae and Auchenipterinae. We highlight here some clades presenting such behavior: Colossomatinae, Cynodontidae, *Megaleporinus*, Prochilodontidae, *Curimata*, *Potamorhina*, *Psectrogaster*, Bryconidae, and *Cynopotamus* (Characiformes); *Gymnorhamphichthys*, *Rhamphichthys*, *Sternopygus*, *Rhabdolichops*, *Adontosternarchus*, and a large clade of Apteronotidae (Gymnotiformes); Cetopsidae, *Conorhynchos*, Pimelodidae, *Entomocorus*, *Epapterus*, *Auchenipterus*, *Auchenipterichthys*, *Ageneiosus*, *Astrodoras*, Wertheimerinae, Rhinodoradinae, and some genera of Doradinae (Siluriformes) (Fig. 20.2; Table 20.2).

Migration directly affects speciation rates due to the homogenization of the gene pool in the distribution range. For example, the large Amazon Basin contains only one species of *Prochilodus* in the central lowlands, and two species of *Semaprochilodus*. *Potamorhina latior* and *P. altamazonica* are distributed along enormous areas of the Amazon basin. Only one species of *Hemisorubim* is present in both Amazon, Orinoco, and La Plata basins. *Salminus brasiliensis*, *Prochilodus lineatus*, and *Pseudoplatystoma corruscans* inhabit vast expanses of the La Plata basin. In the São Francisco, only one genetic population exists for *Prochilodus argenteus* and *P. costatus*. Empirical genetic studies have hypothesized the effect of restricted gene flow over speciation events in Neotropical fishes (Dorini et al. 2020; Melo et al. 2018a).

The occurrence of migratory species in isolated river basins is indicative of dispersal through *river capture* (Albert et al. 2021; Carvalho and Albert 2011). *Megaleporinus obtusidens* has broad distribution over La Plata, São Francisco, and adjacent coastal Atlantic rivers, including the Parnaíba in northeastern Brazil (Britski et al. 2012). The sister relationship of *Prochilodus lineatus* from La Plata and *P. costatus* from the São Francisco river has been hypothesized to have occurred due to Quaternary reactivations of the upper São Francisco crustal faults in the Rio Grande at Serra da Canastra range (Santos et al. 2021). Another hypothesis is the incomplete knowledge status of many species due to taxonomic uncertainty, as it might be the case of the single species of *Hemisorubim* in the Amazon, Orinoco, and La Plata basins.

20.5 Conservation Status of Piracema Fishes

All migratory fishes are directly impacted by hydroelectric dams (Prestes et al. 2022; Hauser et al. 2024). According to the endangered species evaluation of Neotropical freshwater fishes from Brazil, the main threats are agriculture and livestock, river damming, urban expansion, pollution, and overfishing (ICMBIO 2018). Several studies have shown the direct impact of dam construction and establishment of large lentic reservoirs in diverse assemblage of fishes (Agostinho et al. 2008; Pelicice et al. 2015). Similarly, an empirical study has shown the impact of the low water temperature immediately below

Table 20.2 Main migratory otophysan genera, distribution, and references

Genus	Distribution	References
Brachyplatystoma	A-O	Lundberg et al. 2011; Duponchelle et al. 2016
Brycon	A-O-MM-LP-SF	Abe et al. 2014; Lima 2017
Colossoma	A-O	Santos et al. 2007; Farias et al. 2010; Santos et al. 2018
Conorhynchos	SF	Britski et al. 1988; Carvalho and Beheregaray 2018
Curimata	A-O-G-MM	Vari 1989a; Melo et al. 2018b
Ichthyoelephas	MM	López-Casas et al. 2016; Frable et al. 2022
Hemisorubim	A-LP	Agostinho et al. 2004
Hypophthalmus	A-O	Carvalho and Mérona 1986
Megaleporinus	A-O-G-MM-LP-SF	Garavello 1979; Britski et al. 2012; Ramirez et al. 2017
Mylossoma	A-O-LP	Araújo-Lima and Ruffino 2003; Mateussi et al. 2018
Phractocephalus	A	Freitas and Montag 2019; Hahn et al. 2019
Piaractus	A-O-LP	Escobar et al. 2019; Escobar L et al. 2022
Pimelodus	A-O-G-LP-SF	Agostinho et al. 2004
Pinirampus	A-LP	Agostinho et al. 2004
Potamorhina	A-O-LP	Vari 1984; Dorini et al. 2020
Prochilodus	A-O-G-MM-LP-SF-U	Castro and Vari 2004; Sato et al. 2005; López-Casas et al. 2016; Melo et al. 2016b; Melo et al. 2018a
Psectrogaster	A-O-G-LP	Vari 1989b; Melo et al. 2018b
Pseudoplatystoma	A-O-MM-LP-SF	Pereira et al. 2009; Torrico et al. 2009; Hahn et al. 2019
Pterodoras	A-O-LP	Makrakis et al. 2007
Rhaphiodon	A-O-LP	Toledo-Piza 2000; Agostinho et al. 2004
Rhinelepis	LP	Suzuki et al. 2000; Agostinho et al. 2004
Salminus	A-O-LP-SF-U	López-Casas et al. 2016; Machado et al. 2017a
Semaprochilodus	A-O-G	Ribeiro and Petrere 1990; Melo et al. 2016b; Nunes et al. 2023

A Amazon, *O* Orinoco, *G* Guianas, *MM* Magdalena/Maracaibo, *LP* La Plata, *SF* São Francisco, *U* Uruguay

the Três Marias Dam over spawning rate of *Prochilodus* in the São Francisco River (Sato et al. 2005). The construction of dams has posed challenges for conservationists directly working with migratory fishes (Godinho and Kynard 2009).

Several migratory fish species have been included in the Red List assessment of the Brazilian government with the advice of several ichthyologists and local experts. The list includes two species of *Prochilodus* as threatened of extinction: *P. britskii*, endangered (EN), from the Rio Apiacás of the upper Tapajós, and *P. vimboides*, vulnerable (VU), from coastal rivers of eastern Brazil, Paraíba do Sul, and upper Paraná (ICMBIO 2018). Seven species of *Brycon*, generally performing piracema migration for reproduction, are also listed as threatened: *B. devillei* (EN) from Doce and Jequitinhonha, *B. gouldingi* (EN) from Tocantins-Araguaia, *B. insignis* (EN) from coastal rivers of Brazil, *B. nattereri* (VU) from La Plata, São Francisco and upper Tocantins, *B. opalinus* (VU) from Paraíba do Sul and Doce, *B. orbygnianus* (EN) from La Plata, and *B. vermelha* (EN) from the Mucuri basin (ICMBIO 2018). Among the catfishes, the list includes the rare *Kalyptodoras bahiensis* (EN) from the Paraguaçu, *Rhyncodoras xingui* (EN) from Tocantins and Xingu, *Conorhynchos conirostris* (EN) from São Francisco, *Steindachneridion amblyurum*, Critically Endangered (CR) from Jequitinhonha, *S. doceanum* (CR) from Paraíba do Sul, *S. melanodermatum* (EN) from Rio Iguaçu, *S. parahybae* (EN) from Paraíba do Sul, and *S. scriptum* (EN) from upper Paraná and Uruguay. Among gymnotiforms, the red list includes *Sternarchogiton zuanoni* (VU) from the lower Xingu, *Sternarchorhynchus britskii* (EN) from the upper Paraná, among others.

Management plans for migratory freshwater fishes are challenging because their geographic ranges span international borders. Conservationists would ideally be expected to share knowledge of the current status of the same species in different countries (e.g. COP15 of Convention of Conservation of Migratory Species of Wild Animals; https://www.cms.int/cop15). Other solutions involve science-based fishery management plans, sustainable projects (e.g., Mamirauá), fish ladders and other dam-amelioration technologies, and social activism.

References

Abe KT, Mariguela TC, Avelino GS, Foresti F, Oliveira C (2014) Systematic and historical biogeography of the Bryconidae (Ostariophysi: Characiformes) suggesting a new rearrangement of its genera and an old origin of *Mesoamerican ichthyofauna* . BMC Evol Biol 14(1):152–152

Agostinho AA, Gomes LC, Suzuki HI, Júlio JrHF (2004) Migratory fishes of the upper Paraná river basin. In: Carolsfeld J, Harvey B, Ross C, Baer A (Eds.)Eds. Migratory fishes of South America. World Fisheries Trust, Victoria, 19–98

Agostinho AA, Pelicice FM, Gomes LC (2008) Dams and the fish fauna of the Neotropical region: impacts and management related to diversity and fisheries. Braz J Biol 68:1119–1132

Aguirre-Pabón J, Narváez Barandica J, Castro García L (2013) Mitochondrial DNA variation of the bocachico *Prochilodus magdalenae* (Characiformes, Prochilodontidae) in the Magdalena River Basin, Colombia. Aquat Conserv: Mar Freshw Ecosyst 23(4):594–605

Albert JS, Bernt MJ, Fronk AH, Fontenelle JP, Kuznar SL, Lovejoy NR (2021b) Late Neogene megariver captures and the great Amazonian biotic interchange. Glob Planet Change 205:103554

Albert JS, Destouni G, Duke-Sylvester SM, Magurran AE, Oberdorff T, Reis RE, Ripple WJ (2021a) Scientists' warning to humanity on the freshwater biodiversity crisis. Ambio 50(1):85–94

Antoine P-O, Abello MA, Adnet S, Sierra AJA, Baby P, Billet G, Chabain J (2016) A 60-million-year Cenozoic history of western Amazonian ecosystems in Contamana, eastern Peru. Gondwana Res 31:30–59

Araújo-Lima CARM, Ruffino ML (2003) Migratory fishes of the Brazilian Amazon. In: Carolsfeld J, Harvey B, Ross C, Baer A (Eds.)Eds. Migratory fishes of South America. World Fisheries Trust, Victoria, 233–301

Arce-H M, Reis RE, Geneva AJ, Sabaj Pérez MH (2013) Molecular phylogeny of thorny catfishes (Siluriformes: Doradidae). Mol Phylogenet Evol 67(3):560–577

Ashikaga FY, Orsi ML, Oliveira C, Senhorini JA, Foresti F (2015) The endangered species *Brycon orbignyanus*: genetic analysis and definition of priority areas for conservation. Environ Biol Fishes 98(7):1845–1855

Barletta M, Jaureguizar AJ, Baigun C, Fontoura NF, Agostinho AA, Almeida-Val VD, Corrêa MFM (2010) Fish and aquatic habitat conservation in South America: a continental overview with emphasis on neotropical systems. J Fish Biol 76(9):2118–2176

Barradas JRS, Silva LG, Harvey BC, Fontoura NF (2012) Estimating migratory fish distribution from altitude and basin area: a case study in a large Neotropical river. Freshwater Biol 57(11):2297–2305

Barroso RM, Hilsdorf AW, Moreira HL, Cabello PH, Traub-Cseko YM (2005) Genetic diversity of wild and cultured populations of *Brycon opalinus* (Cuvier, 1819) (Characiformes, Characidae, Bryconinae) using microsatellites. Aquaculture 247(1–4):51–65

Barthem RB, de Brito Ribeiro MCL, Petrere JrM (1991) Life strategies of some long-distance migratory catfish in relation to hydroelectric dams in the Amazon Basin. Biol Cons 55(3):339–345

Barthem RB, Goulding M, Leite RG, Cañas C, Forsberg B, Venticinque E, Petry P, Ribeiro MLDB, Chuctaya J, Mercado A (2017) Goliath catfish spawning in the far western Amazon confirmed by the distribution of mature adults, drifting larvae and migrating juveniles. Sci Rep 7(1):41784

Barthem R, Goulding M (1997) The Catfish connection: ecology, . Migration, and conservation of Amazon predators. Columbia University Press

Berdugo GO, Narváez Barandica JC (2014) Genetic diversity and population structure of bocachico *Prochilodus magdalenae* (Pisces, Prochilodontidae) in the Magdalena River basin and its tributaries, Colombia. Genet Mol Biol 37:37–45

Braga-Silva A, Galetti PM (2016) Evidence of isolation by time in freshwater migratory fish *Prochilodus costatus* (Characiformes, Prochilodontidae). Hydrobiologia 765:159–167

Britski HA, Birindelli JLO, Garavello JC (2012) A new species of Leporinus Agassiz, 1829 from the upper Rio Paraná basin (Chraciformes, Anostomidae) with redescription of *L. elongatus* Valenciennes, 1850 and *L. obtusidens* (Valenciennes, 1837). Pap Avulsos Zool 52(37):441–476.

Britski HA, Sato Y, Rosa ABS (1988) Manual de identificação de peixes da região de Três Marias: com chaves de identificação para os peixes da Bacia do São Francisco, 3rd edn. CODEVASF, Brasília

Burns MD, Bloom DD (2020) Migratory lineages rapidly evolve larger body sizes than non-migratory relatives in ray-finned fishes. Proc R Soc B 287(1918):20192615

Cajado RA, Zacardi DM, Arantes CC, Oliveira LS, Silva FK, Andrade MC (2023) Seasonal patterns in the structure of the fish larvae assemblages in the lower Amazon River. Freshwater Biol 68(7):1252–1266

Calcagnotto D, DeSalle R (2009) Population genetic structuring in pacu (*Piaractus mesopotamicus*) across the Parana-Paraguay basin: evidence from microsatellites. Neotrop Ichthyol 7(4):607–616

Calegari BB, Vari RP, Reis RE (2019) Phylogenetic systematics of the driftwood catfishes (Siluriformes: Auchenipteridae): a combined morphological and molecular analysis. Zool J Linnean Soc 187(3):661–773

Carolsfeld J, Harvey B, Ross C, Baer A (2003) Migratory fishes of South America. World Fisheries Trust, Victoria

Carvajal-Vallejos F, Duponchelle F, Desmarais E, Cerqueira F, Quérouil S, Núñez J, Renno J-F (2014) Genetic structure in the Amazonian catfish *Brachyplatystoma rousseauxii*: influence of life history strategies. Genetica 142(4):323–336

Carvalho DC, Beheregaray LB (2018) Conservation genetics of the threatened catfish *Conorhynchos conirostris* (Siluriformes: incertae sedis), an evolutionary relict endemic to the São Francisco River Basin, Brazil. Conserv Genet 19(5):1223–1230

Carvalho JL, de Mérona B (1986) Estudos sobre dois peixes migratorios do baixo Tocantins, antes do fechamento da barragem de Tucurui. Amazoniana 4:595–607

Carvalho TP, Albert JS (2011) The Amazon-Paraguay divide. In: Albert JS, Reis RE (eds) Historical biogeography of Neotropical freshwater fishes. University of California Press, Berkeley and Los Angeles, pp 193–202

García-Castro KL, Márquez EJ (2023) Temporal-scale assessment of population genetics of the freshwater fish *Prochilodus magdalenae* in an area impacted by construction of a dam. Hydrobiologia 851:1513–1531.

Castro RMC, Vari RP (2004) Detritivores of the South American fish family Prochilodontidae (Teleostei: Ostariophysi: Characiformes): a phylogenetic and revisionary study. Smithson Contrib Zool 622:1–189

Chagas ATdA, Carmo AO, Costa MA, Resende LC, Brandão Dias PFP, Martins APV, Kalapothakis E (2015) Description and comparison of two economically important fish species mitogenomes: *Prochilodus argenteus* and *Prochilodus costatus* (Characiformes, Prochilodontidae). mtDNA 27(4):2852–2853

Chapman B, Hulthén K, Brodersen J, Nilsson PA, Skov C, Hansson LA, Brönmark C (2012) Partial migration in fishes: causes and consequences. J Fish Biol 81(2):456–478

Coimbra MR, Dantas HL, Luna MM, Lima MA, Sales M, da Silva BC, Lima AP (2020) High gene flow in two migratory Neotropical fish species, *Salminus franciscanus* and *Brycon orthotaenia*, and implications for conservation aquaculture. Aquat Conserv: Mar Freshw Ecosyst 30(6):1063–1073

Coronel JS, Maes G, Claus S, Van Damme P, Volckaert F (2004) Differential population history in the migratory catfishes *Brachyplatystoma flavicans* and *Pseudoplatystoma fasciatum*

(Pimelodidae) from the Bolivian Amazon assessed with nuclear and mitochondrial DNA markers. J Fish Biol 65(3):859–868

Cunha-Machado AS, Farias IP, Hrbek T, Escobar MD, Alves-Gomes JA, Formiga KM, da Silva Batista J (2022) Genetic differentiation and gene flow of the Amazonian catfish *Pseudoplatystoma punctifer* across the Madeira River rapids prior to the construction of hydroelectric dams. Hydrobiologia 849:29–46

de Brito Ribeiro MCL, Junior MP (1990) Fisheries ecology and management of the Jaraqui (*Semaprochilodus taeniurus* , *S. insignis*) in central Amazonia. Regul Rivers: Res Manag 5(3):195–215

De Lima ÁC, Araujo-Lima CA (2004) The distributions of larval and juvenile fishes in Amazonian rivers of different nutrient status. Freshwater Biol 49(6):787–800

de Magalhães Lopes J, Alves CBM, Peressin A, Pompeu PS (2018) Influence of rainfall, hydrological fluctuations, and lunar phase on spawning migration timing of the Neotropical fish *Prochilodus costatus*. Hydrobiologia 818:145–161

Dingle H, Drake VA (2007) What is migration?. BioScience 57(2):113–121

Dorini BF, Ribeiro-Silva LR, Foresti F, Oliveira C, Melo BF (2020) Molecular phylogenetics provides a novel hypothesis of chromosome evolution in Neotropical fishes of the genus *Potamorhina* (Teleostei, Curimatidae). J Zool Sys Evol Res 58(4):1067–1075

Duponchelle F, Pouilly M, Pécheyran C, Hauser M, Renno JF, Panfili J, García-Dávila C (2016) Trans-Amazonian natal homing in giant catfish. J Appl Ecol 53(5):1511–1520

Escobar LMD, Farias IP, Hrbek T (2022) Genetic comparison of populations of *Piaractus brachypomus* and *P. orinoquensis* (Characiformes: Serrasalmidae) of the Amazon and Orinoco basins. Neotrop Ichthyol 20(3), e220056.

Escobar L MD, Ota RP, Machado-Allison A, Andrade-López J, Farias IP, Hrbek T (2019) A new species of *Piaractus* (Characiformes: Serrasalmidae) from the Orinoco Basin with a redescription of *Piaractus brachypomus*. J Fish Biol 95(2):411–427

Farias IP, Torrico JP, García-Dávila C, Santos MdCF, Hrbek T, Renno JF (2010) Are rapids a barrier for floodplain fishes of the Amazon basin? A demographic study of the keystone floodplain species *Colossoma macropomum* (Teleostei: Characiformes). Mol Phylogenet Evol 56(3):1129–1135

Fernandes CC (1997) Lateral migration of fishes in Amazon floodplains. Ecol Freshwater Fish 6(1):36–44

Fitzgerald DB, Perez MHS, Sousa LM, Gonçalves AP, Py-Daniel LR, Lujan NK, Lundberg JG (2018) Diversity and community structure of rapids-dwelling fishes of the Xingu River: implications for conservation amid large-scale hydroelectric development. Biol Cons 222:104–112

Frable BW, Melo BF, Fontenelle JP, Oliveira C, Sidlauskas BL (2022) Biogeographic reconstruction of the migratory Neotropical fish family Prochilodontidae (Teleostei: Characiformes). Zool Scripta 51(3):348–364

Freitas TMdS, Montag LfdA (2019) Population and reproductive parameters of the red-tailed catfish, *Phractocephalus hemioliopterus* (Pimelodidae: Siluriformes), from the Xingu River, Brazil. Neotrop Ichthyol 17:e190015

Galindo BA, Ferreira DG, Apolinário-Silva C, Terra MC, Aprígio NG, Ota RR et al (2019) Genetic diversity and population structure of *Brycon nattereri* (Characiformes: Bryconidae): a Neotropical fish under threat of extinction. Neotrop Ichthyol 17:e180071

Garavello, JC, (1979). Revisão taxonômica do gênero Leporinus Spix, 1829 (Ostariophysi, Anostomidae), Tese de Doutorado. Universidade de São Paulo, São Paulo

García-Castro KL, Rangel-Medrano JD, Landínez-García RM, Márquez EJ (2021) Population genetics of the endangered catfish *Pseudoplatystoma magdaleniatum* (Siluriformes: Pimelodidae) based on species-specific microsatellite loci. Neotrop Ichthyol 19:e200120

Gayet M, Jégu M, Bocquentin J, Negri FR (2003) New characoides from the upper Cretaceous and Paleocene of Bolivia and the Mio-Pliocene of Brazil: phylogenetic position and paleobiogeographic implications. J Vertebr Paleontol 23(1):28–46

Gayet M, Marshall LG, Sempere T, Meunier FJ, Cappetta H, Rage J-C (2001) Middle Maastrichtian vertebrates (fishes, amphibians, dinosaurs and other reptiles, mammals) from Pajcha Pata (Bolivia). Biostratigraphic, palaeoecologic and palaeobiogeographic implications. Palaeogeogr Palaeoclimatol Palaeoecol 169(1–2):39–68

Godinho AL, Godinho HP (2003) Águas, peixes e pescadores do São Francisco das Minas Gerais. PUC Minas, Belo Horizonte

Godinho AL, Kynard B (2009) Migratory fishes of Brazil: life history and fish passage needs. River Res Appl 25(6):702–712

González-Bergonzoni I, D'Anatro A, Vidal N, Stebniki S, Tesitore G, Silva I, Teixeira de Mello F (2019) Origin of fish biomass in a diverse subtropical river: an allochthonic-supported biomass increase following flood pulses. Ecosystems 22:1736–1753

Gross MR (1987) The evolution of diadromy in fishes. Paper presented at the American Fisheries Society Symposium 1:14–25.

Hahn L, Martins EG, Nunes LD, da Câmara LF, Machado LS, Garrone-Neto D (2019) Biotelemetry reveals migratory behaviour of large catfish in the Xingu River, Eastern Amazon. Sci Rep 9(1):8464

Hashimoto DT, Senhorini JA, Foresti F, Porto-Foresti F (2012) Interspecific fish hybrids in Brazil: management of genetic resources for sustainable use. Rev Aquac 4(2):108–118

Hatanaka T, Henrique-Silva F, Galetti JrPM (2006) Population substructuring in a migratory freshwater fish *Prochilodus argenteus* (Characiformes, Prochilodontidae) from the São Francisco river. Genetica 126:153–159

Hauser M, Doria CR, Pécheyran C, Ponzevera E, Panfili J, Torrente-Vilara G et al (2024) Quantitative impacts of hydroelectric dams on the trans-Amazonian migrations of goliath catfish. Conserv Lett 17(5): e13046

Herrera RGA, Heilpern SA, Couto TB, Victoria-Lacy L, Duponchelle F, Correa SB, Anderson EP (2024) A synthesis of the diversity of freshwater fish migrations in the Amazon basin. Fish Fish 25(1):114–133

Hoorn C, Wesselingh FP, ter Steege H, Bermudez Ma, Mora A, Sevink J, Antonelli A (2010) Amazonia through time: Andean uplift, climate change, landscape evolution, and biodiversity. Science 330:927–931

ICMBIO (2018) Livro Vermelho da Fauna Brasileira Ameaçada de Extinção: Volume VI-Peixes (ICMdCd. Biodiversidade Ed.). ICMBio/MMA, Brasília, DF

Iervolino F, de Resende EK, Hilsdorf AWS (2010) The lack of genetic differentiation of pacu (*Piaractus mesopotamicus*) populations in the Upper-Paraguay Basin revealed by the mitochondrial DNA D-loop region: implications for fishery management. Fish Res 101(1–2):27–31

Jorge PH, Mastrochirico-Filho VA, Hata ME, Mendes NJ, Ariede RB, Freitas MVd, Hashimoto DT (2018) Genetic characterization of the fish *Piaractus brachypomus* by microsatellites derived from transcriptome sequencing. Front Genet 9:46

Joya CD, Landínez-García RM, Márquez EJ (2021) Development of microsatellite loci and population genetics of the catfish *Pimelodus yuma* (Siluriformes: Pimelodidae). Neotrop Ichthyol 19:e200114

Kolmann M, Hughes L, Hernandez L, Arcila D, Betancur-R R, Sabaj M, Ortí G (2020) Phylogenomics of piranhas and pacus (Serrasalmidae) uncovers how dietary convergence and parallelism obfuscate traditional morphological taxonomy. Syst Biol 70(3):576–592

Landínez-García RM, Márquez EJ (2016) Development and characterization of 24 polymorphic microsatellite loci for the freshwater fish Ichthyoelephas longirostris (Characiformes: Prochilodontidae). PeerJ 4:e2419

Landínez-García RM, Marquez EJ (2018) Microsatellite loci development and population genetics in Neotropical fish *Curimata mivartii* (Characiformes: Curimatidae). PeerJ 6:e5959

Landínez-García RM, Márquez EJ (2020) Population genetics of the fish *Brycon henni* (Characiformes: Bryconidae) using species-specific polymorphic microsatellite loci. Rev Biol Trop 68(3):847–861

Lima FC (2017) A revision of the cis-andean species of the genus *Brycon* Müller and Troschel (Characiformes: Characidae). Zootaxa 4222(1):1–189

López-Casas S, Jiménez-Segura L, Agostinho A, Pérez C (2016) Potamodromous migrations in the Magdalena River basin: bimodal reproductive patterns in neotropical rivers. J Fish Biol 89(1):157–171

Lundberg JG, Sullivan JP, Hardman M (2011) Phylogenetics of the South American catfish family Pimelodidae (Teleostei: Siluriformes) using nuclear and mitochondrial gene sequences. Proc Acad Nat Sci Philadelphia 161(1):153–189

Machado CDB, Ishizuka TK, Freitas PDD, Valiati VH, Galetti JrPM (2017a) DNA barcoding reveals taxonomic uncertainty in *Salminus* (Characiformes). Syst Biodivers 15(4):372–382

Machado VN, Collins RA, Ota RP, Andrade MC, Farias IP, Hrbek T (2018) One thousand DNA barcodes of piranhas and pacus reveal geographic structure and unrecognised diversity in the Amazon. Sci Rep 8(1):8387

Machado VN, Willis SC, Teixeira AS, Hrbek T, Farias IP (2017b) Population genetic structure of the Amazonian black flannelmouth characin (Characiformes, Prochilodontidae: *Prochilodus nigricans* Spix and Agassiz, 1829): contemporary and historical gene flow of a migratory and abundant fishery species. Environ Biol Fishes 100(1):1–16

Makrakis M, Miranda L, Makrakis S, Fernandez D, Garcia J, Dias J (2007) Movement patterns of armado, *Pterodoras granulosus*, in the Paraná River Basin. Ecol Freshwater Fish 16(3):410–416

Mateussi NT, Melo BF, Ota RP, Roxo FF, Ochoa LE, Foresti F, Oliveira C (2020) Phylogenomics of the Neotropical fish family Serrasalmidae with a novel intrafamilial classification (Teleostei: Characiformes). Mol Phylogenet Evol 153:106945

Mateussi NT, Oliveira C, Pavanelli CS (2018) Taxonomic revision of the Cis-Andean species of *Mylossoma* Eigenmann Kennedy, 1903 (Teleostei: Characiformes: Serrasalmidae). Zootaxa 4387(2):275–309

Matsumoto CK, Hilsdorf AWS (2009) Microsatellite variation and population genetic structure of a neotropical endangered Bryconinae species *Brycon insignis* Steindachner, 1877: implications for its conservation and sustainable management. Neotrop Ichthyol 7:395–402

McDowall R (1992) Diadromy: origins and definitions of terminology. Copeia 1992(1):248–251

McIntyre PB, Liermann CR, Childress E, Hamann EJ, Hogan JD, Januchowski-Hartley SR, Koning AA, Neeson TM, Oele DL, Pracheil BM (2015) Conservation of migratory fishes in freshwater ecosystems. In: Closs GP, Olden JD, Krkosek M (Eds.) Conservation of freshwater fishes. Cambridge University Press, pp 324–360

Melo BF, Albert JS, Dagosta FCP, Tagliacollo VA (2021) Biogeography of curimatid fishes reveals multiple lowland–upland river transitions and differential diversification in the Neotropics (Teleostei, Curimatidae). Ecol Evol 11(22):15815–15832

Melo BF, Dorini BF, Foresti F, Oliveira C (2018a) Little divergence among mitochondrial lineages of *Prochilodus* (Teleostei, Characiformes). Front Genet 9:107

Melo BF, Ochoa LE, Vari RP, Oliveira C (2016a) Cryptic species in the Neotropical fish genus *Curimatopsis* (Teleostei, Characiformes). Zool Scripta 45(6):650–658

Melo BF, Sato Y, Foresti F, Oliveira C (2013) The roles of marginal lagoons in the maintenance of genetic diversity in the Brazilian migratory fishes *Prochilodus argenteus* and *P. costatus*. Neotrop Ichthyol 11(3):625–636

Melo BF, Sidlauskas BL, Hoekzema K, Frable BW, Vari RP, Oliveira C (2016b) Molecular phylogenetics of the Neotropical fish family Prochilodontidae (Teleostei: Characiformes). Mol Phylogenet Evol 102:189–201

Melo BF, Sidlauskas BL, Hoekzema K, Vari RP, Dillman CB, Oliveira C (2018b) Molecular phylogenetics of Neotropical detritivorous fishes of the family Curimatidae (Teleostei: Characiformes). Mol Phylogenet Evol 127:800–812

Melo BF, Sidlauskas BL, Near TJ, Roxo FF, Ghezelayagh A, Ochoa LE et al (2022) Accelerated diversification explains the exceptional species richness of tropical characoid fishes. Syst Biol 71(1):78–92

Melo BF, Conde-Saldaña CC, Villa-Navarro FA, McMahan CD, Oliveira C (2024) Phylogeographic patterns of *Cyphocharax* from trans-Andean rivers and northward expansion to lower Central America (Teleostei, Curimatidae). J Fish Biol 105(1):314–325

Mendes IS, Melo BF, Damasceno JS, Teixeira DF, Carvalho DC (2022) Phylogeography of hypomasticus copelandii reveals distinct genetic lineages along Atlantic coastal drainages of eastern Brazil (Teleostei, Anostomidae). Diversity 14(1):29

Miranda-Chumacero G, Álvarez G, Luna V, Wallace RB, Painter L (2015) First observations on annual massive upstream migration of juvenile catfish Trichomycterus in an Amazonian River. Environ Biol Fishes 98(8):1913–1926

Moreira-Hara SS, Zuanon JA, Amadio SA (2009) Feeding of *Pellona flavipinnis* (Clupeiformes, Pristigasteridae) in a Central Amazonian floodplain. Iheringia. Sér Zool 99:153–157

Myers GS (1938) Fresh water fishes and West Indian zoogeography, vol 3465. US Government Printing Office, Washington, DC

Myers GS (1949) Usage of anadromous, catadromous and allied terms for migratory fishes. Copeia 1949(2):89–97

Nico L, Jégu M, Andrade M (2018) Family Serrasalmidae-piranhas and pacus. In: Van der Sleen P, Albert J (Eds.) Field guide to the fishes of the Amazon, Orinoco, and Guianas. Princeton University Press, Princeton, NJ, 172–196

Nunes I, Passos K, Ximenes AM, Hrbek T, Farias IP (2023) Spatial and temporal population genetic analysis of *Semaprochilodus insignis* (Prochilodontidae), an overexploited fish from the Amazon basin. PeerJ 11:e15503

Ochoa-Aristizábal AM, Márquez EJ (2024) Genetic insights into *Cyphocharax magdalenae* (Characiformes: Curimatidae): microsatellite loci development and population analysis in the Cauca River, Colombia. PLoS One 19(4):e0302273

Ochoa LE, Pereira LHG, Costa-Silva GJ, Roxo FF, Batista JS, Formiga K, Oliveira C (2015) Genetic structure and historical diversification of catfish *Brachyplatystoma platynemum* (Siluriformes: Pimelodidae) in the Amazon basin with implications for its conservation. Ecol Evol 5(10):2005–2020

Oliveira-Farias KK, Farias RS, Neto MA, de Souza ME, Calado LL, Montes MA, Coimbra MR (2022) Genetic structure and management of the Neotropical migratory fish *Megaleporinus obtusidens* on a highly impacted river basin. Hydrobiologia 849(7):1645–1658

Oliveira RC, Santos MdCF, Bernardino G, Hrbek T, Farias IP (2018) From river to farm: an evaluation of genetic diversity in wild and aquaculture stocks of *Brycon amazonicus* (Spix and Agassiz, 1829), Characidae, Bryconinae. Hydrobiologia 805:75–88

Paixão RV, Ribolli J, Zaniboni-Filho E (2018) Genetic variation of the endangered neotropical catfish *Steindachneridion scriptum* (Siluriformes: Pimelodidae). Front Genet 9:320819

Pelicice FM, Pompeu PS, Agostinho AA (2015) Large reservoirs as ecological barriers to downstream movements of Neotropical migratory fish. Fish Fish 16(4):697–715

Pereira L, Foresti F, Oliveira C (2009) Genetic structure of the migratory catfish *Pseudoplatystoma corruscans* (Siluriformes: Pimelodidae) suggests homing behaviour. Ecol Freshwater Fish 18(2):215–225

Pompeu PdS, Godinho HP (2006) Effects of extended absence of flooding on the fish assemblages of three floodplain lagoons in the middle São Francisco River, Brazil. Neotrop Ichthyol 4:427–433

Prestes L, Barthem R, Mello-Filho A, Anderson E, Correa SB, Couto TBDA, Goulding M (2022) Proactively averting the collapse of Amazon fisheries based on three migratory flagship species. PLoS One 17(3):e0264490

Ramirez JL, Birindelli JL, Galetti JrPM (2017) A new genus of Anostomidae (Ostariophysi: Characiformes): diversity, phylogeny and biogeography based on cytogenetic, molecular and morphological data. Mol Phylogenet Evol 107:308–323

Ramos JVB, Sodré LMK, Orsi ML, Almeida FSd (2012) Genetic diversity of the species *Leporinus elongatus* (Teleostei: Characiformes) in the canoas complex-Paranapanema river. Neotrop Ichthyol 10:821–828

Revaldaves E, Renesto E, Machado MF (1997) Genetic variability of *Prochilodus lineatus* (Characiformes, Prochilodontidae) in the upper Paraná River. Braz J Genet 20:381–388

Ribeiro MCLB, Petrere JrM (1990) Fisheries ecology and management of the jaraqui (*Semaprochilodus taeniurus*, *S. insignis*) in central Amazonia. Regul Rivers: Res Manage 5:195–215

Ribolli J, Hoeinghaus DJ, Johnson JA, Zaniboni-Filho E, de Freitas PD, Galetti PM (2017) Isolation-by-time population structure in potamodromous Dourado *Salminus brasiliensis* in southern Brazil. Conserv Genet 18:67–76

Ribolli J, Melo CMRd, Zaniboni-Filho E (2012) Genetic characterization of the neotropical catfish Pimelodus maculatus (Pimelodidae, Siluriformes) in the Upper Uruguay River. Genet Mol Biol 35:761–769

Ribolli J, Zaniboni-Filho E, Freitas PD, Galetti PM (2018) Genetic evidences of non-reproductive shoaling in the freshwater fish *Salminus brasiliensis*. Hydrobiologia 815:65–72

Rueda EC, Carriquiriborde P, Monzón AM, Somoza GM, Ortí G (2013) Seasonal variation in genetic population structure of sábalo (*Prochilodus lineatus*) in the Lower Uruguay River. Genetica 141(7–9):401–407

Rueda EC, Machado CB, Castro V, Braga-Silva A, Ojeda G, Vargas F, Ortí G (2023) Genetic population structure of *Pseudoplatystoma corruscans* (Siluriformes: Pimelodidae) and evidence of temporal variation in structure. J Fish Biol 102(5):1040–1048

Santos MdCF, Hrbek T, Farias IP (2018) A multilocus approach to understanding historical and contemporary demography of the keystone floodplain species *Colossoma macropomum* (Teleostei: Characiformes). Front Genet 9:263

Santos MdC, Ruffino M, Farias I (2007) High levels of genetic variability and panmixia of the tambaqui *Colossoma macropomum* (Cuvier, 1816) in the main channel of the Amazon River. J Fish Biol 71:33–44

Santos RP, Melo BF, Yazbeck GM, Oliveira RS, Hilário HO, Prosdocimi F, Carvalho DC (2021) Diversification of *Prochilodus* in the eastern Brazilian shield: evidence from complete mitochondrial genomes (Teleostei, Prochilodontidae). J Zool Sys Evol Res 59(5):1053–1063

Sato Y, Bazzoli N, Rizzo E, Boschi MB, Miranda MO (2005) Influence of the Abaeté River on the reproductive success of the neotropical migratory teleost *Prochilodus argenteus* in the São Francisco River, downstream from the Três Marias Dam, southeastern Brazil. River Res Applic 21(8):939–950

Shibuya A (2022) A review of the ecological role of the Neotropical freshwater stingrays (Chondrichthyes: Potamotrygoninae). Food Webs 32:e00244

Sidlauskas BL, Assega FM, Melo BF, Oliveira C, Birindelli JLO (2022) Total evidence phylogenetic analysis reveals polyphyly of *Anostomoides* and uncovers an unexpectedly ancient genus of Anostomidae fishes (Characiformes). Zool J Linnean Soc 194(2):626–669

Silva EA, Stewart DJ (2017) Reproduction, feeding and migration patterns of *Prochilodus nigricans* (Characiformes: Prochilodontidae) in northeastern Ecuador. Neotrop Ichthyol 15(3):e160171

Silva JVd, Hallerman EM, Orfão LH, Hilsdorf AWS (2015) Genetic structuring of *Salminus hilarii* Valenciennes, 1850 (Teleostei: Characiformes) in the rio Paraná basin as revealed by microsatellite and mitochondrial DNA markers. Neotrop Ichthyol 13:547–556

Silva PB, Arantes CC, Freitas CE, Petrere JrM, Ribeiro FR (2021) Seasonal hydrology and fish assemblage structure in the floodplain of the lower Amazon River. Ecol Freshwater Fish 30(2):162–173

Sivasundar A, Bermingham E, Ortí G (2001) Population structure and biogeography of migratory freshwater fishes (*Prochilodus*: Characiformes) in major South American rivers. Mol Ecol 10(2):407–417

Sullivan JP, Lundberg JG, Hardman M (2006) A phylogenetic analysis of the major groups of catfishes (Teleostei: Siluriformes) using rag1 and rag2 nuclear gene sequences. Mol Phylogenet Evol 41(3):636–662

Suzuki H, Agostinho A, Winemiller K (2000) Relationship between oocyte morphology and reproductive strategy in loricariid catfishes of the Paraná River, Brazil. J Fish Biol 57(3):791–807

Tagliacollo VA, Roxo FF, Duke-Sylvester SM, Oliveira C, Albert JS (2015) Biogeographical signature of river capture events in Amazonian lowlands. J Biogeography 42(12):2349–2362

Toledo-Piza M (2000) The neotropical fish subfamily Cynodontinae (Teleostei: Ostariophysi: Characiformes): a phylogenetic study and a revision of *Cynodon* and *Rhaphiodon*. Am Mus Novit 3286:1–88

Torrico JP, Hubert N, Desmarais E, Duponchelle F, Nuñez Rodriguez J, Montoya-Burgos J, Renno JF (2009) Molecular phylogeny of the genus *Pseudoplatystoma* (Bleeker, 1862): biogeographic and evolutionary implications. Mol Phylogenet Evol 51(3):588–594

Turner TF, McPhee MV, Campbell P, Winemiller KO (2004) Phylogeography and intraspecific genetic variation of prochilodontid fishes endemic to rivers of northern South America. J Fish Biol 64:186–201

Vari RP (1984) Systematics of the Neotropical characiform genus *Potamorhina* (Pisces, Characiformes). Smithson. Contr Zool 400:1–36

Vari RP (1989a) Systematics of the Neotropical characiform genus *Curimata* Bosc (Pisces: Characiformes). Smithson. Contr Zool 474:1–63

Vari RP (1989b) Systematics of the Neotropical characiform genus *Psectrogaster* Eigenmann and Eigenmann (Pisces: Characiformes). Smithson Contr Zool 481(400):i-iii +1–43

Wick SL (2021) New early Campanian characiform fishes (Otophysi: Characiformes) from West Texas support a South American origin for known Late Cretaceous characiforms from North America. Cretaceous Res 128:104993

Winemiller KO, Jepsen DB (1998) Effects of seasonality and fish movement on tropical river food webs. J Fish Biol 53:267–296

Winemiller, KO, Jepsen, DB, (2004). Migratory neotropical fish subsidize food webs of oligotrophic blackwater rivers, Food webs at the landscape level. University of Chicago Press, Chicago, IL, pp. 115–132

Winemiller KO, McIntyre PB, Castello L, Fluet-Chouinard E, Giarrizzo T, Nam S, Sáenz L (2016) Balancing hydropower and biodiversity in the Amazon, Congo, and Mekong. Science 351:128–129

Wright S,(1965) The interpretation of population structure by F-statistics with special regard to systems of mating. Evolution 395–420

Ximenes AM, Machado VN, Martinez JG, Farias JG, Farias IP (2017) Using DNA barcodes to confirm new records of Amazon longfin herrings *Pellona castelnaeana* Valenciennes, 1847 and *Pellona flavipinnis* (Valenciennes, 1837) (Clupeiformes: Pristigasteridae) in the Branco River sub-basin. Check List 13(4):391–395

Zaniboni-Filho E, Shulz UH (2003) Migratory fishes of the uruguay river. In: Carolsfeld J, Harvey B, Ross C, Baer A (eds) Migratory fishes of South America. World Fisheries Trust, Victoria, pp 157–194

Zardo ÉL, Behr ER (2015) Population structure and reproductive biology of *Loricariichthys melanocheilus* Reis and Pereira, 2000 (Siluriformes: Loricariidae) in the rio Ibicuí, Brazil. Neotrop Ichthyol 13:371–382

21 Origins of the Richest Continental Fauna: Major Macroevolutionary Events

James S. Albert, Tiago P. Carvalho, William G. R. Crampton, Bruno F. Melo, Victor A. Tagliacollo and Roberto E. Reis

Abstract

Amazonian fishes display the greatest variety of species and ecological traits (*ecotraits*) of any continental vertebrate fauna on Earth. This singular fauna emerged from a fortuitous combination of paleoenvironmental conditions and macroevolutionary events operating across a vast geographic arena representing the whole of continental South America, and over an immense time period of over 100 million years. Like most biotas, Amazonian fishes exhibit a *hollow-curve* distribution of *species richness*, in which most species are members of just a few highly diverse *clades*, and most clades are composed of a single species. About 90% of NFF species are members of just nine clades (6%), which themselves represent just two superorders of teleost fishes; Ostariophysi and Acanthopterygii, while 101 NFF clades (70%) are singletons. The environmental–organismal interactions of these diverse fish *clades* allowed *speciation* rates to surpass *extinction* rates for many millions of years, allowing sustained net *diversification* and, ultimately, high *functional disparity* and regional *diversity*. All the major clades emerged in the super-greenhouse world of the Upper Cretaceous and Paleogene, with the fauna being almost entirely modern by the Early Miocene. The fossil record shows rapid taxonomic turnover after the K-Pg extinction, with some modern genera present by the Eocene, and almost no *extinct* genera since the Late Miocene. *Diversification* was relatively constant throughout the Cenozoic, with a slight acceleration following the Oligocene global cooling and geographic expansion of lowland freshwater *habitats*. *Evolutionary radiations* during the Early Cenozoic were influenced by habitats associated with paleoelevation, while later radiations were tied to formation of lowland river basins and marine regressions, including the Proto-Amazon river and Pebas megawetland system within the Sub-Andean foreland (SAF) basin. Tectonic uplifts of the Andean cordilleras during the Miocene reshaped river drainage patterns, prompting both *vicariance* and *geodispersal*, notably during the Oligocene to Miocene formation of the Bolivian Orocline and the Late Miocene formation of the modern Amazon River and Orinoco River basins. Historical biogeographic analyses support both the *Upland Origins* and *Lowland Origins* hypotheses in different clades, although clades with uplands origins dominate modern diversity in all habitats. Diversification of Amazonian fishes was driven by a variety of organismal, ecological, evolutionary processes, in which most speciation events occurred along geographic and not ecological lines, and with speciation rates exceeding rates of extinction and *adaptation* with nearly uninterrupted diversification under conditions of splendid isolation. These *phylodynamic* results resulted in a fauna with high levels of local *alpha diversity* and regional *beta diversity*, and also high functional redundancy. However, luck may be running out for Amazonian fishes under accelerating pressures from *habitat* destruction and climate changes due to human activities in the

J. S. Albert (✉)
University of Louisiana at Lafayette, Lafayette, LA, USA
e-mail: jsalbe@gmail.com

T. P. Carvalho
Universidad Nacional de Colombia, Bogotá, Colombia
e-mail: carvalho.ictio@gmail.com

W. G. R. Crampton
University of Central Florida, Orlando, FL, USA
e-mail: crampton@ucf.edu

B. F. Melo
American Museum of Natural History, New York, NY, USA
e-mail: brunfmelo@gmail.com

UNICAMP Universidade Estadual de Campinas, Campinas, Brazil

V. A. Tagliacollo
Universidade Federal do Uberlândia, Uberlândia, MG, Brazil
e-mail: vatagliacollo@gmail.com

R. E. Reis
Pontifícia Universidade Católica Rio Grande do Sul, Porto Alegre, RS, Brazil
e-mail: reis@pucrs.br

J. S. Albert and R. E. Reis (eds.), *Ecology and Evolution of Amazonian Fishes*, Fascinating Life Sciences,
https://doi.org/10.1007/978-3-032-09328-8_21

Anthropocene. The waterways and wetlands of tropical America preserve a rich biota that accumulated over deep time, and once damaged or destroyed, this vital ecosystem will never be replaced. As with all the unique forms of life on Earth, extinction is forever.

21.1 Richest Continental Fauna

Given its immense size, tropical climates, and ecological complexity, it is perhaps not surprising that *Greater Amazonia* is the global center of biological diversity, or *biodiversity*, for most continental clades of plants and animals (Balian et al. 2008; Stork et al. 2015; Cardoso et al. 2017; Meseguer et al. 2020; Albert et al. 2021b; Antonelli 2022; Guayasamin et al. 2024; Chap. 1). These clades include vascular plants (i.e., trees, epiphytes, grasses, etc., Antonelli and Sanmartín 2011; Cai et al. 2023; Ter Steege et al. 2023), soil fungi (Tedersoo et al. 2014), arthropods (Kass et al. 2022; Liu et al. 2022), freshwater vertebrates (i.e., fishes; Val et al. 2022; van der Sleen and Albert 2022), and terrestrial vertebrates (i.e., tetrapods; Roll et al. 2017). The density of species in freshwaters is especially pronounced, with approximately 12% of all fish species, and similar values for other aquatic plant and animal taxa, compressed into a tiny 0.003% of the planetary water supply (Lundberg et al. 2000; Levêque et al. 2008).

The larger *Neotropical Freshwater Fish* (NFF) fauna includes at least 6342 valid species distributed throughout the rivers, lakes, and wetlands of tropical South America, Central America, and the Greater Antilles (Matamoros et al. 2015; Reis et al. 2016; Albert et al. 2020). This diversity represents about one-sixth (18%) of all c. 37,200 described fish species globally (Eschmeyer et al. 2025), or about one in 11 (c. 9%) of the described vertebrate species that live on the Earth today. The Neotropics is also the global center of diversity for many biodiversity metrics among freshwater (i.e., continental) fishes, including species richness (van der Sleen and Albert 2022), genetic (Manel et al. 2020), morphological (Su et al. 2019), and ecological (Winemiller 1991; Toussaint et al.) diversity. Neotropical aquatic ecosystems support the food webs of large lakes and rivers that provide a substantial fraction of the animal protein consumed by tens of millions of people (Goulding et al. 2019; Ainsworth et al. 2021; Pelicice et al. 2023). Recent studies have also documented the ecological (Carvalho et al. 2023) conservation (Tagliacollo et al. 2021) and evolutionary (Cassemiro et al. 2023) values of the immense NFF fauna.

Yet, despite these exceptional numbers, the species density of Amazonian fishes, measured as the number of species per unit area (i.e., sub-basin or square km), far exceeds that expected from the usual suspects of potential biodiversity drivers (Oberdorff et al. 2019; Val et al. 2022). Studies using methods and data from macroecology have shown that many classic predictors of *species richness*, including geographic area, precipitation, productivity, habitat heterogeneity, and habitat connectivity, are necessary but not fully sufficient to explain Amazonian biodiversity (Antonelli et al. 2018; Rangel et al. 2018; Rull and Carnaval 2020; Diniz-Filho 2023; Chap. 1). Decades of detailed and arduous studies have revealed that understanding patterns of freshwater fish biodiversity at regional or continental scales require knowledge of the evolutionary processes affecting the origins of phenotypic novelties and lineage diversification through deep time (Smith et al. 2010; Albert et al. 2020; Boschman et al. 2023; Cassemiro et al. 2023). The most important macroevolutionary processes are speciation, extinction, geographic range changes, and adaptation, all occurring under a unique history of paleogeographic events and conditions over millions of years (Jablonski 2000; López-Fernández 2021). Several previous chapters of this book document how *ecotraits* associated with habitat utilization, diet, and life history affect these macroevolutionary processes, culminating in the most diverse aquatic ecosystems on Earth. Further, the evolutionary history of Amazonian fishes, like that of all continental-scale biotas, has been strongly affected by rare but important (i.e., *Black Swan*) paleogeographic events and paleoclimatic conditions (e.g., Fig. 1.10; Albert et al. 2021a).

This chapter merges the concepts and data of the previous chapters to develop a cohesive and multidimensional narrative of the evolutionary history of Amazonian fishes. This synthetic narrative addresses the alternative macroevolutionary hypotheses described in Chap. 1, and concludes by noting the critical importance of ecological and evolutionary principles and information in managing the fate of Amazonian fishes in the Anthropocene (Fig. 21.1).

21.2 Materials and Methods

21.2.1 Four New Datasets

Previous chapters of this book introduce four new datasets of the whole NFF fauna, which in combination form the empirical foundations for the analyses and conclusions developed in this chapter. These datasets are: (1) a syncretic chronology of 20 major paleogeographic, climatic, and geological conditions and events affecting South American biotas over the past 100 million years (Ma or *Mega-annum*), hypothesized to have affected the evolutionary histories of South American fishes (Table 1.1); (2) a time-calibrated phylogeny for all valid NFF species (Fig. 2.1; Tagliacollo et al. 2024); (3) geographic distributions of all NFF species among major *bioregions* and paleobioregions (Figs. 4.2–4.9); and (4) an *ecomatrix* of 42 ecophysiological traits scored for all NFF species (Table 6.1; Albert et al. 2025). It must be reiterated from previous chapters that these four datasets are the product of specific methods and

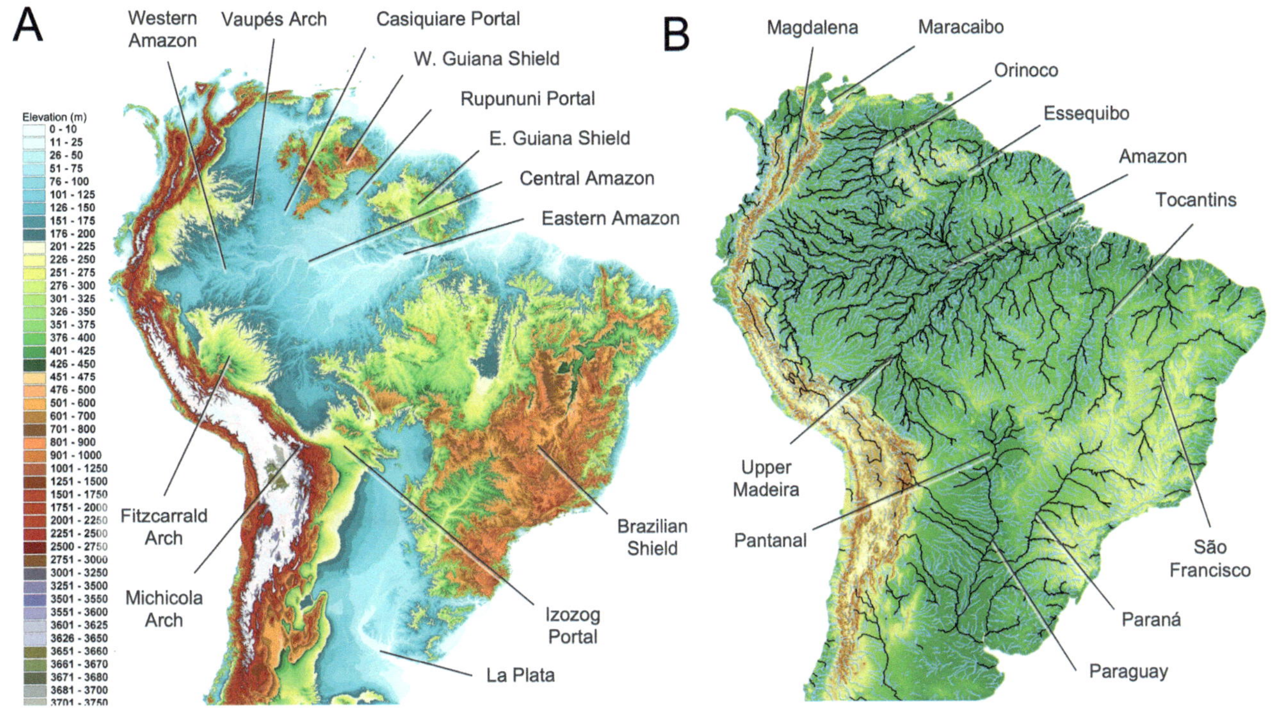

Fig. 21.1 Major geomorphological and hydrological features of tropical South America. (**A**) False-colored elevation map emphasizing differences at lower elevations. Low-elevation (<500 m) watersheds (i.e., portals and arches) divide major river basins. (**B**) River sizes binned by Strahler stream order; blue = SO 1–5, black = SO 6–10. Data from HydroSHEDS v2.0 (Lehner et al. 2021)

empirical datasets, which together produce excellent summaries of our current understanding of the timing, relationships, geographic distributions, and ecological traits of NFF species. However, due to the burgeoning research effort and recent discoveries, there are some discrepancies in these datasets with information in the most currently published literature, and these datasets must be continually updated, enhanced, and refined to reflect our growing knowledge of NFF diversity.

21.2.2 Four New Analyses

Previous chapters of this book also report results of four kinds of analyses on the diversity the NFF fauna: (1) *Ancestral Area Estimation* (AAE) for 558 clades among major bioregions and paleobioregions (Table 4.2); (2) estimated rates of net linage *diversification* (Table 5.1) for the *Big Nine NFF clades* (defined in Chap. 3); (3) *ecospace* analysis using principal components Aaalysis (PCA) of the NFF ecomatrix (Fig. 6.7); and (4) *ecotrace* analyses using ancestral trait estimation (Revell 2025) for all 42 ecotraits; e.g., macrohabitat by *stream order* (Fig. 7.2), water chemistry (Fig. 10.5), whitewater floodplains (várzeas; Fig. 11.2), deep river channels (Fig. 13.5), *trophic position* (Figs. 7.3, 16.2, 17.1, and 18.6), and body size (Fig. 7.4). Results from these kinds of analyses allow investigators to assess the statistical support for alternative models of ecological evolution within a clade (Johnson and Omland 2004; Pontarp et al. 2019). These methods can be used to estimate the environmental conditions associated with ancient diversification events.

21.3 Results

21.3.1 Geographic Origins

Chapter 1 introduced Eigenmann's *Highland Origins hypothesis* (Eigenmann 1906, 1912, 1921) and Smith's *Lowland Origins hypothesis* (Smith 1992) to describe the first-order patterns of evolutionary diversification in NFF clades. AAE was used to study the biogeographic origins and early evolutionary histories of the Big Nine NFF clades (Chap. 4). These analyses reveal slightly more NFF clades with origins in lowland coastal and interior basins (n = 40) than in uplands of the Guiana Shield and Brazilian Shield (n = 31; Fig. 21.2). However, clades with lowland and upland origins have very different subsequent histories of diversification, such that most (5067 = c. 83%) *extant* NFF species are members of the clades with upland origins.

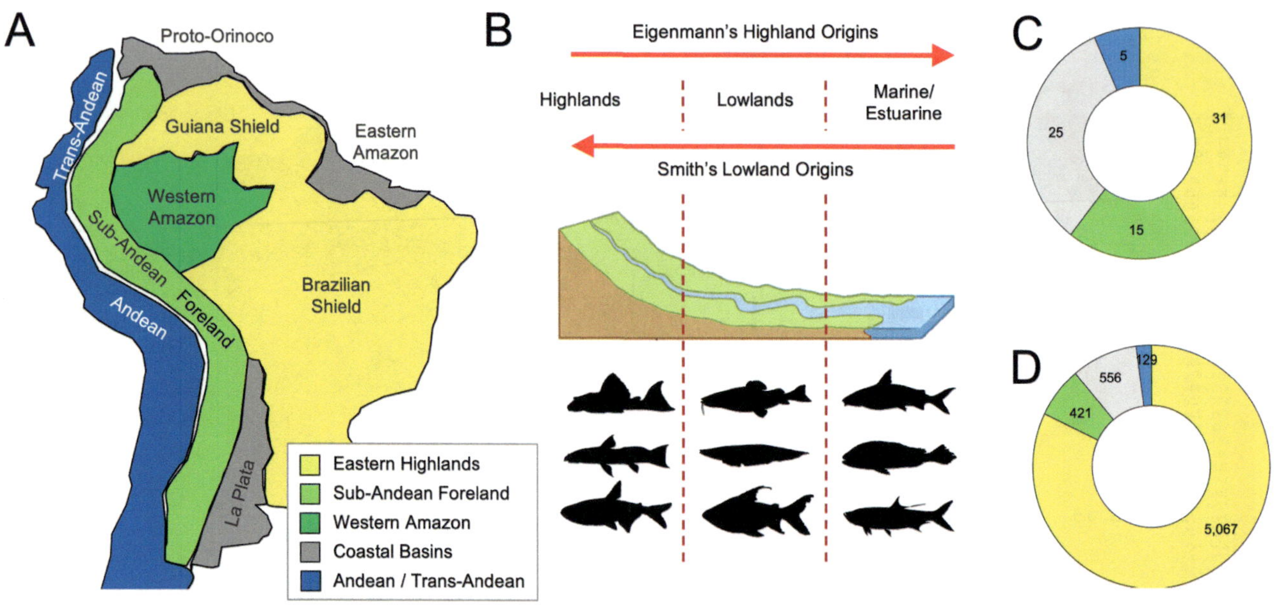

Fig. 21.2 Alternative paleogeographic hypotheses for origins of NFF taxa. (**A**) Map of part of South America illustrating major topographic structures mentioned in the text. (**B**) Opposite directions of hypothesized habitat evolution among NFF taxa. *Eigenmann's Uplands Origins hypothesis* posits downstream colonization from highland ancestors; *Smith's Lowland Origins hypothesis* posits upstream colonization from lowland ancestors, ultimately with marine origins. Longitudinal river profile illustrates major hydrological regions; silhouettes illustrate representative NFF taxa in each elevation zone. (**C**) Estimated geographic origins of 75 South American freshwater fish clades represented by two or more species among major paleobioregions. (**D**) Total number of species (in both uplands and lowlands) that are members of clades with upland or lowland origins

Although AAE is a powerful tool in evolutionary analysis, phylogenetic estimates based on state data (e.g., species geographic range) taken from extant taxa have several known assumptions and limitations. Prominent among these are accurate models of range evolution (McHugh et al. 2024), biogeographic range orthology with areas occupied from common *vicariance* not *coordinated dispersal* (Sanmartín 2012), sufficient representation of trait (e.g., range) disparity among taxa included in the analysis (Bagley and Johnson 2014), and stationarity; i.e., constant rates of range changes among clades and through time (Ferreira et al. 2024). AAE results can also be sensitive to sampling biases from missing taxa due to extinction, or incomplete taxon sampling (e.g., excluded cryptic species or undescribed species), and model misspecification concerning the modes and rates of geographic range expansion (i.e., *dispersal*), or geographic range contraction from local extirpations (Lamm and Redelings 2009; Pirie et al. 2012; Soberón and Osorio-Olvera 2023).

In combination, these factors generally result in loss of spatial resolution in estimated ancestral areas at more ancient nodes of the phylogeny (Wisniewski et al. 2022). The geographic distributions of extant species are always known with higher precision than older nodes because the ancestral species ranges are estimated from information observed in descendent species, with nonzero errors in the estimated topology and branch lengths of the phylogeny due to extinction and model misspecification from incorrect assumptions, omitted variables, or errors in parameter estimation (Caetano et al. 2018; Swiston and Landis 2024). Ultimately, all NFF clades have origins in either South America or perhaps more generally in Western Gondwana during the Upper Cretaceous (c. 100–66 Ma; Maisey 2000; Albert et al. 2011a; Melo et al. 2022) or from another continent (i.e., North America) or the marine realm (Cavin 2017). Fossil taxa can help constrain AAE because they add more taxa with shorter branches from undersampled earlier *time horizons* (Faurby et al. 2024). Most NFF taxa, however, either lack a fossil record or present fragmentary elements, and for these clades, biogeographic data from tip taxa accompanied by information from dated paleogeographic and paleoenvironmental events can help constrain lineage divergence times on a phylogenetic tree, matching allopatric taxa to the formation of vicariance barriers or dispersal corridors (Albert et al. 2006; Ho et al. 2015; Carrillo-Briceño et al. 2023). Paleogeographic data may also constrain estimated changes in ancestral ranges by organisms shifting areas or environments (Hoorn et al. 2010; Albert et al. 2017; Val et al. 2022; Cassemiro et al. 2023).

21.3.2 Ecospace and Ecotrace Analyses

The ecospace analysis of all NFF fishes (Chap. 6) showed that ecotraits associated with all seven of the major ecotrait categories contribute substantially to variance represented in the first two PCs. These ecotraits are related to body size, water chemistry, macrohabitat (e.g., *terra firme* streams, floodplains, river channels), mesohabitat (e.g., benthic, pelagic), microhabitat (e.g., substrate grain), and diet (e.g., *trophic position*; Fig. 21.4A, C). PC1 is dominated by ecotraits associated with body size (ecotrait 1a), stream order (ecotraits 2 and 3), water chemistry (ecotraits 6 and 10), mesohabitat (ecotraits 15 and 16), and behavioral potamodromy (ecotrait 37). PC2 is dominated by ecotraits associated with diet (ecotraits 27 and 31) and microhabitat (ecotraits 21 and 25). However, the first two PCs account for only 18.5% of the total ecospace variance, with the other PCs accounting for the remaining 81.5% of the variance (Fig. 6.6). In other words, most of the variance in NFF ecospace has yet to be fully explored as potential drivers of NFF ecological diversity.

Analysis of ecotrait evolution on the time-calibrated phylogeny of NFF species (Chap. 7) allows us to determine the evolutionary direction (i.e., polarity) of ecotrait variance along the PC axes of the ecospace analysis (Figs. 21.4B, D). The results show origins of all Big Nine NFF clades in low-gradient, upland, clearwater small rivers and streams, with a benthic invertivore diet. Body size and aspects of macrohabitat utilization are strongly structured phylogenetically among NFF species, with large-bodied species modally inhabiting large rivers, small and miniature species largely inhabiting small rivers, and medium-sized species as ecological generalists. The combined results of the ecospace and ecotrace analyses of Fig. 21.3 indicate that all of the Big Nine NFF clades originated about 100 Ma in small rivers or streams (blue circles in Fig. 21.3B) and as epibenthic or benthic invertivores (orange circles in Fig. 21.3D). Members of five clades (Curimatoidea, Gymnotiformes, Aspredinoidea, Pimelodoidea, and Loricarioidei) experienced multiple independent evolutionary transitions to large river channels with Strahler stream orders (SO) 6–10 (red circles in Fig. 21.3B), and members of all Big Nine NFF clades experienced independent transitions to benthic algivory (blue circles in Fig. 21.3D).

The results of the ecospace and ecotrace analyses are generally consistent predictions of the *River Continuum* concept, posting basin-scale gradients in species richness and ecotrait values among aquatic habitats from headwaters to river mainstems (Vannote and Sweeney 1980). These gradients involve predictable associations of local fluvial and hydrological

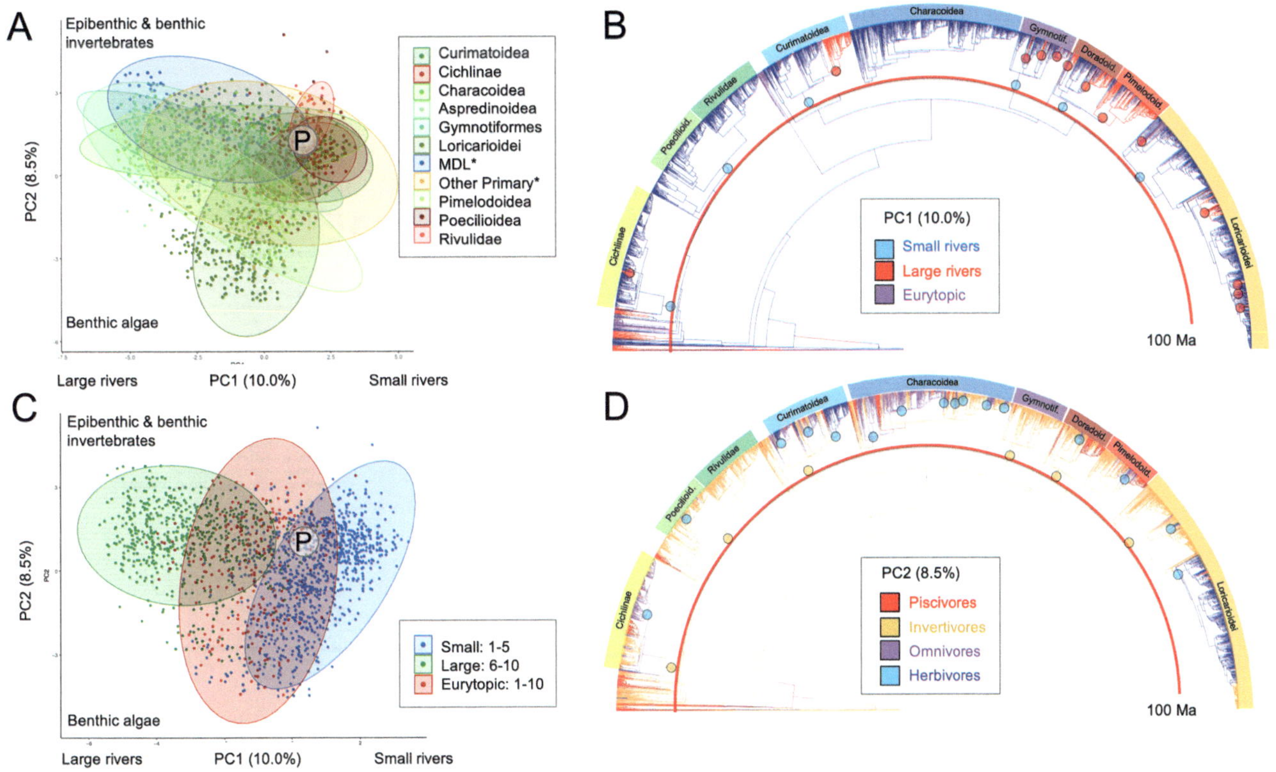

Fig. 21.3 Ecospace (left) and ecotrace (right) analyses of major axes of ecotrait variance among NFF fishes. (**A**) Phylogenetic structure of NFF ecospace formed by PC1 and PC2 (Fig. 6.7). PC values from Chap. 6 with PC1 representing 10.0% and PC2 8.5% of the total ecospace variance. Ecotraits loading strongly on PC1 and PC2 in Table 6.2. Letter P at upper right indicates estimated plesiomorphic (ancestral) values for Big Nine NFF clades at c. 100 Ma. Data for 6342 species. Colored ellipses represent 95% CI for each of the Big Nine NFF clades, plus pooled data for *nonmonophyletic groups; marine-derived clades (MDCs) and other primary freshwater clades. (**B**) Ecotrace analysis of PC1 (Fig. 7.2). Phylogeny of 3167 species (49.6% of NFF total) in 865 genera, 96 families, and 40 orders from Chap. 2. Big Nine NFF clades (colored regions) from Chap. 3. Methods of trait evolution from Chap. 7. Red semicircle at c. 100 Ma marks the time of final separation of South America and Africa and the earliest origin of phylogenetically independent NFF clades. (**C**) Ecospace partitioned by species adapted to macrohabitats along the longitudinal river continuum (Fig. 6.8B). Small rivers have Strahler stream orders (SO) 1–5; large rivers SO 6–10; Eurytopic SO = 1–10. (**D**) Ecotrace analysis of PC2 (Fig. 7.3). Colors indicate trophic level. Ecotraits loading strongly on PC2 are related to diet and vertical position in the water column from eating epibenthic and benthic invertebrates to benthic *algae*

conditions with fish body size and shape (Langerhans and Reznick 2010; Lujan and Conway 2015; Benejam et al. 2018; Cerezer et al. 2023), diet and trophic position in local food webs (Winemiller et al. 2011; Sánchez-Hernández 2023), and life-history traits involved with migration, seasonality, mating, and reproduction (Winemiller et al. 2008).

Local environments in small streams (SO 1–3) have a much smaller habitat volume than comparable reaches of large lowland rivers, but a much larger total stream length when considered as a distinct habitat type across an entire river basin (Fig. 1.4C). Yet, upland streams are also much more spatially fragmented and geographically isolated from one another than are the main stems of larger rivers, thereby promoting barriers to gene flow and potentially leading to higher rates of speciation and extinction (Thomaz et al. 2016). There are also many ecological differences among rivers with different stream orders (Lewis et al. 2006). Many headwater fish species are *rheophilic*, with locomotor, dietary, and respiratory specializations to hold position in swiftly flowing, highly oxygenated water (Lujan and Conway 2015). High-gradient, high-elevation headwaters are also exposed to higher seasonal variation in water flow, temperature, and productivity (Meyer et al. 2007).

21.3.3 Rates of Lineage Diversification

Rates of net diversification summarized in Chap. 5 for 154 NFF clades with origins during the Upper Cretaceous, Paleogene, and Miocene are reported in Fig. 21.4. The most prominent event highlighted in this analysis is a rapid acceleration in net diversification during the Oligocene from 34 to 22 Ma (see also López-Fernández and Albert 2011; Tagliacollo et al. 2015; Melo et al. 2022; Frable et al. 2022). The Oligocene was a time of large fluctuations in global climates, driving dramatic changes in eustatic or global sea levels, and therefore changes to the shorelines of continental margins (Liebrand et al. 2017; De Lira Mota et al. 2023). Large changes in sea levels of up to hundreds of meters resulted in multiple marine regressions into, and transgressions out of, low elevation portions of South America, especially in the Sub-Andean foreland (SAF) and coastal basins (Horton and DeCelles 1997; Mora et al. 2020; Bernal et al. 2023; Horton et al. 2023). The episodic separation and merging of river mouths under fluctuating shorelines have been a perennial geographic driver of freshwater fish diversification along continental margins throughout the Cenozoic (Lundberg et al. 1998; Avise 2000). Shoreline fluctuations

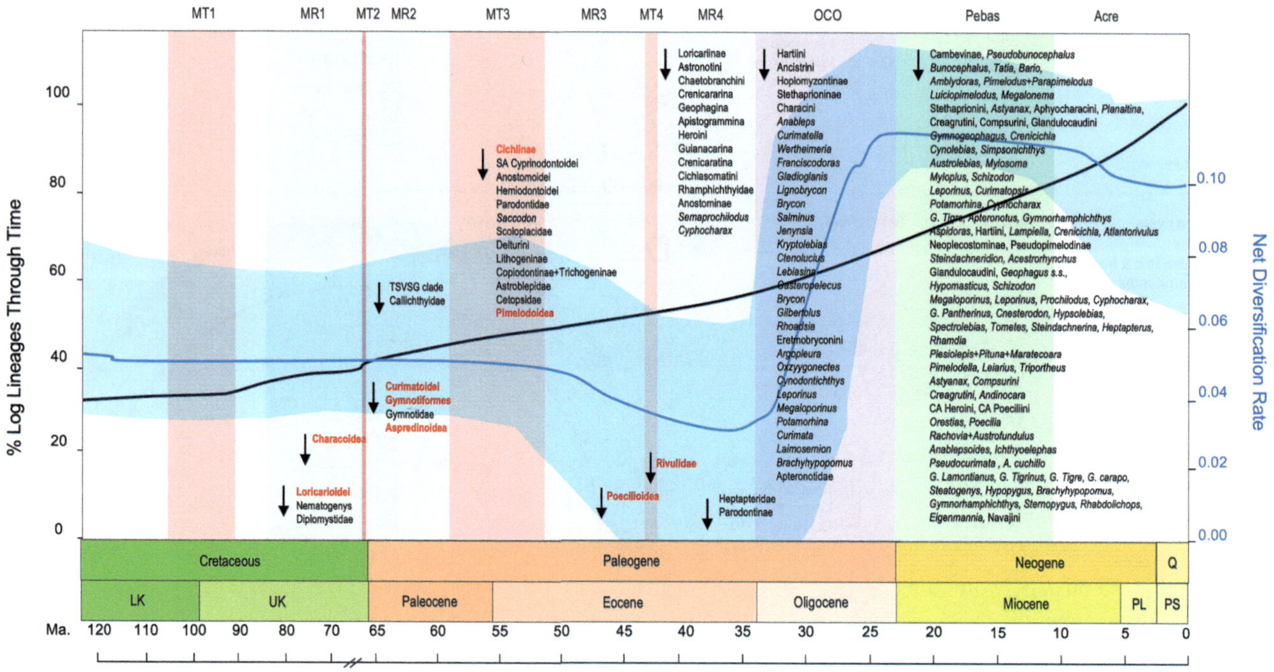

Fig. 21.4 Synthetic chronology of lineage diversification and clade originations among South American fishes. Log lineages through time (LLT, black line) and net diversification rate (median = blue line; shaded blue area = one standard deviation) for extant NFF fishes (n = 6346 species). Taxon names for 154 NFF clades that are 22 Ma or older, arranged by crown clade age (arrows). Big Nine NFF clades in red font. List of taxa, clade ages, and estimated bioregion origin in Table 4.3. Clade ages from time-calibrated phylogeny of Tagliacollo et al. (2024; Chap. 2); net diversification rates from Chap. 5. Paleogeographic events across top as in Fig. 1.10

facilitated geographic dispersal and lineage diversification in many NFF clades through the Pliocene and Pleistocene global climate oscillations (Vari and Weitzman 1990; Huber 1998; Farias and Hrbek 2008; Ochoa et al. 2015; Thomaz et al. 2015, 2017; Tschá et al. 2017; Abreu et al. 2019; Oberdorff et al. 2019; Abreu et al. 2020).

21.4 Major Macroevolutionary Events

The data and analyses reviewed in this chapter are used to develop a synthetic narrative of macroevolutionary diversification in South American freshwater fishes. This narrative merges data from the biogeographic, paleogeographic, ecospace, and ecotrace analyses of earlier chapters. The major paleogeographic conditions and events of South America during Upper Cretaceous and Cenozoic (i.e., the past 100 Ma) are depicted in Fig. 21.5. This narrative is based on the list of biogeographic changes among 558 South American fish clades with crown clade ages more than c. 4 Ma (Table 4.2), including the 154 clades with pre-Miocene origins (>22 Ma) depicted in Fig. 21.4. Each of these 558 macroevolutionary events is time correlated to one of 27 paleogeographic events occurring across ten geological time stages (Table 21.1), with locations spatially constrained to *bioregions* defined by elevation from 100 to 34 Ma (Fig. 21.6A, B) and river basin from 34 to 0 Ma (Fig. 21.6C, D). Given known uncertainties associated with phylogenetic dating, all clade ages and events are reported as median estimates rounded to the nearest Ma.

It is important to note several important limitations in the datasets and analyses used to develop this chronological narrative. The largest sources of systematic error may arise from the estimated dates of internal nodes on the time-calibrated phylogeny, the relatively coarse scale of the ecological variables (i.e., ecotraits) used in the ecospace and ecotrace analyses, and the relatively coarse use of PCA to assess major patterns ecotrait disparity and evolution. Another limitation consists of the lack of multilocus data for several species, as classical multigene phylogenetics has been replaced by modern phylogenomic approaches in Neotropical ichthyology (e.g., Roxo et al. 2019; Alda et al. 2019; Melo et al. 2024). Phylogenetic dating using molecular phylogenetics is limited by uncertainties in mutation rates, an incomplete and possibly biased fossil record, and assumptions of constant evolutionary rates, often resulting in wide error bars (sometimes spanning tens of millions of years) and percentage errors that can exceed 20%, particularly among nodes with deeper divergence times.

Some of the ecotraits pertaining to habitat utilization used in this study, for example, those related to water chemistry and macrohabitat (floodplain, deep channel) were developed specifically to characterize the ecological attributes of Amazonian fishes. Yet, about half of all NFF species scored in the ecomatrix live in other regions of tropical America, from Central America to the Pampas of Argentina, where these particular ecotraits are less useful in characterizing fish functional traits. Although all these ecotraits were scored for all NFF species, a majority of these taxa exhibit a single combination of ecotraits: i.e., inhabiting slowly flowing upland clearwater streams and small rivers, and consuming benthic invertebrates (e.g., insects, microcrustaceans). The total number of functional groups among NFF species is much larger than those scored in the ecomatrix, and many known ecological differences are not represented (e.g., Lujan and Armbruster 2012; Lujan et al. 2013). In other words, the combination of functional traits for each species described in this ecomatrix does not correspond to individual ecological niches.

Tremendous care and effort were invested in assembling the ecomatrix of 42 ecotraits for all NFF species used in this study. These data were scored as adult modal values by 55 taxonomic experts using data from 1315 literature references, and through several rounds of review and revisions (Albert et al. 2025). This ecomatrix is by far the most thorough and data-rich such compilation assembled to date for NFF species. Nevertheless, there are presumably errors in species identifications and ecological data presented in published studies. The alpha taxonomy of NFF species is far from complete, with several hundred new species having been described since the ecomatrix was assembled. Further, among these several thousand fish species, high-quality ecological data are available for only a tiny handful of species. For many NFF species, we have only fragmentary and sometimes incorrect knowledge about characteristic habitats, and seasonal, ontogenetic, and geographic variation in these ecotraits.

The use of principal component (PC) scores to assess taxa in the ecospace and ecotrace analyses has several analytical tradeoffs (Greenacre et al. 2022). By condensing high-dimensional trait data (i.e., 42 ecotraits) into fewer interpretable PC axes, PCA simplifies data visualization and analysis, reduces noise in the data, and standardizes (i.e., normalizes) data at different measurement scales. These patterns help in data visualization for use in exploratory data analysis. Some potential drawbacks of PCA are that it may obscure important nonlinear relationships, underestimate multidimensional trait disparities, and be overly sensitive to data scaling and outliers. All these effects can potentially lead to incomplete or misleading interpretations of ecological disparity. These limitations occur when the data violate the assumptions of linearity, scaling insensitivity of contributing variables, and the influence of data outliers.

21.4.1 Upper Cretaceous (c. 100–66 Ma)

During the Upper Cretaceous, the whole Earth experienced super-greenhouse conditions, with tropical to subtropical climatic conditions extending to high latitudes, no ice caps, and overall high sea levels with large amplitude fluctuations (Burgener et al. 2023). In hindsight, it appears that these geographic and climatic conditions formed a kind of

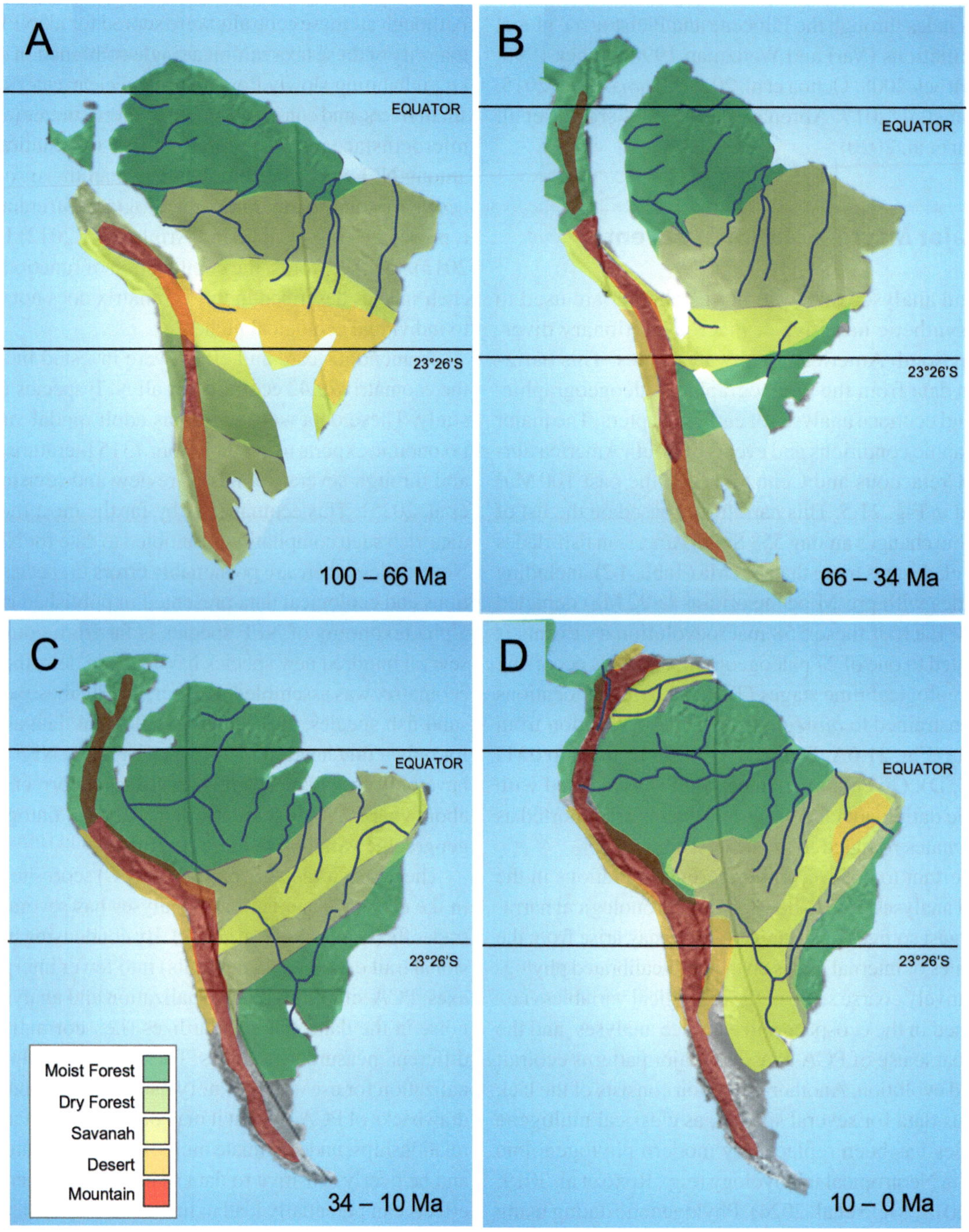

Fig. 21.5 Paleogeographic reconstructions of South America from 100 Ma. to present. (**A**) Upper Cretaceous (100–66 Ma). (**B**) Paleocene–Eocene (66 – 34 Ma). (**C**) Oligocene–Middle Miocene (34–10 Ma). (**D**) Late Miocene–*Recent* (10–0 Ma). Climate (Paleo-Köppen) zones in inset in C. Paleogeographic data (paleoclimates, paleotopography, paleoshorelines, paleo-watersheds) for Upper Cretaceous (**A**) from Burgener et al. (2023); for Cenozoic (**B**–**D**) from Louterbach et al. (2014); Acosta et al. (2022); and Custódio et al. (2023) with additional paleogeographic data from Marcilly et al. (2022), Aminov et al. 2023), Carrillo-Briceño et al. (2023), de Assunção Rodrigues et al. (2023), Salles et al. (2023), and Zapata et al. (2023). Blue curves indicate principal river thalwegs (some details conjectural)

"perfect storm" for the diversification of aquatic taxa, in both near-shore marine and lowland continental (i.e., freshwater) environments of recently separated landmasses (Cavin et al. 2007; Guinot and Cavin 2016; Cavin 2017; Melo and Stiassny 2024). The Upper Cretaceous was embedded within the Mesozoic Marine Revolution, a period of rapid diversification of many taxa that eventually came to dominate Cenozoic marine ecosystems; e.g., diatoms, gastropod mollusks, and teleost fishes (Vermeij 1977; Benton 2009; Cortés and Larsson 2024; Pimiento et al. 2024).

Table 21.1 Summary of major macroevolutionary events for South American freshwater fishes

Period	Horizon	Epoch	Ma	Paleogeographic event	Biotic responses	Bioregions	NFF clades	References
K	1	Santonian	86–83	Global greenhouse	Radiations	CP	2	Roxo et al. 2014; Lujan et al. 2020; Janzen 2023; Tagliacollo et al. 2024
	2	Campanian	83–72	Southern Andes	Vicariance	CP	1	
			83–72	MR1	Radiations	SF	1	Sarmiento-Rojas 2019; Melo et al. 2022; Burgener et al. 2023
	3	Maastric.	72–66	Northern Andes	Vic-Dis-Rad	CP	1	Gayet et al. 2001; Patarroyo et al. 2022
PG	4	Paleocene	66–55	Relicts	Vicariance	CP	1	Alda et al. 2019
			65–58	MR2	Radiations	SF	4	Sullivan et al. 2013; Acosta et al. 2022
	5	Early Eocene	55–48	Relicts	Vicariance	B, G, A	9	Tagliacollo et al. 2024
			52–44	MR3	Radiations	SF	2	Custódio et al. 2023
	6	Late Eocene	42–37	MR4	Radiations	SF	8	Tagliacollo et al. 2024
			42–38	G1	Dispersal	G	5	Tagliacollo et al. 2024
			38	L1	Vicariance	SF→L	2	Karner and Driscoll 1999; Crelier et al. 2025
	7	Oligocene	34	EO cooling	Vicariance	G→A, SF→G	8	Roxo et al. 2014; Carvalho et al. 2018; Melo et al. 2018a, 2018b
			34	A1	Vic-Dis-Rad	SF x B, S→A	7	Ribeiro 2006; Mirande 2019; de Santana et al. 2023
			34–30	T1	Vicariance	CP: T; SFàT	14	Lundberg et al. 1998
			34–30	MR5	Dispersal	G→SF	4	Roddaz et al. 2010; Bernt et al. 2019; Melo et al. 2021; Frable et al. 2022
NG	8	Early Miocene	22–15	L2	Vicariance	A→L, SF→L; W→L	37	Tagliacollo et al. 2015
			22–15	A2	Vicariance	B, G, A, L→ A; B→A	18	dos Reis et al. 2024
			22–15	Dry diagonal	Vic-Dis-Rad	CP: B; G→B	6	Werneck et al. 2012; Meseguer et al. 2022; Hoorn et al. 2023
			20–15	T2	Vic-Dis-Rad	CP: T	19	Horton et al. 2020
			22–10	Pebas	Vic-Dis-Rad	W	13	Bloom and Lovejoy 2017; Cuitiño et al. 2017; Hoorn et al. 2022
	9	Middle Miocene	18	G2	Vicariance	CP: G	2	Fadul et al. 2022; Londoño-Burbano and Britto 2025
	10	Late Miocene	11–8	T3	Vicariance	T, O, W	40	Albert et al. 2006; Montes et al. 2021
			10	GAzBI	Vic-Dis-Rad	W, O, B; G→ B→W	286	Albert et al. 2021; Ribas et al. 2025
			10	L3	Dispersal	A→L	9	Carvalho and Albert 2011
			10–6	A3	Vic-Dis-Rad	L→A, B→A, A × B	15	Ramirez et al. 2023; Vilardo et al. 2023
	11	Pliocene	<5	L4	Dispersal	B→L; W→L	14	Carvalho and Albert 2011
			<5	T4	Dispersal		11	Montes et al. 2021
			<5	A4	Dispersal	A→L; L→A	19	de Queiroz et al. 2021
Total	**1**			**28**			**558**	

Data for 558 clades and 28 paleogeographic events over 11 time horizons from c. 86 to 5 Ma. Color coded by bioregions (Fig. 4.1). Biogeographic events from Figs. 4.2 to 4.9, enumerated in Table 4.2. Abbreviations in Fig. 1.10; *x* vicariance; → dispersal; *Rad (radiation)* dispersal + speciation

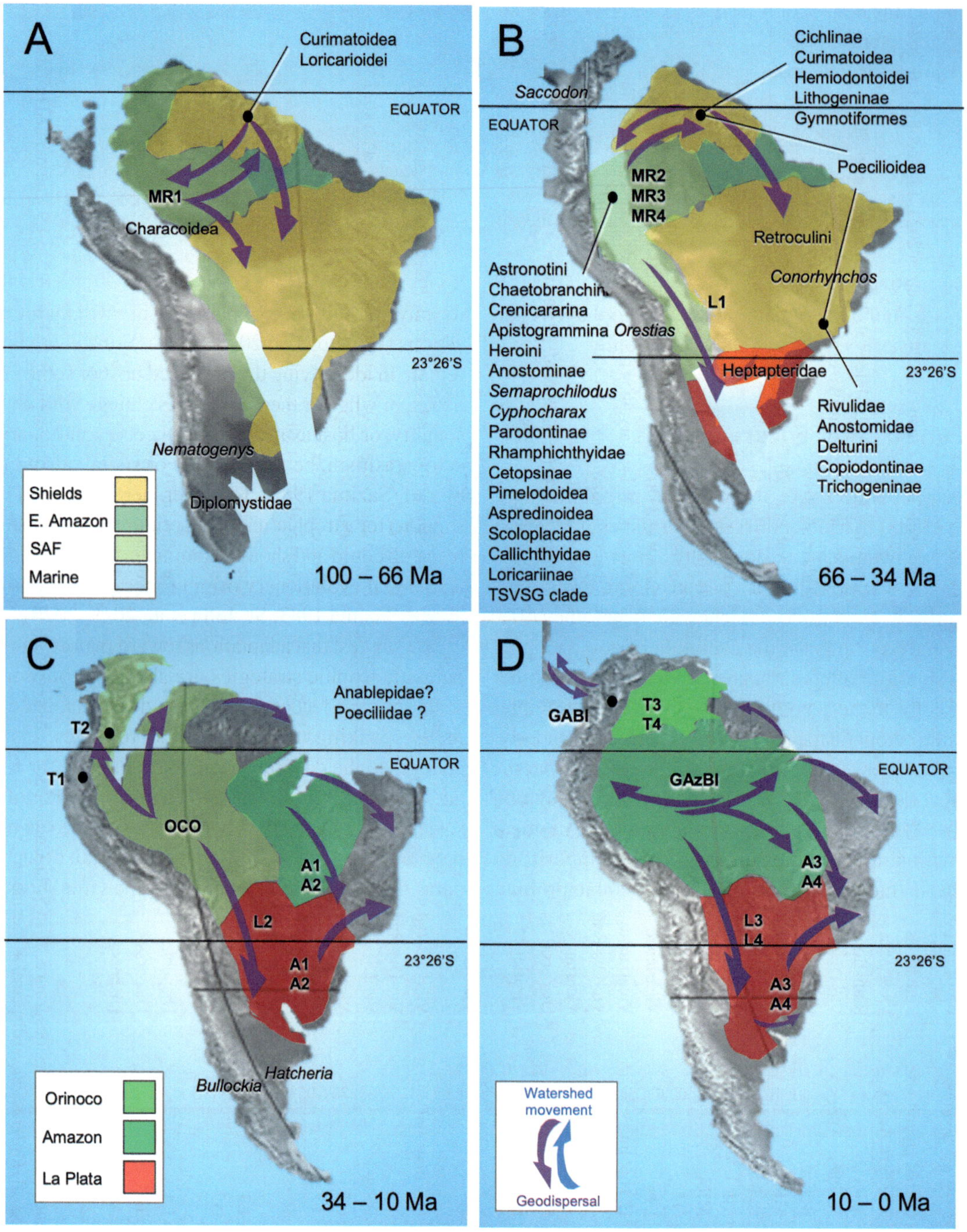

Fig. 21.6 A birds-eye view of macroevolution in South American freshwater fishes. AAE for 558 clades in Table 4.2, including 154 clades in Fig. 21.5. Marine regressions (MR), & other biogeographic events (A, Atlantic; L, La Plata; T, trans-Andean, GABI, GAzBI) in Table 1.1, Fig. 1.10, and Table 4.2? AAE location ambiguous. Paleoclimate (Köppen) zones from Fig. 21.5. Note Upper Cretaceous and early Paleogene origins of upland NFF clades in rainforest refugia of the Guiana Shields (i.e., *Eigenmann's Uplands Origins* hypothesis) and Late Paleogene to Late Neogene origins of lowland NFF clades in SAF and Amazon basins (i.e., *Smith's Lowland Origins* hypothesis). Note also opposite directions of geodispersal (purple arrows) and watershed movement (blue arrows) vectors (Fig. 1.6)

The Upper Cretaceous was also embedded within the Cretaceous Terrestrial Revolution (KTR), otherwise known as the Angiosperm Terrestrial Revolution, which in fact lasted to the end of the Palaeogene c. 34 Ma (Benton et al. 2022; Jaramillo 2023; Thompson and Ramírez-Barahona). The KTR was a time of rapid diversification in many clades of flowering plants (angiosperms) and animals, centered primarily on tropical Western Gondwana, which led to origin of modern rainforest taxa and ecosystems (Labandeira and Wappler 2023; Jouault et al. 2024; Weaver et al. 2024). Major KTR radiations included pollinating holometabolous insects (i.e., Lepidoptera, Hymenoptera, Diptera, and Coleoptera), as well as frugivorous, nectivorous, and insectivorous tetrapods (especially birds, mammals, and squamate reptiles) and web-spinning spiders (Benton et al. 2022). Major clades of continental freshwater fishes also radiated in Western Gondwana during the KTR (Fink and Fink 1981; Lundberg 1993; Cavin et al. 2007, 2010; Arratia 2018; Nogueira et al. 2023; Burns et al. 2024; Melo and Stiassny 2024).

The final continental connections between South America and Africa occurred in the transition to the Upper Cretaceous c. 100 Ma. During this time, sea-levels reached the highest of the whole Phanerozoic Eon (c. 539 – 0 Ma) at +150 m (Pucéat et al. 2003; Gréselle and Pittet, 2005). High-amplitude climate-driven sea-level fluctuations (30–100 m) in the Tethys Sea region episodically flooded the continental interiors, leaving many near-shore marine deposits and fossil-bearing rocks (e.g., Santana Formation in northeastern Brazil), introducing marine taxa into continental freshwaters (Eaton et al. 1997; Keller and Pardo 2004). These pronounced sea level fluctuations also produced episodic vicariance and geodispersal events between the slowly rifting portions of Western Gondwana.

Throughout the Upper Cretaceous and Cenozoic, South America drifted northwest away from Africa, colliding with and overriding portions of the Pacific, Nazca, Cocos, and Caribbean tectonic plates, and driving tectonic uplift of the Andean cordilleras and down warping of the SAF basin. Fossil evidence indicates that some freshwater fish taxa were endemic to portions of the SAF basin during this time, from Venezuela to Bolivia (Gayet et al. 2001). This basin served as the main drainage axis of South America for 66 Ma, until the Late Miocene (c. 10 Ma) formation of the modern tripartite basin configuration; i.e., Orinoco, Amazon, and La Plata basins. Throughout much of the Upper Cretaceous and early Cenozoic, the SAF basin was covered by large interior seaways; i.e., marine transgressions. The later stages of the Upper Cretaceous saw gradual cooling after c. 90 Ma, resulting in a major marine regression from low elevation portions of South America referred to here as MR1 (Figs. 21.4 and 21.6; Table 21.1). The Cenomanian (c. 100–94 Ma), in particular, was a relatively brief but busy interval (of about six million years) that left a strong mark on marine Tethyan and continental Gondwanan paleogeography, and the evolution of their regional ichthyofaunas. Cenomanian localities yield many more fossil fish specimens than any other stages of the Cretaceous (Cavin and Forey 2007; Friedman 2010).

Teleost fishes diversified rapidly during the Upper Cretaceous, eventually coming to dominate both the diversity and ecology of all modern aquatic systems worldwide (Cavin and Forey 2007; Cavin et al. 2007; Cavin 2010). Members of just two teleost subclades currently represent about 88% of all extant global fish species diversity; Otophysi primarily in freshwaters, and Acanthomorpha in both marine and freshwaters. However, only a handful of extant NFF clades trace their origins to the Upper Cretaceous, all to within just 20 Ma before the K-Pg mass extinction event (Lundberg 1993; Albert et al. 2011a; Fig. 21.6). These clades include three of the Big Nine NFF clades; Loricarioidei during the Santonian (86–83 Ma) in the Guiana Shield, Characoidea during the Campanian (83–72 Ma) in the SAF basin associated with the MR1 event, and Curimatoidea in the Maastrichtian (72–66 Ma) also in the Guiana Shield. Several other NFF clades also have origins in Western Gondwana during the latest part of the Upper Cretaceous, including the South American lungfish *Lepidosiren* (Dipnoi), two clades of bony-tongue fishes (Osteoglossiformes) *Arapaima* and *Osteoglossum*, and two catfish clades now restricted to the Southern Cone, i.e., *Nematogenys* (Nematogenyidae) and Diplomystidae (Arratia 1997; Lujan et al. 2020; Melo et al. 2022; Fig. 21.6A). Several other freshwater fish clades with Upper Cretaceous origins are restricted to tropical Central America, including the gars *Atractosteus* (Holostei) and the Chiapas catfish *Lacantunia*.

21.4.2 Paleocene–Eocene (c. 66–34 Ma)

Following the Cretaceous–Paleogene (K-Pg) mass extinction event, Neotropical rainforest plants attained their modern taxonomic composition (Carvalho et al. 2021), and clades of both terrestrial and aquatic vertebrates underwent relatively rapid diversifications during the first ten million years of the Paleocene (c. 66–56 Ma; Friedman 2010; Guinot and Cavin 2016; Alfaro et al. 2018). Super-greenhouse climatic conditions prevailing for another 32 million years, through to the end of the Eocene (c. 34 Ma; Yasuhara et al. 2022). During this time, the already warm background conditions were punctuated by several hyperthermal events that raised global temperatures by 2–6°C and sea levels by up to c. 50 m, each event lasting several hundred thousand years: the Paleocene–Eocene thermal maximum (PETM) c. 56 Ma, Early Eocene thermal maximum (EETM) c. 52–50 Ma, and the Middle Eocene thermal maximum (METM) c. 43 M.

Several extant NFF clades originated and began diversifying in this time period, including the characiform clades Hemiodontoidei (c. 59 Ma), Anostomoidei (c. 57 Ma), and Crenuchidae (c. 63 Ma), Gymnotiformes (c. 63 Ma) and two of its subclades (Gymnotidae c. 62 Ma, Sinusoidea c. 60 Ma), and the siluriform clades Aspredinoidea (c. 65 Ma), Scoloplacidae (c. 59 Ma), Astroblepidae (c. 58 Ma), Callichthyidae (c. 58 Ma), and Cetopsidae (c. 57 Ma). Three other NFF clades have estimated origins just before the K-Pg event, but began diversifying during the Paleocene; Lebiasinidae (c. 71 Ma), Trichomycteridae (c. 68 Ma), and Ctenoluciidae (c. 67 Ma). The estimated ancestral water chemistry tolerances and geographic origins of these clades are Anostomoidea in upland clearwaters of the Brazilian Shield, Hemiodontoidei, Callichthyidae, Scoloplacidae, and Trichomycteridae in uplands clearwaters of the Brazilian and Guiana shields, Gymnotiformes and two of its large subclades (Gymnotidae and Sinusoidea) in upland blackwaters or whitewaters of the Guiana Shield, and Ctenoluciidae,

Lebiasinidae, and Aspredinoidea in lowland whitewaters of the SAF basin. Although the water chemistry tolerances and geographic origins of Crenuchidae and Cetopsidae are not reported in previous chapters, most extant crenuchid species inhabit clearwaters the upland Brazilian Shield, and most extant cetopsid species inhabit large rivers (Cetopsinae) and small rivers (Helogeninae) of the Greater Amazonia.

Many diverse NFF clades, some with more than 100 extant species, originated in the SAF at the time of the pronounced global cooling and marine regressions events MR2 (c. 65–59 Ma), MR3 (c. 51–43 Ma), and MR4 (c. 43–34 Ma; Figs. 21.4 and 21.6). These include two of the Big Nine NFF clades (i.e., Pimelodoidea, Aspredinoidea), several of their subclades (i.e., Astronotini, Chaetobranchini, Crenicararina, Apistogrammina, Heroini, Anostominae, *Semaprochilodus*, *Cyphocharax*, Parodontinae, Rhamphichthyidae, Scoloplacidae, Callichthyidae, Loricariinae, trichomycterid TSVSG clade), and Cetopsidae. Other NFF radiations were associated with the pronounced global warming and MT3 event of the Early Eocene (c. 56–48 Ma), including Cichlinae, Pimelodoidea, and Aspredinidae at c. 55 Ma, Characidae at c. 52 Ma, Rhamphichthyoidea at c. 53 Ma, and Loricariidae at c. 52 Ma. Riverine-adapted Pimelodidae, which inhabit large rivers with Strahler stream orders (SO) 6–10, have estimated origins in the SAF at c. 43 Ma associated with the MT4 event (c. 43 Ma).

During the early Eocene, three siluriform clades became geographically isolated in basins of the Atlantic margin; *Conorhynchos* (stem age c. 53 Ma) in the São Francisco River of the Brazilian Shield, Lithogeninae (stem age c. 52 Ma) in small rivers of the Orinoco basin of the Guiana Shield and adjacent uplands, and Delturinae (stem age c. 48 Ma) in small to medium sized rivers of the Brazilian Shield like the Doce River and Jequitinhonha River (Roxo et al. 2014). However, despite these three large-scale marine transgressions and regressions into the SAF over a period of c. 31 Ma, only two of ten marine-derived NFF clades trace origins to the Eocene: freshwater stingrays (Potamotrygonidae with 23 extant species in four genera), and the Longfin herring clade *Pristigaster* (Pristigasteridae) with two extant species (Bloom and Lovejoy 2017).

Available phylogenetic and biogeographic data indicate origins of the three most diverse NFF clades of Acanthomorpha from overseas dispersal events from Africa to South America during the Eocene; Cichlinae at c. 54 Ma; Poecilioidea at c. 47 Ma; Rivulidae at c. 43 Ma (Tagliacollo et al. 2024; Chap. 2). Previously published dates for the clade age of Cichlinae range from c. 35 to 53 Ma (c. 35 Ma, Friedman et al. 2013; c. 35 Ma, Cassemiro et al. 2023; c. 43 Ma, Irisarri et al. 2018; c. 52 Ma, Rabosky et al. 2013; c. 53 Ma, Betancur-R et al. 2015), for Poecilioidea from c. 42 (Helmstetter et al. 2016) to c. 47 Ma (Cassemiro et al. 2023), and for Rivulidae from c. 43 Ma (Cassemiro et al. 2023) to c. 78 Ma (Rabosky et al. 2013).

Earliest dates for Neotropical Cichlinae from fossils are from phenotypically modern and phylogenetically nested taxa (*†Proterocara argentina, †Gymnogeophagus eocenicus, and †Plesioheros chauliodus*) in the Early Eocene Lumbrera Formation of Argentina (Malabarba et al. 2006, 2010, 2014; Perez et al. 2010). North American killifishes (Cyprinodontoidei) of the families Fundulidae, Cyprinodontidae, Profundulidae and Goodeidae originated in tropical areas of southern North America and Nuclear Central America during the Eocene c. 53 Ma (Hernández-Ávila et al. 2024).

By the start of the Middle Eocene (c. 48 Ma), the NFF fauna had differentiated into almost all of the recognized modern family-level clades, and a majority of the subfamily-level clades. By this time, Characoidei had diversified into all nine families and 12 of the 18 subfamilies (Melo et al. 2022), Curimatoidea into all eight families and 12 of 16 subfamilies (Melo et al. 2018b, 2021; Frable et al. 2022), Gymnotiformes into all five families and nine of ten subfamilies, Pimelodoidea into all five family-level clades (including *Conorhynchos*) and ten subfamilies (Silva et al. 2021a, 2021b, 2024), Aspredinoidea into all three families and four of the six subfamilies, and Loricarioidei into all of the six families and eight of the 12 subfamilies.

Continued cooling in the Late Eocene is associated with 21 fish radiations, including 13 radiations in the centrally located SAF and nine radiations along the continental margins. Radiations associated with MR4 at c. 43–34 Ma include four in Cichlinae (Astronotini + Chaetobranchini, Crenicararina + Geophagina, Apistogrammina, and Heroini), one in Cyprinodontiformes (Rivulidae), two in Siluriformes (Loricariinae, Pimelodoidea), and one in Gymnotiformes (Rhamphichthyidae); and five upland radiations, including three in Cichlinae (Guianacarina, Crenicaratina, and Cichlasomatini) and two in Curimatoidea *Anostomus* + *Pseudanos* + *Petulanos* and *Semaprochilodus*). Radiations in the continental periphery at this time are associated with G1 event at c. 42–38 Ma in five clades (Guianacarina, Crenicaratina, Cichlasomatini, *Anostomus* + *Pseudanos* + *Petulanos*, and *Semaprochilodus*), the L1 event at c. 38 in two clades (Parodontinae and Heptapteridae), and two other biogeographic events located at the continental periphery c. 38 to 37 Ma (Spintherobolidae and *Saccodon*; Melo et al. 2021).

21.4.3 Oligocene-Middle Miocene (34–12 Ma): Fig. 21.6C): 463 Clades

The time period from the Oligocene to Middle Miocene was the most productive stage in NFF paleogeographic history, in terms of the origins of new clades as a proportion of the existing fauna, and also for generating modern geographic and ecographic distributions (Albert et al. 2011a; Albert and Carvalho 2011). The dramatic global cooling across the Eocene–Oligocene

(E-O) boundary (c. 34 Ma) was a seminal event in the global climate history with widespread consequences for biodiversity worldwide (Weppe et al. 2023; Akabane et al. 2024; Rull 2024; Buffan et al. 2025; McCraney et al. 2025). The hypothesized geological driver of the E-O cooling was the opening of the Drake Passage between South America and Antarctica and initial formation of the circumpolar Antarctic current, which thermally isolated Antarctica allowing the initial buildup of the Antarctic ice cap (Sarkar et al. 2019; Martos and Catalán 2024). These geological and climatic events lowered sea levels and caused large marine regressions from the SAF and coastal basins. During the *Oligocene Climatic Oscillations* (OCO, c. 34–22 Ma; Hubbard et al. 1994), cooler and dryer regional climates spread across the Brazilian Shield allowed expansion of xeric open vegetation formations of the *Dry Diagonal*, which separated the more mesic closed-canopy vegetations of the Amazon and Atlantic rainforests (Strömberg 2011; Werneck et al. 2012; Thode et al. 2019).

At least six biogeographic events during the Oligocene and Middle Miocene isolated fish taxa along the continental periphery. The A1 event (c. 34–30 Ma) affected the doradids *Wertheimeria*, *Kalyptodoras*, and *Franciscodoras*, the triportheid *Lignobrycon*, the anablepid *Jenynsia*, and the rivulid *Kryptolebias* (Ribeiro 2006). The A2 event (c. 23–15 Ma) affected the poeciliid *Cnesterodon*, the rivulids *Hypsolebias, Spectrolebias*, and the *Plesiolepis* + *Pituna* + *Maratecoara* clade, and the curimatid *Steindachnerina*. The T1 event (c. 34–30 Ma) associated with uplifts of the northern Andean Cordilleras (Horton et al. 2020) affected 13 clades representing four of the Big Nine NFF clades, including the poecillioid *Oxyzygonectes* in brackish waters of Central America and trans-Andean northern South America (Amorim and Costa 2018), the Mexican rivulid *Cynodonichthys*, some curimatoids in *Leporinus, Megaleporinus, Potamorhina*, and *Curimata*, and some characoids in *Ctenolucius, Lebiasina, Gasteropelecus, Brycon, Gilbertolus, Rhoadsia*, and *Argopleura*. The L2 event (c. 23–15 Ma) associated with formation of the Bolivian Orocline and large *river capture* events (i.e., *mega-river captures* >10,000 sq. km) in the SAF (Tagliacollo et al. 2015) affected 37 clades representing all of the Big Nine NFF clades (Fig. 4.2). The T2 event (c. 20–15 Ma) associated with renewed uplift of the northern Andes affected 19 clades representing seven of the Big Nine NFF clades, only excluding Aspredinoidea and Loricarioidei. The G2 event (c. 18 Ma) isolated the poecilid *Tomeurus* in lowlands basins adjacent to the Guiana Shield.

The Pebas megawetland system was a prominent geomorphological feature affecting in the evolution of Amazonian fishes. The Pebas Formation is a lithostratigraphic unit exposed in portions of the western Amazon of Peru, Colombia, and Brazil (where it is called the Solimões Formation) that ranges in age from the earliest to early–late Miocene (c. 23–10 Ma). The Pebas Formation is composed of grey or blue clays (indicating hypoxic lake bottoms) interspersed with unconsolidated sands (indicating flowing rivers) with some mollusk shell beds and fossiliferous lignite (i.e., soft brownish coal) layers deposited during flooding events (Wesselingh et al. 2001; Salas-Gismondi et al. 2015). Based on study of fossil pollens, ostracods, and mollusks, the Pebas Formation is interpreted to have been a large megawetland system, extending up to one million sq. km, composed of a heterogeneous set of fluvial (riverine), lacustrine (lakes), and mangrove-like areas (Hoorn et al. 2022).

The Pebas megawetlands formed under the combined influences of several distinct geomorphological processes. Dynamic topography, referring to vertical deflections of the lithosphere caused by mantle convection, depressed the western portion of the South American Platform, contributing to marine incursions in the northern SAF as far south as northern Peru, and the formation of freshwater megawetlands in the Western Amazon (Shephard et al. 2010; Sacek 2014; Jaramillo et al. 2017; Bicudo et al. 2019; Sacek et al. 2023). Subsidence of the SAF was enhanced by tectonic uplift of the northern Andean cordilleras, and sediment erosion from these orogens eventually filled the Western Amazon sedimentary basins, overfilling the Purús Arch headwaters, and forming the modern west–east draining transcontinental Amazon river c. 10 Ma (Albert et al. 2021a; Hoorn et al. 2022; Zapata et al. 2023; see also the section below).

Adding to this mix of geomorphological drivers, global warming and high sea-levels during the Middle Miocene thermal maximum (MMTM) drove marine incursions hundreds of kilometers into the northern SAF, introducing multiple marine-derived clades (MDCs) into Amazonian freshwaters. While only two MDCs originated in the 31 million years of the Paleocene to Eocene (see above), eight MDCs originated in the 22 million years of the Oligocene to Miocene (Bloom and Lovejoy 2017). These clade are: a second clade of Longfin herrings (Pristigasteridae) with three species in two genera at c. 44 Ma; Amazon sprats (Dorosomatidae) with two species in the genus *Rhinosardinia* c. 31 Ma; Amazon anchovies (Engraulidae) with 15 species in five genera at c. 32 Ma; two clades of needlefishes (Belonidae), the river needlefishes *Pseudotylusurus* with two species at 31–18 Ma, and the stream needlefishes *Potamorhaphis* + *Belonion* with five species at c. 30–21 Ma; Amazonian drums (Sciaenidae) with 22 species in three genera at c. 22 Ma; Amazonian pufferfishes (Tetraodontidae) with two species of *Spheroides* at c. 12–7 Ma (Araujo et al. 2023); and Amazonian flatfishes (Achiridae) with seven species in two genera at c. 10 Ma (Rodrigues et al. 2023).

Prevailing environmental conditions of the Pebas megawetlands affected ecological and evolutionary processes in all the resident fish clades. At least 13 fish clades trace their origins to the Pebas megawetlands. The callichthyid clade *Corydoras* (sensu Dias et al. 2024) has an estimated origin in clearwaters (Fig. 8.5) of the Western Amazon at c. 26 Ma (Fig. 4.9), with rapid early diversification in the Western Amazon and adjacent basins of Brazilian and Guiana Shields from c. 21 to 10 Ma. Many clades of gymnotiform electric fishes also undertook rapid diversification in blackwaters or whitewater of the Western Amazon during the time of the Pebas megawetlands, including clades representing all five families, and representing clades adapted to both terra firme streams and small rivers with SO 1–5 and channels and floodplains of large lowland rivers with SO 6–10. Within Gymnotidae, these include the upland *Gymnotus Lamontianus* clade, the eurytopic *Gymnotus Tigrinus* and *Gymnotus Carapo* clades, and the obligate riverine *Gymnotus Tigre* clade. Within Rhamphichthyidae, these include the eurytopic *Steatogenys*, the small stream *Hypopygus*, and the eurytopic *Gymnorhamphichthys*. Within Hypopomidae, these include the only species-rich clade *Brachyhypopomus*. Within Sternopygidae, these include the eurytopic *Sternopygus* and *Eigenmannia*, and the obligate riverine *Rhabdolichops*. Within Apteronotidae, these include the obligate riverine species of *Navajini* (Bernt et al. 2019). A recent report from a fauna located in the western Amazon near Iquitos Peru dated as c. 15–12 Ma reported 26 fish species spanning seven taxonomic order, including Potamotrygonidae, Osteoglossidae, Cichlidae, Sciaenidae, five characiform families, five catfish families, and scales of an enigmatic taxon called †*Acregoliath* of unknown phylogenetic affinity (Carrillo-Briceño et al. 2021; Fig. 21.4).

According to the timeframe of the time-calibrated tree of all NFF species developed for this book (Chap. 2), Poeciliinae dispersed to the Greater Antilles and Central America (GACA) in the early Miocene c. 20 Ma, somewhat earlier that estimated by some earlier studies (c. 39 Ma by Reznick et al 2017; c. 50 Ma by Tagliacollo et al. 2017). Under this timeframe, heroine cichlids are estimated to have dispersed to GACA c. 18 Ma, similar to the estimates of c. 15 Ma (Martin and Bermingham 1998; Hulsey et al. 2010), and much younger than the 42 Ma estimated by Tagliacollo et al. (2017).

21.4.4 Late Miocene (11.6–5.3 Ma)

The formation of the modern transcontinental Amazon River was obviously a central event in the evolution of Amazonian fishes. The Amazon river basin, now the largest river basin on Earth by surface area, became assembled through the concerted action of several smaller capture events in the time frame of 10.0–4.5 Ma, which together constitute the largest known river capture episode in the entire geological history of the Earth (Albert et al. 2018a). In combination, these river captures merged the sedimentary basins of the Western and Eastern Amazon river basins across the Purús Arch, and separated the sedimentary basins of the Western Amazon and Orinoco river basins across the Vaupes Arch (Latrubesse et al. 2010; Albert et al. 2018a, 2021b). This pronounced rearrangement of the regional river drainage network allowed a large amount of dispersal between the Western and Eastern Amazon basins, an event referred to as the *Great Amazonian Biotic Interchange* or GAzBI (Albert et al. 2018b). These same river captures of large rivers (SO 6–10) also separated the modern Amazon and Orinoco basins, contributing to the biotic distinctiveness of these megadiverse regions, and thereby elevating the total biodiversity of Greater Amazonia.

Several other major biogeographic events along the continental margins also occurred during this time frame (Fig. 21.6D). The T3 event c. 11–8 Ma involved the initial rise of the Eastern Cordillera of Colombia, which emplaced the final barrier separating the cis-Andean basins (draining to the Atlantic) from the trans-Andean basins (draining to the Caribbean and Pacific basins; Albert et al. 2006). Before the T3 event, geological evidence indicates that the intermontane Cauca and Magdalena valleys drained south and west into the Caribbean-draining Proto-Orinoco Amazon basin (Mora et al. 2020; Montes et al. 2021). Fish fossils from the La Venta Formation in the middle Magdalena basin, and the Cocinetas and Urumaco Formations in the Maracaibo basin, indicate a modern taxonomic composition of lowland riverine Amazonian fish faunas (i.e., *Arapaima, Brachyplatystoma, Lepidosiren, Phractocephalus*) in the Miocene, which have been entirely extirpated from those basin today (Lundberg et al. 1986; Lundberg and Aguilera 2003; Ballen et al. 2022). Comparative analyses of the entire cis- and trans-Andean fish faunas indicate the fauna had already achieved a modern taxonomic composition at the genus and species-complex level before c. 11 Ma (Albert et al. 2006).

A mega river capture in the SAF called the L3 event (c. 10 Ma) affected clades in both small streams, like *Australoheros* (Cichlidae), *Hyphessobrycon* and Diapomini (Characidae), *Microglannis* (Pseudopimelodidae), and *Eigenmannia* (Sternopygidae), and clades in large rivers, like *Brycon* (Bryconidae), *Rhamphichthys* (Rhamphichthyidae), and the *Apteronotus brasiliensis* clade (Apteronotidae). Late Miocene diversification of the hyperdiverse loricariid clade *Hypostomus* (with 152 species) also occurred in the SAF at about the time of the L3 event (c. 12.5 Ma; de Queiroz et al. 2021). Continued climatic cooling during the Late Miocene and Early Pliocene (10–4.5 Ma) promoted desertification in a portion of the continental periphery (da Silva et al. 2018), further separating the Amazon and Atlantic rainforests and resulting in the A3 event c. 10–6 Ma. Species in this event are mostly adapted to small streams and rivers like *Australoheros* (Cichlidae), *Phalloceros* and *Pamphorichthys* (Poeciliidae), *Melanorivulus* (Rivulidae), *Spintherobolus* and *Bryconamericus* (Characidae), *Brachyhypopomus, Eigenmannia*, and *Apteronotus* (Gymnotiformes), and *Heptapterus, Pimelodella*, and *Acentronichthys* (Heptapteridae), with only two clades adapted to larger rivers; i.e., *Pseudoplatystoma* and *Pimelodus* (Pimelodidae). Late Miocene to Pliocene rise of Fitzcarrald arch resulted in the semi-isolation of Upper Madeira and Western

Amazonian basins c. 6.5–4.0 Ma (Albert et al. 2011c; Bishop et al. 2018). The timing of the capture of upper Branco from the Proto-Berbice is poorly constrained by geological data, with estimated dates ranging from the Pliocene to Pleistocene from lineage divergence times; c. 3.6–2.4 Ma for *Hoplias* (Conde-Saldaña et al. 2025); c. 1.9 Ma for *Trichomycterus* (Akin et al. 2024), and the *Gymnotus carapo* clade (Lehmberg et al. 2018).

21.4.5 Plio–Pleistocene (5.3–0.01 Ma)

Pliocene (5.33–2.58 Ma) global cooling and Pleistocene (c. 2.58–0.012 Ma) global Ice Age cycles resulted in an unsteady but progressive contraction and fragmentation of tropical rainforests, formerly widespread across South America as far south as Patagonia (c. 40° S; Palazzesi et al. 2021). The uplift of the Panama isthmus redirected the Circumtropical Current northeast into the north Atlantic forming the Gulf Stream. Global climatic cooling drove the onset of northern hemisphere glaciation and global expansion of tropical-zone savannas and temperate-zone grasslands with concomitant contraction of rainforests (Jaramillo 2023).

Final closure of the Central American seaway and the uplift of the Northern Andes formed a land bridge connecting North and South America, and facilitated the *Great American Biotic Interchange* (GABI) of terrestrial taxa between Central and South America during the Late Miocene–Pliocene (Stehli and Webb 2013). While the GABI strongly affected many terrestrial taxa, it apparently had little effect on the freshwater fishes (Chakrabarty and Albert 2011; Bacon et al. 2015; Matamoros et al. 2015; Tagliacollo et al. 2017; Elías et al. 2023). Relatively few freshwater fishes dispersed after the formation of the Panamanian isthmus; e.g., several colonizations by the hypopomid electric fish *Brachyhypopomus occidentalis* (Picq et al. 2014; De León et al. 2023). Similarly, the Pleistocene megafauna extinctions of large-bodied mammals and reptiles had little effect on Amazonian fishes. Some fish lineages did become extinct in the Orinoco and trans-Andean Maracaibo and Magdalena basins (e.g., *Arapaima, Brachyplatystoma, Lepidosiren, Phractocephalus*), but these extinctions largely predate the GABI, and more likely resulted from reduced habitat area due to climate change and marine incursions (Lundberg et al. 1998; Albert et al. 2006, 2011a; Ballen et al. 2022, 2024).

The L4 event in the SAF at c. 5.0 Ma affected a broad range of clades with varied ecological tolerances, including stream-adapted species of *Cichlasoma* (Cichlinae), *Atlantirivulus, Melanorivulus*, and *Trigonectes* (Rivulidae), *Brachyhypopomus* (Hypopomidae), and *Sternopygus* (Sternopygidae), large-river adapted *Piaractus* (Serrasalmidae), *Triportheus* (Triportheidae), *Acestrorhynchidae* (Acestrorhynchidae), and *Porotergus* (Apteronotidae), and even the piedmont-adapted Creagrutini (Characidae). At about this time (c. 4.5 Ma), a clade of *Jenynsia* in the SAF of the Paraguay basin was separated from congeners in southeastern coastal basins (Amorim and Costa 2019).

By contrast, the T4 and A4 events, both at c. 5 Ma, strongly affected small-stream fishes. The T4 event affected *Satanoperca, Apistogramma, Crenicichla, Bujurquina, Australoheros, Acaronia* (Cichlinae), *Charax, Knodus*, and *Eretmobrycon* (Characidae), and *Gymnotus* (Gymnotidae). The A4 event affected *Gymnogeophagus, Acarichthys, Crenicichla, Acaronia, Aequidens*, and *Cichlasoma* (Cichlidae), *Poecilia* (Poeciliidae), *Metynnis, Serrasalmus*, and *Pygocentrus* (Serrasalmidae), *Parodon* and *Apareiodon* (Parodontidae), *Planaltina* (Characoidea), *Gymnotus* (Gymnotidae), and *Sternopygus* (Sternopygidae).

Major geographical and evolutionary events affecting the diversification of South American fishes during the Pliocene and Pleistocene (5.3–0.01 Ma) were climatically induced fluctuations of major ecozone margins (e.g., forests, savannahs) and shorelines associated with global ice age cycles. Shoreline oscillations resulted in the separation (at high sea levels) and merging (at low sea levels) of river mouths in low-elevation coastal (e.g., Thomaz et al. 2015, 2017; Wendt et al. 2019; Abreu et al. 2019, 2020; Pio and Carvalho 2021) and intracratonic (e.g., Farias and Hrbek 2008; Oberdorff et al. 2019) settings, as well as the formation (at high sea levels) and erosion (at low sea levels) of white-water *várzea* floodplains (i.e., Irion cycles; Albert et al. 2011a; Říčan et al. 2019; Val et al. 2021).

21.5 Fate of Amazonian Fishes

The most important effects of the Holocene (0.01–0.0005 Ma) and Anthropocene (approximately the last five decades) on South American fishes have arisen from human effects, principally land-use change and climate change. The main drivers are deforestation and forest degradation, water pollution, overfishing, river damming, and diversions, and the effects of introduced species (Albert et al. 2021; Val and Wood 2022; Leroy et al. 2023; Guayasamin et al. 2024). Although much of the Amazon remains a wilderness today, many areas are experiencing rapid economic development, and without interventions, the rate of change indicates collapse of many regional fish faunas within the coming few decades (Reis et al. 2016; Albert et al. 2023; Ottoni et al. 2023).

Studies on the ecology and evolution of Amazonian fishes have profound implications for conservation assessments and actions. The fauna was assembled incrementally, clade by clade, trait by trait, across a labyrinth of aquatics habitats extending over a continental arena of tens of millions of square km, and over geological time scales representing tens of millions of years. With a few curious exceptions, the Amazonian fish fauna is not the result of recent or rapid adaptive radiations. Rather, the fauna arose under a fortuitous combination of environmental conditions and paleogeographic events that allowed speciation rates to exceed extinction rates over immense time periods.

Amazonian fishes diversified under a myriad of continually changing environments, primarily driven by river captures and shoreline fluctuations, under the influences of a handful of rare but highly impactful (i.e., *Black Swan*) paleogeographic events (Albert et al. 2018b). The evolutionary responses of each clade to these changing environments depended critically in their own unique combination of ecotraits, including ecophysiological phenotypes associated with body size and metabolic rate, water chemistry tolerances, habitat utilization, dietary tolerances, and behaviors associated with reproduction, sensory biology, and migration (Chap. 6). In combination, these conditions, events, and ecotraits allowed species and novel functional traits to accumulate over an unparalleled duration of 100 million years. The fauna is not the result of one or a few major drivers; i.e., it is a continental not an *adaptive radiation* (Albert et al. 2020; Chap. 22).

The origins of the megadiverse South American fish fauna are unique among the major biotas of the Earth. Among extant continental faunas, only tropical South America retains the immense biodiversity dimensions of the K-Pg global *super greenhouse* (van der Sleen and Albert 2017; Cassemiro et al. 2023). All the other continental and marine biotas on Earth have been substantially diminished by extinctions during the Late Cenozoic global climate cooling, beginning with the Eocene–Oligocene event and accelerating through the Pleistocene Ice Ages cycles. These results highlight a fundamental insight that *the megadiverse Neotropical biota has never, in its lengthy evolutionary history of more than 100 million years, faced such high rates of environmental change as now in the Anthropocene* (Albert et al. 2023).

The twenty-first century poses an existential crisis for this fauna, the most diverse continental fauna on Earth. For the first time in a lengthy evolutionary history, Neotropical aquatic ecosystems face existential destruction from a multitude of pronounced and accelerating anthropogenic activities (Castello et al. 2013; Duponchelle et al. 2021). The greatest challenges come from direct habitat destruction and degradation from dams, agriculture, and overfishing at local scales (10–100 sq. km) and global climate-induced precipitation changes driving droughts, wildfires, and deforestation at regional scales (1000–10,000 sq. km; Albert et al. 2023).

21.6 Future Research Priorities

The macroevolutionary narrative presented in this chapter is based on a "top-down" approach of studies attempting to encompass the whole NFF fauna. This narrative is based on a synthesis of data and concepts from the four new datasets and four new analyses presented in earlier chapters of this book. Each of the events listed in Tables 4.2 and 21.1 can be treated as a hypothesis to be tested by future studies focusing on smaller taxa using much greater data resolution. Each and every event of this narrative needs to be validated and adjusted with more focused analytical studies of targeted clades, using high-density sampling and phylogenomics of taxa at the genus or family level. Such a "bottom-up" approach will require extensive field studies of these animals in their natural habitats. Further, detailed evolutionary studies at the genus and species level can provide crucial insights into biodiversity, speciation, and adaptation, helping resolve taxonomic uncertainties and uncover genetic and ecological drivers of divergence. Building this knowledge into a fauna-wide database will involve community-level studies that benefit from modern identification (e.g., DNA barcoding) tools and from the ability to build trees and use phylogenetic comparative methods. Highest priority must be given to studies of alpha taxonomy (i.e., species delimitation), natural history and autecology, and phylogenomics with greater taxon sampling.

21.7 Conclusions

This final chapter summarizes and synthesizes results and conclusions of the previous chapters regarding the major phylogenetic, ecological, and biogeographic patterns of NFF fishes, and interprets these patterns in light of the syncretic chronology major climatic and geological events affecting South American biotas over the past 100 million years. Most of major NFF clades ranked as orders and families originated in the Upper Cretaceous or Paleogene, and the NFF fauna was almost entirely modern to the genus level by Early Miocene before the Late Miocene formation of the modern transcontinental Amazon River c. 10 Ma. The comparatively limited fossil record of Cenozoic NFF fishes suggests rapid replacement of taxa after the K-Pg mass extinction event, with several highly derived phenotypes ascribed to living genera present by the Eocene c. 50 Ma, a largely modern fauna at the genus level by the Late Miocene c. 10 Ma, and few documented extinctions of genera since the Late Miocene. Most South American fishes exhibited relatively even diversification rates throughout the Cenozoic, with a small acceleration after the Oligocene global cooling event c. 34 Ma, coinciding with the expansion of several million km sq. of freshwater habitats in the Sub-Andean foreland basin. The geomorphological formation of the modern transcontinental Amazon River and associated biotic GAzBI event exerted a modest increase in species diversification.

All Big Nine NFF clades originated in Western Gondwana during the super-greenhouse world of the Upper Cretaceous or Paleogene, either in the uplands of the continental periphery (Loricarioidei, Curimatoidea) or Sub-Andean foreland lowlands following large marine regressions. During the Paleogene (Paleocene to Eocene), numerous lineages originated in the upland Guiana and Brazilian shields and trans-Andean regions, which today are regarded as geographic relicts. Subsequent marine regressions during the Paleogene are associated with the origins of numerous clades in the Sub-Andean foreland lowlands. While earlier evolutionary events during the Upper Cretaceous and Paleogene appear to have been driven more by paleo-elevation patterns, evolutionary events in the Oligocene and Neogene are more associated with modern basin or watershed limits. Subsequent marine regressions in the Oligocene and Miocene, and the formation of the Pebas megawetland system in the Early and Middle Miocene, allowed for further diversification in the Sub-Andean foreland basin. These marine influences introduced multiple MDCs into freshwater Amazonian ecosystems.

Tectonic uplifts of the Northern Andean cordilleras during the Late Miocene (c. 10–5 Ma) separated the cis-Andean and trans-Andean regions of the Sub-Andean foreland basin, and the formation of the modern tripartite continental drainage pattern; i.e., Orinoco, Amazon, and La Plata basins. The origin of the transcontinental Amazon River resulted in merging (or geodispersal) between the Western and Eastern Amazon basins (i.e., Great Amazonian biotic interchange or GAzBI), and separation (or vicariance) between the Amazon and Orinoco basins. From the Late Miocene to Quaternary, NFF fishes experience continued dispersal to, and along, the continental periphery, affording numerous new opportunities for dispersal, vicariance, and evolutionary diversification. The GABI had only minor influences on freshwater fishes, being entirely restricted to trans-Andean (Pacific) regions of the Chocó and southern Central America.

Qualitative results of analyses into the geographic origins of NFF clades found support for both Eigenmann's *Upland Origins* and Smith's *Lowland Origins* hypotheses. Using AAE in a ML framework, we found that these hypotheses account for roughly similar proportions of all NFF clades, with slightly more than half (53%) having origins in the lowland Sub-Andean and coastal basins, somewhat fewer clades (41%) in the upland basins of the Brazilian and Guiana shields, and a small number (7%) in Trans-Andean river basins. However, clades with upland origins vastly dominate the species composition of the modern fauna (83% of the total). Qualitative results of ecomatrix and ecospace analysis discovered many factors contributing to NFF diversification, each of relatively small effect. These results indicate that NFF fauna is a highly heterogeneous assemblage of taxa, each with distinct ecological and environmental requirements, and each with a distinct evolutionary history. The immense diversity of NFF species is accompanied by a myriad of phenotypic specializations, with each clade exhibiting a distinctive suite of physiological, morphological, behavioral, and life-history phenotypes. The phylogenetic clustering of these phenotypic specializations, a phenomenon known as *niche conservatism*, occurs when the rate of speciation is faster than that of adaptation, resulting in functional redundancy with more species than distinct ecological functions. Conversely, *niche evolution* occurs when the anagenetic processes of adaptation and genetic drift modify species traits, resulting in the evolution of new functional phenotypes.

The differential diversification of upland and lowland NFF fish clades may be ascribed to one or more of the four non-exclusive hypotheses advanced by Edward Wilson to explain the existence of hyperdiverse taxa. A null hypothesis against which these hypotheses can be assessed is that lineage diversification patterns arise from chance fluctuations alone; i.e., phylogenetic drift. One hypothesis suggests that small body size, lower vagility, and reduced gene flow contribute to increased diversification. A second hypothesis posits that incumbent taxa, lineages already present in freshwater environments, are more successful than marine invaders because they descend from ancestors with traits preadapted to persist and diversify in riverine habitats. A third hypothesis emphasizes the importance of the appropriate demographic conditions, overlapping with the ideas of the other hypotheses. Finally, a fourth hypothesis proposes that certain key innovations—traits that are preadapted or exapted for freshwater environments—have facilitated rapid diversification among species with upland origins. Results from these analyses reported in this chapter indicate that most diverse clades descended from lineages with long histories of surviving in small, fragmented populations, possessing ecological traits suited to clearwater streams and rivers and feeding on benthic insect larvae. Additionally, certain evolutionary developments in secondary sex characteristics and chromosome structures, particularly among ostariophysans, cyprinodontiforms, and cichlids, may have further enhanced diversification potential.

The conservation status of NFF species varies by region, and many rivers are threatened by anthropogenic activities. Although many Amazonian rivers remain relatively well preserved as compared with those in other regions of the world, many of the aquatic ecosystems in this region are facing rapid environmental deterioration and conservation threats from a variety of human activities. Conserving Amazonian fishes is an obligation and a challenge for human societies. Proactive conservation policies, international cooperation between the multiple countries in the Amazon, effective enforcement measures, and ongoing scientific research are required to preserve the Neotropical fishes, the most species-rich vertebrate fauna on Earth. The survival of many NFF species into the near future is uncertain, especially for species inhabiting headwater streams, many of which have limited distributional ranges.

References

Abreu JMS, Craig JM, Albert JS, Piorski NM (2019) Historical biogeography of fishes from coastal basins of Maranhão State, northeastern Brazil. Neotrop Ichthyol 17(2):e180156

Abreu JMS, Saraiva ACS, Albert JS, Piorski NM (2020) Paleogeographic influences on freshwater fish distributions in northeastern Brazil. J South Am Earth Sci 102:102692

Acosta RP, Ladant JB, Zhu J, Poulsen CJ (2022) Evolution of the Atlantic Intertropical Convergence Zone, and the South American and African monsoons over the past 95-Myr and their impact on the tropical rainforests. Paleoceanogr Paleoclimatol 37(7):e2021PA004383

Ainsworth R, Cowx IG, Funge-Smith SJ (2021) A review of major river basins and large lakes relevant to inland fisheries. Food and Agriculture Organization of the United Nations, Rome, p 314

Akabane TK, Garcia MJ, Kern AK, De Oliveira PE (2024) Eocene–Oligocene vegetation and climate changes in southeastern Brazil. Rev Palaeobot Palynol 327:105142

Akin DR, Weyand CA, Buehler MD, Armbruster JW (2024) Phylogenetics and biogeography of the Guiana Shield Pencil catfishes, genus *Trichomycterus*. Ichthyol Herpetol 112(4):600–614

Albert JS, Lovejoy NR, Crampton WG (2006) Miocene tectonism and the separation of *cis*-and *trans*-Andean river basins: evidence from neotropical fishes. J South Am Earth Sci 21(1–2):14–27

Albert JS, Bart Jr HL, Reis RE (2011a) Species richness and cladal diversity. In: Albert JS, Reis RE (eds) Historical biogeography of neotropical freshwater fishes. University of California, Berkeley, pp 89–104

Albert JS, Val P, Hoorn C (2018b) The changing course of the Amazon River in the Neogene: center stage for neotropical diversification. Neotrop Ichthyol 16(03):e180033

Albert JS, Tagliacollo VA, Dagosta F (2020) Diversification of neotropical freshwater fishes. Annu Rev Ecol Evol Syst 51(1):27–53

Albert JS, Schoolmaster JrDR, Tagliacollo V, Duke-Sylvester SM (2017) Barrier displacement on a neutral landscape: toward a theory of continental biogeography. Syst Biol 66(2):167–182

Albert JS, Carvalho TP, Petry P, Holder MA, Maxime EL, Espino J et al (2011b) Aquatic biodiversity in the Amazon: habitat specialization and geographic isolation promote species richness. Animals 1(2):205–241

Albert JS, Bernt MJ, Fronk AH, Fontenelle JP, Kuznar SL, Lovejoy NR (2021a) Late Neogene megariver captures and the Great Amazonian biotic interchange. Glob Planet Change 205:103554

Albert JS, Destouni G, Duke-Sylvester SM, Magurran AE, Oberdorff T, Reis RE et al (2021b) Scientists' warning to humanity on the freshwater biodiversity crisis. Ambio 50(1):85–94

Albert JS, Carnaval AC, Flantua SG, Lohmann LG, Ribas CC, Riff D et al (2023) Human impacts outpace natural processes in the Amazon. Science 379(6630):eabo5003

Albert JS et al (2025) An ecological trait matrix for neotropical freshwater fishes, the most diverse continental vertebrate fauna. Nat Sci Data 12:1127

Albert JS, Carvalho TP (2011) Neogene assembly of modern faunas. In: Albert JS, Reis RE (eds) Historical biogeography of neotropical freshwater fishes. University of California, Berkeley, pp 119–136

Albert JS, Craig JM, Tagliacollo VA, Petry P (2018a) Upland and lowland fishes: a test of the river capture hypothesis. In: Hoorn C, Perrigo A, Antonelli A (eds) Mountains, climate and biodiversity. John Wiley & Sons Ltd/Wiley-Blackwell, Hoboken, NJ, Chichester, UK, pp 273–294

Albert JS, Hoorn C, Malhi Y, Phillips O, Encalada AC, Steege H (2021b) The multiple viewpoints for the Amazon: geographic limits and meanings. Amazon Assessment Report, Science Panel For the Amazon, United Nations Sustainable Development Solutions Network, New York

Alda F, Tagliacollo VA, Bernt MJ, Waltz BT, Ludt WB, Faircloth BC et al (2019) Resolving deep nodes in an ancient radiation of neotropical fishes in the presence of conflicting signals from incomplete lineage sorting. Syst Biol 68(4):573–593

Alfaro ME, Faircloth BC, Harrington RC, Sorenson L, Friedman M, Thacker CE et al (2018) Explosive diversification of marine fishes at the Cretaceous–Palaeogene boundary. Nat Ecol Evol 2(4):688–696

Aminov J, Dupont-Nivet G, Ruiz D, Gailleton B (2023) Paleogeographic reconstructions using QGIS: introducing Terra Antiqua plugin and its application to 30 and 50 Ma maps. Earth-Sci Rev 240:104401

Amorim PF, Costa WJ (2018) Multigene phylogeny supports diversification of four-eyed fishes and one-sided livebearers (Cyprinodontiformes: Anablepidae) related to major South American geological events. PLOS One 13(6):e0199201

Amorim PF, Costa WJ (2019) Reconstructing biogeographic temporal events in the evolution of the livebearer fish genus *Jenynsia* based on total evidence analysis (Cyprinodontiformes: Anablepidae). Syst Biodiversity 17(2):124–133

Antonelli A (2022) The rise and fall of neotropical biodiversity. Bot J Linean Soc 199(1):8–24

Antonelli A, Sanmartín I (2011) Why are there so many plant species in the neotropics?. Taxon 60(2):403–414

Antonelli A, Ariza M, Albert J, Andermann T, Azevedo J, Bacon C et al (2018) Conceptual and empirical advances in neotropical biodiversity research. PeerJ 6:e5644

Antonelli A, Zizka A, Carvalho FA, Scharn R, Bacon CD, Silvestro D, Condamine FL (2018) Amazonia is the primary source of neotropical biodiversity. Proc Natl Acad Sci 115(23):6034–6039

Araujo GS, Kurtz YR, Sazima I, Carvalho PH, Floeter SR, Vilasboa A et al (2023) Evolutionary history, biogeography, and a new species of *Spheroides* (Tetraodontiformes: Tetraodontidae): how the major biogeographic barriers of the Atlantic Ocean shaped the evolution of a pufferfish genus. Zool J Linnean Soc 199(4):978–993

Arratia G (1997) Brazilian and Austral freshwater fish faunas of South America. A contrast. In: Ulrich H (ed) Tropical biodiversity and systematics. Museum Alexander Koenig, Bonn, pp 179–186

Arratia G (2018) Otomorphs (= otocephalans or ostarioclupeomorphs) revisited. Neotrop Ichthyol 16:e180079

Avise JC (2000) Phylogeography: the history and formation of species. Harvard University, Cambridge

Bacon CD, Silvestro D, Jaramillo C, Smith BT, Chakrabarty P, Antonelli A (2015) Biological evidence supports an early and complex emergence of the Isthmus of Panama. Proc Natl Acad Sci 112(19):6110–6115

Bagley JC, Johnson JB (2014) Phylogeography and biogeography of the lower Central American neotropics: diversification between two continents and between two seas. Biol Rev 89(4):767–790

Balian EV, Segers H, Martens K, Lévéque C (2008) An overview of the results. Freshwater animal diversity assessment. Springer, Netherlands, pp 627–637

Ballen GA, Jaramillo C, Dagosta FC, De Pinna MC (2022) A fossil fish assemblage from the middle Miocene of the Cocinetas Basin, northern Colombia. Pap Palaeontol 8(1):e1405

Ballen GA, Montes C, Carrillo-Briceño JD, Bogan S, Reinales S, de Pinna MC, Jaramillo C (2024) Fossil freshwater fishes from the Pliocene of northern Colombia and the Palaeogeography of northern South America. bioRxiv 2024-12

Benejam L, Tobes I, Brucet S, Miranda R (2018) Size spectra and other size-related variables of river fish communities: systematic changes along the altitudinal gradient on pristine Andean streams. Ecol Indic 90:366–378

Benton MJ (2009) The Red Queen and the Court Jester: species diversity and the role of biotic and abiotic factors through time. Science 323(5915):728–732

Benton MJ, Wilf P, Sauquet H (2022) The Angiosperm Terrestrial Revolution and the origins of modern biodiversity. New Phytol 233(5):2017–2035

Bernt MJ, Tagliacollo VA, Albert JS (2019) Molecular phylogeny of the ghost knifefishes (Gymnotiformes: Apteronotidae). Mol Phylogenet Evol 135:297–307

Betancur-R R, Ortí G, Pyron RA (2015) Fossil-based comparative analyses reveal ancient marine ancestry erased by extinction in ray-finned fishes. Ecol Lett 18(5):441–450

Bicudo TC, Sacek V, de Almeida RP, Bates JM, Ribas CC (2019) Andean tectonics and mantle dynamics as a pervasive influence on Amazonian ecosystem. Sci Rep 9(1):16879

Bishop BT, Beck SL, Zandt G, Wagner LS, Long MD, Tavera H (2018) Foreland uplift during flat subduction: insights from the Peruvian Andes and Fitzcarrald Arch. Tectonophysics 731:73–84

Bloom DD, Lovejoy NR (2017) On the origins of marine-derived freshwater fishes in South America. J Biogeography 44(9):1927–1938

Boschman LM, Carraro L, Cassemiro FA, de Vries J, Altermatt F, Hagen O et al (2023) Freshwater fish diversity in the western Amazon basin shaped by Andean uplift since the Late Cretaceous. Nat Ecol Evol 7(12):2037–2044

Buffan L, Condamine FL, Stutz NS, Pujos F, Antoine PO, Marivaux L (2025) The fate of South America's endemic mammalian fauna in response to the most dramatic Cenozoic climate disruption. Proc Natl Acad Sci 122(20):e2419520122

Burgener L, Hyland E, Reich BJ, Scotese C (2023) Cretaceous climates: mapping paleo-Köppen climatic zones using a Bayesian statistical analysis of lithologic, paleontologic, and geochemical proxies. Palaeogeog Palaeoclim Palaeoecol 613:111373

Burns MD, Knouft JH, Dillman CB (2024) The role of abiotic and biotic factors in the unequal body shape diversification of a Gondwanan fish radiation (Otophysi: Characiformes). Evolution 78(2):253–266

Caetano DS, O'Meara BC, Beaulieu JM (2018) Hidden state models improve state-dependent diversification approaches, including biogeographical models. Evolution 72(11):2308–2324

Cai L, Kreft H, Taylor A, Denelle P, Schrader J, Essl F et al (2023) Global models and predictions of plant diversity based on advanced machine learning techniques. New Phytol 237(4):1432–1445

Cardoso D, Särkinen T, Alexander S, Amorim AM, Bittrich V, Celis M et al (2017) Amazon plant diversity revealed by a taxonomically verified species list. Proc Natl Acad Sci 114(40):10695–10700

Carrillo-Briceño JD, Mora-Rojas L, Hendricks K, Vanegas A, Aguilera O (2023) New clues on the palaeodiversity of the middle Miocene freshwater ichthyofauna from the Tatacoa Desert, Colombia. Geodiversitas 45(10):327–351

Carrillo-Briceño JD, Aguilera OA, Benites-Palomino A, Hsiou AS, Birindelli JL, Adnet S et al (2021) A historical vertebrate collection from the Middle Miocene of the Peruvian Amazon. Swiss J Palaeontol 140:1–30

Carvalho MR, Jaramillo C, de la Parra F, Caballero-Rodríguez D, Herrera F, Wing S et al (2021) Extinction at the end-Cretaceous and the origin of modern neotropical rainforests. Science 372(6537):63–68

Carvalho RL, Resende AF, Barlow J, França FM, Moura MR, Maciel R, Alves-Martins F, Shutt J, Nunes CA, Elias F, Silveira JM (2023) Pervasive gaps in Amazonian ecological research. Curr Biol 33(16):3495–3504

Carvalho TP, Albert JS (2011) The Amazon-Paraguay Divide. In: Albert JS, Reis RE (eds) Historical biogeography of neotropical freshwater fishes. University of California, Berkeley, pp 193–202

Carvalho TP, Arce M, Reis RE, Sabaj MH (2018) Molecular phylogeny of Banjo catfishes (Ostariophysi: Siluriformes: Aspredinidae): a continental radiation in South American freshwaters. Mol Phylogenet Evol 127:459–467

Cassemiro FA, Albert JS, Antonelli A, Menegotto A, Wüest RO, Cerezer F et al (2023) Landscape dynamics and diversification of the megadiverse South American freshwater fish fauna. Proc Natl Acad Sci 120(2):e2211974120

Castello L, McGrath DG, Hess LL, Coe MT, Lefebvre PA, Petry P et al (2013) The vulnerability of Amazon freshwater ecosystems. Conserv Lett 6(4):217–229

Cavin L (2010) The Late Jurassic ray-finned fish peak of diversity: biological radiation or preservational bias?. In: Nelson JS, Schultze H-P, Wilson MVH (eds) Origin and phylogenetic interrelationships of Teleosts. Pfeil, Munich, pp 111–121

Cavin, L, 2017. Freshwater fishes: 250 million years of evolutionary history. Elsevier, London, p 156

Cavin L, Forey PL, Lécuyer C (2007) Correlation between environment and Late Mesozoic ray-finned fish evolution. Palaeogeog palaeoclim palaeoecol 245(3–4):353–367

Cavin L, Tong H, Boudad L, Meister C, Piuz A, Tabouelle J et al (2010) Vertebrate assemblages from the early Late Cretaceous of southeastern Morocco: an overview. J Afr Earth Sci 57(5):391–412

Cavin L, Forey PL (2007) Using ghost lineages to identify diversification events in the fossil record. Biol Lett 3(2):201–204

Cerezer FO, Dambros CS, Coelho MTP, Cassemiro FAS, Barreto E, Albert JS, Wüest RO, Graham CH (2023) Accelerated body size evolution in upland environments is correlated with recent speciation in South American freshwater fishes. Nat Commun 14(1):6070

Chakrabarty P, Albert JS (2011) Not so fast: a new take on the Great American biotic interchange. In: Albert JS, Reis RE (eds) Historical biogeography of neotropical freshwater fishes. University of California, Berkeley, pp 293–305

Conde-Saldaña CC, Melo BF, Roxo FF, Toledo-Piza M, Tagliacollo VA, Oliveira C, Albert JS (2025) Landscape evolution drives continental diversification in Neotropical freshwater fishes of the family Erythrinidae (Teleostei, Erythrinoidea). J Biogeogr 52(6):e15129

Cortés D, Larsson HC (2024) Top of the food chains: an ecological network of the marine Paja Formation biota from the Early Cretaceous of Colombia reveals the highest trophic levels ever estimated. Zool J Linnean Soc 202(1):zlad092

Crelier C, Zumba A, Peifer D, Val P (2025) Mobility of South America's transcontinental drainage divide and shrinkage of the Paraná river basin linked to lithologic and geodynamic controls. Sci Rep 15(1):3134

Cuitiño JI, Dozo MT, del Río CJ, Buono MR, Palazzesi L, Fuentes S, Scasso RA (2017) Miocene marine transgressions: Paleoenvironments and paleobiodiversity. In: Late Cenozoic of Península Valdés, Patagonia, Argentina: an interdisciplinary approach. Springer International Publishing, Cham, pp 47–84

Custódio MA, Roddaz M, Santos RV, Antoine PO, Marivaux L, Stutz NS et al (2023) New stratigraphic and paleoenvironmental constraints on the Paleogene paleogeography of Western Amazonia. J South Am Earth Sci 124:104256

da Silva JMC, Leal IR, Tabarelli M (eds) (2018) Caatinga: the largest tropical dry forest region in South America. Springer, Dordrecht, Netherlands, p 481

De León LF, Arias CF, Sharpe DM, Bravo V, González R, Krahe R, Aguilar C (2023) Unraveling the complex phylogeographic history of freshwater fishes in Lower Central America: a study of the electric fish *Brachyhypopomus occidentalis*. Mol Phylogenet Evol 189:107941

De Lira Mota MA, Dunkley Jones T, Sulaiman N, Edgar KM, Yamaguchi T, Leng MJ et al (2023) Multi-proxy evidence for sea level fall at the onset of the Eocene-Oligocene transition. Nat Commun 14(1):4748

de Santana JFOM, de Araújo-Júnior HI, de Figueiredo FJ, dos Reis Polck MA, Ade MVB (2023) Taphonomy of the fish fauna of the Tremembé Formation (Oligocene of the Taubaté basin), Southeastern Brazil. J South Am Earth Sci 131:104599

Dias AC, Tencatt LF, Roxo FF, Silva GDSDC, Santos SA, Britto MR, et al (2024) Phylogenomic analyses in the complex neotropical subfamily Corydoradinae (Siluriformes: Callichthyidae) with a new classification based on morphological and molecular data. Zool J Linnean Soc zlae053

Diniz-Filho JAF (2023) Diversity patterns in macroecological assemblages. In: Diniz-Filho JA (ed) The Macroecological perspective: theories, models and methods. Springer International Publishing, Cham, pp 251–291

de Queiroz LJ, Meyer X, Cardoso YP, Bahechar IA, Covain R, Parente TE et al (2021) Western Amazon was a center of neotropical fish dispersal, as evidenced by the continental-wide time-stratified biogeographic analysis of the hyper-diverse Hypostomus catfish. bioRxiv, 2021-06

dos Reis RB, Stabile BHM, Frota A, Ferrer J, de Oliveira AV, da Graça WJ (2024) Recent dispersion routes between freshwater ecoregions evidence headwater captures in southern Brazil: a case study using cryptic species of the neotropical freshwater fish genus Cambeva (Siluriformes: Trichomycteridae). Hydrobiologia 852(4):873–890

Duponchelle F, Isaac VJ, Rodrigues Da Costa Doria C, Van Damme PA, Herrera-R GA, Anderson EP et al (2021) Conservation of migratory fishes in the Amazon basin. Aquat Conserv: Mar Freshwater Ecosyst 31(5):1087–1105

Eaton JG, Kirkland JI, Howard Hutchison J, Denton R, O'Neill RC, Michael Parrish J (1997) Nonmarine extinction across the Cenomanian-Turonian boundary, southwestern Utah, with a comparison to the Cretaceous-Tertiary extinction event. Geol Soc Am Bull 109(5):560–567

Eigenmann CH (1921) The nature and origin of the fishes of the Pacific slope of Ecuador, Peru and Chili. Proc Am Philos Soc 60(4):503–523

Eigenmann CH (1906) The fresh-water fishes of South and Middle America. Pop Sci Mon 515–530

Eigenmann CH (1912) The freshwater fishes of British Guiana, including a study of the ecological grouping of species and the relation of the fauna of the plateau to that of the lowlands, vol 5. Carnegie Institute, Pittsburgh, p 578

Elías DJ, McMahan CD, Alda F, García-Alzate C, Hart PB, Chakrabarty P (2023) Phylogenomics of trans-Andean tetras of the genus *Hyphessobrycon* Durbin 1908 (Stethaprioninae: Characidae) and colonization patterns of Middle America. PLoS One 18(1):e0279924

Eschmeyer WN, Fricke R, Van der Laan R (2025) Catalog of fishes: genera, species, references. https://researcharchive.calacademy.org/research/ichthyology/catalog/fishcatmain.asp.

Fadul CM, Oliveira P, Val P (2022) Ongoing landscape transience in the eastern Amazon Craton consistent with lithologic control of base level. Earth Surf Processes Landforms 47(13):3117–3132

Farias IP, Hrbek T (2008) Patterns of diversification in the discus fishes (*Symphysodon* spp. Cichlidae) of the Amazon basin. Mol Phylogenet Evol 49(1):32–43

Faurby S, Silvestro D, Werdelin L, Antonelli A (2024) Reliable biogeography requires fossils: insights from a new species-level phylogeny of extinct and living carnivores. Proc R Soc B 291(2028):20240473

Ferreira PSM, Gonçalves-Souza T, Ouchi-Melo LS, Oliveira-Filho AT, Rodal MJN (2024) Biogeographic history and environmental gradients modulate non-stationary patterns of tropical tree diversity. An Acad Bras Cienc 96(1):e20230063

Fink SV, Fink WL (1981) Interrelationships of the ostariophysan fishes (Teleostei). Zool J Linnean Soc 72(4):297–353

Frable BW, Melo BF, Fontenelle JP, Oliveira C, Sidlauskas BL (2022) Biogeographic reconstruction of the migratory neotropical fish family Prochilodontidae (Teleostei: Characiformes). Zool Scripta 51(3):348–364

Friedman M (2010) Explosive morphological diversification of spiny-finned teleost fishes in the aftermath of the end-Cretaceous extinction. Proc R Soc B 277(1688):1675–1683

Friedman M, Keck BP, Dornburg A, Eytan RI, Martin CH, Hulsey CD et al (2013) Molecular and fossil evidence place the origin of cichlid fishes long after Gondwanan rifting. Proc R Soc B 280(1770):20131733

Gayet M, Marshall LG, Sempere T, Meunier FJ, Cappetta H, Rage JC (2001) Middle Maastrichtian vertebrates (fishes, amphibians, dinosaurs and other reptiles, mammals) from Pajcha Pata (Bolivia). Biostratigraphic, palaeoecologic and palaeobiogeographic implications. Palaeogeog Palaeoclim Palaeoecol 169(1–2):39–68

Goulding MJ, Venticinque E, Ribeiro MLDB, Barthem RB, Leite RG, Forsberg B et al (2019) Ecosystem-based management of Amazon fisheries and wetlands. Fish Fish 20(1):138–158

Greenacre M, Groenen PJ, Hastie T, d'Enza AI, Markos A, Tuzhilina E (2022) Principal component analysis. Nat Rev Methods Primers 2(1):100

Grélselle B, Pittet B (2005) Fringing carbonate platforms at the Arabian Plate margin in northern Oman during the Late Aptian–Middle Albian: evidence for high-amplitude sea-level changes. Sediment Geol 175(1-4):367–390

Guayasamin JM, Ribas CC, Carnaval AC, Carrillo JD, Hoorn C, Lohmann LG et al (2024) Evolution of Amazonian biodiversity: a review. Acta Amazonica 54(spe1):e54bc21360

Guinot G, Cavin L (2016) 'Fish' (Actinopterygii and Elasmobranchii) diversification patterns through deep time. Biol Rev 91(4):950–981

Helmstetter AJ, Papadopulos AS, Igea J, Van Dooren TJ, Leroi AM, Savolainen V (2016) Viviparity stimulates diversification in an order of fish. Nat Commun 7(1):11271

Hernández-Ávila SG, Hoagstrom CW, Matamoros WA (2024) Historical biogeography of North American killifishes (Cyprinodontiformes) recapitulates geographical history in the Gulf of México watershed. Zool J Linnean Soc 202(2):zlae105

Ho SY, Tong KJ, Foster CS, Ritchie AM, Lo N, Crisp MD (2015) Biogeographic calibrations for the molecular clock. Biol Lett 11(9)

Hoorn C, Lohmann LG, Boschman LM, Condamine FL (2023) Neogene history of the Amazonian flora: a perspective based on geological, palynological, and molecular phylogenetic data. Annu Rev Earth Planet Sci 51(1):419–446

Hoorn C, Boschman LM, Kukla T, Sciumbata M, Val P (2022) The Miocene wetland of western Amazonia and its role in neotropical biogeography. Bot J Linean Soc 199(1):25–35

Hoorn C, Wesselingh FP, Ter Steege H, Bermudez MA, Mora A, Sevink J et al (2010) Amazonia through time: Andean uplift, climate change, landscape evolution, and biodiversity. Science 330(6006): 927–931

Horton BK, DeCelles PG (1997) The modern foreland basin system adjacent to the Central Andes. Geology 25(10):895–898

Horton B, Baker PA, Gomez S, Parra M, Jackson L, Banks CL, Gutierrez EG, Vallejo C, Silva CG, Wahnfried I, Kiefer G, Sawakuchi AO, Fritz S (2023) Drainage evolution in the Amazon Foreland Basin during uplift of the Andes. Geol Soc Am. https://doi.org/10.1130/abs/2023AM-390363

Horton BK, Parra M, Mora A (2020) Construction of the Eastern Cordillera of Colombia: insights from the sedimentary record. In: Gómez J, Mateus–Zabala D (eds) The Geology of Colombia, vol 3 Paleogene – Neogene. Servicio Geológico Colombiano, Publicaciones Geológicas Especiales 37, Bogota, p 507

Hubbard RNLB, Boulter MC, Manum SB (1994) Cenozoic dinoflagellate Palaeoecology elucidated, and used For Marine-Terrestrial biological correlation. In: Cenozoic plants and climates of the Arctic. Springer, Berlin Heidelberg, pp 57–72

Huber JH (1998) Comparison of Old World and New World Tropical Cyprinodonts. Societe Francais de D'Ichtyologie, Paris

Hulsey CD, Hollingsworth PR, Fordyce JA (2010) Temporal diversification of Central American cichlids. BMC Evol Biol 10:1–12

Irisarri I, Singh P, Koblmüller S, Torres-Dowdall J, Henning F, Franchini P (2018) Phylogenomics uncovers early hybridization and adaptive loci shaping the radiation of Lake Tanganyika cichlid fishes. Nat Commun 9:3159

Jablonski D (2000) Micro-and macroevolution: scale and hierarchy in evolutionary biology and paleobiology. Paleobiology 26(S4):15–52

Janzen F (2023) Evolutionary biogeography of catfishes (Siluriformes, Actinopterygii): the influence of habitat and landscape on gene flow

and genetic diversification. Doctoral Dissertation, Université d'Ottawa/University of Ottawa.
Jaramillo C (2023) The evolution of extant South American tropical biomes. New Phytol 239(2):477–493
Jaramillo C, Romero I, D'Apolito C, Bayona G, Duarte E, Louwye S et al (2017) Miocene flooding events of western Amazonia. Sci Adv 3(5):e1601693
Johnson JB, Omland KS (2004) Model selection in ecology and evolution. Trends Ecol Evol 19(2):101–108
Jouault C, Condamine FL, Legendre F, Perrichot V (2024) The Angiosperm Terrestrial Revolution buffered ants against extinction. Proc Natl Acad Sci 121(13):e2317795121
Karner GD, Driscoll NW (1999) Tectonic and stratigraphic development of the West African and eastern Brazilian Margins: insights from quantitative basin modelling. Geol Soc London, Special Publications 153(1):11–40
Kass JM, Guénard B, Dudley KL, Jenkins CN, Azuma F, Fisher BL et al (2022) The global distribution of known and undiscovered ant biodiversity. Sci Adv 8(31):eabp9908
Keller G, Pardo A (2004) Age and paleoenvironment of the Cenomanian–Turonian global stratotype section and point at Pueblo, Colorado. Mar Micropaleontol 51(1–2):95–128
Labandeira CC, Wappler T (2023) Arthropod and pathogen damage on fossil and modern plants: exploring the origins and evolution of herbivory on land. Annu Rev Entomol 68(1):341–361
Lamm KS, Redelings BD (2009) Reconstructing ancestral ranges in historical biogeography: properties and prospects. J Syst Evol 47(5):369–382
Langerhans RB, Reznick DN (2010) Ecology and evolution of swimming performance in fishes: predicting evolution with biomechanics. Fish Locomotion: Eco-ethol Perspect 200:248
Latrubesse EM, Cozzuol M, da Silva-Caminha SA, Rigsby CA, Absy ML, Jaramillo C (2010) The Late Miocene paleogeography of the Amazon Basin and the evolution of the Amazon River system. Earth-Sci Rev 99(3-4):99–124
Lehmberg ES, Elbassiouny AA, Bloom DD, López-Fernández H, Crampton WG, Lovejoy NR (2018) Fish biogeography in the "Lost World" of the Guiana Shield: Phylogeography of the weakly electric knifefish *Gymnotus carapo* (Teleostei: Gymnotidae). J Biogeography 45(4):815–825
Lehner B, Roth A, Huber M, Anand M, Grill G, Osterkamp N et al (2021) HydroSHEDS v2.0 - Refined global river network and catchment delineations from TanDEM-X elevation data. In: EGU General Assembly Conference Abstracts (pp EGU21-9277)
Leroy B, Bellard C, Dias MS, Hugueny B, Jézéquel C, Leprieur F et al (2023) Major shifts in biogeographic regions of freshwater fishes as evidence of the Anthropocene epoch. Sci Adv 9(46):eadi5502
Levêque C, Oberdorff T, Paugy D, Stiassny MLJ, Tedesco PA (2008) Global diversity of fish (Pisces) in freshwater. In: Balian V, Lévêque C, Segers H, Martens K (eds) Freshwater animal diversity assessment. Springer, Netherlands, pp 545–567
Lewis Jr WM, Hamilton SK, Saunders III JF (2006) Rivers of northern south America. Ecosystems of the World: Rivers 219–256
Liebrand D, de Bakker AT, Beddow HM, Wilson PA, Bohaty SM, Ruessink G et al (2017) Evolution of the early Antarctic ice ages. Proc Natl Acad Sci 114(15):3867–3872
Liu S, Xia S, Wu D, Behm JE, Meng Y, Yuan H et al (2022) Understanding global and regional patterns of termite diversity and regional functional traits. IScience 25(12):105538
Londoño-Burbano A, Britto MR (2025) Species delimitation and historical biogeography of *Sturisoma* Swainson, 1838 (Loricariidae: Loricariinae): hidden diversity along the Amazon River. Mol Phylogenet Evol 203:108248
López-Fernández H (2021) Neotropical riverine cichlids: adaptive radiation and macroevolution at continental scales. In: Abate ME, Noakes DL (Eds.) The behavior, ecology and evolution of Cichlid fishes. Springer, Dordrecht, Netherlands, 135–173
López-Fernández H, Albert JS (2011) Paleogene radiations. In: Albert JS, Reis RE (eds) Historical biogeography of neotropical freshwater fishes. University of California, Berkeley, pp 105–118
Louterbach M, Roddaz M, Bailleul J, Antoine PO, Adnet S, Kim JH et al (2014) Evidences for a Paleocene marine incursion in southern Amazonia (Madre de Dios sub-Andean zone, Peru). Palaeogeog palaeoclim palaeoecol 414:451–471
Lujan NK, Armbruster JW, Werneke DC, Teixeira TF, Lovejoy NR (2020) Phylogeny and biogeography of the Brazilian–Guiana Shield endemic *Corymbophanes* clade of armoured catfishes (Loricariidae). Zool J Linnean Soc 188(4):1213–1235
Lujan NK, Armbruster JW (2012) Morphological and functional diversity of the mandible in suckermouth armored catfishes (Siluriformes: Loricariidae). J Morphol 273(1):24–39
Lujan NK, Conway KW (2015) Life in the fast lane: a review of rheophily in freshwater fishes. In: Riesch R, Tobler M, Plath M (eds) Extremophile fishes: ecology, evolution, and physiology of Teleosts in extreme environments. Springer, Cham, Switzerland, pp 107–136
Lujan NK, Roach KA, Jacobsen D, Winemiller KO, Vargas VM, Ching VR, Maestre JA (2013) Aquatic community structure across an Andes-to-Amazon fluvial gradient. J Biogeogr 40(9):1715–1728
Lundberg JG, Aguilera O (2003) The late Miocene *Phractocephalus* catfish (Siluriformes: Pimelodidae) from Urumaco, Venezuela: additional specimens and reinterpretation as a distinct species. Neotrop Ichthyol 1(2):97–110
Lundberg JG, Machado-Allison A, Kay RF (1986) Miocene characid fishes from Colombia: evolutionary stasis and extirpation. Science 234(4773):208–209
Lundberg JG, Marshall LG, Guerrero J, Horton B, Malabarba MCSL, Wesselingh F (1998) The stage for Neotropical fish diversification: a history of tropical South American rivers. Phylog Classif Neotrop Fishes 27:13–48
Lundberg JG (1993) African–South American freshwater fish clades and continental drift: problems with a paradigm. In: Biological relationships between Africa and South America. Yale University, New Haven, p 648
Lundberg JG, Kottelat M, Smith GR, Stiassny ML, Gill AC (2000) So many fishes, so little time: an overview of recent ichthyological discovery in continental waters. Ann Mo Bot Gard 26–62
Maisey JG (2000) Continental break up and the distribution of fishes of Western Gondwana during the Early Cretaceous. Cretac Res 21(2–3):281–314
Malabarba MC, Zuleta O, Papa CD (2006) *Proterocara argentina*, a new fossil cichlid from the Lumbrera Formation, Eocene of Argentina. J Vertebr Paleontol 26(2):267–275
Malabarba MC, Malabarba LR, Papa CD (2010) *Gymnogeophagus eocenicus*, n. sp. (Perciformes: Cichlidae), an Eocene cichlid from the Lumbrera Formation in Argentina. J Vertebr Paleontol 30(2):341–350
Malabarba MC, Malabarba LR, López-Fernández H (2014) On the Eocene cichlids from the Lumbrera Formation: additions and implications for the Neotropical ichthyofauna. J Vertebr Paleontol 34(1):49–58
Manel S, Guerin PE, Mouillot D, Blanchet S, Velez L, Albouy C, Pellissier L (2020) Global determinants of freshwater and marine fish genetic diversity. Nat Commun 11(1):692
Marcilly CM, Torsvik TH, Conrad CP (2022) Global Phanerozoic sea levels from paleogeographic flooding maps. Gondwana Res 110:128–142
Martin AP, Bermingham E (1998) Systematics and evolution of lower Central American cichlids inferred from analysis of cytochrome b gene sequences. Mol Phylogenet Evol 9(2):192–203
Martos YM, Catalán M (2024) The Drake Passage asthenospheric and oceanic gateway. Earth-Sci Rev 252:104731
Matamoros WA, McMahan CD, Chakrabarty P, Albert JS, Schaefer JF (2015) Derivation of the freshwater fish fauna of Central America

revisited: Myers's hypothesis in the twenty-first century. Cladistics 31(2):177–188
McCraney WT, Thacker CE, Faircloth BC, Harrington RC, Near TJ, Alfaro ME (2025) Explosion of goby fish diversity at the Eocene-Oligocene transition. Mol Phylogenet Evol 207:108342
McHugh SW, Donoghue MJ, Landis MJ (2024) A phylogenetic model of established and enabled biome shifts. bioRxiv.
Melo BF, Stiassny ML (2024) Phylogenomic and anatomical evidence for the Late Cretaceous diversification of African characiform fishes, including a new family, under the influence of the Trans-Saharan Seaway. Evol J Linnean Soc 3(1):kzae030
Melo BF, Dorini BF, Foresti F, Oliveira C (2018a) Little divergence among mitochondrial lineages of *Prochilodus* (Teleostei, Characiformes). Front Genet 9:107
Melo BF, Albert JS, Dagosta FC, Tagliacollo VA (2021) Biogeography of curimatid fishes reveals multiple lowland–upland river transitions and differential diversification in the neotropics (Teleostei, Curimatidae). Ecol Evol 11(22):15815–15832
Melo BF, Sidlauskas BL, Hoekzema K, Vari RP, Dillman CB, Oliveira C (2018b) Molecular phylogenetics of neotropical detritivorous fishes of the family Curimatidae (Teleostei: Characiformes). Mol Phylogenet Evol 127:800–812
Melo BF, Sidlauskas BL, Near TJ, Roxo FF, Ghezelayagh A, Ochoa LE et al (2022) Accelerated diversification explains the exceptional species richness of tropical characoid fishes. Syst Biol 71(1):78–92
Melo BF, Ota RP, Benine RC, Carvalho FR, Lima FC, Mattox GM et al (2024) Phylogenomics of Characidae, a hyper-diverse neotropical freshwater fish lineage, with a phylogenetic classification including four families (Teleostei: Characiformes). Zool J Linnean Soc 202(1):zlae101
Meseguer AS, Antoine PO, Fouquet A, Delsuc F, Condamine FL (2020) The role of the neotropics as a source of world tetrapod biodiversity. Global Ecol Biogeogr 29(9):1565–1578
Meseguer AS, Michel A, Fabre PH, Escobar OAP, Chomicki G, Riina R et al (2022) Diversification dynamics in the neotropics through time, clades, and biogeographic regions. eLife 11:e74503
Meyer JL, Strayer DL, Wallace JB, Eggert SL, Helfman GS, Leonard NE (2007) The contribution of headwater streams to biodiversity in river networks. JAWRA J Am Water Resour Assoc 43(1):86–103
Mirande JM (2019) Morphology, molecules and the phylogeny of Characidae (Teleostei, Characiformes). Cladistics 35(3):282–300
Montes C, Silva CA, Bayona GA, Villamil R, Stiles E, Rodriguez-Corcho AF et al (2021) A middle to late Miocene trans-Andean portal: geologic record in the Tatacoa Desert. Front Earth Sci 8:587022
Mora A, Villagómez D, Parra M, Caballero VM, Spikings R, Horton BK et al (2020) Late Cretaceous to Cenozoic uplift of the northern Andes: Paleogeographic implications. Geol Colombia 3:ca-4
Nogueira AF, Oliveira C, Langeani F, Netto-Ferreira AL (2023) Phylogenomics, evolution of trophic traits and divergence times of hemiodontid fishes (Ostariophysi: Characiformes). Mol Phylogenet Evol 186:107864
Oberdorff T, Dias MS, Jézéquel C, Albert JS, Arantes CC, Bigorne R et al (2019) Unexpected fish diversity gradients in the Amazon basin. Sci Adv 5(9):eaav8681
Ochoa LE, Pereira LHG, Costa-Silva GJ, Roxo FF, Batista JDS, Formiga K (2015) Genetic structure and historical diversification of catfish *Brachyplatystoma platynemum* (Siluriformes: Pimelodidae) in the Amazon basin with implications for its conservation. Ecol Evol 5(10):2005–2020
Ottoni FP, South J, Azevedo-Santos VM, Henschel E, de Bragança PHN (2023) Freshwater biodiversity crisis: multidisciplinary approaches as tools for conservation. Front Environ Sci 11:1155608
Palazzesi L, Vizcaíno SF, Barreda VD, Cuitiño JI, del Río CJ, Goin F et al (2021) Reconstructing Cenozoic Patagonian biotas using multi-proxy fossil records. J South Am Earth Sci 112:103513
Patarroyo GD, Kochhann KG, Ceolin D, Guerra RM, Alegret L, Bom MH (2022) Paleoenvironmental changes recorded at a late Maastrichtian marine succession of northern South America. J South Am Earth Sci 119:104015
Pelicice FM, Agostinho AA, Azevedo-Santos VM, Bessa E, Casatti L, Garrone-Neto D et al (2023) Ecosystem services generated by neotropical freshwater fishes. Hydrobiologia 850(12):2903–2926
Perez PA, Malabarba MC, del Papa C (2010) A new genus and species of Heroini (Perciformes: Cichlidae) from the early Eocene of southern South America. Neotrop Ichthyol 8:631–642
Picq S, Alda F, Krahe R, Bermingham E (2014) Miocene and Pliocene colonization of the Central American Isthmus by the weakly electric fish *Brachyhypopomus occidentalis* (Hypopomidae, Gymnotiformes). J Biogeography 41(8):1520–1532
Pimiento C, Kocáková K, Mathes GH, Argyriou T, Cadena EA, Cooper JA et al (2024) The extinct marine megafauna of the Phanerozoic. Camb Prism Extinction 2:e7
Pio NL, Carvalho TP (2021) Evidence on the paleodrainage connectivity during Pleistocene: Phylogeography of a hypoptopomatine endemic to southeastern Brazilian coastal drainages. Neotrop Ichthyol 19:e200128
Pirie MD, Humphreys AM, Antonelli A, Galley C, Linder HP (2012) Model uncertainty in ancestral area reconstruction: a parsimonious solution?. Taxon 61(3):652–664
Pontarp M, Bunnefeld L, Cabral JS, Etienne RS, Fritz SA, Gillespie R et al (2019) The latitudinal diversity gradient: novel understanding through mechanistic eco evolutionary models. Trends Ecol Evol 34(3):211–223
Pucéat E, Lécuyer C, Sheppard SMF, Dromart G, Reboulet S, Grandjean P (2003) Thermal evolution of Cretaceous Tethyan marine waters inferred from oxygen isotope composition of fish tooth enamels. Paleoceanography 18:1–12
Rabosky DL, Santini F, Eastman J, Smith SA, Sidlauskas B, Chang J, Alfaro ME (2013) Rates of speciation and morphological evolution are correlated across the largest vertebrate radiation. Nat Commun 4(1):1958
Ramirez JL, Machado CB, de Mello Affonso PRA, Galetti JrPM (2023) Speciation in coastal basins driven by staggered headwater captures: dispersal of a species complex, *Leporinus bahiensis*, as revealed by genome-wide SNP data. Syst Biol 72(5):973–983
Rangel TF, Edwards NR, Holden PB, Diniz-Filho JAF, Gosling WD, Coelho MTP et al (2018) Modeling the ecology and evolution of biodiversity: biogeographical cradles, museums, and graves. Science 361(6399):eaar5452
Reis RE, Albert JS, Di Dario F, Mincarone MM, Petry P, Rocha LA (2016) Fish biodiversity and conservation in South America. J Fish Biol 89(1):12–47
Revel, LJ (2025) Ancestral state reconstruction of phenotypic characters. Evol Biol:1–25
Reznick DN, Furness AI, Meredith RW, Springer MS (2017) The origin and biogeographic diversification of fishes in the family Poeciliidae. PLoS One 12(3):e0172546
Ribas CC, Sawakuchi AO, de Almeida RP, Pupim FN, Rego MA, Batista R, Knowles LL (2025) The role of rivers in the origin and future of Amazonian biodiversity. Nat Rev Biodiversity 1(1):14–31
Ribeiro AC (2006) Tectonic history and the biogeography of the freshwater fishes from the coastal drainages of eastern Brazil: an example of faunal evolution associated with a divergent continental margin. Neotrop Ichthyol 4:225–246
Říčan O, Říčanová Š, Dragová K, Piálek L, Almirón A, Casciotta J (2019) Species diversity in *Gymnogeophagus* (Teleostei: Cichlidae) and comparative biogeography of cichlids in the Middle Paraná basin, an emerging hotspot of fish endemism. Hydrobiologia 832:331–354
Roddaz M, Hermoza W, Mora A, Baby P, Parra M, Christophoul F et al (2010) Cenozoic sedimentary evolution of the Amazonian foreland

basin system. In: Amazonia, landscape and species evolution: a look into the past. vol 5. Blackwell-Wiley, Hoboken, pp 61–88

Rodrigues AES, Ready JS, Pereira da Silva LG, Silva KDF, Ferreira Silva DJ, Sabaj MH et al (2023) The role of Miocene-Pleistocene environmental change in diversification of the genus *Hypoclinemus Chabanaud*, 1928. bioRxiv, 2023-12

Roll U, Feldman A, Novosolov M, Allison A, Bauer AM, Bernard R et al (2017) The global distribution of tetrapods reveals a need for targeted reptile conservation. Nat Ecol Evol 1(11):1677–1682

Roxo FF, Albert JS, Silva GS, Zawadzki CH, Foresti F, Oliveira C (2014) Molecular phylogeny and biogeographic history of the armored neotropical catfish subfamilies Hypoptopomatinae, Neoplecostominae and Otothyrinae (Siluriformes: Loricariidae). PLoS One 9(8):e105564

Roxo FF, Ochoa LE, Sabaj MH, Lujan NK, Covain R, Silva GS et al (2019) Phylogenomic reappraisal of the neotropical catfish family Loricariidae (Teleostei: Siluriformes) using ultraconserved elements. Mol Phylogenet Evol 135:148–165

Rull V (2024) Oligocene revolution. In: Origin and evolution of Caribbean Mangroves: a time-continuum ecological approach. Springer Nature Switzerland, Cham, pp 85–124

Rull, V, Carnaval, AC (2020). Neotropical diversification: patterns and processes. Springer, Berlin, (e pp. 13–49

Sacek V (2014) Drainage reversal of the Amazon River due to the coupling of surface and lithospheric processes. Earth Planet Sci Lett 401:301–312

Sacek V, Mutz SG, Bicudo TC, de Almeida RP, Ehlers TA (2023) The Amazon paleoenvironment resulted from geodynamic, climate, and sea-level interactions. Earth Planet Sci Lett 605:118033

Salas-Gismondi R, Flynn JJ, Baby P, Tejada-Lara JV, Wesselingh FP, Antoine PO (2015) A Miocene hyperdiverse crocodylian community reveals peculiar trophic dynamics in proto-Amazonian mega-wetlands. Proc R Soc B 282(1804):20142490

Salles T, Husson L, Rey P, Mallard C, Zahirovic S, Boggiani BH et al (2023) Hundred million years of landscape dynamics from catchment to global scale. Science 379(6635):918–923

Sánchez-Hernández J (2023) Fresh perspectives on the river continuum concept require trophic ecology approaches focused on food web structure and energy mobilization routes. J Anim Ecol 92:957–964

Sanmartín I (2012) Historical biogeography: evolution in time and space. Evol: Educ Outreach 5(4):555–568

Sarkar S, Basak C, Frank M, Berndt C, Huuse M, Badhani S, Bialas J (2019) Late Eocene onset of the proto-Antarctic circumpolar current. Sci Rep 9(1):10125

Sarmiento-Rojas LF (2019) Cretaceous stratigraphy and paleo-facies maps of northwestern South America. In: Geology and Tectonics of Northwestern South America: the Pacific-Caribbean-Andean Junction. Springer International Publishing, Cham, pp 673–747

Shephard GE, Müller RD, Liu L, Gurnis M (2010) Miocene drainage reversal of the Amazon River driven by plate–mantle interaction. Nat Geosci 3(12):870–875

Silva GS, Roxo FF, Melo BF, Ochoa LE, Bockmann FA, Sabaj MH et al (2021a) Evolutionary history of Heptapteridae catfishes using ultraconserved elements (Teleostei, Siluriformes). Zool Scripta 50(5):543–554

Silva GS, Melo BF, Roxo FF, Ochoa LE, Shibatta OA, Sabaj MH, Oliveira C (2021b) Phylogenomics of the bumblebee catfishes (Siluriformes: Pseudopimelodidae) using ultraconserved elements. J Zool Sys Evol Res 59(8):1662–1672

Silva GS, Rocha MS, Melo BF, Reia L, Roxo FF, Sabaj MH, Oliveira C (2024) Phylogenomics of the catfish family Pimelodidae with focus on the genus *Pimelodus* support the recognition of Sorubiminae and Pimelodinae (Teleostei, Siluriformes). Zool Scripta 53(5): 541–554

Smith GR (1992) Phylogeny and biogeography of the Catostomidae, freshwater fishes of North America and Asia. In: Mayden RL (ed) Systematics, historical ecology, and North American freshwater fishes. Stanford University Press, Stanford, pp778–826

Smith GR, Badgley C, Eiting TP, Larson PS (2010) Species diversity gradients in relation to geological history in North American freshwater fishes. Evol Ecol Res 12(6):693–726

Soberón J, Osorio-Olvera L (2023) A dynamic theory of the area of distribution. J Biogeogr 50(6):1037–1048

Ter Steege H, Pitman NC, Do Amaral IL, de Souza Coelho L, de Almeida Matos FD, de Andrade Lima Filho D et al (2023) Mapping density, diversity and species-richness of the Amazon tree flora. Commun Biol 6(1):1130

Stehli FG, Webb SD (eds) (2013) The Great American biotic interchange, vol 4. Springer Science & Business Media, New York, pp 532

Stork NE, McBroom J, Gely C, Hamilton AJ (2015) New approaches narrow global species estimates for beetles, insects, and terrestrial arthropods. Proc Natl Acad Sci 112(24):7519–7523

Strömberg CA (2011) Evolution of grasses and grassland ecosystems. Annu Rev Earth Planet Sci 39:517–544

Su G, Villéger S, Brosse S (2019) Morphological diversity of freshwater fishes differs between realms, but morphologically extreme species are widespread. Global Ecol Biogeogr 28(2):211–221

Sullivan JP, Muriel-Cunha J, Lundberg JG (2013) Phylogenetic relationships and molecular dating of the major groups of catfishes of the neotropical superfamily Pimelodoidea (Teleostei, Siluriformes). Proc Acad Nat Sci Philadelphia 162(1):89–110

Swiston SK, Landis MJ (2024) Testing relationships between multiple regional features and biogeographic processes of speciation, extinction, and dispersal. Syst Biol 74(2):282–300

Tagliacollo VA, Roxo FF, Duke-Sylvester SM, Oliveira C, Albert JS (2015) Biogeographical signature of river capture events in Amazonian lowlands. J Biogeography 42(12):2349–2362

Tagliacollo VA, Duke-Sylvester SM, Matamoros WA, Chakrabarty P, Albert JS (2017) Coordinated dispersal and pre-isthmian assembly of the Central American ichthyofauna. Syst Biol 66(2):183–196

Tagliacollo VA, Dagosta FCP, Pinna MD, Reis RE, Albert JS (2021) Assessing extinction risk from geographic distribution data in neotropical freshwater fishes. Neotrop Ichthyol 19:e210079

Tagliacollo VA, Tan M, Reis RE, Gaia R, Carrijo V, Ranuzi M et al (2024) Time-calibrated phylogeny of neotropical freshwater fishes. Front Bioinf 4:1433995

Tedersoo L, Bahram M, Põlme S, Kõljalg U, Yorou NS, Wijesundera R et al (2014) Global diversity and geography of soil fungi. Science 346(6213):1256688

Thode VA, Sanmartín I, Lohmann LG (2019) Contrasting patterns of diversification between Amazonian and Atlantic forest clades of neotropical lianas (*Amphilophium*, Bignonieae) inferred from plastid genomic data. Mol Phylogenet Evol 133:92–106

Thomaz AT, Christie MR, Knowles LL (2016) The architecture of river networks can drive the evolutionary dynamics of aquatic populations. Evolution 70(3):731–739

Thomaz AT, Malabarba LR, Knowles LL (2017) Genomic signatures of paleodrainages in a freshwater fish along the southeastern coast of Brazil: genetic structure reflects past riverine properties. Heredity 119(4):287–294

Thomaz AT, Malabarba LR, Bonatto SL, Knowles LL (2015) Testing the effect of palaeodrainages versus habitat stability on genetic divergence in riverine systems: study of a neotropical fish of the Brazilian coastal Atlantic Forest. J Biogeography 42(12):2389–2401

Thompson JB, Ramírez-Barahona S (2023) No phylogenetic evidence for angiosperm mass extinction at the Cretaceous–Palaeogene (K-Pg) boundary. Biol Lett 19(9):20230314

Toussaint A, Charpin N, Brosse S, Villéger S (2016) Global functional diversity of freshwater fish is concentrated in the neotropics while functional vulnerability is widespread. Sci Rep 6(1):22125

Tschá MK, Bachmann L, Abilhoa V, Boeger WA (2017) Past connection and isolation of catchments: the sea-level changes affect the distribution and genetic variability of coastal freshwater fishes. Estuarine Coastal Shelf Sci 190:31–39

Val AL, Wood CM (2022) Global change and physiological challenges for fish of the Amazon today and in the near future. J Exp Biol 225(10):jeb216440

Val P, Lyons NJ, Gasparini N, Willenbring JK, Albert JS (2022) Landscape evolution as a diversification driver in freshwater fishes. Front Ecol Evol 9:975

Val P, Figueiredo J, de Melo G, Flantua SG, Quesada CA, Fan Y et al (2021) Geological history and geodiversity of the Amazon. In: Amazon Assessment Report. pp 1–41

van der Sleen P, Albert JS (2022) Patterns in freshwater fish diversity. In: Reference Module in Earth Systems and Environmental Sciences, vol 263. pp 894–907

van der Sleen P, Albert JS (eds) (2017) Field guide to the fishes of the Amazon, Orinoco, and Guianas, vol 115. Princeton University, Princeton, p 464

Vannote RL, Sweeney BW (1980) Geographic analysis of thermal equilibria: a conceptual model for evaluating the effect of natural and modified thermal regimes on aquatic insect communities. Am Nat 115(5):667–695

Vari RP, Weitzman SH (1990) A review of the phylogenetic biogeography of the freshwater fishes of South America. In: G. Peters and R. Hutterer (eds). Vertebrates in the Tropics. Alexander Koenig Zoological Research Institute and Zoological Museum, Bonn, pp 381–393

Vermeij GJ (1977) The Mesozoic marine revolution: evidence from snails, predators and grazers. Paleobiology 3:245–258

Vilardo PJ, Katz AM, Costa WJEM (2023) Phylogeny and historical biogeography of neotropical catfishes *Trichomycterus* (Siluriformes: Trichomycteridae) from eastern Brazil. Mol Phylogenet Evol 186:107836

Weaver LN, Kelson JR, Holder RM, Niemi NA, Badgley C (2024) On the role of tectonics in stimulating the Cretaceous diversification of mammals. Earth-Sci Rev 248:104630

Wendt EW, Silva PC, Malabarba LR, Carvalho TP (2019) Phylogenetic relationships and historical biogeography of *Oligosarcus* (Teleostei: Characidae): examining riverine landscape evolution in southeastern South America. Mol Phylogenet Evol 140:106604

Weppe R, Condamine FL, Guinot G, Maugoust J, Orliac MJ (2023) Drivers of the artiodactyl turnover in insular western Europe at the Eocene–Oligocene Transition. Proc Natl Acad Sci 120(52):e2309945120

Werneck FP, Gamble T, Colli GR, Rodrigues MT, Sites JrJW (2012) Deep diversification and long-term persistence in the South American 'dry diagonal': integrating continent-wide phylogeography and distribution modeling of geckos. Evolution 66(10): 3014–3034

Wesselingh FP, Räsänen ME, Irion G, Vonhof HB, Kaandorp R, Renema W et al (2001) Lake Pebas: a palaeoecological reconstruction of a Miocene, long-lived lake complex in western Amazonia. Cainozoic Res 1(1/2):35–68

Winemiller KO (1991) Ecomorphological diversification in lowland freshwater fish assemblages from five biotic regions. Ecol Monograph 61(4):343–365

Winemiller KO, Zeug SC, Robertson CR, Winemiller BK, Honeycutt RL (2011) Food web structure of coastal streams in Costa Rica revealed by dietary and stable isotope analyses. J Trop Ecol 27:463–476

Winemiller KO, Agostinho AA, Caramaschi ÉP (2008) Fish ecology in tropical streams. In: Dudgeon D (ed) Tropical stream ecology. San Diego Academic Press/Elsevier, London, pp 107–146

Wisniewski AL, Lloyd GT, Slater GJ (2022) Extant species fail to estimate ancestral geographical ranges at older nodes in primate phylogeny. Proc R Soc B 289(1975):20212535

Yasuhara M, Huang HHM, Reuter M, Tian SY, Cybulski JD, O'Dea A et al (2022) Hotspots of Cenozoic tropical marine biodiversity. In: Hawkins SJ et al. (eds) Oceanography and marine biology. CRC Press/Taylor & Francis, Boca Raton, pp 243–300

Zapata S, Calderon-Diaz L, Jaramillo C, Oboh-Ikuenobe F, Piedrahita JC, Rodríguez-Cuevas M et al (2023) Drainage and sedimentary response of the Northern Andes and the Pebas system to Miocene strike-slip tectonics: a source to sink study of the Magdalena Basin. Basin Res 35(5):1674–1717

Glossary

Commonly Used Terms Used in Continental Ecology and Evolution. This glossary is offered to aid readers in navigating rich literature on biodiversity, with the goal of simplifying and clarifying language whenever possible. Related terms are in *italic* font and indicated by *Cf.* Note that some distinctions and nuances in these definitions are not universally observed among evolutionary biologists, but for clarity, these terms are used consistently throughout this book. All terms are nouns unless otherwise specified.

***Adaptation* (noun):** *Ecotraits* and other *phenotypes* that arise due to the process of adaptation. (verb): Improving ecophysiological function(s) of an *ecotrait* or other *phenotype*, generally from *natural selection. Cf. Exaptation, Genetic drift.*

***Adaptive radiation* (noun or verb):** *Macroevolutionary diversification*, usually under geographically restricted conditions (c. thousands of square km), with *speciation* occurs along ecological lines in *sympatry*, and processes of *speciation* and *adaptation* are coupled as a result of *natural selection. Cf. Continental radiation.*

***Algae*:** Heterogenous assemblage of plants, protists, and bacteria that use chlorophyll-α as their primary photosynthetic pigment, and which lack traditional terrestrial plant characteristics such as roots, leaves, and stems.

***Algivore*:** An animal that primarily feeds on or consumes algae, either on the benthos (benthic) or in the water column (planktonic). *Cf. Trophic position.*

***Allochthonous* (adj):** Energy and nutrients that enter an environment from elsewhere (e.g., leaf litter and runoff).

***Allopatric* (adj):** Geographic separation of *populations* or *species. Cf. Sympatric.*

***Allotopic* (adj):** Ecological separation of populations or *species* among different *habitats. Cf. Syntopic.*

***Alpha diversity*:** Species richness of a local *assemblage. Cf. Beta diversity.*

***Amazon or Amazonia*:** A geographic region with several distinct, although overlapping meanings, in different contexts. (1) Amazon River Basin (or Amazon Basin) as a hydrological drainage unit, (2) Amazonas sedimentary basin as a geological unit, (3) Amazon Rainforest as a *biodiversity* province, (4) Amazonas state or department as one of several administrative or political units.

***Ancestral area estimation (AAE)*:** Statistical method to estimate the geographic origin of a *clade* using methods of Maximum Likelihood or Bayesian Analysis.

***Assemblage*:** Ecological unit composed of all *species* of a local environment or *habitat*, without regard to *ecotrait* or other *phenotype* functions or interactions. *Cf. Community.*

***Autochthonous* (adj):** Energy and nutrients that originate within the environment itself, e.g., photosynthetic outputs from *algae.*

***Bell curve*:** A common statistical distribution to describe differences within a *population* or *species*, characterized by a symmetrical, bell-shaped curve. Also known as a *normal distribution* in which the peak represents the most probable outcome (i.e., *mode*) and the width represents *standard deviation. Cf. Hollow curve.*

Benthos: The bottom of a water body, e.g., river, lake, sea. Benthic (Adj.).

***Beta diversity*:** Change or turnover in taxonomic composition (i.e. names) of *species* among local *assemblages* (*habitats*, areas) or through time (e.g., seasons, years). *Cf. Alpha diversity.*

***Biodiversity*:** Diversity (numbers and variety) of genes, *traits*, *species*, *habitats*, landscapes and other biological entities in an area or portion of the biosphere. *Cf. Variation.*

***Biome*:** Biogeographical unit is defined by climate (i.e. temperature and precipitation) and a characteristic vegetation *physiognomy*. Biomes may span more than one continent. Neotropical examples include tropical rainforest, tropical dry forest, tree savannah, and grass savannah. *Cf. Biorealm, Bioregion, Biota.*

***Biorealm*:** A biogeographic realm, the broadest biogeographic division of Earth's land surface, based on distributional patterns of terrestrial organisms. The Earth is divided into seven *biorealms*, including the *Neotropical Realm*. Realms are subdivided into *bioregions*, which are further subdivided into *ecoregions. Cf. Bioregion.*

***Bioregion*:** Subdivision of a *biogeographic realm* whose limits are defined by topographic and biological features,

J. S. Albert and R. E. Reis (eds.), *Ecology and Evolution of Amazonian Fishes*, Fascinating Life Sciences,
https://doi.org/10.1007/978-3-032-09328-8

such as mountain ranges and *biome* margins. A *bioregion* is smaller than a *biogeographic realm* and larger than an *ecoregion. Cf. Ecoregion, Ecozone.*

Biota: All animal (e.g., fauna) and plant (e.g., flora) *species* living in an area of Earth's surface. *Cf. Biorealm, Biome, Bioregion, Ecozone.*

Black Swan events: Rare but impactful historical events or conditions that structure current conditions, but which are not usually observable from study of current conditions alone; e.g., one-in-ten-thousand-year-storm-events, Ice ages, Hyperthermal events, VEI8 volcanic eruptions, Richter scale 11 earthquakes, 10 km diameter asteroid impacts, formation and breakup of supercontinents. *Cf. Hollow curve.*

Clade age: Age of a crown group, meaning the time of origin of the most recent common ancestor of all *extant species* of a *clade*. Contrasted with the stem group ancestor of all known *taxa* in a clade including extinct (fossil) *species*. Stem groups are older than their corresponding crown groups, and possibly older than *phenotypes* shared by crown group *taxa*.

Clade: A complete group of related *species*, representing a complete branch of the Tree of Life. A higher taxon (e.g., genus, family) or *monophyletic group*, including an ancestral *species* and all of its descendant *species. Cf. Assemblage.*

Community: An ecological unit composed of all *species* and intraspecific and interspecific interactions (e.g., predation, competition, parasitism, etc.) of a local environment or *habitat. Cf. Assemblage.*

***Continental radiation* (noun or verb)**: *Macroevolutionary diversification* at a continental spatial scale (millions of square km) and geological time frames of tens of millions of years, in which speciation is mostly along geographic lines in *allopatry*, and processes of *speciation* and *adaptation* are decoupled, such that speciation usually outpaces adaptation, resulting in functional redundancy of *species. Cf. Adaptive radiation.*

***Coordinated dispersal* (noun or verb)**: Refers to the influence of similar paleogeographic constraints on the distribution of *taxa* in space.

Deterministic: Events or outcomes are determined by preceding causes or natural laws, meaning that given the same initial conditions, the same results will always occur. Implies processes that are inevitable, fated, destined, predetermined, or preordained, with no randomness. *Cf. Stochastic.*

Detritus: Organic-rich non-living matter, varying from fallen tree branches and leaf litter to shed insect casings and fish fecal matter.

***Dispersal* (verb)**: Movements and establishment of organisms outside a *species* geographic range. Geographic range expansion of a clade. *Cf. Dispersion, Migration.*

Dispersal filter: Any factor that limits or biases the movement and colonization of organisms among areas or *habitats*. Filters are physical barriers, environmental conditions, or biotic interactions that partially restrict biotic *dispersal. cf. Portal.*

***Dispersion* (verb)**: Movements of organisms within an established *species* geographic range. *Cf. Dispersal.*

***Dissolved Organic Carbon* (*DOC*)**: Amount of carbon found in organic molecules dissolved in water. DOC is a major component of Earth's carbon cycle and plays a vital role in aquatic *ecosystems. Cf. Dissolved Organic Matter, Particulate Organic Matter.*

***Dissolved Organic Matter* (*DOM*)**: Portion of organic matter in water that passes through a filter with a pore size of 0.2–0.7 μm. DOM is a complex mix of molecules released by living organisms and derived from the breakdown of organic materials. *Cf. Dissolved Organic Carbon.*

Diversification: Change in the number of *species* within a *clade* as a result of different rates of *speciation* and *extinction*. Called *net diversification* when estimated from a phylogeny of *extant species*. A clade expands when the *speciation* rate exceeds the *extinction* rate.

Diversity: Genetic and phenotypic differences among *species* are often characterized by skewed or lognormal frequency distributions and are subject to processes of speciation and extinction. *Cf. Variation.*

Drainage basin: Land surface area where precipitation from rain and snowmelt drains into a specific river and its tributaries. Also called a catchment area. *Cf. Sedimentary basin, Watershed.*

Ecomatrix: A matrix of ecological data summarizing physiological and behavioral information of *ecotraits* among *species* of a *biota*.

Ecophysiology: Study of interactions between organismal physiology and physical and biological environmental factors.

Ecoregion: Subdivision of a *bioregion*, with one or more *ecosystems*, characterized by a distinct *assemblage, community* or *species*. A geographically defined area characterized by similar *ecosystems*, environmental conditions, and type, quality, and quantity of environmental resources. *Cf. Bioregion, Ecozone.*

***Ecospace analysis* (noun or verb)**: A multivariate statistical analysis (e.g., principal component analysis or PCA), which is a quantitative representation of phenotypes in a multidimensional space with each dimension corresponding to a combination of traits that covary together, and which are independent of other such trait combinations.

Ecospace: Graphical representation of ecological or *functional traits* (e.g., *ecotraits*) resulting from a multivariate statistical analysis (e.g., PCA). *Cf. Morphospace.*

Ecosystem: Biotic and abiotic components of a *species assemblage or community*, including energy and material flows and biogeochemical interactions.

***Ecotrace analysis* (noun or verb)**: Estimate of evolutionary history of an *ecotrait* on a *phylogeny*, assuming heritability of *trait* values among *species*, tree topology and branch-length data, and a model of trait evolution (e.g., Brownian Motion). Estimates include the number and proportions of transitions for qualitative states, amounts and proportions of evolutionary change for quantitative traits, and overall rates of trait evolution.

Ecotrait: Any *trait* or organismal feature that affects ecological functions; e.g., *habitat* utilization, diet, *trophic position*, or life-history attributes related to predation, *dispersal*, mating or reproduction. *Cf. Functional trait, Trait.*

Ecozone: A large geographical region having a distinct biodiversity of plant (flora) and animal (fauna) *species*. *Cf. Bioregion, Ecoregion.*

Elevational Diversity Gradient. *Biodiversity* patterns in which *species richness*, *taxonomic composition* and ecological productivity change with altitude, typically decreasing in values as elevation increases. *Cf. Beta diversity, Latitudinal Diversity Gradient; Mid-Domain Effect.*

Epigenome: Chemical modifications of DNA and associated proteins that influence how genes are expressed without altering the DNA sequence of *genome*. *Cf. Genome, Phenome.*

Evolutionary radiation. General term to describe any *macroevolutionary diversification* without specifying tempo, modes or underlying processes, *Cf. Adaptive radiation, Continental radiation.*

Evolutionary relict: A taxon (*species* or *clade*) that remains from an earlier geologic time when environmental conditions were different than at present. A distribution area that is a small fraction of a formerly wider geographic range. Cf. *Peripheral relict.*

Eurytopic **(adj)**: Ability of organisms to inhabit or tolerate a wide range of *habitats* and environmental conditions. *Cf. Stenotopic.*

Exaptation: Noun: *Ecotraits* and other phenotypes with different function(s) from preexisting previous function(s). Verb: Changing ecophysiological function(s) of a *phenotype* away from that of previous function(s), either by *natural selection* or *genetic drift*. *Cf. Adaptation.*

Extant: Referring to living *species* that live in the *Recent* time horizon. *Cf. Extinct*

Extinct: *Species* or *clades* that have not survived to the *Recent* time horizon. *Cf. Extinct.*

Extinction **(noun or verb)**: Termination of a *species* lineage or a *clade* by the death of its last member. *Cf. Diversification, Speciation.*

Functional disparity: *Biodiversity* metric assessed by the quantitative amount of functional differences among organisms. *Cf. Genetic diversity, Ecospace, Functional trait, Species richness.*

Functional trait: A *trait* that influences organismal performance, survival, development, growth, or reproduction. Can affect *community* assembly, *species* coexistence, and *biodiversity* loss. *Cf. Ecotrait, Functional disparity.*

Gene flow: Transfer of genes among populations of a *species* by *migration* or *dispersal*. *Cf. Genetic drift.*

Genetic diversity: Variety of genes and differences in DNA sequences among members of a clade or *ecosystem*. *Cf. Functional disparity, Genetic variation, Phylogenetic diversity, Species richness.*

Genetic drift **(noun or verb):** *Stochastic* changes in gene frequencies within a species due to chance and entropy. *Cf. Exaptation, Genetic flow, Natural selection.*

Genetic variation: Differences in DNA sequences among members of a *species*. *Cf. Genetic diversity.*

Genome: Complete set of genes (i.e. DNA sequences) containing organismal information needed for an organism to develop and function. Under this definition, the *epigenome* is considered part of *phenome* but not *genome*. *Cf. Epigenome, Phenome.*

Geodispersal **(noun or verb)**: Merging of biotic regions by landscape evolution potentially allows biotic *dispersal* and gene flow. *Cf. Vicariance.*

Geodiversity: Geophysical features of Earth's surface and subsurface (e.g., mountains, canyons, plateaus, rivers, wetlands, aquifers), including materials (e.g., minerals, sediments, fossils, soil and water chemistry), and processes (e.g., tectonics, weathering, erosion, *river capture*, soil formation, sedimentation). *Cf. Biodiversity.*

Grazer: Organism that feeds on the outer layers of algae/detritus/periphyton without completely removing the food source. *Cf. Trophic position.*

Greater Amazonia: Area of *Amazon* and Orinoco river basins and Guiana Shield, extending over an area of about 8.4 million km^2 of northern South America east of the Andes. *Cf. Amazon.*

Habitat: Location and features of a local environment where organisms naturally live and grow. Includes physical, chemical, and biological conditions. Important freshwater *habitat* parameters include: temperature, dissolved oxygen, pH, turbidity (clarity), depth, velocity, dissolved ions (conductivity), daily and annual fluctuations, amount and type of vegetation cover, substrate type and grain size, and accessibility to migration routes.

Historical ecology: Study of the evolutionary history of ecological associations (i.e., mutualisms, parasitism) and effects of speciation and adaptation on clades and community structure.

Hollow curve: A commonly observed frequency distribution in natural phenomena is that many items are rare and a few items are common. Also known as a *Power-law distribution*, *Pareto distribution* (economics), *Zipf's Law* (linguistics), or the 80/20 rule, in which roughly 80% of effects result from 20% of causes. *Cf. Bell Curve, Black Swan event.*

Homology: Similarity of organismal traits and phenotypes due to inheritance from a common ancestor. *Cf. Homoplasy.*

Homoplasy: Similarity of organismal traits and phenotypes due to convergence, parallelism, loss, or chance. *Cf. Homology.*

Horton's Law: Describes relationships between number, length, and area of streams in a drainage basin, stating number of streams of a given *stream order* decreases following an inverse geometric series.

***Insectivore*:** Animal that primarily consumes insects, either as aquatic larvae or terrestrial adults. *Cf. Trophic position.*

***Intermediate-rate capture hypothesis (ICH)*:** Most diverse *ecosystems* are found in landscapes where *river capture* events occur at an intermediate rate optimal for driving diversification. *Cf. Stochastic speciation hypothesis, Topographic controls hypothesis.*

***Latitudinal Diversity Gradient (LDG)*:** Global pattern of higher *species richness* at lower latitudes (closer to the equator) and lower richness at higher latitudes (closer to the poles). A fundamental and widely observed feature of *biodiversity* in both terrestrial and marine *ecosystems*. *Cf. Beta diversity, Mid-Domain Effect.*

***Macroecology*:** Study of broad-scale ecological patterns and processes, focusing on relationships between organisms and their environment across large spatial and temporal scales. *Cf. Macroevolution.*

***Macroevolution*:** Study of large-scale evolutionary changes, at or above the *species* level, including origin of new *species*, *clades*, and body plans (e.g., animal phyla), formation of new ecological phenomena (e.g., *biomes*, *ecosystems*), long-term evolutionary trends (e.g., supercontinent and global climate cycles, origins of terrestrial or abyssal marine clades and biotas), and rare but high-impact *Black Swan events* (e.g., mass extinctions, adaptive radiations). *Cf. Macroecology, Microevolution.*

***Map*:** A diagram or illustration of spatially arranged data, generally on a planar (2D) surface; e.g., Earth, sky, body, brain, etc. In mathematics, correspondence of elements in one set to elements in the same set or another set.

***Mega-river capture hypothesis (MCH)*:** Qualitatively distinct consequences of large *river capture* events >10,000 sq. km or 3,861 sq. miles) on ecological and evolutionary processes of *evolutionary diversification* in riverine and riparian *taxa*. *Cf. Intermediate-rate capture hypothesis, Topographic controls hypothesis.*

***Microevolution*:** Evolution within a *species* is the subject of study for population genetics. *Cf. Genetic drift; Macroevolution; Natural Selection.*

***Microphytobenthos*:** Microscopic algae that inhabit the thin layer of substratum between sediments and the water column (also benthic microalgae).

***Microvore*:** Animal that primarily consumes small animals of less than 1.0 mm along the largest body dimension. *Cf. Trophic position.*

***Mid-Domain Effect (MDE)*:** Tendency of *species richness* to peak at middle of a geographical domain. This pattern arises from a simple geometric constraint that the overlap of *species*' ranges tends to be greatest in the middle of a region when *species* are randomly distributed. *Cf. Elevational diversity gradient, Latitudinal diversity gradient.*

Migration: Regular (e.g., seasonal, annual) movements of organisms within parts of an established *species* geographic range. *cf. Dispersal, Dispersion.*

***Morphospace*:** Mathematical representation of shape diversity among *taxa* resulting from a multivariate statistical analysis (e.g., PCA) where each data point represents an individual *taxon*, and each axis represents a mathematically distinct aspect of organismal shape. *Cf. Ecospace.*

***Mosaic evolution*:** A pattern where different parts or systems of an organism evolve at different rates and times, rather than all changes occurring simultaneously. Results in organisms that exhibit a variety of primitive and derived traits.

***Natural selection*:** Differential survival and reproduction of individual organisms within a *species* due to differences in heritable *traits* or *phenotypes*. A mechanism of that changes in heritable traits characteristic of a *population* or *species* over multiple generations. *Cf. Adaptation, Genetic drift, Microevolution.*

***Paraspecies*:** A *species* that gave rise to another species without itself going extinct. *Cf. Peripheral relict.*

***Particulate Organic Matter (POM)*:** Fraction of organic matter retained by a filter, meaning it's not dissolved. Includes materials like plant litter, animal feces, and microbial detritus. POM plays a crucial role in soil health and the carbon cycle. *Cf. Dissolved Organic Matter, Dissolved Organic Carbon.*

***Peripheral freshwater fishes*:** Freshwater fishes with little or high tolerance to salty marine waters and Neogene or Quaternary origins in continental freshwater *habitats*. Also called Marine Derived Lineages. *Cf. Primary freshwater fishes, Secondary freshwater fishes.*

***Peripheral relict*:** Taxon (*species* or *clade*) with low diversity, a small and geographically isolated distribution near margins of an evolutionary arena (e.g., continent, ocean basin), and with an early-branching phylogenetic position as sister taxon to a highly diverse clade with broad geographic distribution. *Cf. Evolutionary relict, Paraspecies.*

***Periphyton*:** Heterogenous assemblage of benthic algae, detritus, and microbial decomposers (biofilm) overlying submerged surfaces.

***Phenome*:** Complete set of all organismal traits, attributes and properties, encompassing both physical traits and physiological and behavioral functions, resulting from the interaction of genes and environment. *Cf. Genome, Phenotype.*

***Phenotype*:** Organismal trait, attribute or property that characterizes individual organisms of a *species*, and which affects their ecological and evolutionary processes. Encompasses any aspect of *phenome* throughout the hierarchy of organismal organization, including structure and function of macromolecules, organelles, cells and organs, as well as physiological, morphological, functional, behavioral and ecological attributes that characterize *species* and arise from *genome*, including phenotypic plasticity and *ecosystem* engineering. *Cf. Phenome, Trait.*

Phylodynamics**:** Study of phylogenetic tree parameters (e.g., branch-length frequencies, branching symmetry) to estimate evolutionary processes.

Phylogenetic diversity**:** *Biodiversity* metric assessed by the summed branch lengths of all *species* in a *phylogeny*. *Cf. Functional disparity, Genetic diversity, Species richness.*

Phylogenetic lability**:** Propensity for evolutionary change in trait values of *phenotypes* and *ecotraits*. *Cf. Homology, Homoplasy, Phylogenetic niche conservatism.*

Phylogenetic niche conservatism**:** Tendency of *species, phenotypes* and *ecotraits* to resemble their close evolutionary relatives by inheritance from a common ancestor. *Cf. Homology, Phylogenetic niche evolution.*

Phylogenetic niche evolution**:** Evolutionary changes in *phenotypes* and *ecotraits* that affect ecological functions, e.g., breadth and quality of *habitat* utilization, diet, and life history attributes. *Cf. Homology, Homoplasy, Phylogenetic niche conservatism.*

Phylogeny**:** The evolutionary history of a *trait, species*, or *clade*. A graphical representation, usually as a branching diagram, depicting evolutionary relationships among *species* or higher *taxa*. Also known as a phylogenetic tree or evolutionary tree.

Physiognomy**:** Overall structure and physical appearance of a plant community based on height, spacing and shape of dominant *species*.

Phytoplankton**:** Subset of plankton that are photosynthetic.

Piscivore**:** Animals that primarily consume fish, either in part (by biting) or whole (by swallowing). *Cf. Trophic position.*

Plankton**:** Organisms that drift in the water column. *Cf. Trophic position.*

Portal**:** A low-elevation (<500 m) *watershed* divide between adjacent river basins that allows partial or episodic hydrological and biotic connections. *Cf. Dispersal filter.*

Primary freshwater fishes**:** Freshwater fishes with little or no tolerance to salty marine waters and Mesozoic origins in continental freshwater *habitats*. *Cf. Secondary freshwater fishes, Peripheral freshwater fishes.*

Rapoport's Rule**:** A biogeographic pattern in which the range size of *species* tends to increase with increasing latitude, meaning *species* at higher latitudes generally have larger geographic ranges compared to those at lower latitudes.

Recent: The current or present time horizon with all living or *extant species*.

River capture**:** A perennial geomorphological (i.e., landscape evolution) process operating on all continental surfaces (except frozen Antarctica), which regularly divides and reunites portions of river drainage networks.

River capture hypothesis (RCH)**:** A model of biological *diversification* in which river network rearrangements serve to both separate and merge portions of adjacent river basins. The RCH posits that changing the connection and fragmentation of aquatic biotas affects the rates of *dispersal, speciation* and *extinction*. Merging river segments promotes dispersal and gene flow, reducing rates of speciation and extinction. River capture rates differ among waterways with different *stream orders*, being faster in smaller low-order rivers than in larger high-order rivers. *Cf. Intermediate-rate capture hypothesis, Topographic controls hypothesis.*

River continuum**:** Model describing how physical and biological characteristics of a river system change along a river's longitudinal axis from headwaters to mouth.

Riverscape**:** Geographic features found on and along a river. Most drainage basins exhibit distinct riverscapes in their upper (waterfalls, rapids), middle (riffles, alluvial fans, inland deltas) and lower (e.g., meanders, oxbow lakes, levees, floodplains, estuaries) reaches.

Secondary freshwater fishes**:** Freshwater fishes with some tolerance to salty marine waters and Paleogene origins in continental freshwater *habitats*. *Cf. Primary freshwater fishes, Peripheral freshwater fishes.*

Sedimentary basin**:** Geological depression in Earth's crust where sediments accumulate and are preserved. *Cf. Drainage basin.*

***Shoreline fluctuation* (noun or verb):** A regular geomorphological process operating on all continental surfaces (except frozen Antarctica) in which eustatic (i.e., global) sea-level changes associated with Plio-Pleistocene climate cycles divide (during warmer interglacial intervals) and reunite (during colder glacial intervals) the lower reaches of coastal and lowland river basins.

***Speciation* (noun or verb):** The process by which ancestral *species* give rise to descendant *species*. *Cf. Extinction.* Theoretically, speciation can occur in *sympatry* by populations diverging under different ecological conditions, in *allopatry* due to geographic separation, or in *parapatry* due to evolution along a geographic or ecological gradient.

Species **(*noun;* abbrev. *sp.*, singular; *spp.*, plural):** A fundamental unit of biodiversity, biogeography, evolution and ecology. A *species* taxon is an individual entity (e.g., *Gymnotus carapo*), while a *species* category is the least inclusive level of taxonomic hierarchy. *Cf. Species concept.*

Species concept**:** A scientific theory on the nature of *species* entities. Three widely-held *species* concepts in Neotropical ichthyology are based on a single criterion: *Biological, Ecological*, and *Phylogenetic species* concepts are based on shared breeding (i.e. reproduction), niche utilization (e.g., *habitat*, diet, life history), and phenotypic/genotypic traits, respectively. The *Evolutionary species* concept hypothesizes that each *species* taxon is a distinct lineage of ancestors and descendants, without specifying a single criterion.

Species richness**:** A count of the number of *species* in a geographic area, *habitat*, or *clade*. *Cf. Phylogenetic diversity.*

***Stenotopic* (adj):** Ability of organisms to inhabit or tolerate a narrow range of *habitats* and environmental conditions. *Cf. Eurytopic.*

***Stochastic speciation hypothesis (SSH)*:** Hypothesis that speciation results from the regular accumulation of genetic incompatibilities by *genetic drift* in *allopatry*. *Cf. Intermediate-rate capture hypothesis*.

***Stochastic*:** Events or outcomes characterized by a random probability distribution, in which events may be unpredictable or predicted probabilistically, but are not inevitable, fated, destined, predetermined, or preordained by initial conditions. *Cf. Deterministic*.

***Stream order*:** Classification of rivers based on hierarchical tributary relationships. Under the Strahler system, headwaters (with no upstream tributaries) are designated first-order streams down to their first confluence with another first-order stream, where they merge into a second-order stream. The *Amazon* River, below its confluence with Negro River, is the only 10th-order waterway on Earth.

***Sympatric* (adj):** Presence of two or more *species* in same geographic region. *Cf. Allopatric*

***Syntopic* (adj):** Presence of two or more *species* within the same *habitat*. *Cf. Allotopic*.

***Taxon* (singular; *Taxa*, plural):** A group of organisms with a formal Latin name. A taxon may or may not be a *clade* (i.e. monophyletic); e.g., Actinopterygii, Fishes.

Taxonomic composition: Names of *taxa* in a *clade* or *bioregion*. *Cf. Species richness*.

***Taxonomy*:** The scientific practice of naming natural objects and entities. Biological taxonomy involves naming *species* and *clades*.

***Terra firme*:** Areas of lowland Amazonian rainforest are located above the level of the annual floodplain.

***Tobler's First Law of Geography*:** A core principle of spatial analysis highlighting the concept of *spatial autocorrelation* that "everything is connected to everything else, but near things are more connected than distant things."

***Topographic controls hypothesis (TCH)*:** Topographic features and processes, like tectonic uplift, erosion, and base-level fall, influence rate and timing of *river capture* events. *Cf. Intermediate-rate capture hypothesis*.

***Trait*:** Any feature of an organismal *phenotype* at any level in the hierarchy of organismal organization, including molecular, cellular, morphological, physiological, or behavioral phenotypes. *Cf. Phenome, Phenotype*.

***Trophic position*:** Position an organism occupies within a local food web. Quantified as the number of trophic steps from primary producers (i.e., plants) at the base of the web.

***Variation*:** Genetic and phenotypic differences within a *species* are often characterized by a normal frequency distribution and are subject to processes of *natural selection* and *genetic drift*. *Cf. Diversity*.

***Vicariance*:** Separation of biotic regions by landscape evolution allows for biotic separation and speciation. *Cf. Geodispersal*.

***Watershed*:** Hydrological margins of a river *drainage basin* or *catchment*. *Cf. Sedimentary basin*.

***Xylivore*:** Animal that primarily feeds on or consumes xylem in wood. *Cf. Trophic position*.

Index

J. S. Albert and R. E. Reis (eds.), *Ecology and Evolution of Amazonian Fishes*, Fascinating Life Sciences,
https://doi.org/10.1007/978-3-032-09328-8

D

P

T

U

V

W

GPSR Compliance

The European Union's (EU) General Product Safety Regulation (GPSR) is a set of rules that requires consumer products to be safe and our obligations to ensure this.

If you have any concerns about our products, you can contact us on ProductSafety@springernature.com

In case Publisher is established outside the EU, the EU authorized representative is:

Springer Nature Customer Service Center GmbH
Europaplatz 3
69115 Heidelberg, Germany

Batch number: 10731667

Printed by Printforce, the Netherlands